MODERN
GEOTECHNICAL
ENGINEERING

Other Companion Books by the Author

1. Soil Engineering in Theory and Practice
   Volume 2, Geotechnical Testing and Instrumentation
   (Second Enlarged Edition 1989)
2. Soil Engineering in Theory and Practice
   Volume 1, Fundamentals and General Principles
   (Third thoroughly revised and enlarged edition, 1990)
3. Basic Soil Mechanics and Foundations, 1990
   (A new text for undergraduates, professional examinations
   and competitions)

NOVEL IDEAS

By HIS GRACE
and GRACE alone,
Novel ideas may trickle
like crude diamonds,
Gather them immediately
and polish at leisure.

Author

MODERN GEOTECHNICAL ENGINEERING

THIRD EDITION

Professor Dr. Alam Singh

Geotechnical Consultant
A-24 Shastri Nagar, Jodhpur
(Ex-Professor of Civil Engineering
and Head of Civil Engineering Department
University of Jodhpur
Jodhpur)

Foreword by
H. Y. Fang
Professor of Civil Engineering & Director
Geotechnicol Engineering Division, Lehigh University
Bethlehem, PA 18015 USA

CBS Publishers & Distributors Pvt. Ltd.

New Delhi • Bengaluru • Chennai • Kochi • Kolkata • Mumbai
Hyderabad • Nagpur • Patna • Pune • Vijayawada

ISBN: 81-239-0121-6

Second Enlarged Edition: 1990
Third Edition: 1992
Reprint: 1999, 2001, 2003, 2008, 2009 (Twice),
        2010, 2012, 2014, 2016

Published by:
Satish Kumar Jain for CBS Publishers & Distributors Pvt. Ltd.,
4819/XI Prahlad Street, 24 Ansari Road, Daryaganj, New Delhi - 110002
delhi@cbspd.com, cbspubs@airtelmail.in • www.cbspd.com
Ph.: 23289259, 23266861, 23266867 • Fax: 011-23243014

Corporate Office: 204 FIE, Industrial Area, Patparganj, Delhi - 110 092
Ph: 49344934 • Fax: 011-49344935
E-mail: publishing@cbspd.com • publicity@cbspd.com

Branches:
• *Bengaluru:* 2975, 17th Cross, K.R. Road, Bansankari 2nd Stage,
  Bengaluru - 70 • Ph: +91-80-26771678/79 • Fax: +91-80-26771680
  E-mail: cbsbng@gmail.com, bangalore@cbspd.com
• *Chennai:* No. 7, Subbaraya Street, Shenoy Nagar, Chennai - 600030
  Ph: +91-44-26681266, 26680620 • Fax: +91-44-42032115
  E-mail: chennai@cbspd.com
• *Kochi:* Ashana House, 39/1904, A.M. Thomas Road, Valanjambalam,
  Ernakulum, Kochi • Ph: +91-484-4059061-65
  Fax: +91-484-4059065 • E-mail: cochin@cbspd.com
• *Kolkata:* 6-B, Ground Floor, Rameshwar Shaw Road, Kolkata - 700014
  Ph: +91-33-22891126/7/8 • E-mail: kolkata@cbspd.com
• *Mumbai:* 83-C, Dr. E. Moses Road, Worli, Mumbai - 400018
  Ph: +91-9833017933, 022-24902340/41 • E-mail: mumbai@cbspd.com

Representatives:

• Hyderabad: 0-9885175004   • Nagpur: 0-9021734563
• Patna: 0-9334159340   • Pune: 0-9623451994
• Vijayawada: 0-9000660880

Printed at: Neekunj Print Process, Delhi

FOREWORD

A textbook on geotechnical engineering must serve dual purposes which are, providing the basic concepts of the nature and properties of soil and rock and present up-to-date information on how these concepts are used in analysis of various geotechnical problems relating to ground failure mechanisms and foundation design. From this aspect, Professor Singh admirably provides both of the requirements. In addition, this book includes depositional and environmental effects on soil and rock and soil-water interaction. The book also emphasises the understanding of clay mineralogy in basic geotechnical properties such as sensitivity, thixotropy, swelling, dispersibility and collapsibility.

Recent developments in ground improvement techniques such as chemical, physico-chemical soil stabilization, low-cost energy saving lime piles and stone columns, geotextiles and geomembranes and dynamic consolidation are introduced. Even though the descriptions of these subjects are brief they are concisely cited with proper references for those who are interested in further investigations.

The Modern Geotechnical Engineering text prepared by Professor Singh not only serves as a basic textbook, it should also be valuable to the practising engineer. I recommend it as a useful and well-timed addition to the geotechnical library.

March, 1987

H. Y. Fang
Professor & Director
Geotechnical Engineering Div.
Fritz Engineering Laboratory
Lehigh University
Bethlehem, PA 18015 (USA)

PREFACE
(Third Edition)

The Third enlarged edition has a new chapter entitled 'Soil Engineering Experiments' which describes the laboratory tests normally performed at the undergraduate level. The test description is in the format of a laboratory manual. The following tests are covered: moisture content determination, specific gravity determination, particle size determination, consistency limits determination, permeability determination, consolidation test, compaction tests, field density and voids ratio determination, shear strength determination, California Bearing Ratio determination, North Dakota Cone Test and swelling pressure determination.

With these additions, the utility and scope of the book as 'text' is expected to be much enlarged for the students.

Jodhpur, 1992

(Alam Singh)

PREFACE
(Second Edition)

Encouraged by wide acclaimation of the book as a *'real* modern text of Geotechnical Engineering', the second enlarged edition is now presented, which has new chapters on Machine Foundations, Design of Wells and Rings, and Fortran Computer Programs. Discussing briefly preliminary concepts of vibration and machine foundation requirements, lumped parameter method of design is illustrated by solved examples. Design of wells include worked examples for wells surrounded by clay but resting on rock, soft rock, sand and clay, according to the recent design concepts advanced by the IIT, Bombay. A Russian Engineers method is given for design of ring foundations often used for tower-type structures. Looking to the advent of computerization in present day technology development, typical programs for solving geotechnical problems are added.

Enriched with the above additional matter, the book is expected to be and all-the-more preferred text for students and useful to teachers an professional engineers.

Illustrations 269, Examples 187, Tables 111.

Jodhpur
July, 1989

ALAM SINGH

PREFACE
(First Edition)

This book is designed primarily to provide a clear-cut, contemporary and stimulating text in a convenient form for engineering students who are beginning the study of Geotechnical Engineering, a speciality of Civil Engineering. It covers more or less completely the under-graduate course and partly the post-graduate course of this subject usually prescribed in the Universities/Institutes and for professional examinations in India.

The general style of the book has developed from the author's experience in teaching the subject for over three decades and from his professional advisory work as well as from other sources. An attempt has been made to provide quite modern and up-to-date coverage of the science and art of Geotechnical Engineering which are changing rapidly. The emphasis throughout is on the practical and admittedly empirical knowledge of soil behaviour required by the geotechnical engineers for the design of foundations, embankments and other soil-related structures rather than on sophisticated analysis, although the material covered may be sophisticated at times. Soil being a unique engineering material, requires considerable laboratory and field testing for understanding its behaviour and hence brief description of soil tests is given, where necessary, to develop in the young engineers a 'feel' for soils and soil behaviour, so essential for the successful practice of geotechnical engineering.

Including a wealth of worked-out examples, many of them being design problems, the book is almost 'self-teaching'. This aspect of the book also frees the teacher in a formal course from the necessity of working examples during lectures and thus allows him to concentrate more on basic principles and applications of the points in question. Apart from the students and teachers, the professional geotechnical engineers will also find the book full of practice oriented information useful to them.

The extensive references quoted in the text and listed at the end of the book may be used by the readers for in-depth review and/or research work. The SI system of units is used in the text.

(viii)

To sum up, the author undertook this work with aspirations of designing an introductory, enjoyable and informative book which may be lucid for 'novice' up-to-date in knowledge for 'connoisseur' and a compendium for the profession. How far he has succeeded, is left to the readers' judgment and he would always welcome suggestions for improvement.

Illustrations 252, Examples 173, Tables 108, References 225.

Jodhpur
March, 1987
Alam Singh

ACKNOWLEDGEMENT

For condensing extensive developments of geotechnical engineering, the author had to make use of the experience and publications of other engineers, researchers, organisations and institutions. Care has been taken to acknowledge the source and give specific credit for all the material so presented in the book and any omission is purely out of inadvertance.

As a matter of great privilege, the author wishes to put on record his sincere gratitude to Professor H. Y. Fang, Professor of Civil Engineering & Director, Geotechnical Engineering Division, Lehigh University, Bethlehem, PA, USA, who could find time to scan the book before its release and so pleasingly write a Forweword.

The recommendatory remarks of Professor Fang, an established academician and consulting geotechnical engineer, are not only exhilaratory to the author, but also an assurance of worth-reading material. The author would feel rewarded if he is able to serve the engineering education and profession to some extent through his present work.

ALAM SINGH

CONTENTS

(xiii)

1

Introduction to soils and rocks

1.1 SOIL AND ROCK

Naturally occurring deposits of the earth's crust are classified by engineers into "soil" and "rock" with an arbitrary division based on strength, related physical properties and use. Soil, in an engineering sense, is the relatively loose mass of mineral and organic materials and sediments found above the bedrock, which can be relatively easily broken down into its constituent mineral or organic particles. The strength for a deposit to be qualified as 'soil' is arbitrarily taken by some engineers to be such that the deposit is loose enough to be excavated without blasting and to be penetrated in borings by ordinary soil sampling equipment. Soil is the natural product of weathering of rocks and decomposition of organic matter. It is a 'particulate' material, which means that a 'soil mass' is an accumulation of individual particles that are bonded together by mechanical or attractive means, the strength of the bonds being a small fraction of the mineral particles. The particles may range from colloidal size to small boulders (>300 mm).

Rock is considered to be a natural aggregate of mineral grains bonded together and possessing rigid internal bonds whose strength is of the same order as that of mineral grains proper. Rock is generally meant to refer to 'bedrock' or to large and hard fragments of bedrock. The term "bedrock" includes any of the solid rocks that make up the earth crust. Where exposed at the surface, bedrock forms "outcrops". Loose deposits covering the bedrock are called "superficial deposits" or "surface deposits" (Hunt 1972), which essentially constitute "soil" in the engineering sense.

Some natural materials which may be geologically called as rocks should be treated as soils. These materials are (CGS 1978-I): (i) soft or weakly cemented rocks with unconfined compressive strength less than 1000 kPa, and (ii) cemented sands and gravels having discontinuous cementing.

Although, both soil mass and rock mass consist of assemblage of mineral grains and flakes, the presence of discontinuities in a rock mass has a major effect on its properties. For example, a rock mass, in comparison to soil, possesses a lower degree of freedom with regard to movement due to discontinuities which are planar or nearly planar. Similarly, the insitu permeability of a rock mass is an orientation dependent property. Water develops an isotropic pore pressure in a soil mass but has an anisotropic effect (oriented joint water pressure) in a rock mass.

1.2 GEOTECHNICAL ENGINEERING

Geotechnical engineering is a speciality of Civil Engineering which deals with the properties, behaviour and use of earth materials (soil and rock) in engineering works. The successful practice of geotechnical engineering requires integration of knowledge from several fields, such as geology, material science and testing, mechanics and hydraulics. Soil and rock are naturally occurring, highly complex materials with variable ingredients and variable properties. Theoretical predictions of their behaviour have to be based on a number of simplifying and idealised assumptions. A lot of experience, imagination and judgment is, therefore, needed in the practice of geotechnical engineering.

1.3 WEATHERING PROCESSES

Weathering and erosion of rocks and minerals leads to the formations of soils. Weathering processes are of three type: mechanical, chemical and biological.

1. **Mechanical processes.** Mechanical weathering means the breakdown of rock masses into smaller fragments and subordinate sizes of particles without chemical changes, There are many mechanical processes of disintegrating rocks such as splitting action of frost, roots and fire, expansion and contraction due to temperature changes and breakdown and abrasion during transport as sediment, etc.

2. Chemical processes. Chemical weathering is the decomposition of parent minerals and their transformation into new compounds such as clay particles (clay colloids) silica, carbonates and iron oxides. Chemical weathering depends on the presence of water, temperature and the dissolved materials in water. The processes involved are: oxidation, hydration, hydrolysis, carbonation and leaching.

3. Biological processes. Bacterial and other microorganisms which induce chemical changes in their surroundings (specially by contributing organic acids) play an important role in further weathering of soils. The breakdown of organic matter in soils is accomplished almost entirely by the microorganisms.

1.4 WEATHERED ROCK MASS

The stage of weathering of a rock mass may be described relative to the distribution of weathered rock materials within it and the effect of weathering on discontinuities. The classification of weathered rock mass is expressed by grades and terms as given in Table 1.1. Their field recognition and engineering properties also follow (Fookes and Horswill 1970, Little 1969).

Table 1.1

ENGINEERING GRADE CLASSIFICATION OF WEATHERED ROCK MASS

Grade	Term
VI	Residual soil
V	Completely weathered
IV	Highly weathered
III	Moderately weathered
II	Slightly weathered
I	Fresh

VI Residual soil. This is the topmost layer. All rock material is converted to soil. There is no recognizable rock fabric. There is a large change in volume but the soil has not been significantly transported. Surface layer may contain humus and plant roots.

It is unsuitable for important foundations and is unsuitable on slopes when cover is destroyed. Selection is necessary before using as fill.

V Completely weathered. All rock material is decomposed and/or disintegrated to soil but the original fabric is still largely intact.

It is friable and can be excavated by hand or ripping without use of explosives. It is unsuitable for foundations of concrete dams or large structures but may be suitable for foundations of earth dams and fills. High cuttings at steep angles are unstable. Protection is required to avoid erosion.

IV Highly weathered. More than half of the rock material is converted to soil. Discoloured rock is present either as a discontinuous framework or as corestones. The rock mass is partially friable.

The engineering properties are similar to those of grade V. It is unlikely to be suitable for foundations of concrete dams. Erratic presence of boulders make it unreliable foundation for large structures.

III Moderately weathered. Less than half of the material is coverted to soil. Discoloured rock is present either as a continuous framework or as corestones. The intack rock is noticeably weaker, as determined in the field, than the fresh rock. The rock mass is not friable.

It is excavated with difficulty without the use of explosives and mostly crushes under bulldozer tracks. It is suitable for foundations of small concrete structures and rockfill dams and may be suitable for semi-pervious fills. Stability in cuttings depend on structural features, especially joint attitudes.

II Slightly weathered. The rock may be slightly discoloured, particularly adjacent to discontinuities which may be open. The strength of intack rock approaches that of fresh rock.

It requires explosives for excavation. It is suitable for concrete dam foundations. The mass is highly permeable through open joints; often more permeable than the zones above and below. The material is questionable as concrete aggregate.

I Fresh rock. There is no visible sign of rock material weathering; there may perhaps be slight discolouration on major discontinuity surfaces.

Staining indicates water percolation along joints. Individual pieces may be loosened by blasting or stress relief and support may be required in tunnels and shafts.

1.5 SOIL DEPOSITS BASED ON ORIGIN

Soils can be divided into two broad groups — transported or residual — depending on the method of deposition. "Transported soils" are the soils which have been moved from their original place of formation to other. places of deposition. Transporting agencies may be glaciers, water, wind, or gravity, acting either singularly or in combination. The method of transportation and deposition has a marked influence on the properties and behaviour of the resulting soil deposition. Soils remaining at the location of their original formation are the "residual soils".

1. **Glacial deposits.** They include materials transported and redeposited by glacial ice or by melt waters flowing from glaciers. The size composition may vary from boulders to clays. Differentiation is made by using the following terminology:

(a) *Moraine*—glacial material deposited by ice rather than by its melt waters.

(b) *Drift*—superficial material of rock debris of any sort transported by glaciers and deposited either directly from the ice or from the melt water.

(c) *Till or boulder clay*—unstratified, heterogeneous mixtures of clay, silt, sand, gravel and boulders, directly deposited by ice, without transportation or sorting by water.

(d) *Glaciofluvial deposits*—materials moved by glaciers and subsequently sorted and deposited with stratification by streams of melt water. Fine materials deposited in temporary glacial lakes are known as the "laminated or varved clays". They consist of alternating thin layers of medium grey silt and darker silty clay or clay. A pair of such layers is called "varve" and these are regarded as annual deposits indicating seasonal differences in sedimentation.

2. **River deposits.** The materials transported and redeposited by action of water is known as "alluvium". Typical river deposit consists of fine grained material of recent origin overlying coarser

strata dating from earlier stage of river development. The deposits usually have pronounced stratification. The grain size varies from very coarse fraction (boulders or cobbles) to gravel, sand, silt and some clay. Alluvial deposits are usually poorly graded.

3. **Lacustrine and marine deposits.** Fine grained materials like silts and clays are deposited in layers when flowing water comes to rest, as in lakes, estuaries and deltas. Soils deposited from suspension in quiet, fresh water lakes are the "lacustrine soils" and those deposited in salt water are the "marine soils". These deposits may contain appreciable amounts of organic matter and exhibit medium to very high compressibility. If rich in calcium carbonate, the deposit is called "marl".

4. **Wind deposits.** Soil materials transported and deposited by wind are called "aeolian deposits"; dune sand and loess are the two typical deposits.

(a) *Dune sand*. Dune is a deposit of sand formed by wind action. Wind transports sand particles (> 0.05 mm) either by rolling along the surface or by lifting within a few metres of the ground only to be deposited a short distance away as soon as the wind slackens. Dunes generally occur in sandy desert areas and along sea shore or lakes having sandy beaches. Continual shifting or migration in the direction of prevailing wind is an important characteristic of sand dunes and various techniques are adopted to stop this shifting, although not always successful.

Dune sands are composed of relatively uniform. rounded to subrounded particles of fine to medium sand. Non-plastic fine fraction (minus 75 micron) in typical Rajasthan dune sand is found to vary from 4 to 15 percent and the angle of shearing resistance varies from 32 to 35 degrees; the critical void ratio (Chapter 10) ranges from 83 to 72 percent under normal stress varying from 20 to 500 kPa (0.2 to 5 kg/cm^2). (Singh and Maheshwari 1977).

(b) *Loess*. Loess is the most uniform in gradation of all principal soil types, consisting of 50—90% particles of silt size. Loess deposition is presumed to have occurred in the semi-arid grass-lands, the grasses having left behind fossil root-holes, which are typical of loess. The result is a high vertical porosity and cleavage combined with an extremely loose, metastable structure. The permeability in appreciably higher in the vertical than in the horizontal direction. When dry, loess is reasonably strong and incompressible due to the presence of a binder which may be predominantly calcareous or clayey. A vertical cut is stable. On saturation, however, loess deposits become soft and may loose their stability; such materials are termed "collapsible soils". The liquid limit of loess averages about 30, plastic index ranges from about 4 to 9 (average 6) and the angle of shearing resistance varies from 30 to 34 degrees (Bell 1978).

5. Residual soils. Residual soils are the product of rock weathering which remain insitu with little or no alteration by transport, The deposits almost invariably become more compact, rockier and less weathered with increasing depth. The composition of residual materials may be very variable from large fragments to gravel, sand, silt, clay and colloids. Complete weathering leads to the formation of clay of a type depending on the parent rock and the weathering process, plus varying amounts of resistant silica product; e.g. the expansive "black cotton soil" results from the decomposition insitu of basalt and similar rocks, under alkaline conditions which are generally the result of poor drainage. Similarly, in tropical and semitropical climates, with heavy rains, high temperatures and good drainage, some materials from weathered igneous rocks (and also metamorphic, Moh 1969) are leached out and high iron and alumina contents are left behind. The process is known as "laterization", and the soil as "lateritic".

6. Organic soils. Organic soils are formed insitu by growtn and decay of plants. Accumulations of decomposed or partially decomposed plant matter in swamps or marshes are termed "muck" or "peat". Muck, geologically older than peat, is almost fully decomposed and relatively dense. Peat is partially decomposed vegetation and is normally spongy and light. Peats are dark brown or black, loose and highly compressible. The void ratio is high (normally 5—15, but may be as high as 25; Bell 1978) and the moisture content is 50 to 1000 percent or more (Mitchell 1976).

7. Agricultural soils. The weathered uppermost layers oɪ the surface deposits which are of importance to the growth and support of plant life constitute the agricultural soils. The thickness of these layers does not generally exceed one metre, although in some places roots may go deeper. The development of soil is usually in the form of layers or "horizons", with distinctive colour and granular composition. The uppermost layer, called "A-horizon" is usually darker in colour, contains decayed organic matter (*humus*) and is fertile. It is subjected to maximum bacterial activity and is altered by natural processes such as leaching and washing away of materials. The "B-horizon" which lies below the A-horizon represents the layer of accumulation or deposition of the material washed down. The third layer, "C-horizon", lying below the B-horizon contains soil that is unaltered by weathering subsequent to deposition or formation. 'A' and 'B' horizons constitute the "top soil" which overlies the C-horizon or the "sub-soil". The development of 'A' and 'B' horizon in the soil profile is of primary importance, which is determined by climate, vegetation, parent material, topography and time.

1.6 SOIL PARTICLES

1. Nomenclature and size. Soil particles may be either 'inorganic' or 'organic'; the former are derived from weathering of rocks and the latter from plant or animal remains. Inorganic soil particles may vary in size within very wide limits from boulders to colloidal clay particles. A broad division can be made into "coarse" and "fine" particles depending on whether they are visible to the naked eye or not, which is roughly the 75 micron size (one micron $= 1 \mu m = 1 \times 10^{-3}$ mm). The coarse particles are further designated as "boulder" "cobble" "gravel" and "sand" and the fine particles as "silt" and "clay". The nomenclature of particles and their size limits are given in Table 1.2.

Table 1.2
SOIL PARTICLES AND SIZE LIMITS

Nomenclature		Size limits (mm)	
(1)	(2)	(3)	(4)
Boulder	Over 300	Over 200	—
Cobble	75—300	60—200	—
Gravel	4.75—75	2—60	Over 2
Sand	0.075—4.75	0.06—2	0.05—2
Silt	Less than 0.075	0.002—0.06	0.002—0.05
Clay	Less than 0.075	Less than 0.002	Less than 0.002

Column 2 gives the size limit according to the Unified Soil Classification System (Wagner 1957) and also adopted as the Indian Standard (IS:1948). Both silt and clay particles are smaller than 75 micron size and they are to be identified essentially by their behaviour. Column 3 is according to the British Soil Classification System (CP:2001, BS:1377) and column 4 according to the U.S. Department of Agriculture (Hillel 1971). Sand, silt and clay sizes according to the International Soil Science Society are also the same as given in column 3 (Hillel 1971).

2. **Composition.** The coarse particles — boulder, cobble, gravel and sand — have the same mineralogical composition as the parent rock from which they are derived by weathering processes or by mechanical disintegration. Silt particles, though classified as 'fines' have the same mineralogical composition as the coarse particles and they rarely break down to less than 2 micron size. Silt particles are mainly the product of grinding action between larger rock fragments during their transportation by glacier, water or wind. The least plastic varieties, sometimes called "rock flour", generally consist of more or less equi-dimensional grains of quaitz. The plastic types are formed by chemical weathering. All true clay particles mainly consist of 'clay minerals' which are the product of chemical weathering. These particles are very small, their main dimension being seldom more than 0.002 mm, and frequently very much less. They are capable of developing cohesion and plasticity.

3. **Shape and surface.** Soil particles are commonly of either of the two shapes, "bulky" or "flaky". When the length, width and thickness of the particles is of the same order of magnitude (i.e. one dimension is no more than 2 or 3 times larger or smaller than another dimension), the shape is 'bulky'. Coarse particles, and also silt particles, are bulky particles. Their degree of rounding at edges and corners is described in terms of angularity; the particle may be "angular", "subangular", "sub-rounded" or "rounded". Angular: corners and edges sharp and unworn. Subangular: corners worn off, angles not worn off. Subrounded: corners and angles worn off, flat surfaces remain. Rounded: worn to almost spherical shape.

Clay particles are commonly 'flaky' or in the form of flat

plates, although needles, tubes or rods may occur. The ratio of the length (diameter) of the plates to their thickness may vary between limits as wide as 10:1 and 250:1 (Kezdi 1974). They may be made of many sheets on top of one another. They are mostly found in sizes less then 2 μm or easily break down to this size. It may, however, be emphasized that it is the mineral type and not the small size which is primarily responsible for the cohesion and plasticity of clay particles. Where a soil particle, although broken down to clay size (less than 2 μm), does not possess clay properties, it is composed of 'non-clay' minerals.

The surface of soil particles (fine particles) is the seat of many physico-chemical phenomena which affect the nature and behaviour of the soil. The magnitude of the surface of particles in relation to the mass is expressed by the "specific surface", i.e. the surface area per unit mass. The smaller the particle in size and flater in shape, the greatei is its specific surface. Thus, sand particles of about 1 mm in diameter have a specific surface around 0.002 m²/g, and kaolin and bentonite (two types of clays) have specific surfaces of 80 m²/g and 1300—1390 m²/g respectively (Priklonskii 1949).

4. Behaviour. The behaviour of bulky particles is governed primarily by gravity forces or mass energy. Grading and state oi packing have an important influence on the properties of a soil which is composed mainly of bulky particles. The shape of particles also has some effect on the properties, but the mineral composition is of insignificance. On the other hand, the shape, size, specific surface and mineralogical composition all influence the engineering behaviour of clay particles. Electro-chemical forces rather than gravity forces, are predominant in determining their properties

Organic matter has properties which are undesirable in engineering, e.g. its large water absorption capacity (upto 5 times its own mass), large volume changes on loading and expulsion of water, low shear strength and deletarious reaction with other engineering materials such as cement, etc.

Certain soils may also have considerable quantity of cementing material such as calcite, iron oxide or silica, deposited on the surface of the soil particles. Cementing increases the shear strength and reduces the compressibility of the soil.

1.7 CLAY MINERALOGY

A "mineral" is an inorganic chemical compound formed in nature. As a solid it may occur in an amorphous state or in a crystalline state. A "crystal" is a homogeneous body bounded by smooth plane surfaces. Soil particles are largely composed of mineral crystals. Molecules of minerals are composed of atoms of chemical elements. The atoms in a crystal are arranged in a definite orderly manner to form a three dimensional net-work, called a "lattice". An atom consists of a small nucleus having a positive electromagnetic charge around which a definite number of negatively charged electrons rotate. The electrons rotate in orbits of different radii forming the so-called electron shells.

Many compounds loose their identity, in solution by separating into "ions". The ions consist of only one element of the compound or of two or more elements which are not electrically balanced. Atoms get transformed into ions by the gain or loss of electrons. The positively charged ions are called "cations" and the negatively charged ions are called "anions". On removal from solution the cations and anions unite to form the original solid compound. Many elements do not form ions, yet they unite to form compounds. Solutions of non-ion forming elements or compounds in water are poor conductors of electric current.

1. Atomic and molecular bonds. Forces which bind atoms and molecules to build up the structure of substances are primarily of electrical nature. They may be broadly classified into "primary bonds" and "secondary bonds". Primary bonds combine the atoms into molecules. Secondary bonds link atoms in one molecule to atoms in another; they are much weaker than the primary bonds. Primary bonds are the ionic bond and the covalent bond. Secondary bonds are the hydrogen bond and the Van der Waals bond.

(a) *Ionic bond.* The ionic bond is the simplest and strongest of the bonds which holds atoms together. This bond is formed between oppositely charged ions by the exchange of electrons. Atoms held together by ionic bonds form 'ionic compounds', e.g. common salt (sodium chloride), and a majority of clay mineral crystals fall into this group.

Ionic bonding causes a separation between centres of positive and negative charge in a molecule, which tends the molecule to orient in an electric field forming a 'dipole'. Dipole is the arrangement of two equal electro-static charges of opposite sign. A dipolar (or simply polar) molecule is one which is neutral but in which the centres of positive and negative charges are separated such that the molecule behaves like a short bar magnet with positive and negative poles.

(b) *Covalent bond.* The covalent bond is formed when one or more bonding electrons are shared by two atoms so that they serve to complete the outer shell for each atom. The covalent bond exists generally in non-electrolytes (elements which do not form ions, such as carbon) and between elements of negative valence.

Covalent bond usually occurs in combination with ionic bond, e.g. the atomic bonding in silica ($Si O_2$) (which is the most abundant constituents of most soils) is about half ionic and half covalent (Mitchell 1976). Covalent and ionic bonds are of approximately the same strength. Covalent bonding may also produce dipolar molecules.

(c) *Hydrogen bond.* A hydrogen bond is formed when the hydrogen atom in one molecule or crystal can attract two electronegative atoms (anions). For example, it can link the oxygen from a water molecule to the oxygen on the clay particle surface. Hydrogen bonding between two oxygen atoms (anions) is responsible for some of the weaker bonds between crystal layers (e.g. in the clay mineral kaolinite), for holding water at the clay surface and for bonding organic molecules to the clay surface.

(d) *Van der Waals bonds.* These are forces of attraction between uncharged molecules, and are of three types. The first type is the force of attraction between the oppositely charged ends of permanent dipoles on orientation. For examples, water molecules (which are permanent dipoles) are attracted together as oriented dipoles shown in Fig. 1.1. Secondly, the attractive force may also result between permanent dipoles and dipoles induced by them in adjacent molecules which are orginally non-polar. This is known as the "induction effect". Thirdly, even molecules which are non-polar can behave as instantaneous, fluctuating dipoles due to the constant spinning of electrons, (as at any time there may be more electrons on one side of an atomic nucleus than the other). Attraction between such molecules is known as the "dispersion effect". Van der Waals forces are thus contributed by the three effects — orientation, dispersion and induction, the maximum contribution being by orientation and the minimum by induction.

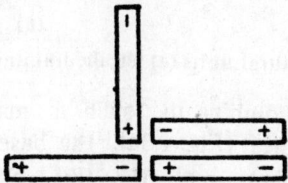

Fig 1.1   Oriented dipoles

The relative strengths of the various kinds of bonds are given in Table 1.3 (Kezdi 1974).

Table 1.3

RELATIVE STRENGTHS OF BONDS (KEZDI 1974)

Bond	Relative strength
Ionic and covalent bonds	40—400
Hydrozen bond	10—20
Van der Waals bonds	1—10

2. **Basic structural units**. The clay minerals are a group of complex alumino-silicates, i.e. oxides of aluminium and silicon with smaller amounts of metal ions substituted within the crystal. The atomic structures of clay minerals are built up of two basic units — silica tetrahedral units and aluminium (or magnesium) octahedral unit, held together by ionic bonds.

The silica unit consists of a silicon ion (Si^{4+}) surrounded by four oxygen ions (O^{2-}) arranged in the form of a tetrahedron (Fig. 1.2a). The octahedral unit has an aluminium ion (Al^{3+}) or a magnesium ion (Mg^{2+}) enclosed by six hydroxyl radicals (OH) or oxygens arranged in the form of an octahedron (Fig. 1.2b).

In some cases other cations (e.g. Fe) are present in place of Al and Mg.

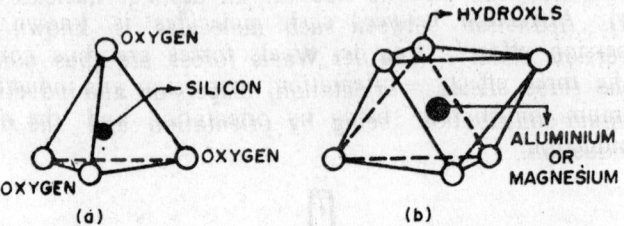

Fig. 1.2   Basic structural units (a) tetrahedral unit (b) octahedral unit

The basic units combine in such a manner as to form a *sheet*. In the silica sheet (Fig. 1.3a), the bases of the tetrahedrals are all in the same plane and the tips all point in the same direction. Each of the three oxygens at the base is shared by two silicons of adjacent units. This sheet has a thickness of 4.63 A° in clay minerals (Grim 1968) (A° = Angstrom = 10^{-10}m). The unit has a negative charge of 1.0. Neutrality can be obtained by

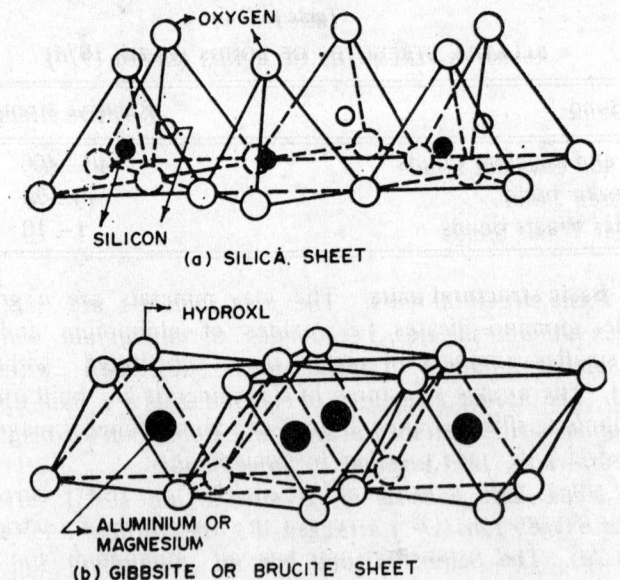

Fig. 1.3   Sheet structure of tetrahedral and octahedral units (Grim 1959)
(a) silica sheet   (b) gibbsite or brucite sheet

replacement of oxygen by hydroxyls or by union with a positively charged sheet of different composition. Combination of octahedral units forms an octahedral sheet, (Fig. 1.3b) which is called a "gibbsite" sheet if the central cation of the unit is aluminium or a "brucite' sheet if the central cation is magnesium. The sheet is 5.05 A° thick in clay minerals.

For describing clay mineral structure, the sheets are symbolically represented as shown in Fig. 1.4. A water layer is represented by 00000.

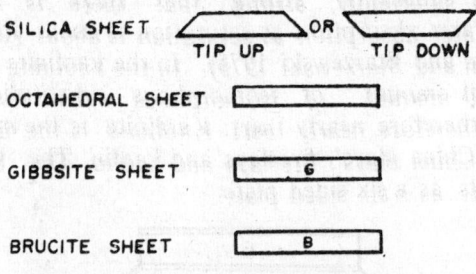

Fig. 1.4 Symbolic representation of sheets

3. **Clay minerals and their structure**. Clay minerals are classified on the basis of the component layers. In the formation of minerals the sheets of basic structural units get stacked one above the other. The stack of sheet is known as a "unit-cell" "unit cell layer" or a "basic layer" which may be composed of 2, 3, or 4 sheets. Bonding between the sheets is of the primary type (ionic and covalent) and thus very strong. The bonds holding the unit cell layers may be of several types. In the naturally occurring clay minerals, the phenomenon of 'isomorphous substitution' frequently occurs. Isomorphous (meaning 'same form') substitution consists in the substitution of the central cations of the basic units by cations other than those in the normal structure. Thus aluminium may take the place of silicon (in a tetrahedral unit), magnesium may replace aluminium and ferrous ion (Fe^{2+}) may replace magnesium; the crystal structure remaining unchanged. When the replacing ion is of lower valency, e.g. Al^{3+} replacing Si^{4+} and Mg^{2+} replacing Al^{3+}, the crystal is left with a net negative charge. The stacking of sheets into basic layers, the bonding between layers and the isomorphous substitution account for the different minerals. The most common

clay minerals can be divided into three main groups—kaolinite, montmorillonite and illite.

(a) *Kaolinite*. The basic layer of kaolinite consists of a gibbsite sheet joined to a silica sheet through the unbalanced oxygen atoms at the apexes of the silica sheet (primary bonding), as shown in Fig. 1.5. The kaolinite crystal consists of many such basic layers (may be 70—100 or more), one on top of the other. The linkage between two successive basic layers is both by the hydrogen bonds and the Van der Waals forces. This bonding is sufficiently strong that there is no inter layer swelling. Water absorption at saturation is about 90% of the dry mass (Wilun and Starzewski 1975). In the kaolinite mineral there is very small amount of isomorphous substitution, and the mineral is therefore nearly inert. Kaolinite is the most important mineral in 'China clays', fireclays and kaolin. The basic kaolinite crystal exists as a six sided plate.

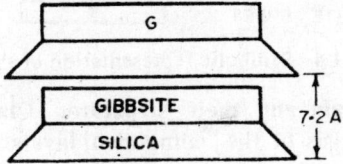

Fig. 1.5 Symbolic structure of kaolinite

Another important mineral in the kaolinite group is *halloysite* which has the similar basic structure as that of kaolinite, with the difference that it can occur in a hydrated form with a layer of water molecules between the basic layers, and which causes halloysite to exist in tubular shape. The hydrated form found in nature loses its water layer easily on heating and does not rehydrate on wetting. On drying this mineral behaves differently from kaolinite.

(b) *Montmorillonite*. The basic layer of montmorillonite consists of a gibbsite sheet between two silica sheets. Isomorphous substitution occurs mainly in the gibbsite sheet, with magnesium or iron replacing aluminium leading to charge deficiency which is made up by the absorption of exchangeable cations between the basic layers and on the surfaces of particles. Bonding between the successive layers is by Van der Waals forces and by the interlayer cations. The symbolic structure is shown in Fig 1.6.

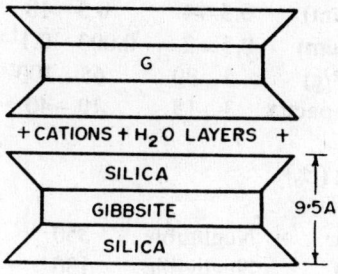

Fig. 1.6 Symbolic structure of montmorillonite

The bond between basic layers is relatively weak and depends on the type of exchangeable cations. Water can easily penetrate between the layers pushing them apart and causing very large volume changes within the crystal itself. The basal spacing between the layers is variable, ranging from about 9.6 A° to complete separation. Water absorption of montmorillonite particles at saturation is 300—700 percent. The montmorillonite clays have a high activity and high liquid limit. Montmorillonites are the main minerals in bentonites and fullers' earth and also in black cotton soil. The particles are platy in shape.

(c) *Illite.* The basic structure is similar to that of montmorillonite, i.e. the basic layer comprises a gibbsite sheet sandwitched between two silica sheets. There is a pronounced substitution of aluminium for silicon in the silica sheet, which results in a larger net negative charge than in case of montmorillonite. The charge deficiency is balanced partly by the non-exchangeable potassium cations. The interlayer bonding by potassium is sufficiently strong and the mineral swells much less than montmorillonite. The basic layer thickness is 10 A° and the particles are platy.

Some of the properties of clay minerals are given in Table 1.4.

Table 1.4

PROPERTIES OF CLAY MINERALS (GROMKO 1974)

Property	Kaolinite	Illite	Montmorillonite
Particle diameter (μm)	0.5—4	0.5—10	0.5—10
Particle thickness (μm)	0.5—2	0.003—0.1	9.5 A°
Specific surface (m²/g)	5—30	65—100	600—800
Cation exchange capacity (meq/100 g)	3—15	10—40	80—150
Maximum swelling (%) for surcharge pressure			
9.5 kPa (0.98 t/m²)	Negligible	350	1500
19 kPa (1.96 t/m²)	Negligible	150	350

4. Water and ion adsorption. Surfaces of clay minerals are termed 'active' as they are capable of attracting and holding water molecules and ions by electro-chemical forces. Clay minerals carry an unbalanced negative charge and are, therefore, *polar*, i.e. they are capable of producing electro fields around them without any contributory external cause. Unbalanced charge arises mainly due to two sources — internally unbalanced negative charge due to isomorphous replacement of one ion for another in the crystal lattice and the surface unbalanced charge due to incomplete charge neutralization of terminal atoms on lattice edges. The charge distribution is generally negative at the mineral faces and positive at the edges and there are localized areas of high and low charge. These charges are balanced externally by electro-chemical attraction, or "adsorption", of polar molecules (e.g. water dipoles) and exchangeable ions (mostly cations) from the aqueous environment.

A number of mechanisms are involved in adsorption of water to clay minerals, such as hydrogen bonding, dipole attraction, osmotic attraction, hydration of cations, etc. As clay mineral surfaces usually comprise a layer of either oxygens or hydroxyls, hydrogen bonding of water molecules occurs, which is probably the main force of attraction (Yong and Warkentin 1975). Water molecules being dipolar get attracted to the negatively charged mineral surface electrostatically, with their positive poles directed towards the surface. Since cations are attracted to clay surface, they also hold some of the water as their water of hydration. This mechanism is most important at low moisture contents. There is a concentration of cations near the clay surface and water molecules tend to diffuse towards the surface (attraction by osmosis) in an attempt to equalize concentration. Van der Waals forces probably also contribute to attraction of water. The exact arrangement of the attracted water molecules is disputed, but it is certain that they are tightly bound against the crystal face in a fairly regular pattern, or in an oriented manner with the degree of orientation decreasing with distance from the centre.

Within the layer of bound water, a number of ions (mostly cations) are also embedded. Nearer the surface, the greater is the concentration of cations and lesser are the anions. The negative force field of the particle attracts the cations and repels the anions. The swarm of cations consists partly of a layer more or less fixed in the proximity of the particle surface and partly of a diffuse distribution extending to some distance away from the surface. The diffuse layer of cations together with the negatively charged surface of the particles is called a "diffuse double layer" or an "electrical double layer". The water of the double layer which is attracted by physico-chemical forces around the soil

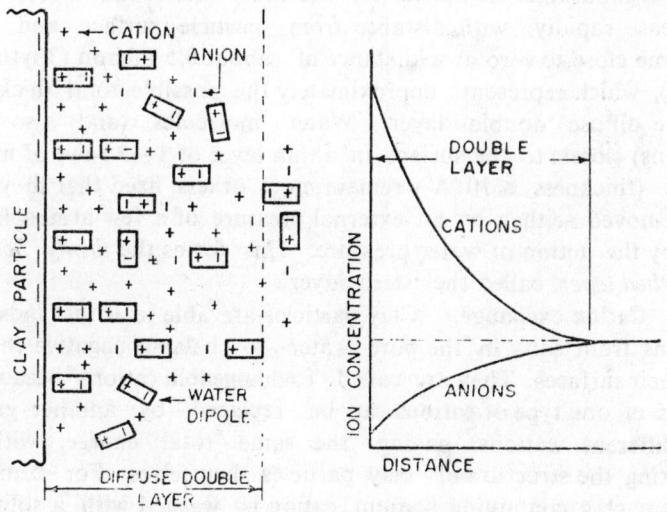

Fig. 1.7 Diffuse double layer and ion concentration

particle is known as the   "double layer water" or "adsorbed water". The diffuse double layer and ion concentration are symbolically shown in Fig. 1.7.

The thickness of the electrical double layer depends on the concentration of ions and the type of cations. Increasing ion concentration in the soil water increases the number of cations near the particle surface and, therefore, reduces the layer thickness needed to neutralize surface charge. Monovalent cations (e.g. Na^+) lead to thicker layer than the divalent cations (Ca^{2+}), because twice as many of the former are needed to balance a given charge. The thickness of double layer water for a typical kaolinite particle (10,000 by 1000 A°) was 400 A° and for a typical montmorillonite particle (1000 by 10 A°) 200 A° (Lambe 1960).

The properties of water close to the surface (i.e. adsorbed water) differ from those of free water. The density of adsorbed water in the first molecular layer may be of the order of 1.4 t/m^3 and the viscosity may be 100 times greater than that of free water (Yong and Warkentin 1975). The forces of attraction at the particle surface are very strong and may reach the value of several thousands of kg/cm^2 for the first molecular layer. They decrease rapidly with distance from particle surface and may become close to zero at a distance of about 0.5 micron (Tsytovich 1976), which represents approximately the possible total thickness of the diffuse double layer. Water molecules (and also the cations) closest to the surface in a thin layer of 1—3 rows of mole-cules (thickness $\leqslant 10$ A°) remain more or less fixed that they can be removed neither by an external pressure of a few atmospheres nor by the action of water pressure. This forms the *firmly bonded adsorbed layer*, called the 'stern layer'.

5. **Cation exchange.**   Clay particles are able to attract (adsorb) cations from salts in the pore water to balance negative charge on their surfaces. They are called 'exchangeable cations' because a group of one type of cations can be replaced by another group of different cations having the same total charge, without affecting the structure of clay particles themselves. For example, when a clay containing sodium cation is washed with a solution of calcium chloride, each calcium ion (Ca^{2+}) replaces two sodium ions (Na^+) and the sodium is washed out in the solution. The

reaction is indicated as below:

$$Na_2\text{-Clay} + CaCl_2 \Leftrightarrow Ca\text{-Clay} + 2NaCl$$

This process is called "cation exchange" or "base exchange". The quantity of exchangeable cations held by a clay is known its 'cation exchange capacity' and is usually expressed in milligram equivalent or milliequivalent per 100 g of the mineral (meq %).

The most commonly found exchangeable cations in soils, in decreasing order of abundance, are calcium (Ca^{2+}), magnesium (Mg^{2+}), sodium (Na^+) and potassium (K^+). Acid soils contain aluminium and hydrogen. The case with which a cation can be replaced by another depends mainly on the valency, relative concentration of the different types and the cation size. Normally, trivalent cations are held more tightly than divalent and divalent more tightly than monovalent. Ordinarily, small cations tend to replace large cations. The cations can be arranged approximately in the following order of increasing replacing power:

$$Li^+ < Na^+ < K^+ < Mg^{2+} < Ca^{2+} < Al^{3+}$$

For natural soils the cation exchange capacity varies between 0 to 40 meq per 100 g. Typical values for clay minerals are given in Table 1.4. The kind of adsorbed cations at the clay mineral surfaces and their exchange have a far reaching influence on the properties of clays. For example, monovalent cations such as sodium increase the activity of clay, its swelling, etc. Organic matter also contributes to the cation exchange capacity of organic and surface soils.

6. Interaction of particles. Clay particles approaching one another in a suspension interact through adsorbed water layer and through diffuse layer of cations. In some cases, the interaction is through direct particle contact. Forces of attraction and repulsion play an important role in interaction.

(a) *Attraction.* Attraction develops due to Van der Waals forces, but these are short range forces decreasing very rapidly with increasing distance of separation. They manifest when the particles are brought closer together by drying or by consolidation. Net attraction also results when the particles have a separation of less than 15 A° due to a more or less uniform distribution of exchangeable ions in the interparticle space rather than the ions being separated into two diffuse layers, one associated with each surface.

(b) *Repulsion.* Repulsion is due to water adsorption and interaction of two positively ionised diffuse layers. Adsorption of water molecules on the surface forces adjacent particles apart. This is manifested in swelling of clay at low moisture contents where adsorbed water is strongly held. At interparticles separation exceeding about 15 A°, diffuse ion layers form, with a resulting net repulsion. Repulsion will be greatest with monovalent ions in the diffuse layer and with distilled water as pore water.

(c) *Net force.* The interparticle net force varies with the particle separation and the ion concentration in the soil water. At very small separations, the Van der Waals forces of attractions are always the larger. As separation increases, repulsive forces increase and the Van der Waals forces become insignificant. Lower the ion concentration and lower the cation valency, the thicker is the diffuse double layer and greater will be the repulsion. The net force between two particles can be represented as shown in Fig. 1.8.

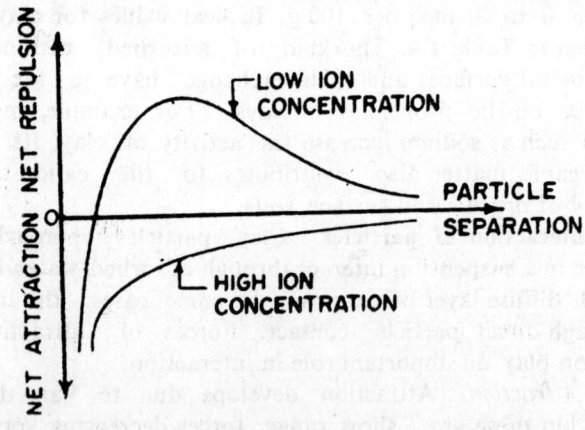

Fig. 1.8 Interparticle net force in suspension

Attraction and repulsion in clay soils are thus variable but the maximum repulsion is greater than the maximum attraction. Attraction can manifest only if the conditions do not favour repulsion.

(d) *Flocculation and dispersion*.   Particles approaching face-to-face develop greater repulsion than in the edge-to-face approach. As a result, particles settling in a suspension may tend to develop the edge-to-face arrangement, as shown in Fig. 1.9, and the particles are said to "flocculate".

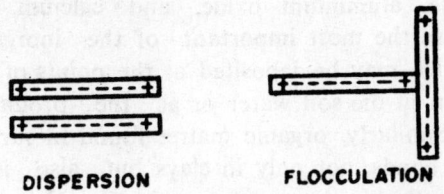

DISPERSION   **FLOCCULATION**

Fig. 1.9   Face-to-face and edge-to-face arrangements

The tendency of flocculation or otherwise is greatly affected by the pH value of the medium (*Note*. $pH = 1/\log_{10} H^+$ concentration: $pH < 7$ means acidic and high H^+ concentration and $pH > 7$ means alkaline and low H^+ concentration).

A low pH (acid medium) promotes a positive edge to negative surface interaction, leading to flocculation. Ion concentration (presence of salt) reduces double layer thickness and thus increases attraction and flocculation.

If the net force during the course of sedimentation is repulsion, the particles will settle in a 'dispersed' state with absence of edge-to-face contacts. The forces of attraction between particles in suspension cannot be easily varied. The forces of repulsion (and hence dispersion) can, however, be easily increased by creating conditions such as high pH, low salt concentration, monovalent exchangeable cations, and high water contents to increase interparticles distance. On the other hand, if repulsion is decreased, e.g. by adding salt, the particles will again flocculate. Flocculation and dispersion of clay particles in suspension are thus reversible processes.

(e) *Interparticle bonds*.   The most important forces holding soil particles together at field moisture content are the bonds which gradually develop at points of contact between particles. The nature of these bonds is not quite certain, but they are probably due to orientation of adsorbad water and cementation due to non-clay materials.

Although the water molecules on the particle surface are firmly bonded and cannot be displaced, it is probable that the cations embedded in the adsorbed layer are driven away from the points of contact allowing the water molecules to develop a stronger and more regular structure than that existing else-where in the adsorbed layer (Scott 1974). This orientation of adsorbed water becomes the cause of bonding.

Iron oxide, aluminium oxide, and calcium or magnesium carbonates are the most important of the inorganic cementing materials which may be deposited at the points of contact, either from solution in the soil water or as the product of chemical weathering, Similarly, organic matter found in surface soils forms interparticle bonds, not only in clays but also in-between clay particles and other bulky particles such as sand, as well as between sand particles themselves.

1.8 COHESIVE AND COHESIONLESS SOILS

By 'cohesion' is meant the shearing resistance inherent in a soil which does not require any normal pressure or other outside influence for its development. It is the property which holds the particles together in a soil mass mainly due to interparticle molecular attraction and bonds. A soil in which interparticle attraction and adsorbed water work together to produce a mass that holds together and deforms plastically at varying moisture contents is called a "cohesive soil". It is the clay minerals which impart cohesive property to a soil, cohesion increasing with increasing proportion of clay minerals in a soil. Hence these soils are also known as "clay soils" or "clayey soils".

A soil which does not exhibit cohesion is termed 'cohesionless' or "non-cohesive". Cohesionless soils possess no shearing resistance, except as developed by normal pressure between their particles. Soils composed of bulky particles are cohesionless regardless of the fineness of the particles. These soils are also known as "granular soils". Composite soils are mixtures of cohesive and cohesionless soils.

2
Basic properties

2.1 PHASE RELATIONSHIPS

Soil mass is composed of matter existing generally in three states or phases — solid, liquid, and gaseous. Soil particles form the solid phase, water forms the liquid phase, and air (water vapour or gases) the gaseous phase. These phases are symbolically represented in Fig. 2.1. Water and air constitute the voids or pore space of a soil mass.

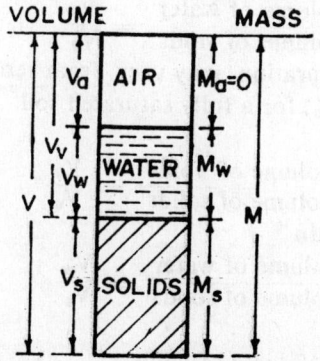

Fig. 2.1 Phase diagram of soil mass

1. Density and unit weight. Density ρ is the ratio of mass M to volume V, and the unit weight γ is the ratio of weight W (a force) to volume V:

$$\rho = \frac{M}{V} \text{ and } \gamma = \frac{W}{V} = \frac{Mg}{V} \tag{2.1}$$

or $\qquad \gamma = \rho g \tag{2.2}$

where g = gravitational acceleration $\simeq$ 9.8 m/s^2

Density is usually expressed in g/cm^3 (g/ml) or t/m^3 (Mg/m^3) and the unit weight is expressed in kN/m^3 (1 g/cm^3 = 1 t/m^3 =9.8 kN/m^3)

For practical purposes the density of water ρ_w = 1 g/cm^3 (1 t/m^3) and the unit weight of water γ_w = 9.8 kN/m^3.

2. **Density of solids (mean particle density)**

$$\rho_s = \frac{\text{Mass of solid particles}}{\text{Volume of solid particles}} = \frac{M_s}{V_s} \qquad (2.3)$$

3. **Relative density (specific gravity).** The relative density of *solid soil particles* G is the ratio of the density of solids to the density of water:

$$G = \rho_s / \rho_w = \gamma_s / \gamma_w \qquad (2.4)$$

or $\qquad \rho_s = G\rho_w$ and $\gamma_s = G\gamma_w \qquad\qquad$ (2.5)

where $\qquad \gamma_s$ = unit weight of solids = $W_s / V_s \qquad$ (2.6)

4. **Moisture content (water content)**

$$w = \frac{\text{Mass of water}}{\text{Mass of solids}} = \frac{M_w}{M_s} \qquad (2.7)$$

5. **Degree of saturation**

$$S_r = \frac{\text{Volume of water}}{\text{Volume of voids}} = \frac{V_w}{V_v} \qquad (2.8)$$

The degree of saturation may vary from zero for a completely dry soil to 1 (or 100%) for a fully saturated soil.

6. **Void ratio**

$$e = \frac{\text{Volume of voids}}{\text{Volume of solids}} = \frac{V_v}{V_s} \qquad (2.9)$$

7. **Water void ratio**

$$e_w = \frac{\text{Volume of water}}{\text{Volume of solids}} = \frac{V_w}{V_s} \qquad (2.10)$$

8. **Porosity**

$$n = \frac{\text{Volume of voids}}{\text{Total volume of soil}} = \frac{V_v}{V} \qquad (2.11)$$

9. **Air content**

$$n_a = \frac{\text{Volume of air}}{\text{Total volume of soil}} = \frac{V_a}{V} \qquad (2.12)$$

Alternative terms used for air content are "air-filled porosity" and "percentage air voids".

From the definition of void ratio, if the volume of solids is 1 unit, the volume of voids will be 'e' units and the total volume of soil will be '1+e' units. The mass of solids is then $G\rho_w$ and, from the definition of moisture content, the mass of water is $wG\rho_w$ and the volume of water is wG.   Similarly, if voids are represented by 'n' units, the total volume is 1 unit and the volume of solids is '1−n' units.   The corresponding phase diagrams are shown in Fig. 2.2.

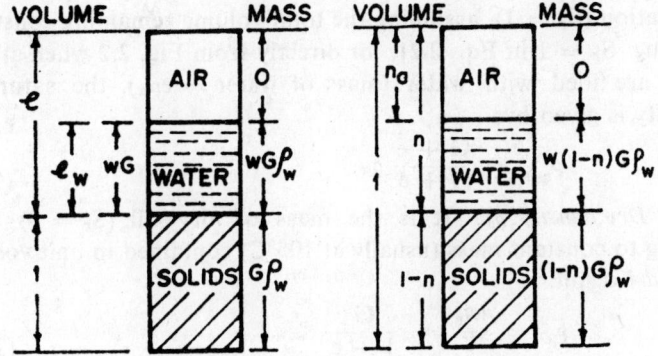

Fig. 2.2 Phase diagrams in terms of e and n

From Fig. 2.2, the following inter-relations can be established.

$$e = \frac{n}{1-n} \tag{2.13}$$

$$n = \frac{e}{1+e} \tag{2.14}$$

$$e_w = wG \tag{2.15}$$

$$S_r = \frac{e_w}{e} = \frac{wG}{e} \tag{2.16}$$

or $\qquad e = \dfrac{wG}{S_r} \tag{2.17}$

At full saturation, $S_r = 1$
and $\qquad e = w_{sat}G \tag{2.18}$
where $\quad w_{sat}$ = moisture content at full saturation

$$n_a = \frac{e - e_w}{1+e} = \frac{e(1-S_r)}{1+e} = n(1-S_r) \tag{2.19}$$

10. Densities. (a) *Bulk density* ρ. It is the mass of soil (solid particles + any contained water) per unit volume including voids.   The bulk density of a partly saturated soil can be expressed from Fig. 2.2 as follows:

$$\rho = \frac{M}{V} = \frac{G(1+w)}{1+e} \rho_w \qquad (2.20)$$

or from Eq. 2.16:

$$\rho = \frac{G + eS_r}{1+e} \rho_w \qquad (2.21)$$

(b) *Saturated density* ρ_{sat}. It is the bulk density at full saturation ($S_r = 1$), assuming the total volume remaining constant. Putting $S_r = 1$ in Eq. 2.21, or directly from Fig. 2.2 when all the voids are filled with water (mass of water $= e\rho_w$), the saturated density is given by:

$$\rho_{sat} = \frac{G + e}{1+e} \rho_w \qquad (2.22)$$

(c) *Dry density* ρ_d. It is the mass of dry soil ($S_r = 0$) after drying to constant mass (usually at $105°C$) contained in unit volume of *undried* soil.

$$\rho_d = \frac{M_s}{V} = \frac{G}{1+e} \,^w \qquad (2.23)$$

and

$$e = \frac{G\rho_w}{\rho_d} - 1 \qquad (2.24)$$

(d) *Submerged density* ρ'. When a soil is below water table (submerged), the solid soil particles (volume 1 unit, mass $G\rho_w$) are subjected to buoyant effect of water. This upthrust equals the mass of the volume of water displaced (ρ_w). Hence,

$$\rho' = \frac{G\rho_w - \rho_w}{1+e} = \frac{G-1}{1+e} \rho_w \qquad (2.25)$$

or

$$\rho' = \rho_{sat} - \rho_w \qquad (2.26)$$

Considering the phase diagram in terms of 'n' (Fig. 2.2b), ρ_d and ρ_{sat} can also be expressed as follows. The volume of soil is 1 unit.

$$\rho_d = (1 - n) G\rho_w \qquad (2.27)$$

and

$$\rho_{sat} = (1 - n) G\rho_w + n\rho_w = \rho_d + n\rho_w \qquad (2.28)$$

11. Unit weights. Similar expressions can be obtained for unit weights.

Bulk unit weight

$$\gamma = \frac{G(1+w)}{1+e} \gamma_w \qquad (2.29)$$

or

$$\gamma = \frac{G + eS_r}{1+e} \gamma_w \qquad (2.30)$$

Saturated unit weight $\gamma_{sat} = \dfrac{(G + e)}{1 + e} \gamma_w$ (2.31)

Dry unit weight $\gamma_d = \dfrac{G}{1 + e} \gamma_w$ (2.32)

Submerged unit weight $\gamma' = \dfrac{G - 1}{1 + e} \gamma_w$ (2.33)

or $\gamma' = \gamma_{sat} - \gamma_w$ (2.34)

12. Inter-relations

(a) Between ρ, ρ_d and w

$$\rho = \frac{M_s + M_w}{V} = \frac{M_s + wM_s}{V} = \frac{M_s}{V}(1 + w)$$

$$= \rho_d (1 + w) \tag{2.35a}$$

or $\rho_d = \rho/(1 + w)$ (2.35b)

Similarly, $\gamma_d = \gamma/(1 + w)$ (2.36)

(b) Between ρ_d, n_a (S_r), w and G

$$V = V_a + V_w + V_s$$

$$1 = \frac{V_a}{V} + \frac{wM_s}{V\rho_w} + \frac{M_s}{V\rho_s}$$

$$= n_a + \frac{w\rho_d}{\rho_w} + \frac{\rho_d}{G\rho_w}$$

or $\rho_d = \dfrac{G(1 - n_a)}{1 + wG} \rho_w$ (2.37)

At zero air voids, $n_a = 0$, $(w = w_{sat})$

and $\rho_d = \dfrac{G}{1 + w_{sat}G} \rho_w$ (2.38a)

The above relation (Eq. 2.38a) can also be obtained directly from Eq. 2.23 by putting $e = w_{sat} G$. It is referred to as the 'zero air voids dry density'.

If the soil is not fully saturated, $e = wG/S_r$ and

$$\rho_a = \frac{G\rho_w}{1 + e} = \frac{G\rho_w}{1 + wG/S_r} \tag{2.38b}$$

(c) Between ρ, ρ_d, ρ_{sat} and S_r

From Eq. 2.21: $\rho = \dfrac{G\rho_w}{1 + e} + S_r \dfrac{e\rho_w}{1 + e}$

$$\rho = \rho_d + S_r (\rho_{sat} - \rho_d) \tag{2.39}$$

13. Density index. In case of coarse grained soils, the insitu state of compaction is expressed in terms of density index I_D defined as:

$$I_D = \frac{e_{max} - e}{e_{max} - e_{min}} \qquad (2.40)$$

where e = natural void ratio

   e_{max} = maximum void ratio (loosest state)

   e_{min} = minimum void ratio (densest state)

Since the volume of soil of a given dry mass is proportional to $(1+e)$, density index can also be expressed as:

$$I_D = \frac{V_{max} - V}{V_{man} - V_{min}} \qquad (2.41)$$

where V = volume in the natural state

   V_{max} = maximum volume of soil of the same dry mass as in the natural state

   V_{min} = minimum volume of the same soil

Density index can be expressed in terms of density (or unit weight) also

$$I_D = \frac{1/\rho_{d\ min} - 1/\rho_d}{1/\rho_{d\ min} - 1/\rho_{d\ max}} \qquad (2.42)$$

where ρ_d = dry density in the natural state

   $\rho_{d\ min}$ = minimum dry density

   $\rho_{d\ max}$ = maximum dry density

The density index of a natural soil deposit very strongly affects its engineering behaviour. The density index may vary from zero in its loosest possible state ($e = e_{max}$) to 1 (100%) in its densest possible state ($e = e_{min}$). Qualitatively, it is described by terms 'very loose', 'loose', 'medium dense', 'dense' and 'very dense'. The insitu density index may be estimated from empirical correlations established with the standard penetration test SPT (Chapter 15) and the static cone penetration test CPT (Chapter 15) as given in Table 2.1 (Mitchell and Katti 1981). The table also gives corresponding values of dry unit weight and angle of internal friction.

Table 2.1

DESCRIPTION OF DENSITY INDEX AND CORRELATIONS
(MITCHELL AND KATTI 1981)

State of compaction	Density index (%)	SPT blows*	CPT resistance* (kg/cm²)	Dry unit weight (kN/m³)	Friction angle (degree)
Very loose	0 - 15	<4	<50	<14	<30
Loose	15 - 35	4 - 10	50 - 100	14 - 16	30 - 32
Medium dense (compact)	35 - 65	10 - 30	100 - 150	16 - 18	32 - 35
Dense	65 - 85	30 - 50	150 - 200	18 - 20	35 - 38
Very dense	85 - 100	>50	>200	>20	>38

*At an effective vertical overburden pressure of 100 kPa.

Examples 2.1 — 2.16

2.1 The dry density ρ_d of a soil is 1.8 g/cm³. What is its dry unit weight γ_d ? Find the bulk density ρ if the unit weight is 19.6 kN/m³,

$$\gamma_d = \rho_d g = 1.8 \times 9.8 = 17.64 \text{ kN/m}^3$$
$$\rho = \gamma/g = 2.0 \text{ g/cm}^3$$

2.2 The moisture content of a saturated soil w_{sat} is 35% and the relative density 'G' of its particles is 2.7. Find the void ratio 'e' and the porosity 'n'. What will be the degree of saturation S_r and the air content n_a if the moisture content 'w' gets reduced to 5% on drying ?

$$e = w_{sat} G = 0.945$$
$$n = \frac{e}{1 + e} = 0.486 = 48.6\%$$
$$S_r = \frac{wG}{e} = 0.143 = 14.3\%$$
$$n_a = n(1 - S_r) = 0.42 = 42\%$$

2.3 The mass M of an undisturbed sample is 1250 g and its volume V is 630 cm³. On oven drying the sample weighs 1102 g (M_s). The average relative density G of soil is 2.68. Find the bulk density ρ , natural moisture content w, dry density ρ_d, void ratio e, degree of saturation S_r and the air content n_a. What will be the density and unit weight on full saturation ?

$$\rho = \frac{M}{V} = 1.98 \text{ t/m}^3$$

$$w = \frac{M_w}{M_s} = \frac{M - M_s}{M_s} = 0.134 = 13.4\%$$

$$\rho_d = \frac{\rho}{1 + w} = 1.75 \text{ t/m}^3 \left(\text{or } \rho_d = \frac{M_s}{V} \right)$$

$$e = \frac{G\rho_v}{\rho_d} - 1 = 0.53$$

$$S_r = wG/e = 0.68$$

$$n_a = \frac{e}{1 + e} (1 - S_r) = 0.11$$

$$\rho_{sat} = \frac{G + e}{1 + e} \rho_w = 2.1 \text{ t/m}^3, (\rho_w = 1 \text{ t/m}^3)$$

$$\gamma_{sat} = \rho_{sat} \times 9.8 = 20.58 \text{ kN/m}^3$$

2.4 The dry unit weight γ_d of a soil having 15% moisture content is 17.5 kN/m³. Find bulk unit weight γ, saturated unit weight γ_{sat} and submerged unit weight γ'. Assume $G = 2.7$.

$$\gamma = \gamma_d (1 + w) = 20.13 \text{ kN/m}^3$$

$$e = \frac{G\gamma_w}{\gamma_d} - 1 = 0.51 \quad (\gamma_w = 9.8 \text{ kN/m}^3)$$

$$\gamma_{sat} = \frac{(G + e) \gamma_w}{1 + e} = 20.83 \text{ kN/m}^3$$

$$\gamma' = \frac{(G - 1) \gamma_w}{1 + e} = 11.03 \text{ kN/m}^3$$

or $\gamma' = \gamma_{sat} - \gamma_w = 11.03 \text{ kN/m}^3$

2.5 The bulk unit weight γ of a soil is 19.6 kN/m³, moisture content is 20% and the degree of saturation S_r is 80%. What will be the moisture content and unit weight on full saturation?

$$S_r = e_w/e = wG/w_{sat}G = w/w_{sat}; \; w_{sat} = 0.25$$

$$\gamma_{sat} = \gamma \frac{1 + w_{sat}}{1 + w} = 20.4 \text{ kN/m}^3$$

2.6 The unit weights of a soil in the dry and saturated states are 17.2 kN/m³ and 19.9 kN/m³ respectively. What will be the unit weight if the degree of saturation is 60% ?

$$\gamma = \gamma_d + S_r (\gamma_{sat} - \gamma_d) = 18.82 \text{ k N/m}^3$$

2.7 Find the dry density ρ_d and saturated density ρ_{sat} of a soil whose porosity n is 35% and the relative density G is 2.74.

$$\rho_d = (1 - n)\, G\rho_w = 1.78 \text{ g/cm}^3$$
$$\rho_{sat} = \rho_d + n\, \rho_w = 2.13 \text{ g/cm}^3$$

2.8 The insitu moisture content w_1 of a soil is 15% and the bulk density ρ_1 is 1.85 t/m³. Due to rains the moisture content w_2 increases to 20%. Find the bulk density ρ_2 and the bulk unit weight.

$$\rho = \rho_d\, (1 + w)\ ;\ \rho \propto (1 + w)$$
$$\rho_2 = \rho_1 \frac{1 + w_2}{1 + w_1} = 1.93 \text{ t/m}^3$$
$$\gamma = \rho_2\, g = 18.91 \text{ kN/m}^3$$

2.9 The void ratio e of a soil sample is 0.7 and the relative density G is 2.72. Find the moisture content w and the bulk unit weight γ, if the degree of saturation is 80%.

$$w = e\, S_r/G = 20.6\%$$
$$\gamma = \frac{G + e\, S_r}{1 + e}\, \gamma_w = 18.9 \text{ kN/m}^3$$

2.10 A soil is compacted at a moisture content w of 12% with 10% air content. Find the dry unit weight. The value of G is 2.65.

$$\gamma_d = G\, (1 - n_a)\, \gamma_w/(1 + w\, G) = 17.7 \text{ kN/m}^3$$

2.11 Find the dry density ρ_d and air content n_a of a soil compacted to a bulk density ρ of 1.98 g/cm³ at a moisture content of 14%. The value of G is 2.72.

$$\rho_d = \rho/(1 + w) = 1.74 \text{ g/m}^3$$
$$\rho_d = G\, (1 - n_a)\, \rho w/(1 + wG)\ ;\ n_a = 12\%$$

2.12 Find the moisture content w_{sat} necessary to fully saturate a soil having a dry density of 1.7 t/m³. The value of G is 2.65.

$$\rho_d = G\rho_w\, /\, (1 + e) = G\rho_w\, /\, (1 + w_{sat}G)\ ;\ w_{sat} = 21\%$$

2.13 Soil, 2000 m³ in volume (V_1), is excavated from a borrow pit where the natural moisture content w_1 is 15% and the unit weight γ is 17 kN/m³. The soil is used in an embankment compacted at a porosity n_2 of 30%. What will be the volume of compacted embankment, assuming no change in dry mass and moisture content of soil ? Relative density G is 2.68. Find also the degree of saturation and air content of the borrow pit and the embankment.

$$\gamma_d = \gamma/(1 + w_1) = 14.78 \text{ kN/m}^3$$

$$e_1 = \frac{G\gamma_w}{\gamma_d} - 1 = 0.777, \ n_1 = e_1/(1 + e_1) = 0.437$$

$$e_2 = n_2/(1 - n_2) = 0.429$$

$$V \propto (1 + e) \ ; \ V_2 = 2000 \ \frac{(1 + e_2)}{(1 + e_1)} = 1608 \text{ m}^3$$

$$S_{r1} = w_1 G/e_1 = 51.7\%$$
$$S_{r2} = w_1 G/e_2 = 93.7\%$$
$$n_{a1} = n_1 (1 - S_{r1}) = 21\%$$
$$n_{a2} = n_2 (1 - S_{r2}) = 1.9\%$$

2.14 A borrow area soil has a natural moisture content w_1 of 8% and a bulk density of 1.75 t/m³. Relative density of soil is 2.65. The soil is used for an embankment to be compacted at 15% moisture content w_2 to a dry density ρ_d of 1.8 t/m³. Find the amount of water to be added to 1.0 m³ of borrow soil. How many cubic metres of excavation is required for 1.0 m³ of compacted embankment ?

ρ_d (field) = 1.75/1.08 = 1.62 t/m³

Mass of dry soil in 1 m³ of borrow soil = 1.62 t

Mass of water = $w_1 M_s$ = 0.13 t

Mass of water needed at w_2 of 15% = $w_2 M_s$ = 0.243 t

Water to be added to 1 m³ of borrow soil = 0.113 t

Mass of dry soil in 1 m³ of embankment = 1.8 t

$$\text{Required excavation} = \frac{1}{1.62} \times 1.8 = 1.11 \text{ m}^3$$

2.15 The natural dry density ρ_d of a sandy deposit is 1.78 t/m³. Laboratory tests give the maximum dry density ρ_{dmax} of 1.9 t/m³ and the minimum dry de·· v ρ_{dmin} of 1.66 t/m³. Find the density index I_D. The value o· ·· .. 2.67.

$$e = \frac{G\rho_w}{\rho_d} - 1 = 0.5 \ ; \ e_{max} = 0.608 \ ; \ e_{min} = 0.405$$

$$I_D = (e_{max} - e)/(e_{max} - e_{min}) = 0.53 = 53\%$$

or $\quad I_D = \left(\frac{1}{\rho_{dmin}} - \frac{1}{\rho_d} \right) \bigg/ \left(\frac{1}{\rho_{dmin}} - \frac{1}{\rho_{dmax}} \right) = 0.53$

2.16 A sample of sand has a volume V of 1000 cm³ in its natural state. On drying and compaction by vibration its minimum volume V_{min} is 850 em³. When gently poured in a measuring cylinder the maximum volume V_{max} is 1350 cm³. Find the density index.

$$I_D = (V_{max} - V)/(V_{max} - V_{min}) = 0.7 = 70\%$$

2.2 RELATIVE DENSITY DETERMINATION

The mean relative density (specific gravity) of soil particles can be determined by using either a 1 litre gas jar (for coarse grained soils) or a 50 ml density bottle (for fine grained soils).   A glass bottle of about 0.8 litre capacity fitted with a screwed conical cap (known as pycnometer) can also be used in place of the gas jar.

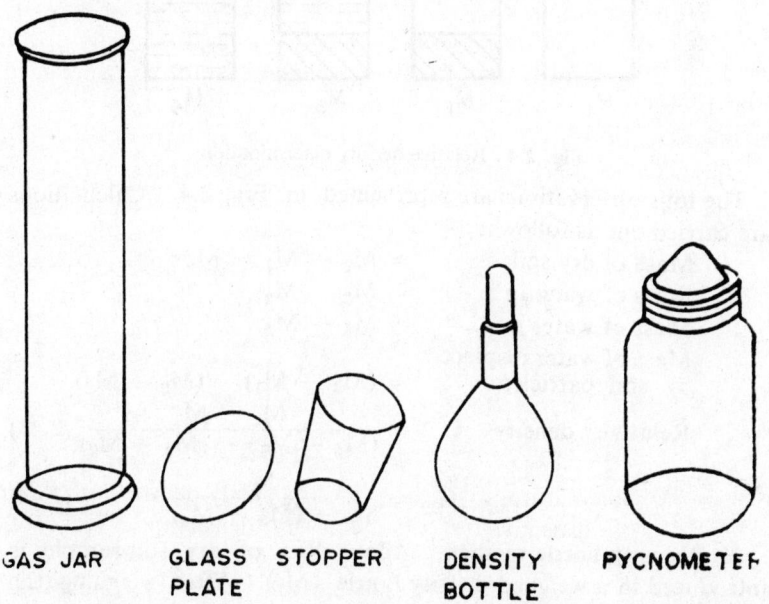

GAS JAR   GLASS STOPPER   DENSITY   PYCNOMETEF
           PLATE   BOTTLE

Fig. 2.3

1. Gas jar method. The gas jar is provided with a rubber stop-per and a ground glass plate to close its open end, as necessary. The empty jar and glass plate are first weighed (M_1). 200—400 g of dry soil is introduced into the jar.   The jar, plate and soil are weighed (M_2).   About 500 ml of water is poured into the jar.   In-serting the stopper into the jar, it is allowed to rest for sometime

(4 hours). The jar is then first shaken by hand and then in an end-over-end shaking apporatus for 20—30 minutes. After thorough shaking, the stopper is removed. Any soil adhering to the stopper or the top of jar is carefully washed into the jar. Water is added to fill the jar to the brim. Placing the glass plate on top (taking care not to trap air), the jar and contents are weighed (M_3). Lastly, the jar is completely emptied, thoroughly washed out, filled to the brim with water, covered with glass plate and weighed (M_4).

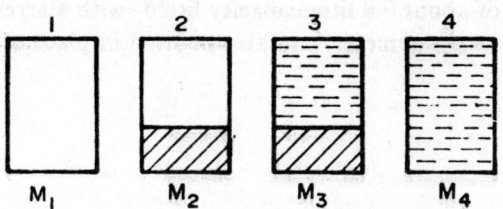

Fig. 2.4 Relative density determination

The four observations are represented in Fig. 2.4. Calculations are carried out as follows:

Mass of dry soil $\qquad = M_2 - M_1 = M_s$

Mass of water in 3 $\qquad = M_3 - M_2$

Mass of water in 4 $\qquad = M_4 - M_1$

Mass of water displaced
by soil particles $\qquad = (M_4 - M_1) - (M_3 - M_2)$

Relative density G $\quad = \dfrac{M_2 - M_1}{(M_4 - M_1) - (M_3 - M_2)}$ (2.43)

or

$$G = \frac{M_s}{M_s - (M_3 - M_4)} \qquad (2.44)$$

2. Density bottle method. Adout 10 g of dry soil sample is introduced in a weighed density bottle (M_1). After weighing the bottle containing soil (M_2), it is partly filled with distilled water. The bottle, without the stopper, is kept in a vacuum desiccator and the contents are subjected for a few hours or preferably over-night to partial vacuum (about 20—25 mm of mercury) to remover air from soil water mixture. The bottle is removed from the desiccator and more water is added to soil. The bottle is again subjected to partial vacuum. When no more air is evolved from the soil, the bottle is removed from the desiccator, and completely filled with water. Inserting the stopper. the bottle is kept in a constant temperature bath (if available) for about 1 hour. The bottle is wiped dry and weighed (M_3). The bottle is cleaned out and filled with water alone. After keeping in the constant temperature bath, the bottle is wiped dry and weighed (M_4). The relative density is calculated from Eq. 2.43 or 2.44.

Entrapped and dissolved air is a great source of error which results in a lower calculated value of relative density. For certain soils, kerosene, in place of water, gives better results. The relative density is then calculated as:

$$G = G_1 G' \tag{2.45}$$

where G_1 = relative density of the liquid
used in place of water

G' = relative density of soil obtained
by using the liquid

The relative density may be reported at a standard temperature of 27°C by the following relation.

$$G_{27} = KG \tag{2.46}$$

where G_{27} = relative density at 27°C

G = relative density at test temperature T°C

$$K = \frac{\text{Density of water at T°C}}{\text{Density of water at 27°C } (= 0.99654)}$$

3. **Typical values.** The relative density of soil particles depends on their mineralogical composition. Most of the commoner primary minerals have relative densities in the range of 2.55 to 2.75 with a mean value of 2.65. The clay minerals generally have higher values in the range of 2.70 to 2.85 with a mean value of 2.75. Organic soils have a low and quite variable relative density such as in the range of 2.2 to 2.64.

Examples 2.17—2.19

2.17 The following observations are obtained for determining the relative density (specific gravity) G of soil by the gas jar method. Find the value of G.

Mass of empty jar (M_1)   = 480 g

Mass of jar + soil (M_2)   = 726.8 g

Mass of jar + soil + water (M_3) = 1667.6 g

Mass of jar + water (M_4)   = 1512.5 g

$$G = \frac{M_2 - M_1}{(M_4 - M_1) - (M_3 - M_2)} = 2.69$$

2.18 A pycnometer containing 0.27 kg of dry soil (M_s) is filled with water and its mass (M_3) is found to be 1.952 kg. The pycnometer filled with water alone weighs 1.784 kg (M_4). Determine the relative density.

$$G = \frac{M_s}{M_s - (M_3 - M_4)} = 2.65$$

2.19 A density bottle containing 9.5 g of dry soil (M_s) and filled with distilled water weighs 78.325 g (M_3). The mass of density bottle filled with water alone is 72.343 (M_4). Determine the relative density.

$$G = \frac{M_s}{M_s - (M_3 - M_4)} = 2.7$$

2.3 MOISTURE CONTENT DETERMINATION

1. Direct measurement. (a) *Oven drying method.* Drying a wet soil sample in a thermostatically controlled, electric oven is the standard laboratory method. A soil sample 30—50 g for fine grained soils and 250—2500 g for coarse grained soils, is taken in a container of known mass (M_1) and weighed with the container (M_2). The container with soil is kept in the oven and the soil is allowed to dry at a temperature of 105—110°C for a sufficient period (usually 24 hours). After drying, the container and the soil are allowed to cool and weighed (M_3). The moisture content is calculated as a percentage of the dry soil mass, from the formula:

$$w = \frac{M_w}{M_s} = \frac{M_2 - M_3}{M_3 - M_1} \times 100 \qquad (2.47)$$

Soils containing gypsum or significant amount of organic matter are dried at a lower temperature (60 — 80°C) and possibly for a longer time. Though very accurate, the oven drying method is time consuming.

(b) *Infrared heat method.* The apparatus comprises a infrared lamp and a torsion balance. A small amount of wet soil (about 5 g) is placed on a drying pan under the infrared lamp and heated at 110°C. The torsion balance is calibrated to give the percentage reduction of mass on drying. It is a rapid method requiring 5 to 10 minutes only. The disadvantage is that it can be used only for fine soils, as only a small sample can be tested.

The test gives the moisture content (m) on *wet mass basis*, defined as:

$$m = \frac{M_w}{M} \times 100 \qquad (2.48)$$

The moisture contents on dry mass basis (w) and wet mass basis (m) are inter-related as follows:

$$w = \frac{m}{1 - m} \text{ and } m = \frac{w}{1 + w} \qquad (2.49)$$

(c) *Sand bath method*. It is a field method requiring only about one hour's drying period. A sand bath (a basin containing about 50 mm thick sand layer) is heated over a gas burner or stove. Wet soil sample, kept in a container, is placed on the sand bath and heated until dry. The sample is stirred with a palette knife (or a rod) during heating to assist drying. Observations are similar to those of the oven drying method and the moisture content is calculated from Eq. 2.47. The method suffers from the disadvantage of possible overheating.

2. **Indirect measurement.** (a) *Calcium carbide method*. A weighed sample of wet soil and calcium carbide are mixed in a closed container, known as the speedy moisture tester (Fig. 2.5). Calcium carbide (CaC_2) reacts with soil moisture to form acetylene gas (C_2H_2):

$$CaC_2 + H_2O \rightarrow Ca(OH)_2 + C_2H_2 \qquad (2.50)$$

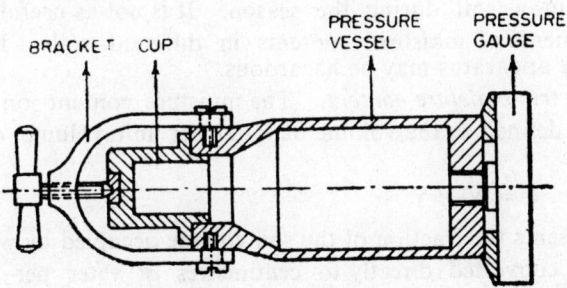

Fig. 2.5 Speedy moisture tester

The resulting pressure of the acetylene gas is measured with a pressure gauge provided in the apparatus. Pressure is directly related to the moisture content on wet mass basis which can be converted to dry mass basis by Eq. 2.49. It is a rapid method and works best with granular soils.

(b) *Nuclear method.* The apparatus, known as a neutron moisture gauge, consists of two principal parts—a *probe* and a *scaler* or *count rate meter.* The probe contains a source of fast neutrons, e.g. a 2—5 millicurie mixture of radium and beryllium or a mixture of americiun and beryllium. The probe is lowered into a metallic access tube inserted vertically into the soil. Fast neutrons are emitted radially into the soil, where they encounter and colloid elastically with various atomic nuclei, gradually losing some of their kinetic energy. The loss of energy is primarily due to collision with hydrogen atoms existing mainly as part of soil water molecules. The slowed neutrons, called 'thermalized neutrons', scatter randomly into the soil forming a cloud around the probe. Some of these return to the probe where they are picked up by a special detector, housed in the probe itself, which is sensitive only to slow neutrons. Receipt of neutrons by the detector is converted into electric pulses which are counted or indicated over a measured time interval by the scaler. The count rate is related to the moisture content on volume basis. Each instrument requires a calibration curve.

This is a very rapid method of measuring moisture content in undisturbed soil in the field. The calibration curve may vary with the soil. The method is best suited to follow moisture changes in a soil during the season. It is not as useful for spot measurements of moisture contents in different soils. Improper use of the apparatus may be hazardous.

Volumetric moisture content. The moisture content on volume basis θ is defined as the volume of water per unit volume of moist soil:

$$\theta = V_w / V \tag{2.51}$$

θ represents the fraction of the soil volume occupied by water and it can be converted directly to centimetres of water per metre of soil. This expression is often more useful for irrigation and drainage calculations. The neutron moisture gauge indicates volumetric moisture content. θ may also be expressed in grammes of water per cm^3 of soil. θ and w are inter-related as follows:

$$\theta = V_w / V = \frac{M_w / \rho_w}{M_s / \rho_d} = w \frac{\rho_d}{\rho_w} \tag{2.52}$$

(c) *Penetration resistance method.* The moisture content of compacted soils can be determined by correlating it with the penetration resistance of the Proctor needle apparatus as described in Chapter 21.

2.4 DENSITY INDEX DETERMINATION

The natural void ratio is determined from the known values of w, ρ and G of a soil. The soil is dried and its minimum void ratio or maximum dry density may be determined by vibratory compaction. A simple method is to compact soil in a CBR mould or a Proctor mould (Chapter 21). The compactor may be a vibrating hammer (BS : 1377) or it can be improvised by an electric needle vibrator commonly used in concrete works. A steel plate (tamper) is attached to the vibrating hammer or the vibrator. The soil is compacted in three layers in the Proctor mould and in five layers in the CBR mould, each layer being vibrated under firm pressure of the vibrator for one minute. The density is obtained from the mass and volume of the compacted soil and the void ratio is calculated from Eq. 2.24. Alternative method of compaction is the use of a vibrating table and moulds (IS: 2720—14).

The maximum void ratio may be obtained by pouring oven dry soil through a filter funnel (with opening large enough to permit free fall of soil) into a standard mould maintaining the bottom of the funnel at 10—15 mm above the soil surface. Coarser soils ($>$ 10 mm) can be deposited loosely in the mould by placing in thin layers using a hand scoop. From the known volume and measured mass of soil its minimum dry density is calculated.

2.5 PARTICLE SIZE DISTRIBUTION

By particle size distribution is meant the range of sizes of particles in a soil and the percentage (by mass) of particles which occur within a certain size range. This information is obtained by two methods — 'sieve analysis' and 'sedimentation analysis', the former being generally used for coarse grained soils and the latter for fine grained soils. In many cases a combination of both the methods may be necessary.

1. Sieve analysis. The set of sieves may consist of 75 mm, 37.5 mm, 19 mm, 9.5 mm, 4.75 mm, 2.0 mm and 1.0 mm sieves, and 600, 425, 300, 212, 150 and 75 (or 63) micron sieves with lids and receivers. Not all the sieves may be required for a soil. It is desirable to include 75 mm, 19 mm, 4.75 mm, 2 mm, 425 micron and 75 (or 63) micron sieves in the set as they are required to subdivide gravel into coarse gravel (75—19 mm) and fine gravel (19—4.75 mm) and sand into coarse sand (4.75—2 mm), medium sand (2 mm—425 micron) and fine sand (425—75 micron).

Wet or dry procedures may be adopted for sieving. Wet procedure is preferable.

(a) *Wet sieving.* The representative dry sample is first divided into two portions by sieving on the 19 mm sieve. The fraction retained on the sieve is subdivided into different fractions by sieving on the 75, 37.5 and 19 mm sieves. A sub-sample (about 2 kg) is taken from the fraction passing the 19 mm sieve and it is soaked in water for about 1 hour. (Sodium hexametaphosphate may be added at the rate of 2 g/l of water). The soaked material is washed, a little at a time, through a 2 mm sieve nested in a 75 micron sieve, until wash water is virtually clear. The wash water containing silt and clay may be allowed to run to waste. The washed material retained on 2 mm and 75 micron sieves is transferred to suitable containers and dried in an oven.

When dry, all the material is dry sieved through 10 mm and 4.75 mm sieves and the mass retained on each sieve is weighed. The fraction passing the 4.75 mm sieve (or a smaller sub-sample of this fraction, in case it is large) is sieved through the set of 2 mm, 1 mm, and 600, 425, 300, 212, 150 and 75 micron sieves. Material retained on each sieve is weighed.

(b) *Dry sieving.* Dry sieving may be adopted for granular soil containing little or no fines (< 75 micron). A dry soil sample is first separated into plus 4.75 mm fraction and minus 4.75 mm fraction by sieving on a 4.75 mm sieve. A coarse sieve analysis is run on the plus 4.75 mm gravel fraction starting with the largest size sieve appropriate to the maximum size of particles present. A fine sieve analysis (from 2 mm to 75 micron size) is run on the minus 4.75 mm fraction.

(c) *Particle size distribution curve.* The mass retained on each sieve is weighed and expressed as the percentage of the total mass of the sample. The "cumulative per cent" retained for each sieve is then calculated. Cumulative percent for a sieve is the percentage retained if whole of the sample were sieved through that particular sieve only. It is thus the total of the percentages retained on the particular sieve and on all the coarser sieves above it. By subtracting the 'cumulative per cent retained' from 100, the 'cumulative percent finer' (also termed simply 'percent finer N') is calculated. The results are plotted on a semi-logarithmic chart with percent finer 'N' as ordinate on a linear scale and particle size 'D', taken equivalent to the sieve aperture, on logarithmic scale. As the particle size may vary between very wide limits, it would not be practical to plot the particle size on a conventional linear scale and hence, a logarithmic scale is adopted. The resulting curve is known as a 'particle size distribution curve' or a 'grading curve'.

2. Sedimentation analysis. The sedimentation analysis is based on the Stokes' law which governs the velocity at which spherical particles settle in a suspension, the larger the particle the greater is the settling velocity and vice versa. The Stokes' law may be stated as follows:

$$v = \frac{g (G - 1) D^2}{18 \eta} \tag{2.53}$$

where v = settling velocity (cm/s)
    D = diameter of particle (mm)
    g = gravitational acceleration (9.8 m/s^2)
    G = relative density of particle
    η = viscosity of water (poise, i.e. dyne-sec/cm^2)

If H′ is the depth of settlement (cm) of a particle in t minutes, the settling velocity is given by:

$$v = H'/60t \quad (cm/s) \tag{2.54}$$

and the Stokes' law may be expressed as:

$$D = \sqrt{\frac{0.3\eta}{g (G - 1)}} \times \sqrt{\frac{H'}{t}} \tag{2.55}$$

or $$D = K\sqrt{H'/t} \tag{2.56}$$

where $$K = \sqrt{\frac{0.3\eta}{g (G - 1)}} \tag{2.57}$$

By substituting $g = 9.8$ m/s^2, $G = 2.7$ and $\eta = 0.00855$ poise (at 27°C), an ap proximate expression for the Stokes' law (Eq. 2.53) is obtained as:

$$v \simeq 108\, D^2 \quad (cm/s) \tag{2.58}$$

The SI unit for viscosity is $mN\text{-}s/m^2$ which is converted to poise ($1mN\text{-}s/m^2 = 0.01$ poise) for use in the above equations.

Stokes' law assumes that (i) the particles are spherical, rigid and and smooth, (ii) the particles are sufficiently large so that Brownian movement does not influence their rate of fall, that is, the particles are much larger than water molecules, (iii) the settling velocity is sufficiently low so that the viscosity of the liquid is the only resistance to settlement, that is, the flow around the particles is laminar, and (iv) the particles settle independently of one another.

Soil particles are rigid but they are not spherical and smooth. A large percentage of fine particles are platelets which do not settle in the same manner and at the same rate as smooth spheres. The particle size calculated from the Stokes' law is, therefore, called the 'equivalent diameter'. The law is found to remain valid for the range of 0.0002 mm to 0.2 mm in diameter. Particles smaller than 0.0002 mm are affected by the Brownian movement, and turbulence accompanies the settlement of particles larger than about 0.2 mm in diameter. In the sedimentation analysis it is usual to separate out the sand and coarser particles first so that the assumption of laminar flow holds. The assumption of independent fall of particles is satisfied by keeping a low suspension concentration (less than 5 %).

Sedimentation analysis is carried out by pipette and hydrometer methods, the former is considered the standard laboratory method. The hydrometer method is much simpler and the results are quite accurate for practical purposes.

3. **Pipette method.** The apparatus consists of a sampling pipette of about 10 ml capacity (Fig. 2.6), a 500 ml sedimentation tube (boiling tube), constant temperature bath, and glass weighing bottles, stop watch and balance.

Independent settling of particles requires that their aggregation should be destroyed. If the soil contains cementing material, such as organic matter and carbonates, it is first treated with hydrogen peroxide to oxidise organic matter and then with dilute hydrochloric acid to remove carbonates. The treated soil is washed with distilled water to remove acid and dried. This pretreatment is not given to a soil free of cementing materials.

A dry soil sample (sufficient to yield about 15 g fraction passing the 75 micron sieve) is taken in a beaker and covered with dispersing agent solution (50 ml) and distilled water. The dispersing agent solution commonly used is prepared by dissolving 33 g of sodium hexametaphosphate, 'calgon', and 7 g of sodium carbonate in one litre of distilled water.   Alternative dispersing agents, e.g. sodium oxalate, may also be used.   A dispersing agent increases interparticle repulsion by introducing monovalent ex-changeable cations and thus decreases the tendency of particle flocculation during sedimentation.   The soil water slurry is then stirred in a mechanical stirrer (to further increase dispersion), after adding some more water.   It is then washed on a 75 micron sieve to separate out sand fraction.   The suspension passing the sieve is transferred to the sedimentation tube and the volume is made up with distilled water to the 500 ml mark.

The tube containing the soil suspension is shaken thoroughly by inverting several times, keeping the open end closed with palm of hand.   It is then kept upright in the constant temperature bath and the stop watch is started.   Samples of the suspension are taken by means of the pipette from a depth of 10 cm below the surface of the suspension at suitable time intervals, (after 2 minutes of the start of sedimentation and further at 4, 8, 15, 30 minutes and 1, 2, 4, 8 and 24 hours).   Each sample is dried and weighed to get the day mass of solids per ml of suspension at various sampling times.

According to the Stokes' law particles settle according to their size, the coarser ones settling faster.   Let H' be the sampling depth and let D_1 be the diameter of the particles which settle through H' during sampling time t_1.   After t_1 no particle of dia-meter D_1 and also larger diameter remains in suspension at depth H'.   All Particles finer than D_1 are still in suspension at depth H', at the same concentration as in the original suspension.   This is because particles of the same size settle with the same velocity throughout the whole height of the suspension.   Particles (finer than D_1) that are able to settle lower than the sampling level are con-tinuously replaced by the same quantity of similar particles arriving from above keeping their concentration constant.   The dry mass of solids M_{D1} per ml of suspension corresponding to sampling time t_1

also contains the mass of dry dispersing agent equal to m/V, where m is the mass of dispersing agent added to total suspension of volume V, (m = 2 g if 50 ml of dispersing agent solution is added). The percentage N_1 of particles finer than D_1 is thus given by:

$$N_1 = \frac{M_{D1} - m/V}{M_s/V} \times 100 \qquad (2.59)$$

where M_s = total dry mass of soil sample taken for the test

V = volume of suspension (500 ml)

The particle size D_1 (mm) is calculated from Eq. 2.55. Sampling at different time intervals thus leads to the determination of different particle sizes that are able to settle below the sampling level and the percentages finer than these particular sizes.

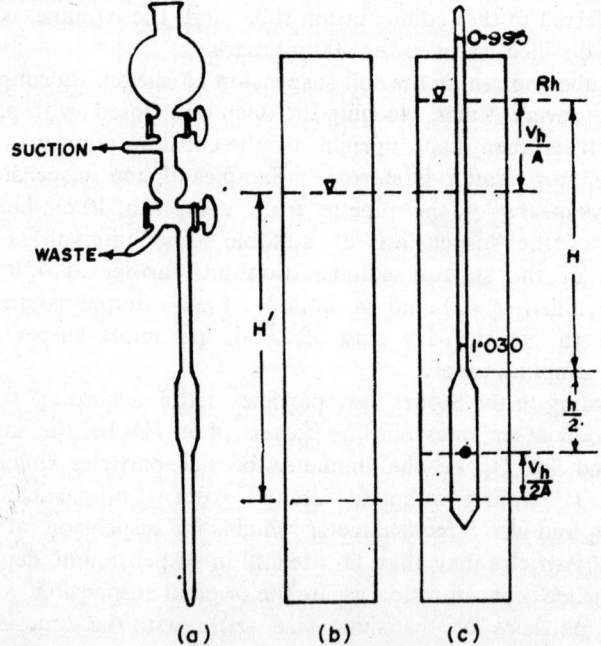

Fig. 2.6 Showing the sampling pipette and effective depth of hydrometer

4. **Hydrometer method.** The apparatus consists of a hydrometer calibrated at 27°C to read density of suspension (usual range 1.030 g/ml — 0.995 g/ml), two 1000 ml graduated glass measuring

cylinders without lip, set of 4.75 mm, 2 mm, 425 micron and 75 micron sieves, stop watch and balance.

(a) *Calibration of hydrometer*. The effective depth H' at which the hydrometer indicates the density is taken to be the centre of its bulb. Because of the displacement of suspension (Fig. 2.6) on immersion of hydrometer, each set of hydrometer and cylinder requires calibration to know the effective depth.

The volume of the hydrometer V_h is measured either by immersing it in a partly filled measuring cylinder or weighing the hydrometer in grams. The mass in grams approximately gives the volume in ml (density of hydrometer is approximately unity). The inside sectional area 'A' (cm^2) of the cylinder, in which the hydrometer is to be used, is found by measuring the distance between two graduations (say 100—900 ml marks) and dividing the volume included between the graduations by the distance (cm) between them. The height of the hydrometer bulb 'h' (bottom to neck) and the distances from the neck to each of the major graduations R_h on the stem are measured. The effective depth H' (Fig. 2.6) is calculated from the relation:

$$H' = H + \frac{1}{2}\left(h - \frac{V_h}{A}\right) \tag{2.60}$$

A calibration curve is plotted between H' and R_h.

(b) *Hydrometer corrections*. The hydrometer is calibrated to read the density of distilled water as 1.000 g/ml at 27°C. When a hydrometer is immersed in a soil suspension a meniscus is formed along the stem of the hydrometer (water rises due to surface tension) and the bottom of the meniscus, which is the surface level of the suspension, cannot be read as the suspension is opaque. The top of the meniscus is read on the stem, hence a 'meniscus correction' ($+C_m$) is necessary. The addition of dispersing agent to soil suspension increases its density which requires 'dispersing agent correction' ($-C_d$). Similarly, if the test temperature is other than the calibration temperature, a 'temperature correction' ($\pm C_t$) is required.

To know the corrections, distilled water having the same amount of dispersing agent as is to be used in the test is filled in one of the cylinders and the hydrometer is floated into it. The top and bottom of the meniscus are read and the difference is the 'meniscus correction'. If distilled water is taken at 27°C without dispersing agent and the bottom of the meniscus is read, the hydrometer reading is 1.000. Hence the difference between 1.000 and the actual hydrometer reading at top of meniscus in distilled water having dispersing agent at test temperature gives the 'composite correction' ($\pm C$) due to combined effects of meniscus, dispersing agent and temperature. When composite correction is determined, it is not necessary to determine the dispersing agent correction separately. The overall corrected hydrometer reading R is thus expressed as:

$$R = R'_h \pm C \qquad (2.61)$$
and
$$R_h = R'_h + C_m \qquad (2.62)$$
where R'_h = observed hydrometer reading at top of meniscus

R_h = hydrometer reading corrected for meniscus only

(c) *Sedimentation test.* 50 — 150 g of dry soil sample (pretreated, if necessary) passing a 4.75 mm sieve is taken; the sample should be sufficient to yield about 30 g of fraction passing the 75 micron sieve. The sample is placed in a beaker and soaked with 125 ml of dispersing agent solution and distilled water (for at least 2 hours). The soil water slurry is washed through a 2 mm sieve and the wash water containing soil fraction finer than 2 mm is transferred to the cup of the mechanical stirrer. After stirring (1-2 minutes), the suspension is passed through the 75 micron sieve into the 1000 ml cylinder. Material retained on the sieve is washed and the wash water transferred to the cylinder. Distilled water is added to fill the cylinder upto the 1000 ml mark.

The material retained on the 2 mm and 75 micron sieves is dried. The plus 75 micron material may be subdivided into medium and fine sand fractions by sieving on the 425 micron sieve. If the plus 2 mm fraction is not much, the soil water slurry can be transferred directly from the beaker into the mechanical stirrer. If the mechanical stirrer is not available the soil is well stirred in the cup after soaking period, washed through 75 micron sieve and the suspension is transferred to the cylinder.

Closing the open end of the cylinder with a suitable rubber stopper, it is shaken end-over-end and then allowed to stand starting the stop watch simultaneously. The hydrometer is inserted and top of the meniscus is read at 0.5, 1, 2 and 4 minutes. The hydrometer is taken out of the suspension slowly, rinsed with distilled water and allowed to float in the second cylinder containing distilled water and dispersing agent. Temperature of the suspension is recorded. Further readings of the suspension are taken at 8, 15, 30 minutes and at 1, 2, 4, 8 and 24 hours, after the start of the test. After each reading, the hydrometer is taken out of the suspension, rinsed and floated in the second cylinder. Insertion and withdrawal must be done carefully, taking 10 seconds for each operation. Temperature of the suspension and composite correction are periodically noted.

(d) *Calculations.* It is usual to record the hydrometer reading by subtracting 1 and multiplying the remaining digits by 1000. Thus 1.030 is recorded as 30 and 0.995 is recorded as —5. Let R_1 be corrected hydrometer reading at elapsed time t_1 minutes. The density of suspension is $1 + R/1000$ g/ml. Let M_{D1} be the mass (g) of solids in one ml of suspension at depth H' after time t_1 and which are finer than the particles of diameter D_1 that have settled below the depth H'. Adding the mass of solids and water in 1 ml of suspension, the density of suspension is also given as:

$$\text{Density of suspension} = M_{D1} + \left(1 - \frac{M_{D1}}{\rho_s} \right) \rho_w \quad \text{g/ml}$$

$$\text{or} \quad 1 + \frac{R}{1000} = 1 + M_{D1} \left(\frac{G-1}{G} \right)$$

$$\text{and} \quad M_{D1} = \frac{G}{G-1} \times \frac{R}{1000} \tag{2.63}$$

The percentage N_1 of particles finer than D_1 still remaining in suspension after time t_1 is given by:

$$N_1 = \frac{M_{D1}}{M_s/V} \times 100 = \frac{100\,G}{M_s\,(G-1)} R \ (\%) \tag{2.64}$$

where M_s = mass of dry soil taken for the test
             V = volume of suspension (1000 ml)

The hydrometer reading R_h corrected for meniscus equals $R'_h + C_m$. The effect depth H' corresponding to R_h is read from the calibration curve. The equivalent diameter D_1 (mm) is calculated from Eq. 2.56:

$$D_1 = K \sqrt{H'/t_1}$$

The results of sedimentation analysis are combined with the sieve analysis to give a composite 'particle size distribution curve' for a soil having particles ranging from coarse to fine sizes.

5. Analysis of grading curve. The particle size distribution curve (grading curve) is used to determine percentage contents of particle sizes necessary for classification of soils, and to define the grading of the soil. The grading is defined in terms of the 'effective size', the 'uniformity coefficient' and the 'coefficient of curvature' Three sizes are determined from the grading curve (Fig. 2.7):

(1) D_{10} = particle diameter at which 10% of the soil mass is finer, i.e. 90% is coarser than D_{10}. This diameter is called the 'effective size', or the 10% finer size.

(2) D_{30} = particle diameter at which 30% of the soil mass is finer, called the 30% finer size.

(3) D_{60} = particle diameter at which 60% the soil mass is finer, called the 60% finer size.

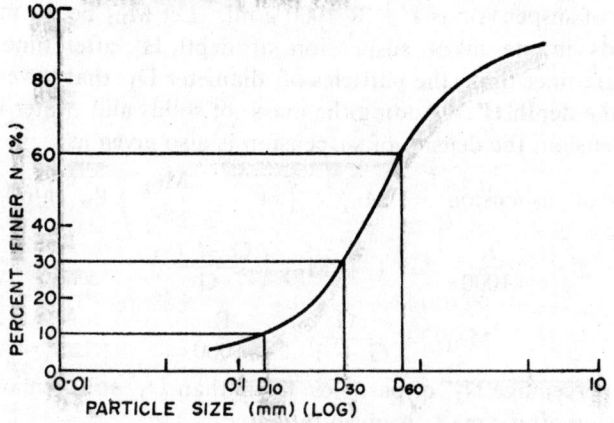

Fig. 2.7 Analysis of grading curve

The uniformity coefficient C_u and coefficient of curvature C_c are defined as:

$$C_u = D_{60} / D_{10} \qquad (2.65)$$

$$C_c = \frac{D^2_{30}}{D_{60} \times D_{10}} \qquad (2.66)$$

Low values of C_u imply a uniform close grading. If all the particles are of the same size, C_u is unity. A granular soil is termed 'well graded' if C_c lies between 1—3 and C_u is greater than 6 in case of sands and greater than 4 in case of gravels. When these requirements are not met, the soil is termed 'poorly graded'.

Examples 2.20—2.23

2.20 Find the sampling time so that particles finer than 0.006 mm and 0.002 mm may be sampled by a pipette from a depth of 10 cm. The value of G is 2.72 and the viscosity η is 0.01 poise. Gravitational acceleration g is 9.8 m/s².

$$v = \frac{g(G-1)D^2}{18\eta}, \quad v_1 = 3.37 \times 10^{-3} \text{ cm/s}$$

$$\text{for } D = 0.006 \text{ min}$$

$$v_2 = 3.7 \times 10^{-4} \text{ cm/s}$$

$$\text{for } D = 0.002 \text{ min}$$

$$t_1 = H'/60\,v_1 = 49.5 \text{ min}$$

$$t_2 = 450.5 \text{ min} = 7 \text{ hr } 30.5 \text{ min}$$

2.21 A soil sample with gradation from 0.075 mm to 0.002 mm is dropped on the surface of still water, 1 m deep. The viscosity of water is 0.009 poise. Find the time in which the first and the last particles of soil will be able to settle to the bottom of water. The value of G is 2.68.

$$v = \frac{g(G-1)D^2}{18\eta} = 101.63\,D^2$$

$$v_1 = 0.572 \text{ cm/s}, \qquad t_1 = 2 \text{ min } 55 \text{ s}$$

$$v_2 = 4.065 \times 10^{-4} \text{ cm/s}, \quad t_2 = 68 \text{ hr } 20 \text{ min } 2.5 \text{ s}$$

2.22 The corrected hydrometer reading at commencement of test (t = 0) is 1.020. At 60 minutes the corrected reading is 1.008 and the effective depth of hydrometer is read from the graph as 18 cm. The relative density of soil is 2.7 and the viscosity of water is 0.01 poise. Find the total mass of soil in the suspension. Find the particle size corresponding to the 60 minute reading and the percent finer than this diameter.

$$M_D = \frac{G}{G-1} \times \frac{R}{1000}$$

At $\quad t = 0, \dfrac{M_s}{1000} = \dfrac{G}{G-1} \times \dfrac{R_0}{1000}$

$R_0 = 20$ and therefore $M_s = 31.76$ g

At $\quad t = 60$ min, $R_1 = 8$

$$M_{D1} = \frac{G}{G-1} \times \frac{R_1}{1000} = 0.0127 \text{ g}$$

$$N_1 = \frac{M_{D1}}{M_s/V} \times 100 = 40\%, \ (V = 1000 \text{ cm}^3)$$

Alternately $N_1 = \dfrac{R_1}{R_0} \times 100 = 40\%$

$$D_1 = \sqrt{\frac{0.3\,\eta\,H'}{g\,(G-1)t}} = 0.0073 \text{ mm}$$

2.23 The grading curve of a soil gives the effective size as 0.14 mm, 30 percent finer size D_{30} as 0.38 mm and 60 percent finer size D_{60} as 0.80 mm. Find the uniformity coefficient C_u and the coefficient of curvature C_c (Fig. 2.7).

$$C_u = D_{60} / D_{10} = 5.7$$

$$C_c = \frac{D^2_{30}}{D_{60} \times D_{10}} = 1.29$$

2.6 CONSISTENCY AND PLASTICITY

The term "consistency" is used to describe the physical state, i.e. the degree of coherence between particles, of a soil at a given moisture content. Consistency is the resistance to deformation of soil and is obviously related to the force of attraction between particles or aggregates of particles. "Plasticity" is the ability of a soil to change shape (deform) continuously under an applied stress, and to retain the new shape on removal of the stress. Only finer particles of soils, clays and to some extent silts, exhibit plastic behaviour. Consistency and plasticity are directly related to moisture content, but different soils may have different consistency at the same moisture content and they exhibit plastic behaviour over varying range of moisture content.

At high moisture contents, fine grained soils form suspensions and behave like fluids. As the moisture content is gradually reduced, the flow properties change to those of paste-like materials and a stage is reached when a slight disturbing force is necessary to make the soil-water mixture flow. Upto this stage the soil is said to be in a "liquid state". On further drying, the soil can be moulded and develops the plastic behaviour. This is the "plastic state". As the moisture content decreases still further, the plasticity is lost and soil starts crumbling on application of pressure; the soil is then said to be in the "semi-solid state". At still lower moisture contents the soil takes on the properties of a solid or attains the "solid state". Thus depending on its moisture content a fine grained soil can exist in any of the four states of consistency — liquid, plastic, semi-solid and solid states. The moisture contents at the boundaries between the adjacent states are termed 'consistency limits'. There are three consistency limits—"liquid limit w_L" at which soil changes from liquid to plastic state, "plastic limit w_P" at which soil changes from plastic to semi-solid state and "shrinkage limit w_S" at which the solid state begins. Being first studied by Swedish scientist Atterberg in 1911, the consistency limits are also called 'Atterberg-limits'.

Changes in moisture content are accompanied by change in total volume of soil as shown in Fig. 2.8. It may, however, be understood that the transition from one stage to another does not occur abruptly. It is a gradual change and the dividing moisture content limits are based on more or less arbitrarily set criteria.

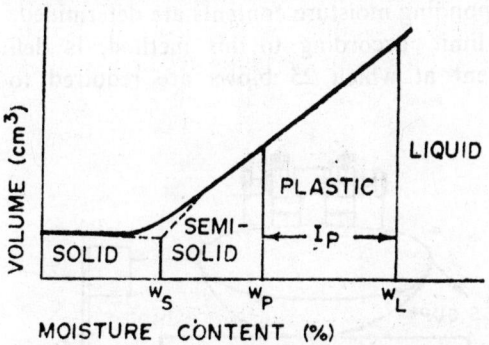

Fig. 2 8 States of consistency and consistency limits

1. Liquid limit. Liquid limit w_L is the minimum moisture content at which the soil can flow under a specified small disturbing force, the disturbing force being defined by the method of testing. Two methods of testing are described, the Casagrande apparatus method and the cone penetrometer method.

(a) *Casagrande apparatus method.* The Casagrande liquid limit apparatus (Fig. 2.9) consists of a brass cup and a carriage mounted on a hard rubber base. Grooving tools are used to cut out a groove in the soil paste kept in the cup. The cup is adjusted to drop from a height of 10 mm on the base when the handle is rotated. An air dry soil sample (about 200 g) passing the 425 micron sieve is mixed with distilled water on a glass plate or in a dish to form a stiff paste which is left for a suitable maturing time. A small portion of the paste is kept in the cup and levelled off with a spatula to give a maximum depth of 10 mm. A clean, straight groove is cut by a grooving tool through the paste. The groove is V-shaped and 2 mm wide at the bottom. The handle is turned at 2 revolutions per second and the number of blows (drops of the cup) are counted until the two parts of the soil come in continuous contact at the bottom of the groove along a distance of about 13 mm. A sample is taken from the closed portion of the groove for moisture content determination. The remaining soil is removed from the cup, mixed with the main sample on the glass plate and the moisture content is changed. The cup is cleaned and the test is repeated. It is preferable to proceed from the drier to the wetter condition of soil. The number of blows for each test are counted and the corresponding moisture contents are determined.

The liquid limit, according to this method, is defined as the moisture content at which 25 blows are required to close the groove.

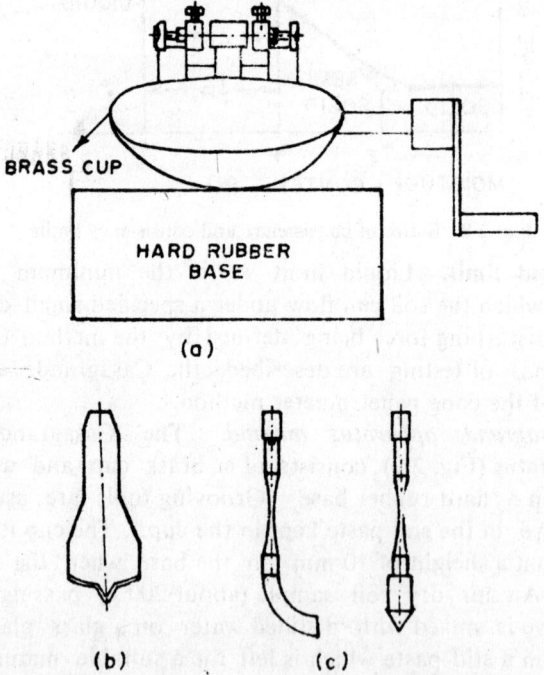

Fig. 2.9 Casagrande liquid limit apparatus
Grooving tools: (b) Casagrande type (type A) (c) ASTM type (type B) (IS : 9259)

A semi-log plot is obtained between moisture content as ordinate on the linear scale and the corresponding number of blows as abscissa on the logarithmic scale. The resulting curve drawn as the best fitting straight line is called the 'flow curve' (Fig. 2.10). The moisture content corresponding to 25 blows is read as the liquid limit.

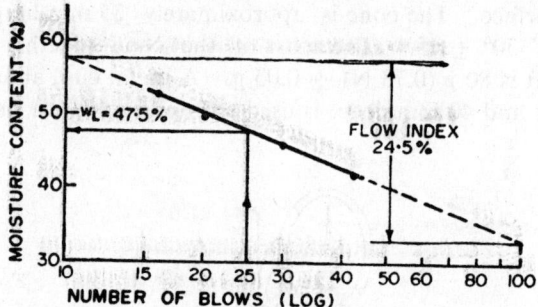

Fig. 2.10 Flow curve and liquid limit determination

The slope of the flow curve is termed the 'flow index I_F' which can be calculated as the difference between the moisture contents at 1 and 10 blows (or at 10 and 100 blows). Flows index is expressed as:

$$I_F = \frac{w_1 - w_2}{\log (n_2/n_1)} \tag{2.67}$$

where w_1 = moisture content at n_1 blows

w_2 = moisture content at n_2 blows

Flow curves plotted on a double log graph (log w, log n) are observed to give parallel straight lines (Lambe 1951), represented by the general equation:

$$w_L = w \left(\frac{n}{25} \right)^e \tag{2.68a}$$

where   w = moisture content (%) at n blows

The value of index 'e' may be adopted as 0.1 (Kapre and Kulkarni 1972). The moisture content 'w' should range in the 15 to 35 blows. Eq. 2.68a affords a 'one-point method' of determining the liquid limit by the Casagrande apparatus. An alternative expression for one point method of determining liquid limit is as follows

(Nagaraj and Jayadeva 1981):

$$w_L = \frac{w}{1.3215 - 0.23 \log n} \qquad (2.68b)$$

where   w = moisture content at n blows, n ranging from 15 to 35.

(b) *Cone penetrometer method.* The penetrometer (Fig. 2. 11) is fitted with a stainless steel or duralumin cone having a smooth, polished surface. The cone is approximately 35 mm long and has an angle of 30° ± 1°. The mass of the cone together with its sliding shaft is 80 g (0.78 N) ± 0.05 g. A metal cup, about 55 mm in diameter and 40 mm deep, is used to contain the test sample.

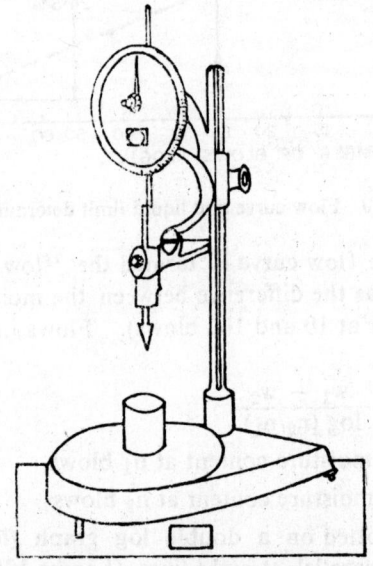

Fig 2.11   Cone penetrometer

An air dry soil sample (about 250 g) passing the 425 micron sieve is mixed with distilled water. The soil paste is filled in the metal cup and the surface struck off level. The cone is lowered to just touch the surface of soil and then released for a period of 5 seconds. The penetration is measured. The cone is lifted and cleaned and the depression in the soil surface is filled up by adding a little more wet soil. The test is repeated. If the difference between the two measured penetrations is less than 0.5 mm, the tests are considered valid. The average penetration is noted and the moisture content of the soil is determined. The test is repeated at least 4 times with increasing moisture contents. The moisture contents used in the tests should be such that the penetrations obtained lie within a range of 15 to 35 mm.

To obtain the liquid limit, cone penetration is plotted against moisture content (both on linear scales) to give the best fitting straight line (Fig. 2.12). The moisture content corresponding to a cone penetration of 20 mm is taken as the 'cone penetration limit 'w_e' of the soil (BS: 1377) which for all practical purposes is the same as the liquid limit determined by the Casagrande apparatus.

The cone penetrometer method gives a more consistent estimate of the liquid limit than the Casagrande apparatus, with greater repeatability and less operator susceptibility (Sherwood and Ryley 1970).

As a 'one-point method' using the cone penetrometer the liquid limit can be determined from the following relationship (Nagaraj and Jayadev 1981, IS: 2720—5, 1985):

$$w_L = \frac{w}{0.77 \log D} \qquad (2.68c)$$

or
$$w_L = w/(0.65 + 0.0175D) \qquad (2.68d)$$

where $w =$ moisture content corresponding to penetration D, D ranging from 16 to 26 mm

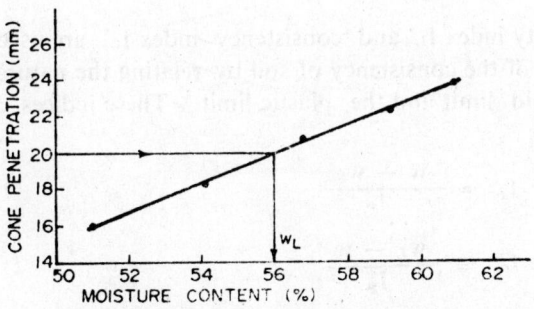

Fig. 2.12 Liquid limit by cone penetrometer

2. **Plastic limit.** The plastic limit w_P is the minimum moisture content at which the soil can be deformed plastically. As standard-

ized, it is taken to be the minimum moisture content at which the soil can be rolled into a thread 3 mm thick.

About 30g of soil (minus 425 micron) is mixed with just enough water so that it may be rolled into a ball. About 5 g of wet soil is shaped into a ball and rolled into a thread on the glass plate with tips of the fingers of one hand. When a diameter of 3—4 mm is reached, the soil is remoulded again into a ball. Rolling and remoulding process is repeated until the thread starts just crumbling at a diameter of 3 mm; the soil is then said to be at its plastic limit and its moisture content is determined. The test is repeated on two more samples to get the average value of plastic limit.

The range of moisture content over which a soil is in the plastic condition is known its 'plasticity index I_P', which is equal to the numerical difference between the liquid and plastic limits:

$$I_P = w_L - w_P \tag{2.69a}$$

The plasticity index indicates the percentage of water a soil absorbs in changing from the semi-solid state to the liquid state; the finer the soil, the greater is its plasticity index.

The plasticity index can be known approximately from the following relation (Nagaraj and Jayadeva 1983)

$$I_P = 0.74 (w_L - 8) \tag{2.69b}$$

A soil is termed 'non-plastic' (NP) if it cannot be rolled into threads to determine the plastic limit, or if its plasticity index is zero.

'Liquidity index I_L' and 'consistency index I_C' are used to get an indication of the consistency of soil by relating the natural moisture to the liquid limit and the plastic limit. These indices are defined as follows:

$$I_L = \frac{w - w_P}{I_P} \tag{2.70}$$

$$I_C = \frac{w_L - w}{I_P} \tag{2.71}$$

where   w = natural moisture content
The two indices are inter-related as:

$$I_C = 1 - I_L \tag{2.72}$$

A soil is at liquid limit when $I_L = 1$ ($I_C = 0$) and at plastic limit when $I_L = 0$ ($I_C = 1$)

The ratio of plasticity index I_P to flow index I_F is termed the 'toughness index I_T'

$$I_T = I_P / I_F \tag{2.73}$$

The toughness index is indicative of the toughness or resistance to deformation at the plastic limit. I_T varies from 0 to 3 for most of the clays, but may be as high as 5. $I_T < 1$ indicates that the soil is friable at the plastic limit (Means and Parcher 1963).

3. **Liquid and plastic limits in terms of shear strength.** On the basis of experimental evidence it seems reasonable to assign a unique strength to all soils at their liquid limit. Comprehensive studies by Youssef et al (1965) on clays over the range of liquid limit of 30—200% indicate that range of undrained shear strength (Chapter 9) at liquid limit is from 24 g/cm² to 13 g/cm²(2.4—1.3 kPa) with a mean value of about 17 g/cm² (1.7 kPa). This mean value of 1.7 kPa (17 g/cm²) may be adopted as the present best estimate of the undrained shear strength of a soil when at the liquid limit (Wroth and Wood 1978). It is further evident that the shear strength at the plastic limit is about 100 times that at the liquid limit (Skempton and Northey 1953) and a value of 1.7 kg/cm² (170 kPa) may be adopted as the best estimate. On the basis of these findings Wroth and Wood (1978) redefined the plastic limit as the moisture content at which the soil strength is 100 times that at the liquid limit.

The plasticity index may thus be defined as the change of moisture content producing a 100-fold change in strength of the soil. If the cone penetrometer (Fig. 2.11) is fitted with two cones of different masses M_1 and M_2 (total sliding masses) and the corresponding penetrations for the same soil are plotted on a semi-log plot, a set of parallel lines in obtained as shown in Fig. 2.13. The plasticity index I_P can be obtained from the expression:

$$I_P = \triangle w \ \frac{\log 100}{\log (M_1 / M_2)} \tag{2.74}$$

where $\triangle w$ = moisture content separation
One of the cones is of the standard mass of 80 g (M_2) and the other may be of 240 g (M_1) so that $M_1/M_2 = 3$ and Eq. 2.74 may be written as:

$$I_P = 2 \triangle w / \log_{10} 3 \tag{2.75}$$

From the typical results of Fig. 2.13, $\triangle w = 0.124$ and

$I_F = 2 \times 0.124/\log 3 = 0.52$ (or 52%). This is comparable with a value of $I_P = 54\%$ obtained by the conventional method.

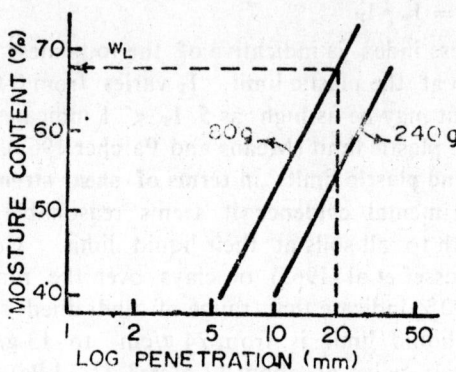

Fig. 2.13   Results of cone penetration tests with sliding masses of 80 g and 240 g (Wroth and Wood 1978)

Assuming the undrained shear strength s_n as 17 g/cm^2 and 1.7 kg/cm^2 respectively at the liquid limit and the plastic limit, the following correlations are proposed by Wroth and Wood (1978):

$$C_o' = \tfrac{1}{2} I_P G \qquad (2.76)$$

or $\qquad C_o' = \triangle wG/\log 3 \qquad (2.77)$

and $\qquad s_n = 1.7 \exp(-4.6\ I_L) \text{ kg/cm}^2 \qquad (2.78)$

or $\qquad s_u = 170 \exp(-4.6\ I_L) \text{ kPa} \qquad (2.79)$

where $\qquad C_c' =$ compression index of remoulded soil
(Chapter 14)

$\qquad\qquad s_u =$ undrained remoulded strength

4. Shrinkage and shrinkage limit. When a saturated soil is dried, a curved air-water interface (meniscus) develops in each void at the soil surface. This produces a lower pressure (capillary pressure) on the convex inner side in the soil moisture and a corresponding compression in the soil structure. Water is drawn from inside the soil due to pressure difference on the two faces of the menisci. Water evaporates and the soil decreases in volume. As long as the capillary forces exceed the resistance of soil particles to come closer, i.e. to shrink, the soil remains saturated and the volume change equals the volume of water lost. With further evaporation, the capillary menisci retreat into the narrowing voids exerting an increasing capillary pressure which is accompanied by an increase in inter-particle stresses and a reduction in adsorbed water layer thickness at the points of contact.   Eventually, a stage is reached when particle interaction restricts shrinking and further loss of water is partly replaced by air.   Moisture content at this stage is known as the "shrinkage limit".   The phenomenon of drying of soil by evaporation accompanied by shrinkage (and possibly cracking) is known as 'desiccation'.

When the shrinkage limit is reached, further drying causes no appreciable reduction in volume, the lost water being replaced by air drawn into the voids.   Shrinkage limit w_s is defined as the maximum moisture content below which the soil ceases to decrease in volume on further drying.   Below the shrinkage limit, the mass of soil goes on decreasing until it is fully dried.

Consider a saturated soil mass initially at volume V and mass M which is gradually dried to attain volume V_d and mass M_s in the dried state (Fig. 2.14).   During the process of drying, when soil reaches shrinkage limit its volume becomes V_d, (the same as will ultimately be in the fully dried state) and the soil remains still saturated.   The moisture content at stage marked 2 in Fig. 2.14 is thus the shrinkage limit which can be calculated as follows:

Mass of water at stage 1 $= M - M_s = M_w$

Volume reduction from stage 1 to 2 $= V - V_d$

Mass reduction from stage 1 to 2 $= (V - V_d)\, \rho_w$

Mass of water at stage 2 $= (M - M_s) - (V - V_d)\, \rho_w$

Hence, shrinkage limit w_s $= \dfrac{(M - M_s) - (V - V_d)\, \rho_w}{M_s}$

$$(2.80)$$

or $w_s = \left[w - \dfrac{(V - V_d)\, \rho_w}{M_s} \right]$ (2.81)

where w = moisture content (%) of the initial saturated sample of volume V

Defining shrinkage limit as the moisture content just sufficient to saturate a dry sample at constant volume, an alternative expression

for shrinkage limit can be obtained by calculating the water needed
to saturate the sample at stage 3 in Fig. 2.14.

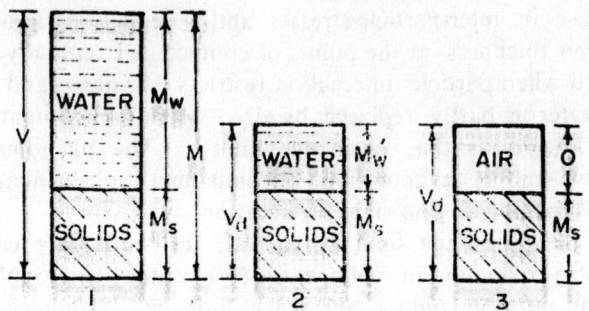

Fig. 2.14 Shrinkage limit calculation

$$w_s = \frac{(V_d - M_s / \rho_s) \rho_w}{M_s} = \frac{\rho_w}{\rho_d'} - \frac{1}{G} \qquad (2.82)$$

where ρ_d' = dry density of a *dry* sample

Substituting $\rho_d' = G\rho_w / (1 + e')$

$$w_s = \frac{e'}{G} \qquad (2.83)$$

where e' = void ratio of a dry sample (at minimum volume
obtained by drying)

If shrinkage limit is determined from Eq. 2.80, an approximate
value of G can be indirectly determined from Eq. 2.82.

(a) *Shrinkage limit determination.* A shrinkage dish of stainless steel or porcelain about 30—40 mm in diameter and 15 mm
deep is weighed and its volume V is measured by filling mercury
and weighing. Inside of the dish is greased. A soil sample passing
the 425 micron sieve is mixed with distilled water to form a creamy
paste which is filled in the dish in level with the rim. Dish containing soil is weighed to get initial wet mass of soil M. It is then
dried in an oven and weighed to get the dry mass M_s. The volume
V_d of the dry soil pat is obtained by displacement of mercury.
Mercury is filled in a glass or stainless steel cup over which a glass
plate with prongs is pressed. The dry soil pat is placed on the
mercury surface and submerged under mercury by pressing with the
glass plate with prongs. The mass of displaced mercury divided by

its density (13.55 g/ml) gives the dry volume V_d. The shrinkage limit is calculated from Eq. 2.80.

(b) *Shrinkage index I_s*. The numerical difference between liquid limit and shrinkage limit is termed the shrinkage index (Ranganathan and Satyanarayana 1965).

$$I_s = w_L - w_s \qquad (2.84)$$

I_s, so defined, can be used to find swell potential of soil (Section 2.9).

According to IS:2720—6. shrinkage index I_s' is the numerical difference between plastic limit and shrinkage limit.

$$I_s' = w_P - w_s \qquad (2.85)$$

(c) *Shrinkage ratio SR*. It is defined by the expression:

$$SR = \frac{\Delta V / V_d}{\Delta w} \qquad (2.86)$$

where ΔV = change in volume of soil for change of moisture content Δw

Upto the shrinkage limit, $\Delta V = (\Delta w M_s)/\rho_w$

Hence $$SR = \frac{M_s}{V_d} \cdot \frac{1}{\rho_w} = \frac{\rho_d'}{\rho_w} \qquad (2.87)$$

(d) *Volumetric shrinkage VS*. It is the change in volume, expressed as percentage of dry volume, which occurs when a soil is dried from a given moisture content w_1 upto the shrinkage limit.

$$VS = \frac{\Delta V}{V_d} \times 100 \qquad (2.88)$$

Combining with Eq. 2.86,

$$VS = (\Delta w) SR = (w_1 - w_s) SR \qquad (2.89)$$

(e) *Linear shrinkage LS*. Linear shrinkage is the decrease in one dimension of a soil sample, expressed as a percentage of the initial dimension when moisture content is reduced from a given value to the shrinkage limit (IS:2809). A soil paste prepared by mixing distilled water to soil sample passing the 425 micron sieve is filled in a shrinkage mould (SI:2720—20) having the form of a semi-cylindrical trough 140 mm in length and 12.5 mm in internal radius. The soil paste is first dried in air, then at a temperature of 60—65°C (in an oven) until shrinkage has largely ceased and finally completely dried at 105—110°C. The mean length of the dry specimen is measured. The linear shrinkage LS is calculated from the formula:

$$LS = \frac{L_i - L_d}{L_i} \times 100 \qquad (2.90)$$

where L_i == initial length of specimen

L_d = length of dry specimen

5. Activity. The plasticity of soil is closely related to the amount of water that is attracted to the surfaces of soil particles. Finer the particle, greater the specific surface and greater is the amount of water which is likely to be attracted. With this in mind Skenpton (1953) related the plasticity index to the percentage by weight of clay size particles (finer than 2 μm; 1 μm = 0.001 mm) and called it the "activity A".

$$\text{Activity 'A'} = \frac{\text{Plasticity index}}{\% < 2\mu\text{m}} \qquad (2.91)$$

Typical values of activity for some clay minerals are as follows: kaolinite—0.40; illite—0.90; Ca-montmorillonite—1.5; Na-montmorillonite—6.0. The higher the activity, the more important is the influence of the clay fraction. On the basis of activity a soil is classified as 'inactive' when $A < 0.75$, 'normal' when $A = 0.75$ —1.25, and 'active' when $A > 1.25$.

Studies (Seed, Woodword and Lundgren 1962) reveal that while there is a linear relationship between plasticity index I_p and percent clay fraction C (< 2 μm) it does not always pass through the origin. Thus, activity can be redefined as:

$$A = \frac{I_P}{C - n} \qquad (2.92)$$

where $n \approx 5$ for natural soils

6. Degree of consistency. The degree of consistency is described qualitatively by the terms 'very soft', 'soft', 'firm (or medium)', 'stiff', 'very stiff' and 'hard'. Approximate criteria for field identification and their rough estimate on the basis of soil strength are given in Table 2.2. Different criteria may not yield identical results.

Table 2.2

QUALITATIVE AND QUANTITATIVE CRITERIA OF DESCRIBING CONSISTENCY

Degree of con- sistency	Field identification	Consistency index (I_c)	Unconfined compressive strength[1] (kPa)	Static cone resistance[2] (kPa) (1×10^2)
Very soft	Exudes between fingers when squeezed in hand	$\leqslant 0$	< 20	
Soft	Moulded by light finger pressure	0—0.25	20—40	< 10
Firm	Can be moulded by strong finger pressure	0.25—0.5	40—75	10—20
Stiff	Cannot be moulded by fingers, can be indented by thumb	0.5—0.75	75—150	20—50
Very stiff	Can be indented by thumb nail	0.75—1.0	150—300	50—100
Hard	Indented with diffi- culty by thumb nail	1.0. $w \leqslant w_P$	> 300	> 100

1. See Chapter 9
2. See Chapter 15. 1 kPa $\simeq$ 0.01 kg/cm^2

7. Use of consistency limits. The consistency limits and related indices are very useful for soil identification and classification. The limits are often used in specifications for soil compaction and in semi-empirical methods of design. The liquid and plastic limits depend on both the type and amount of clay, but the plasticity index depends to a first approximation only on the amount of clay present. Their inter-relation can thus give information both about the type and amount of clay. Compressibility of soil increases markedly with increasing plastic limit, whereas strength decreases. On the other hand, the strength of soil increases with increasing plasticity index. Shrinkage limit is helpful in estimating the moisture content changes that may take place after construction of a foundation. The value of liquidity index (or consistency index) is indicative of the stress history of the soil (Chapter 14).

Examples 2.24 — 2.30

2.24 A liquid limit test by the Casagrande apparatus gave the following results:

Number of blows   16   20   30   43
Moisture content (%) 52.5 49.5 46.0 41.8

Plot the flow curve and find the liquid limit and the flow index.

From Fig. 2.10, liquid limit $= 47.5\%$

$$n_1 = 10, w_1 = 57.5\%; n_2 = 100, w_2 = 33\%$$

$$I_F = \frac{w_1 - w_2}{\log (n_2 / n_1)} = w_1 - w_2 = 24.5\%$$

2.25 Find the liquid limit by the 'one point method' if the moisture content at 30 blows is 46%.

$$w_L = w \left(\frac{n}{25}\right)^{0.1} = 46.8\%$$

2.26 A cone penetrometer test was performed on a soil sample with the following results:

Penetration (mm)   16.0   18.2   20.8   22.2   24.0
Moisture content(%)   51   54.1   56.8   59.2   60.9

Determine the liquid limit.

From Fig. 2.12, $w_L = 56\%$

2.27 A soil has a liquid limit of 65%, plastic limit of 24% and shrinkage limit of 11%. Find the void ratios at these limits, if the value of G is 2.72.

$$e_L = w_L G = 1.768$$
$$e_P = w_P G = 0.653$$
$$e_s = w_s G = 0.299$$

2.28 A saturated sample of soil weighing 39.4 g (M) has a volume of 20.4 cm³ (V). On drying it weighs 30.2 g (M_s) and its volume is determined by displacement of mercury as 16.0 cm³ (V_d). Find the shrinkage limit. The relative density G is 2.7. Find also the shrinkage ratio SR and volumetric shrinkage VS.

$$w_s = \frac{(M - M_s) - (V - V_d)\rho_w}{M_s} = 15.9\%$$

Alternately $\rho_d' = M_s/V_d = 1.887$ g/cm³

$$w_s = \frac{\rho_w}{\rho_d'} - \frac{1}{G} = 15.9\%$$

or $\quad e' = \dfrac{G\rho_w}{\rho_d'} - 1 = 0.43$

$$w_s = e'/G = 15.9\%$$

$$SR = \frac{M_s}{V_d} \cdot \frac{1}{\rho_w} = \frac{\rho_d'}{\rho_w} = 1.89$$

$$w_1 = \frac{M - M_s}{M_s} = 0.305$$

$$VS = (w_1 - w_s)\ SR = 27.5\%$$

Alternately: $VS = \dfrac{\Delta V}{V_d} \times 100 = \dfrac{V - V_d}{V_d} \times 100 = 27.5\%$

2.29 A soil sample was initially at its liquid limit of 40%. On drying it weighs 26.8 g (M_s) and has a volume of 14.5 cm³. Find the shrinkage limit, shrinkage index and volumetric shrinkage, assuming G as 2.7.

$$\rho_d' = M_s / V_d = 1.848\ \text{g/cm}^3$$

$$w_s = \frac{\rho_w}{\rho_d'} - \frac{1}{G} = 17.1\%$$

$$I_s = w_L - w_s = 22.9$$

$$SR = \rho_d' / \rho_w = 1.848$$

$$VS = (w_L - w_s)\ SR = 42.3\%$$

2.30 A dry sample of soil weighs 35.5 g and has a volume of 18.2 cm³. Find the shrinkage limit, if G is 2.75.

$$\rho_d' = M_s / V_d = 1.95\ \text{g/cm}^3$$

$$w_s = 14.9\%$$

2.7 FABRIC AND STRUCTURE

The arrangement of particles and pore spaces and inter-particle forces in soil (and rock) masses are described in terms of "fabric" and "structure", which are used by some interchangeably. Making a distinction between these two terms, the following terminology is adopted:

(a) *Fabric.* Arrangement of individual particles, particle groups, and pore spaces in a soil (or rock) mass.

(b) *Structure.* Gradation and arrangement of particles, porosity and pore-size distribution, bonding agents, and specific inter-particle forces in a soil mass.

(c) *Mass structure.* Larger scale inter-relationship of textural features of the entire formation of a given soil (or rock) mass. (*Texture* refers to particle size composition.)

The arrangement of individual particles in granular soils is also referred to as "packing", instead of fabric. On the basis of degree of magnification required to study the particle arrangement, fabric may be broadly classified as "microfabric" and "macrofabric". Microfabric requires at least an optical microscope for study. Macrofabric is distinguishable with the naked eye or a hand lens.

In general, soil particles (specially the clay particles) tend to form groups or group units having definable physical boundaries. An identiable group or assemblage of particles is called a "fabric unit". The smallest group of particles consisting of two or more particles acting as a unit which can be visually observed in the ultra-microscopic level using electron microscopy is called "domain". Several domains may combine to form a "cluster" or a "floc" which becomes possible to be observed under the optical microscope. Clusters, in turn, combine to form "peds" or "aggregates" that can be identified visually with the naked eye. In other words, peds are the macroscopic fabric units. Brewer (1964) defined ped as an individual soil aggregate consisting of a cluster of primary particles, and separated from adjoining peds by surfaces of weaknesses.

1. Granular soil fabric. (a) *Single grain fabric*. The depoition and settlement of granular particles (gravels, sands and silts > 0.02 mm) is largely influenced by their individual weight as the particle surface forces are negligible. They settle individually and independently of one another. The weight of a particular particle causes it to settle and role to a position of equilibrium amongst the other particles. The resulting arrangement (Fig. 2.15) of single particles is called the "single grain fabric".

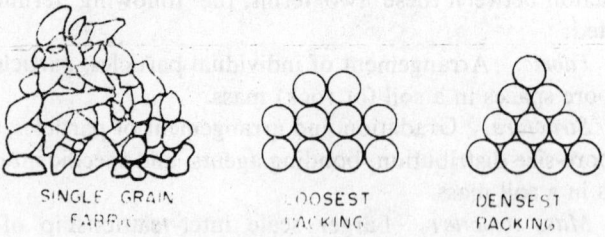

SINGLE GRAIN
FABRIC

LOOSEST
PACKING

DENSEST
PACKING

Fig. 2.15 Single grain fabric of granular soil and packing of equal spheres

The fabric (or packing) of granular soils is very strongly influenced by particle shape and size distribution. The particles may form a loose or dense deposit. For comparison, it may be of interest to know that equal spheres in their loosest and densest packings (Fig. 2.15) have void ratios 0.91 and 0.35 respectively (porposity 47.64 and 25.95). Well graded sands and silts usually attain denser packing and have lower values of void ratio in both the loosest and densest states in comparison to equal spheres. On the other hand, irregular particle shapes result in a tendency towards lower densities and higher void ratios. State of packing has an important bearing on the stability and deformation of granular soils.

(b) *Honeycomb fabric.* Certain silty deposits are observed to have void ratio exceeding the upper limit of about 1.0 for single grain fabric of bulky particles. It is thought that grains of silt and rock flour start settling out of still water more or less as single grains. As they approach near the bottom and start coming in contact with one another, their behaviour can be to some extent influenced by surface forces of magnitude larger than their submerged weight. This prevents the particles from rolling down immediately into positions of equilibrium among the particles already deposited. They remain attracted to one another and form miniature arches bridging over relatively large void spaces. The resulting arrangement (Fig. 2.16) is called a 'honeycomb' fabric (Terzaghi 1925). Honeycomb fabric may also develop under certain conditions in silt and clay mixtures. This type of fabric is usually metastable and may collapse suddenly or liquefy under the action of rapidly applied stress.

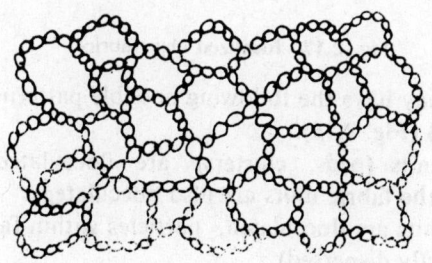

Fig. 2.16   Honeycomb fabric

2. Clay soil fabric. The total fabric of clay soils includes the arrangement of individual particles in fabric units and the arrangement of groups of particles, i.e. of individual fabric units and sub-units. Typical arrangements have been proposed by Lambe (1953, 1958), van Olphen (1963), Smart (1969), Young and Shearan (1973), and Collins and McGown (1974). To simplify fabric description, two basic types of arrangements can be thought of as *flocculated* and *dispersed* (or *oriented*). Flocculated means the edge-to-edge or edge-to-face association of individual particles, or of groups of particles. Dispersed means a more or less parallel or face-to-face association of individual particles or of groups of particles. There may be an infinite number of intermediate stages between the above two basic types. Flocculated and dispersed fabrics are shown in Fig. 2.17.

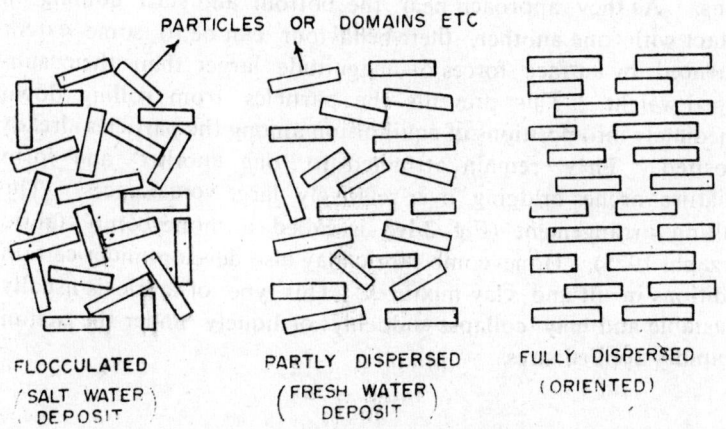

PARTICLES OR DOMAINS ETC

FLOCCULATED
(SALT WATER
DEPOSIT)

PARTLY DISPERSED
(FRESH WATER
DEPOSIT)

FULLY DISPERSED
(ORIENTED)

Fig. 2.17 Idealized clay fabrics

Total fabric may have the following possible patterns (Young and Warkenten 1975) (Fig. 2.18).

(a) Fabric units (peds, clusters) are flocculated, individual particles within the fabric units are also flocculated.

(b) Fabric units are flocculated, particles within fabric units are dispersed (or partly dispersed).

(c) Fabric units are dispersed, particles within fabric units are flocculated.

(d)   Fabric units are dispersed, particles within fabric units are also dispersed.

The four types of total fabric are shown in Fig. 2.18.

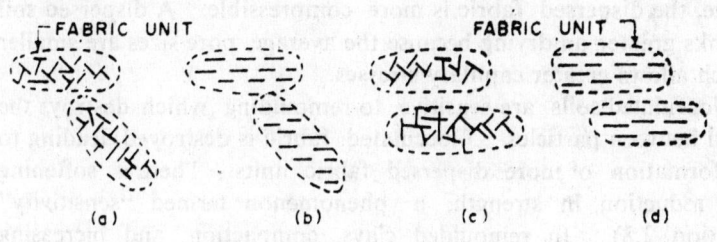

(a)　　　　　(b)　　　　(c)　　　　　(d)

Fig. 2.18 Types of total fabric

The type of fabric likely to develop in clay suspensions and deposits depends largely on inter-particle forces and the composition of the water in which deposition takes place. The presence and proportion of bulky grains also influence final arrangement of the clay particles.   Inter-particle forces of attraction and repulsion have been discussed in Section 1.8. A flocculated fabric results when attraction predominates between particles and between fabric units and a dispersed fabric results when there is net repulsion. As dissolved salts decrease inter-particle repulsion (by depressing diffuse ion layer), attractive forces become dominant and thereby the tendency of flocculation increases.   Clays deposited in salt water (marine environment) develop a characteristic flocculated fabric with typical edge-to-face arrangement (Fig. 2.17).   The fabric is more open and considerable free water gets trapped in the large voids.   Organic acid from plant decay in shallow ponds and marshes also induces a high degree of flocculation.   On the other hand, fresh water deposits from rivers and in lakes where ion concentration is low are likely to be partially flocculated or even dispersed.   In this case, only a few particles are in direct contact. most of them being separated by the adsorbed water layers.

The engineering behaviour of a soil depends very much on its fabric and structure.   Flocculated soils have high void ratios, low density, and quite probably high moisture content.   They are, however, relatively strong and resistant to external forces because of inter-particle attraction. Compared at equal void ratio, a flocculated soil is more permeable than a dispersed soil. This is because the flow channels (pores) are larger in size, although fewer in number, in flocculated fabric. A flocculated fabric is more compressible than a dispersed fabric under high pressures. In the low pressure range, the dispersed fabric is more compressible. A dispersed soil shrinks greater on drying because the average pore sizes are smaller which allows greater capillary stresses.

Flocculated soils are sensitive to remoulding which destroys the bond between particles. Flocculated fabric is destroyed leading to the formation of more dispersed fabric units. There is softening and reduction in strength, a phenomenon termed "sensitivity" (Section 2.8). In remoulded clays, compaction and increasing moisture content induce preferred orientation (dispersion). Clays compacted dry of optimum moisture content (Chapter 21) have a flocculated fabric. Partial orientation of particles takes place if compacted at moisture content above the optimum; the greater the moulding moisture content, the more dispersed would be the fabric.

3. **Composite soil fabric.** A wide variety of fabrics can develop in composite soils depending on the relative amounts of the clay particles and bulky particles, the type of binder, and the method of deposition. In case clay particles are sufficient enough to occupy a volume more that about twice that of bulky particles, the latter may not be able to attain a grain-to-grain contact and they appear to float in a cohesive matrix. The arrangement may be termed "cohesive matrix fabric" Clay particles, in themselves, may exist in flocculated or dispersed fabric units with localised departures from the pattern in the vicinity of bulky particles. With increased proportion of bulky particle, a grain-to-grain contact gets established forming a frame work, which, the author terms as "bulky-grain framed fabric". The frame-work may be loose and honeycombed, or it may be quite rigid and incompresible depending on the packing of bulky particles and the inter-particle binder. Inter-particle bonding may be through clay particles or other cementing agents. Voids of the bulky frame-work are filled up usually by flocculated clay fabric units.

4. **Mass structure.** The types of mass structure of soils and their field identification are as follows:
    (1) *Homogeneous.* Essentially of one type of texture and exhibiting the same physical properties.

(2) *Heterogeneous.* A mixture of different types.

(3) *Stratified.* Alternating layers of different soil textures.

(4) *Laminated.* Stratified with thin layers, usually less than 6 mm.

(5) *Varved.* Special case of laminated structure, composed of thin laminae of varying colour indicating seasonal deposition of clay, and silt or sand alternately.

(6) *Fissured.* Presence of shrinkage cracks, frequently filled with fine sand or silt; breaks into polyhedral fragments along fissures.

(7) *Intact.* No fissures.

(8) *Slickensided.* Contains fracture planes of weaknesses which are smooth and glossy in apprearance: a variety of fissured structure.

(9) *Lensed.* One type of soil containing lenses (pockets) of other materials.

(10) *Friable.* Can be powdered or broken easily with hands, usually dry.

(11) *Weathered (Crumbly).* Usually has crumb or columnar structure, breaks into small blocks and crumbs on drying; pertains to cohesive soils.

2.8 SENSITIVITY AND THIXOTROPY

Sensitivity is a measure of the change in consistency or strength of a clay soil on remoulding. In terms of shear strength, sensitivity 'S_t' is defined as the ratio of the unconfined compressive strength q_u (Chapter 9) of the undisturbed soil to the remoulded or disturbed strength of soil reconstituted at the same natural density and moisture content:

$$S_t = \frac{q_u \ (\text{undisturbed})}{q_u \ (\text{remoulded})} \qquad (2.93)$$

Vane shear strength or Swedish fall-cone test strength (Chapter 9) may also be used as an alternative to the unconfined compressive strength for defining sensitivity in case of soft soils having low remoulding strength.

The sensitivity of most of the clays ranges from 2 to 4, for marine clay from 1.6 to 26 and for peat from 1.5 to 10 (Winterkorn and

Fang 1975).   Clay may be classified as given in Table 2.3 on the basis of their sensitivity (Skempton and Northey 1952, Bjerrum 1954).

Table 2.3

CLASSIFICATION OF CLAY ON THE BASIS OF SENSITIVITY

Sensitivity	Classification
<2	Insensitive
2—4	Medium sensitive
4—8	Sensitive
8—16	Very sensitive
16—32	Slightly quick
32—64	Medium quick
>64	Quick

The decrease in strength on remoulding of soil is principally due to destruction of soil structure and inter-particle bonds that exist in the natural soil.   Also the structure of the adsorbed water is broken up. Subsequent to remoulding, and with passage of time, however, the strength increases, though not necessarily back to the original value. This phenomenon of loss in strength on remoulding and gain in strength on standing, with no change in volume or moisture content is termed as "thixotropy".   The strength regain is called 'thixotropic regain' or 'rest-hardening'.

Several factors contribute to sensitivity of soil, such as metastable fabric, cementation, thixotropic bonding, weathering, leaching and ion exchange.   Flocculated clayey soils possess a metastable fabric which on remoulding has a tendency for the volume to decrease leading to reduction in effective stresses under which the fabric is in equilibrium and hence the strength is less.   Some soils have inter-particle bonding through cementing agents (free carbonates, iron oxide, alumina and organic matter) which are destroyed on remoulding leading to a loss of strength.   Thixotropy is the property of certain suspensions of very fine particles to stiffen (form a *gel*) while at rest and to soften or liquefy (form a *sol*) upon remoulding.   This phenomenon is different from flocculation (coagulation) in that during flocculation separate flocks form with no bond between them, whereas in thixotropic hardening all the particles form a continuous frame-work to develop a thixotropic structure (Wilun

and Starzewski 1975).   This thixotropic bonding of particles can be
destroyed by remoulding, shock or vibration resulting to strength
loss.   Knowing that interparticle repulsion reduces strength, it can
be inferred that the development of sensitive and quick clays requires
an increase in inter-particle repulsion   through expansion of the
double layers.   Decrease in salt concentration through leaching
increases double layer thickness.   Mere leaching of salts may not be
sufficient to develop highly sensitive and quick clays.   The ratio of
monovalent to   total cations   should   increase through selective
removal of divalent cations, possibly by organic matter (Soderblom
1969).   Weathering processes   also contribute to sensitivity by
causing changes in the types and relative proportions of ions in soil
water.

Thixotropic strength regain following remoulding depends on the
original structure, activity of the clay minerals and the degree of
disturbance.   Remoulding induces orientation and dispersion.   Loss
in strength is greater for a flocculated clay than for a partly oriented
clay at the same void ratio and moisture content, since a partly
oriented clay possesses a measure of dispersion to begin with.
Rest-hardening reverses the process so as to restore inter-particle
force equilibrium (i.e. to increase attraction).   The percentage
increase in strength will be more for an initially partly dispersed
fabric than for a flocculated fabric, although, quantitatively the total
regain might be greater for the latter.   Thixotropic effects are
greater for active clays that absorb large quantities of water, such as
montmorillonites, than for less active or inactive clay types, such
as kaolinite.

2.9 SWELLING

Swelling is the increase in volume of a fine grained soil on wetting,
which may occur in two ways — intercrystalline and intracrystalline
(Grim 1962).   In relatively dry clays, particles are held by capillary
forces.   On wetting, these forces get relaxed and the soil expands.
The intake of moisture content is restricted to the external crystal
surfaces and the voids between the crystals.   This is known as
intercrystalline swelling which takes place in any type of clay
irrespective of its mineralogical composition and the process is
reversible.

Intracrystalline swelling is mainly due to the absorption of moisture into the lattice structure of the clay minerals. The amount of swelling depends on a number of factors such as type and amount of clay minerals and their arrangement on orientation, valence of exchangeable cations, pore water salt concentration, and cementing bonds between particles. Monovalent exchangeable cations, e.g. sodium, cause greater swelling than divalent calcium cations. An increase in pore water salt concentration decreases swelling. Relative swelling of some clay minerals is given in Table 1.4. For compacted clays, swelling depends on the density, moisture content at compaction and the method of compaction. Denser and drier the clay the greater is the swelling on wetting. Static compaction allows more swell than impact or dynamic compaction. Surcharge loads reduce swelling. Identification criteria for swelling soils are described in Chapter 3.

1. **Differential free swell determination.** Two oven dry soil samples, 10 g each, passing a 425 micron sieve are placed separately in two 100 ml graduated glass cylinders. Distilled water is filled in one cylinder and kerosene (a non-polar liquid) in the other cylinder upto the 100 ml mark. The final volume of soil is read after 24 hours (or more). The "differential free swell S_d" is calculated from the relation :

$$S_d = \frac{\text{soil vol. in water} - \text{soil vol. iu kerosene}}{\text{soil volume in kerosene}} \times 100 \quad (2.94)$$

The differential free swell is used as an indication of the degree of expansion of a soil (IS: 2911—3). Differential free swell is also termed 'free swell index' (IS: 2720—11).

2. **Swell pressure determination.** The pressure exerted by a swelling soil against a confining surcharge when water is available for volume increase is known as 'swell pressure P_s'. The swell pressure is usually determined by testing a laterally confined specimen for which a number of devices are described in literature. The author's device* (Singh 1966, 1981) consists of a swelling pressure cylindrical mould 50 cm^2 in sectional area and 6 cm high, in which a 24 cm thick test specimen is placed in-between two porous stones. Water is allowed to enter the specimen through both the top and

---

*For detailed description of apparatus and test procedure please see author's book Soil Engineering in Theory and Practice, Vol. 2, Geotechnical Testing and Instrumentation, Asia Publishing House, Bombay, 1981.

bottom faces. Swell pressure is measured as the restraint (by means of a load cell or a proving ring) which has to be applied through the load frame to keep the volume constant without permitting any expansion.

3. Swell potential determination. The swell potential S_p of a soil is a measure of the ability and degree to which a soil might swell if its environment were to be changed in some definite way. Swell potential, i.e. the amount of swell, can be inferred from the Atterberg limits and indices, differential free swell, and colloid content particles usually smaller than $1 \mu m$ or $2 \mu m$. A direct measurement of swell potential is made in terms of the percentage swelling of a specimen on wetting.

According to Seed et al (1962), swell potential is determined on a specimen compacted at the optimum moisture content in the Proctor mould by the IS light compaction (Chapter 21). For convenience of specimen preparation, the 100 mm diameter mould is made of three parts so that the central part, 25 mm high, could serve as a confining ring during a swell test. The soil in this ring is trimmed from the compacted sample, covered on top and bottom with porous stones, and then allowed to swell by giving it access to water while at the same time subjected to a surcharge of 6.9 kPa (1 lb/in², 0.07 kg/cm²). Swell potential S_p is defined as the percentage swell of laterally cofined sample on soaking under 6.9 kPa (1 lb/in²) surcharge after being compacted to maximum density at optimum moisture content according to the Standard Proctor Compaction test (IS light compaction test).

The following correlation is found to exist between swell potential S_p, activity 'A' (Eq. 2.91 or 2.92) and percent clay C ($<2 \mu m$):

$$S_p = 3.6 \times 10^{-5} \times A^{2.44} \times C^{3.44} \qquad (2.95)$$

For natural soils with clay content 8 to 65%, S_p can be related to the plasticity index I_p, with an accuracy of $\pm$ 33 percent, as follows (Seed et al 1962):

$$S_p = 2.16 \times 10^{-3} (I_p)^{2.44} \qquad (2.96)$$

The swell potential of some soils, specially the black cotton soils of India, can better be obtained by the following relation obtained by Ranganathan and Satyanarayan (1965):

$$S_p = \frac{1}{6.3} (I_s)^{1.17} \qquad (2.97)$$

where I_s = shrinkage index = $w_L - w_s$

Another relationship derived from extensive tests on undisturbed samples is as follows (Chen 1975):

$$S_p = B \, e^{A(I_p)} \tag{2.98}$$

in which $A = 0.0838$ and $B = 0.2558$

The author (Singh 1981) defined the swell potential as the percentage change in thickness of a confined specimen from the condition of full swell pressure to the condition of nominal pressure of 5 kPa (0.05 kg/cm²). A laterally confined 24 mm thick specimen (50 cm² in section) is allowed to develop maximum swell pressure on submergence at practically no volume change. The pressure is then reduced in decrements to the nominal value of 5 kPa (0.05 kg/cm²) and the increase in thickness is recorded, from which S_p is calculated.

2.10 COLLAPSE

Collapse (or subsidence) is the phenomenon of relatively large decrease in bulk volume of certain soils on their becoming saturated or by saturation and loading acting together. Such soils are known as "collapsible" or "metastable" soils. Collapsible soils may be residual, water deposited or wind deposited (Dudley 1970). They are composed mainly of bulky grains, often in the silt to fine sand range. They occur usually in arid and semi-arid regions. Most of the collapsible soils have liquid limit below 45 and plasticity index below 25, usually much lower even down to the nonplastic condition (Dudley 1970). Loess is a typical example of collapsible soil.

For collapse to occur a soil must have a partly unstable, open fabric (i.e. large voids ratio) with moisture content less than saturation. There must be a temporary source of strength to stabilize the fabric when dry. The predominant source of strength appears to be derived from capillary tensions in the pore water; clay binder and other cementing agents may also play some role. These temporary sources of strength are reduced by addition of water and the fabric collapses causing considerable subsidence. Identification criteria for collapsible soils are given in Chapter 3.

2.11 DISPERSIBILITY

Certain clays and clayey soils called 'dispersive' soils erode by a process in which the individual clay particles go into suspension in practically still water. This process is quite different from the normal erosion of ordinary clays which requires considerable velocity in the eroding water. Dispersive soils are highly erodible even under low flow velocity. Erosion of dispersive soils has been the cause of failure of a number of well constructed low, homogeneous dams and other earth structures. Dispersive erosion (piping) in dams usually commences from the upstream face, which is fundamentally different from the piping of cohesionless soils and which develops starting at the discharge face (Chapter 7).

Soil dispersibility* (susceptibility to colloidal erosion) can be studied by the pinhole test, crumb test and hydrometer dispersion test standardized by Sherard et al (1972, 1976a, 1976b). The pinhole test is the most reliable single test for identifying dispersive soils. In this test distilled water is allowed to flow through a 1.0 mm diameter hole drilled through a compacted specimen at velocities of about 0.3 m/s to 3 m/s. The water becomes muddy and the hole erodes rapidly in dispersive clays. For non-dispersive clays, the water remains clear and there is no erosion. In the crumb test, a 6—10 mm crumb of soil at natural moisture content is dropped into a beaker containing about 150 ml of distilled water and the tendency of the clay particles to go into suspension is observed. In the hydrometer dispersion test, the percentage of particles finer than 5 μm (0.005 mm) is determined by hydrometer analyses on samples with and without dispersing agents in the suspension. 'Percent dispersion' is calculated as the ratio of the minus 5 μm particles without dispersing agent to that measured with a dispersing agent. Soils with more than 50 percent dispersion are generally dispersive. The main factor governing dispersibility is the percentage of sodium in the pore water in relation to other cations. Dispersive soils have a higher content of sodium cations.

---

*For detailed test procedures. please see author's book Soil Engineering in Theory and Practice, Vol. 2, Geotechnical Testing and Instrumentation, Asia Publishing House, Bombay, 1981.

3
Soil classification and identification

The object of soil classification is to divide soils into a limited number of groups, on the basis of a few key characteristics — the grading and the plasticity. These characteristics play a major role in determining the engineering performance of the soil and provide a quick means of identification. A classification system is also meant to provide a common language for the exchange of information and experience about various types of soil.

3.1 UNIFIED SOIL CLASSIFICATION

The Unified Soil Classification (USC) system originally proposed in the United States (Wagner 1957) is a widely adopted system. The system is so simple that fairly accurate classification can be made even by soil examination in the field. Soil is identified and allocated to an appropriate group on the basis of grading and plasticity, after excluding "boulders" (> 300 mm) and "cobbles" (300 mm —75 mm). Each group is represented by a group symbol consisting of a primary and a secondary descriptive letter. The descriptive letters and the basic soil components are given in Table 3.1. The essential features of the USC—system are given in Tables 3.2 and 3.3, and the plasticity chart to be used for classifying fine grained soils is shown in Fig. 3.1. Soil *fines* mentioned in Table 3.2. refer to the fraction smaller than 75 μm (micron) sieve. C_u is the uniformity coefficient (Eq. 2.65) and C_c is the coefficient of curvature (Eq. 2.66).

The 'A' line on the plasticity chart (Fig. 3.1) has the following equation:

$$I_p = 0.73 \, (w_L - 20 \%) \qquad (3.1)$$

Inorganic clays plot above the 'A' line, and the inorganic silts and organic soils plot below the 'A' line.   A soil falling above 'A' line with I_p between 4 and 7 represents a borderline case and is given a dual symbol, e.g. ML—CL.

Table 3.1

DESCRIPTIVE LETTERS AND BASIC SOIL COMPONENTS IN THE UNIFIED
SOIL CLASSIFICATION SYSTEM

Primary letter	Soil component	Size and description	Secondary letter	Description
G	Gravel	Bulky, hard rock particles (75 mm—4.75 mm) *Coarse*: 75 mm—19 mm *Fine*: 19 mm—4.75 mm	W P M	Well graded Poorly graded Non-plastic fines
S	Sand	Bulky, hard rock particles (4.75 mm—75 μm) *Coarse*: 4.75 mm—2.0 mm) *Medium*: 2.0 mm—425 μm *Fine*: 425 μm—75 μm	C L H	Plastic fines Low plasticity ($w_L < 50\%$) High plasticity ($w_L > 50\%$)
M	Silt	Particles smaller than 75 μm, identified by plasticity		
C	Clay	Particles smaller than 75 μm, identified by plasticity		
O	Organic matter	Organic matter in various sizes and stages of decomposition		
P_t	Peat	Highly organic soils		

Note. 63 μm sieve may be adopted in place of 75 μm sieve.

Table 3.2

UNIFIED SOIL CLASSIFICATION OF COARSE GRAINED SOILS (MORE THAN HALF OF MATERIALS IS LARGER THAN 75 μm SIEVE SIZE)

Primary division	Group symbols	Description	Fines (%)	Laboratory criteria		Remarks
				Grading/plasticity		
Gravels (more than 50% of coarse fraction of gravel size)	GW	Well graded gravels, sandy gravels, with little or no fines	0—5	$C_u > 4$ $1 < C_c < 3$		Use dual symbols if fines are 5—12% or if $I_p = 4$—7. For example:
	GP	Poorly graded gravels, sandy gravels, with little or no fines	0—5	Not satisfying GW requirements		GW — GM
	GM	Silty gravels, silty sandy gravels	>12	Below 'A' line or $I_p < 4$		SM — SC
	GC	Clayey gravels, clayey sandy gravels	>12	Above 'A' line or $I_p > 7$		
Sands (more than 50% of coarse fraction of sand size)	SW	Well graded sands, gravelly sands, with little or no fines	0—5	$C_u > 6$ $1 < C_c < 3$		
	SP	Poorly graded sands, gravelly sands, with little or no fines	0—5	Not satisfying SW requirements		
	SM	Silty sands	>12	Below 'A' line or $I_p < 4$		
	SC	Clayey sands	>12	Above 'A' line or $I_p > 7$		

Note. Soil fraction smaller than 75 μm size is termed 'fines'.

Table 3.3

UNIFIED SOIL CLASSIFICATION OF FINE GRAINED AND ORGANIC SOILS (MORE THAN HALF OF MATERIAL IS SMALLER THAN 75 μm SIEVE SIZE)

Primary division	Group symbol	Description	Field criteria		
			Dilatancy	Dry strength	Toughness
Silts and clays (liquid limit less than 50%)	ML	Inorganic silts, silty or clayey fine sands, with slight plasticity	Quick to slow	None to slight	None
	CL	Inorganic clays, silty clays, sandy clays, gravelly clays of low plasticity, lean clays	None to very slow	Medium to high	Medium
	OL	Organic silts and organic silty clays of low plasticity	Slow	Slight to medium	Slight
Silts and clays (liquid limit more than 50%)	MH	Inorganic silts of high plasticity	Slow to None	Slight to medium	Slight to medium
	CH	Inorganic clays of high plasticity, fat clays	None	High to very high	High
	OH	Organic clays of high plasticity	None to very slow	Medium to high	Slight to medium
Highly organic soils	Pt	Peat and other highly organic soils	Organic colour, odour, spongy feel, frequently fibrous		

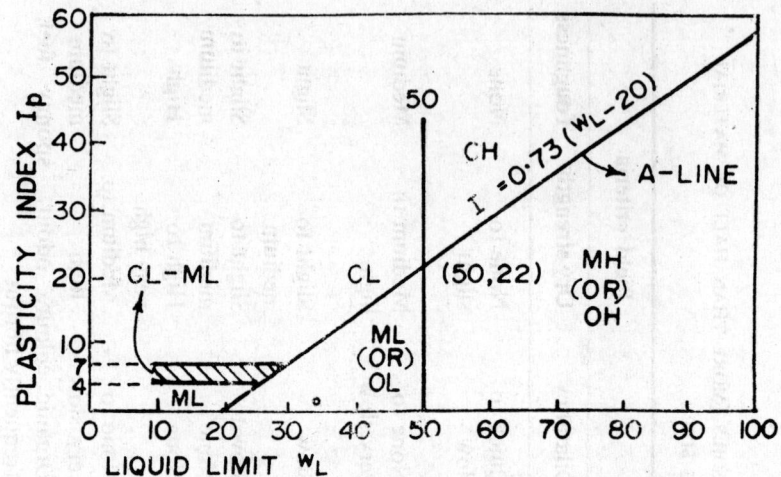

Fig 3.1 Plasticity chart (USC system) for laboratory classification of fine-grained soils

1. Field identification and classification. A soil sample is spread on a flat surface and, based on observation, is classified as "coarse grained" if more than half of the material is visible to the naked eye, otherwise it is classified as "fine grained". Coarse grained soil is classified as "gravel" or "sand" depending on whether more than half of the coarse fraction is of gravel size (larger than about 5 mm) or of sand size. Gravels and sands are classified as 'clean' (i.e. with little or no fines) or 'with appreciable fines'. Clean gravels and sands are either well graded or poorly graded, the former having wide range in grain size and substantial amount of all intermediate particle sizes, and the latter has predominantly one size or a range of sizes with some intermediate sizes missing. Gravels and sands with appreciable fines are classified according to the nature of the fines as for fine grained soil described below.

For fine grained soils or for the fine fraction of coarse grained soils the following tests are performed after removing particles larger than about 0.5 mm (425 μm sieve size).

(a) *Dilatancy test.* A small pat of moist soil is prepared by adding enough water to make it soft but not sticky. The pat is placed in the open palm of one hand and shaken horizontally by striking against the other hand several times. If water appears on the surface of the pat giving a glossy appearance, the pat is said to have a 'positive reaction'. On pressing between the fingers, the water and glossiness disappear from the surface, the pat becomes stiff and eventually cracks or crumbs. The phenomenon of appearance of water on shaking and disappearance on squeezing, followed by cracking, is termed "dilatancy". The rapidity of appearance and disappearance of water is observed. Very fine clean sands give the quickest positive reaction. Inorganic silts and rock flour show a moderately quick reaction and plastic clays no reaction.

(b) *Dry strength test.* The strength of a completely dry pat of soil (oven dried or sun dried) is assessed by breaking and crumbling between the fingers. The dry strength depends on the nature and the proportion of colloidal clay particles; the strength increases with increasing plasticity. Dry strength is recognised as follows:

Slight: Can be easily broken between fingers

Medium: Can be broken between fingers with considerable pressure

High: Can barely be broken under palm of hand with the aid of body weight

Very high: Cannot be broken under palm of hand

(c) *Toughness test.* Toughness is the resistance to moulding at the plastic limit. A soil is brought to plastic limit and rolled into threads to break at about 3 mm in diameter. The broken pieces are lumped together and subjected to kneading until the lump crumbles. The tougher the thread and stiffer the kneaded lump just before it crumbles, the greater is the activity of colloidal clay fraction in the soil, which indicates a clay of high plasticity. Weakness of the thread at the plastic limit and quick loss of coherence of the lump below the plastic limit indicate either an inorganic clay of low plasticity or an organic clay. Non-plastic soils cannot be rolled into 3 mm diameters at any moisture content. Highly organic clays have a very weak and spongy feel at the plastic limit.

(d) *Other tests.* Fine sands are felt gritty, silts rough and clays smooth. Wet clay sticks to fingers and does not wash off readily. Silt is washed off easily, or brushed off, if dry. A slightly moist lump of highly plastic soil produces a shining surface on being rubbed with a knife blade. Calcium carbonate, if present, may also indicate high dry strength, which should be distinguished by treating soil with a little dilute hydrochloric acid.

Field identification criteria based on dilatancy, dry strength and toughness are given in Table 3.3. The broad identification between fine sand, silt and clay is given in Table 3.4.

Table 3.4

BROAD IDENTIFICATION BETWEEN FINE SAND, SILT AND CLAY

Fine sand	Silt	Clay
Individual particles visible	Some particles visible	No particles visible
Exhibits dilatancy	Exhibits dilatancy	No dilatancy
Easy to crumble when dry	Easy to crumble when dry	Hard to crumble when dry
Gritty feel	Rough feel	Smooth feel
Non-plastic	Non-plastic to partly plastic	Plastic

(e) *Organic silt and clay.* Organic soils usually have darker colours (grey, brown or almost black) and a distinctive odour; odour can be made more noticeable by heating a wet sample.

Note. In case organic soils cannot be easily recognised, liquid limit of the soil is determined before and after oven drying. A decrease in liquid limit by over 25 percent on oven drying indicates organic soils.

(f) *Peat.* Highly organic soils are characterized by undecayed vegetable matter giving the soil a fibrous texture. They are spongy, dull brown to black in colour and have a characteristic odour.

2. Comparative soil properties on plasticity chart. The variation of some of the soil properties depending on the plotted position on the plasticity chart (Fig. 3.1) is given in Table 3.5.

Table 3.5

VARIATION OF SOIL PROPERTIES WITH LIQUID LIMIT AND PLASTICITY INDEX

Property	At equal w_L, with increasing I_p	At equal I_p, with increasing w_L
Dry strength	Increases	Decreases
Toughness	Increases	Decreases
Compressibility	About the same	Increases
Permeability	Decreases	Increases

3. Soil description. As a complement to the appropriate group symbol, a general description of the soil is also given which usually includes information about colour, particle size, shape and gradation, compactness and consistency, plasticity, dry strength, mass structure and any other pertinent geological or local feature.

3.2 INDIAN SOIL CLASSIFICATION

The Indian Soil Classification (ISC) system adopted by the Indian Standards Institution (IS: 1498) is essentially the USC system with modification in the classification of the fine grained soils only.   The basic soil components are the same as given in Table 3.1.   Silt and clay are the fractions smaller than 75 μm sieve (or the 63 μm sieve) to be identified by plasticity.   However, a division may be reported as 'silt size' from 75 μm to 2 μm and 'clay size' smaller than 2 μm. Classification of coarse grained soil is the same as given in Table 3.2. Fine grained soils are divided into three categories of plasticity: low plasticity with $w_L < 35\%$ (symbol L), medium plasticity with $w_L = 35—50\%$ (Symbol I), and high plasticity with $w_L > 50\%$ (symbol H).   The ISC plasticity chart is shown in Fig. 3.2 and the groups of fine grained soils are given in Table 3.6.

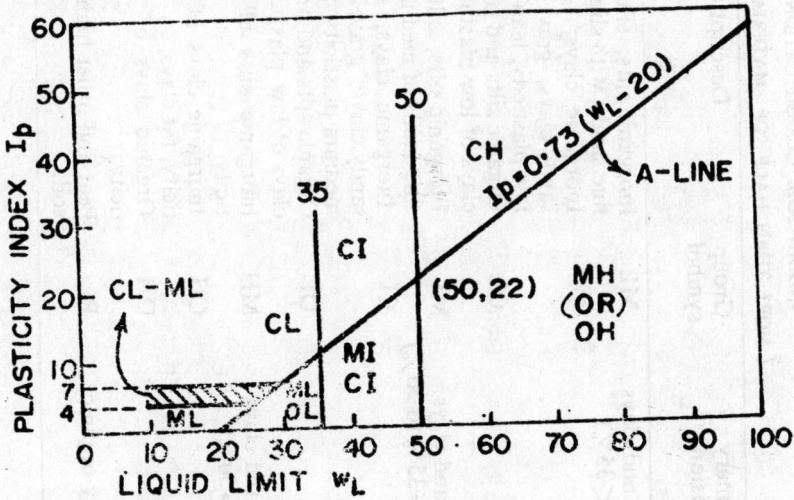

Fig. 3.2 Plasticity chart for laboratory classification of fine-grained soils
(Indian Soil Classification system)

Table 3.6

INDIAN SOIL CLASSIFICATION OF FINE GRAINED AND ORGANIC SOILS

(MORE THAN HALF OF MATERIAL IS SMALLER THAN 75 μm SIEVE SIZE)

Primary division	Group symbol	Description	Field criteria		
			Dilatancy	Dry strength	Toughness
Silts and clays ($w_L < 35\%$)	ML	Inorganic silts, silty or clayey fine sands, with slight plasticity	Quick	None to slight	None
	CL	Inorganic clays, silty clays, sandy clays, gravelly clays of low plasticity, lean clays	None to very slow	Medium	Medium
	OL	Organic silts and organic silty clays of low plasticity	Slow	Slight	Slight
Silts and clays ($w_L = 35\%$ to 50%)	MI	Inorganic silts, silty or clayey fine sands of medium plasticity	Quick to slow	Slight	None
	CI	Inorganic clays, silty clays, sandy clays, gravelly clays of medium plasticity, lean clays	None	Medium to high	Medium
	OI	Organic silts and organic silty clays of low plasticity	Slow	Slight to medium	Slight
Silts and clays ($w_L > 50\%$)	MH	Inorganic silts of high plasticity	Slow to none	Slight to medium	Slight to medium
	CH	Inorganic clays of high plasticity, fat clays	None	High to very high	High
	OH	Organic clays of high plasticity	None to very slow	Medium to high	Slight to medium
Highly organic soils	P_t	Peat and other highly organic soils	Organic colour, odour, spongy feel, frequently fibrous		

1. Black cotton soils. Black cotton soils (IS: 1498) are inorganic clays of medium to high plasticity and form a major soil group in India. The majority of the soils, when plotted on the plasticity chart, lie along a band above the 'A' line. A few of them are also found to lie below the 'A' line.

Black cotton soils derive their name because of their colour (black or blackish grey) and their being conducive to cotton cultivation. They are also called 'Regur soils'. The clay mineral is predominantly montmorillonite or a combination of montmorillonite and illite. They are susceptible to high seasonal volume changes on wetting and drying.

3.3 IDENTIFICATION AND CLASSIFICATION OF EXPANSIVE SOILS

There are three groups of methods of recognising expansive soils: (1) mineralogical identification, (2) indirect methods, such as the index properties, soil suction and activity, and (3) direct measurement. Methods of minerological identification are important for exploring the basic properties of clays, but are impractical and uneconomical for practising engineers. The other two groups of methods are generally used, out of which the direct measurement offers the most useful data.

Potentially expansive soils are usually recognised in the field by their fissured or shattered (and slicken sided) condition or by obvious structural damage caused by such soils to existing buildings. Digging tools leave a high polish on expansive clays and an extensive crazing occurs as the clay dries on the sunny side of a trench. The "potential expansion PE" (also termed potential swell or degree of expansion) is a convenient term used to classify expansive soils, viz. to ascertain 'how bad' potentially expansive soils are.

Table 3.7

POTENTIAL EXPANSION FROM SHRINKAGE LIMIT AND LINEAR SHRINKAGE (ALTMEYER 1955)

Shrinkage limit (%)	Linear shrinkage (%)	Potential expansion classification (degree of expansion)
>12	0—5	Non-critical
10—12	5—8	Marginal
<10	>8	Critical

Table 3.7 to 3.10 give the various criteria proposed for classifying expansive soils.

Table 3.8

USBR CLASSIFICATION SYSTEM (HOLTZ 1959)

Colloid content percent minus 0.001 mm	Plasticity index	Shrinkage limit (%)	Probable expansion, percent total vol. change	Potential expansion (degree)
<15	<18	>15	<10	Low
13—23	15—28	10—16	10—20	Medium
20—31	25—41	7—12	20—30	High
>28	>35	<11	>30	Very high

Note. Data based on vertical loading of 1.0 psi (6.89 kPa).

The three soil properties, colloid content, I_p and w_s, are not to be used separately, but all three must be considered to arrive at the estimated degree of expansion from air dry to saturated conditions.

Table 3.9

CHEN'S METHOD OF CLASSIFICATION (CHEN 1965)

Percentage passing 75 μm sieve	Liquid limit (%)	Standard penetration resistance (No. of blows)	Probable expansion, percent total vol. change	Swelling pressure (kg/cm^2)	Degree of expansion
<30	<30	<10	<1	0.5	Low
30—60	30—40	10—20	1—5	1.5—2.5	Medium
60—95	40—60	20—30	3—10	2.5—9.8	High
>95	>60	>30	>10	>9.8	Very high

Note. Data based on vertical loading of 1000 psf (48 kPa)

Table 3.10

USAEWES CLASSIFICATION SYSTEM (SNETHEN 1979)

Liquid limit (%)	Plasticity index	Initial suction (kPa)	Potential expansion (%)	Potential expansion classification
<50	<25	<144	<0.5	Low
50—60	25—35	144—383	0.5—1.5	Marginal
>60	>35	>383	>1.5	High

Note. Potential expansion is given for a confined sample with vertical pressure equal to overburden pressure, expressed as a percentage of sample height. Suction is a more definitive measure of potential expansion than moisture content and index properties.

Table 3.11
IS CLASSIFICATION SYSTEM (IS: 1498)

Liquid limit (%)	Plasticity index	Shrinkage index	Free swell (%)	Degree of expansion	Danger of severity
20—35	<12	<15	<50	Low	Non-critical
35—50	12—23	15—30	50—100	Medium	Marginal
50—70	23—32	30—60	100—200	High	Critical
70—90	>32	>60	>200	Very high	Severe

A potential expansion chart originally proposed by Williams (1958) with modification by Van der Merwe (1975) is shown in Fig. 3.3. The chart has been widely used to determine the potential total heave.

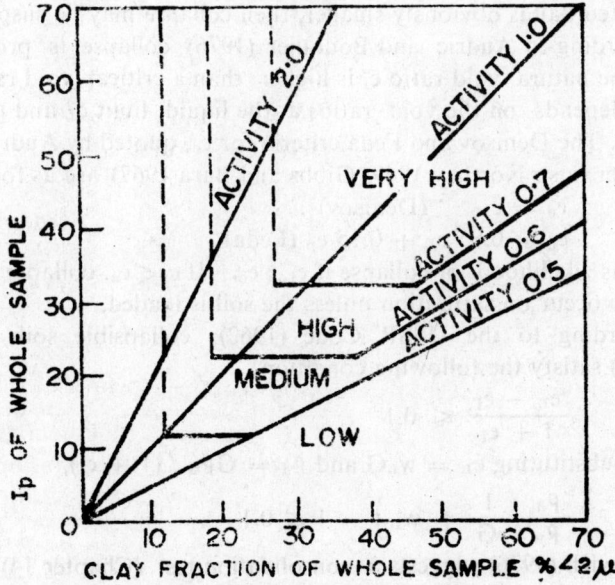

Fig. 3.3 Potential expansion classification chart (Williams 1958, Van der Merwe 1975)

While estimating expansion in the design of foundations, it is necessary to consider the following factors; (1) natural moisture content or rather the degree of saturation, (2) climate and (3) possibility of surface drainage being altered after construction of building. If the moisture content of the soil is at the shrinkage limit, maximum heave could occur on wetting, but if the soil is at its plastic limit the heave will be much less.

3.4 IDENTIFICATION OF COLLAPSIBLE SOILS

Soil deposits most likely to collapse are (Clemence and Finbarr 1981, Clemence 1983): (1) loose fills, (2) altered wind blown sands, (3) hill wash of loose consistency, and (4) decomposed granite or orther acid igneous rocks.

A very simple field test for recognising collapsible soil is the "sausage test" (Clemence 1983). A hand size of the soil to be tested in broken into two pieces and each is trimmed to approximately equal volumes. One of them is then wetted and moulded in the hands to form a damp ball. Comparing the two volumes, if the wetted ball is obviously smaller, then collapse may be suspected.

According to Audric and Bouquier (1976) collapse is probable when the natural void ratio e_i is higher than a critical void ratio e_o which depends on the void ratios at the liquid limit e_L and plastic limit e_p. The Denisov and Feda criteria for e_o, quoted by Audric and Bouquier (also Northey 1969, Gibbs and Bara 1967) are as follows:

$$e_o = e_L \quad \text{(Denisov)} \tag{3.2}$$

$$e_o = 0.85\, e_L + 0.15\, e_p \text{ (Feda)} \tag{3.3}$$

There is likelihood of collapse if $e_i > e_o$. If $e_i < e_o$, collapse is not likely to occur on saturation unless the soil is loaded.

According to the USSR Code (1962), collapsible soils (with $S_r \leqslant 0.6$) satisfy the following condition:

$$\frac{e_L - e_i}{1 + e_i} \leqslant 0.1 \tag{3.4}$$

or, on substituting $e_L = w_L G$ and $\rho_d = G\rho_w / (1 + e_i)$,

$$\frac{\rho_d}{\rho_w} \left(\frac{1}{G} + w_L \right) - 1 \leqslant 0.1 \tag{3.5}$$

Tsytovich (1976) suggests a consolidation test (Chapter 14) to be performed on an undisturbed specimen at natural moisture content and to record the thickness 'H' on consolidation under a pressure 'p' equal to the overburden pressure plus the external pressure likely to be exerted on the soil. The specimen is then submerged under the same pressure and the final thickness H' recorded. "Relative subsidence I_{subs}" is found from the expression:

$$I_{subs} = \frac{H - H'}{H} \qquad (3.6)$$

where   H = sample thickness on consolidation under pressure 'p'

Structurally unstable soils in which $I_{subs} \geqslant 0.02$ are considered to be collapsible. It is found that I_{subs} does not remain constant but increases with an increase in external pressure. Wilun and Starzewski (1975) suggest a standard pressure of 300 kPa (3 kg/cm²) for determining I_{subs}. In general, I_{subs} can be related to pressure p ($p \leqslant 2.5$ kg/cm²) as follows:

$$I_{subs} = C_1 + C_2 p \qquad (3.7)$$

The coefficients C_1 and C_2 can be found from the results of tests on two undisturbed specimens at two different pressures. Great care is needed in specimen preparation and placement in the consolidation cell, otherwise unrealistic results are obtained.

In the "collapse potential" test suggested by Knight (1963), a sample of initial height H_i (natural void ratio e_0) fitted in the consolidation ring is progressively loaded to about 200 kPa. At the end of this loading, the sample is flooded with water and the change in height $\triangle H$ is recorded. The test is then carried on to its maximum loading unit. The 'collapse potential CP' is then defined as:

$$CP = \frac{\triangle H}{H_i} = \frac{\triangle e}{1 + e_0} \qquad (3.8)$$

This collapse potential is used as a guide to the collapse which may be encountered. Suggested values are given in Table 3.12 (Jennings and Knight 1957).

Table 3.12

COLLAPSE POTENTIAL VALUES (JENNINGS AND KNIGHT 1957)

Collapse potential (%)	Severity of problem
0—1	No problem
1—5	Moderate trouble
5—10	Trouble
10—20	Severe trouble
>20	Very severe trouble

3.5 IDENTIFICATION OF DISPERSIVE SOILS

A reliable identification of dispersive soils can be made on the basis of the pin hole test (Sherard et al 1976a, Singh 1981). The test is run under 50, 180, 380 and 1020 mm heads and the soil is classified as given in Table 3.13.

Table 3.13

CLASSIFICATION OF SOIL ON THE BASIS OF PINHOLE TEST

(SHERARD ET AL 1976a)

Classification symbol	Type of soil
D1 and D2	Dispersive soils: fail rapidly under 50 mm head
ND4 and ND3	Intermediate soils: erode slowly under 50 mm or 180 mm head
ND2 and ND1	Nondispersive soils: no colloidal erosion under 380 mm or 1020 mm head

Another method of identification (von Harmse 1975) is to first determine the pH of a 1:2.5 soil/water suspension. If the pH is above 7.8, the soil may contain enough sodium to disperse the mass. Then determine: (1) total exchangeable bases, i.e. K^+, Ca^{2+}, Mg^{2+} and Na^+ (milliequivalents per 100 g of air dried soil), (2) cation exchange capacity (CEC) of soil (milliequivalents per 100 gm of air dried soil). The Exchangeable Soil Percentage ESP is calculated from the relation:

$$ESP = \frac{Na}{CEC} \times 100 \, (\%) \tag{3.9}$$

If the ESP is above 8 percent and the ESP + EMgP is above 15, dispersion will take place. EMgP is given by:

$$EMgP = \frac{Mg}{CEC} \times 100 \, (\%) \tag{3.10}$$

Sherard et al (1972) report that soils with ESP = 7 to 10 are moderately dispersive in combination with reservoir waters of low dissolved salts. Soils with ESP greater than 15 have serious piping potential.

4
Rock classification and characteristics

Rock is the consolidated, coherent and relatively hard portion of the earth's crust. It is a naturally formed, solidly bonded mass of mineral matter which cannot be readily broken by hands nor will disintegrate on its first drying and wetting cycle.

4.1 GEOLOGICAL CLASSIFICATION

Rocks, in the geological sense, are classified into three broad groups based on their mode of formation:

1. **Igneous rocks.** They are crystalline or glassy rocks which have been formed mainly by solidification from molten or liquid material (megma). Example: granite, diorite and basalt.

2. **Sedimentary rocks.** They are made up of rock or mineral particles which have been transported, or which have been precipitated in water. Sediments have generally well defined layers. Example: sandstone, limestone and shale.

3. **Metamorphic rocks**. They are crystalline rocks resulting from the transformation of existing rocks, (which may be of igneous, sedimentary or metamorphic origin) by partial or complete recrystallisation under the action of heat or pressure. Example: quartzite, schist and gneiss.

4.2 STRUCTURAL FEATURES

Some of the important geological features (and their terminology) which have a significant influence on rock behaviour are as follows:

1. Rock material. Rock material means the consolidated aggregate of mineral particles forming solid material between structural discontinuities.

2. Discontinuity. Structural or geomechanical discontinuity is simply a fracture which separates solid blocks of a rock mass. Discontinuities are in the form of joints, bedding planes, fissures, faults, shear zones, cleavage planes, and solution cavities. These features constitute planes of weakness which reduce the strength of rock mass appreciably.

3. Intact rock. It is the rock material which is free of the larger-scale structural features (major discontinuities), and which can be sampled and tested in the laboratory.

4. Major discontinuity. Major discontinuities or major structures are structural discontinuities which are sufficiently well developed and continuous such that shear failure along them would involve little or no shearing of intact rock material.

5. Rock mass. Rock mass represents the insitu rock formation of volume as may be concerned in a given geomechanical problem. It may be defined as the aggregate of regular or irregular blocks of rock material. These blocks are separated by structural features such as bedding planes, joints, cavities, fissures and other discontinuities. Rock mass is thus characterized by an immense anisotropy and structural discontinuity. Rock material is the 'matrix' of the solid phase of the rock mass. This solid part of the rock mass typically obtained as a drill core is known as the 'rock substance'. "Regolith" or "mantle rock" refers to the loose fragments of rock and soil that act as a cover for bedrock. The soil above the bedrock is called "overburden".

4.3 CLASSIFICATION BASED ON STRENGTH

The classification of rocks with regard to strength, based on the recommendations of the International Society of Rock Mechanics (Brown 1981, CGS 1985), is given in Table 4.1.

Table 4.1
CLASSIFICATION OF ROCKS WITH REGARD TO STRENGTH

| Strength | Range of uncon- | Field |
Grade classification	fined compressive strength (MPa)	identification method
R0 Extremely weak	<1	Indented by thumb nail
R1 Very weak	1—5	Crumbles under firm blows of geological hammer; can be peeled with a pocket knife
R2 Weak rock	5—25	Can be peeled by a pocket knife with difficulty; shallow indentations made by a firm blow with point of geological hammer
R3 Medium strong	25—50	Cannot be scrapped or peeled with a pocket knife; specimen can be fractured with a single firm blow of geological hammer
R4 Strong	50—100	Specimen requires more than one blow of geological hammer to fracture
R5 Very strong	100—250	Specimen requires many blows of geological hammer to fracture
R6 Extremely strong	>250	Specimen can only be chipped by the geological hammer

4.4 CLASSIFICATION BASED ON DISCONTINUITIES

The overall strength of a rock mass is reduced due to the presence of discontinuities. The degree of such reduction is governed by the spacing and orientation of the discontinuities. The classification of rock mass with respect to the spacing of discontinuities (joints and bedding) is given in Table 4.2 (Deere 1968, CGS 1978, IRC: 78).

Table 4.2

CLASSIFICATION OF ROCK MASS BASED ON SPACING OF

DISCONTINUITIES

Average spacing	Joints description	Bedding description	Rock mass grading
>3 m	Very wide	Very thick	Solid
1—3 m	Wide	Thick	Massive
0.3—1 m	Moderately close	Medium	Blocky/seamy
5—30 cm	Close	Thin	Fractured
<5 cm	Very close	Very thin	Crushed

4.5 ROCK QUALITY DESIGNATION

The concept of rock quality designation RQD, introduced by Deere et al (1967), offers a general method of classifying rock mass and obtaining the quality of the rock at a site based on the relative amount of fracturing and alteration. The RQD is a more general measure than fracture frequency and is based indirectly on both the degree of fracturing and the amount of weathering in rock. The RQD is defined as the collective length of those core sticks (of NX, i.e. 57.2 mm or larger diameter) in excess of 10 cm, expressed as a percentage of total core length drilled.

$$\text{RQD (\%)} = 100 \ \frac{\text{Length of core in pieces 10 cm and longer}}{\text{Length of run}}$$

$$(4.1)$$

Classification of rock quality is given in Table 4.3.

Table 4.3

CLASSIFICATION OF ROCK QUALITY

RQD (%)	Rock quality
100—90	Excellent
90—75	Good
75—50	Fair
50—25	Poor
25—0	Very poor

4.6 ROCK MASS QUALITY INDEX

A very sophisticated method of classifying rocks has been proposed by Barton et al (1975) of the Norwegian Geotechnical Institute (NGI). They introduced the concept of "rock mass quality index Q", which is defined in terms of six parameters: the RQD or an equivalent estimate of joint density, the number of joint sets, the roughness of the most unfavourable joint set, the degree of alteration or filling of the joint set, the degree of water seepage, and the stress reduction factor which accounts for the loading on a tunnel. They provided a rock mass description and ratings for each of the six parameters which enables the derivation of the NGI quality index Q. The numerical value of Q ranges from 0.001 for exceptionally poor quality sqeezing ground, to 1000 for exceptionally good quality rock which is practically unjointed. The rock quality classification is as follows (Hoek and Brown 1980): Exceptionally good (Q = 1000—400); Extremely good (400—100); Very good (100—40), Good (40—10); Fair (10—4); Poor (4—1); Very poor (1—0.1); Extremely poor (0.1—0.01); and exceptionally poor (Q = 0.01—0.001).

4.7 ROCK MASS RATING

Bieniawski (1976) of the South African Council for Scientific and Industrial Research (CSIR) used "rock mass rating RMR" to classify rock quality, which is as follows (Hoek and Brown 1980): Very good (RMR = 100—80); Good (80—60); Fair (60—40); Poor (40—20); and Very poor (20—0). The NGI quality index Q and the South African CSIR rock mass rating RMR are related as follows (Bieniawski) 1978):

$$RMR = 9 \log_e Q + 44 \qquad (4.2)$$

4.8 HARDNESS

The following hardness criteria (IRC: 78) may be used for engineering description of rocks:

1. **Very hard** : Cannot be scratched with knife or a sharp pick. Breaking of hand specimens requires several hard blows of a geologist's pick.

2. **Hard** : Can be scratched with knife or pick only with difficulty. Hard blows of hammer required to detach hand specimen.

3. **Moderately hard** : Can be scratched with knife or pick. Gouges or grooves 6 mm deep can be excavated by hard blow of the point of geologist's pick. Hand specimens can be detached by moderate blow.

4. **Medium hard** : Can be grooved or gouged 1.5 mm deep by firm pressure on knife or pick point. Can be excavated in small chips to pieces about 25 mm maximum size by hard blows of a pick point.

5. **Soft** : Can be gouged or grooved readily with knife or pick point. Can be excavated in chips to pieces several centimetres in size by moderate blows of pick point. Small thin pieces can be broken by finger pressure.

6. **Very soft** : Can be carved with knife, Can be excavated readily with pick point. Pieces 2.5 cm or more in thickness can be broken by finger pressure. Can be scratched readily by finger nail.

5

Potential, suction and pressure

5.1 SOIL WATER POTENTIAL

Soil water, like other bodies in nature, can possess free energy in different forms and quantities. "Free energy" is the capacity to do work. The energy status of water is primary importance in determining its state and movement through soil. It is not the absolute amount of energy but rather the relative magnitude of energy of water at different regions or times within the soil which governs the behaviour of water. The standard reference state for comparing energy is taken to be that of a hypothetical pool of pure and free water, at atmospheric pressure, at the same temperature as that of soil water, and at a given and constant elevation. By "free water" is meant the water at atmospheric pressure under a horizontal air-water interface. The air-water interface (boundary surface of water) at which the air pressure is atmospheric is called the "free surface". The relative free energy of soil water is described as its "potential". The energy may also be expressed in terms of equivalent 'head' or 'pressure'.

A number of force fields may act on soil water which cause its potential to differ from that of the 'reference pool'. The force fields are generated from the attraction of the soil solid matrix for water, as well as from the presence of solutes and the action of gravity and other external gas pressure. The "total potential" of soil water can be defined as the work required to transfer reversibly a unit quantity of water from the reference pool to the point in the soil. It is given a negative sign convention when the soil tends to absorb water and a positive sign when it tends to expel water. The total potential can be subdivided into components on the basis of the forces responsible for the

differences in energy between the soil water and that of the reference
pool. The various components of the total potential are described
below (Aitchison et al 1965, Richards 1974, Yong and Warkenton
1975).

   1.   **Matrix or capillary potential.** When a capillary tube is placed
in water (Fig. 5.1), the water rises inside the tube to a certain level
above its surrounding free surface. This phenomenon, known as
'capillarity', occurs due to water surface tension and the attraction
(adhesion) of water to the walls of the tube. The height of capillary
rise depends on the diameter of the tube; the smaller the tube
diameter, the greater is the capillary rise.

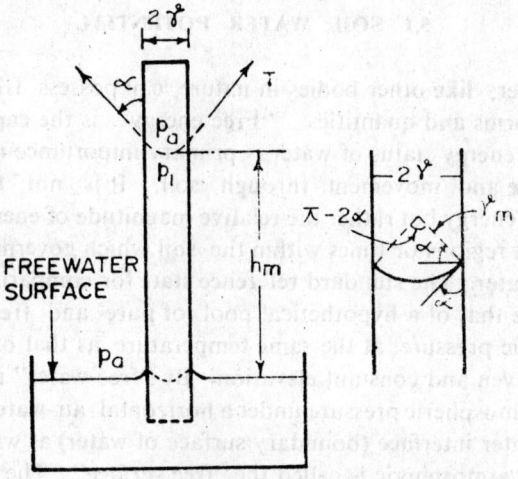

Fig. 5.1 Capillary rise

   As the capillary tube is dipped in water, a meniscus is formed
inside the tube as a result of the contact angle α of water with the walls
of the tube. This angle is acute in case of water on glass and the
meniscus becomes concave towards the air implying a smaller
water pressure p_1 (absolute pressure) under the meniscus
than the atmospheric pressure p_a above the meniscus. For
this reason, water is driven up the tube from its initial location (shown
dotted in Fig. 5.1) by the greater pressure p_a of the free water outside
the tube at the same level, until the pressure difference $(p_a - p_1)$ is
relieved by the weight of the water column in the tube.

The water-air interface (meniscus) in the tube is in equilibrium under the action of a molecular free energy at the surface, giving rise to "surface tension T", which is assumed to act as a force per unit length of the meniscus perimeter. Considering the meniscus as a 'free body' membrane and equating the forces vertically,

$$(p_a - p_1) \pi r^2 = 2\pi r \, T \cos \alpha$$

or
$$p_m = p_a - p_1 = \frac{2T \cos \alpha}{r} \qquad (5.1)$$

where p_m = 'pressure difference' between the capilliary water (under the meniscus) and the atmosphere

r   = radius of the capillary tube

The radius of capillary tube r is related to the radius of the meniscus r_m (Fig. 5.1) as follows:

$$r = r_m \cos \alpha \qquad (5.2)$$

The pressure difference p_m (Eq. 5.1) can be expressed as :

$$p_m = 2T/r_m \qquad (5.3)$$

Eq. 5.3 shows that the pressure difference is dependent only on the meniscus radius and it varies inversely with it.

The water held in the tube below the meniscus induces a compressive stress in the walls of the tube, the total compressive force being equal to the weight of water column.

Similar to the retention of water in a capillary tube, water can also be held by surface tension around the point of contact of two spheres (or soil particles). (Fig. 5.2) This phenomenon of capillarity occurs in partly saturated soils. The meniscus surface is curved in two directions perpendicular to each other and the resulting pressure difference p_m due to surface tension forces in two directions is given by the following equation of capillarity :

$$p_m = T \left(\frac{1}{r_1} + \frac{1}{r_2} \right) \qquad (5.4)$$

where r_1 and r_2 are the radii of curvature of a point on the meniscus in two orthogonal principal planes. For a circular pore, $r_1 = r_2$ and $p_m = 2T/r$.

The height of capillary rise h_m can be obtained by equating the weight of the water column (Fig. 5.1) to the resultant of the surface tension.

$$h_m \pi r^2 \rho_w g = 2\pi r T \cos \alpha$$

or
$$h_m = \frac{2T \cos \alpha}{r \rho_w g} \qquad (5.5)$$

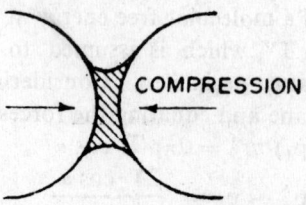

Fig. 5.2 Curved menisci around points of contact of spheres

The pressure difference p_m, also called the "pressure deficiency" or the "capillary pressure", in water is assigned a negative sign indicating that it is below atmospheric. The contact angle α for clean glass and water is zero, and the meniscus becomes spherical with its radius r_m equal to the radius 'r' of the tube. The equations of capillarity can be then re-written as:

$$p_m = -\frac{2T}{r} \tag{5.6}$$

$$h_m = \frac{2T}{r\rho_w g} = \frac{p_m}{\rho_w g} \tag{5.7}$$

and

$$p_m = -h_m \rho_w g \tag{5.8}$$

Substituting $T = 72.75$ dynes/cm* (value at 20°C), $\rho_w = 1$ g/cm³ and $g = 980$ cm/s², a simplified formula for capillary rise is obtained as:

$$h_m \simeq \frac{0.15}{r(cm)} \quad (cm) \tag{5.9}$$

If the soil is considered to be composed of inert soil particles, the voids between particles may be regarded to act roughly as capillary tubes. The contact angle between water and particles is generally assumed zero.

Assuming an average void diameter of 0.01 mm ($r = 0.0005$ cm), the capillary rise from Eq. 5.9 would be 300 cm. Laboratory

---

* 1 dyne/cm $= 10^{-3}$ N/m $= \frac{1}{980}$ gram/cm.

If T is expressed in gram/cm units, gravitational accelerating 'g' will not occur in Eq. 5.7.

Dyne $=$ force which imparts to a free mass of 1 g an acceleration of 1 cm/s².
Newton $=$ force which imparts to a mass of 1 kg an acceleration of 1 m/s².
1 dyne $= 10^{-5}$ N.

experiments, however, give a considerably smaller capillary rise than the calculated values. This is because the capillary tube concept is not wholly valid for soils. Soil has cellular voids inter-connected through openings of various sizes and they cannot be expected to behave as bundles of capillaries of different sizes. Surface activity of particles and their environment also play a major role in influencing the movement of water.

The capillary potential ψ_m, better called the "matrix potential", is a soil matrix property. It may be defined as the work required to move a unit mass of water from the free water surface of zero potential to the point in the soil. It takes into account the combined effects of the water-air interface forces and the forces of adsorption and swelling.

Fig. 5.3 shows a soil column with free water on either side. The air pressure on the free water surface is atmospheric. Water at such a surface is said to have 'zero pressure potential'. The soil column is saturated in line with the free water surface. The surface of the zone of saturation of soil is called the "water table" as the potential of soil water is also zero at the water table. Soil also holds water upto a certain height above the water table or the free water surface.

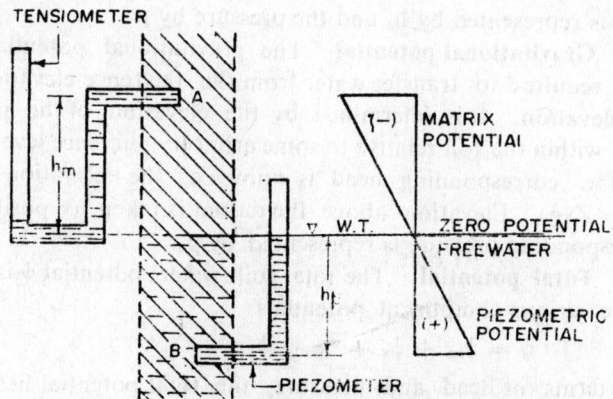

Fig. 5.3 Matrix potential and piezometric potential
(Negative and positive pressure potentials)

Soil water above the free water surface is at a pressure lower than atmospheric and is said to be at negative pressure potential, called the *matrix potential*. The equivalent matrix head h_m can be measured by a 'tensiometer'.

2. Piezometric potential. The piezometric potential ψ_p, or submergence potential, is the work required to transfer a unit mass of water to a point below the water table. Below the water table, the soil water is at a hydrostatic pressure greater than atmospheric and its pressure potential is considered positive. The equivalent piezomatric head h_p at the given point in soil can be measured by a 'piezometer', as shown in Fig. 5.3. The equivalent pressure may be represented by 'p'.

The matrix potential and the piezometric potential are often taken together and called the "pressure potential"; the former is considered as the negative pressure potential and the latter as the positive pressure potential. Soil water may exhibit either of the two potentials, but not both simultaneously.

3. Osmotic or solute potential. The presence of solutes (dissolved substances) in soil water lowers its potential energy. In particular, solutes lower the vapour pressure. The osmotic potential ψ_s is the work required to transfer water from the reference pool of pure water to a pool of soil solution at the same elevation and temperature. The osmotic potential is negative, i.e. water tends to flow in the direction of increasing concentration. The equivalent head is represented by h_s and the pressure by p_s.

4. Gravitational potential. The gravitational potential is the work required to transfer water from the reference elevation to the soil elevation. It is determined by the elevation of the pertinent point within the soil relative to some arbitrary reference level (datum) and the corresponding head is known as the "elevation head Z" ($\psi_g = Zxg$). Elevation above the datum is taken as positive. The corresponding pressure is represented by p_z.

5. Total potential. The total soil water potential ψ is the sum of the various component potentials.

$$\psi = \psi_m + \psi_s + \psi_p + \psi_g \qquad (5.10)$$

In terms of head and pressure, the total potential head H and the total pressure P are similarly expressed as the sum of their corresponding components.

$$H = h_m + h_s + h_p + Z \qquad (5.11)$$

$$P = p_m + p_s + p + p_z \qquad (5.12)$$

When there is equilibrium and no movement of water, there can be no variation in ψ, H or P within the soil.

Potential is expressed as energy per unit mass (units* : erg/g, joule/kg), head is expressed in terms of *equivalent height* of a vertical water column (units: cm, m) and the pressure is expressed as the *force per unit area* (units: dyne/cm², kPa, bar, t/m², atm). The potential ψ, head H and pressure P are inter-related as follows:

$$\psi = Hg = \frac{P}{\rho_w} \tag{5.13}$$

or $$H = \frac{\psi}{g} = \frac{P}{\rho_w g} \tag{5.14}$$

and $$P = H\rho_w g = \psi \rho_w \tag{5.15}$$

where ρ_w = density of water

g = gravitational acceleration

If P is expressed in t/m² or g/cm², Eq. 5.15 is written as:

$$P = H\rho_w \text{ or } H = \frac{P}{\rho_w} \tag{5.16}$$

The units are convertable as follows: 1 bar = 10^2 kPa = 10^6 dynes/cm² = 10^6 erg/g = 10^2 joules/kg = 10.20 t/m² = 0.99 atm = 10.20 m head of water.

Examples 5.1 — 5.2

5.1 Water is observed to rise to a height h_m of 0.4 m above water table in a fine sandy soil. What are the equivalent matrix potential ψ_m and pressure p_m ?

$$\psi_m = h_m g = 0.4 \text{ m} \times 9.8 \frac{m}{s^2} \times \frac{kg}{kg} = 3.92 \text{ J/kg}$$

or $$\psi_m = 40 \text{ cm} \times 980 \frac{cm}{s^2} \times \frac{g}{g} = 3.92 \times 10^4 \text{ erg/g}$$

$$p_m = h_m \rho_m g = 0.4 \text{ m} \times 1 \frac{t}{m^3} \times 9.8 \frac{m}{s^2}$$

$$= 3.92 \frac{kN}{m^2} = 3.92 \text{ kPa} = 0.0392 \text{ bar}$$

or $$p_m = 40 \text{ cm} \times 1 \frac{g}{cm^3} \times 980 \frac{cm}{s^2} = 3.92 \times 10^4 \text{ dynes/cm}^2$$

or $$p_m = h_m \rho_w = 0.4 \text{ m} \times 1 \frac{t}{m^3} = 0.4 \text{ t/m}^2$$

---

* Erg = work done by a force of 1 dyne to move by 1 cm; 1/4 erg = 1 dyne x 1 cm Joule = work done by a force of 1 Newton to move by 1 metre; Joule J = 1 N x 1 m. 1 J = 10^7 erg. 1 J/kg = 10^4 erg/g.

5.2 Determine the equivalent head and pressure for a soil water potential of 10 J/kg.

$$H = \frac{\psi}{g} = \frac{10}{9.8} = 1.02 \text{ m}$$

$$P = \psi \rho_w = 10 \times 1 = 10 \text{ kPa} = 0.1 \text{ bar}$$

5.2 SOIL WATER SUCTION

The force or energy of retention of water in soil can be specified as the soil water potential (Section 5.1). The term 'potential' (total and components) is preferred to be used generally in theoretical treatments of soil water. In practical usage, the term "soil water suction", or simply "soil suction" is used. When an unsaturated soil comes in contact with a pool of free water, it absorbs water; the soil exerts a suction, called the 'soil suction', on the free water. It indicates that the soil water is in equilibrium with a pressure less than atmospheric. Soil suction values are expressed by *positive* figures. Thus soil suction and potential are numerically the same but opposite in sign and the analogous definitions are as follows:

   1. Matrix suction S_m. It is the negative gauge pressure related to the external gas pressure on the soil water (normally atmospheric), to which a solution identical in composition with the soil water must be subjected in order to be in equilibrium through a permeable (both to water and solutes) membrane with the soil water. This results from the interaction of water with the capillary and adsorptive force fields emanating from solid surfaces. It is the equivalent component of matrix potential ($S_m = -h_m$ or $-p_m$).?

   2. Osmotic suction S_s. It is the negative gauge pressure to which a pool of pure water must be subjected in order to be in equilibrium through a semi-permeable (i.e. permeable to water only) membrane with a pool containing a solution identical in composition with the soil water. This results from the interaction of water with the force fields emanating from solutes. It is the equivalent component of osmotic potential ($S_s = -h_s$ or $-p_s$).

   3. Total suction S. It is the negative gauge pressure (relative to atmospheric pressure on the soil water) to which a pool of pure water at the same elevation and temperature must be subjected in order to be in equilibrium through a semipermeable membrane with the soil water. Total suction S is thus equal to the sum of matrix suction S_m and osmotic suction S_s.

The total potential is related to a fixed point in space (e.g. the water table level) and thus includes the gravitational or elevation component, whereas, suction refers to the same elevation as the point in consideration. The total soil water potential P (expressed in pressure units) and the total suction S (in equivalent height of water column) are inter-related as follows from Eq. 5.11 and 5.14:

$$- \frac{P}{\rho_{w}g} = - H = - (h_m + h_s + Z) = S_m + S_s - Z$$

or

$$S = S_m + S_s = - \frac{P}{\rho_{w}g} + Z \qquad (5.17)$$

where $Z =$ height above datum

There being a wide range in suction values of soil, it is convenient to express suction on a logarithmic scale by the use of "pF" index defined as $\log_{10}$ of suction measured in cm of water (Schofield 1935). A pF of 1 represents, thus, a suction of 10 cm head of water, a pF of 2 is a suction of 100 cm head of water and so forth. Due to the logarithmic effect, pF = 0 does not quite correspond to zero cm head of water. Bar or centibar (1 centibar = 0.01 bar) is the other commonly used unit for suction. 1 bar = 1020 cm of water $\approx$ pF 3. 1 centibar = 10.2 cm of water. 1 millibar = 1.02 cm of water.

The matrix suction in clean sands seldom exceeds a pF of 1.7 (0.5 m of water), but in dry clays it may be of the order of pF = 6. The osmotic suction may be of the order of pF = 3. ($\approx$ 1 bar or 1 atm) between the layers of montmorillonite mineral but seldom exceeds a pF of 2 in the water between the particles. Osmotic suction may, therefore, be of small significance in many cases. In the absence of osmotic effects (soluble salts), soil water suction equals matrix suction only, otherwise it is the sum of matrix and osmotic suctions.

The suction (or potential) concepts provides a fundamental approach to the study of water flow and moisture changes in soils. Water flow is controlled by potential gradients; e.g. water may be observed in practice to flow from a sand of *low* moisture content to a clay of *high* moisture content because of high suction clay. Effective stresses in partially saturated soils are also best studied in terms of soil water suction or the negative pore pressures.

5.3 MEASUREMENT OF SUCTION

Soil suction can be measured for laboratory samples and also insitu. The two laboratory methods described here are based on the principle of applying a known force (suction) to the soil water and measuring the resultant change in the moisture content   Three methods are further described for measuring suction in the field.

1. Suction plate apparatus. The suction plate apparatus (Fig. 5.4) has a porous ceramic plate in contact with water on its underside. The size of the pores in the plate are such that no air passes through when the pressure difference across the saturated plate is approximately 1 bar ($\approx$ 10 m of water, 1 atmospheric).   A small soil specimen is placed firmly on the plate. Suction (or negative pressure) can be applied either through a vacuum line or by changing the elevation of a measuring burette.   When the moisture content of the specimen has reached equilibrium, its moisture content is determined by weighing.   By testing several specimens, allowing

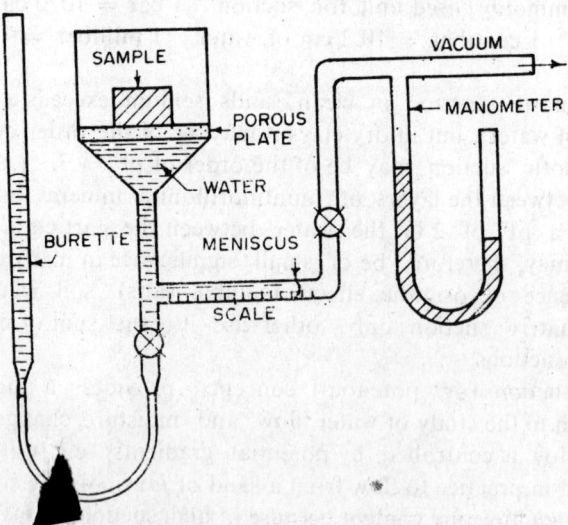

Fig. 5.4 Suction plate apparatus

them to attain equilibrium under different applied suctions and then determining their moisture contents, a relationship can be established between suction and moisture content. Alternatively, the change in moisture content of a particular specimen can be directly observed by measuring the changing position of the meniscus in the burette or the horizontal tube leading to vacuum. The meniscus in the horizontal tube moves towards the sample as water is drawn by suction in the soil. The meniscus can be returned to the original position by the application of a vacuum which is measured by the manometer and gives directly the value of the soil suction.

It is not possible to apply a negative pressure greater than minus one atmosphere gauge pressure (zero absolute pressure) and the method is restricted, in practice, to a (matrix) suction range of 0 to 0.8 bar ($\approx$ 800 cm of water, 0.8 atm, 2.9 pF, 80 kPa).

2. Pressure membrane apparatus. A wet soil specimen is placed in a pressure chamber (cell) on a cellulose membrane supported on a porous bronze plate which is in contact with water at atmospheric zero (Fig. 5.5). Instead of applying a suction to the soil water, as

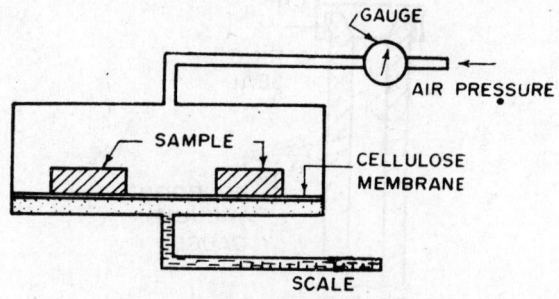

Fig. 5.5 Pressure membrane apparatus

in the suction plate, a positive air pressure is applied above the specimen. The moisture content of the specimen is allowed to attain equilibrium and the soil suction then equals the applied air pressure. The pressure is released, the specimen is removed and its moisture content is determined.

The limit of matrix suction obtainable depends on the design of the chamber (i.e. its safe working pressure) and the air entry value (i.e. pressure at which air passes through the saturated plate) of the membrane. Ceramic plates alone can hold pressures upto about

15 bar, but cellulose membranes can be used with pressures exceeding 100 bar (pF > 5).

3. Tensiometer. A tensiometer gives a direct measurement of matrix suction in the field. It consists of a porous ceramic cup (Fig. 5.6) connected through a clear, heavy wall plastic tube to either a very sensitive vacuum gauge or a mercury manometer, with all parts filled with water. When the cup is placed in soil, the water of the tensiometer (which is initially at atmospheric pressure) comes into contact with soil water through the ceramic cup. The soil water, being generally at subatmospheric pressure, exerts a suction on the water in the tensiometer, and this is measured with a gauge or a manometer. Time to time changes in the suction of soil water can be observed by a tensiometer left in the soil for a long period of time. This way tensiometer are widely used to indicate when irrigation is required. In practice, the maximum suction which can be satisfactorily measured by most of the tensiometers is about 0.8 bar. A long response time needed for equilibrium, specially in clayey soils, is a draw back and a source of error.

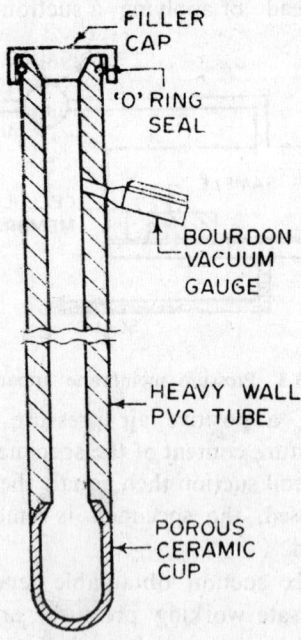

FILLER CAP

'O' RING SEAL

BOURDON VACUUM GAUGE

HEAVY WALL PVC TUBE

POROUS CERAMIC CUP

Fig. 5.6 Tensiometer

4. Psychrometer. A thermocouple pyschrometer is a precise and very convenient instrument for measuring total soil suction (matrix plus osmotic) both in the laboratory and in the field. It has a good response time (normally less than 1-2 days) and covers the range of soil suction encountered in practice (pF 0 to 5, bar 0 to 100) (Richards 1971).

At equilibrium, the absolute relative humidity H (%), i.e. ratio of the water vapour pressure in air above a soil surface to the vapour pressure over free pure water, is related to free energy or total soil suction 'S' (cm water) of soil water by the following thermodynamic equation (Richards 1968).

$$S = R T \log_e (H/100) \qquad (5.18)$$

where   R = universal gas constant (cm head/gm/°K)

T = absoulute temperature (°K)

A psychrometer works on the principle of measuring the relative humidity within the soil which is related to its suction.

A chromel-constantan thermocouple* end is fixed inside the probe

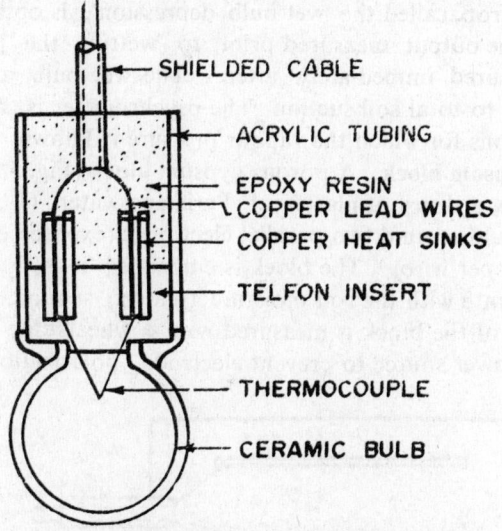

SHIELDED CABLE

ACRYLIC TUBING

EPOXY RESIN
COPPER LEAD WIRES

COPPER HEAT SINKS

TELFON INSERT

THERMOCOUPLE

CERAMIC BULB

Fig. 5.7 Peltier psychrometer

---

* A thermocouple formed by joining two dissimilar metals has the property of liberation or absorption of heat at the point of union depending on the direction of current.

of the psychrometer (Richards 1971). The lower end of the probe
in which the thermocouple is fixed is made of a small ceramic bulb
about 1 cm diameter (OD 2 cm) as shown in Fig. 5.7 (Rawlins and
Dalton 1967, Yong and Warkenton 1975). As an alternative, the
lower end of the probe may have a stainless steel mesh (Richards
1974).

The psychrometer is inserted into the soil and the 'dry bulb' or
'zero' reading is taken by reading the output from the thermocouple
in the humid atmosphere in equilibrium with the soil.   A current
is then passed through the thermocouple in such a direction that the
junction is cooled below the dew point, transforming thus the thermocouple into a 'wet bulb'.   A minute amount of water gets con-
densed on the junction.   As this water evaporates, the temperature
of the junction (wet bulb) gets lowered.   The rate of evaporation
is inversely related to vapour pressure or relative humidity in the
probe.   A second measurement is then taken of the   thermocouple
output which corresponds to the wet bulb temperature.   The temperature drop, called the 'wet bulb depression', is obtained by subtracting the output measured prior to wetting the junction from
that measured immediately after.   The wet bulb depression is
correlated to total soil suction. The psychrometer is calibrated with
salt solutions for which the vapour pressure is known.

5.   **Gypsum block.** A porous gypsum block (Fig. 5.8) is prepared
by pouring a slurry of plaster of Paris and water (1 : 1 by weight)
into a mould around two parallel electrodes (exposed ends of multi-
strand copper wire). The block is embedded is soil where it tends
to equilibrate with the soil moisture (matrix) suction.   The electric
resistance of the block is measured with a wheatstone bridge using
an A-C power source to prevent electrode polarization.   Knowing

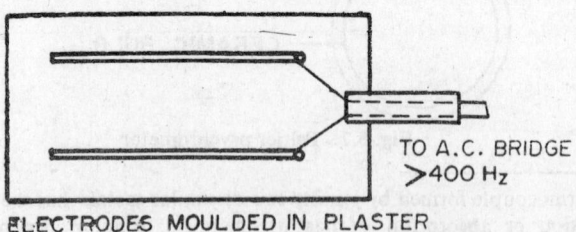

TO A.C. BRIDGE
>400 Hz

ELECTRODES MOULDED IN PLASTER

Fig. 5.8   Gypsum block with electrodes

the resistance, soil suction or moisture content can be determined from the calibration curve of the block.

Gypsum blocks may be calibrated by embedding them in wet soil in the field or in a garden pot. The resistance of the blocks is periodically measured as the soil dries, determining at the same time the suction or moisture content of the samples taken near the blocks. A more precise calibration can be carried out by placing the blocks in soil in a pressure membrane apparatus.

Gypsum block is a simple device for measuring suction or moisture content in field. The blocks are left embedded into the soil to measure suction and moisture variations with time and at different depths. They may last for about 5 years in well drained soils, but in saline, highly organic and wet soils, the life may be reduced to one season (Dastane 1972). Their life may be prolonged by impregnating them with alcohol soluble nylon resin. Their useful range of soil suction (i.e. for a sensitive change in moisture content) is 0.1 to 10 bar (Richards 1974). They are found to be more dependable in the drier than in the wetter range (Johnson 1962).

5.4 EQUILIBRIUM SUCTION AND MOISTURE

Infiltration and evaporation of water through ground surface are prevented when the surface is covered with a seal such as by the construction of an impervious pavement. Under a seal the moisture conditions within the soil (and also in the pavement materials) tend towards stable values and are referred to as the "equilibrium moisture conditions". For each depth there results particular values of suction and moisture content. They are respectively referred to as the 'equilibrium suction' and the 'equilibrium moisture content'. Although relatively constant at a particular point, the equilibrium moisture condition may vary over the cross section of the pavement due mainly to edge effects; and there may also be seasonal variations. Since the strength of soil and granular pavement materials is in most instances markedly affected by moisture content and suction, it is often necessary to estimate the final condition. The period for stabilization of moisture conditions in subgrades may vary from three months to more than six years but two years may be taken as a reasonable guide.

To predict the equilibrium moisture conditions in fine grained subgrades three cases are distinguished (Aitchison et al 1965, NAASRA 1972): (a) regions having shallow water table, i.e. where the depth of water table is less than 6 m in clays, 3 m in sandy clays and silts, and 1 m in sands; (b) regions having deep water table and a total annual rainfall greater than 250 mm; (c) regions having deep water table but a total annual rainfall less than 250 mm.

1. Equilibrium suction. The equilibrium suction within the soil under the central region of the pavement may be predicted as follows:

(a) In regions of shallow water table, the position of the water table regardless of the climate, is the main factor governing the moisture condition.   The soil suction profile is given by:

$$S_m = z_w - z \qquad (5.19)$$

where S_m = matrix suction at depth z

z_w = depth of water table from the surface

The highest position of the water table is taken for design purposes. This conditions is also adopted where the water table exists at some time during a year due to impedded drainage, flood plains, etc.

(b) In regions of deep water table and annual rainfall greater than 250 mm, the predicted equilibrium suction at subgrade level is approximately equal to the value of suction measured in the uncovered soil below the depth of seasonal fluctuations (say 3 to 4.5 metres). Suction is measured either in situ or in the laboratory on undisturbed samples obtained from the field.

(c) In regions of deep water table and annual rainfall less than 250 mm, the predicted equilibrium suction at subgrade level is equal to the value of suction measured in the uncovered soil at the same depth. Here it is assumed that the effects of seasonal fluctuations in the uncovered soil are negligible.

An alternative procedure to predict suction for proposed subgrades can be to extrapolate from the suction measured under existing roads, provided similar moisture conditions exist. This procedure is also adopted for predicting equilibrium moisture content as described below.

2. Equilibrium moisture content. Samples taken from the subgrade of an existing road are tested to measure their moisture content w, optimum moisture content w_o (Chapter 21) and plastic limit w_p.   The moisture content is expressed as a ratio of either w_0 or w_p; this ratio is termed "relative moisture content". In granular soils, w_0 is used in preference to w_p for determining the relative moisture content.   The relative moisture content is assumed to be constant for any other soil in the area. In other words, to allow for variations in soil characteristics the design moisture content for the new situation may be derived from the relation:

$$\frac{\text{Design } w}{w_0} \text{ (proposed subgrade)} = \frac{\text{Measured } w}{w_0} \text{ (existing subgrade)} \qquad (5.20)$$

In the above relation, optimum moisture content w_0 may be replaced by the plastic limit w_p.

In regions of shallow water table, the moisture content of the soil half a metre above the water table can be taken as a guide for field purposes using the constancy of the ratio w/w_0 or w/w_p for any adjustments. In regions of deep water table and annual rainfall greater than 250 mm, the moisture content at a depth below that affected by seasonal changes is adopted for field use.   Where the rainfall is less than 250 mm and the water table is deep, the ultimate moisture content of the subgrade will be similar to uncovered soil at the same depth.   (Aitchison et al 1965).

The prediction of the equilibrium moisture content, however, appears to be of concern mainly in areas having an annual rainfall greater than 650 mm or where the pF is less than about three. In drier areas, the pF is generally greater than three and a high subgrade strength value usually occurs.   The main concern in drier areas would appear to be moisture changes in soils with high shrink-swell properties (NAASRA 1972).

3. **Moisture in granular soils.**   Present methods of predicting moisture conditions in granular subgrades and pavement materials are limited.   One way would be to extrapolate from existing roads by using the relative moisture content ratio. In many cases the likely moisture conditions may be decided only by judgement.

5.5   EFFECTIVE PRESSURE AND PORE PRESSURE

The 'total' normal pressure σ on a plane underlying a saturated mass of soil is the sum of two components: owing to the load transmitted by the solid particles of the soil structures and to the pressure in water filling the pore space. The former is termed the "effective pressure" and the latter the "pore water pressure" or briefly "pore pressure". Thus:

$$\sigma = \sigma' + u \qquad (5.21)$$

where σ = total pressure (or stress)

σ' = effective pressure (or stress)

u = pore pressure

or $\sigma' = \sigma - u$ $\qquad (5.22)$

Eq. 5.22 is the effective pressure equation first conceived by Terzaghi (1925). The difference between the total pressure and the pore pressure has been termed the "effective pressure" as it controls the strength and deformation (compression) of soil. The definition of effective pressure and its controlling effect on soil behaviour are described by the 'principle of effective stress' given by Terzaghi (1936) which, in its simplest form, can be stated as follows:

1. The effective pressure is equal to the total pressure minus the pore pressure for a saturated soil.
2. The effective pressure controls certain aspects of soil behaviour, notably strength and deformation.

Of the terms in the effective pressure equation (Eq. 5. 22), only the total pressure σ and pore pressure u can be directly measured. The effective pressure is not directly measurable and is simply a deduced quantity.

If W' is the weight of soil solids in a submerged soil mass resting on the plane of area 'A' and A_v is the sectional area of pores on this plane, the forces are equated as follows:

$$\sigma A = W' + u A_v \qquad (5.23)$$

$$\sigma = \frac{W'}{A} + u \left(\frac{A_v}{A} \right) \qquad (5.24)$$

The pressure component W'/A transmitted by the solid phase of the soil is called the effective pressure or effective stress σ' and since $A \approx A_v$, Eq. 5.24 gets transformed to the effective pressure equation 5.22:

$$\sigma' = \sigma - u$$

The effective pressure may be viewed as the force transmitted from particle to particle through their points of contact divided by the *gross* area of the plane under consideration. It does not represent the true contact pressure between particles but is, simply, a pressure difference $(\sigma - u)$ giving a *measure* of the force or loading transmitted by the soil structure. Being transferrable through particle contacts, the effective pressure is also termed "inter-granular pressure".

The pore water pressure acts equally in every direction on any point within the soil mass. It also acts equally on the entire surface of any particle but is assumed not to change the volume of the particle. The pore pressure does not cause particles to be pressed together and cannot in itself cause volume change in soil or resist shear; hence it is also called "neutral pressure".

1. Computation of effective pressure. The computation of effective pressure requires the separate determination of the total pressure σ and of the pore pressure u. The effective pressure is then found as the difference $\sigma' = \sigma - u$.

Consider the typical cases where the watar table is: (a) at the ground surface, (b) above the ground surface, and (c) below the ground surface, as shown in Fig. 5.9.

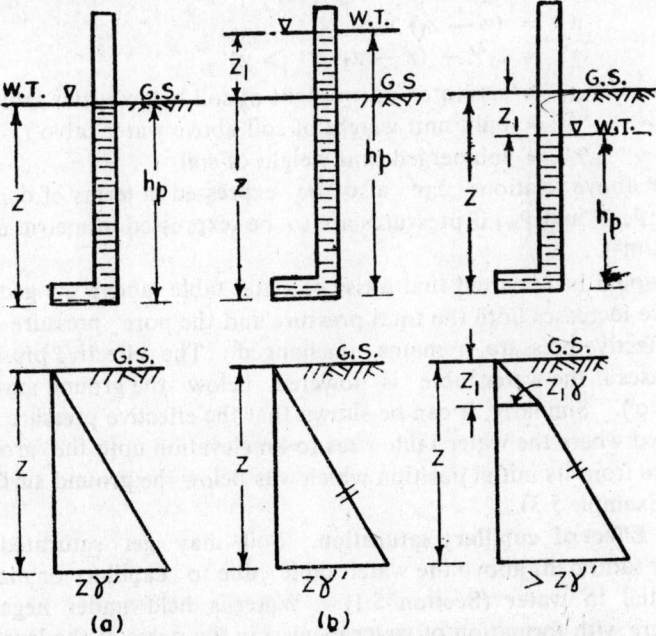

Fig. 5.9 Effective pressure for different positions of water table

The total vertical pressure at depth z below the ground surface is equal to the weight of all the material (solids + water) per unit area above that depth. The pore water pressure is the hydrostatic pressure equivalent to the water column in the piezometer inserted at depth z, that is,

$$u = h_p \gamma_w \quad (\text{or } h_p \rho_w) \tag{5.25}$$

The pressures for the three cases are found as follows:

(a) Water table at ground surface,

$$\sigma = z\gamma_{sat}$$
$$u = h_p\gamma_w = z\gamma_w$$
$$\sigma' = \sigma - u = z(\gamma_{sat} - \gamma_w) = z\gamma' \tag{5.26}$$

(b) Water table above ground surface,

$$\sigma = z_1\gamma_w + z\gamma_{sat}$$
$$u = (z_1 + z)\gamma_w$$
$$\sigma' = z\gamma'$$

(c) Water table below ground surface.

$$\sigma = z_1\gamma + (z - z_1)\gamma_{sat}$$
$$u = (z - z_1)\gamma_w$$
$$\sigma' = z_1\gamma + (z - z_1)\gamma' > z\gamma'$$

where
γ_{sat} = saturated unit weight of soil below water table
γ = bulk unit weight of soil above water table
γ' = submerged unit weight of soil

The above relations can also be expressed in terms of density (ρ_{sat}, ρ, ρ' and ρ_w) if pressures are to be expressed in metric units (e.g. t/m^2).

It would be apparent that a rise of water table above the ground surface increases both the total pressure and the pore pressure but the effective pressure remains unchanged. The effective pressure increases if the water table is lowered below the ground surface (case 'c'). Similarly, it can be shown that the effective pressure gets reduced where the water table rises to an elevation upto the ground surface from its initial position which was below the ground surface. (See Example 5.3).

2. Effect of capillary saturation. Soil may get saturated (or nearly saturated) above the water table due to capillary or matrix potential in water (Section 5.1). Water is held under negative pressure with formation of water menisci in the pores at the level of top of capillary saturation. This water induces compression in the soil mass and thereby increases effective pressure. Being under negative pressure, the capillary water does not contribute to the hydrostatic pressure below the water table.

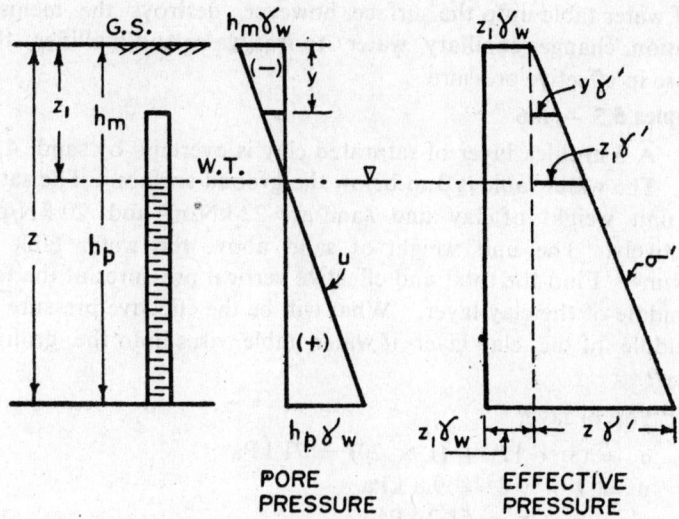

PORE
PRESSURE

EFFECTIVE
PRESSURE

Fig. 5.10 Effect of capillary saturation

Consider the case where soil is saturated by capillary rise above the water table upto the ground surface (Fig. 5.10). The only pressure acting just below the menisci level is the negative pore pressure ($u = -h_m \gamma_w = -z_1 \gamma_w$) which is equivalent to an effective pressure of $\sigma' = z_1 \gamma_w$ transmitted to the soil particles. At a certain depth 'y' below the menisic level in the capillary zone, the total and effective pressures are:

$$\sigma = y \gamma_{sat}, \quad u = -(z_1 - y) \gamma_w$$
$$\sigma' = \sigma - u = y \gamma' + z_1 \gamma_w \tag{5.27}$$

At the level of the water table, $\sigma = z_1 \gamma_{sat}$ and $u = 0$, hence:

$$\sigma' = z_1 \gamma_{sat} = z_1 \gamma' + z_1 \gamma_w \tag{5.28}$$

Similarly, at depth z below the ground surface,

$$\sigma = z_1 \gamma_{sat} + (z - z_1) \gamma_{sat}$$
$$u = h_p \gamma_w = (z - z_1) \gamma_w$$
and
$$\sigma' = z \gamma_{sat} - (z - z_1) \gamma_w = z \gamma' + z_1 \gamma_w \tag{5.29}$$

The distribution of effective pressure at various levels within the soil is shown in Fig. 5.10. It is thus seen that capillarity increases uniformly the effective pressure equivalent to capillary rise, i.e. by $z_1\gamma_w$, from the menisci level upto the point under consideration. A rise of water table upto the surface, however, destroys the menisci formation, changes capillary water to free water and nullifies the increase in effective pressure.

Examples 5.3 — 5.6

5.3 A 5 m thick layer of saturated clay is overlain by sand 4 m deep. The water table is 3 m below the ground surface. The saturated unit weight of clay and sand are 22 kN/m³ and 20 kN/m³ respectively. The unit weight of sand above the water table is 17 kN/m³. Find the total and effective vertical pressures at the top and middle of the clay layer. What will be the effective pressure at the middle of the clay layer if water table rises upto the ground ·surface?

(a)   Top of layer.

$\sigma = (3 \times 17) + (1 \times 20) = 71$ kPa

$u = 1 \times 9.8 = 9.8$ kPa

$\sigma' = \sigma - u = 61.2$ kPa

(b)   Middle of layer.

$\sigma = (3 \times 17) + (1 \times 20) + (2.5 \times 22) = 126$ kPa

$u = 3.5 \times 9.8 = 34.3$ kPa

$\sigma' = \sigma - u = 91.7$ kPa

(c)   Middle of layer; water table at G.S.

$\sigma = (4 \times 20) + (2.5 \times 22) = 135$ kPa

$u = 6.5 \times 9.8 = 63.7$ kPa

$\sigma' = 71.3$ kPa

(σ' decreases with rise of water table.)

5.4 What is the effective pressure at the middle of clay layer in Ex. 4.3 if sand is saturated with capillary water to a height of 1 m above the water table ?

$\sigma = (2 \times 17) + (2 \times 20) + (2.5 \times 22) = 129$ kPa

$u = 3.5 \times 9.8 = 34.3$ kPa

$\sigma' = 94.7$ kPa

5.5 The strata in the flat bottom of a valley consist of 3 m of coarse sand overlying 10 m of clay. Below the clay is a gravel layer in which the water is under artesian pressure corresponding to a standpipe level of 4 m above ground level. The water table in the top sand layer is 1 m below the surface. The unit weights of soil are as follows: sand above water table $= 17$ kN/m^3, saturated sand below water table $= 20$ kN/m^3, and saturated clay $= 21$ kN/m^3.

Find the total pressure, pore pressure and effective pressure at 'a) 1 m below ground level, (b) top of clay layer and (c) bottom of clay layer. What will be the effective pressure at the bottom of clay layer assuming that the water level in the gravel layer is lowered by 3 m using relief wells, but the water level in top sand remains unchanged?

(a)   At 1 m below ground level.

$\sigma = 1 \times 17 = 17$ kPa

$u = 0$

$\sigma' = 17$ kPa

(b)   At top of clay.

$\sigma = (1 \times 17) + (2 \times 20) = 57$ kPa

$u = 2 \times 9.8 = 19.6$ kPa

$\sigma' = 37.4$ kPa

(c)   At bottom of clay.

$\sigma = (1 \times 17) + (2 \times 20) + (10 \times 21) = 267$ kPa

$u = 17 \times 9.8 = 166.6$ kPa

$\sigma' = 100.4$ kPa

(d)   At bottom of clay with artesian pressure lowered by 3 m.

$u = (17 - 3) \times 9.8 = 137.2$ kPa

$\sigma' = 267 - 137.2 = 129.8$ kPa

5.6 A layer of saturated sand extends to a depth of 10 m below the ground surface, and is underlain by a 8 m thick layer of clay. The water table is at the ground surface. The saturated densities of sand and clay are respectively 1.99 t/m^3 and 2.05 t/m^3. The dry density of sand is 1.65 t/m^3.

Find the effective pressure at the middle of clay layer. What will be the effective pressures if water table is lowered to 3 m below ground surface by drainage and (a) the degree of saturation of sand above lowered water table remains 20 percent, (b) the upper 3 m of sand remains saturated by capillary water?

(a) Water table at ground surface.

$\sigma = (10 \times 1.99) + (4 \times 2.05) = 28.1$ t/m²

u $= 14 \times 1 = 14$ t/m²

$\sigma' = 14.1$ t/m³

(b) Lowered water table, $S_r = 20\%$.

Density of sand above water table is given by:

$\rho = \rho_d + S_r (\rho_{sat} - \rho_d) = 1.718$ t/m³

$\sigma = (3 \times 1.718) + (7 \times 1.99) + (4 \times 2.05) = 27.28$ t/m²

u $= 11$ t/m²

$\sigma' = 16.21$ t/m²

(c) Lowered water table, 3 m sand saturated by capillary water.

$\sigma = (10 \times 1.99) + (4 \times 2.05) = 28.1$ t/m²

u $= 11$ t/m²

$\sigma' = 17.1$ t/m²

6

Permeability and groundwater flow

6.1 FLOW CONDITIONS

A soil mass has interconnected pores through which water may move. Water movement or moisture transfer in soils may be considered under two general conditions: (a) the saturated state where all the voids are essentially filled with water and (b) the partly saturated state where both water and air are present. In many groundwater and seepage problems, flow may be assumed to be in a saturated state and it is governed by the energy potential or 'head' of water.

Considering saturated flow through a uniformly porous soil mass of length L in Fig. 6.1, the "positive pressure head" or the "piezometric head" at the entry face is represented by the height of

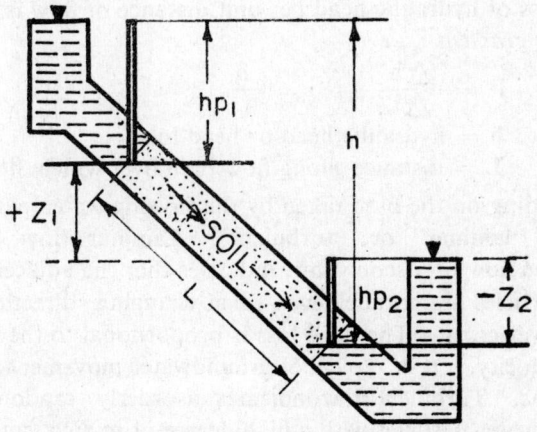

Fig. 6.1 Determination of hydraulic head and gradient

water column hp_1. Similarly, the piezometric head at the exit face is hp_2. Assuming the down-stream water level as datum, the "elevation head" at entry is $+Z_1$ and at exit $-Z_2$. The "total head" is the sum of the piezometric head and the elevation head; the velocity head for flow through soils is negligible. Thus the total heads H_1 and H_2 at the entry and exist faces are respectively:

$$H_1 = hp_1 + Z_1 \text{ and } H_2 = hp_2 - Z_2 = 0$$

It is the 'difference' in total heads which causes the flow. The total head difference will be referred to as the "hydraulic head h". It is also known as the 'head loss' or the 'head drop'. Any elevation can be selected for datum as a base of elevation heads. The absolute magnitude of the elevation head is unimportant, as it is the difference in elevation head which is required and this difference of elevation heads between any two points is the same regardless of the position of the datum. The advantage of taking the down-stream free water surface as datum is that total head at exit becomes zero and the elevation of water in a piezometer at any point in the soil measured above the datum gives directly the hydraulic head or the head loss which will occur when the water flows from that point in soil upto the exit end. The hydraulic head h at any point in the soil water can thus be expressed as:

$$h = h_p \pm Z \tag{6.1}$$

where h_p = piezometric head

Z = elevation head

The loss of hydraulic head per unit distance of flow is called the *hydraulic gradient* i, i.e.

$$i = \frac{\Delta h}{\Delta L} \text{ or } i = \frac{h}{L} \tag{6.2}$$

where h = hydraulic head or head loss

L = distance along flow path over which head loss is h

Depending on the path taken by the flowing water particles, flow is termed "laminar" or "turbulent". Laminar flow, also called streamline flow or viscous flow, indicates that the adjacent paths of water particles are parallel, even when changing direction, and the paths never cross. The head loss is proportional to the first power of the velocity. It is typical of groundwater movement under most conditions. Turbulent flow indicates disorderly random paths for moving water particles with a high degree of mixing, and in which the head loss is approximately proportional to the second power of the velocity. Flow velocity has a direct bearing on whether a flow remains laminar or turbulent.

6.2 DARCY'S LAW FOR FLOW

In the mid-eighteenth century, H. Darcy (1856), working in Paris, studied experimentally the flow of water through sands. For laminar flow through saturated soil, Darcy established experimentally that the *rate of flow* q. defined as the volume of water flowing per unit time across a sectional area 'A' of soil, was proportional to the hydraulic gradient. Mathematically,

$$q = k i A \tag{6.3}$$

or $$v = \frac{q}{A} = k i \tag{6.4}$$

where v = flow velocity
        k = constant of proportionality

The constant of proportionality 'k' in Eq. 6.3 or 6.4 is known as the Darcy "coefficient of permeability" or the "hydraulic conductivity". Darcy stated that k is almost independent of the hydraulic gradient i. It has the dimensions of a velocity. The coefficient of permeability, or simply the "permeability", may be defined as the rate of flow of water under laminar flow conditions through a unit cross-sectional area of a porous medium under a unit hydraulic gradient and standard temperature conditions (usually 27°C in India). In other words, coefficient of permeability (or permeability) is the flow velocity under a unit hydraulic gradient.

The flow velocity v in Eq. 6.4 is an 'apparent' velocity being equal to the average rate of flow of water across a unit *gross* area in the soil. The actual or true velocity of flow, called the *"seepage velocity* v_s", is that with which water flows through the soil voids. If A_v is the area of void space and 'A' the gross area of soil perpendicular to the direction of flow, the rate of flow may be equated as:

$$q = v A = v_s A_v \tag{6.5}$$

or $$v_s = v \frac{A}{A_v} = \frac{v}{n} = v \frac{1 + e}{e} \tag{6.6}$$

where n = porosity
        e = void ratio

The porosity n in Eq. 6.6 should be the 'effective' porosity, i.e. the ratio of the actual volume of pores through which water is seeping to the total volume; usually it is somewhat less than the true porosity.

Permeability (or hydraulic conductivity) is the property of soil which permits appreciable movement of water through it when saturated and actuated by hydrostatic pressure of the magnitude normally encountered in natural subsurface water. Permeability is characteristic of a given soil medium and its range of magnitude is extremely wide; it can be lower than 10^{-8} cm/s for clays and exceed 1 cm/s for gravels. Typical values of permeability of natural soils are given in Table 6.1. The commonly used physical unit for permeability is cm/s or m/s. In practical calculations, the units m/year and m/day are also used (1 cm/s $= 3.15 \times 10^5$ m/year; 1 m/year $= 3.17 \times 10^{-6}$ cm/s; 1 m/s $= 86.4 \times 10^3$ m/day; 1 m/day $= 11.57 \times 10^{-6}$ m/s). Soils may be rated as practically *impervious* when k is less than 10^{-6} cm/s, *semi-pervious* for k from 10^{-6} cm/s to 10^{-4} cm/s, and *pervious* for k greater than 10^{-4} cm/s.

Table 6.1
TYPICAL VALUES OF PERMEABILITY

Soil type	Coeff. of permeability (cm/s)
Clean gravel	$10^2 - 1.0$
Clean sands, clean sand and gravel mixtures	$1.0 - 10^{-3}$
Very fine sands, organic and inorganic silts, mixtures of sand silt and clay, glacial till, stratified clay deposits, etc.	$10^{-3} - 10^{-7}$
Homogeneous, intact clays	$10^{-7} - 10^{-9}$

6.3 VALIDITY OF DARCY'S LAW

Validity of the Darcy law implies that (a) the flow velocity v is directly proportional to hydraulic gradient i, and (b) the relationship between v and i is linear through the origin. The condition of linearity between v and i has been verified for a variety of soils — sands, silts and many clays. In sands and silts, linearity exists so long as the flow remains approximately laminar. The flow velocity is affected by the hydraulic gradient as well as the void opening, the latter being a function of particle size and arrangement. Flow in pervious soils tends to change from laminar to turbulent as the average particle size increases and medium sand $(D \not> 2\ mm)$ seems to be the most pervious soil through which flow remains laminar. There is ample experimental evidence to establish the validity of the Darcy law in silts, fine sands and medium sands under low hydraulic gradients. The upper limit of hydraulic gradient in sands should not exceed unity or preferably 0.5. Coarser the soil, the lower is the hydraulic gradient required for laminar flow. At higher gradients and in coarse grained soils $(D > 1\ mm)$ where flow becomes turbulent, the non-linear form of the Darcy law may be as follows:

$$v = k(i)^n \qquad (6.7)$$

Where n is the exponent of turbulence; experiments indicate 0.65 $< n < 1.0$ (Tuma and Hady 1973). Since the present engineering practice is not to use a non-linear relationship between v and i, it is suggested that only the linear form of the Darcy law be used in analysis and that the permeability be measured under hydraulic gradients of the same magnitude as anticipated in the field.

Some of the flow studies with low hydraulic gradients in saturated fine grained soils with prominent adsorption complex and clay-water forces suggest the existence of a "threshold" or initial gradient which has to be exceeded before flow can take place (Miller and Low 1963). Increasing gradients, however, yield a linear relationship between v and i, which on extension does not pass through the origin. Some further studies do not confirm the existence of a threshold gradient (Olsen 1965, 1969, Mitchell and Younger 1967, Chan and Kenney 1973). Experimental evidences are thus inconclusive. The initial, apparent deviation from the Darcy law for flow in clays is probably due to a combination of experimental artifacts, particle migrations and other changes in the soil fabric that take place as flow occurs (Mitchell 1976). Once the steady state of flow gets established, the law becomes equally valid for clays as for silts and sands.

6.4 FACTORS AFFECTING PERMEABILITY

The permeability in the Darcy law depends on the properties of both the flowing water and the soil. An insight into the factors influencing permeability can be obtained from the following theoretical Kozeny-Carman equation for flow through porous media (Kozeny 1927, Carman 1956):

$$k = \frac{1}{C'T^2S^2} \left(\frac{e^3}{1+e} \right) \frac{\rho_w}{\eta} \qquad (6.8)$$

where C' = pore shape factor

T = tortuosity factor = $\dfrac{\text{effective flow path}}{\text{thickness of test sample}}$

S = specific surface = surface area per unit volume of particles, cm^2/cm^3

ρ_w = density of water

η = viscosity of water

For sand and silt size particles, $C'\, T^2$ may be taken approximately equal to 5 (Carman 1956, Perloff and Baron 1976). An alternative expression for Eq. 6.8 is as follows:

$$k = C D_o^2 \left(\frac{e^3}{1+e} \right) \frac{\rho_w}{\eta} \qquad (6.9)$$

where C = composite shape factor

D_o = representative particle size

The Kozeny-Carman equation (Eq. 6.8 or Eq. 6.9) describes the behaviour of cohesionless soils reasonably well, but is inadequate for clays. The factors affecting permeability may be briefly discussed as follows:

1. Density and viscosity of water. Permeability varies directly with the density and inversely with the viscosity of water:

$$k \propto \frac{\rho_w}{\eta} \qquad (6.10)$$

Sincer both ρ_w and η vary with temperature, k is also dependent on the temperature. The effect of temperature variation is considerable on η but negligible on ρ_w. Neglecting density variation, it is usual to report k at a standard temperature of 27°C using the expression:

$$k_{27} = k \frac{\eta}{\eta_{27}} \qquad (6.11)$$

where   k and η = permeability and viscosity at test temperature

$\quad\quad\quad$ k_{27} and η_{27} = permeability and viscosity at 27°C

2. Void ratio. The marked influence of void ratio (or porosity) on permeability as demonstrated by Eq. 6.8 has been experimentally varified. The following empirical relations can be used for coarse grained soils:

$$k \propto e^2 \tag{6.12}$$

and $\quad\quad k \propto \dfrac{e^3}{1+e} \tag{6.13}$

A semi-log relationship also exists between 'k' and 'e'. A plot of log k (logarithmic scale) versus 'e' (linear scale) is approximately a straight line both for coarse grained and fine grained soils (Taylor 1948, Lambe and Whitman 1973).

3. Size, shape and arrangement of soil particles. These are the most significant factors affecting k since they decide the void ratio, size and shape of pores and the pore-size distribution. A coarse soil has larger pore sizes and hence greater k than that of a fine grained soil, even though the void ratio is frequently greater for the fine grained soil; the pore sizes in a soil mass are roughly of the same size as the particles themselves. As indicated by Eq. 6.8 and 6.9, k varies inversely with the square of the specific surface or directly with the square of the particle size. Permeability is roughly proportional to square of the harmonic mean of the particle diameters (Kezdi 1974). Smaller particles in a soil mass have a greater influence on k than the larger ones, as a small percentage of fines can clog the pores of an otherwise coarse soil. At equal void ratio, a soil composed mainly of bulky grains is more permeable than a soil composed of flakyones.

Natural soil formations are generally non-uniform in structure, being more-or-less stratified. The average permeability along stratification is always greater than that in the direction perpendicular to stratification. Even in man-made deposits, the horizontal permeability tends to be larger than the vertical. This is because the usual methods of placement and compaction tend to build up stratification into embankments and fills. Wind-blown sands and silts are often more permeable in the vertical direction because of tubular voids believed to be left by grass roots.

The effect of soil fabric on permeability is important specially in case of clayey soils. Compared at equal void ratio, a flocculated fabric is more permeable than a dispersed one, because the pores are larger in size, though fewer in number, in a flocculated fabric. Within a well dispersed clay itself, flow parallel to particle orientation will be more than that in a direction normal to the orientation. A flocculated fabric is more-or-less insensitive to direction of flow. Equally important are the effects of shrinkage cracks, fissures, joints, warping of layers and shear zones.

4. **Degree of saturation.** The Darcy law presumes a fully saturated state of soil Air which may be originally entrapped in the soil pores or may get liberated and deposited from flowing water causes a great obstruction to flow. The permeability is observed to vary directly with the cube of the degree of saturation (Mitchell 1976). If the degree of saturation reduces to less than about 85 percent, much of the air becomes continuous through the voids and the Darcy law no longer holds good (Scott 1974). A 100 percent saturated state hardy exists even in natural groundwater flow and drainage, however, the Darcy law still remains approximately valid for degree of saturation greater than about 85 percent.

5. **Adsorption complex and clay water interaction.** Fine particles of clay are surrounded by films of adsorbed water whose properties differ essentially from those of normal water. Forces of adsorption and development of diffuse ion-layer around the clay particles creat immobilized hydrodynamic layers of water, thereby reducing the effective pore space available for seepage. The thickness of these immobilized hydrodynamic layers depends on the interaction characteristics of the soil-water system. Permeability of a fine grained soil can be altered by changing its electrolyte concentration. Out of the common exchangeable ions, sodium is the one which gives the lowest permeability to a clay. Sodium montmorillonite is one of the least permeable soil minerals.

6.5 DETERMINATION OF PERMEABILITY

The permeability can be determined in three ways: by laboratory tests, by field tests and by empirical formulae.

1. Laboratory tests. Direct determination of permeability of representative soil specimens is made in an apparatus called the 'permeameter'. Two types of test set-ups are used: 'constant head permeability test' and the 'falling head (or variable head) permeability test'. The constant head test requires the measurement of the rate of flow through a soil sample under a constant head and, hence, it is suitable for coarse grained pervious soils where the quantity of flow is measurable in a reasonable test time. The test can be used for permeability down to about 10^{-3} cm/s.

For soils of low permeability the rate of flow through the sample is too small to be accurately measured in a constant head permeameter. Moreover, the test results may become unreliable due to evaporation. In such cases, the falling head permeameter is more suitable. In this test, the decrease in the hydraulic head causing the flow is measured in a given time interval and the quantity of flow through sample need not be measured. The test can be used for values of k from 10^{-1} cm/s to about 10^{-7} cm/s.

The permeability of fine grained soils ($k < 10^{-5}$ cm/s) can better be inferred indirectly from the results of the consolidation test (Chapter 14).

Laboratory tests can be performed under controlled conditions. They are used to establish relationships between void ratio and permeability (e.g. 'log k' versus 'e' curve) from which k can be interpolated at the desired void ratio.

2. Field tests. Laboratory tests have an inherent limitation of testing only a small sample which is rarely truly representative of large soil masses whose properties often show large local variations. Most of the soil masses are anisotropic in character. In a sequence of layered deposit, the permeability of the individual beds will vary, also the average horizontal permeability is usually greater than that in the vertical direction. The permeability of even a fairly uniform deposit tends to increase with depth because the overburden pressure may lead to densification. Moreover, it is difficult to collect undisturbed samples of coarse grained soils for laboratory testing.

Field permeability tests have the advantage of determining the average permeability of a large mass of soil insitu and should be considered most reliable, specially in case of coarse grained soils below the groundwater table. The field tests are, however, generally more expensive and frequently more time consuming. They are conducted in cases where the question of permeability seems to be of major importance. A number of test methods are available, depending, among other things, on whether the soil to be tested is below or above the groundwater table.

3. Empirical formulae. A number of empirical formulae have be suggested to estimate the permeability of soil. These formulae are generally expressed as functions of void ratio and particle size. A simple formulae, proposed by Hazen (1911) and frequently used for loose clean sands and gravels is:

$$k = C_1 (D_{10})^2 \quad cm/s \tag{6.14}$$

where D_{10} is the effective particle size or the 10 percent finer size expressed in millimeter and C_1 is an empirical coefficient varying from about 0.9 to 1.2 (Cedergren 1977). A value of 1.0 is often used for C_1. The Hazen formulae was developed by testing clean filter sands in a loose state. Even minute quantities of fines may greatly diminish the permeability of sands.

Empirical formulae are inadequate to correctly reflect the often considerable influence on permeability of such soil characteristics as the grading or the amount of fines, the shape and other surface characteristics of the particles. They should be looked upon as approximations only and, as a rule, the permeability of most soils should be determined by direct test methods.

6.6 LABORATORY MEASUREMENTS OF PERMEABILITY

Laboratory tests can be performed on both undisturbed and remoulded samples of soils. Undisturbed samples are obtained by sampling tubes and core cutters, trimmed to test size and placed in the permeameter. Remoulded specimens are generally prepared in the permeameter mould itself. A variety of permeameters are available or can be fabricated with either downward or upward flow. A commonly used permeameter has a cylindrical mould of the same size as used for compaction studies, viz, a 1000 cm³ mould (Chapter 21). The Jodhpur Permeameter designed by the author (Singh 1964, 1981), has a 300 cm³ mould (7.98 cm diameter, 6 cm height, 50 cm² sectional area) in which soil can be compacted by a 2.5 kg Dynamic Ramming Tool (DRT). A set of collars and plugs is provided for preparing statically compacted specimens. A 300 cm³ core cutter is used for taking undisturbed specimens which can be directly mounted between the top cap and the perforated base of the permeameter.

1. **Specimen preparation and mounting.** Using the Jodhpur Permeameter, the 300 cm³ mould is fixed upside down to the compaction plate and a collar is attached to the other end. Soil is compacted in two layers by the 2.5 kg DRT at the desired moisture content and density. The two-layer compaction with 15 blows to each layer is nearly equivalent to the IS Light Compaction (Chapter 21). After compaction the mould is fixed between the perforated base and the top cap. For preparing statically compacted specimen, collars are attached to either ends of the mould. A weighed quantity of soil at desired moisture content is pressed into the mould by end plugs working through the collors. An undisturbed specimen is obtained by the 300 cm³ cutter which is directly clamped between the top cap and the perforated base. Wire gauzes or filter papers are placed at the top and bottom faces of the specimen before fixing it between the top plate and the base plate.

2. **Constant head test.** A typical set up of the permeameter for the constant head test is shown in Fig. 6.2a. The constant head tank is a large diameter glass tube (20 to 60 mm in diameter and about 1 m long) with stopper and air intake tube at top and outlet nozzle at bottom. The tank is initially filled *completely* with water without any air space. As soon as water starts flowing out of the tank and air enters the air intake tube, the pressure at the bottom end of the air intake tube becomes atmospheric (constant), and it remains so as long as the water level remains above the bottom of the tube (Singh 1981). The constant hydraulic head causing the flow is the elevation difference between the bottom level of the air intake tube and the free water level in the bottom tank. The head can be increased or decreased by raising or lowering the air intake tube.

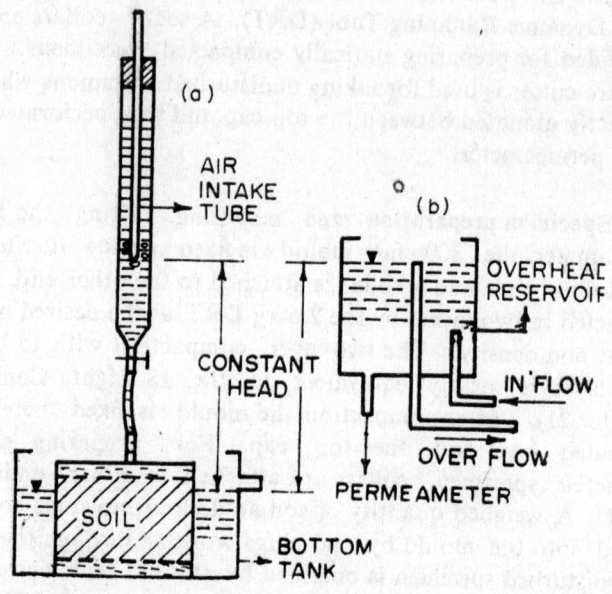

Fig. 6.2 Typical set-ups of constant head permeameters

In the alternative set up of Fig. 6.2 b, a reservoir with an inlet and overflow serves as a constant head tank. The permeameter has a lucite mould for containing the specimen. Filters of coarse material are placed at top and bottom of the specimen to prevent fines being washed out. While testing less pervious fine sands and silty sands, more reliable results can be obtained by measuring the actual head loss over a fixed length in the specimen by inserting piezometer tubes. This also eliminates the error due to some head loss which occurs in the filters or due to the formation of filter skins by fines which may be carried along with the flowing water.

Flow through the specimen is maintained long enough so that the flow rate becomes constant. Onces the steady state has been attained, the quantity of flow Q during a time interval 't' is collected and measured. By the Darcy law:

$$q = \frac{Q}{t} = k\,i\,A = k\,\frac{h}{L}\,A \qquad (6.15)$$

where   A = cross sectional area of the specimen

h = constant head loss over length L of the specimen

Hence $k = \dfrac{Q\,L}{A\,t\,h}$ (6.16)

3. **Falling head test.** The constant head tank (of Fig. 6.2) is replaced by a standpipe fixed vertically to the top of the permeameter mould, as shown in Fig. 6.3. The diameter of the standpipe is so adjusted to the permeability of the soil being tested that the rate of fall of the water level in the standpipe is neither excessively slow nor too fast for accurate measurements. Water is filled in the standpipe and the stop watch is started when the water level touches a marked top point of the standpipe. The time required for the water level to drop to one or more lower points is recorded.

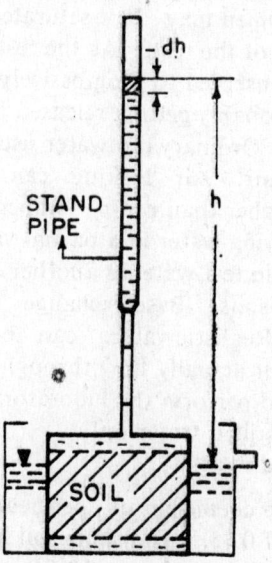

Fig. 6.3 Set-up of falling head permeameter

The hydraulic head h at any moment is taken to be the difference between the water levels in the standpipe and in the bottom tank. Let the level drop by (−dh) in time (dt). Then the rate of flow is given by:

$$q = -\frac{a\,dh}{dt} = k\,\frac{h}{L}\,A \qquad (6.17)$$

Let h_1 be the head at the start of the test ($t = 0$) and h_2 be the head after time interval 't'. Transposing the terms and integrating between the limits h_1 and h_2 for time 0 to t,

$$\frac{Ak}{aL}\int_0^t dt = \int_{h_1}^{h_2}\frac{-dh}{h} = \int_{h_2}^{h_1}\frac{dh}{h}$$

or $\qquad k = \dfrac{aL}{At}\log_e\dfrac{h_1}{h_2} = \dfrac{2.3\,aL}{At}\log_{10}\dfrac{h_1}{h_2} \qquad (6.18)$

where $\qquad$ a = area of standpipe

$\qquad\qquad$ A = area of specimen

4. Precautions. Entrapped air in the specimen is a great source of error. The specimen may be saturated by applying partial vacuum at the start of the test. As the test is run, the permeability may frequently be observed to progressively decrease. When this is the case, air is probably getting released from the test water and causing air locking. Ordinary tap water usually contains a considerable amount of air. Air locking can be eliminated by using distilled water at higher than room temperature. Water can also be de-aired by spraying water in a partial vacuum.

Presence of salts in test water is another source of error while testing fine grained soils. Base exchange may considerably alter the permeability. Realistic values can be obtained by using the soil water which will actually flow through the soil in the field. It is, however, usual to perform the laboratory tests under standard conditions using distilled water only.

Examples 6.1 — 6.9

6.1 Estimate the coefficient of permeability k_2 of a sandy soil at a void ratio e_2 of 0.75. The same soil has a permeability k_1 of 2×10^{-2} cm/s at a void ratio e_1 of 0.64.

$\qquad$ (a) $\quad k \propto e^2$; $\quad k_2 = k_1\,\dfrac{e_2^2}{e_1^2} = 2.75 \times 10^{-2}$ cm/s

$\qquad$ (b) $\quad k \propto \dfrac{e^3}{1+e}$; $\quad k_2 = 3.02 \times 10^{-2}$ cm/s

6.2 The permeability of a soil at a test temperature of 35°C is 1.8×10^{-4} cm/s. What will be the permeability at a temperature of 27°C.   The viscosities of water at 35°C and 27°C are respectively 7.21 $\times$ 10^{-3} poise and 8.55 $\times$ 10^{-3} poise.   Neglect the effect of change in density of water.

$$k_{27} = k \frac{\eta}{\eta_{27}} = 1.52 \times 10^{-4} \text{ cm/s}$$

6.3   The effective size D_{10} of a filter sand is 0.075 mm.   Estimate the approximate value of the coefficient of permeability.

$$k = C_1 (D_{10})^2 = 1 \times (0.075)^2 = 5.63 \times 10^{-3} \text{ cm/s}$$

6.4   During a constant head permeameter test, a flow Q of 160 cm^3 is measured in 5 minutes under a constant head of 15 cm. The specimen is 6 cm long and has a sectional area of 50 cm^2. The porosity n_1 of the specimen is 42%. Determine the permeability, the flow velocity v and the seepage velocity v_s. Estimate k_2 for n_2 = 35%.

$$k = \frac{Q L}{A t h} = 4.267 \times 10^{-3} \text{ cm/s}$$

$$v = k \frac{h}{L} = 1.067 \times 10^{-2} \text{ cm/s}$$

$$v_s = \frac{v}{n_1} = 2.54 \times 10^{-2} \text{ cm/s}$$

$$n_1 = 0.42, e_1 = 0.724$$

$$n_2 = 0.35, e_2 = 0.538$$

$$k \, \alpha \, \frac{e^3}{1 + e} \; ; \; k_2 = 1.96 \times 10^{-3} \text{ cm/s}$$

6.5   A specimen, 10 cm in diameter and 30 cm long, is tested in a constant head permeameter.   A flow of 235 cm^3 is measured in 5 minutes when the constant head difference between tapping points 20 cm (L) apart is 6.5 cm (h).   Determine the coefficient of permeability.

$$q = \frac{235}{5 \times 60} \text{ cm}^3/\text{s} \, , \; i = \frac{6.5}{20}$$

$$k = \frac{q}{A i} = 3.07 \times 10^{-2} \text{ cm/s}$$

6.6   In a falling head permeability test on a specimen 6 cm high and 50 cm^2 in cross sectional area, the water level in the standpipe, 0.8 cm^2 in sectional area, dropped from a height of 60 cm to 20 cm in 3 minutes 20 seconds.   Find the permeability.

$$k = \frac{2.3 \, a \, L}{A \, t} \log_{10} \frac{h_1}{h_2} = 5.27 \times 10^{-4} \text{ cm/s}$$

6.7 A falling head permeability test is performed on a specimen of clean uniform sand 100 mm in diameter and 150 mm long. The diameter of the standpipe is 5 mm. The water level in the standpipe is 1500 mm above the overflow which drops by 500 mm in 120 seconds. Find the permeability. How much additional time will the water level in the standpipe take to drop by another 500 mm?

(a) $A = 7854$ mm^2, $a = 19.63$ mm^2

$L = 150$ mm, $h_1 = 1500$ mm, $h_2 = 1000$ mm

$$k = \frac{2\cdot3\,a\,L}{A\,t}\ \log_{10}\frac{h_1}{h_2} = 1.265 \times 10^{-3}\ \text{mm/s}$$

(b) $h_1 = 1000$ mm, $h_2 = 500$ mm

$t = 205$ seconds

6.8 A glass tube, fitted with a screen at bottom, contains saturated sand upto a height of 8 cm above the screen (Fig. 6.4). The tube is filled with water upto a height of 40 cm above the bottom and the water is allowed to flow through the sand specimen. Find the permeability if the water level drops by 20 cm in 160 seconds.

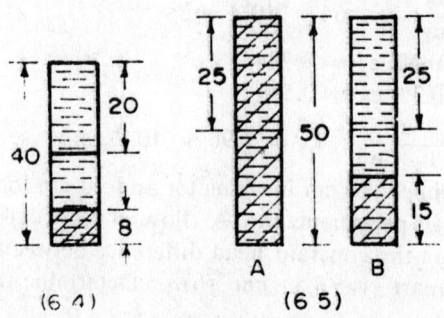

Figs. 6.4 & 6.5 Examples 6.8 & 6.9

$$k = \frac{2.3\,a\,L}{A\,t}\ \log_{10}\frac{h_1}{h_2}$$

$a = A$, $h_1 = 40$ cm, $h_2 = 20$ cm

$k = 3.46 \times 10^{-2}$ cm/s

6.9 Two glass cylinders A and B, each 45 cm^2 in inside cross sectional area and 50 cm long, are fitted with perforated screens at bottom and are open at top (Fig. 6.5). Saturated, uniform sand at a porosity of 40 percent is placed in cylinder 'A' upto its top and in cylinder 'B' upto a height of 15 cm above the screen. Both the cylinders are completely flooded with water upto their tops, and water is allowed to drain out through the sand. In cylinder A water level drops by 25 cm in 7 minutes 25 seconds. How much time will the water level take to drop by the same height in cylinder B ? Assume the sand to be completely drainable and to have the same permeability in both the cylinders. What is the value of the permeability ?

(a) Cylinder 'A' represents flow under a constant hydraulic gradient of unity ($h/L = 1$).

Volume of sand drained $V = 24 \times 45 = 1125$ cm^3
Volume of water drained $= n V = 450$ cm^3

$$q = k i A ; \quad \frac{450}{445} = k \times 1 \times 45$$

$$k = 2.247 \times 10^{-2} \text{ cm/s}$$

(b) Cylinder 'B' represents flow under a variable head.

$h_1 = 50$ cm, $h_2 = 25$ cm, $L = 15$ cm, $a = A$

$$k = \frac{2.3 \, aL}{A \, t} \, \log_{10} \frac{h_1}{h_2}$$

$t = 7$ min 42 s

6.7 FIELD MEASUREMENT OF PERMEABILITY

A widely used field method of measuring permeability insitu is the 'well pumping test', in which water is pumped into or out of a well. The 'pumping out' test is feasible in coarse grained soils only (k not less than about 10^{-3} or 10^{-4} cm/s). A water bearing porous formation which can yield water to a well is known as an 'aquifer'. When the water table exists *within* the aquifer, it is called an 'unconfined aquifer.' When the aquifer is surrounded by formations of less permeable or impermeable material, it is called a 'confined aquifer'. Here the groundwater is under pressure and it would rise in a well or piezometer above the bottom of the superimposed confining formation, hence it is also known as an 'artesian aquifer'.

1. **Pumping-out test in an unconfined aquifer.** A test well is sunk through the aquifer upto the underlying impervious formation. It is provided with perforations over the full length which is below the water table. Such a well is called the 'perfect well'. (*Note.* An 'imperfect' well is in which the length of strainer is less than the thickness of the aquifer or which does not penetrate the lower impervious boundary). A series of piezometers or observation wells are installed around the test well along radial lines. The piezometers often need to penetrate only to the lowest level that the water table will reach in a test. Each radial line has at least 2 piezometers. Water is pumped out from the test well at some constant rate until a steady state is attained, i.e. the water level in the test well and in the piezometers stays more or less constant under a constant rate of pumping.

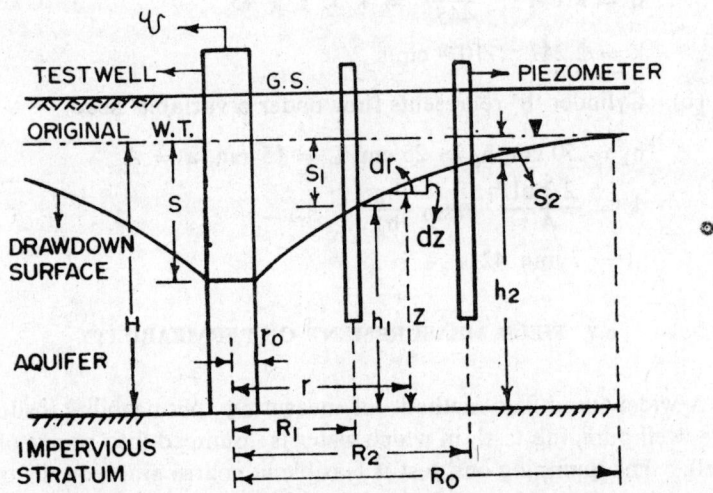

Fig. 6.6 Pumping out test in an unconfined aquifer

Fig. 6.6 shows a section through the test well and two observation wells (piezometers). The radius of the test well is r_0 and the piezometers are located at distances R_1 and R_2 from the test well. Pumping generates a radial flow of water towards the test well, and as a result, drawdown of the water table takes place. When the steady state is established the respective water table levels in the two piezometers are at heights of h_1 and h_2 above the bottom of the aquifer. Let R_0 be the maximum radius of influence, i.e. the distance upto which the 'drawdown surface' extends and at which the drawdown is negligible.

The analysis is based on the following assumptions (Dupuit 1983).

(1) The water bearing stratum is homogeneous, isotropic, horizontal and extends an infinite distance in all directions with a constant thickness.

(2) The water table is of infinite extent in horizontal directions. In its original state the groundwater is at rest.

(3) The test well is a 'perfect' well.

(4) The flow is horizontal and the flow velocity at every point of a vertical flow section is the same. The hydraulic gradient is constant at any given radius and is equal to the *slope* of the water surface. This assumption introduces large errors near the test well but is reasonaly accurate at moderate distance. As indicated by the curved drawdown surface, the hydraulic gradient varies with distance from the test well.

During steady state flow conditions, the rate of flow q towards the well must be the same at all radii. Then, at radius r the hydraulic gradient is dz/dr and the flow rate is given by:

$$q = k\, i\, A = k\, \frac{dz}{dr}\, 2\pi r\, z \qquad (6.19)$$

where A = area of cylindrical surface of radius r and height z.

Separating the variables and integrating for the two piezometers,

$$\frac{2\pi k}{q} \int_{h_2}^{h_1} z\, dz = \int_{R_2}^{R_1} \frac{dr}{r}$$

Hence $\quad k = \dfrac{q}{\pi\,(h_2{}^2 - h_1{}^2)}\, \log_e \dfrac{R_2}{R_1} \qquad (6.20)$

or $\quad k = \dfrac{2.3\, q}{\pi\,(h_2{}^2 - h_1{}^2)}\, \log_{10} \dfrac{R_2}{R_1} \qquad (6.21)$

If S_1 and S_2 are the drawdowns at the respective piezometers at distances R_1 and R_2 and H is the thickness of aquifer. Eq. 6.21 may be expressed as:

$$k = \frac{2.3\, q}{\pi\,(2H - S_1 - S_2)\,(S_1 - S_2)}\, \log \frac{R_2}{R_1} \qquad (6.22)$$

If the ratio R_2/R_1 is kept equal to 2.72, Eq. 6.21 reduces to:

$$k = \frac{q}{\pi (h_2{}^2 - h_1{}^2)} \tag{6.23}$$

If there is only one piezometer at distance R_1 from the test well and integration is carried out from one piezometer to the circumference of the test well, an approximate formula is obtained as:

$$k = \frac{2.3 \, q}{\pi (2H - S - S_1)(S - S_1)} \log \frac{R_1}{r_0} \tag{6.24}$$

where S = drawdown at the test well

r_0 = radius of the test well

Still another approximate formula (Eq. 6.25) is obtained by considering the limits of integration for the complete drawdown surface from the test well upto the radius of influence R_0. Thus k can approximately be determined by observing the drawdown S at the test well alone and assuming a suitable value of R_0.

$$k = \frac{2.3 \, q}{\pi (2H - S) S} \log \frac{R_0}{r_0} \tag{6.25}$$

The radius of influence R_0 depends on many factors such as the thickness and permeability of the aquifer, rate of pumping, and the conditions of inflow and drainage, etc. It may vary from about 50 m in fine sands to about 200 m in gravelly soils.

The *specific yield* q_s, or the rate of flow per unit length of the drawdown is given by:

$$q_s = q/S \quad (m^3/s/m) \tag{6.26}$$

The rate of flow (m^3/s) occurring through a vertical strip of the aquifer of unit width (1 m) and thickness H (m) under unit hydraulic gradient is called the *transmissibility* T, which is related to permeability k (m/s) as follows:

$$T = kH \quad (m^3/s/m) \tag{6.27}$$

2. Pumping-out test in a confined aquifer. Fig. 6.7 shows a test well sunk through an aquifer confined at both top and bottom by practically impervious strata. The water table is not within the aquifer and the piezometric surface is every where above the top of the aquifer.

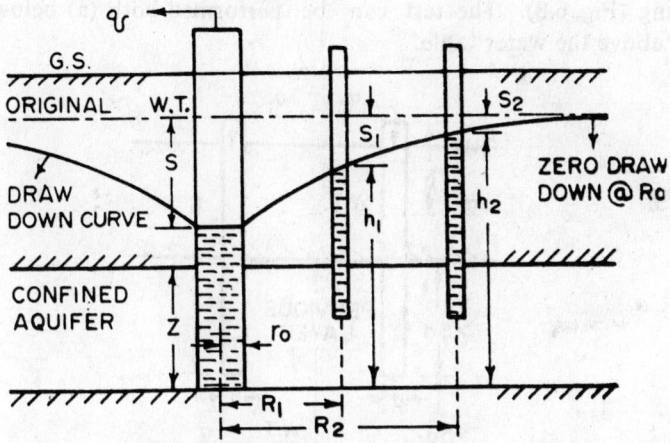

Fig. 6.7 Pumping-out test in a confined aquifer

For steady conditions, the flow rate q at radius 'r' is given by:

$$q = k \, i \, A = k \, \frac{dz}{dr} \, 2\pi r \, Z \qquad (6.28)$$

where Z = thickness of confined aquifer
On integration, as before,

$$k = \frac{2.3 \, q}{2\pi \, Z \, (h_2 - h_1)} \, \log_{10} \frac{R_2}{R_1} \qquad (6.29)$$

or $$k = \frac{2.3 \, q}{2\pi Z \, (S_1 - S_2)} \, \log_{10} \frac{R_2}{R_1} \qquad (6.30)$$

where S_1 = drawdown in the first piezometer at distance R_1
S_2 = drawdown in the second piezometer at distance R_2
Similarly, the formula for the test well alone is:

$$k = \frac{2.3 \, q}{2\pi \, Z \, S} \, \log \frac{R_0}{r_0} \qquad (6.31)$$

where S = drawdown in the test well
3. Constant head borehole test. The borehole test, also known as the 'open-end pipe test' (USBR 1973, 1974), consists in drilling a hole upto the test level into the pervious stratum and sinking a pipe casing upto the bottom of the hole. The lower open end of the casing should not be less than '5d' from either the top or bottom of the stratum, where 'd' is the internal diameter of the

casing (Fig. 6.8). The test can be performed both (a) below and (b) above the water table.

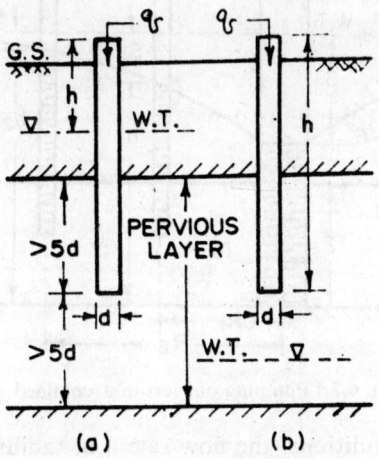

Fig. 6.8 Constant head borehole test

The hole is carefully cleaned out to the bottom of the casing. For the hole extending below the groundwater level, the casing is kept filled with water to minimize squeezing of soil into the bottom of the casing. Water is allowed to enter the casing through a measuring device to maintain a constant level. Water thus flows through the bottom of the cased hole into the pervious stratum under a constant head 'h' (gravity head) which is taken to be equal to the difference in elevations of the water level in the casing and the groundwater level for a test performed below the water table. For test above the water table, the hydraulic head is the height of water column in the casing. If desired, water may be fed under pressure. The additional pressure head, in terms of height of water, is then added to the gravity head to get the total hydraulic head 'h'. The permeability is calculated from the following relationship developed by electrical analogy tests:

$$k = \frac{q}{2.75 \, d \, h} \tag{6.32}$$

where q = constant rate of flow into the hole

4. **Variable head borehole test.** In a variable head test the rate of flow from the stratum into the borehole is measured by observing

the time 't' for the water level in the hole, relative to the water table, to change from a value h_1 to a value h_2. Hvorslev (1951) published formulae to determine permeability in a number of borehole situations. Two typical examples are given below.

To determine permeability at shallow depths below the water table, a cased hole of internal diameter 'd' is drilled upto a short distance (not exceeding 1.5 m) below the water table in a stratum assumed to be of infinite depth, as shown in Fig. 6.9a. The permeability is calculated from:

$$k = \frac{\pi\, d}{11\, t} \log_e \left(\frac{h_1}{h_2}\right) \qquad (6.33)$$

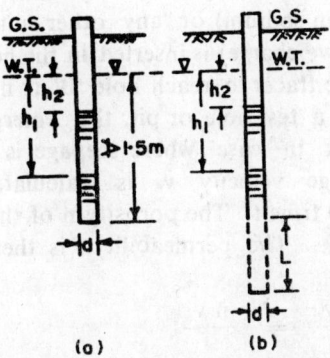

Fig. 6.9 Variable head borehole tests

Permeability determination at greater depths below water table is made by drilling a borehole with a cased length and an uncased or perforated extension of length L, where $L > 4d$ (Fig. 6.9b). The permeability is given by:

$$k = \frac{d^2}{8\, L\, t} \log_e \left(\frac{2L}{d}\right) \lg_e \left(\frac{h_1}{h_2}\right) \qquad (6.34)$$

5. Method based on seepage velocity. In situations of sloping water table, the permeability of coarse grained soil can be obtained from insitu measurement of speepage velocity v_s. Two uncased boreholes, or test pits, are excavated at two points 'A' and 'B' along the direction of flow, as shown in Fig. 6.10.

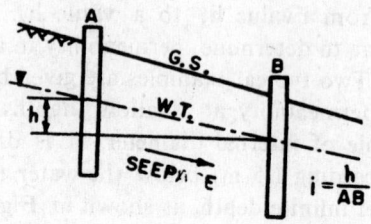

Fig. 6.10 Permeability by measurement of seepage velocity

The difference in the steady state water levels in the boreholes divided by the distance AB gives the hydraulic gradient i. A dye (such as fluorescein sodium) or any other suitable tracer (an electrolyte or radioactive charge) is inserted in the hole 'A' and the time t for the dye or the tracer to reach hole 'B' is measured (Cedergren 1977). Instead of a test hole or pit, the emergent point B may also be on a bank in case where seepage is emerging from a bank. The seepage velocity v_s is calculated by dividing the distance AB by the time t. The porosity n of the soil is determined from density tests. The permeability is then calculated from Eq. 6.35.

$$k = \frac{v}{i} = \frac{n \, v_s}{i} \qquad (6.35)$$

Examples 6.10 — 6.19

6.10 A field pumping-out test is made in sand extending to a depth of 13 m and underlain by impervious clay. Observation wells are drilled at distances of 5 m and 10 m from the centre of the test well. The water in the test well and in the observation wells stands originally at the same level, 1 m below the top of the test well. A steady state is established in about 24 hours when the rate of flow is 3.5 l/s. The drawdowns at the two observation wells are observed to be 1.2 m and 0.25 m respectively. Determine the coefficient of permeability.

$$q = 3.5 \text{ l/s} = 3.5 \times 10^{-3} \text{ m}^3/\text{s}$$
$$R_1 = 5 \text{ m}, \ h_1 = 12 - 1.2 = 10.8 \text{ m}$$
$$R_2 = 10 \text{ m}, \ h_2 = 12 - 0.25 = 11.75 \text{ m}$$
$$k = \frac{2.3 \, q}{\pi \, (h_2{}^2 - h_1{}^2)} \ \log \frac{R_2}{R_1} = 3.6 \times 10^{-5} \text{ m/s}$$

Alternatively,

$$k = \frac{2.3\ q}{\pi\ (2H - S_1 - S_2)\ (S_1 - S_2)}\ \log\ \frac{R_2}{R_1}$$

$H = 13 - 1 = 12$ m, $S_1 = 1.2$ m, $S_2 = 0.25$ m

$k = 3.6 \times 10^{-5}$ m/s

6.11 A test well, 0.6 m in diameter, is sunk through an aquifer 15 m thick upto the underlying impermeable stratum. The original water table is at the ground surface. At steady state when the discharge is 7.5×10^{-3} m³/s, the drawdown at the test well is 6 m. Assuming the radius of influence as 150 m, calculate the permeability. What will be the permeability if the radius of influence is assumed as 300 m ?

$q = 7.5 \times 10^{-3}$ m³/s, $r_o = 0.3$ m, $R_o = 150$ m

$H = 15$ m, $S = 6$ m

$$k = \frac{2.3\ q}{\pi\ (2H - S)\ S}\ \log\ \frac{R_0}{r_0} = 1.03 \times 10^{-4}\ \text{m/s}$$

When $R_o = 300$ m, $k = 1.14 \times 10^{-4}$ m/s

The permeability at $R_0 = 300$ m is only about 11 percent greater than the value obtained for $R_o = 150$ m. Since R_o is always very large in comparison to r_o, the value of log (R_o/r_o) will vary only over a narrow range with the variation of R_o.

6.12 A test well, 0.5 m in diameter penetrates through a saturated aquifer, 8 m thick, overlying an impervious layer. The steady discharge of the well is 18.72 m³/hr. The drawdown at a distance of $R_1 = 15$ m from the centre of the test well is found to be 1.8 m. What will be the drawdown at a distance of $R_2 = 50$ m, if the permeability is 3.8×10^{-4} m/s ? Estimate approximately the drawdown at the test well also, Find the specific yield and transmissibility.

$q = 18.72$ m³/hr $= 5.2 \times 10^{-3}$ m³/s

$H = 8$ m, $R_1 = 15$ m, $S_1 = 1.8$ m, $R_2 = 50$ m

$$k = \frac{2.3\ q}{\pi\ (2H - S_1 - S_2)\ (S_1 - S_2)}\ \log_{10}\ \frac{R_2}{R_1}$$

$S_2 = 1.39$ m

For drawdown at test well :

$$k = \frac{2.3\ q}{\pi\ (2H - S - S_1)\ (S - S_1)}\ \log_{10}\ \frac{R_1}{J_0}$$

$$S = 2.25 \text{ m}$$
$$q_s = q/S = 2.31 \times 10^{-3} \text{ m}^3/\text{s/m}$$
$$T = kH = 3.04 \times 10^{-3} \text{ m}^3/\text{s/m}$$

6.13 A 5 m thick aquifer is confined at both top and bottom by practically impervious strata. The top impervious stratum is 6 m thick. A test well is sunk upto the lower impervious stratum. Initially the water level stands in the test well at a depth of 1.5 m below the ground surface. Two piezometers are radially installed at distances of 12 m and 60 m from the test well. The piezometeric surface stands at 3.3 m and 1.9 m below the ground surface in the two respective piezometers when a steady discharge is established. If the permeability is 2.45×10^{-4} m/s, find the yield of the well.

$$Z = 5 \text{ m}, R_1 = 12 \text{ m}, S_1 = 3.3 - 1.5 = 1.8 \text{ m}$$
$$R_2 = 60 \text{ m}, S_2 = 1.9 - 1.5 = 0.4 \text{ m}$$
$$k = \frac{2.3 \, q}{2 \, \pi \, z \, (S_1 - S_2)} \log \frac{R_2}{R_1}$$
$$q = 6.7 \times 10^{-3} \text{ m}^3/\text{s}$$

6.14 The maximum drawdown produced by steady pumping from a perfect artesian well is 3 m. The thickness of the aquifer is 10 m. Find the yield of the well, if the permeability of the aquifer is 5×10^{-3} m/s. The radius of the well is 0.15 m and the maximum radius of influence of the well may be assumed as 200 m.

$$S = 3 \text{ m}, Z = 10 \text{ m}, k = 5 \times 10^{-3} \text{ m/s}$$
$$r_o = 0.15 \text{ m}, R_o = 200 \text{ m}$$
$$q = \frac{2 \, \pi \, Z \, S \, K}{2.3 \log (R_o/r_o)} = 131 \times 10^{-3} \text{ m}^3/\text{s}$$

6.15 Calculate the rate of flow into the artesian well of Ex. 6.14 if

    (a) the maximum drawdown is increased to 6 m,

    (b) the radius of the well is increased to 0.3 m but the drawdown remains 3 m.

      (a) $S = 6$ m, $q = 262 \times 10^{-3}$ m^3/s

      (b) $r_o = 0.3$ m, $S = 3$ m, $q = 145 \times 10^{-3}$ m^3/s

Percent increase in rate of flow $= \dfrac{145 - 131}{131} \times 100 = 10.7\%$

Increasing the drawdown by 100 percent produces an increase of 100 percent in the rate of flow, but a 100 percent increase of the radius of well increases the rate of flow by only about 10.7 percent.

6.16 A constant head borehole test is performed in an open-end casing of 10 cm diameter drilled into a pervious stratum lying below the groundwater level. Water is observed to flow into the cased hole at a rate of 1.2 l/min when the water level in the casing is maintained constant at an elevation of 2 m above the water table. Find the permeability.

$$d = 0.1 \text{ m}, h = 2 \text{ m}, q = 1.2 \times 10^{-3} \text{ m}^3/\text{min}$$

$$k = \frac{q}{2.75 \, d \, h} = 2.18 \times 10^{-3} \text{ m/min}$$

6.17 To determine the permeability, a cased hole of 7.5 cm diameter is drilled 1.5 m below the water table. Water is pumped out to lower the water level in the casing by 1.2 m below the water table. Water is observed to rise in the casing to an elevation of 0.5 m below the water table in 15 minutes. Determine the permeability.

$$d = 7.5 \text{ cm}, h_1 = 1.2 \text{ m}, h_2 = 0.5 \text{ m}, t = 900\text{s}$$

$$k = \frac{\pi}{11} \frac{d}{t} \log_e \frac{h_1}{h_2} = 2.08 \times 10^{-3} \text{ cm/s}$$

6.18 A borehole, 8 cm in diameter, with a cased length and a perforated extension of 40 cm is drilled in a pervious stratum to a depth of 4 m below the water table. Water is bailed out of the borehole and the rise of water is observed to take place from a depth of 1.0 m to 0.5 m below the water table in 20 minutes. Estimate the permeability. What will be the rise in water level in the next 20 minutes.

(a) $d = 8$ cm, $L = 40$ cm, $h_1 = 100$ cm
$h_2 = 50$ cm, $t = 1200$ s

$$k = \frac{d^2}{8 \, L \, t} \log_e \left(\frac{2L}{d} \right) \log_e \left(\frac{h_1}{h_2} \right)$$

$$= 2.66 \times 10^{-4} \text{ cm/s}$$

(b) $h_1 = 50$ cm
Rise of water in 20 minutes = 25 cm

6.19 To estimate the hydraulic conductivity of a bed of coarse gravel, dyne is inserted in a borehole and its time to travel to a borehole 50 m away on the down slope is observed to be 56 minutes. The head difference of the sloping water table in the two holes is 0.5 m. Find the permeability, assuming the porosity of gravel to be 30 percent.

$$L = 50, \quad h = 0.5 \text{ m}, \quad i = h/L = 0.01$$

$$t = 56 \text{ min}, \quad v_s = L/t = 0.893 \text{ m/min}$$

$$k = \frac{n \, v_s}{i} = 3.86 \times 10^4 \text{ m/day}$$

6.8 GROUNDWATER FLOW

Three cases of groundwater flow are considered: unconfined horizontal flow, unconfined inclined flow and the artesian flow.

1. Unconfined horizontal flow. A uniform water bearing stratum resting on a horizontal, permeable layer is shown in Fig. 6.11. Water table is within the aquifer and the flow is assumed to be steady, parallel and horizontal. The height of water column in the two observations wells (piezometers) spaced at a distance L along the line of flow is respectively h_1 and h_2.

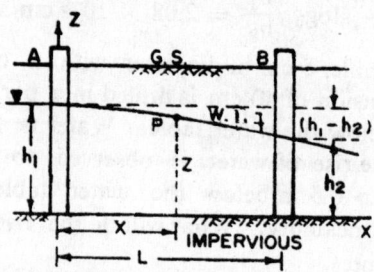

Fig. 6.11 Unconfined horizontal groundwater flow

Consider a flow section at a distance 'x' from the piezometer 'A' to have a height 'z'. Based on Dupuit's assumption, the hydraulic gradient is represented by the slope of the free water table, i.e. the tangent dz/dx and the rate of flow, q per unit width (normal to the plane of the figure) can be expressed as:

$$q = - k \frac{dz}{dx} z \quad \text{(per unit width)}$$

where x, z = coordinates of point P on the free surface of flowing water

Separating the variables and intergrating for the two observation wells:

$$q \int_{o}^{L} dx = -k \int_{h_1}^{h_2} z \, dz$$

or $\qquad q = k \dfrac{h_1^2 - h_2^2}{2L}$ \hfill (6.36)

or $\qquad q = k \dfrac{h_1 - h_2}{L} \times \dfrac{h_1 + h_2}{2}$ \hfill (6.37)

Eq. 6.37 suggests that an alternative solution of the problem is to simply express the rate of flow as:

$$q = k i \overline{A} \qquad\qquad (6.38)$$

where $\qquad i = \dfrac{h_1 - h_2}{L}$

and $\qquad \overline{A} =$ average cross section of the aquifer per unit width

$$= \dfrac{h_1 + h_2}{2}$$

hence $\qquad q = k \dfrac{h_1 - h_2}{L} \times \dfrac{h_1 + h_2}{2} = k \dfrac{h_1^2 - h_2^2}{2L}$

2. Unconfined inclined flow. A uniform flow of groundwater is shown in Fig. 6.12 occurring through a uniform pervious soil overlying an inclined impervious formation. The thickness (or depth) of the flowing sheet of water is constant 'd' throughout, and the grade of the groundwater surface is the same as that of the underlying impervious boundary. Let it be further assumed that the width of the groundwater stream is also constant. Under these conditions the hydraulic gradient is constant at every section of the flow stream and equals h/L, where h is the difference in levels of two points on the water surface in the direction of flow separated by a distance L measured *along* the water surface i.e. $i = \sin \alpha$, where α is the slope angle of the free water surface. (*Note.* Dupuit assumed the hydraulic gradient as the tangent dz/dx, Fig. 6.11.) Further, the flow is 'uniform' which means that the velocity is the same in both the magnitude and direction from point to point along the flowing stream.

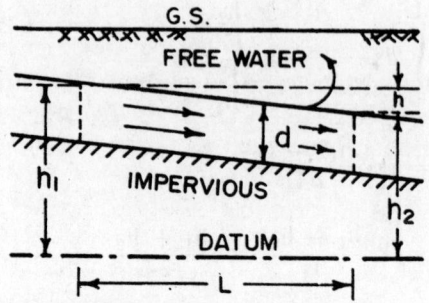

Fig. 6.12 Unconfined, uniform inclined flow

The rate of flow q per unit width is given by:

$$q = k\,i\,A = k\,\frac{h}{L}\,d = k\,d\,\sin\alpha \qquad (6.39)$$

where $\quad i = \sin\alpha = $ constant

3. Artesian groundwater flow. Flow through a confined aquifer is shown in Fig. 6.13.

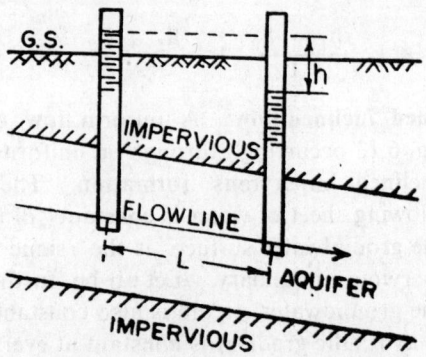

Fig. 6.13 Artesian groundwater flow

The artesian groundwater is under sufficient head to cause it to rise above the zone of saturation at that place if opportunity were afforded to do so. The hydraulic gradient is given by h/L, where h is the difference in levels of the top of the water columns in two piezometers separated by a distance L *along* the flow path. The flow velocity v is given by: $v = k\,i$

Examples 6.20 — 6.21

6.20 Unconfined, horizontal groundwater flow is occurring through a bed of medium sand, 15 m thick, overlying a horizontal impermeable layer (Fig. 6.11). The respective groundwater levels in the two piezometers, spaced 300 m apart along the direction of flow are observed to be 3 m and 3.25 m below the ground surface. Estimate the rate of flow per metre width (normal to the direction of flow), if the coefficient of permeability is 4×10^{-3} m/s.

$$h_1 = 15 - 3 = 12 \text{ m}, \quad h_2 = 15 - 3.25 = 11.75 \text{ m}$$

$$L = 300 \text{ m}$$

$$q = k \frac{h_1^2 - h_2^2}{2L} = 3.96 \times 10^{-5} \text{ m}^3/\text{s} = 3.42 \text{ m}^3/\text{day}$$

6.21 Determine the rate of flow per day per unit width of a uniform, parallel flow along the incline of Fig. 6.12. The thickness of the water sheet is 5 m. The top free surface of the flowing water drops by 0.25 m in a distance of 100 m along the incline. The permeability is 4×10^{-3} m/s.

$$i = \sin \alpha = 0.25/100 = 0.0025$$

$$d = 5 \text{ m}$$

$$q = k \, d \sin \alpha = 5 \times 10^{-5} \text{ m}^3/\text{s} = 4.32 \text{ m}^3/\text{day}$$

6.9 PERMEABILITY OF STRATIFIED SOILS

Where a soil profile consists of a number of strata having different permeability, the equivalent or average permeability of the soil is different in directions parallel to, and normal to, the strata.

Consider the soil profile shown in Fig. 6.14 consisting of three layers whose properties are indicated by the suffixes 1, 2 and 3 respectively. For flow parallel to the layers (Fig. 6.14a), the hydraulic gradient in each layer is the same $(i = h/L)$ and the total flow rate is the sum of the flow rates in all the three layers.

$$q = q_1 + q_2 + q_3 \tag{6.40}$$

or

$$k_x \, i \, H = k_1 \, i \, H_1 + k_2 \, i \, H_2 + k_3 \, i \, H_3 \tag{6.41}$$

hence

$$k_x = \frac{k_1 H_1 + k_2 H_2 + k_3 H_3}{H} \tag{6.42}$$

where k_x = equivalent or average permeability in direction parallel to the layers

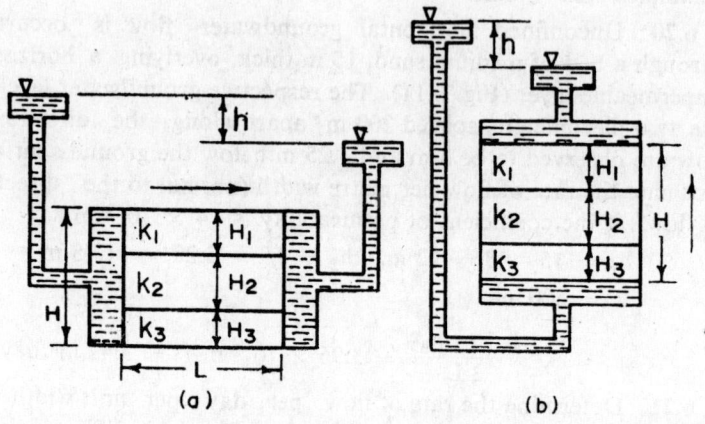

Fig. 6.14 Flow through multilayered system

For flow normal to the layers (Fig. 6.14b) the flow rate must be the same in all layers for steady state flow, and as the flow area 'A' is constant, the flow velocity across each layer is also the same.

$$v = v_1 = v_2 = v_3 \tag{6.43}$$

The total head loss h is equal to the sum of the losses in the three layers.

$$h = h_1 + h_2 + h_3 \tag{6.44}$$

$$i H = i_1 H_1 + i_2 H_2 + i_3 H_3 \tag{6.45}$$

$$\frac{vH}{k_z} = \frac{v_1 H_1}{k_1} + \frac{v_2 H_2}{k_2} + \frac{v_3 H_3}{k_3} \tag{6.46}$$

$$\frac{H}{k_z} = \frac{H_1}{k_1} + \frac{H_2}{k_2} + \frac{H_3}{k_3} \tag{6.47}$$

hence

$$k_z = \frac{H}{\dfrac{H_1}{k_1} + \dfrac{H_2}{k_2} + \dfrac{H_3}{k_3}} \tag{6.48}$$

where k_z = equivalent permeability for flow normal to the layers

Eqs. 6.42 and 6.48 can be re-written to include any desired number of layers. It would be found that the equivalent permeability for flow parallel to the strata is always greater than that for flow normal to the strata, i.e. k_x is always greater than k_z. Here it is assumed that each individual layer is isotropic, which may not be the case in actual practice. The individual layers may be anisotropic and have a greater permeability in a direction parallel to the bedding planes, which are often horizontal.

Examples 6.22 — 6.26

6.22 A sand deposit it made up of three horizontal layers of equal thicknesses. The permeability of the top and bottom layers is 2×10^{-4} cm/s and that of the middle layer is 3.2×10^{-2} cm/s. Find the equivalent permeability in the horizontal and vertical directions and their ratio.

$$k_x H = k_1 H_1 + k_2 H_2 + k_3 H_3$$
$$\frac{H}{k_z} = \frac{H_1}{k_1} + \frac{H_2}{k_2} + \frac{H_3}{k_3}$$
$$H_2 = H_3 = H_1, H = 3 H_1, k_1 = k_3 = 2 \times 10^{-4} \text{ cm/s}$$
$$k_x = 1.08 \times 10^{-2} \text{ cm/s}, k_z = 2.99 \times 10^{-4} \text{ cm/s}$$
$$k_x/k_z = 36.1$$

6.23 Alternating layers of coarse and fine material of equal thickness constitute the strata at a site. The permeability of the coarse material (k_1) is 100 times that of the fine material (k_2). Find the ratio of the average horizontal and vertical permeabilities.

$$k_1 = 100 \, k_2$$
Consider two layers only, $H = 2 H_1, H_1 = H_2$
$$k_x = 50.5 \, k_2 \, , \quad k_z = 1.98 \, k_2$$
$$k_x/k_z = 25.5$$

6.24 A glacial clay deposit consists of a series of thin layers of silt, 5 mm thick, after every 1.5 m thick layer of clay. The silt is 100 times more permeable than the clay. Assuming both materials to be hydraulically isotropic, find the ratio of the horizontal and vertical permeabilities.

Consider only two adjacent layers.
$$H_1 = 150 \text{ cm}, H_2 = 0.5 \text{ cm}$$
$$k_2 \text{ (silt)} = 100 \, k_1 \text{ (clay)}$$
$$k_x = 1.329 \, k_1, \quad k_z = 1.003 \, k_1$$
$$k_x/k_z = 1.325$$

6.25 Water is flowing under an average hydraulic gradient of 0.1 in the vertical direction through two horizontal layers of sand, 2 m and 3 m thick, whose respective permeabilities are 5×10^{-3}.

cm/s and 2×10^{-2} cm/s. Find the flow velocity, gradient and head loss in each layer.

$$k_z = 9.1 \times 10^{-3} \text{ cm/s}$$

Velocity in each layer is the same $= k_z i_a = 9.1 \times 10^{-4}$ cm/s

$$i_1 = v/k_1 = 0.182, \; h_1 = i_1 H_1 = 0.364 \text{ m}$$
$$i_2 = v/k_2 = 0.0455, \; h_2 = i_2 H_2 = 0.1365 \text{ m}$$
$$h_1 + h_2 \simeq 0.5 \text{ m}, \; h = i_a H = 0.5 \text{ m}$$

6.26 Water is flowing under a constant head of 0.4 m through soils A and B of different permeabilities as shown in Fig. 6.15. The permeability of soil A is 4.5×10^{-2} cm/s. If 37.5 percent of the head is lost in flowing through soil A, find the flow velocity and the permeability of soil B.

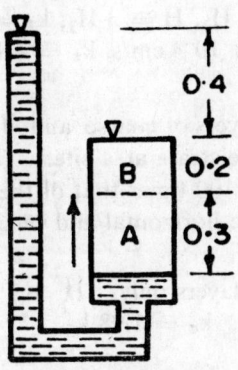

Fig. 6.15   Example 6.25

Head lost in soil A $= 0.4 \times 0.375 = 0.15$ m
$$i_A = 0.15/0.3 = 0.5$$
$$v_A = k_A i_A = 2.25 \times 10^{-2} \text{ cm/s} = v_B = \text{constant}$$
Head lost in soil B $= 0.25$ m
$$i_B = 1.25, \; v_B = k_B i_B; \; k_B = 1.8 \times 10^{-2} \text{ cm/s}$$
Alternatively
$$i_a \text{ (average)} = 0.4/0.5 = 0.8$$
$$k_z = v/i_a = 2.8125 \times 10^{-2} \text{ cm/s}$$
$$\frac{H}{k_z} = \frac{H_1}{k_A} + \frac{H_2}{k_B}; \; k_B = 1.8 \times 10^{-2} \text{ cm/s}$$

6.10 PERMEABILITY OF ROCKS

Like soils, rocks are also porous media and may be permeable to water. A distinction is, however, necessary between water flow through 'rock material or rock matrix' and through 'rock mass'. The rock material generally exhibits a very low permeability, around 10^{-7} to 10^{-14} cm/s, depending on the nature of the rock and is thus virtually impervious. The Darcy law governs the flow of water through rock material. Water flow through rock material is of little interest as practical flow problems involve large rock masses. The rock masses always contain joints and cracks which may make a formation highly permeable to the flow of water.

The permeability of rock masses is established almost entirely by the discontinuity or fracture pattern. Slight changes in the width of fissures greatly alter the permeability of jointed rocks (Londe 1972). The rock mass has always a very complex geometry in which the discontinuities are spread in an anisotropic and heterogeneous manner. A correct assessment of the permeability of a rock mass can be made only by insitu testing.

1. **Packer tests.** The permeability of an individual bed of rock can be determined by a packer test performed in a drill hole ($\simeq 5$ cm diameter), below the casing. The test can be performed during drilling or, more often, after the hole has been drilled. In the former situation, a top rubber packer is placed just inside or below the casing and water is introduced under pressure in the uncased newly drilled section of the hole, as shown in Fig, 6.16a. When the hole can stand without casing, it is drilled to its final depth and flushed to remove sediments. Two packers on a pipe or drill stem are set, as shown in Fig. 6.16b, to seal off a selected test length. The bottom of the pipe holding the packers is kept plugged and it has a perforated portion between the packers. It is found convenient to perform tests from the bottom of the hole upwards.

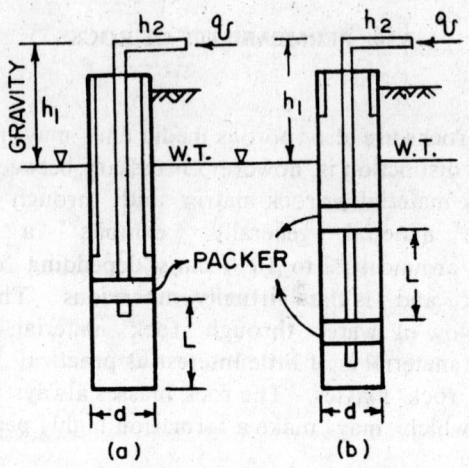

Fig. 6.16
(a) Single packer test (b) Two packer test

The coefficients of permeability are calculated from the following formulae (USBR 1973,1974):

$$k = \frac{q}{2\pi L h} \log_e \frac{2L}{d}; \quad L \geqslant 5d \qquad (6.49)$$

$$k = \frac{q}{2\pi L h} \sin h^{-1} \frac{L}{d}; \quad 5d > L \geqslant \frac{d}{2} \qquad (6.50)$$

where q = constant rate of flow into the hole
L = test length
d = diameter of hole
sin h⁻¹ = arc hyperbolic sine
h = differential head = h_1 (gravity) + h_2 (pressure)

According to the USBR (1973, 1974). Eqs. 6.49 and 6.50 are most valid when the thickness of the tested stratum is at least 5 L and are more accurate for tests below groundwater table than above. If the test is performed above the water table, the gravity head h_1 is measured from the swivel upto the middle of the test length L. The tests can also be used for testing soils.

2. Directional permeability. Structural discontinuities render a rock mass hydraulically anisotropic. The packer tests described above give only the total value of the permeability, without any allowance being made for anisotropy. A test procedure has been perfected by Louis (1974) for measuring the directional hydraulic conductivities. In case of a rock mass having, for example, three sets of fractures, pumping tests are performed in three different directions; the direction of drilling in order to test one of the sets is chosen to be parallel to the direction of the other two joint sets.

Example 6.27

A constant head, single packer test is performed during drilling through a fractured rock below the ground water level. The drill hole is 5 cm in diameter and test length below the packer is 2 m. Water is fed at the rate of 18×10^{-3} m^3/min under gravity head with groundwater level being 2.5 m below the swivel. Find the permeability.

$$h = 2.5 \text{ m}, \ L = 2 \text{ m}, \ d = 0.05 \text{ m}$$

$$k = \frac{q}{2 \pi L h} \log_e \frac{2 L}{d} = 2.51 \times 10^{-3} \text{ m/min.}$$

7

Seepage and flow net

7.1 ONE DIMENSIONAL FLOW

Seepage is the percolation or slow movement of water through soil or rock. It may be one dimensional (unidirectional), two dimensional in one plane, or three dimensional. A simple example of one dimensional flow is illustrated in Fig. 7.1 where water is flowing in the horizontal direction through a sample of sand contained within a tube of square cross section. If a dye is injected in sand on certain points on the upstream face where water is entering the sample, coloured streaks will form indicating the path taken by the percolating water. The path which a particle of water follows in its course of seepage under laminar flow conditions is called a "flow line" or a 'stream line'. An infinite number of flow lines is possible; however, only a limited number is adopted to represent a given flow. Only two flow lines are shown in Fig. 7.1. There is also a limiting flow line at each horizontal boundary of the sample; these are known as the 'boundary flow lines'. These four flow lines divide the area into three *flow channels* of equal dimensions. A "flow channel" is the space bounded by two adjacent flow lines. All the flow lines are straight because the confining boundaries are parallel. In most flow problems, however, the flow lines are curved.

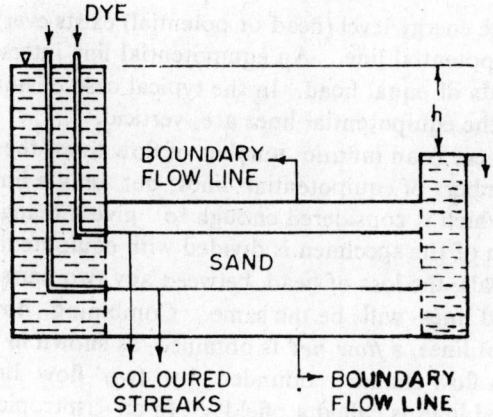

Fig. 7.1 Illustration of flow lines

As water moves along the flow lines, it experiences a continuous loss of head. The line AB in Fig. 7.2 represents the piezometric surface upto which water will rise if small piezometers are inserted at several positions along the length of the sample. If several piezometers are inserted at different elevations on any vertical section, say C or D, water rises to exactly the same level. The line along which water rises to the same elevation in piezometric tubes is called the "equipotential line" or the 'piezometric line'. It indicates

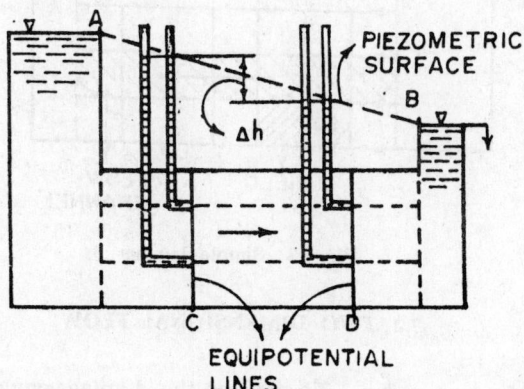

Fig. 7.2 Illustration of equipotential lines

that the same energy level (head or potential) exists everywhere along a given equipotential line. An equipotential line intersects the flow lines at points of equal head. In the typical case of horizontal flow of Fig. 7.2, the equipotential lines are vertical.

Just as there is an infinite number of flow lines, there is also 'an infinite numbers of equipotential lines; but only a limited number is selected which is considered enough to give a readable pattern. If the length of the specimen is divided with equipotential lines at equal intervals, the loss of head between any two pairs of adjacent equipotential lines will be the same. Combining flow lines with equipotential lines, a *flow net* is obtained, as shown in Fig. 7.3. The portion of a flow channel bounded by two flow lines and two equipotential lines is called a "field".   In an   isotropic soil the flow lines and equipotential lines intersect at right   angles, meaning that the flow direction is normal to the equipotential lines. The simplest pattern of orthogonal lines is one of squares. Thus the simple flow net of Fig. 7.3 is composed of square fields. Typical lines of equal pressure (piezometric head) are also shown in the figure.

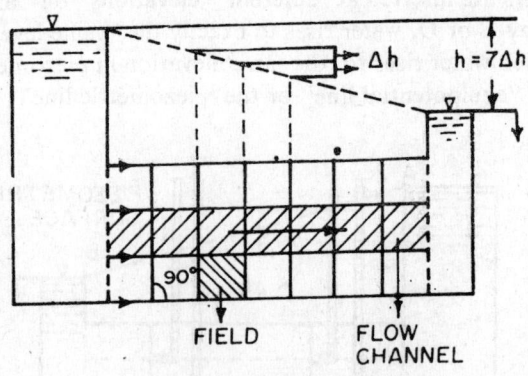

Fig. 7.3 Simple flow net

7.2 TWO DIMENSIONAL FLOW

1. Laplace equation. To develop the Laplace equation for flow of water through porous media, the following assumptions are made:

1. The soil is homogeneous and isotropic with respect to hydraulic conductivity (permeability).
2. The soil it saturated and the capillary effects are negligible.
3. Both the soil and water are incompressible. No compression or expansion takes place during the flow.
4. Flow is laminar and the Darcy law is valid.

Considering the general case of seepage in two dimensions only, let dx, dy, dz be the differential dimensions of a soil element (Fig. 7.4) in the x, y and z directions respectively, with flow taking place in the x-z plane only. Let v_x be the component of the flow velocity in the horizontal direction (x direction) and let i_x be the hydraulic gradient component in this direction. Let v_z and i_z be the corresponding values for the vertical direction. As the flow is assumed two dimensional, the velocity in the direction normal to the plane of the figure is zero.

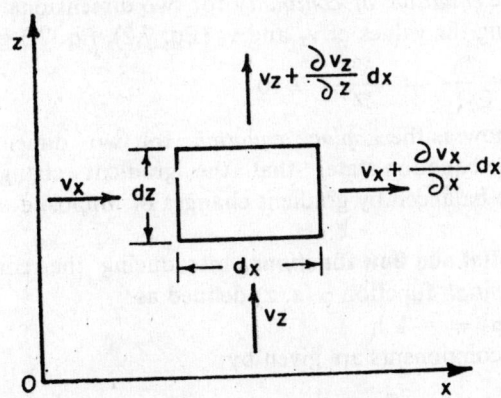

Fig. 7.4 Two dimensional flow through a soil element

According to the Darcy law:

$$i_x = -\frac{\partial h}{\partial x} \quad \text{and} \quad i_z = -\frac{\partial n}{\partial z} \tag{7.1}$$

$$v_x = -k\frac{\partial h}{\partial x} \quad \text{and} \quad v_z = -k\frac{\partial h}{\partial z} \tag{7.2}$$

where h = hydraulic head decreasing in the directions of v_x and v_z

The rates of change of velocity in the 'x' and 'z' directions are $\partial v_x/\partial x$ and $\partial v_z/\partial z$ respectively.

The value of water entering the element per unit time is:

$$v_x \; dz \; dy + v_z \; dx \; dy$$

and the volume of water leaving the element per unit time is:

$$(v_x + \frac{\partial v_x}{\partial x} \; dx) \; dz \; dy + (v_z + \frac{\partial v_z}{\partial z} \; dz) \; dx \; dy$$

As soil and water have been assumed incompressible and there are no volume changes, the quantity of water entering the element must be equal to the quantity leaving. Equating the above two expressions:

$$\frac{\partial v_x}{\partial x} \; dx \; dy \; dz + \frac{\partial v_z}{\partial z} \; dx \; dy \; dz = 0$$

whence
$$\frac{\partial v_x}{\partial x} + \frac{\partial v_z}{\partial z} = 0 \qquad (7.3)$$

Eq. 7.3 is the *equation of continuity* for two dimensional flow.

Substituting the values of v_x and v_z (Eq. 7.2), Eq. 7.3 gives:

$$\frac{\partial^2 h}{\partial x^2} + \frac{\partial^2 h}{\partial z^2} = 0 \qquad (7.4)$$

Eq. 7.4 is know as the *Laplace equation* for two dimensional flow. In words, the equation states that the gradient changes in the x direction are balanced by gradient changes of opposite sign in the z direction.

2. Potential and flow functions. Introducing the concept of the *veloctiy potential* function φ (x, z) defined as:

$$\varphi = - k \; h \qquad (7.5)$$

the velocity components are given by:

$$v_x = \frac{\partial \varphi}{\partial x} \quad \text{and} \quad v_z = \frac{\partial \varphi}{\partial z} \qquad (7.6)$$

and the Laplace equation may be expressed as:

$$\frac{\partial^2 \varphi}{\partial x^2} + \frac{\partial^2 \varphi}{\partial z^2} = \nabla^2 \varphi = 0 \qquad (7.7)$$

Then if a second function ψ (x,z) is introduced such that

$$v_x = \frac{\partial \psi}{\partial z} \quad \text{and} \quad v_z = - \frac{\partial \psi}{\partial x} \qquad (7.8)$$

it may be seen that

$$\frac{\partial^2 \psi}{\partial x^2} + \frac{\partial^2 \psi}{\partial z^2} = 0 \qquad (7.9)$$

Thus the function ψ, called the *flow function* or the *stream function*, also satisfies the Laplace equation. The two functions φ and ψ which satisfy the Laplace equation are, infact, both parts of the solution of the equation. Their physical significance is examined below.

The potential function φ (x,z) gives the distribution of φ in the x, z plane. Consider a curve along which φ (x, z) = constant = φ_1, as shown in Fig. 7.5. To obtain the tangent on a point on the curve, the total differential of the function is written as:

$$d\varphi = \frac{\partial \varphi}{\partial x}dx + \frac{\partial \varphi}{\partial z} dz$$

$$= v_x \, dx + v_z \, dz$$

Then for constant φ, $d\varphi = 0$ and the slope of the tangent is given by:

$$\frac{dz}{dx} = - \frac{v_x}{vz} \tag{7.10}$$

The curve along which φ is constant is called the "equipotential line". It is a line of constant head 'h' or energy.

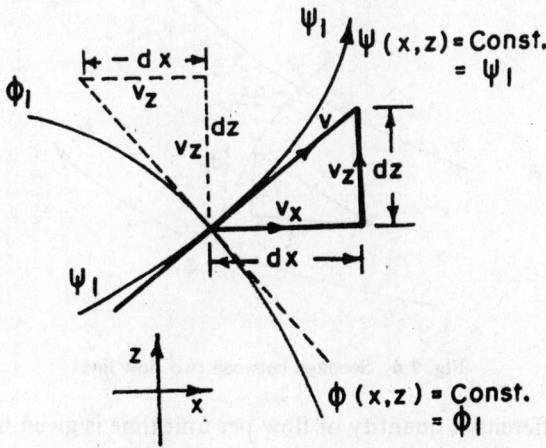

Fig. 7.5 Orthogonal relationship between an equipotential line and a flow line

Considering a second curve defined by a constant flow function, the total differential of the function ψ (x, z) is:

$$\partial\psi = \frac{\partial\psi}{\partial x}\ dx + \frac{\partial\psi}{\partial z}\ dz$$

$$= -v_z\ dx + v_x\ dz$$

As the function $\psi\ (x, z) = \text{constant} = \psi_1$, $d\psi = 0$ and the slope of the tangent is given by:

$$\frac{dz}{dx} = \frac{v_z}{v_x} \tag{7.11}$$

The curve of constant ψ is called the "flow line" or the "stream line". Comparing Eqs. 7.10 and 7.11 it is apparent that the product of the gradients of an equipotential line and a flow line is (-1), i.e. the two lines intersect each other at right angles.

The solution of the Laplace equation for the potential and flow functions thus takes the form of two families of orthogonal curves, the equipotential lines and the flow lines. The form of these curves is governed by the boundary conditions of the problem.

3. Rate of flow. Consider two flow lines (Fig. 7.6) for which the values of the flow functions are ψ_1 and ψ_2.

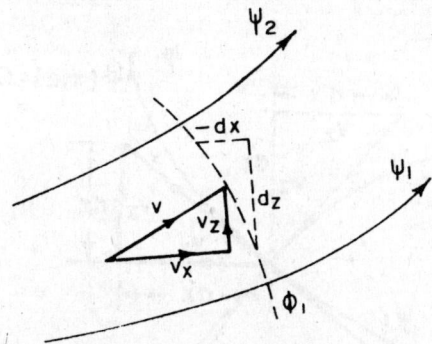

Fig. 7.6 Seepage between two flow lines

The differential quantity of flow per unit time is given by:

$$dq = v_x\ dz - v_z\ dx \tag{7.12}$$

The differential distance dx has been expressed negative in Eq. 7.12 because, for the orientation shown in the figure, a positive change in z corresponds to a negative change in x. The flow rate $\triangle q$ between the two flow lines is obtained on integration as:

$$\Delta q = \int_{\psi_1}^{\psi_2} (v_x \, dz - v_z \, dx)$$

$$= \int_{\psi_1}^{\psi_2} \left(\frac{d\psi}{dz} \, dz + \frac{d\psi}{dx} \, dx \right) = \int_{\psi_1}^{\psi_2} d\psi$$

or $\qquad \Delta q = \psi_2 - \psi_1$ $\hfill$ (7.13)

Thus the rate of flow through the 'channel' between the two flow lines is constant and is simply the difference between the two values of ψ.

7.3 FLOW NET CONDITIONS

A flow net is a graphical representation of the solution of the Laplace equation for a flow problem. It is a plot of flow lines and equipotential lines used in the study of seepage phenomena. The form of a flow net is governed by the internal and boundary conditions.

1. Internal conditions. It has already been shown that a flow line intersects an equipotential line at right angles. Since an equipotential line represent a line of equal head, no flow can occur along it. In other words, flow cannot have a resolved component along an equipotential line and the resultant flow line must cross the equipotential line orthogonally.

Consider two dimensional flow through three fields, 1, 2 and 3, of the section of a flow net, as shown in Fig. 7.7. The rates of flow through the fields per unit width normal to the plane of the figure are given by:

$$\Delta q_1 = k \frac{\Delta h_1}{l_1} b_1, \ \Delta q_2 = k \frac{\Delta h_2}{l_2} b_2 \ \text{and} \ \wedge q_3 = k \frac{\Delta h_3}{l_3} b_3$$

$$\hfill (7.14)$$

where Δh_1, Δh_2 and Δh_3 are the head losses or the 'potential drops' in the respective fields, and $l_1 b_1$, $l_2 b_2$ and $l_3 b_3$ are the average lengths (along the flow line) and average widths (along the equipotential line) of the three fields.

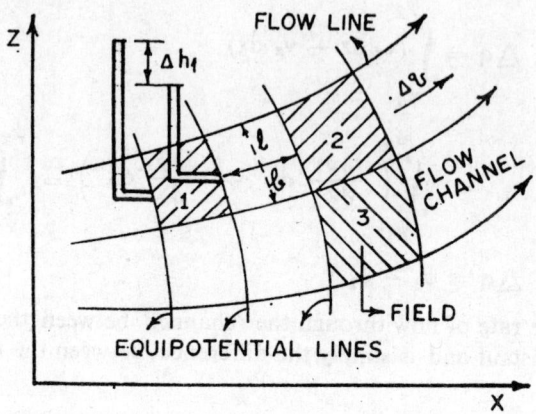

Fig. 7.7 Section of a flow net

Since the fields 1 and 2 are located on the same flow channel, $\triangle q_1 = \triangle q_2$. Similarly, since the boundary equipotential lines are common to fields 2 and 3, $\triangle h_2 = \triangle h_3$. If it is further possible to draw the flow net so that $b_1/l_1 = b_2/l_2 = b_3/l_3 = \text{constant}$, the following relationships can be obtained from Eq. 7.14:

$$\triangle q_1 = \triangle q_2 = \triangle q_3 \tag{7.15a}$$

and

$$\triangle h_1 = \triangle h_2 = \triangle h_3 \tag{7.15b}$$

Thus, if the b/l ratio is kept the same for all the fields, there will be the same rate of flow through each field and the head lost across each field will also be equal. Any value may be given to b/l, but it is convenient to arrange that b = l, so that each field has the same mean dimension in both directions and the fields assume the shape of *approximate squares*. They are also called *curvilinear squares* as they have curved sides which meet at right angles at the corners. It may, however, be noted that with equal head lost in each field, the flow velocity and hydraulic gradients through the fields will be inversely proportional to their dimensions, the smaller the size of the field, the greater are the values of the velocity and the hydraulic gredients.

The 'fundamental properties' imparted by the above mentioned internal conditions to a flow net may be summarized as follows:

1. Flow lines and equipotential lines intersect one another at right angles.

2. The fields of a net are essentially approximate squares.
3. Each flow channel (field) of a net transmits the same quantity of seepage.
4. The potential drop through each field is the same.
5. The spacing of lines at any point in the net is inversely proportional to the hydraulic gradient and to the seepage velocity.

2. Boundary conditions. Seepage problems may be grouped under 'confined flow' and 'unconfined flow'. In the case of confined flow, all the boundaries of the flow domain are initially defined, whereas in the case of an unconfined flow, the position of at least one boundary of the flow domain is not initially known and its location is determined by the solution itself. Typical examples of the two types of flows and boundary conditions are shown in Fig. 7.8.

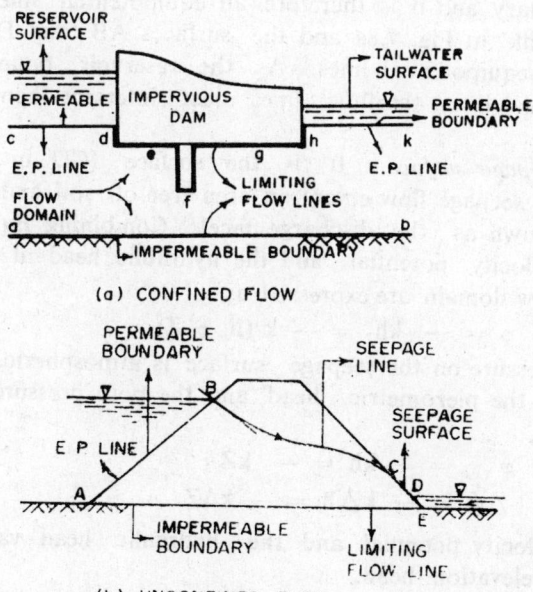

Fig. 7.8 Examples of confined and unconfined flow and types
of boundary conditions

There are four possible types of boundary conditions which need consideration.
1. Impervious boundary.
2. Submerged permeable boundary of structure.
3. Seepage surface
4. Seepage line or free surface

The first two types pertain to both confined and unconfined flow problems; the last two types apply only to unconfined flow problems. These boundaries are discussed below. There can be yet another type of boundary — a boundary between materials of different hydraulic conductivities; this is described in Section 7.5.

(a) *Impervious boundary.* Since there can be no flow across this boundary, ψ is constant and the boundary is a flow line. In Fig, 7.8, ab, defgh and AE represent the limiting flow lines.

(b) *Submerged permeable boundary of structure.* If piezometers are inserted along a submerged permeable boundary, water will rise to the same level. The head 'h' is constant on the boundary and it is therefore an equipotential line. Surfaces 'cd' and 'hk' in Fig. 7.8a and the surfaces AB and DE in Fig. 7.8b are equipotential lines. As the reservoir boundaries are equipotential lines, the flow lines must intersect them at right angles.

(c) *Seepage surface.* It is the surface (CD in Fig. 7.8b) where the seepage flow enters an area free of soil and water; it is also known as the 'discharge face'. Combining Eqs. 7.5 and 6.1 the velocity potential and the hydraulic head at any point in the flow domain are expressed as.

$$\varphi = - kh = - k (h_p \pm Z) \tag{7.16}$$

The pressure on the seepage surface is atmospheric, assumed zero, and the piezometric head and the pore pressure are zero. Therefore,

$$\varphi = - kh = - kZ \tag{7.17a}$$

and

$$\triangle\phi = - k\triangle h = - k\Delta Z \tag{7.17b}$$

The velocity potential and the hydraulic head vary linearly with the elevation head.

If $\triangle\phi$ or $\triangle h$ is to be kept constant, as required for the construction of a flow net, the equipotential lines must meet the seepage surface at constant vertical intervals (See Fig. 7.18).

The seepage surface is, however, neither a flow line nor an equipotential line since there are components of the flow velocity both normal and tangential to the seepage surface; the squares along such a boundary are incomplete.

(d) *Seepage line.* The upper free water surface of the zone of seepage (BC in Fig. 7.8b) is called a "seepage line" or a "phreatic line". It is the top flow line. In contrast to the confined flow problem, the position of the seepage line is not predetermined. The two boundary conditions on the seepage line are:

(i) that ψ is constant, i.e. the seepage line is a flow line; it is the uppermost flow line in the flow domain, separating the saturated region of flow from that part in which flow does not occur, and

(ii) that the pressure is atmospheric, $h_p = 0$, so that

$$\Delta\phi = -k\Delta h = -k\Delta Z,$$ as on the seepage surface.

Since the seepage line is a flow line, it must, unless limited by some other condition, be normal to the equipotential line AB (Fig. 7.8b) and tangential to the seepage surface CD. Also, as the velocity potential and the hydraulic head vary linearly with elevation head, the head loss Δh between equipotential lines is constant equal to ΔZ for a flow net composed of curvilinear squares. This is shown in Fig. 7.9.

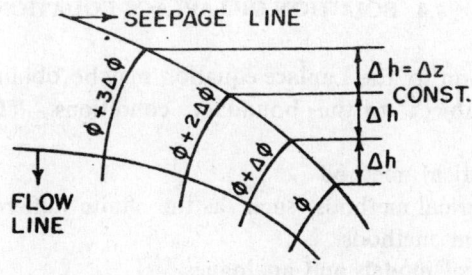

Fig. 7.9 Conditions for a seepage line

Various entrance and emergence conditions for the seepage line are shown in Figs. 7.10 and 7.11. (For the significance of the dimension 'a' see Figs. 7.14 and 7.15)

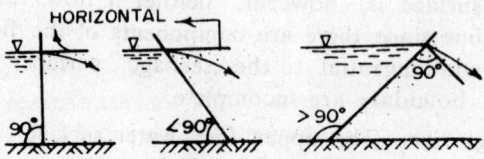

Fig. 7.10 Inclination of seepage line at entry from a submerged permeable boundary

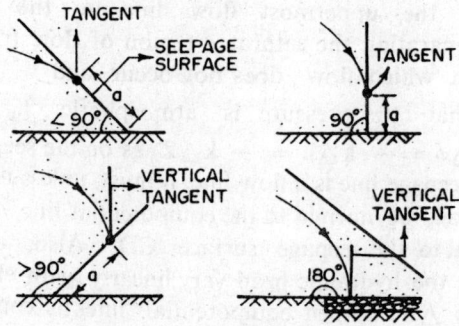

Fig. 7.11 Inclination of seepage line at point of discharge

7.4 SOLUTION OF LAPLACE EQUATION

The solution to the Laplace equation may be obtained by various methods subject to the boundary conditions. These methods include:

1. Analytical method
2. Numerical methods, such as the finite difference and finite element methods
3. Physical models and analogues
4. Graphical method

Analytical solutions of the Laplace equation may be workable in certain cases if the boundary conditions are relatively simple. Such solutions are not possible in more involved cases which constitute the majority of the problems. In the finite difference method, an approximate solution to the function φ may be obtained by replacing the Laplace equation by the finite difference approximation at node points on a square grid covering the flow zone (Southwell 1946, Harr 1962).   In the finite element method, the seepage zone is subdivided into small elements within which an approximate form of the solution φ (x, z) is assumed (Zienkiewicz 1971). This method is exceedingly versatile and is of increasing importance with the general use of computers.   Models and analogues are used to find an experimental solution to many seepage problems of practical interest.   A direct seepage test may be performed in a small scale model e.g. a sand model constructed in a flume with parallel glass plates, in which the pressure head at various points can be actually measured by piezometers. The flow lines can be traced by suitable injecting a dye. Analogues are based on the analogy between the steady state water flow through porous media and other physical processes, such as heat or current conduction. The electrical analogy method is more useful and commonly used. The commonest and the most effective method for practical use is the graphical method. The graphical and electrical analogy methods are described further.

7.5 GRAPHICAL METHOD

The graphical method consists in *sketching a trial flow net* and observing whether it satisfies the solution requirements and the boundary conditions.   As discussed earlier, the basic requirements of a flow net for seepage through isotropic soil are:

1. Flow lines and equipotential lines must intersect at right angles.
2. The fields formed by the intersection of the flow lines and the equipotential lines must be curvilinear squares.
3. Certain entrance and exit requirements must be met.

By trial and error, the quality of the flow net is improved to meet the above requirements and conditions. When the trial flow net is made to satisfy the requirements of orthogonality and the boundary conditions and the fields are curvilinear squares, it represents a plot of the *unique solution* to the problem, and then

(i) head drop between adjacent equipotential lines is the same, and

(ii) rate of flow through adjacent flow channels is also the same.

1. Confined flow. Considering first the confined flow where all seepage boundaries are defined in advance of the flow net construction, the step-by-step procedure is as follows.

1. Prepare a scale drawing of the flow region showing all boundary flow lines and equipotential lines, as shown typically for a simple problem in Fig. 7.12.

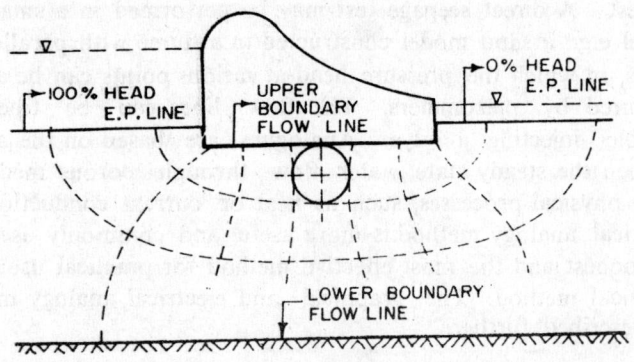

Fig 7.12 Construction of flow net, confined flow

2. Sketch the first trial flow line such that it is approximately consistent with the boundary conditions and represents a reasonable estimate of the flow path. Note that the flow line is smooth and it should intersect the equipotential lines including the boundary equipotential lines at right angles.

3. Divide the first flow channel into approximate square fields by drawing equipotential lines starting from the boundary flow line. The first and last fields and also those near the sharp edges of the structure may be of irregular, shapes; they are called the 'singular squares'. The number of fields should be a full integer; if not, the first trial flow line is adjusted and the fields redrawn to be a whole number.

4. Extend the equipotential lines obtained by dividing the first flow channel and draw the second flow line cutting the extended equipotential lines at right angles and forming square fields. The second flow channel is thus obtained.

5. Continue the extension of the flow net, keeping in view the requirement of orthogonality and of curvilinear squares. If the first flow line has been correctly chosen, the last flow line must coincide with the lower boundary line. If it is not so, as is often the case in the first trial, ascertain the nature of change required in the first flow line by visually examining the entire flow net backwards. Redraw the first flow line and repeat the whole construction. Thus obtain by trial and error the correct flow net.

The flow net can also be constructed with the last flow channel as a 'partial' flow channel having b/l ratio less than 1.0, but a constant value.

6. Sketching could also be started by drawing a few flow lines (two to four are usually sufficient) before starting to sketch equipotential lines. Alternatively, a plausible family of equipotential lines may be drawn first. Usually it is expedient to start with an integer number of equipotential drops, dividing the head by a whole number.

The following practical hints and general suggestions are greatly helpful to beginners in flow net construction (Casagrande 1973, Cedergren 1977).

1. Draw the boundaries of the problem on the *back side* of a sheet of tracing paper and construct the flow net on the front side. Erasures and corrections can thus be carried out without disturbing the boundaries.

2. Three to five flow channels may be sufficient for the first trial. Too many lines may distract the attention from essential features. Remember that the flow net can always be subdivided to any required degree.

8. Keep in view the overall shape while working on details. Do not refine a small portion of the flow net before other parts have been fairly well drawn. Corrections to the flow net should be attempted by modifying one line at a time and then changing the entire flow net in accordance with this modification. The next series of corrections should then be taken up.

4. Use a pair of dividers to measure the mean square dimensions. The precision of the 'squares' in a flow net can be improved by the 'circle method' developed by Leliavsky (1955) from the basic concept that four curves, intersecting at right angles, must

all be tangent to a common circle. Thus it should be possible to place a circle in a curvilinear square as shown in Fig. 7.12.

5. Make smooth transitions around corners. Also use gradual transitions from small to large squares.

6. As a practice, try to absorb in mind the appearance of well-constructed flow nets and reproduce them without looking at the available solution.

It is best that a flow net has an integral number of flow channels and equipotential drops. At times it may not be the case. Although there is no objection having fractional number of flow channels and equipotential drops, one of them should always be made a whole number. Suppose there is a whole number of flow channels but there results a 'partial' potential drop some where in the flow net in order to insure curvilinear squares everywhere else. Then the b/l ratio for all the curvilinear rectangles within that partial equipotential drop must be kept constant. Alternatively, if there happens to be a partial flow channel, the b/l ratio of the fields throughout the length of the partial flow channel must be constant.

2. Unconfined flow. In contrast to the confined flow problem, sketching the flow net for an unconfined flow may be quite tedius because the unknown boundary flow line has to be located simultaneously with the drawing of the flow net. The determination of the position of this boundary flow line thus becomes one of the major objectives of the analysis. Taking the example of unconfined flow through an earth dam, the 'seepage line' (phreatic line) is the unknown boundary flow line. Casagrande (1937) developing a solution obtained by Kozeny has suggested graphical methods for estimating the seepage line in a homogeneous, isotropic dam. It has been found that the seepage line is very nearly parabolic in shape except near the points of entrance and exit. (A parabola is a curve such that any point on it is equidistant from both a fixed point, called the 'focus', and a fixed straight line, called the 'directrix'.) Two cases are considered:

1. The dam rests on a base which is permeable at its downstream end; there is an internal filter drain and no seepage surface.

2. The dam rests on an impermeable base and the filter drain is omitted; discharge occurs on a seepage surface.

(a) *Homogeneous dam with filter drain.* For the typical earth dam shown in Fig. 7.13, the focus of the Kozeny basic parabola lies at point F, the upstream edge of the filter. Based on an extensive study of the earth dam problems, Casagrande (1937) recommended that the initial point of the parabola should be taken at C, where AC $\approx$ 0.3 AB.

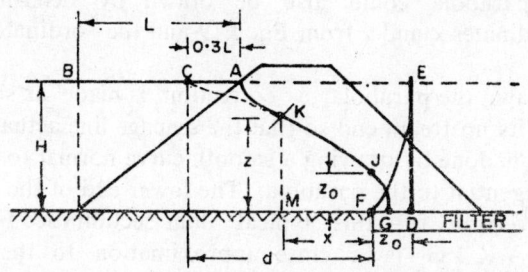

Fig. 7.13 Casagrande's construction for the seepage line in a dam with filter

The parabola can be constructed graphically as follows:
1. With centre C and radius CF, draw an arc to cut the water line extension at point E. Draw the vertical ED which is the 'directrix'.
2. Bisect the distance FD to get G, the 'vertex'.
3. Establish point J, by drawing FJ parallel to DE and equal to FD.

The basic parabola can be drawn with the three known points C, J and G.
4. If an additional point, say point K, is required, draw the vertical line KM. With focus F as centre and the distance DM as radius, draw an arc to cut the line KM at the required point K.

The equation of the parabola is obtained by taking F as the origin. Let x and z be the coordinates of a point, say K, on the parabola and let z_0 be the ordinate of the parabola at point x = 0. Equating the distances MD and KF,

$$x + z_0 = \sqrt{x^2 + z^2} \qquad (7.18)$$

or

$$x = \frac{1}{2 z_0} (z^2 - z_0^2) \qquad (7.19)$$

Eq. 7.19 is the equation of the Kozeny basic parabola. Substituting the coordinates of the initial point C as $x = d$ and $z = H$ in Eq. 7.18, the ordinate z_0 can be expressed as:

$$z_0 = \sqrt{d^2 + H^2} - d \qquad (7.20)$$

The focal length a_0 where the parabola meets the filter (x-axis) is $z_0/2$. The parabola could also be drawn by determining the related coordinates x and z from Eq. 7.19 and the ordinate z_0 from Eq. 7.20.

Having drawn the parabola, a correction is made as shown in Fig. 7.13, to its upstream end so that the seepage line actually starts from A; this is done by drawing a smooth curve normal to the slope at A and tangential to the parabola. The lower end of the parabola needs no correction for this typical dam section (See Fig. 7.11). The curve A K J G is a close approximation to the seepage line.

(b) *Homogeneous dam without filter drain.* When a homogeneous dam rests on an impermeable base and is not provided with special, internal drainage measure, the seepage line cuts the downstream slope above the base forming a 'seepage surface' or the 'discharge face'. The basic parabola is drawn as before, but with the toe of the dam taken as the focus (Fig. 7.14). The upper end of the curve is corrected as in Fig. 7.13.

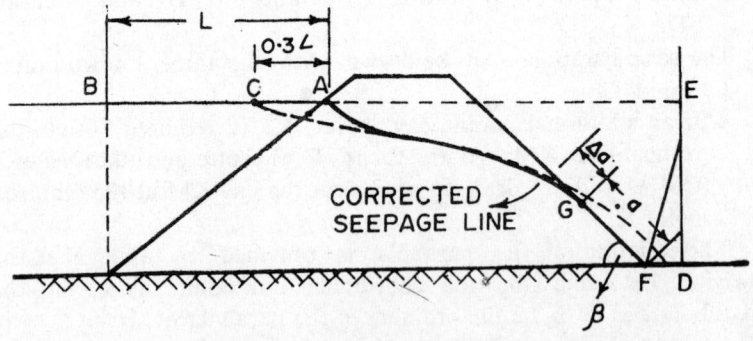

Fig. 7.14 Casagrande's construction for the seepage line in a dam without filter

The parabola cuts the downstream face a distance Δa above the actual seepage surface. The actual point of emergence G of the seepage line can be determined from the relationship established by Casagrande between 'a' and 'Δa' in terms of β the slope of the discharge face. This relationship, plotted in Fig. 7.15, can be used for steep and overhanging discharge faces ($60° < β ≤ 180°$). Typical values of the corrections are given in Table 7.1.

Table 7.1

CASAGRANDE'S DOWNSTREAM CORRECTION FOR SEEPAGE LINE

β	30°	60°	90°	120°	150°	180°
Δa / (a + Δa)	0.36	0.32	0.26	0.18	0.10	0

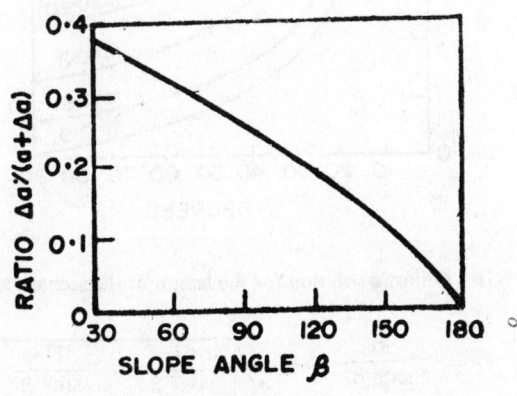

Fig. 7.15 Casagrande's correction for seepage line at exit

(c) *Seepage line for shallow slope.* The point of emergence G of the seepage line can also be determined from Gilboy's curves (Gilboy 1933) shown in Fig. 7.16. The corrected seepage line is obtained by drawing a smooth curve tangential to the parabola and to the seepage surface at point G.

For shallow discharge slopes where $β ≤ 60°$; Casagrande (1937) developed a method for locating the seepage line, which is based on an earlier work of Schaffernak and Iterson (Perloff and Baron 1976). The point of emergence G of the seepage line is located at a distance 'a' along the slope above the toe point F, given by:

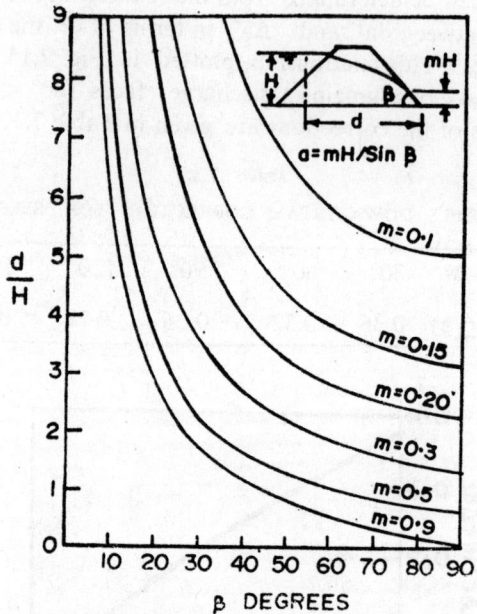

Fig. 7.16 Gilboy's solution for the length of the seepage surface

$$a = \frac{d}{\cos \beta} - \sqrt{\frac{d^2}{\cos^2 \beta} - \frac{H^2}{\sin^2 \beta}} \qquad (7.21a)$$

or
$$a = \sqrt{d^2 + H^2} - \sqrt{d^2 - H^2 \cot^2 \beta} \qquad (7.21b)$$

The parabolic seepage line is constructed graphically as follows (Fig. 7.17):

1. Locate point G by determining the distance 'a' from Eq. 7.21.
2. Extend the horizontal through point C to meet the D/S slope line (upward extension of the seepage surface) at point T.
3. Divide the distances CT and GT into an arbitrary but equal number of equal parts (e.g. 4 parts in Fig. 7.17).
4. Join points 1, 2 and 3 to point G. Draw horizontal lines through I, II and III to intersect the respective lines drawn through points 1, 2 and 3.

5. Draw the parabola through the above points of intersection
    (step 4). Make the U/S correction to the seepage line, as
    described earlier (Fig. 7.13).

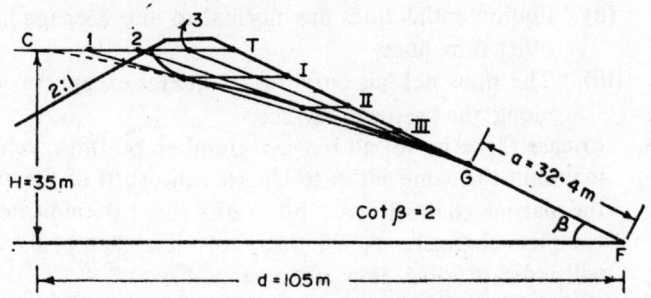

Fig. 7.17 Construction of seepage line for shallow discharge slopes
(β ≤ 60°)   (Example 7.9)

(d) *Flow net construction.* Having located the seepage line
(even approximately), the flow net construction may be taken up
as follows:

1. Divide the hydraulic head h_1 (the diffeience of the reservoir
    level and the tail water level) into a convenient number of
    equal parts or increments $\triangle h$. Draw a series of light
    horizontal lines, called the *head* lines, at vertical intervals of
    $\triangle h$ to intersect the seepage line and the seepage surface, as
    shown in Fig. 7.18.   The points of intersections indicate the
    positions of the equipotential lines.

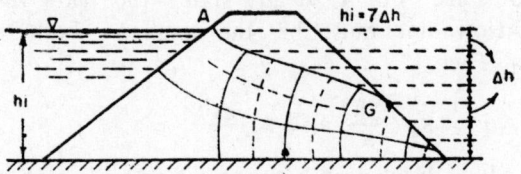

Fig. 7.18 Construction of flow net, unconfined flow

2. Select a convenient number of flow channels, say three or
    four.

3. Draw the equipotential lines and the flow lines keeping in view the requirements:
  (i) At the upstream slope where water enters the dam, all the flow lines are normal to the slope line.
  (ii) Equipotential lines are normal to the seepage line and other flow lines.
  (iii) The flow net has curvilinear squares except, of course, along the seepage surface.
4. In case there is not an *integral* number of flow channels, maintain the same width-to-length ratio (b/l) of the fields in the partial channel. A b/l ratio of 1.0 indicates one complete channel. A b/l ratio of less than 1.0, say 0.5, will indicate a 0.5 flow channel.

As a first trial in Fig. 7.18, h_l is divided into 7 equal parts ($h_l = 7\triangle h$) and a rough flow net can be drawn consisting of 7 potential drops and 1 full channel and one partial channel equivalent to 0.5 full channel. The net can be refined by adopting $h_l = 14\triangle h$ which will give 14 potential drops and approximately 3 flow channels. Usually a few trials will be necessary to get a flow net of desired accuracy.

3. Boundary effect of soils. When water flows from one soil to other soil of different permeability, the flow net gets transformed on crossing the boundary. Considering the simplest case of horizontal flow through two soils with a vertical boundary, where the flow lines are horizontal and the equipotential lines are vertical, the flow net has square fields on the left side of the boundary (Fig 7.19). On crossing the boundary the spacing of the equipotential lines changes according to the relative permeability of the two soils. For a steady state, the flow rate between any two flow lines must be the same on both sides of the boundary. Then,

$$\triangle q = k_1 \frac{\triangle h}{a} a = k_2 \frac{\triangle h}{l} a \qquad (7.22)$$

where $\triangle h$ = head drop between any two adjacent equipotential lines

k_1, k_2 = coefficients of permeability of soils
  $(k_2 > k_1)$
a, l = dimensions of fields as shown in Fig. 7.19

From Eq. 7.22

$$\frac{a}{l} = \frac{k_1}{k_2} \tag{7.23}$$

Thus the square fields get transformed to rectangles on crossing the boundary.

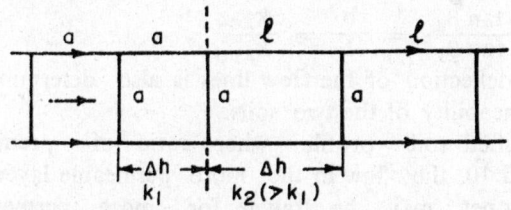

Fig. 7.19 Horizontal flow across a vertical boundary of two soils

If the boundary is not perpendicular to the direction of flow, the flow lines get deflected at the boundary, as shown in Fig. 7.20.

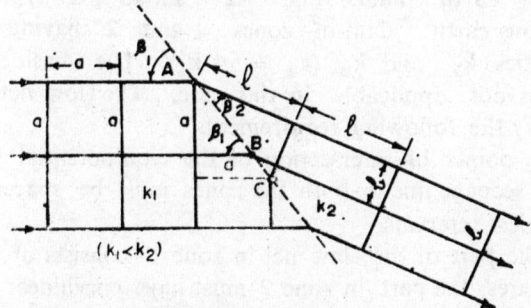

Fig. 7.20 Deflection of flow lines at a boundary which is not normal to direction of flow

To satisfy the continuity condition as before, the rate of flow must be the same in both the soils.

Hence,

$$\wedge q = k_1 \frac{\wedge h}{a} a = k_2 \frac{\wedge h}{l} b \tag{7.24}$$

where a, b, l = dimensions of fields as shown in Fig. 7.20

or

$$\frac{b}{l} = \frac{k_1}{k_2} \quad (k_2 > k_1) \tag{7.25}$$

Thus the square fields get transformed to rectangles on crossing the boundary. If square fields are adopted downstream of the boundary, there must be rectangular fields upstream of the boundary having b/l ratio of k_2/k_1.

By geometry in Fig. 7.20,

$$AB = \frac{a}{\sin \beta_1} = \frac{b}{\sin \beta_2}$$

and

$$AC = \frac{a}{\cos \beta_1} = \frac{1}{\cos \beta_2}$$

so that

$$\frac{\tan \beta_2}{\tan \beta_1} = \frac{b}{1} = \frac{k_1}{k_2} \tag{7.26}$$

Thus the deflection of the flow lines is also determined by the relative permeability of the two soils.

In a stratified soil profile where ratio of premeability of layers exceed 10, the flow in the more permeable layer controls; i.e. the flow net may be drawn for more permeable layer assuming the less permeable layer to be impervious (NAVFAC, 1982). The head on the interface thus obtained is imposed on the less pervious layer for construction of the flow net within it.

4. Zoned earth dam. Fig. 7.21 shows a typical non-homogeneous earth dam of zones 1 and 2 having different permeabilities k_1 and k_2 ($k_2 = 4\ k_1$). The basic parabola concept is not applicable in this case. The flow net, however, must satisfy the following requirements.

1. The points of intersection of the equipotential lines with the seepage line in both the zones must be spaced at equal vertical intervals.

2. If the part of the flow net in zone 1 consists of curvilinear squares, the part in zone 2 must have curvilinear rectangles with $b/1 = k_1/k_2 = 1/4$.

3. The flow lines at the interzone boundary must deflect according to Eq. 7.25.

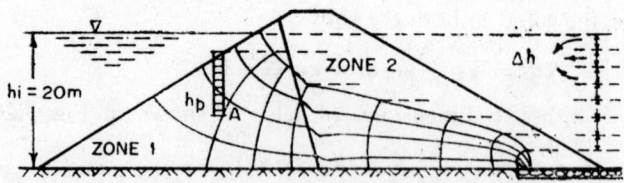

Fig 7.21 Flow net construction for a zoned dam
(With adaptation from Cedergren 1977)

To construct the flow net (Cedergren 1977), a trial seepage line is drawn by guess in both the zones. The hydraulic head h_1 is divided into a convenient number of equal parts $\triangle h$ and horizontal 'head' lines are drawn at intervals of $\triangle h$ across the downstream part of the section. A trial flow net is constructed with square fields in zone 1 and rectangles in zone 2. The b/l ratio of the rectangles should be kept approximately constant ($=0.25$, Eq. 7.25) by adjusting the seepage line. The flow net in zone 1 of Fig. 7.21 consists of 3.5 flow channels as can be counted between any two adjacent equipotential lines.

5. **Anisotropic soil condition.** Many natural soil deposits and most of the compacted soils, although homogeneous, are anisotropic with respect to permeability, the horizontal permeability being much greater than the vertical. Let k_x and k_z be the coefficients of permeability in the x and z directions of the two dimensional flow shown in Fig. 7.4. The velocity components (Eq. 7.2) are expressed as:

$$v_x = - k_x \frac{\partial h}{\partial x} \text{ and } v_z = - k_z \frac{\partial h}{\partial z} \qquad (7.27)$$

Substituting in Eq. 7.3, the equation of continuity can be written as:

$$k_x \frac{\partial^2 h}{\partial x^2} + k_z \frac{\partial^2 h}{\partial z^2} = 0 \qquad (7.28a)$$

or

$$\frac{\partial^2 h}{(k_z/k_x) \partial x^2} + \frac{\partial^2 h}{\partial z^2} = 0 \qquad (7.28b)$$

If $k_x \neq k_z$, Eq. 7.28 may still be expressed in the Laplace form by substituting:

$$x_r = x \sqrt{k_z/k_x} \qquad (7.29)$$

so that

$$\frac{\partial h}{\partial x_r} = \frac{\partial h}{\sqrt{k_z/k_x} \cdot \partial x} \qquad (7.30)$$

and

$$\frac{\partial^2 h}{\partial x^2_r} = \frac{\partial^2 h}{(k_z/k_x) \partial x^2}$$

Eq. 7.28 thus transforms to the Laplace form:

$$\frac{\partial^2 h}{\partial x^2_r} + \frac{\partial^2 h}{\partial z^2} = 0 \qquad (7.31)$$

Eq. 7.31 is the Laplace equation for the anisotropic seepage condition.

(a) *Flow net construction.* To construct an orthogonal flow net for the anisotropic soil, it is necessary to reduce the dimensions of the seepage zone in the x-direction according to the scale factor

of Eq. 7.29.   This will transform a given anisotropic flow   region
into a fictitious isotropic region in which   the   Laplace equation
(Eq. 7.31) is valid.   The true flow net for the natural, anisotropic
region can be obtained by applying the inverse of the scale factor.

A typical anisotropic flow section is   shown in   Fig. 7.22 where
$k_x = 2.25 \ k_z$ in the foundation soil under the dam. The   section
is redrawn, using the original vertical scale, but with a transformed
horizontal scale   in   which the reduced   horizontal dimension
$x_r = x\sqrt{k_z/k_x} = x/1.5$. An orthogonal flow   net is constructed
for the transformed section as shown in Fig. 7.23.   The true flow
net is obtained by transforming back the section of the flow zone
as well as the flow net to the original horizontal and vertical
scales, i.e. by increasing all horizontal dimensions by the scale
factor of $\sqrt{k_x/k_z} = 1.5$. The true flow net is shown in Fig. 7.22.
The flow net is no more orthogonal, the actual directions of flow
are not normal to the equipotential lines.

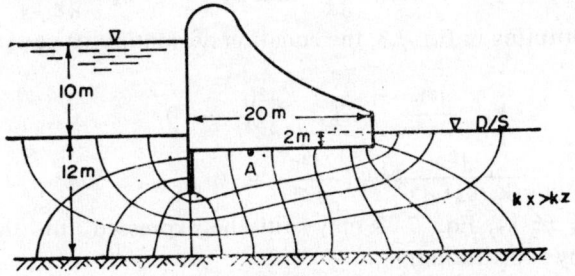

Fig. 7.22   Typical anisotropic flow section
with true flow net, $k_x = 2.25 \ k_z$

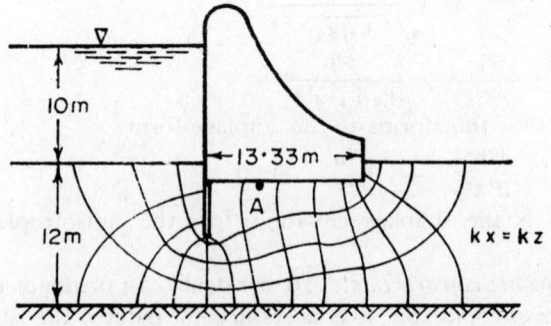

Fig. 7.23   Transformed isotropic flow section and flow net

(b) *Equivalent permeability.* Consider a typical flow net field through which flow is essentially in the x-direction. The field drawn to transformed and natural scales is shown in Fig. 7.24, the transformation being in the x-direction. The flow rate can be expressed in terms of either an equivalent permeability k' (transformed section) or k_x (natural section), i.e.

$$\triangle q = k' \frac{\triangle h}{a} a = k_x \frac{\triangle h}{a \sqrt{k_x/k_z}} a \qquad (7.32)$$

$$k' = \sqrt{k_x k_z} \qquad (7.33)$$

where $\triangle h$ = head drop

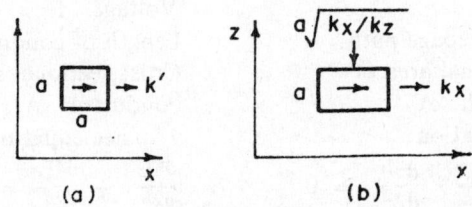

Fig. 7.24 Flow net fields on transformed and natural scales

It may be noted that the actual hydraulic gradients for anisotropic conditions are to be determined in flow nets redrawn at natural horizontal scale because the actual distance over which a given head drop occurs can be measured only on a true section.

7.6 ELECTRICAL ANALOGUE

The two dimensional flow of electricity through a conducting medium is governed by the Laplace eqation similar to the one for two dimensional seepage flow in an isotropic medium. The solution of a two dimensional seepage problem can therefore be obtained from studying the form of electrical flow through a conducting medium geometrically similar to the seepage flow domain. Salt solution in a shallow tray and coated conducting paper are the two commonly used types of conducting media. The analogy between seepage and electrical flow is given in Table 7.2.

Table 7.2

ANALOGY BETWEEN FLOW OF WATER AND FLOW OF ELECTRICITY

Flow of water	Flow of electricity
Darcy's law:	Ohm's law:
$v = - k \dfrac{\partial h}{\partial L}$	$j = - \sigma \dfrac{\partial E}{\partial L}$
$q = k \dfrac{h}{L} A$	$I = \sigma \dfrac{E}{L} A$
Flow velocity v	Current density j
Flow rate q	Current I
Hydraulic conductivity k (Permeability)	Specific conductivity σ (Reciprocal of resistivity)
Head h	Voltage E
Length of seepage path L	Length of conductor L
Cross sectional area of seepage path A	Cross sectional area of conductor A
Laplace equation:	Laplace equation:
$\dfrac{\partial^2 h}{\partial x^2} + \dfrac{\partial^2 h}{\partial z^2} = 0$	$\dfrac{\partial^2 E}{\partial x^2} + \dfrac{\partial^2 E}{\partial z^2} = 0$

1. Electrical analogy tray. The model is assembled in a shallow tray with flat bottom and constructed of non-conducting material. The boundary equipotential lines are simulated by copper

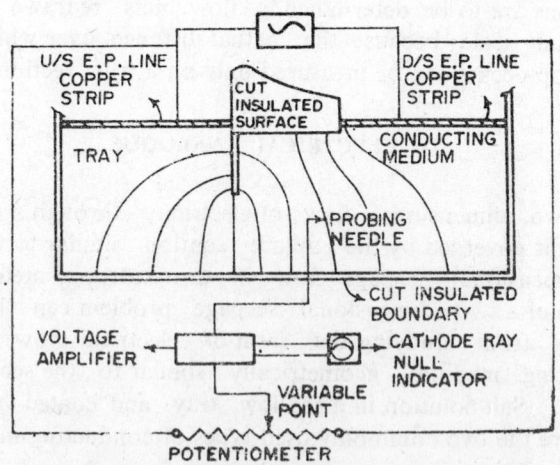

Fig. 7.25 Set-up of an electrical analogy model in a tray

strips and the boundary flow lines by non-conducting material such as perspex. The strips are secured in position by non-conducting adhesives such as plasticene (modelling clay), beewax or other synthetic adhesive. A week salt-water solution, filled in the flow zone of the model, serves as a conducting material. A typical set-up is shown in Fig. 7.25.

To plot a particular equipotential line the variable point is set at the desired potential on the potentiometer. The probing needle is moved within the conducting medium of the tray until a null point is obtained. The probing point (null point) in the tray and the contact point at the potentiometer then have the same potential. Keeping the variable point at the particular potential, a number of null points are obtained in the tray by moving the probing needle to different positions. On joining the null points the particular equipotential line is traced. The variable point is then set at some other potential to get another equipotential line. Having traced the complete set of equipotential lines, orthogonal flow lines conforming to the boundary conditions are drawn as in the graphical method.

For a problem where hydraulic conductivity varies in different parts of the seepage zone, one way of simulation of greater conductivity is to increase the depth of electrolyte by depressing the bottom of the tray. The other method is to provide water-tight partitioning strips and use the electrolyte of different conductivities in different zones. The strips are made of perpex, over which closely spaced U-shaped copper wire clips are placed to conduct current from one face of the partition to the other (Singh et al 1972, Singh 1981).

2. **Conducting paper.** A graphite-impregnated paper provides a continuous conducting medium which can be conveniently used for the electrical analogue. The paper can be easily cut to the shape of the seepage zone. The paper is cut-through at all impervious boundaries. An electrical potential is applied to the equipotential boundaries, using silver paint to ensure contact with low resistance. A set of equipotential lines are obtained as in the electrical analogy tray. Using a probe attached to a graphite-pointed pencil, the equipotential lines can be drawn directly on the paper. The flow lines are then sketched in, to complete the orthogonal net.

The resistivity of the paper is found generally to differ in the two dimensions; the transverse resistivity (across the roll) being greater than the longitudinal resistivity. The model dimensions are, therefore, modified as described earlier under 'anisotropic soil condition'. In many cases, however, the difference in resistivities is not much and, even otherwise, the correction may not be necessary.

3. Location of seepage line. The seepage line can be experimentaly determined by the electrical analogy tray and the conducting paper. In case of tray, a trial seepage line is made of plasticene and its position is shifted until the potential along it corresponds to the elevation head (Singh 1981). Similarly, the conducting paper is cut along an assumed seepage line, erring on the safe side, and potentials are determined. The boundary is trimmed off, as necessary, until the equipotentials along it are at equal vertical distance apart.

7.7 SEEPAGE FORCE

Considering horizontal flow of water through a soil sample of cross sectional area A and length L, as shown in Fig. 7.26, the 'boundary water forces', P_1 and P_2, at the entry and exit faces are: $P_1 = h_1 \gamma_w A$ and $P_2 = h_2 \gamma_w A$. The resultant force $F = P_1 - P_2$, which must be consumed in friction, is termed the seepage force. Thus,

$$F = (h_1 - h_2) \gamma_w A = h \gamma_w A \qquad (7.34)$$

where h = head drop (hydraulic head)

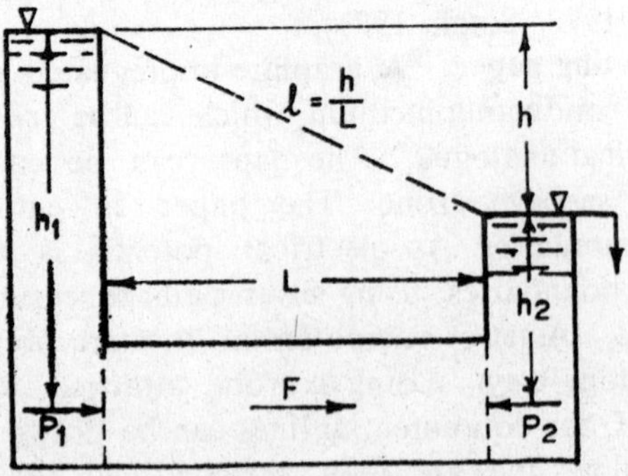

Fig. 7.26 Seepage force in a soil

The seepage force is the force applied by the moving water to the soil (or rock) grains through frictional drag. It is the energy of seeping water transferred to soil as an effective pressure. For a given head drop and a given area, the energy loss (Eq. 7.34) is constant irrespective of the distance over which it is consumed. Thus the seepage force is proportional to the head drop (hydraulic head or differential head) and the cross sectional area of the soil. In an isotropic soil seepage force always acts in the direction of flow.

The seepage force may also be expressed in relation to hydraulic gradient i (= h/L) by writing Eq. 7.34 as follows:

$$F = \frac{h}{L} L A \gamma_w = i V \gamma_w \qquad (7.35)$$

where   V = volume of soil

It is found convenient to express the seepage force per unit volume 'j' as:

$$j = i \gamma_w \qquad (7.36)$$

The seepage force per unit of cross sectional area of soil is called the "seepage pressure p_s", which may be expressed as:

$$p_s = h \gamma_w \qquad (7.37)$$

or $$p_s = \frac{h}{L} L \gamma_w = i L \gamma_w \qquad (7.38)$$

where   i = hydraulic gradient

L = length over which h or i is dissipated

It may be noted that a seepage force or seepage pressure always occurs when there is a head drop $\triangle h$ across a flow path of length L and will vary from point to point within the length of soil under consideration.

Depending on the direction of flow, seepage increases or decreases the vertical effective pressure of a soil. The effective pressure increases if the flow is downwards and it decreases if the flow is upwards, Thus,

$$\sigma' = L \gamma' \pm p_s = L \gamma' \pm i L \gamma_w \qquad (7.39)$$

where   L = depth below the surface of a submerged soil where σ' is desired

7.8 QUICK CONDITION

When the effective pressure is zero for a cohesionless soil, the soil particles just touch with no frictional resistance available. This state is termed as a "quick" or "quicksand" *condition*. At a quick condition,

$$\sigma' = \sigma - u = 0 \qquad (7.40)$$

The effective pressure obviously is zero when the pore pressure equals the total pressure. There are two common situations where this equality may arise.

(1) An upward water flow of such magnitude that the total upward water force equals the total soil weight, e.i. the upward seepage pressure equals the downward pressure due to submerged weight of soil.

(2) A shock on certain loose soil causing a sudden decrease in soil volume with the result that the effective pressure is transferred to the pore pressure (Chapter 9).

1. Critical hydraulic gradient. Considering upward flow through a soil sample of length L and cross sectional area A, as shown in Fig. 7.27, the hydraulic gradient necessary to develop a quick condition may be obtained by equating the *total* upward and downward pressures at the base of the sample. Thus,

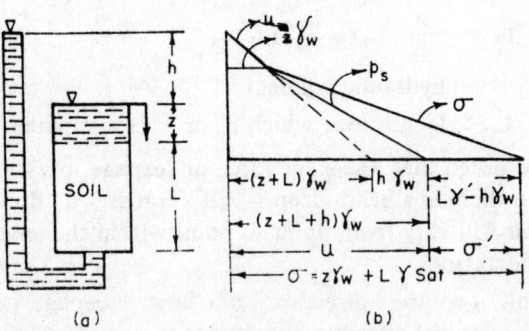

Fig. 7.27

$$(L + z + h) \, \gamma_w \; = L \, \gamma_{sat} + z \, \gamma_w$$

and simplifying

$$h \, \gamma_w = L \, (\gamma_{sat} - \gamma_w) = L \, \gamma'$$

or

$$\frac{h}{L} \; = i \; = \frac{\gamma'}{\gamma_w}$$

The gradient causing a quick condition is called the "critical hydraulic gradient i_c":

$$i_c = \frac{\gamma'}{\gamma_w} = \frac{G-1}{1+e} \tag{7.41}$$

The same expression can be obtained by considering the seepage pressure and the downward submerged pressure of soil. In an upward flow the resultant effective pressure is given by Eq. 7.39 as:

$$\sigma' = L\,\gamma' - p_s = L\,\gamma' - i\,L\,\gamma_w \tag{7.42}$$

At the quick condition, $\sigma' = 0$

or $i = i_c = \gamma'/\gamma_w = \dfrac{G-1}{1+e}$

The resultant body force on a soil element can thus be obtained in two ways: in the first method by working with total (saturated) weight plus boundary water forces; in the second, with submerged weight plus seepage force. The two methods lead to the same result.

It may be noted that the term 'quick' or 'quick sand' *refers to a condition* and not to a material. The two factors necessary for a soil to become 'quick' are: the strength of soil must be proportional to effective pressure and the effective pressure must reduce to zero. It is the cohesionless soil whose strength varies with the effective pressure and the strength becomes zero when $\sigma' = 0$. The upward gradient needed to cause a quick condition in an unloaded cohesionless soil is equal to γ'/γ_w. Since the ratio γ'/γ_w is usually close to unity, the critical upward gradient is approximately equal to one. This phenomenon can be readily observed in sands both in the laboratory and in the field locations such as on the land side of levees during floods, on the downstream side of an earth dam and in excavations where water tends to flow upwards. In a quick state the sand has practically no supporting power and the individual grains remain suspended in the upward flowing water,— a condition visibly resembling 'boiling', and the phenomenon is also termed as "sand boil". Clays may have strength even at zero normal stress, so that the quick condition may not necessarily result when the hydraulic gradient equals the critical value given by Eq. 7.41.

2. Seepage uplift of cohesive stratum. In case of cohesive soils, interparticles attractive forces produce a condition in which a mass of soil rather than individual grains may be lifted due to upward head difference. Consider the example of excavation into a clay deposit underlain by coarse sand within a cofferdam, as shown in Fig. 7.28. Water level inside the cofferdam is kept lowered at the excavation line by pumping out. The permeability of clay is so low that there is hardly any seepage and there are no seepage forces to consider.

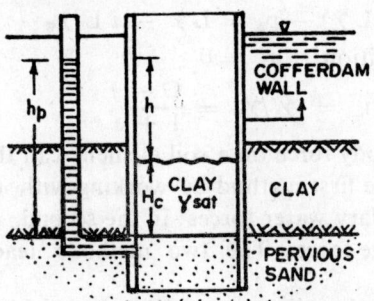

Fig 7.28 Excavation into a clay deposit
subjected to uplift pressure

Neglecting any side friction, the critical thickness H_c at which the clay will be on the verge of uplift (safety factor $F = 1$) is obtained by equating the downward and upward pressures as the bottom plane of the clay.

$$H_c \, \gamma_{sat} = h_p \, \gamma_w = (h + H_c) \, \gamma_w \qquad (7.43)$$

where γ_{sat} = saturated unit weight of clay (may not always be fully saturated)

h_p = piezometric head at the bottom of clay

h = head difference

Rearranging Eq. 7.43:

$$H_c = \frac{h_p \, \gamma_w}{\gamma_{sat}} \qquad (7.44)$$

or $$H_c = \frac{h \, \gamma_w}{\gamma_{sat} - \gamma_w} \qquad (7.45)$$

For any thickness H ($>$ H_c), the factor of safety against uplift is given by:

$$F = \frac{H \, \gamma_{sat}}{h_p \, \gamma_w} \qquad (7.46)$$

Examples 7.1 — 7.5

7.1 The likely range of void ratio for a sand deposit is from 0.4 to 0.85. The specific gravity of sand is 2.67. What is the range of critical hydraulic gradient?

$$i_c = \frac{G - 1}{1 + e}$$

$e = 0.4,$ $i_c = 1.19$

$e = 0.85,$ $i_c = 0.90$

7.2 Two soils (1) and (2) are placed one above the other in a constant head permeameter as shown in Fig. 7.29. The void ratios and specific gravities are $e_1 = 0.6$, $e_2 = 0.7$, $G_1 = 2.67$, $G_2 = 2.70$. If 25 percent of the hydraulic head is lost by upward flow through soil (1), examine the possibility of quick condition of instability. What is the critical hydraulic gradient at which the instability occurs and what would be the corresponding hydraulic head?

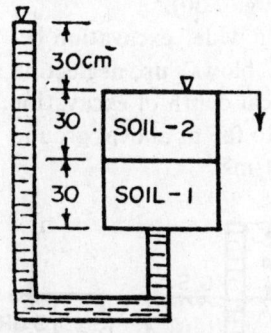

Fig. 7.29 Example 7.2

Head loss: Soil (1) = 7.5 cm, Soil (2) = 22.5 cm

Hydraulic gradient (h/L): Soil (1) = 0.25, Soil (2) = 0.75. As the gradient in soil (2) is three times larger than that in soil (1), the instability of soil (2) must precede the instability of soil (1).

Critical gradient for soil (2): $i_{c_2} = 1.0$

Critical head loss for soil (2): $h_{o2} = i_{c2} L_2 = 30$ cm

Since the head loss in soil (2) is equivalent to 75 percent of the total head loss through both the samples, the total hydraulic head above the surface of soil (2) necessary for quick condition is $30 \times 100/75 = 40$ cm.

Quick condition is thus not possible in the above system, unless the hydraulic head is increased from 30 cm to 40 cm. Even when $h = 40$ cm, soil (1) will not become quick.

7.3 The clay deposit under excavation inside a cofferdam is underlain by coarse sand. A piezometer located in sand at the bottom of clay indicates a pressure head of 11.5 m of water. What should be the minimum thickness of clay to be left intact so that the bottom does not blow up? The saturated unit weight of clay is 21.5 kN/m³. Assume the excavation is kept pumped dry.

$$H_0\, \gamma_{sat} = h_p\, \gamma_w; \quad \gamma_w = 9.8\ kN/m^3$$
$$H_0 = 5.24\ m$$

7.4 The strata in the flat bottom of a valley consist of 3 m of gravel overlying 10 m of clay. Beneath the clay is fissured sandstone of relatively high permeability. The water in the sandstone is under artesian pressure corresponding to a piezometer level of 4 m above ground level (Fig. 7.30).

To what depth can a wide excavation be made into the clay before the bottom blows up, neglecting side shear resistance? What will be the critical depth of excavation, if relief wells reduce the artesian pressure to 0.5 m above ground level? The saturated density of clay is 2.1 t/m³.

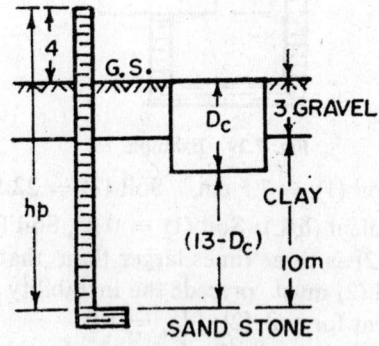

Fig. 7.30 Example 7.4

$$(13 - D_0)\, \rho_{sat} = h_p\, \rho_w; \quad h_p = 17\ m$$
$$D_0 = 4.9\ m$$
When $h_p = 13.5\ m, \quad D_0 = 6.57\ cm$

7.5 A 8 m thick clay layer is underlain by coarse sand containing artesian water under a piezometric head of 10 m of water. To what depth must the piezometric head in the sand be lowered so that a 5 m deep excavation can be made in clay with a factor of safety of 1.2 ? The density of clay is 2.2 t/m³.

$$\text{F.S.} = \frac{3 \times 2.2}{h_p \times 1} ; \quad h_p = 5.5 \text{ m}$$

Reduce the pressure to 2.5 m below ground level.

7.9 USE OF FLOW NET

A flow net may be used to estimate seepage, uplift pressure or hydraulic gradient within a flow domain.

1. Seepage rate. If N_d in the total number of potential drops (number of fields) in a flow channel and h_l is the hydraulic head causing the flow,

Potential drop in each field $\Delta h = h_l/N_d$

Hydraulic gradient in a field of average dimension $l_1 = \Delta h/l_1$. Seepage rate through the field per unit width normal to flow plane,

$$\Delta q = k \frac{\Delta h}{l_1} l_1 = k \Delta h$$

$$= k \frac{h_i}{N_d}$$

Seepage rate for the complete flow net in an isotropic soil is given by:

$$q = k h_l \frac{N_f}{N_d} \tag{7.47}$$

or $$q = k h_i \frac{M-1}{N-1} \tag{7.48}$$

where   M = number of flow lines

N = number of equipotential lines

For an earth dam built on an impervious foundation the seepage rate can also be derived from the property that the flow is parabolic near the lower end and the flow lines and equipotential lines are confocal parabolis (Kozeny). At point F, $h = z_o$, the distance of F from directrix; and $N_f = N_d$ ($= 6$ in Fig. 7.31). From Eq. 7.47,

$$= k z_o \frac{6}{6} = k z_o \qquad (7.49)$$

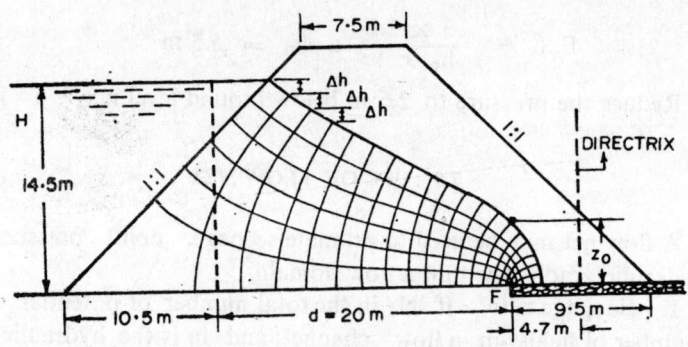

Fig. 7.31　Flow net for earth dam (Example 7.8)

For shallow discharge slopes ($\beta \leqslant 60°$), the flow rate can be determined by the following equation (Casagrande 1937):

$$q = k \, a \sin^2 \beta \qquad (7.50a)$$

or according to Schaffernak and Iterson (Carter 1983) by the following equation (Eq. 7.50b),

$$q = k \, a \sin \beta \cos \beta \qquad (7.50b)$$

where 'a' is given by Eq. 7.21, or by the Gilboy solution (Fig. 7.16):

$$a = \frac{mH}{\sin \beta} \qquad (7.51)$$

and thus $\quad q = k \, m \, H \sin \beta$ ⠀⠀⠀⠀⠀⠀⠀⠀⠀⠀⠀⠀(7.52)

For an anisotropic flow section, equivalent coefficient of permeability k' ($= \sqrt{k_x \, k_s}$, Eq. 7.33) should be used for finding the seepage rate by Eqs. 7.47—7.50).

2. Uplift pressure. Water flowing through the foundation of a structure (e.g. masonry dom) exerts uplift pressure against the base of the structure. The hydrostatic pressure or the 'uplift pressure' at any point is given by:

$$u = h_p \, \gamma_w \qquad (7.53a)$$

or $\qquad u = h_p$ (in terms of head of water) ⠀⠀⠀⠀⠀$(7.53b)$

and the piezometric head h_p may be written as follows from Eq. 6.1:

$$h_p = h - Z \qquad (7.54)$$

where Z = elevation or position head above datum (rear water level) considered positive upwards

The hydraulic head 'h' at any point within the flow net located after 'n' potential drops is given by:

$$h = h_i - n \triangle h \qquad (7.55)$$

where   n = number of fields (may be fractional) *upstream* of the point

3. Pore pressure net. A pore pressure net represents lines of equal water pressures or piezometric heads.   To plot a pressure net the three quantities h_p, h and Z of Eq. 7.54 are expressed as percentage of the initial hydraulic head h_i.   Suppose a point of $h_p = 20\%$ h_i is to be located on an equipotential line of h $= 70\%$ h_i; from Eq. 7.54, $Z_1 = +50\%$ h_i.   Similarly, a point of $h_p = 20\%$ h_i on an equipotential line of h $= 80\%$ h_i, will have $Z_2 = +60\%$ h_1.   Three points A, B and C of the same $h_p = 20\%$ h_i are shown in Fig. 7.32, which on joining form a line of the pore   pressure net of $h_p = 20\%$ h_i.   The pressure net of various 'equal pore pressure lines', can thus be plotted.

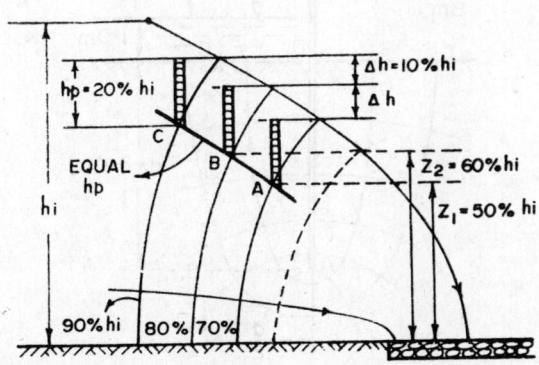

Fig. 7.32   Plotting of a pore pressure net

4. Hydraulic gradient. The hydraulic gradient across any field of average dimensional 'l' is given by $\triangle h/l$.   The hydraulic gradient at the downstream end of the flow lines where the percolating water leaves the soil mass is called the *exit gradient* i_e, which is important from stability considerations as discussed in the next section.

7.10 PIPING AND HEAVE

1. Piping. Piping is the subsurface erosion or 'boiling' caused by flowing water. For example, water flowing through or under a dam may carry with it to the surface at the downstream toe some of the finer material when the seepage pressure is equal to, or greater than, the effective pressure between the soil particles. Such action may result in excessive leakage or even failure, because with the increased porosity of the soil due to removal of fines, the velocity of water increases and in turn more and more larger-sized particles   are also removed.   The erosion starting at the exit end advances back-wards forming pipes or channels through a dam or its foundation. For water percolating through the foundation of a dam, the flow is upwards near the toe and the fields of the flow net are also smaller. The exit gradient may thus reach the critical value (Eq. 7.41) much earlier   near the toe rather than at distances away from the toe, giving start to erosion by piping or forming 'sand boils'.

2. Heave failure. High upward hydraulic gradient in soil adjacent to the downstream face of a sheet pile wall of a cofferdam,

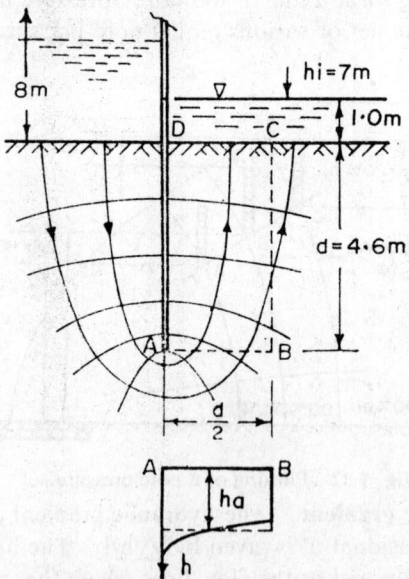

Fig. 7.33 Heave failure adjacent to sheet piles
Example 7.10

or adjacent to the cutoffs provided near the down-stream end of a dam, may cause failure by "heave" or "blowout" of the soil material. Terzaghi (1943) has shown that heave failure is likely to occur within a distance of about d/2 from the sheet piling, where d is the depth of soil above the level of instability. For single row of sheet piles, such as shown in Fig. 7.33, the depth of instability is almost exactly the same as the depth of penetration of the sheet piles below the downstream ground surface (Terzaghi 1943).

The variation of hydraulic head on the base AB of the soil mass ABCD likely to heave can be obtained from the epuipotential lines, but for purposes of analysis it is sufficient to determine the average hydraulic head h_a as follows:

$$h_a = \frac{n_1}{N_d} \times h_i \tag{7.56}$$

where N_d = total number of fields in a flow channel

n_1 = average number of fields *downstream* of the base AB

The average hydraulic gradient i_a is then given by:

$$i_a = h_a/d \tag{7.57}$$

and the safety factor F against heaving may be expressed as:

$$F = i_c/i_a \tag{7.58}$$

The possibility of erosion or boiling starting at the upper surface CD may also be examined by determining the exit gradient i_e ($i_e = \triangle h/l$, where l = average dimension of field adjacent to the piling). Then the safety factor is:

$$F = i_c/i_e \tag{7.59}$$

The two safety factors given by Eqs. 7.58 and 7.59 are not likely to differ appreciably (Craig 1983).

The stability of the soil mass ABCD likely to heave can be examined by two method: (a) using total pressures. or (b) using seepage and effective pressures.

(a)   Total pressure analysis:

Downward total pressure on AB = γ_{sat} d

Average hydraulic head on AB = h_a

Elevation head on AB = $-$ d

Average upward water pressure on AB = $(h_a + d) \gamma_w$

Resultant downward pressure = γ_{sat} d $-$ $(h_a + d) \gamma_w$

$= \gamma'd - h_a \gamma_w$

(b) Effective pressure analysis:

Downward effective pressure on AB $= \gamma'$ d

Average hydraulic gradient through ABCD $= h_a/d$

Average upward seepage pressure on AB $= i \, L \, \gamma_w$

$$= \frac{h_a}{d} \, d \, \gamma_w = h_a \, \gamma_w$$

Resultant downward pressure $= \gamma'$ d $- h_a \, \gamma_w$

(c) Condition for heave:

Heave will occur when the resultant downward pressure is zero, i.e. when γ' d $= h_a \, \gamma_w$ and the safety factor may be expressed as:

$$F = \frac{\gamma' \, d}{h_a \, \gamma_w} = \frac{i_c}{i_a} \qquad (7.58)$$

3. **Control measures**. Piping and heave can be controlled by
    1. Increasing the seepage path.
    2. Reducing seepage,
    3. Controlling the exit of seepage and reducing uplift pressure,
    4. Providing additional stabilizing load.

The seepage path through a dam foundation may be increased by providing sheet pile walls and other thin cutoffs, cutoff trenches, upstream impervious blankets, and by widening the base of the dam. Longer the seepage path, smaller will be the hydraulic gradient. Cutoff trenches, sheet piling, foundation grouting and upstream impervious blankets also reduce seepage through the foundation. Embankment zoning (providing impervious zone or a core wall) reduces seepage within an earth dam. Downstream drainage blankets, toe drains and drainage trenches are used to control the exit of seepage and allow its safe escape without inducing piping. Pressure relief wells near the downstream toe of a dam are the effective means of releasing high pressures in the soil water and thus reducing the gradients to a safe value below the critical gradient. Boiling and heave can also be prevented by covering the area with blankets of heavy, coarse graded material, called "protective filters" which are permeable enough for water to escape without appreciable resistances, but not sufficiently coarse to permit the base material, to be washed into the voids of the filter by the seepage water.

Examples 7.6 — 7.10

7.6 The depth of water on the U/S side of a zoned earth dam (shown in Fig. 7.21) is 20 m. The coefficients of permeability of zones 1 and 2 are respectively 1.5×10^{-7} m/s and 6.0×10^{-7} m/s. Determine the quantity of seepage per unit length through the

dam. What is the pore water (uplift) pressure at point 'A' in zone 1 ?
The flow net is shown in Fig. 7.21. $N_f = 3.5$, $N_d = 8$.
Considering zone 1,

$$q = k_1\, h_i\; \frac{N_f}{N_d} = 1.5 \times 10^{-7} \times 20 \times \frac{3.5}{8}$$

$$= 13.1 \times 10^{-7} \text{ m}^3/\text{s}$$

For point 'A', $h = 20 - \triangle h = 17.5$ m and $Z = 7.5$ m

$h_p = h - Z = 10$ m (pore pressure).

h_p can be obtained by direct measurement also.

7.7 The horizontal and vertical permeabilities of the foundation soil of a masonry dam, shown in Fig. 7.22, are respectively $k_x = 2.25 \times 10^{-7}$ m/s and $k_z = 1 \times 10^{-7}$ m/s. The full reservoir water level is 10 m above the D/S ground surface. Plot the flow net and determine the seepage loss. Find the uplift pressure at point 'A', located after 6.5 potential drops beneath the base of dam. Find also the exit gradient.

(a)   Scale factor for transformation is the x direction is given by:

$$\sqrt{k_z/k_x} = 1/1.5$$

and the equivalent isotropic permeability is:

$$k' = \sqrt{k_x\, k_z} = 1.5 \times 10^{-7} \text{ m/s}$$

The section is drawn to the transformed scale in Fig. 7.23 and a flow net with square fields is drawn. The true flow net is shown in Fig. 7.22.

$$N_f = 4, \quad N_d = 12, \quad h_i = 10 \text{ m}$$

Quantity of seepage per unit length is given by:

$$q = k'\, h_i\; \frac{N_f}{N_d} = 5 \times 10^{-7} \text{ m}^3/\text{s}$$

(b)   For point A, $Z = -2$ m (D/S ground surface as datum)

$$h = h_i - n\,\Delta h = 10 - 6.5 \times \frac{10}{12} \approx 4.58 \text{ m}$$

$$h_p = h - Z = 6.58 \text{ m} \quad \text{(uplift pressure)}$$

(c)   Average length of field near the toe = 1.8 m

$$i_e = \Delta h/1.8 = 0.46$$

7.8   For the homogeneous earth dam of Fig. 7.31, $k = 3 \times 10^{-6}$ cm/s. Plot the flow net and find the seepage rate. Depth of water in the reservoir above the D/S filter is 14.5 m.

The flow net is plotted in Fig. 7.31.

$$N_f = 6, \quad N_d = 19, \quad h_i = 14.5 \text{ m}$$

$$q = k \, h_i \frac{N_f}{N_d} = 13.74 \times 10^{-8} \text{ (m}^3/\text{s)}/\text{m}$$

Alternatively,

$$z_0 = 4.7 \text{ m (as measured)}$$

$$q = k \, z_0 = 3 \times 10^{-8} \times 4.7 = 14.1 \times 10^{-8} \text{ (m}^3/\text{s)}/\text{m}$$

Because of the graphical nature the two methods may not give the same result. The basic parabola may be a poor estimate of the flow lines for some geometries ($d < H$) (Dunn et al 1980).

7.9 An earth dam, built on an impervious foundation with side slopes 2 horizontal to 1 vertical, retains water to a height of 35 m, with a free board of 2 m. The base width is 154 m. Plot the top seepage line and estimate the flow rate if the permeability is 2.5×10^{-6} cm/s.

The seepage line is constructed graphically in Fig. 7.17.

$$d = 105 \text{ m}, \quad H = 35 \text{ m}, \quad \cot \beta = 2$$

Eq. 7.21: a = 32.4 m

Alternatively, from Gilboy solution,

$$d/H = 3, \quad \beta = 26.5^0, \quad m = 0.44$$

$$a = mH/\sin \beta = 34.5 \text{ m}$$

$$q = ka \sin^2 \beta = 2.5 \times 10^{-8} \times 32.4 \times (0.446)^2$$

$$= 16.1 \times 10^{-8} \text{ (m}^3/\text{s)}/\text{m}$$

7.10 A sheet pile wall penetrated to a depth of 4.6 m below the ground surface retains water to a height of 8 m above the ground surface. The downstream water level is 1 m above ground. A part flow net, shown in Fig. 7.33, consists of 12 potential drops in a channel adjacent to the piling. The saturated unit weight of soil is 20 kN/m^3. Determine the safety factor against failure by heaving adjacent to the downstream face of the piling. Also examine the safety against boiling starting at the exit of water.

Mass of soil ABCD likely to heave is 4.6 m by 2.3 m in section. (See Fig. 7.33)

Average number of fields downstream of AB = 4.2

$$h_a = \frac{n_1}{N_d} h_i = \frac{4.2}{12} \times 7 = 2.45 \text{ m}$$

$i_a = h_a/4.6 = 0.533$

$i_c = \gamma'/\gamma_w = 10.2/9.8 = 1.04$

$F = i_c/i_a = 1.95$

Safety against boiling at exit:

$\Delta h = h_i/N_d = 0.583$

Average length of field at exit = 1.3 m

$i_e = 0.448$

$F = i_c/i_e = 2.32$

8

Stress, strain and constitutive concepts

8.1 PRINCIPAL AND OCTAHEDRAL STRESSES

For any system of forces acting at a point there will be three orthogonal (i.e. mutually perpendicular) planes on which there are zero shear stresses, and the stresses are wholly normal. These are called the "principal planes" and the normal stresses acting on them are called the "principal stresses". The largest of these stresses is called the "major principal stress σ_1" the smallest is called the "minor principal stress σ_3", and the third one is the "intermediate principal stress σ_2". The direction of the principal stress are called the 'principal axes'.

The principal axes of three dimensional stress system are shown in Fig. 8.1. A line OD making an equal angle of $\cos^{-1}(1/\sqrt{3})$ with each axis defines the locus of all points for which $\sigma_1 = \sigma_2 = \sigma_3$ and is known is the "space diagonal". The plane perpendicular to the space diagonal is called the "octahedral plane"; it makes equal intercepts on the three axes. For any point on this plane, the mean principal stress is constant and is called the "octahedral normal stress τ_{oct}":

$$\tau_{oct} = \tfrac{1}{3}(\sigma_1 + \sigma_2 + \sigma_3) \tag{8.1}$$

The octahedral shear stress τ_{oct} can be obtained as:

$$\tau_{oct} = \tfrac{1}{3}[(\sigma_1 - \sigma_2)^2 + (\sigma_2 - \sigma_3)^2 + (\sigma_3 - \sigma_1)^2]^{1/2} \tag{8.2}$$

Eq. 8.2 shows that no shear stresses are produced when $\sigma_1 = \sigma_2 = \sigma_3$. This is a hydrostatic stress condition. Shear stresses are produced when $\sigma_1 > \sigma_2$ or $\sigma_1 > \sigma_3$.

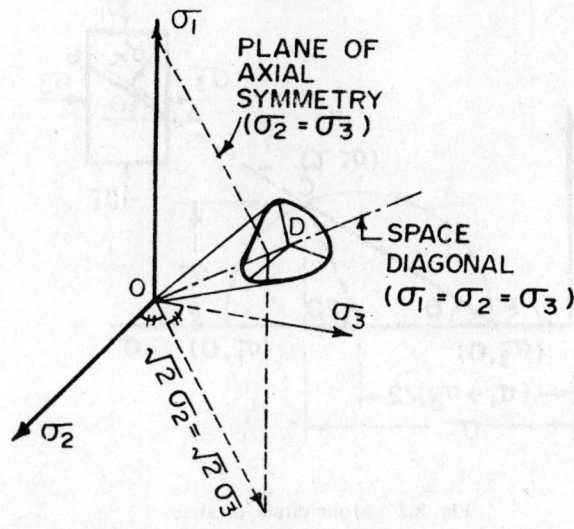

Fig. 8.1 Three dimensional stress space

In geotechnical engineering, it has been convenient to assume either that $\sigma_2 = \sigma_3$ or that $\sigma_2 = 0$, procuding a "plane stress" condition. Field results indicate that the assumption of $\sigma_2 = \sigma_3$, while not strictly correct, does not introduce any large error and does greatly simplify the elastic analysis. If $\sigma_2 = \sigma_3$, the stress space becomes a stress plane.

8.2 MOHR'S STRESS CIRCLE

Consider a soil element (Fig. 8.2) in a two dimensional state of stress with σ_1 and σ_3 acting on the principal planes (element faces). The analysis could be made for any two of the principal stresses, but the major and minor values are generally chosen, as here. The normal and shear stresses (σ and τ) on any plane AB inclined at an angle θ to the major principal plane can be obtained easily by constructing a circle with a radius $(\sigma_1 - \sigma_3)/2$ and its centre at $[(\sigma_1 + \sigma_3)/2, 0]$ in the σ-τ plane, as shown in Fig. 8.2. It is known as the "Mohr circle of stress". It represents the state of stress at *a point at equilibrium* and it applies to any material, not just soil.

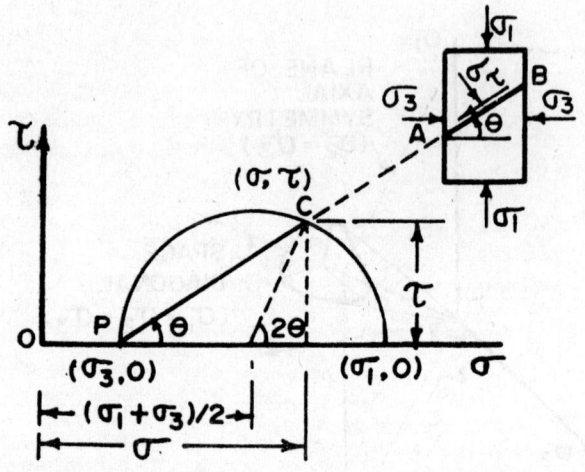

Fig. 8.2   Mohr circle of stress

The stresses σ and τ on the plane AB are obtained by drawing a line through the point P on the circle, parallel to AB, to intersect the circle at point C whose coordinates yield the values of σ and τ as:

$$\sigma = \frac{\sigma_1 + \sigma_3}{2} + \frac{\sigma_1 - \sigma_3}{2} \cos 2\theta \qquad (8.3)$$

$$\tau = \frac{\sigma_1 - \sigma_3}{2} \sin 2\theta \qquad (8.4)$$

The point P on the circle is a unique point called the "pole" or the "origin of planes".   It has a very useful property: any straight line drawn through the pole intersects the Mohr circle at a   point which represents the state of stress on a plane inclined at the same orientation in space as the line. Conversely, if the state of stress, σ and τ, on some plane is known, a line drawn parallel to that plane through the coordinates of σ and τ on the Mohr circle, would pass through the pole F; the pole can thus be located. Once the pole is known, the stresses on any plane can   readily be found by simply drawing a line from the pole parallel to that plane.

8.3   STRESS PATH AND p-q DIAGRAM

Instead of plotting a Mohr circle to represent the state of stress at a point in equilibrium, it may sometimes be more convenient to represent that state of stress by a 'stress point' on the circle, which has the coordinates of:

$$p = \frac{\sigma_1 + \sigma_3}{2} \ , q = \frac{\sigma_1 - \sigma_3}{2} \qquad (8.5)$$

This is the point of maximum shear stress, as shown in Fig. 8.3. The successive states of stress undergone by a test specimen or a typical element in the field during loading or unloading can thus be represented either by a series of Mohr circles or by their stress points, the former may be confusing and it is simpler to adopt the system of stress points only. The 'locus' of the stress points is called a "stress path" and its plot is known as the "p-q diagram," as shown in Fig 8.3b. Both p and q can be defined either in terms of total stresses or effective stresses. Thus in terms of effective stresses, $p' = p - u$ and $q' = q$, where 'u' is the excess hydrostatic or pore water pressure.

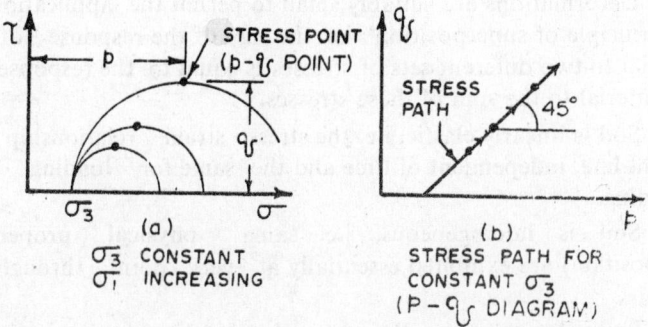

Fig. 8.3   Stress points and stress path for successive Mohr circles

The behaviour of a soil element depends on the stress path for the history and for the stressing the element receives in future. Lambe has developed the stress path method into a practical engineering tool for the solution of stability and deformation problems (Lambe 1967, Lambe and Marr 1979).

8.4 CONSTITUTIVE RELATIONS

Relations or equations which are established to define the mechanical behaviour (stress-strain relationships and consequent effects due to their changes) of soil (or any other material) are referred to as the 'constitutive relations or laws', because they reflect the constitution of the material. The constitutive relations for soils (and for all real materials), if could be defined, would be very complicated. For example, the mechanical response of soil to imposed stress, will depend on the type and magnitude of imposed stress, time duration of its action, past stress and strain history of soil, and possibly other environments. The incorporation of all these relevant factors in the mathematical description of the constitutive relations is complex and not always solvable. Consequently, the practice is to idealize the constitutive relations to a form whose solutions could be obtained and provide reasonable and useful results.

Idealization involves the simplifying assumptions like the ones given below and others, which are used in suitable combinations or separately.

1. Deformations are suitably small to permit the application of the 'principle of superposition', i.e. the sum of the response of the material to two different sets of stresses is equal to the response of the material to the sum of these stresses.

2. Soil is linearly elastic, i e. the stress strain relationship is a straight line, independent of time and the same for loading and unloading.

3. Soil is homogeneous, i.e. same physical properties (composition) are exhibited essentially at every point throughout the mass.

4. Soil is isotropic, i.e. the material properties (e.g. elasticity) at a point are the same in all directions.

5. Soil is linearly viscoelastic, i.e. there is a time-dependent component of response, in addition to a time-independent elastic response.

An example of the constitutive relations for an isotropic, homogeneous, linear-elastic material is the generalized Hooke's law which for a three dimensional principal-stress system   may   be expressed as:

$$\epsilon_1 = \frac{1}{E} [\Delta\sigma_1 - \mu (\Delta\sigma_2 + \Delta\sigma_3)]$$

$$\epsilon_2 = \frac{1}{E} [\Delta\sigma_2 - \mu (\Delta\sigma_3 + \Delta\sigma_1)] \qquad (8.6)$$

$$\epsilon_3 = \frac{1}{E} [\Delta\sigma_3 - \mu (\Delta\sigma_1 + \Delta \sigma_2)]$$

where ϵ_1, ϵ_2 and ϵ_3 are the major, intermediate and minor principal strains, E is the 'Young's modulus of elasticity', and μ is the 'Poisson's ratio'. 'Young's modulus' is the ratio of stress to resultant strain. 'Poisson's ratio' is the ratio of strain orthogonal to the stress of interest to the strain colinear with that stress, e.g.

$$\mu = \frac{\text{horizontal strain}}{\text{vertical strain}} = \frac{\epsilon_h}{\epsilon_v} \qquad (8.7)$$

Typical values of μ for silts and sands range from 0.2 for loose materials to 0.4 for dense materials. Values for saturated clays vary from about 0.4 to 0.5. The theoretical maximum for a saturated clay undergoing no volume change when stressed (undrained) is 0.5. A value of $\mu=0$ may also be used for dry soils as a computational convenience; this is because μ is difficult to obtain for soils.

Soil exhibits linear stress-strain characteristics only at very low strain amplitude. Principal soil deformations are state changes caused by relative particle motion, since soil is a particulate mass and not a continuum. Only a small amount of soil deformation is elastically recoverable. Moreover, soil is commonly anisotropic rather than isotropic due to formation by sedimentation. Because of non-linear (elasto-plastic, plastic) and time-dependent response of real soils, analytically derived constitutive laws have their own limitations. Field tests or laboratory tests under simulated field conditions are often required to establish the constitutive relations. Many of them are simply empirical or semi-empirical based on phenomenological description of soil behaviour.

8.5 RHEOLOGICAL AND MATHEMATICAL MODELS

Mechanical response of soil can also be represented in some cases by rheological and mathematical models.

1. Rheological Model. Rheology is the study of time-dependent deformations of materials. For the analysis, the real (physical) soil system is replaced by an ideal mechanical model, called the 'rheological model', composed of springs, dashpots and friction elements in various combinations. The three basic rheological models of practical interest are shown in Fig. 8.4. These models characterize the stress-strain relationship (constitutive law) in terms of the material constants known from experiments.

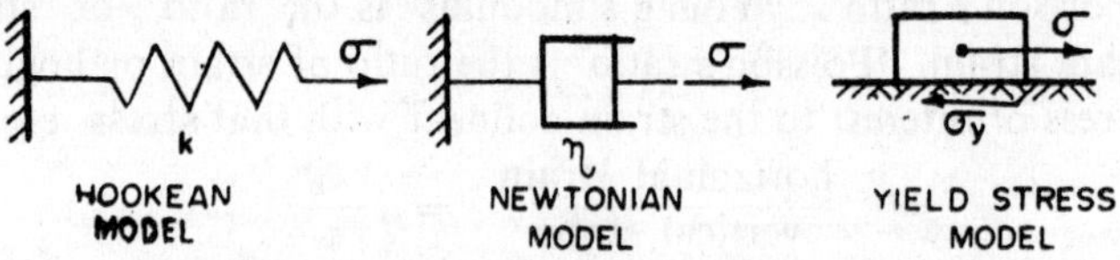

HOOKEAN
MODEL

NEWTONIAN
MODEL

YIELD STRESS
MODEL

Fig. 8.4 Basic rheological models

(a) *Hookean model.* The Hookean model or spring element represents a perfect elastic response of soil independent of time, i.e. stress is a linear function of strain, $\sigma = k\epsilon$, where k is the constant of proportionality, also known as the 'modulus of elasticity'. In the model representation, k is the spring constant.

(b) *Newtonian model.* The Newtonian model or dash represents a perfect viscous response of soil, i.e. stress is a linear function of the rate of change in strain with respect to time, $\sigma = \eta (\epsilon + \epsilon_0)/t$, where η is the constant of viscosity and ϵ_0 is the strain at t = 0. In the model, η is the dashpot constant.

(c) *Yield stress model.* In this case, the stress σ can generate a strain only if $\sigma > \sigma_y$, where σ_y is a certain minimum stress necessary for causing strain (or slip). σ_y is called the 'yield stress'. In the model, σ_y is the frictional resistance.

To represent the complex behaviour of most of the soils, several 'composite models' are constructed by combining the above mentioned simple models.

2. Mathematical model. *Hyperbolic function.* The non-linear stress-strain response of soils can be represented with a high degree of accuracy by a rectangular hyperbola having the following equation (Kondner 1963):

$$y = x / (a + b x) \tag{8.8}$$

where 'a' and 'b' are material constants. For stress-strain relationship, 'y' is replaced by normal stress σ and 'x' by axial strain ϵ yielding:

$$\sigma = \epsilon / (a + b\ \epsilon) \qquad (8.9)$$

When Eq. 8.9 is plotted for an assumed set of values of a and b, the curve obtained is shown in Fig. 8.5. From the property of the rectangular hyperbola, a transformed plot of ϵ/σ versus ϵ is a straight line, as seen from Fig. 8.5. The equation of the transformed plot is given by:

$$\epsilon/\sigma = a + b\ \epsilon \qquad (8.10)$$

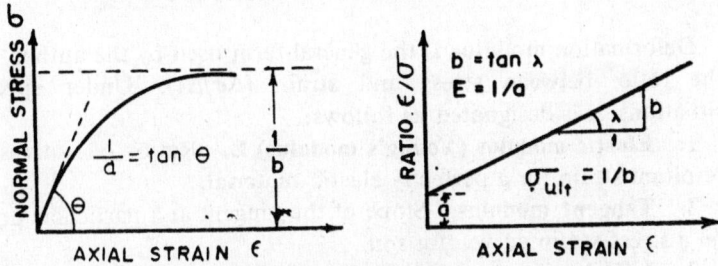

Fig. 8.5 Hyperbolic representation of stress-strain

The ultimate value of the stress can be obtained by taking the limit of Eq. 8.9 as ϵ becomes very large, or

$$\sigma_{ult} = \lim_{\varepsilon \to \infty} \sigma = \frac{1}{b} \qquad (8.11)$$

Thus the reciprocal of the slope of the straight line (i.e. 1/b) gives the ultimate strength.

Differentiating Eq. 8.9 with respect to the strain and evaluating the derivative at ϵ equal to zero yields:

$$\left(\frac{d\sigma}{d\epsilon} \right)_{\epsilon\ \text{z}0} = \frac{1}{a} \qquad (8.12)$$

Thus, the inverse of the intercept of the straight line on the ϵ/σ axis is a measure of the "initial tangent modulus" (See Section 8.6) of the material tested (Kondner 1963).

To represent the soil response in the conventional triaxial compression test, the hyperbolic stress-strain relation takes the form:

$$(\sigma_1 - \sigma_3) = \epsilon / (a + b\ \epsilon) \tag{8.13}$$

where $\quad \sigma_1 - \sigma_3 =$ deviator stress

The parameters a and b represent functions of the rate of axial strain, preconsolidation pressure, rebound stress and soil under consideration. The reciprocal of b gives the ultimate compressive strength $(\sigma_1 - \sigma_3)_u$ of the soil which is not exactly equal to the failure strength $(\sigma_1 - \sigma_3)_f$, but it is proportional to the failure strength:

$$(\sigma_1 - \sigma_3)_u \geqslant (\sigma_1 - \sigma_3)_f$$

8.6 DEFORMATION MODULUS

Deformation modulus is the general term used by the author for the ratio between stress and strain $(\Delta\sigma/\Delta\epsilon)$. Under specific situations, it is designated as follows:

1. **Elastic modulus (Young's modulus) E.** Ratio of stress to resultant strain for a perfectly elastic material.

2. **Tangent modulus.** Slope of the tangent at a particular point on a stress-strain curve of a soil.

3. **Initial tangent modulus.** Slope of the tangent to the stress-strain curve of a soil at its origin.

4. **Secant modulus.** Slope of a straight line joining any two separate points on the stress-strain curve. As the two points come closer, the secant modulus tends to become equal to the tangent modulus.

5. **Compressibility modulus (Constrained modulus) E_v.** Ratio of axial stress (increment) to axial strain for laterally confined or restrained compression (i.e. $\epsilon_x = \epsilon_y = 0$) within the stress range of interest:

$$E_v = \frac{\sigma_z}{\epsilon_z} = \frac{\Delta\sigma_z}{\Delta H/H_i} = \frac{1}{m_v} \tag{8.14}$$

where $\quad \Delta H =$ vertical deformation in consolidation cell of a specimen H_i thick under stress $\Delta\sigma_z$

$\qquad m_v =$ coefficient of volume change (Chapter 14),

6. **Total deformation modulus (General deformation modulus) E_0.** Stress-strain ratio for laterally unrestrained compression of soil within the stress range of interest.

$$E_o = \beta\, E_v, (\beta < 1 \text{ depending on } \mu) \qquad (8.15)$$

7. **Bulk modulus** E_b. (for isotropic compression). Ratio of compressive stress σ_o to volumetric strain $\Delta V/V$.

$$\frac{\Delta V}{V} = \frac{3\sigma_o}{E} (1 - 2\mu)$$

and $\qquad E_b = \dfrac{\sigma_c}{\Delta V/V} = \dfrac{E}{3(1 - 2\mu)} \qquad (8.16)$

8. **Strain modulus**. Inverse of deformation modulus.

i.e. $\dfrac{\Delta \epsilon}{\Delta \sigma}$

9

Shear strength of soil

9.1 COULOMB-TERZAGHI EQUATION

The 'shear strength of a soil in any direction is the maximum shear stress that can be applied to the soil in that direction. When this maximum has been reached, the soil is regarded as having *failed*. The shear strength of soil is basically made up of:

1. Frictional component due to interlocking of particles and the friction between them when subjected to normal stress; and

2. Cohesive component due to "cohesion", i.e. mutual attraction that exists between the fine particles of some soils and thus tends to hold them together in a solid mass without the application of external forces.

The shear strength phenomenological law was first discovered by Coulomb (1773) who expressed shear strength 's' (shear stress at failure) as a linear function of the (total) normal stress σ on the potential surface of sliding:

$$s = c + \sigma \tan \phi \qquad (9.1)$$

where c and ϕ are called the "shear strength parameters" of the soil and are now described as the "cohesion intercept" (or the 'apparent cohesion') and the "angle of shearing resistance" (angle of internal friction) respectively. The Coulomb strength envelope (Eq. 9.1) is shown plotted as a straight line in Fig. 9.1. The friction component (denoted by tan ϕ) increases with normal stress but the cohesion component (denoted by c) remains constant. If there is no normal stress, the friction disappears.

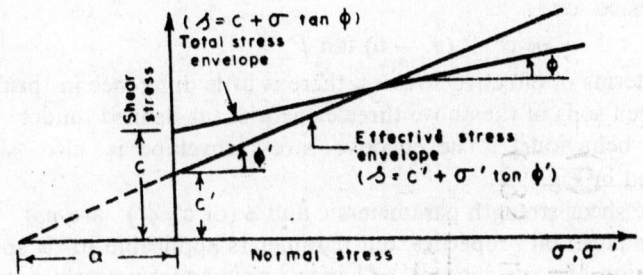

Fig. 9.1 Coulomb strength envelopes in terms
of total and effective stresses

Soils may be divided into three categories on the basis of their
total strength properties.

1.   Frictional or cohesionless (granular) soils.   They possess no
cohesion (c = 0), but derive their   shear   strength   entirely   from
intergranular friction.   The strength envelope   passes   through the
origin.   Example: dry sands or completely submerged free-draining
sands.

2.   Cohesive soils.   They exhibit cohesion, but if no   change in
moisture content occurs during the test, they appear to possess   no
friction ($\phi \approx 0$).   The strength enevelope is   virtually   horizontal.
Example: virtually saturated clays.

3.   Cohesive-frictional soils.   They are the   intermediate   soil
types which possess both cohesion and friction (c, ϕ soils).

Experience has since   shown   that   shear strength can only be
expressed in terms of the total stress in a few exceptional cases.   In
accordance with Terzaghi's fundamental concept   that shear   stress
in a soil can be resisted only by the   skeleton   of   soild   particles,
shear strength is expressed as a function of *effective* normal   stress
σ' and the Coulomb equation is restated in the form:

$$s = c' + \sigma' \tan \phi' \qquad (9.2)$$

where c' and ϕ' are the   shear   strength   parameters   in   terms   of
effective   stress.   For all practical   purposes, it may   be assumed
that

$$\sigma' = \sigma - u \qquad (9.3)$$

where   u   = pore pressure

The Coulomb-Terzaghi shear strength equation can thus be expressed as:

$$s = c' + (\sigma - u) \tan \phi' \qquad (9.4)$$

In terms of effective stresses. there is little difference in principle between soils of the above three categories (as defined under total stress behaviour). The effective stress envelope is also shown plotted in Fig. 9.1.

The shear strength parameters c and ϕ (or c', ϕ') are *not* basic or fundamental properties, but parameters applicable to a specific condition of a specific soil. They are only approximately constant and not the unique values for any particular soil. Defined as such, the shear strength parameters have no physical meaning. The angle ϕ (or ϕ') is not a true angle of friction; it is simply the slope of the strength envelope. The parameter c (or c') simply represents that part of the shear strength which is independent of the normal stress. For explicitness of the total stress parameters c and ϕ (in comparison to c' and ϕ') they are represented by the symbols c_u and ϕ_u or c_T and ϕ_T. In general, c_u and ϕ_u depend on the conditions (drainage conditions) operative in the test and the normal stress range.

9.2 MOHR-COULOMB FAILURE THEORY

The Mohr theory (1900) for failure states that materials fail 'when the shear stress on the failure plane at failure reaches some unique function of the normal stress on that plane', or

$$\tau_t = f(\sigma_f) \qquad (9.5)$$

The shear stress at failure τ_t is called the '*shear strength*' of the material. The Mohr failure criterion can also be expressed in terms of principal stresses $(\sigma_1 > \sigma_2 > \sigma_3)$ at failure as follows:

$$(\sigma_1 - \sigma_3) = f(\sigma_1 + \sigma_3) \qquad (9.6)$$

Eq. 9.6 implies that, at failure, the radius of the Mohr circle is some unique function of the mean principal stress $(\sigma_1 + \sigma_3)/2$. The failure condition may therefore be defined in terms of an envelope called the *Mohr failure envelope*, which is tangential to all such failure circles (Fig. 9.2). This envelope expresses the functional relationship between τ_t and σ_f at failure.

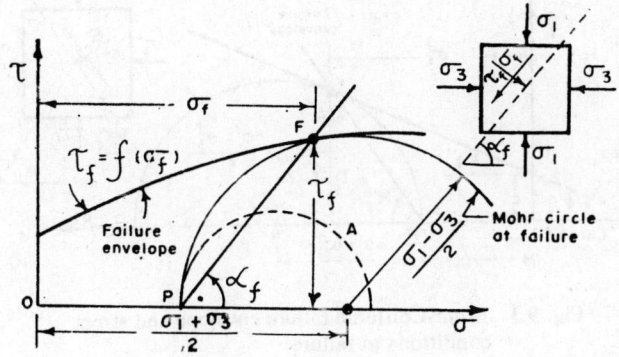

Fig. 9.2 Mohr failure criterion

It may be noted that any Mohr circle, such as A, lying below the failure envelope represents a stable condition. Failure occurs only when the combination of τ and σ is such that the circle is *tangent* to the failure envelope. Circles lying above the failure envelope cannot exist, as the material would fail before reaching that state of stress. Since the failure envelope gives the stress conditions on the failure plane at failure, the point of tangency of the failure envelope with the Mohr circle determines the inclination of the failure plane α_f.

The failure envelope defined by Eq. 9.5 or 9.6 is a curved line. In practice, the Mohr envelope for a soil is commonly found to be approximately straight over a considerable range of normal stress and, therefore as a fairly close approximation, the Mohr failure envelope is assumed to be a straight line over some given stress range and identical with the Coulomb envelope (Eq. 9.1 or 9.2). The combination of the Coulomb concept of linear relationship between shear stress and normal stress at failure and the Mohr's condition of tangency between the Mohr circle and the failure envelope at the time of failure is known as the 'Mohr-Coulomb failure theory'. A Mohr-Coulomb failure envelope (or strength envelope) is thus a straight line defined by Coulomb's equations (Eq. 9.1 or 9.2) and it is tangential to the Mohr circle at failure, as shown in Fig. 9.3.

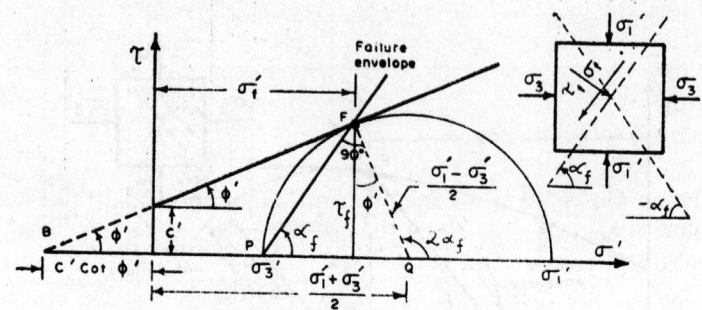

Fig. 9.3   Mohr-Coulomb failure envelope and stress
conditions at failure

The coordinates of the tangent point F are σ_f' and τ_f,
where:

$$\sigma_t' = \tfrac{1}{2}(\sigma_1' + \sigma_3') + \tfrac{1}{2}(\sigma_1' - \sigma_3')\cos 2\,\alpha_t \tag{9.7}$$

$$\tau_f = \tfrac{1}{2}(\sigma_1' - \sigma_3')\sin 2\,\alpha_f \tag{9.8}$$

The theoretical angle α_t between the failure plane and the major
principal plane can be easily obtained from triangle QFB as:

$$2\,\alpha_t = 90 + \phi'$$

or $\qquad \alpha_f = 45 + \dfrac{\phi'}{2}$ $\hspace{3cm}$ (9.9)

From Fig. 9.3 the relationship between the principal stresses at
failure and the shear strength parameters can be obtained as
follows:

$$\sin \phi' = \frac{(\sigma_1' - \sigma_3')/2}{c'\cot\phi' + (\sigma_1' + \sigma_3')/2}$$

or $\quad (\sigma_1' - \sigma_3') = 2\,c'\cos\phi' + (\sigma_1' + \sigma_3')\sin\phi'$ (9.10)

Also $\quad \sigma_1' = \sigma_3'\dfrac{1 + \sin\phi'}{1 - \sin\phi'} + 2\,c'\dfrac{\cos\phi'}{1 - \sin\phi'}$ (9.11)

or $\qquad \sigma_1' = \sigma_3'\tan^2\alpha_t + 2\,c'\tan\alpha_t$ (9.12)

where $\quad \alpha_t = 45 + \phi'/2$

Eq. 9.10 (or Eq. 9.12) is also referred to as the Mohr-Coulomb
failure criterion in terms of principal stresses. The Mohr-Coulomb
failure theory can thus be summarised as follows:

1. Failure of soil is possible only under a *critical combination* of
shear stress and normal stress. This combination is defined by the

straight line relationship of Eq. 9.2 (in terms of effective stress) or Eq. 9.1 (in terms of total stress). This relationship is termed the 'failure envelope or the strength envelope' of the soil. The particular plane where this critical combination is first reached becomes a 'failure plane'

2. In a general three dimensional stress state the unique relationship between the major and minor principal stresses at failure is expressed by Eq. 9.10. (Effect of σ_2 on failure, if any, is neglected.)

3. At failure, one of the points on the Mohr circle drawn for major and minor principal stresses must satisfy the failure criterion Eq. 9.1 or 9.2; i.e. the circle must be tangent to the failure envelope.

4. Failure occurs on a plane whose inclination is given by Eq. 9.9.

For a cohesionless soil (c = 0), Eq. 9.10 yields

$$\sin \phi' = \frac{\sigma_1' - \sigma_3'}{\sigma_1' + \phi_3'} \tag{9.13}$$

and

$$\frac{\sigma_1'}{\sigma_3'} = \frac{1 + \sin \phi'}{1 - \sin \phi'} = \tan^2 (45 + \frac{\phi'}{2})$$

$$= \tan^2 \alpha_f \tag{9.14}$$

or

$$\frac{\sigma_3'}{\sigma_1'} = \frac{1 - \sin \phi'}{1 + \sin \phi'} = \tan^2 (45 - \frac{\phi'}{2}) \tag{9.15}$$

The following useful relationships can also be derived from the Mohr failure diagram.

$$\tau_f = \frac{\sigma_1' - \sigma_3'}{2} \cos \phi' \tag{9.16}$$

and $\qquad \sigma_f' = \sigma_3' (1 + \sin \phi')$, (for c = 0) $\tag{9.17}$

or $\qquad \sigma_f' = \sigma_1' (1 - \sin \phi')$, (for c = 0) $\tag{9.18}$

Eqs. 9.14 and 9.15 are called the 'obliquity relations' because the maximum inclination (slope) or 'obliquity' of the Mohr failure envelope occurs where c is equal to zero. (The relations are, of course, only valid where c = 0). The coordinates of the point of tangency F (σ_f', τ_f) are the stresses on the plane of maximum obliquity in the soil element. In other words, the ratio τ_f/σ_f' is a maximum on this plane. This maximum ratio is termed the 'obliquity' (Taylor 1948).   Failure occurs only when the obliquity (slope) equals ϕ'.

The Mohr criterion implies that the intermediate principal stress σ_2' has no influence on the shear strength of soil. Since by definition σ_2' lies somewhere between σ_1' and σ_3', the Mohr circles for the three stresses would look like those shown in Fig. 9.4. It is obvious that σ_2', irrespective of its magnitude, can have no influence on the conditions at failure for the Mohr failure criterion.   The intermediate principal stress σ_2' probably does have an influence in real soils, but it is neglected in the Mohr-Coulomb failure theory.

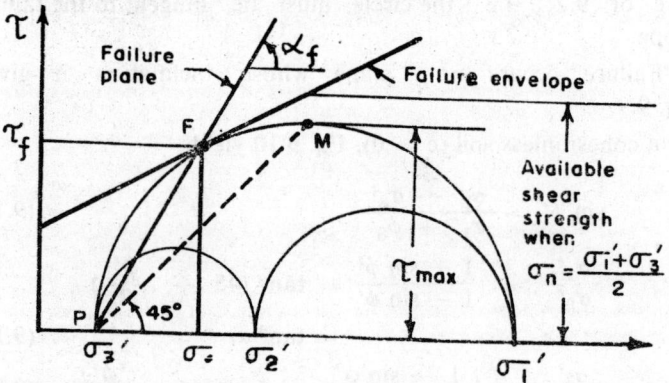

Fig. 9.4 Mohr circles for a three dimensional state of stress and plane of maximum shear stress

It may also be noted that the maximum shear stress τ_{max} in a soil element acts on the plane inclined at 45° and is given by:

$$\tau_{max} = \frac{\sigma_1' - \sigma_3'}{2} > \tau_f \qquad (9.19)$$

Then why should failure not occur at the 45° plane? This is because on this plane, the available shear strength of a soil as defined by the strength envelope is greater than τ_{max} (Fig. 9.4). On this plane ($\alpha = 45°$), the obliquity is less than the maximum value since the ratio of τ_{max} to $(\sigma_1' + \sigma_3')/2$ is less than τ_f/σ_f' (Refer example 9.6).

The Mohr-Coulomb failure criterion because of its simplicity, is widely used in practice, although it is by no means the only possible failure criterion for soils.

1. Modified failure envelope — K_f line. If a number of states of stress are known, each producing shear failure in soil, each state of stress will be represented by a Mohr circle at failure and the common tangent to the various circles, called the failure (or strength) envelope, will yield the shear strength parameters. An alternative is to plot a p-q diagram as described by Eq. 8.5. It is easier to draw the best-fit line through a series of points than tangent to a series of circles. The line through the p-q points (at failure) is called the "modified failure envelope" or a "K_f Line", as shown in Fig. 9.5.

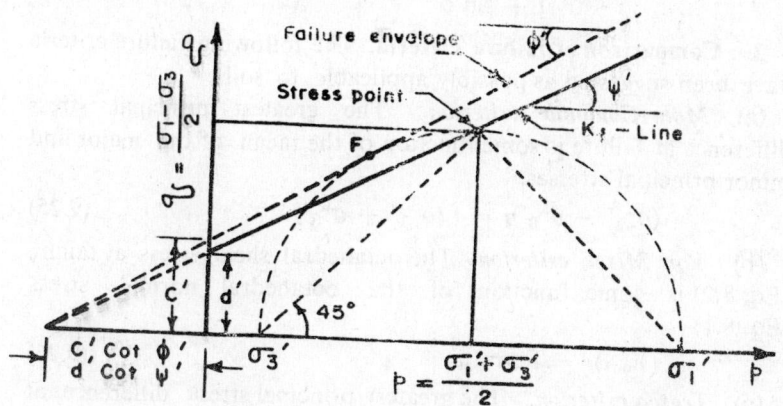

Fig. 9.5 Relationship between the Mohr-Coulomb failure envelope and the K_f line

The K_f line is represented by:

$$q_f = d' + p_f \tan \psi' \tag{9.20}$$

or
$$\frac{\sigma_1' - \sigma_3'}{2} = d' + \frac{\sigma_1' + \sigma_3'}{2} \tan \psi' \tag{9.21}$$

where d' and ψ' are the 'modified shear strength parameters'. By comparing Eqs. 9.10 and 9.21, the parameters c' and ϕ' are then given by:

$$c' = \frac{d'}{\cos \phi'} \tag{9.22}$$

and $\sin \phi' = \tan \psi'$
$$\tag{9.23}$$

From a given stress point on the K_f line, lines drawn at angles of 45° to the horizontal, as shown in Fig. 9.5, intersect the p-axis at points representing σ_1' and σ_3' for that stress point. It should also be noted that $(\sigma_1' - \sigma_3')/2 = (\sigma_1 - {}_3)/2$. Defining the principal stress ratio at failure $(\sigma_3'/\sigma_1')_f$ by the factor K_f, the following relation can be obtained for a cohesionless soil:

$$K_f = (\frac{\sigma_3'}{\sigma_1'})_t = \frac{1 - \tan \psi'}{1 + \tan \psi'}$$

$$= \frac{1 - \sin \phi}{1 + \sin \phi'} = \tan^2 \left(45 - \frac{\phi'}{2} \right) \qquad (.24)$$

2. Comparison of failure criteria. The following failure criteria have been suggested as possibly applicable to soils.

(a) *Mohr-Coulomb criterion.* The greatest principal stress difference at failure is some function of the mean of the major and minor principal stresses.

$$(\sigma_1' - \sigma_3')_t = f(\sigma_1' + \sigma_3') \qquad (9.25)$$

(b) *Von Mises criterion.* The octahedral shear stress at failure (Eq. 8.2) is some function of the octahedral normal stress (Eq. 8.1)

$$(\tau_{oct})_t = f(\sigma_{oct}) \qquad (9.26)$$

(c) *Tresca criterion.* The greatest principal stress difference at failure is some function of the octahedral normal stress.

$$(\sigma_1' - \sigma_3')_t = f(\sigma_{oct}) \qquad (9.27)$$

In a three dimensional stress space, any failure criterion for the material will define its 'failure surface' surrounding the 'space diagonal' (See Section 8.1). The failure surface encloses all points representing stable stress states. If a stress state represented by a point outside the failure surface is imposed, the soil yields. The octahedral plane, which makes equal intercepts on the three axes (σ_1', σ_2' and σ_3' axes) will be projected as equilateral triangular surface $A_1 A_2 A_3$ (Fig. 9.6) when viewed along the space diagonal. The triangle $A_1 A_2 A_3$ forms one face of a regular octahedron centred at 0, the origion of stress; point D, defined as the intersection point between the space diagonal and the octahedral plane, coincides with point 0. (See Fig. 8.1 also). The lines DA_1 DA_2 and DA_3 represent the conditions of triaxial compression tests

(Section 9.6) in which $\sigma_1' > \sigma_2' = \sigma_3'$, etc. Lines DB_1, DB_2 and DB_3 represent triaxial extension ($\sigma_1' < \sigma_2' = \sigma_3'$, etc).

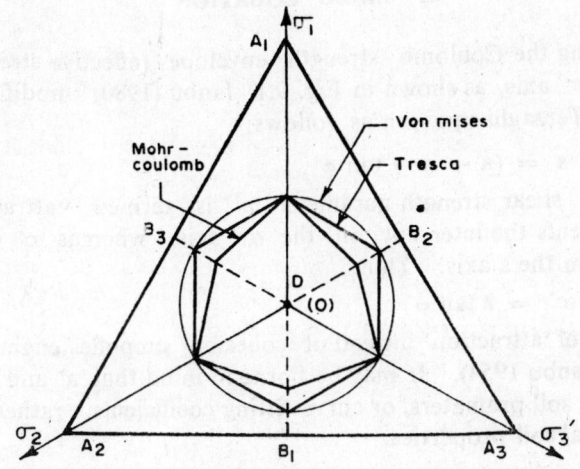

Fig. 9.6 Failure loci on an octahedral plane

The three failure criteria (Eqs. 9.25, 9.26, 9.27) can be compared by plotting the intersections of their failure surfaces with an octahedral plane. The failure loci for these criteria are shown plotted in Fig 9.6. The following observations may be made (Scott 1974) about the failure criteria (Bishop 1966, Schofield and Wroth 1968).

1. The failure loci for real soils are observed to lie between those for the Mohr-Coulomb and von Mises criteria, i.e. the soil strength will be underestimated by the Mohr-Coulomb criterion and overestimated by the von Mises and Tresca criteria.

2. The Mohr-Coulomb criterion correctly represents the peak strength of dense sand in triaxial extension and somewhat underestimates its strength in plane strain. The other two criterion grossly overestimate the strength of dense sands.

3. Failure loci for some soft clays lie rather closer to the von Mises criterion.

For the above reasons and because of simplicity the Mohr-Coulomb criterion is almost invariably used in practice (Scott 1974).

9.3 JANBU EQUATION

Extending the Coulomb strength envelope (effective stress) to meet the σ' axis, as shown in Fig. 9.1, Janbu (1980) modified the Coulomb-Terzaghi equation as follows:

$$s = (a + \sigma') \tan \phi' \qquad (9.28)$$

where the shear strength parameter 'a' is termed "attraction" and represents the intercept on the σ' axis; whereas c' is the intercept on the s axis. Thus,

$$c' = a \tan \phi' \qquad (9.29)$$

The use of 'attraction' instead of 'cohesion' simplifies engineering formulii (Janbu 1954).   It may be borne in mind that 'a' and tan ϕ' are merely soil pramaters, or curve fitting coefficients, rather than fundamental soil properties.

9.4 SHEAR STRENGTH MEASUREMENT AND DRAINAGE CONDITIONS

Shear strength is measured both in the laboratory and in the field. Laboratory tests are performed on representative soil samples and must be done in a way that simulates field conditions. The commonly used laboratory tests are: shear box test, triaxial compression test, unconfined compression test, fall cone test and laboratory vane shear test. Field tests include the in-situ shear box test and vane shear test. The field strength of both granular and cohesive soils is also measured through empirical relation-ships with penetrometer tests.

There are three basic types of shear test procedures depending on sample drainage conditions and which can be adopted in case of shear box and triaxial tests. These are commonly designated by a two-letter symbol; the first letter refers to what happens *before shear* (i.e. whether the sample is consolidated) and the second letter refers to the drainage conditions *during shear*. The tests are designated as given in Table 9.1.

Table 9.1
TYPES OF SHEAR TESTS DEPENDING ON DRAINAGE CONDITIONS

Designation (Before shear-During shear)	Symbol	Also called
Unconsolidated-Undrained	UU	Undrained; or quick
Consolidated-Undrained	CU	Consolidated-quick
Consolidated-Drained	CD	Drained; or slow

1. **Unconsolidated-undrained test.** The sample is subjected to a specified initial stress (normal stress in case of shear box and all-round pressure in case of triaxial test) and without waiting for any consolidation, shearing process is started without permitting any drainage of water. Thus no drainage or dissipation of pore pressure is permitted during the entire test. The results of such a test can be used for field problems where critical stresses develop in a saturated soil mass too rapidly for any moisture content to change appreciably. Examples: bearing capacity of foundations on saturated homogeneous clays immediately after construction; earth pressure against bracings in temporary excavations; stability of side slopes of cuttings in clays immediately after excavation; short term stability of embankments and earth dams during construction.

2. **Consolidated-undrained test.** Consolidation or drainage of the sample is permitted prior to actual shearing process under a specified initial stress. Sample is then sheard allowing no drainage. Test results are applicable to field situations where the soil has consolidated under the foundation pressure during construction or under its own weight, and which is then followed by quick increase in loads resulting in rapid change of critical stresses during which no further change in moisture content can take place. Examples: stability of consolidated earth dams and slopes of cohesive soils under conditions of 'rapid drawdown' of water, where water has no time to drain out of the voids.

3. **Consolidated-drained test.** Drainage and full consolidation of the sample is permitted throughout the test so that no pore pressure is developed at any stage of the test. The results are applied to field problems where the stresses develop within the soil mass sufficiently slowly for all changes in moisture content to take

place. Examples: final bearing capacity of soil where the foundation is erected more slowly than the soil consolidates; strength of sandy soils, or clay embankment in which drainage channels are embedded.

9.5 SHEAR BOX TEST

The shear box test (also called a direct shear test) is the simplest, the oldest, and the most straightforward method for measuring the 'immediate' or short-term shear strength of soils in terms of total stresses. In this test a sample of soil is placed in a metal box, square or circular in section, which is split in half horizontally. The usual, small size shear box is 6 cm square; a. 10 cm square shear box has also been introduced (Head 1982). The large shear box is 30 cm square. The small circular box has a diameter of 79.8 mm giving a sectional area of 50 cm² (author). The sample thickness in the small boxes is 2 to 2.5 cm. The lower half of the box can slide relative to the upper half when pushed by a hand operated or motorised drive unit, while a yoke supporting a load hanger provides the normal pressure. The principle of the test is shown in Fig. 9.7.

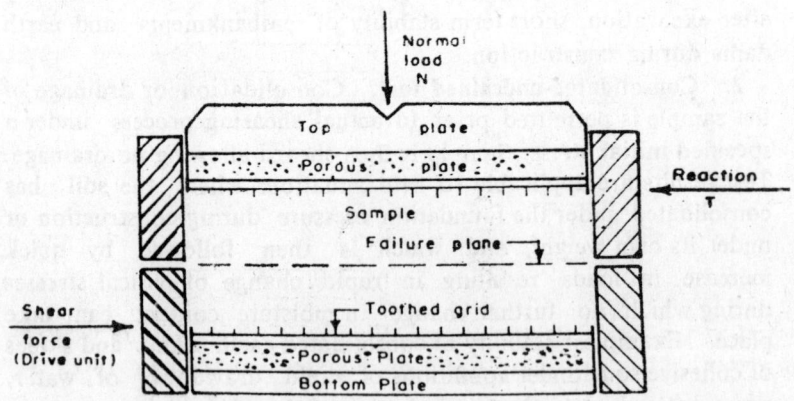

Fig. 9.7 Principle of shear box test

After applying a normal load, the shearing stress is gradually applied by causing the two halves of the box to move relative to each other. The shear force T together with the corresponding shear displacement ΔL are both measured so that a load/displacement curve can be drawn. Normally the vertical movement of the top surface of the sample, which indicates the change in volume, is also measured, and it enables changes in density and void ratio during shear to be evaluated. The test is repeated with different normal stresses on number of identical samples. The value of shear stress at failure is plotted against the normal stress for each test. The shear strength parameters are then obtained from the best line fitting the plotted points.

For performing a UU test, plain toothed grids (without perforations) are used at the top and bottom faces of the samples. Shear force is applied immediately after applying the normal load. For a CU test, perforated grids are used and the sample is first allowed to fully consolidate under the normal load; it is then sheared quickly. For a CD test, the sample is first consolidated under the normal load and then sheared very slowly so that further drainage can take place during shear. As no excess pore pressure is allowed to develop in a CD test, the total, applied normal and shear stresses represent the corresponding effective stresses also.

The main limitations and the advantages of the shear box test are listed below.

1. Limitations. (1) It is difficult to control the drainage conditions, i.e. it would be very difficult to prevent the escape of pore water from a highly permeable soil if an undrained test is required.

(2) As pore water pressure cannot be measured, only the total stresses are known; they are equal to effective stresses only if the pore pressure is zero.

(3) The shear failure plane is predetermined and may not necessarily be the weakest one.

(4) Shear stress distribution on the failure plane is non-uniform (stress is maximum at the edges and minimum at centre); failure occurs progressively from the edges towards the centre.

(5) Most of the distortion occurs on a plane of limited but unknown, thickness. The actual stress pattern is complex and

the directions of principal planes rotate as the shear displacement increases.

(6) The contact area between the soil in the two halves of the box decreases as the test proceeds. (It affects the shear stress and normal stress in equal proportions, and the effect on the failure envelope is usually negligible, Head 1982).

2. Advantages. (1) The test is simple and relatively fast especially for granular materials.

(2) The basic principle is easily understood which can be extended to gravelly soils and other materials containing large particles, which would be more expensive to test by other means.

(3) Quick drainage of pore water in sample is usually easy to achieve because of the thin depth of the sample.

(4) Friction between rocks and angle of friction between soils and some other engineering materials can be easily measured.

(5) Recompacted test samples are not difficult to be prepared.

Examples 9.1 — 9.3

9.1 The following results are obtained from a series of drained shear box test on 36 cm² samples of silty clay. Plot the strength envelope and find the shear strength parameters c' and ϕ'. What is the value of 'attraction' (Janbu)?

Normal stress (kPa)	20	40	60
Maximum shear stress (kPa)	19.5	28.6	38.0

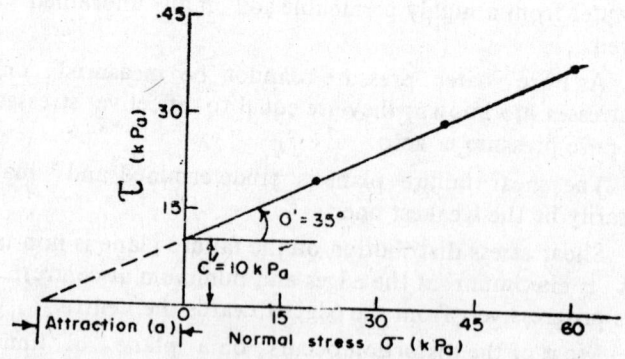

Fig. 9.8 Example 9.1

From the strength envelope of Fig. 9.8,

$c' = 10$ kPa, $\phi' = 25°$, a = 21.4 kPa

9.2 A direct shear test is run on a medium sand under the normal stress of 60 kPa. The maximum shear stress at failure is measured as 37.5 kPa. Draw the Mohr circle at failure and determine the magnitude and direction of the principal stresses in the failure zone. What is the orientation of the plane of maximum shear stress at failure?

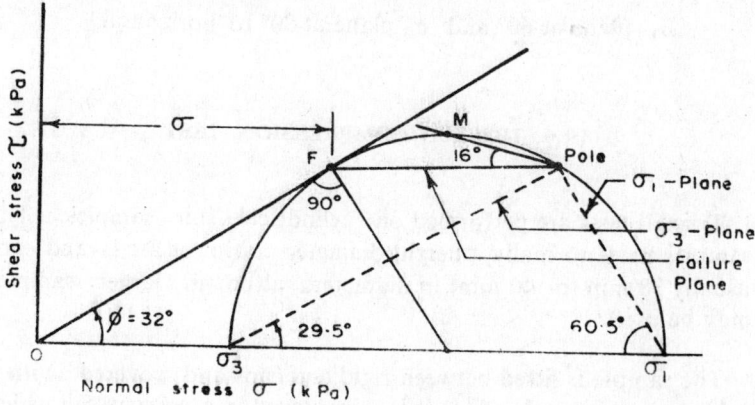

Fig. 9.9 Example 9.2

Plot the failure point F (60, 37.5) and failure envelope through F and O. Draw the Mohr circle by trial with centre at σ axis and tangential to the failure envelope at F. Draw a horizontal line from F to get the pole. The principal planes are then drawn.

$\sigma_1' = 127.6$ kPa, σ_1 plane is inclined at 60.5° to horizontal

$\sigma_3' = 39.2$ kPa, σ_3 plane is inclined at 29.5° to horizontal

M is the point of maximum shear stress and its plane is inclined at about 16° to horizontal.

Analytically $\tan\phi' = s/\sigma = 37.5/60 = 0.625$, $\phi' = 32°$

$\sigma_1' = \sigma' / (1 - \sin \phi') = 127.6$ kPa

$\sigma_3' = \sigma' / (1 + \sin \phi') = 39.2$ kPa

Note. In a shear box the potential failure plane (horizontal plane) is a principal plane only initially (no shear stress), but as the shearing stress is applied and at failure, by definition, it cannot be a principal plane. Hence, a rotation of principal planes must occur in the test. The rotation depends on the slope of the failure envelope.

9.3 Sand sample in shear box test: $\sigma' = 4$ kg/cm², s $= 2.31$ kg/cm². Find ϕ', σ_1', σ_3' and orientation of principal planes.

$$\phi' = 30°, \sigma_1' = 8 \text{ kg/cm}^2, \sigma_3' = 2.67 \text{ kg/cm}^2$$

σ_1 plane at 60° and σ_3 plane at 30° to horizontal

9.6 TRIAXIAL COMPRESSION TEST

Triaxial tests are performed on cylindrical soil samples. The samples have normally a height diameter ratio of 2 : 1 and are usually 38 mm (or 40 mm) in diameter, although larger samples may be used.

The sample is fitted between rigid end caps and covered with a rubber (latex) membrane. It is then placed in a perspex cell which is filled with water (Fig. 9.10). The sample is subjected to a confining (all around) pressure σ_3 by applying pressure to the water in the cell. An additional axial stress σ_d ($= \sigma_1 - \sigma_3$) is then applied by loading the sample through a ram bearing on the top end cap. The ram is forced down at a constant rate, recording the load on the plunger, until the sample fails, The test is repeated on other samples under other confining pressures and the results are interpreted by plotting Mohr circles for the stress conditions of each sample at the time of failure. Recognizing that the horizontal and vertical planes are the principal planes, the minor principal stress σ_3 equals the confining pressure (pressure in cell water) and the major principal stress σ_1 is the additional axial stress plus the confining pressure (water pressure acts on the top of sample as well). Mohr circles are drawn for different tests and a common tangent to them is the failure envelope which gives the shear strength parameters c and ϕ.

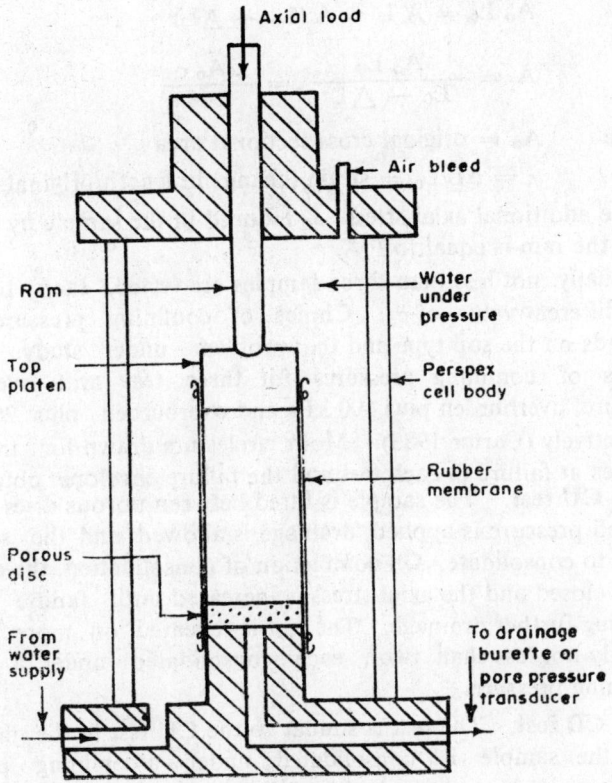

Fig. 9.10 Triaxial test cell

1. UU test. The sample is fitted between solid end caps, so that no change in moisture content is possible. The confining cell pressure is raised to the required value. Without allowing the sample to consolidate (drain) under the confining pressure, the axial load is applied without allowing any drainage of sample. The axial load is usually applied at a rate of 2% strain per minute and is increased until either a definite peak has been passed or an axial strain of 20% has been reached.

The cross-sectional area 'A' of the sample at any strain is calculated assuming the sample deforms as a right cylinder and the volume remains unchanged.

$$A_o L_o = A L = A (L_o - \triangle L)$$

or
$$A = \frac{A_o L_o}{L_o - \triangle L} = \frac{A_o}{1 - \varepsilon} \qquad (9.30)$$

where A_o = original cross sectional area

$\varepsilon = \triangle L / L_o$ = strain (change in length/original length)

The additional axial stress σ_d induced in the sample by the load P on the ram is equal to P/A.

Usually, not less than three samples are tested, each test being at a different value of σ_3. Choice of confining pressures used depends on the soil type and the problem under study. Typical values of confining pressures for three test are: overburden pressure, overburden plus 100 kPa and overburden plus 200 kPa, respectively (Carter 1983). Mohr circles are drawn for the total stresses at failure in each test and the failure envelope obtained.

2. **CU test.** The sample is fitted between porous discs. When the cell pressure is applied, drainage is allowed and the specimen is left to consolidate. On completion of consolidation, the drainage tap is closed and the axial stress is increased until failure without allowing further drainage. The test is repeated on more samples (usually not less than two), each consolidated under a different confining pressure.

3. **CD test.** The test is similar to the CU test in the first stage when the sample is consolidated under a confining pressure. Drainage is also allowed when the axial load is applied and the rate of shear is kept low enough to allow excess pore pressures to dissipate. This permits effective stress parameters to be determined.

The corrected area 'A' at failure for stress calculation is given by:

$$A = \frac{V_o \mp \Delta V}{L_o - \Delta L} \qquad (9.31)$$

or
$$A = A_o \frac{1 \mp \Delta V/V_o}{1 - \Delta L/L_o} \qquad (9.32)$$

where $\Delta V/V_o$ = volumetric strain

4. **Pore pressure measurement.** In CU and UU test at failure only the total stresses are known. The effective stresses may be

determined by measuring the pore pressure developed within the sample and deducting it from the total stresses. The sample is set up between porous discs which are connected to an apparatus for measuring pore pressure.

5. Advantages and limitations.

(1) The triaxial test is much more complicated than the shear box but also much more versatile. Drainage conditions during test can be controlled and a variety of test condition is possible. By measuring pore pressure, effective stresses can also determined in the undrained tests. (2) The sample is not constrained to induce failure on a predetermined surface. Failure plane can occur anywhere and the test may reveal a surface of weakness relating to some natural feature of the soil structure. (3) The applied stresses represent a closer approximation to those which occur in situ than are the conditions imposed in a shear box test. (4) The applied stresses are principal stresses and close control is possible on stresses and strains. The stress paths to failure are under reasonable control and the field stress paths can be modelled in the laboratory. (5) Tests on small diameter samples (38 mm) give unrealistically high strength of stiff fissured clays. To obtain realistic results large diameter samples should be tested. Similarly, soils containing gravel-size particles require large diameter samples. (6) Samples of cohesionless soil such as sands can be difficult to prepare and are perhaps more conveniently tested in a shear box. (7) Consolidation and drainage of cohesive samples in a triaxial cell take a much longer time compared to that in a shear box.

9.7 PORE PRESSURE PARAMETERS

A knowledge of pore pressure is required if effective stress parameters (c', ϕ') are to be used in strength and deformation problems of soil. If a drained condition is applicable then no excess pore pressures are generated. The changes in pore pressure due to change in total stresses under undrained conditions can be expressed in terms of 'pore pressure parameters'.

Consider a cylindrical soil element as in triaxial compression test. Let the minor principal (or radial) total stress increase by $\Delta\sigma_3$ and let the major vertical total stress increase by $\Delta\sigma_1$. These stress increments will generate an excess pore pressure Δu, greater than the existing pore pressure. The total pore pressure Δu may be split in two components: $\Delta u_b =$ pressure generated due to an isotropic change of stress $\triangle \sigma_3$ and $\triangle u_a =$ pressure generated due to a uniaxial change of stress $(\triangle \sigma_1 - \triangle \sigma_3)$, i.e. due to change in deviater stress. Thus by the principle of superposition,

$$\Delta u = \Delta u_b + \Delta u_a \qquad (9.33)$$

Assuming that pore pressure changes are simple functions of stress changes,

$$\Delta u_b = B \Delta \sigma_3 \text{ and } \Delta u_a = \overline{A} (\triangle \sigma_1 - \triangle \sigma_3) \qquad (9.34)$$

where B and $\overline{A}$ = empirical pore pressure parameters determined experimentally.

The total pore pressure change Δu can thus be expressed as follows (Simons and Menzies (1977):

$$\triangle u = B \triangle \sigma_3 + \overline{A} (\triangle \sigma_1 - \triangle \sigma_3) \qquad (9.35)$$

Skempton (1954) derived and expressed the pore pressure equation as:

$$\Delta u = B [\triangle \sigma_3 + A (\triangle \sigma_1 - \triangle \sigma_3)] \qquad (9.36)$$

i.e.

$$\overline{A} = A \times B \qquad (9.37)$$

The pore pressure parameters 'B' and 'A' may be measured in a triaxial compression test. To measure B, a sample is set up under any value of cell pressure and the pore pressure is measured. Under undrained conditions the confining pressure is then increased by an amount $\triangle \sigma_3$ and the *change* in pore pressure $\triangle u_b$ measured, enabling the calculation of B from Eq. 9.34. To further measure A, the sample is then loaded axially again under undrained condition and the pore pressure increase $\triangle u_a$ measured.

For practical purposes, B = 1 for saturated soils (and $\overline{A}$ = A), and B = 0 for dry soils. Like parameter B, the parameter A also is not constant. It depends on factors such as strain, initial stress system (isotropic or anisotropic), stress history (over consolidation ratio) and type of stress change (loading or unloading). Typical values of A may range from greater than one for soils whose structure collapses on loading (highly sensitive clays) to less than zero for soils which dilate on loading (heavily overcondolidated clays). The value of A at failure is called A_f, Typical values of A_f are (Skempton 1954): highly sensitive clays ($+$ 0.75 to $+$ 1.5), normally consolidated clays ($+0.5$ to $+1$), compacted sandy clays ($+0.25$ to $+0.75$), lightly over consolidated clays (0 to $+0.5$), compacted clay-gravels (-0.25 to 0.25), and heavily over consolidated clays (-0.5 to 0).

9.8 UNCONFINED COMPRESSION TEST

This test imposes uniaxial compression conditions on a soil sample and is thus a special case of the UU triaxial test with the confining or cell pressure equal to zero (atmospheric pressure). It is used for determining the undrained shear strength s_u of saturated cohesive soils for which $\phi_T = 0$ and the failure envelope is horizontal. As $\sigma_3 = 0$, the Mohr circle passes through the origin of stress ($0, 0$ point) and c_T is one half the axial stress (σ_1) at failure. The axial stress at failure in an unconfined compression test is also called the 'unconfined compressive strength q_u' of the soil. Thus,

$$s_u = c_T = \frac{\sigma_1}{2} = \frac{q_u}{2} \qquad (9.38)$$

The test is performed on a cylindrical sample with a height: diameter ratio of 2:1. The usual size is 38 mm diameter, although larger samples of 50, 75 and 100 mm diameters are also used. Without any covering or lateral support, the sample is placed between the platens of a mechanical load frame. Axial load is applied to give a rate of strain between 1 to 2% of sample height per minute; (the rate of deformation is about 1.5 mm/min for a 38 mm diameter sample). Load and deformation readings are taken and loading continued until either three or more consecutive readings of the load dial gauge show a decreasing or a constant load, or a strain of 20% (15 mm compression of a 38 mm diameter sample) has been reached. Typical time to failure is 5 to 15 minutes. The axial stress is calculated on the basis of deformed cross-sectional area as given by Eq. 9.30. The peak axial stress or that corresponding to 20% strain is termed the 'unconfined compressive strength q_u', half of which is termed the 'undrained shear strength s_u'. A Mohr circle may be plotted as shown in Fig. 9.11.

While testing an undisturbed sample, if the remoulded strength is also to be measured for the determination of sensitivity (Section 9.11), the failed sample is enclosed in a small polythene bag, together with a little more of the soil at the same moisture content, and remoulded thoroughly by squeezing and kneading with fingers. A test sample is formed by working the remoulded soil into a 38 mm diameter tube or split mould. It is then tested to get the remoulded strength.

Advantages and limitations. (1) The unconfined compression test is a quick, simple and widely used test. It has a significant cost advantage over a triaxial test due to simpler testing requirements. (2) The test is convenient and quite suitable for studying sensitivity of clays. (3) It is only applicable to soils which can stand unsupported and are fairly impervious so that undrained conditions effectively exist on the test duration. (4) The sample must be fully saturated so that '$\phi_T = 0$' condition may be assumed. In a partly saturated sample, there will be compression of air voids causing a decrease in void ratio and an increase in strength. (5) The sample must be free from fissures, silt seams, varves or other defects, this means that the test is suitable for intact homogeneous clays only. Fissures manifest in low strength values. (6) The test under-estimates in-situ strength because sampling inevitably causes a remoulding disturbance and decrease in strength. The test should not be used for the estimation of undrained modulus.

Example 9.4

An unconfined compression test is conducted on a clay sample 38 mm in diameter and 80 mm high. The load at failure is 4.25 kg (41.65 N) and the axial deformation is 12 mm. Calculate the

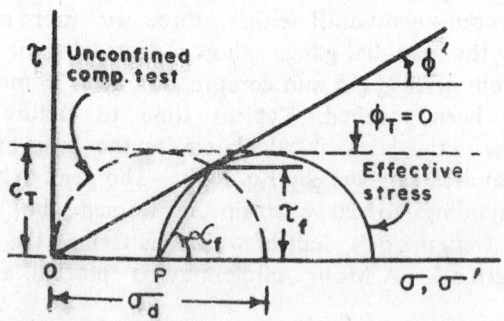

Fig. 9.11   Example 9.4

unconfined compressive strength q_u and the undrained shear strength s_u. What is the error in assuming $s_u = c_T = \sigma_d/2$, if actual $\phi' = 30°$?

$$A_o = 1134 \text{ mm}^2, \epsilon = \triangle L/L_o = 0.15, A = 13.34 \text{ cm}^2$$

$$q_u = 4.25/13.34 = 0.32 \text{ kg/cm}^2 = 31.22 \text{ kPa}$$

$$s_u = c_T = q_u/2 = 15.6 \text{ kPa}$$

If $\phi' = 30°$, $\alpha_f = 45 + \phi'/2 = 60°$

$$\tau_f = \frac{\sigma_1 - \sigma_3}{2} \sin 2\alpha_f = \frac{\sigma_d}{2} \sin 120° = 0.433 \sigma_d$$

$$s_u = c_T = 0.5 \sigma_d$$

$$\text{Error} = \frac{0.5 - 0.433}{0.433} \approx 15\%$$

The undrained strength ($s_u = c_T$) is about 15% greater than the actual shear stress on the failure plane at failure. The error decreases for smaller ϕ'. In actual practice, sample disturbance, anisotropy and the assumption of plane strain conditions for design purposes, tend to reduce the undrained shear strength and induce compensating errors so that the difference between s_u and τ_f becomes negligible (Holtz and Kovacs 1981, Ladd et al 1977).

9.9 VANE SHEAR TEST

The vane shear test (IS:4434) is used to determine the insitu undrained shear strength s_u of clay and is particularly suitable for use in soft or sensitive clays which are difficult to sample or whose properties are significantly affected by normal sampling methods.

The apparatus consists of a four-bladed vane on the end of a rod, as shown in Fig. 9.12. The height of the vane is usually twice its width (diameter), typical dimensions are 150 mm high by 75 mm wide for soft clays and 100 mm high by 50 mm wide for firm clays. A boring is made to the depth at which the test is to be performed. The vane is inserted into the soil at the bottom of the hole and slowly rotated (0.1° per second), using a special instrument to measure the torque.

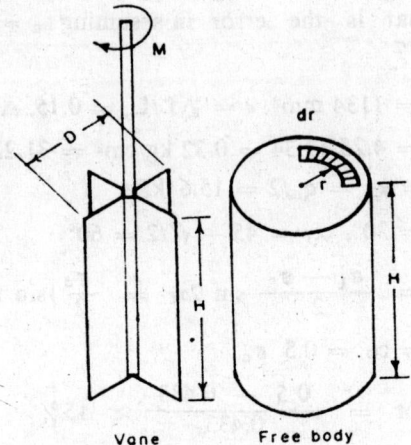

Vane   Free body

Fig. 9.12 Vane shear test

Assuming s_u is constant over the cylinder of soil sheared by the vane, the torque M required to shear the soil is calculated as:

$$M = (\pi D H \times 0.5 D s_u) + 2 \int_{r=0}^{r=0.5D} (2\pi r \, dr \times r \, s_u)$$

$$= \left(\frac{\pi D^2 H}{2} + \frac{\pi D^3}{6} \right) s_u \qquad (9.39)$$

or $s_u = M/\pi D^2 (H/2 + D/6)$ \qquad (9.40)

where   H = height of vane

D = diameter of vane

After measuring the maximum torque in the undisturbed state, the remoulded strength may also be determined. The vane is rotated rapidly through a few revolutions so as to remould the soil in the sheared zone. Without further delay, the vane is rotated (0.1°/s) to get the remoulded strength. Using a smaller vane (24 mm × 12 mm), the test is also performed in the laboratory (IS:2750-30) to find the undisturbed and remoulded strength of soft clays. The sample can be tested in the sample tube itself by clamping the sample tube under the vane of the laboratory model.

Example 9.5

A vane, 75mm in diameter and 150 mm in height, was pressed into soft clay in a borehole. The torque was applied and gradually increased to 50 Nm when failure took place. Determine the undrained shear strength.

$$s_u = M/\pi D^2 \left(\frac{H}{2} + \frac{D}{6} \right)$$

Using the metre as the unit of length,

$$s_u = \frac{50}{\pi (75)^2 \left[\frac{150}{2} + \frac{75}{6} \right] \times 10^{-9}} = 3.23 \times 10^4 \text{ N/m}^2$$

$$= 32.3 \text{ kPa}$$

9.10 SHEAR STRENGTH OF COHESIONLESS SOILS

The shear strength of granular materials (sands and gravels) can be thought to be made up, in general, of two components: frictional resistance due to relative sliding and rolling of particles and the structural resistance due to particle interlocking. The interlocking structural resistance is of special significance in dense granular soils where a shear displacement is possible only if the particles do not merely slip along the plane of shear but also move upwards.

The behaviour of cohesionless soils during shear can be typically illustrated by performing consolidated drained (CD) shear box tests, or preferably, triaxial tests with volume change measurements on saturated 'loose' and 'dense' sands under equal vertical pressures in the shear box or equal confining pressures in the triaxial cell. As shown in Fig. 9.13, there is a definite peak in the stress-strain curve for the dense sample, followed by a decrease in the deviator stress with increasing strain until it levels out to a 'residual' or 'ultimate' value. Failure may be defined either by the 'peak stress' point A or the 'ultimate stress' point B. These two failure criteria would produce failure envelopes similar to those shown in Fig 9.13c. The choice of failure criteria depends on the field conditions that are to be simulated. However, the peak stress criterion is used for most applications. Loose sand does not exhibit a peak stress point and failure condition is generally defined by the stress condition at the maximum deviator stress or by the stress condition at a given strain, such as 15 or 20 percent.

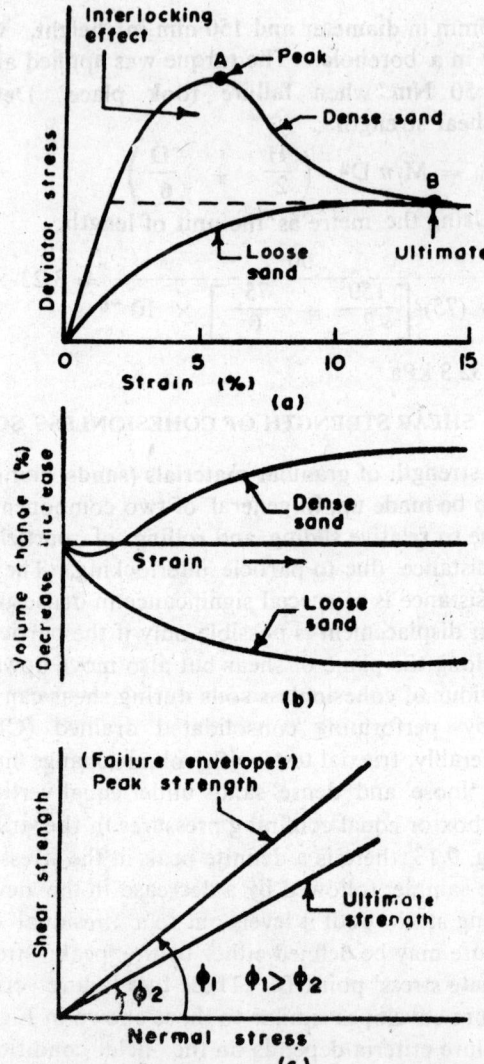

Fig. 9.13 Stress strain and volume change behaviour of dense and loose sands during shear and their failure envelopes

The shear resistance of loose sand is due essentially to the friction between the particles. Dense sand has, in addition to the frictional component, an interlocking component as a result of dense packing of particles. At strains exceeding the peak point, the interlocking effect is overcome and the curve for the dense sample at high strains tends to the same level as the loose sample.

1. Critical void ratio. As shown in Fig. 9.13, the particles in a loose sample tend to seek a more compact arrangement on application of the shearing stress and the volume tends to decrease. A dense sample decreases in volume slightly at first, then expands or "dilates". This is because the particles must raise out of their positions (or fracture) to pass by one another, thus leading to an increase in volume. In-between the loose and dense states there must lie a density at which the sample changes very little in volume during the shearing process. The density at which no change in volume is brought about upon the application of shear is called the "critical density". The void ratio of a sand at its critical density is called the "critical void ratio". The ultimate values of the void ratio for the loose and dense sands (under equal confining pressures or vertical stresses) tend to be the same and approach the critical value.

The critical void ratio depends chiefly on the uniformity of the material. The more uniform the sand, the higher is its critical void ratio. It also depends on the confining pressure (or the vertical stress in a shear box test). The greater the confining pressure, the lower is the critical void ratio. In other words, a higher pressure produces a stronger tendency for volume contraction in response to shear strain. Hence, the definition of loose and dense cohesionless soils depends on both the confining pressure and the void ratio or the density index. A large enough confining pressure will cause a sand even with $I_D = 70\%$ to behave as if it were loose. For a given confining pressure in a triaxial test (or the vertical stress in a shear box), the critical void ratio can be determined by testing a series of samples of varying initial void ratio. Similarly for a given void ratio, a confining pressure, called the "critical confining pressure" may be obtained. It is the pressure at which zero volumetric strain occurs at failure for a given void ratio.

2. Drained and undrained strengths. Most of the granular materials being highly pervious, only the drained strength is normally relevant in practice and it can be determined either by drained triaxial tests or shear box tests on saturated samples. Dry samples can also be tested. The characteristics of dry and saturated states are the same provided all stress conditions are given in terms of effective stresses and there is no excess pore pressure in the case of saturated state.

During undrained shear, saturated, loose sample tends to decrease in volume and develop a positive pore pressure, which causes a reduction in the effective stress. The total stress failure envelope will thus lie to the right of the effective stress envelope, i e. $\phi_T < \phi'$. Reverse is the behaviour in case of dense sample under low and moderate confining pressures when the sample *dilates* (expands) developing negative pore pressure. The total stress envelope lies to the left of the effective stress envelope, i.e. $\phi_T > \phi'$. When the confining pressure becomes sufficiently large, however, the dense sample behaves as if it were loose.

The behaviour of partly saturated cohesionless soils depends on the amount of water present. Small amounts of moisture can introduce an apparent cohesion due to capillary action, and this should be neglected for design purposes.

3. Factors influencing ϕ. The angle of shearing resistance of cohesionless soils is influenced by factors such as void ratio or density index, particle size, angularity and roughness, particle size distribution, water, and stress state. Void ratio, related to the density of soil, is perhaps the most important single factor influencing the shear strength. For a given granular soil, the lower the void ratio (or higher the dry density), the higher are the shear strength and the angle of shearing resistance. Particle size, at constant void ratio, does not seem to influence ϕ significantly (Holtz and Kovacs 1981). Greater the angularity and roughness, the greater will be ϕ. A better graded soil has also a larger ϕ. Wetting (saturation) is found to lower ϕ' by 1° or 2° only compared to the value in the dry state of a sand. For a well graded material the angle of shearing resistance is related to density index approximately as follows (Zeevaert 1983):

$$\phi = 20° I_D + 26° \tag{9.41}$$

Some of the field deformations correspond to "plane strain" conditions, i.e. particle movements are two dimensional restricted in one plane and no strains can occur in direction normal to that plane. This occurs whenever structures have one plan dimension very much larger than the other, e.g. strip footings, retaining walls and long earth dams. It is observed typically that ϕ in plane strain is larger than ϕ in triaxial shear by 4° to 9° for dense sands and 2° to 4° for loose sands (Ladd et al 1977). The following equations can be used to give a conservative estimate of ϕ_p (plane strain) from the triaxial value of ϕ (Lade and Lee 1976):

$$\phi_p = 1.5 \phi - 17° \quad (\phi > 34°) \tag{9.42a}$$

$$\phi_p = \phi \quad (\phi \leqslant 34°) \tag{9.42b}$$

Pre-stressing or over consolidation is not found to significantly affect ϕ, but it strongly affects the deformation modulus of granular materials.

4. Angle of repose. This is the angle with the horizontal of the steepest stable slope for a given dry granular material. The angle of repose represents the angle of internal friction of the granular material at its loosest state.

5. Liquefaction. Liquefaction is the sudden large decrease of the shearing resistance of a saturated cohesionless soil. It is caused by a collapse of the structure by shock or other type of strain (cylic or dynamic loading) and is associated with a sudden but temporary build-up of pore pressure. A saturated sand with a void ratio higher than the critical void ratio, i.e. a loose sand, is liable to liquefaction when it is shocked by earth tremor or by sudden loading or more water flowing into it. The sand becomes temporarily 'quick' and a flow slide is caused. This is a phenomenon of fine sands and silty sands. They have a rather low permeability and the excess, possitive pore pressure generated under rapid or cyclic loading conditions cannot dissipate, which can lead to a complete loss of strength. Liquefaction is not observed in gravels and is difficult to develop in fine silty sands or medium to coarse sands. A similar phenomenon is caused by the seepage pressure of water percolating through the soil, and is called the "quick sand condition" (Chapter 7).

Examples 9.6 — 9.9

9.6 A CD triaxial test is performed on a sand sample with a cell pressure of 80 kPa. The sample fails at a deviator stress of 160 kPa. Plot the Mohr circle at failure and determine ϕ', assuming $c' = 0$. What is the inclination of the failure plane? Determine also (a) shear stress (shear strength) τ_f on the failure plane, (b) maximum obliquity and orientation of the plane of maximum obliquity, (c) maximum shear stress τ_{max} at failure and the angle of the plane on which τ_{max} acts, and (d) factor of safety on the plane on which τ_{max} acts.

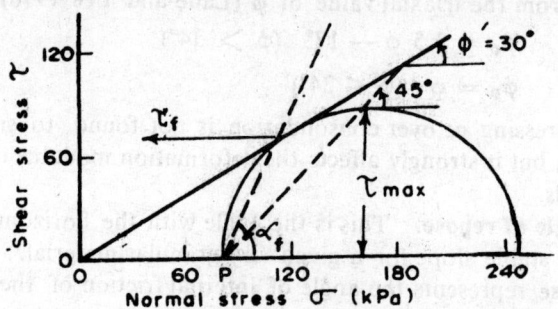

Fig. 9.14 Example 9.6

(a) $\sigma_1 = \sigma_d + \sigma_3 = 240$ kPa

From Mohr cricle (Fig. 9.14), $\phi' = 30°$

Also analytically, $\phi' = \arcsin \dfrac{\sigma_1 - \sigma_3}{\sigma_1 + \sigma_3} = 30°$

$$\alpha_f = 45 + \frac{\phi'}{2} = 60°$$

$$\sigma_f = \frac{\sigma_1 + \sigma_3}{2} + \frac{\sigma_1 - \sigma_3}{2} \cos 2\alpha_f$$

$$= \frac{\sigma_1 + \sigma_3}{2} - \frac{\sigma_1 - \sigma_3}{2} \sin \phi' = 120 \text{ kPa}$$

$$\tau_f = \sigma_f \tan \phi' = 69.28 \text{ kPa}$$

For a 'c = 0' soil, σ_f can also be found from Eq. 9.17 on 9.18, and τ_f can be found from Eq. 9.16.

(b) Maximum obliquity $= \dfrac{\tau_f}{\sigma_f} = \dfrac{69\,28}{120} = 0.58$ at

$$\alpha_f = 60°$$

(c) $\tau_{max} = \dfrac{\sigma_1 - \sigma_3}{2} = 80$ kPa on a plane inclined at 45° to the horizontal.

Obliquity at this orientation $= \dfrac{\tau_{max}}{\sigma} = \dfrac{80}{160} = 0.5$

Available τ_a at 45° plane $= \sigma \tan \phi' = \dfrac{\sigma_1 + \sigma_3}{2} \tan \phi'$

$= 160 \tan 30° = 92.38 \text{ kPa} > \tau_{max}$

(d) Safety factor on 45° plane $= \tau_a/\tau_{max} = 1\,16$

(Safety factor on 60° plane $= \tau_a/\tau_f = 69.28/69.28 = 1$)

9.7 A sand sample tested in a CD triaxial test with confining pressure σ_3 of 100 kPa fails when the maximum principal stress ratio σ_1'/σ_3' reaches 3.85. Determine ϕ' and the deviator stress σ_d at failure.

In the CD test $\sigma_3' = $ cell pressure

$$\frac{\sigma_1'}{\sigma_3'} = \frac{1 + \sin \phi'}{1 - \sin \phi'} = \tan^2 \left(45 + \frac{\phi'}{2}\right) = 3.85$$

$\varphi' = 36°$

$$\sigma_d = \sigma_1' - \sigma_3' = \sigma_3 \left(\frac{\sigma_1}{\sigma_3} - 1\right) = 285 \text{ kPa}$$

Knowing σ_1' and σ_3', ϕ' can also be detemined graphically from the Mohr circle at failure.

9.8 Same sand as in Ex. 9.7. A CU triaxial test with pore pressure is performed. Total cell pressure $\sigma_3 = 100$ kPa, pore pressure at failure $u_f = 40$ kPa, σ_1'/σ_3' at failure $= 3.85$. Determine σ_1' and σ_d at failure and ϕ_T in terms of total stress.

In the first part of the test the sample is consolidated under $\sigma_3 = 100$ kPa as in Ex. 9.7. It will have the same void ratio as in Ex. 9.7, hence ϕ' may be assumed the same as in Ex. 9.7; i.e. $\phi' = 36°$.

$$(\sigma_3')_f = \sigma_3 - u_f = 60 \text{ kPa}$$

As shown in Fig. 9.15, draw effective stress failure envelope with $\varphi' = 36°$. By trial, draw the Mohr circle with $\sigma_3' = 60$ kPa. Measure diameter of circle: $\sigma_d = 171$ kPa and $\sigma_1' = 231$ kPa.

The failure circle in terms of total stress will have the same diameter $\sigma_d = \sigma_1 - \sigma_3 = \sigma_1' - \sigma_3'$. Draw the circle with $\sigma_3 = 100$ kPa and then the failure envelope. $\phi_T = 27.4°$.

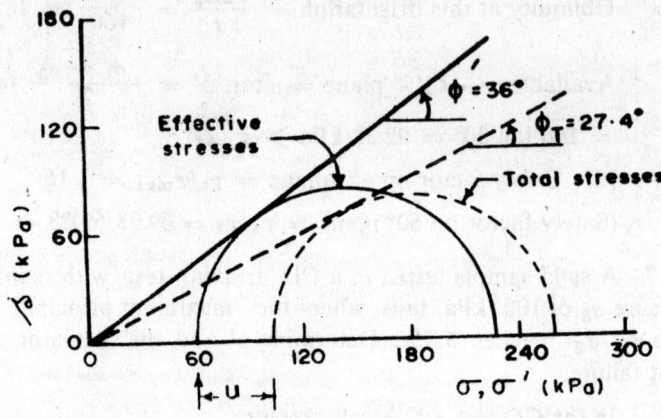

Fig. 9.15 Example 9.8

Analytically: $\sigma_d = \sigma_3' \left(\dfrac{\sigma_1'}{\sigma_3'} - 1\right) = 171$ kPa

$\sigma_1' = \sigma_d + \sigma_3' = 231$ kPa, $\sigma_1 = \sigma_1' + u_t = 271$ kPa

$\sin \varphi_T = \dfrac{\sigma_1 - \sigma_3}{\sigma_1 + \sigma_3}$, $\phi_T = 27.4°$

Alternately, use $\sigma_1/\sigma_3 = \tan^2 \left(45 + \dfrac{\phi_T}{2}\right)$

9.9 A sand sample when tested in a CU triaxial test under cell pressure $\sigma_3 = 100$ kPa and with pore pressure measurement showed the following results: $(\sigma_d)_f = 171$ kPa, $u_t = 40$ kPa, $\phi_T = 27.4°$, $\phi' = 36°$. (Same sand as in Ex. 9.8). Find u_t, if the sample were tested with cell pressure of 200 kPa. Also find σ_d at failure.

Draw total stress and effective stress failure envelopes with $\phi_T = 27.4°$ and $\phi' = 36°$ respectively. With $\sigma_3 = 200$ kPa draw by trial a total stress circle tangent to ϕ_T envelope. Measure $\sigma_d = 342$ kPa. Shift circle to left to be tangent to the ϕ' envelope. Measure $\sigma_3' = 120$ kPa. $u_t = 80$ kPa.

Analytically: From first test, $\sigma_1 = 100 + 171 = 271$ kPa and $\sigma_1/\sigma_3 = 2.71$.

In the second test, $\sigma_1 = 2.71 \sigma_3 = 2.71 \times 200 = 542$ kPa $\sigma_d = \sigma_1 - \sigma_3 = 542 - 200 = 342$ kPa

From first test, $\sigma_3' = 60$ kPa, $\sigma_1' = 231$ kPa, $\sigma_1'/\sigma_3' = 3.85$

$$\sigma_3' = \sigma_d / (\frac{\sigma_1'}{\sigma_3'} - 1) = 120$$

$$u_t = 200 - 120 = 80 \text{ kPa}$$

Also check: $u_t = \sigma_1 - \sigma_1' = 542 - 3.85 \times 120 = 80$ kPa

9.11 SHEAR STRENGTH OF COHESIVE SOILS

1. **Stresss history and soil types.** The stress history of cohesive soils greatly affects their shear strength parameters. Based on stress history soils are classified as 'normally consolidated' and 'overconsolidated' or 'preconsolidated'

(a) *Normally consolidated soil.* It is the soil which, at no time in its history, has been subjected to effective pressures greater than the existing effective overburden pressure, i. e. the present effective pressure is the maximum value to which the soil has ever been subjected.

(b) *Overconsolidated soil.* It is the soil which, during its history, has been subjected to effective pressures greater than its existing effective overburden pressure.

The ratio of the preconsolidation pressure σ_p' (maximum past pressure) divided by the present overburden pressure σ_o' is called the "overconsolidation ratio OCR".

2. **Simulation of consolidation state.** A normally consolidated state will exist in a shear test so long as the confining pressure in a triaxial cell (or the normal pressure in a shear box) is equal to or greater than the maximum past pressure for the soil. If the confining pressure in the triaxial cell (or the normal pressure in the shear box) remains lower than the maximum past pressure, an overconsolidated state exists. Undisturbed samples obtained from naturally consolidated deposits in the field are usually tested under confining pressures (or normal pressures) equal to or greater than the existing overburden pressure in case of normally consolidated soil and under pressures lower than the maximum past pressure in case of overconsolidated soil. When soil is sampled from the field, stresses are released and reconsolidation under confining pressure equal to overburden

pressure does not reproduce exactly the field consolidated state. Laboratory consolidation under an equal all-round pressure in the triaxial cell represents isotropic consolidation, whereas it is anisotropic in the field. Special test procedures are, however, available for producing anisotropic consolidation in the laboratory also.

Overconsolidated samples can also be prepared in the laboratory by first consolidating the samples under a certain confining pressure in the triaxial cell (or a certain normal pressure in the shear box) and then allowing the samples to swell under a reduced (or zero) pressure. This consolidation pressure will be the preconsolidation pressure σ_p' for the sample.

3. **Normally consolidated saturated clay.** (a) *CD and CU tests.* For performing both these tests in a triaxial cell, a series of samples are consolidated under increasing confining pressures as the first part of the tests. The samples are then failed with full drainage (zero pore pressure) in a CD test and with no drainage in a CU test by applying and increasing the additionl axial load. The cell pressure σ_3 for a sample in the second part of the test (when deviator stress is applied) is kept the same as was used for its consolidation in the first part. The failure envelope for the CD test corresponds to effective stresses and that for the CU test corresponds to total stresses, as pore pressure develops in the latter undrained part of the test. These two envelopes are shown in Fig. 9.16. If pore pressures are measured in the undrained part of the CU test, effective stresses at failure can be calculated. The effective stress circle for a normally consolidated case is shifted to the left, towards the origin, because the sample develops positive pore pressure ($A_f > 0$) during shear and $\sigma' = \sigma - u$. Note that both the total stress and effective stress circles have the *same diameters* because of the definition of failure at maximum $(\sigma_1 - \sigma_3) = (\sigma_1' - \sigma_3')$. The effective stress CU envelope coincides with the CD envelope, as shown in Fig. 9.6. It is experimentally observed that both the total stress and effective stress envelopes for CU and CD tests pass (or almost pass) through the origin for normally consolidated clays (also for remoulded clays) giving zero cohesion intercept (c' and $c_T \approx 0$.) The total stress angle ϕ_T is less than the effective stress angle ϕ', and often it is about one-half of ϕ'.

Values of ϕ' range from more than 30^0 for clays of low plasticity index to 15^0 for clays of higher plasticity index (Dunn et al 1980). An empirical correlation between ϕ' and I_D of NC clays is (Kenney 1959):

$$\sin \phi' = 0.814 - 0.234 \log (I_D) \tag{9.43}$$

The equations for the total stress (CU test) and the effective stress envelopes (CD test) for normally consolidated clay are:

$$s = \sigma \tan \phi_T \tag{9.44}$$

$$s = \sigma' \tan \phi' \tag{9.45}$$

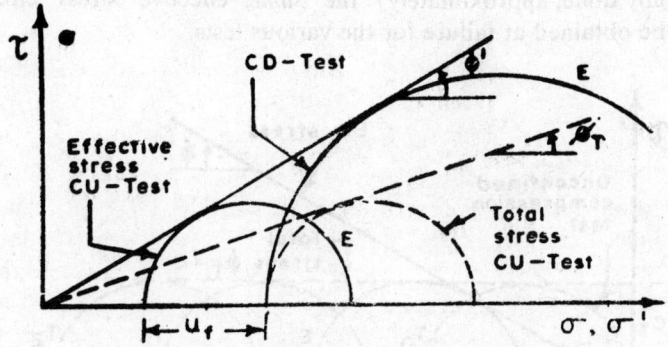

Fig. 9.16 Total and effective stress envelopes for CD and CU tests on normally consolidated clay ($A_f \rightleftharpoons$ positive)

(b) *UU test.* When a sample is taken from insitu, the pore pressure, called the 'residual pressure', is negative, which results from stress release during sampling. If the sample is placed in a triaxial cell and confining pressure applied with the drainage valves closed, a positive pore pressure, equal to confining pressure, is induced in the sample. If the applied confining pressure σ_3 is equal to the pressure to which the sample had been consolidated (overburden pressure σ_0'), the pore pressure within the sample becomes zero. If $\sigma_3 > \sigma_0'$, the pressure will be positive; if $\sigma_3 < \sigma_0'$, the pore pressure will be negative. Any increase in confining pressure is entirely carried by the pore water because the soil is fully saturated ($B = 1$) and thus the effective stresses within the sample remain unchanged. Assuming all samples to be identical, a number of UU tests, each with a different value of confining pressure, therefore, result in equal values of deviator stress at failure, i.e. all the Mohr circles at failure will have the same diameter. The results are plotted in terms of total stresses, as shown in **Fig 9.17.** The failure envelope is horizontal ($\varphi_T = 0$) and represents the undrained strength $s_u = c_u$. In case of fissured clays, the fissures get opened to some extent on sampling resulting in lower strength and the failure envelope is curved at low values of confining pressures. Only when the confining pressure becomes high enough to close the fissures again does the strength become constant. If pore pressures are measured during the tests, although it is not commonly done, approximately the *same* effective stress circle would be obtained at failure for the various tests.

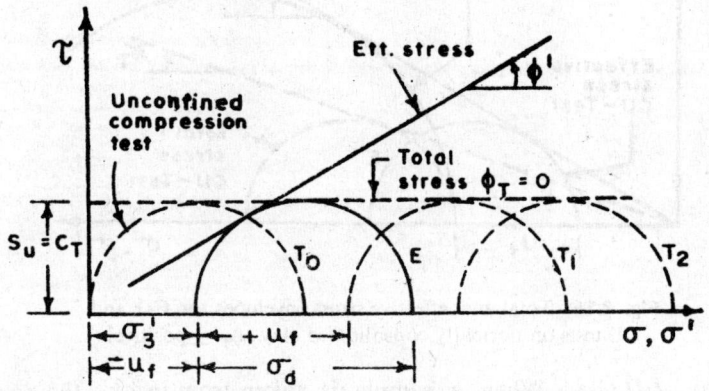

Fig. 9.17 Typical failure envelope for UU tests on normally consolidated clay

The unconfined compression test circle is also shown in Fig. 9.17. It is a special case of UU test without using a confining pressure. Since $\sigma_3 = 0$, the negative pressure (maximum possible→103 kPa) acts as a confining agent and may be approximately equal to the effective overburden pressure.

In practical applications, failure envelope for clay soils will nearly always have a cohesion intercept because: (i) near surface soils may be desiccated and show over consolidation characterisitics (See the following sub-section), (ii) compacted clay appears to be over consolidated, (iii) capillary pressure in clay above the water table causes apparent cohesion, and (iv) for many short term loading conditions, the UU envelope is generally applicable which has a cohesion intercept.

4. Overconsolidated saturated clay. (a) *CU and CD tests.* An overconsolidated clay tends to expand like dense sand during shear and the failure envelope shows the cohesion intercept. In a CU test drainage or intake of water is not permitted during the application of the deviator stress. Due to tendency for expansion of sample, the pore pressure decreases or may even become negative. If the clay is *highly* overconsolidated (OCR $>$ 4 to 8, Terzaghi and Peck 1967) for which $A_f < 0$, the pore pressure would become negative and the effective stress circle would shift to the right of the total stress circle (as shown in Fig. 9.18) and hence the undrained strength will be greater than the drained strength. The shift of the effective stress circle to the right sometimes means that ϕ' is less than ϕ_T. Thus c_T (cu) $>$ c ' but ϕ' may be slightly greater or smaller than ϕ_T.

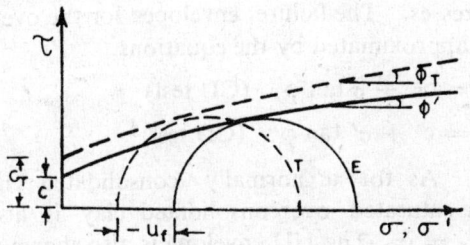

Fig. 9.18 Failure envelopes for overconsolidated clay

It may noted that the failure envelopes for overconsolidated clay are actually non-linear. Typical failure envelopes over a wide range of stresses spanning the preconsolidation stress σ_p' are shown in Fig. 9.19. The 'break' in the total stress envelope (point B) occurs roughly at about $2\sigma_p'$ for typical clays (Hirschfield 1963, Holtz and Kovacs 1981). If the failure envelopes of the normally consolidated range are extended backwards they pass through the origin giving c $= 0$

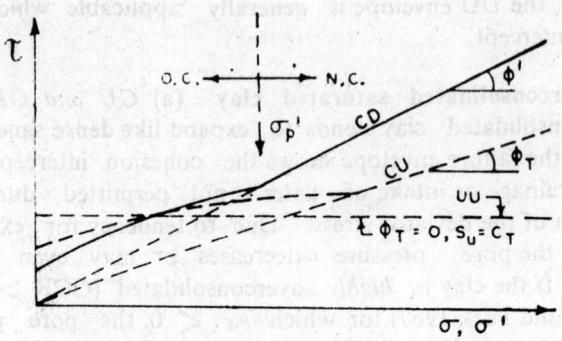

Fig. 9.19 Failure envelopes over a range of stresses spanning the preconsolidation stress σ_p'

The value of c' for an overconsolidated clay does not usually exceed 30 kPa (Craig 1978). According to Ladd (1971), for natural overconsolidated non-cemented clays with a preconsolidation stress of less than 500 to 1000 kPa, c' will probably be less than 5 to 10 kPa at low stresses. The failure envelopes for the overconsolidated range can be approximated by the equations:

$$s = c_T + \sigma \tan \phi_T \quad \text{(CU test)} \tag{9.46}$$

$$s = c' + \sigma' \tan \phi' \quad \text{(CD test)} \tag{9.47}$$

(b) *UU test.* As for a normally consolidated clay, the UU envelope for a saturated overconsolidated clay is also horizontal, $\phi_T = 0$ and $s_u = c_T$. The UU envelope is also shown in Fig. 9.19.

5. **Stress-strain and volume changes.** Typical curves of stress and volume changes versus axial strain for normally consolidated (NC) and highly overconsolidated (OC) remoulded or compacted clays obtained from CD tests are shown in Fig. 9.20. A NC clay behaves like a loose sand, but usually has a greater volume decrease, and exhibits a peak strength. An OC clay expands like dense sand, but with lower modulus and less volume expansion than the sand. There is a general trend towards lower stress-strain modulus and greater volumetric strain with increased soil plasticity. Even though tested at the same effective confining pressure, an OC sample will have a greater strength than a NC sample.

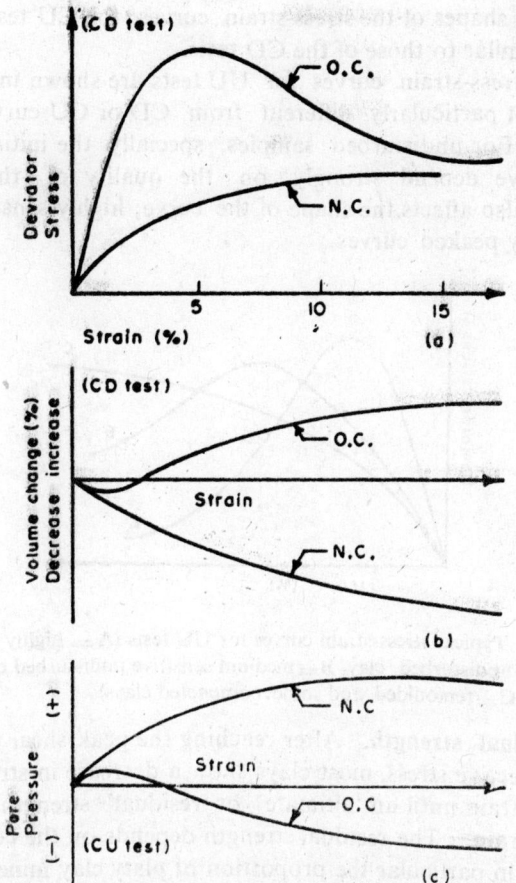

Fig. 9.20 Typical stress, strain, volume and pore pressure
curves for clays — CD and CU tests

During CU tests, the samples after consolidation under the
confining pressure are maintained at a constant volume when the
deviator stress is applied; the volume changes are thus zero.
Because of the tendency of dilation (expansion), a saturated OC
sample will develop negative pore pressure increasing the undrained
strength than the drained strength. A NC sample on other hand
will have positive pore pressure and show a lower strength.
Typical pore pressure changes curves are also shown in Fig. 9.20c.

The general shapes of the stress-strain curves for CU tests remain, however, similar to those of the CD tests.

Typical stress-strain curves for UU tests are shown in Fig. 9.21. They are not particularly different from CD or CU curves for the same soil. For undisturbed samples, specially the initial portions of the curve depend strongly on the quality of the sample. Sensitivity also affects the shape of the curve; highly sensitive clays have sharply peaked curves.

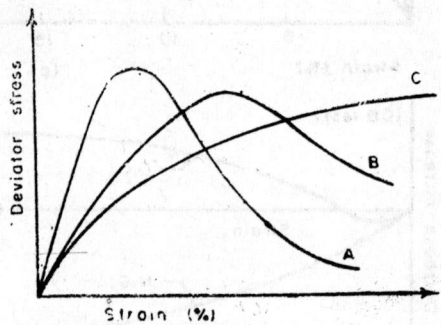

Fig. 9.21   Typical stress-strain curves for UU tests (A = highly sensitive undisturbed clay, B = medium sensitive undisturbed clay, C = remoulded and some compacted clays)

6. Residual strength. After reaching the peak shear strength in terms of effective stress, most clays show a decrease in strength with increasing strain until an 'ultimate' or 'residual' strength is reached at a large strain. The residual strength depends on the composition of the clay, in particular the proportion of platy clay minerals. It is independent of past stress history. The decrease in strength is considered to be due to the increase in moisture content accompanying volume increase during shear and the reorientation of the platy particles parallel to the failure plane. With respect to effective stresses, the cohesion intercept c'_r is zero and ϕ_r' is typically in the range of 9° to 30° (Lee et al 1983). ϕ_r' is invariably less than peak φ' of NC clays (unlike loose sands where peak and residual strengths are equal).

7. Variation of undrained strength with depth. The field undrained strength s_u for both the normally consolidated and the over-consolidated clays is observed to increase with depth. For NC clays,

s_u is almost directly proportional to the effective consolidation pressure σ_0' and the ratio s_u/σ_0' has been empirically related by Skempton to the plasticity index I_p (%) as follows :

$$s_u/\sigma_0' = 0.11 + 0.0037\ I_p \tag{9.48}$$

Near the surface, the strength gets affected by weathering and seasonal water content changes. Also, desiccation close to the surface produces the effect of overconsolidation, even though the present overburden pressure has never been exceeded.

8. **Partially saturated and compacted clays.** The shear strength of partially saturated soils is a function of the relevant effective stress. The effective stress may be expressed as follows (Bishop 1960):

$$\sigma' = \sigma - [\chi\ u_w + (1 - \chi)\ u_a] \tag{9.49}$$

or $$\sigma' = \sigma - u_e \tag{9.50}$$

where u_w = pore water pressure

u_a = pore air pressure

χ = empirical parameter

u_e = equivalent pore pressure

The shear strength τ_f may then be expressed as:

$$\tau_f = c' + (\sigma - u_e) \tan \varphi' \tag{9.51}$$

The parameter χ is determined experimentally. It is related primarily to the degree of saturation S_r. For a fully saturated soil ($S_r=1$) and $\chi = 1$: for a dry soil ($S_r = 0$) and. $\chi = 0$. The soil structure and the way in which a particular degree of saturation is brought about also influence χ to a lesser extent. The above equations are, however, not convenient for use in practice due to the presence of parameter χ.

The total stress parameters can be determined by UU tests on undisturbed samples or on samples compacted to the field density and moisture content. The general behaviour is similar to a moderately overconsolidated soil. The failure envelope· is usually curved. The undrained strength increases with increasing cell pressure, but the strength increase becomes progressively smaller as the void ratio is reduced due to compression of the pore air. When the cell pressure reaches a high enough value that all the pore air passes into solution in pore water, the soil behaves as a saturated soil. The failure envelope becomes horizontal with $\varphi_T = 0$.

Examples 9.10—9.16

9.10 In a CU triaxial test on a normally consolidated clay ($c' = c_T = 0$), $(\sigma_d)_f$ and u_f are respectively 140 kPa and 110 kPa. The sample was consolidated under $\sigma_3 = 200$ kPa. Find σ_1'/σ_3', ϕ', ϕ_T and α_f.

$$\sigma_1 = \sigma_3 + \sigma_d = 340 \text{ kPa}, \quad \sigma_1' = \sigma_1 - u_f = 230 \text{ kPa}$$

$$\sigma_3' = \sigma_3 - u_f = 90 \text{ kPa}, \quad \sigma_1'/\sigma_3' = 2.56$$

$$\sigma_1/\sigma_3 = 1.7$$

$$\sin \phi' = \frac{\sigma_1' - \sigma_3'}{\sigma_1' + \sigma_3'}; \quad \phi' = 25.9° \simeq 26°$$

$$\sin \phi_T = \frac{\sigma_1 - \sigma_3}{\sigma_1 + \sigma_3}; \quad \phi_T = 15°$$

$$\alpha_f = 45 + \phi'/2 = 58°$$

Graphically: Draw total stress circle with $\sigma_3 = 200$ kPa and $\sigma_1 = 340$ kPa. Draw effective stress circle with $\sigma_3' = 90$ kPa and $\sigma_1' = 230$ k Pa

Draw failure envelopes passing through origin of stress and tangent to the circles.

Measure $\phi_T = 15°$, $\phi' = 26°$

9.11 The sample in Ex. 9.10 was saturated ($B = 1$). Find the pore pressure parameter A_f at failure.

$$\text{As } B = 1, \quad A_f = \frac{\Delta u - \Delta \sigma_3}{\Delta \sigma_1 - \Delta \sigma_3}$$

Since σ_5 is held constant during a CU test, $\Delta \sigma_3 = 0$.

and $\Delta \sigma_1 = \sigma_d = \sigma_1 - \sigma_3 = 140$ kPa

$$A_f = \frac{u_f}{\Delta \sigma_1} = \frac{110}{140} = 0.785$$

9.12 UU triaxial tests on saturated clay samples, 38 mm dia and 76 mm high, gave the following results. Find total stress parameters ϕ_T and c_T

σ_3 (kPa)	Axial load (N)	Axial deformation (mm)
200	195	9.5
400	194	10.1
600	199	10.3

$L_0 = 76$ mm, $A_0 = 1134$ mm^2

σ_3 (kPa)	$\dfrac{\Delta L}{L_0}$	A (mm^2)	σ_d (kPa)	σ_1 (kPa)	From failure envelope
200	0.125	1296	150	350	
400	0.133	1308	148	548	$\phi_T = 0$
600	0.136	1312	152	752	$c_T = 75$ kPa

9.13 Two CD triaxial tests on saturated clay (sample $L_0 = 76$ mm, $D = 38$ mm, $A_o = 1134$ mm^2, $V_o = 86$ cm^3) gave the following results. Find effective stress parameters graphically.

σ_3 (kPa)	Axial load (N)	ΔL (mm)	ΔV (cm^3)
200	431	8.4	7.5
400	858	9.2	8.6

Area at failure $A = A_o \dfrac{1 - \Delta V/V_o}{1 - \Delta L/L_o}$

σ_3' (kPa)	A (mm^2)	σ_d (kPa)	σ_1' (kPa)	Failure envelope passes through origin
200	1165	370	570	$c' = 0$
400	1161	739	1139	$\phi' = 29°$

9.14 CU triaxial tests with pore pressure measurements on saturated clay. Determine c' and φ' by plotting (a) Mohr circles (b) modified failure envelope (p-q diagram). Given :

σ_3 (kPa)	σ_d (kPa)	u_f (kPa)
350	251	30
300	431	65
450	610	100

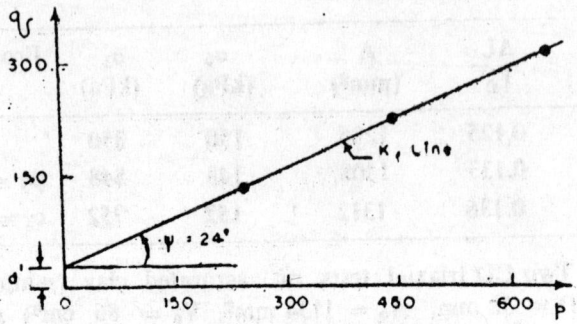

Fig. 9.22   Example 9.14

σ_3', (kPa)	σ_1' (kPa)	$q = \dfrac{\sigma_1 - \sigma_3}{2}$ (kPa)	$p = \dfrac{\sigma_1' + \sigma_3'}{2}$ (kPa)
150	371	125	245
300	666	215	450
450	960	305	655

From failure envelope: $\varphi' = 26°$, $c' = 20$ kPa
From modified failure envelope (K_f-line): $\psi' = 24°$
$d' = 17.5$ kPa, hence $\phi' = 26°$ and $c' = 19.5$ kPa

·9.15 An embankment of cohesive soil is constructed to a height of 20 m at a density of 1.9 g/cm³. The effective stress parameters as determined by CU tests with pore pressure measurements are $c' = 18$ kPa and $\varphi' = 22°$. What will be the shear strength of the soil on a horizontal plane at the base of the embankment if an average pore pressure of 150 kPa is indicated there by the piezometers.

$$\sigma = 1.9 \times 9.8 \times 20 = 372 \text{ kPa}$$
$$s = c' + (\sigma - u) \tan \phi' = 108 \text{ kPa}$$

9.16 An embankment constructed to a height of 3 m of cohesive soil ($c' = 25$ kPa and $\phi' = 20°$, $\rho = 1.7$ g/cm³) is left over for pore pressure dissipation. Its height is subsequently raised rapidly to 5.5 m with negligible dissipation of pore pressure. The pore pressure parameters are measured from triaxial tests as

$A = 0.6$ and $B = 0.9$.   Find the shear strength of the soil at the base of the embankment just after construction of the second phase, assuming that the lateral pressure at   any point is one-half of the vertical pressure.

$\Delta\sigma_1$ due to addition of 2.5 m of fill

$$= 2.5 \times 1.7 \times 9.8 = 42 \text{ kPa}$$

$\Delta\sigma_3 = \Delta\sigma_1/2 = 21$ kPa

$\Delta u = B[\Delta\sigma_3 + A(\Delta\sigma_1 - \Delta\sigma_3)] = 30$ kPa

Original $\sigma_1 = 3 \times 1.7 \times 9.8 = 50$ kPa

Final effective pressure $\sigma' = \sigma_1 + \Delta\sigma_1 - \Delta u = 62$ kPa

$s = 47.6$ kPa

10

Shear strength of rock

10.1 STRENGTH OF INTACT ROCK

It has been demonstrated by Barton (1978) that the peak strength envelope obtained by performing triaxial compression tests on intact rock samples is usually curved as shown in Fig. 10.1. The shear strength increases with an increase in the confining pressure σ_3 until at a very high pressure the envelope reaches a point of zero gradient, point C in Fig 10.1. This represents the maximum possible shear strength of the rock. This stress condition of the envelope at zero gradient is defined by Barton as the 'critical state' for initially intact rock. For each rock there will be a 'critical effective confining

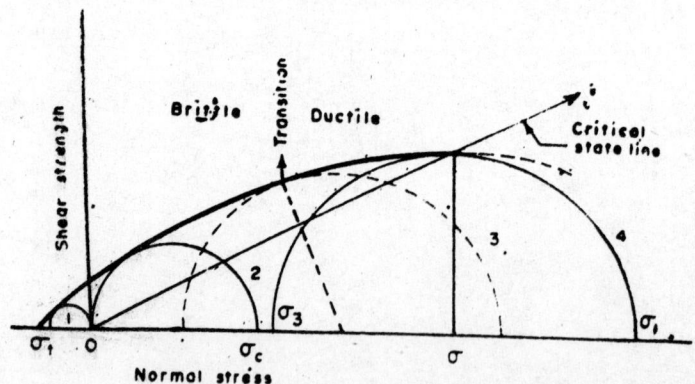

Fig. 10.1 Typical strength envelope for intact rocks 1. Uniaxial tension, 2. Uniaxial compression, 3. Brittle-ductile transition, 4. Critical state (Read tangential points on circles 2 and 4 as T and C)

pressure' above which shear strength cannot be made to increase. The major and minor principal effective stresses σ_1 and σ_3 associated with failure at the critical state are in the ratio of 3 to 1, i.e. $\sigma_1 = 3\sigma_3$. The line joining the origin of stress point to the point of zero gradient on the strength envelope (point C) is known as the 'critical state line'. The slope angle of this line φ_0 is equal to 26.6° (i.e. $\tau/\sigma = 1/2$.) The effective normal stress σ mobilized on the orthogonal conjugate failure surfaces at the critical state is found to be equal to the confined compressive strength of the rock $(\sigma_1 - \sigma_3)$.

The peak shear strength s of intact rock can be described for all practical purposes by the following empirical law proposed by Barton (1978):

$$s = \sigma \tan \left[50 \log_{10} \left(\frac{\sigma_1 - \sigma_3}{\sigma} \right) + \phi_0 \right] \qquad (10.1)$$

where $\quad \varphi_0 = 26.6°$

Eq. 10.1 is valid from the critical state circle (point C) down to stress level equivalent to point T in Fig. 10.1. It will probably also be valid for negative values of σ_3, provided σ remains on the positive side of the axis. Assuming a linear envelope between the uniaxial compression and uniaxial tension circles, the cohesion intercept c can be obtained from the geometrical relations of Mohr theory as:

$$c = \tfrac{1}{2} (\sigma_c \times \sigma_t)^{0.5}$$

where $\quad \sigma_0 =$ uniaxial (unconfined) compressive strength

$\quad \sigma_t =$ uniaxial tensile strength

Assuming isotropic rock properties, most rocks have σ_0/σ_t in the range 5—20. If for convenience values of 9 and 16 are assumed, c will range from $1/6 \, \sigma_0$ to $1/8 \, \sigma_c$. These values are conservative estimates, when the curvature of the real envelopes are considered.

An empirical strength criterion for intact rock proposed by Hoek and Brown (1980) is further described in Section 10.4.

10.2 STRENGTH OF ROCK MASS

The shear strength of a rock mass and its deformability are largely influenced by the discontinuity pattern, its geometry and how well it is developed. The roughness of the discontinuity surface, type and amount of filling and direction of shearing relative to discontinuity are also important factors. The stress levels are important in the sense whether it is a case of simple frictional shear or the *fracture* of interlocking surface projections on rock discontinuities is also involved. Shearing along a single discontinuity or along a family of parallel discontinuities is described in this section. The behaviour of actual rock surfaces involving both frictional strength and fracture strength are described in other sections.

1. Shearing along planar discontinuities. Consider the failure under direct shear of samples of rock obtained from the same block of rock which contains a through-going discontinuity such as a bedding plane. The bedding plane is cemented and is assumed to be absolutely planar, having no surface undulations or roughness. A particular normal stress σ is applied across the discontinuity surface and an increasing shear stress τ is applied parallel to the discontinuity. A typical curve between shear stress and shear displacement for a sample is shown in Fig. 10.2. If peak strength values obtained from tests carried out at different normal stress values are plotted against normal stress, an approximately linear curve will be obtained, similar to the Coulomb strength envelope for soils (Eq. 9.1), defining the 'peak' shear strengt s by the equation (Hoek and Bray 1977):

$$s = c_p + \sigma \tan \phi_p \qquad (10.3)$$

where ϕ_p = peak fiction angle

   c_p = cohesive strength of cementing material

If the 'residual' shear strength is plotted against the normal stress, the following linear relationship is obtained, which shows that all

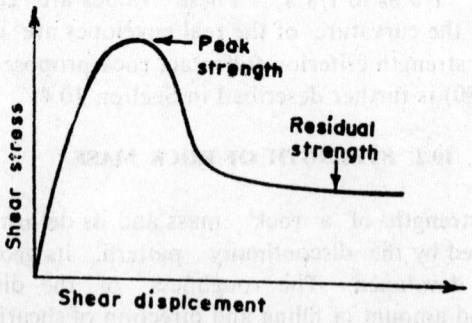

Fig. 10.2 Shear stress and displacement curve for shear along a bedding plane

the cohesive strength of the cementing material has been lost (Hoek and Bray 1977):

$$s = \sigma \tan \phi_r \qquad (10.4)$$

where ϕ_r = residual friction angle (usually lower than ϕ_p)

Presence of water in a discontinuity in rock will reduce the shear strength due to a reduction of the effective normal stress, $\sigma' = \sigma - u$. The influence of water on the cohesive and frictional properties of the rock discontinuity depends on the nature of the filling or cementing material. In most hard rocks (as in many sandy soils and gravels), these properties are not significantly altered by water, but shales, mudstones and similar materials (clays) exhibit significant changes due to changes in moisture content.

2. **Shearing on an inclined plane.** If the discontinuity surface is inclined at an angle 'i' to the shear stress direction (Fig. 10.3), the shear and normal stresses (τ_i and σ_i) acting on the failure surface are given by (Hoek and Bray 1977):

$$\tau_i = \tau \cos^2 i - \sigma \sin i \cos i \qquad (10.5)$$
$$\sigma_i = \sigma \cos^2 i + \tau \sin i \cos i \qquad (10.6)$$

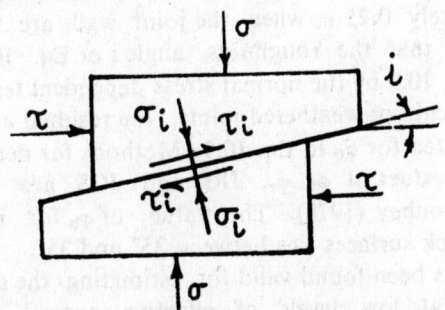

Fig. 10.3 Shearing on an inclined plane

Assuming that the discontinuity surface has zero cohesive strength, its shear strength s_i is given by:

$$s_i = \sigma_i \tan \varphi \qquad (10.7)$$

Substituting Eqs. 10.5 and 10.6 into Eq. 10.7, the shear strength (maximum *applied* shear stress) can be expressed as:

$$s = \sigma \tan (\varphi + i) \qquad (10.8)$$

Eq. 10.8 was confirmed by Patton (1966a, 1966b) in a series of tests on models with regular surface projections.

The strength components φ and i appearing in Eq. 10.8 are usually termed the "basic friction angle ϕ_b" and the "effective roughness" or "i-value" for the rock joint. The basic friction angle can be obtained from residual shear tests on flat unweathered rock surfaces (Barton and Choubey 1978). Unfortunately it is difficult to estimate the average roughness angle i for a given joint surface (without performing shear tests). It depends on the order of projections which is taken into consideration.

10.3 BARTON EMPIRICAL LAW

Barton (1978) proposed the following empirical equation for deriving the peak shear strength s along rough joint surfaces:

$$s = \sigma \tan [\varphi_b + JRC \log_{10} (\frac{JCS}{\sigma})] \qquad (10.9)$$

where $\sigma =$ effective normal stress
JRC is a "joint roughness coefficient" ranging from 0 to 20, from the smoothest to the roughest surface. JCS is the "joint wall compressive strength" which is equal to unconfined compressive strength σ_0 of the rock if the joint is unweathered, but may reduce to approximately 0.25 σ_c when the joint walls are weathered. It may be noted that the roughness angle i of Eq. 10.8. has been replaced in Eq. 10.9 by the normal stress dependent term containing JRC. For the case of weathered joints, the residual angle ϕ_r ($<\phi_b$) can be substituted for φ_b in Eq. 10.9. Methods for determining and estimating the values of φ_b. φ_r, JRC and JCS are described by Barton and Choubey (1978). The value of φ_b for most smooth unweathered rock surfaces lies between 25° and 35°.

Eq. 10.9 has been found valid for estimating the shear strength of rock masses at low levels of effective normal stress (range 0.1—2.0 MN/m², or 1-20 kg/cm²) and provided that the joints are unweathered. This is the range appropriate to most rock engineering problems. In this range the frictional strength of a joint is related to the fracture strength of the intack rock by means of the unconfined compressive strength σ_0. The dimensionless ratio σ_0/σ controls the frictional strength. The shear strength envelopes at low σ levels are steeply inclined. However, in view of the safety requirements the maximum allowable value of arctan s/σ is taken as 70° and any possible 'cohesion' intercept is neglected. If the unconfined compressive strength of the rock is low, or the normal stress is high, the ratio σ_0/σ reduces towards unity and the resultant shear strength *theoretically* approaches the value given by:

$$s = \sigma \tan \phi_b \qquad (10.10)$$

At high levels of effective normal stress the frictional strength is related to the fracture strength by means of the confined compressive strength represented by $(\sigma_1 - \sigma_3)$ at fracture (i.e. $JCS = \sigma_1 - \sigma_3$) and Eq. 10.9 is written as (Barton 1978):

$$s = \sigma \tan \left[\phi_b + JRC \log_{10} \left(\frac{\sigma_1 - \sigma_3}{\sigma} \right) \right] \qquad (10.11)$$

where $\sigma_1 =$ axial stress at failure

$\sigma_3 =$ effective confining pressure

The increase in JCS is thought to be caused by the increasing *contact area* and consequently improved confining effect as the normal stress is increased. The dimensionless ratio $(\sigma_1 - \sigma_3)/\sigma$ (of which σ_0/σ is a special case, i. e. $\sigma_3 = 0$) varies relatively little over a wide range of σ_3 and results in a limited range of frictional strength at high stress level compared to the wide range of frictional strength exhibited at low stress levels (Barton 1978).

At very high stress levels the shear stress required to fracture intact rock is no greater than the shear strength of the resulting fault. The fracture behaviour of rocks *changes from brittle to ductile* (See Fig. 10.1) with increasing confining pressure, the transition pressure being higher for stronger rocks. For silicate rocks, Mogi (1966) defined the brittle-ductile transition by the condition:

$$\sigma_1 - \sigma_3 = 3.4 \, \sigma_3 \qquad (10.12)$$

that is, if the confined compressive strength is greater than $3.4 \, \sigma_3$, the behaviour will be brittle, and vice versa. It is observed that the transition pressure of weaker marbles and lime-stones is appreciably lower, the transition boundary being curved (concave downward) (Barton 1978).

10.4 HOEK AND BROWN EMPIRICAL EQUATIONS

Hoek and Brown (1980) have developed the following empirical relationship between the principal stresses at failure:

$$\frac{\sigma_1}{\sigma_c} = \frac{\sigma_3}{\sigma_c} + \sqrt{\frac{m \sigma_3}{\sigma_c} + k} \qquad (10.13)$$

where σ_c = uniaxial compressive strength of intact rock material

 m, k = constants depending on the properties of rock and the extent to which it had been broken before being subjected to failure stresses

Eq. 10.13 may be rewritten in the form:

$$\sigma_1^* = \sigma_3^* + \sqrt{m \sigma_3^* + k} \qquad (10.14)$$

where σ_1^* and σ_3^* are the values of principal stresses normalised with respect to σ_c.

For intact rock material, $k = 1.0$, and the normalised form of the strength criterion becomes:

$$\sigma_1^* = \sigma_3^* + \sqrt{m \sigma_3^* + 1.0} \qquad (10.15)$$

By putting $\sigma_3 = 0$ in Eq. 10.13, the uniaxial compressive strength of the rock is obtained as:

$$\sigma_{ck} = \sqrt{k \sigma_c^2} \qquad (10.16)$$

For intact material, $k = 1.0$ and $\sigma_{ck} = \sigma_c$ as required.

For previously broken rock, $k < 1$ and for a completely granulated specimen or rock aggregate, $k = 0$.

Putting $\sigma_1 = 0$ in Eq. 10.13 and solving the resulting equation for $\sigma_3 = \sigma_t$, the uniaxial tensile strength is obtained as:

$$\sigma_t = \frac{\sigma_c}{2} (m - \sqrt{m^2 + 4k}) \qquad (10.17)$$

A wide range of experimental data for intact rock and rock mass confirm that the failure envelopes are generally nonlinear. Using the stress components as shown in Fig. 10.4, the strength criterion can also be expressed in terms of the normal stress and the peak shear strength s as follows:

$$\sigma = \sigma_3 + \frac{\tau_m^2}{\tau_m + m\sigma_c/8} \qquad (10.18)$$

and

$$s = (\sigma - \sigma_3) \sqrt{1 + \frac{m \sigma_c}{4 \tau_m}} \qquad (10.19)$$

where $\tau_m = \frac{1}{2} (\sigma_1 - \sigma_3)$ \qquad (10.20)

From Fig. 10.4, the normalised values of σ and s ($\sigma^* = \sigma/\sigma_c$ and $s^* = s/\sigma_o$) can be related as:

$$s^* = A \; (\sigma^* - \sigma_t^*)^B \tag{10.21}$$

where σ_t^* is the normalised tensile strength of the rock given by:

$$\sigma_t^* = \tfrac{1}{2} \; (m - \sqrt{m^2 + 4k}) \tag{10.22}$$

and A and B are constants depending on the value of m.

In developing the above criterion, the influence of the intermediate principal stress σ_2 has been assumed to be negligible. The criterion has been expressed in terms of total or applied stresses. The effective stress form of the criterion can be obtained by substituting effective stresses, but the parameters m and k will not necessarily be the same for both the total and effective stress cases (Hoek and Brown 1980).

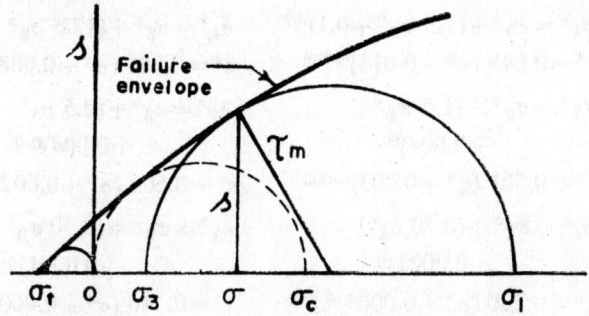

Fig. 10.4   Failure envelope showing relationship between stress components
(Hoek and Brown 1980)

As a rough guide in preliminary design calculations, Hoek and Brown have suggested values of the parameters m, k, A and B for use in the normalised strength criterion (Eqs. 10.14 and 10.21). The uniaxial compressive strength σ_o for use in these equations should be determined on approximately 50 mm (2.0 inches) diameter samples of fresh intact rock material at field moisture conditions. Table 10.1 gives the approximate strength relationships for only two types of rock with respect to rock quality. Such relationships for other types of rocks are also available (Hoek and Brown 1980). These relationships are intended to be used in estimating the overall strength of rock masses in which the discontinuity spacing is small on the scale of the problem and in which complex modes of rock mass failure might be expected to occur. They should not be used where failure is likely to occur by slip on one or two discontinuities as, e.g. in wedge failure of slopes.

Table 10.1

APPROXIMATE STRENGTH CRITERIA FOR INTACT ROCK AND JOINTED ROCK MASSES (HOEK AND BROWN 1980)

Rock quality	Sandstone and quartzite	Igneous and metamorphic rocks
A	$\sigma_1{}^* = \sigma_3{}^* + (15\ \sigma_3{}^* + 1.0)^{0.5}$ $s^* = 1.044\ (\sigma^* + 0.067)^{0.692}$	$\sigma_1{}^* = \sigma_3{}^* + (25\ \sigma_3{}^* + 1.0)^{0.5}$ $s^* = 1.220\ (\sigma^* + 0.040)^{0.705}$
B	$\sigma_1{}^* = \sigma_3{}^* + (7.5\ \sigma_3{}^* + 0.1)^{0.5}$ $s^* = 0.848\ (\sigma^* + 0.013)^{0.702}$	$\sigma_1{}^* = \sigma_3{}^* + (12.5\ \sigma_3{}^* + 0.1)^{0.5}$ $s^* = 0.988\ (\sigma^* + 0.008)^{0.712}$
C	$\sigma_1{}^* = \sigma_3{}^* + (1.5\ \sigma_3{}^*$ $+ 0.004)^{0.5}$ $s^* = 0.501\ (\sigma^* + 0.003)^{0.695}$	$\sigma_1{}^* = \sigma_3{}^* + (2.5\ \sigma_3{}^*$ $+ 0.004)^{0.5}$ $s^* = 0.603\ (\sigma^* + 0.002)^{0.707}$
D	$\sigma_1{}^* = \sigma_3{}^* + (0.30\ \sigma_3{}^*$ $+ 0.0001)^{0.5}$ $s^* = 0.280\ (\sigma^* + 0.0003)^{0.688}$	$\sigma_1{}^* = \sigma_3{}^* + (0.50\ \sigma_3{}^*$ $+ 0.0001)^{0.5}$ $s^* = 0.346\ (\sigma^* + 0.0002)^{0.700}$
E	$\sigma_1{}^* = \sigma_3{}^* + (0.8\ \sigma_3{}^*$ $+ 0.00001)^{0.5}$ $s^* = 0.162\ (\sigma^* + 0.0001)^{0.672}$	$\sigma_1{}^* = \sigma_3{}^* + (0.13\sigma_3{}^*$ $+ 0.00001)^{0.5}$ $s^* = 0.203\ (\sigma^* + 0.0001)^{0.686}$
F	$\sigma_1{}^* = \sigma_3{}^* + (0.015\ \sigma_3{}^* + 0)^{0.5}$ $s^* = 0.061\ (\sigma^*)^{0.546}$	$\sigma_1{}^* = \sigma_3{}^* + (0.025\ \sigma_3{}^* + 0)^{0.5}$ $s^* = 0.078\ (\sigma^*)^{0.556}$

A = Intact rock samples — laboratory size rock samples free from structural defects (African CSIR rating RMA 100 +; NGI index Q 500, See Chapter 4)

B = Very good quality rock mass — tightly interlocking undisturbed rock with unweathered joints spaced at 3 m ± (RMA 85, Q 100)

C=   Good quality rock mass — fresh to slightly weathered rock, slightly disturbed with joints spaced at 1-3 m (RMA 65, Q 10)

D=   Fair quality rock mass—several sets of moderately weathered joints spaced at 0.3—1 m (RMA 44, Q 1.0)

E=   Poor quality rock mass — numerous weathered joints spaced at 30-500 mm with some gouge filling/clean waste rock (RMA 23, Q 0.1)

F=   Very poor quality rock mass — numerous heavily weathered joints spaced less than 50 mm with gouge filling/waste rock with fines (RMA 3, Q 0.01)

11

Lateral pressure and retaining structures

11.1 RETAINING STRUCTURES

A retaining structure is a permanent or temporary structure, (e.g. retaining wall, sheet piling, bulk head, basement wall, and bracing, etc.) used to provide lateral support to soil or other material. A 'retaining wall' is a wall built to hold back soil or other solid material. Retaining walls may be of the following types: (a) gravity or massive retaining walls, (b) reinforced concrete walls (cantilever, counterfort or buttressed walls), and (c) crib walls. A 'crib wall' is essentially a gravity wall constructed of rectangular interlocking precast concrete or timber members to form a cellular structure, laid no top of each other and filled with soil or rock pieces. 'Sheet piling' or 'sheet pile wall' is a wall of sheet piles, which may be a contilever wall or anchored back at one or two levels. 'Bulk head' is also a sheet pile which may be 'free' standing but is more usually anchored back near to the top and is used in water front construction. 'Bracing' represents a system of wallings, struts and other members used to provide temporary support in excavation works.

11.2 LATERAL EARTH PRESSURE

Lateral earth pressure is the force exerted by a soil mass on its side retained by a structure, for instance, a retaining wall. The magnitude of the lateral earth pressure is known to vary considerably with the displacement of the retaining wall and nature of the soil.

Depending on the magnitude and nature of displacement or yield of soil, lateral pressures may be grouped into three states: (i) at-rest, (ii) active and (iii) passive.

The "at-rest" lateral pressure is the value of the pressure existing in a soil deposit in its natural state that has not been subject to lateral yielding. The "active" pressure is the minimum value of the lateral pressure which is attained when a soil deposit tends to expand horizontally, e.g. a wall moving away from its backfill. The maximum value of the earth pressure is the "passive" pressure occurring when a soil mass tends to compress, e.g. a thrust block moving against the soil.

11.3 STATES OF EQUILIBRIUM

1. **At rest state.** The state of stress in a natural or artificially placed soil deposit which is not subject to lateral yield is called the "at-rest state" of equilibrium. The vertical and horizontal effective pressures (σ_v' and σ_h') are then related as follows:

$$\sigma_h' = K_0 \, \sigma_v' \tag{11.1}$$

where K_0 is the 'coefficient of earth pressure at rest'.

A theoretical expression of K_0 for an isotropic, homogeneous and elastic material can be obtained from the condition of zero lateral strain ($\epsilon_h = 0$). From Eq. 8.6,

$$\epsilon_h = \frac{1}{E} \left[\sigma_h' = \mu \, (\sigma_v' + \sigma_h') \right] \tag{11.2}$$

or $$\sigma_h' = \frac{\mu}{1 - \mu} \sigma_v' = K_0 \, \sigma_v' \ (\text{for } \epsilon_h = 0) \tag{11.3}$$

where $\mu =$ Poisson's ratio

The field value of K_0 depends on the manner of formation of a soil deposit, the density index (angle of shearing resistance) for sands, and the stress history (overconsolidation ratio) for clays. The following values are suggested as a guide (Lee et al 1983): normally consolidated clay 0.4 to 0.7; overconsolidated clay 1 to 4; compacted clay 1 to 2; heavily machine compacted clay 2 to 4; loose sand 0.45 to 0.5; dense sand 0.35. Tamping the sand in layers may increase K_0 to about 0.8 (Dunn et al 1980). For normally consolidated soils (both sands and clays), K_0 can be empirically correlated with ϕ' (Jaky 1948, Bishop 1958):

$$K_0 \approx 1 - \sin \phi' \tag{11.4}$$

The plasticity index $I_p(\%)$ has a secondary effect on K_o and Alpan (1967) suggested the following relation for normally consolidated clays:

$$K_o = 0.19 + 0.233 \log I_p \qquad (11.5)$$

Overconsolidation increases the value of K_o. If pore water pressure is present in soil, the total lateral pressure would be:

$$\sigma_h = \sigma_h' + u \qquad (11.6)$$

2. Limiting equilibrium states. When *every* part of a soil mass is just on the verge of failure the soil is said to be in a state of "limiting equilibrium" or "plastic equilibrium". The failure criterion describing this state in a two-dimensional element is the Mohr-Coulomb criterion (Eq. 9.12). A combination of stress states leading to shear failure can be developed either by decreasing or increasing the lateral stresses. Considering a soil element below a horizontal surface in a dry, cohesionless soil deposit, the horizontal pressure σ_h' is initially at its 'at-rest' value ($\sigma_h' = K_o \sigma_v'$) and the corresponding Mohr circle is shown dotted in Fig. 11.1. If the state of plastic equilibrium is to be reached by lateral stretching and thereby decreasing the horizontal pressure while the vertical pressure is held constant, the horizontal pressure must drop to the lower limiting value σ_a' which is the minor principal stress value defined by a Mohr circle tangential to the failure envelope towards the left side of the intial stress state in Fig. 11.1. Similarly, the horizontal

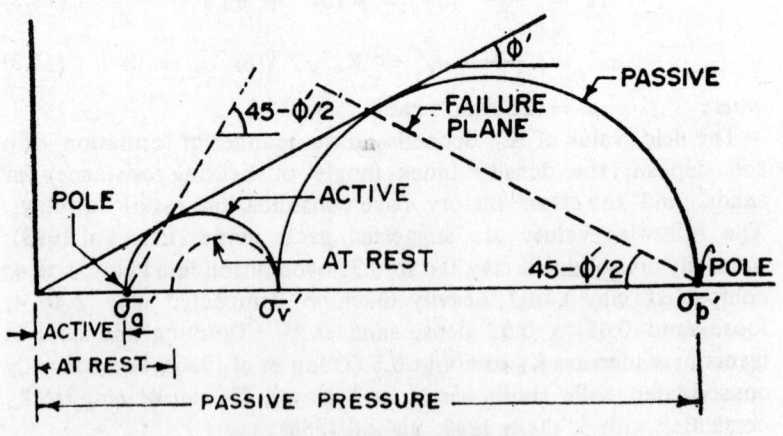

Fig. 11.1 States of equilibrium

pressure must increase to an upper limiting value σ_p' for the state of plastic equilibrium to be reached by lateral compression and the corresponding failure circle lies to the right of the initial stress state circle. σ_p' is now the major principal stress. The two states of plastic equilibrium are respectively called the "Rankine active state" and the "Rankine passive state", after the British engineer Rankine who in 1857 noted the relationship between the active and passive conditions.

In the plastic equilibrium states, there will be a network of failure planes (known as *slip line field*) equally inclined to the principal planes. In the active state, the failure planes are inclined at an angle of $45 + \phi'/2$ to the horizontal and in the passive state at $45 - \phi'/2$.

3. **Strains for Rankine states.** On the basis of triaxial test analysis on dense sands, Lambe and Whitman (1973) conclude that: (i) very little horizontal strain, less than -0.5%, is required to attain the active state, (ii) little compressive strain, about 0.5%, is necessary to develop one-half the maximum passive resistance, and (iii) much more horizontal strain, about 2%, is necessary for a full passive state. For loose sands the first two conclusions remain valid, but the strain required for full passive state may be as high as 15%. The above conclusions are based on triaxial test data, and the magnitudes for field conditions are not likely to be the same.

11.4 RANKINE ANALYSIS

Rankine derived the lateral pressures in a soil deposit assuming it to be in a state of plastic equilibrium. The state of stress along the interface between the soil (i.e. backfill) and the retaining structure (wall) is assumed identical with the stress state within the soil mass away from the wall. In other words, the presence of a wall does not modify the state of stress in its vicinity and the state remains as if the soil mass were semi-infinite, (homogeneous and istropic). For this condition to be satisfied, only a vertical wall with a *smooth* back (with no friction or adhesion on the soil wall interface) supporting a cohesionless soil with a horizontal backfill surface is considered. In the usual situation, shear stresses are developed along the soil wall interface and the Rankine solution is not applicable. The inaccuracies are generally small for the active case and are on the conservative side, i.e. the lateral pressure on the back of the wall is over-estimated. The error from neglecting the wall friction for the passive case may be quite large, and the passive pressure is under-estimated.

1. Active pressure — cohesionless soil. Considering a soil element at a depth z behind a smooth, vertical wall supporting a cohesionless backfill with horizontal surface in level with the top of wall, the vertical pressure on the element is $\sigma_v = \gamma z$.

Let the wall move forward away from the backfill such that a state of plastic equilibrium is reached in the soil mass. The horizontal pressure reaches the minimum value, called the 'active pressure p_a' which is related to σ_v as follows:

$$\sigma_h = p_a = K_a \sigma_v = K_a \gamma z \qquad (11.7)$$

where K_a is called the "coefficient of active pressure". As the backfill surface is horizontal, there can be no lateral transfer of weight and no shear stresses exist on horizontal and vertical planes. The vertical and horizontal stresses (σ_v and σ_h) are therefore, principal stresses, the former (σ_v) being the major principal stress σ_1 and the latter (σ_h) the minor one (σ_3) at the time of plastic equilibrium. These stresses also satisfy the Mohr-Coulomb criterion (Eq. 9.12):

$$\sigma_1' = \sigma_3' \tan^2 \alpha_f \quad (c = 0)$$

or

$$\sigma_3' = \sigma_1' \cot^2 \alpha_f \quad (\alpha_f = 45 + \phi'2) \qquad (11.8)$$

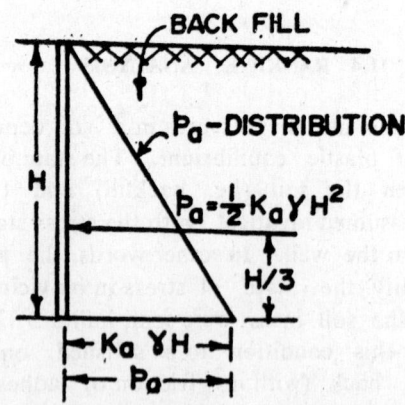

Fig. 11.2 Rankine active pressure distribution, dry cohesionless backfill

Comparing Eqs. 11.7 and 11.8, K_a is equal to the ratio of the principal stresses:

$$K_a = \frac{\sigma_3'}{\sigma_1'} = \cot^2 \alpha_f = \frac{1 - \sin \phi'}{1 + \sin \phi'} \qquad (11.9)$$

Eq. 11.7 states that p_a varies linearly with depth, as shown in Fig. 11.2, and the total lateral force P_a per unit length of the wall of height H will be:

$$P_a = \tfrac{1}{2} K_a \gamma H^2 \qquad (11.10)$$

which will act at a point $H/_3$ above the base of the wall.

(a) *Effect of uniform surcharge.* If a uniformly distributed surcharge pressure of intensity q per unit area acts over the entire surface of the soil mass, the vertical pressure σ_s at any depth is increased to $(\gamma z + q)$ and which causes an *additional* lateral pressure of

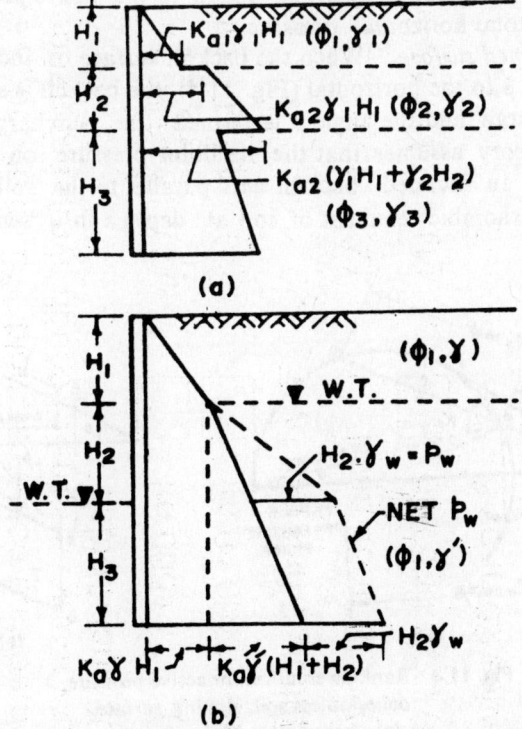

(a)

(b)

Fig. 11.3 Pressure distribution due to (a) stratum change and (b) submergence

a uniform intensity K_aq behind the wall. The total additional pressure due to surcharge is thus K_aq H acting at mid-height H/2.

(b) *Effect of stratum change.* If the backfill is stratified such that K_a and γ are not constant with depth, the pressure will not increase linearly but will change abruptly at the strata interfaces. The pressure distribution is obtained by using appropriate values of K_a for each strata. For a particular layer the weight of the overlying layers is considered as a surchage. A typical pressure diagram is shown in Fig. 11.3a.

(c) *Effect of submergence.* For a submerged backfill, as shown in Fig. 11.3b, K_a is applied to the *effective* vertical pressure only. The effective active pressure distribution is computed on the basis of bulk unit weight γ above the water table and of submerged unit weight γ' below the water table. The *net* water pressure *below* the water table must be added to the active pressure to obtain the total horizontal pressure.

(d) *Inclined surface.* When the backfill surface is inclined, say at an angle β to the horizontal (Fig. 11.4), the backfill is said to be 'with surcharge' and the angle β is termed the 'surcharge angle'. Rankine theory assumes that the resultant pressure on the wall supporting an inclined backfill acts parallel to the soil surface. Consider a rhombic element of soil at depth z in a semi-infinite

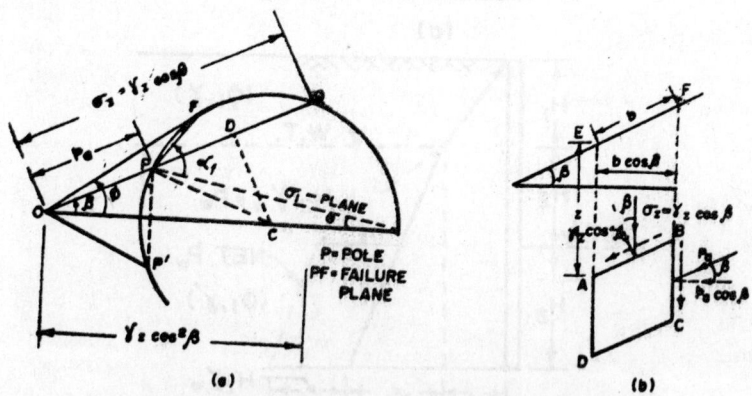

(a)   (b)

Fig. 11.4 Rankine solution for active pressure,
cohesionless soil, sloping surface
(a) active state Mohr circle
(b) stresses on an element

mass, as shown in Fig. 11.4b. Since at any constant depth below the surface, all such elements are similar, the forces on the vertical surfaces EA and FB are equal and parallel to the ground surface. The vertical pressure σ_z on AB and the lateral pressure p_a on BC are each inclined at β to the normals on these faces and are therefore not the principal stresses; they are the 'conjugate stresses; (i.e. the direction of one pressure is parallel to the plane on which the other pressure acts).

The conjugate stresses p_a and σ_z are represented by OP' (= OP) and OQ on the active state Mohr circle from which the conjugate stress ratio K (= p_a/σ_z) is derived as follows (Fig. 11.4a):

$$\frac{p_a}{\sigma_z} = \frac{OP}{OQ} = \frac{OD - DP}{OD + DQ}$$

Now

$$OD = OC \cos \beta$$

$$DP = DQ = (PC^2 - DC^2)^{0.5}$$
$$= (FC^2 - DC^2)^{0.5}$$
$$= OC (\sin^2 \phi - \sin^2 \beta)^{0.5}$$
$$= OC (\cos^2 \beta - \cos^2 \phi)^{0.5}$$

Therefore

$$K = \frac{p_a}{\sigma_z} = \frac{\cos \beta - \sqrt{(\cos^2 \beta - \cos^2 \phi)}}{\cos \beta + \sqrt{(\cos^2 \beta - \cos^2 \phi)}} \quad (11.11)$$

Also

$$\sigma_z = \frac{\gamma z \, b \cos \beta}{b} = \gamma z \cos \beta \quad (11.12)$$

Thus the active pressure acting *parallel* to the slope is given by:

$$p_a = K \sigma_z = \gamma z \cos \beta \frac{\cos \beta - \sqrt{(\cos^2 \beta - \cos^2 \phi)}}{\cos \beta + \sqrt{(\cos^2 \beta - \cos^2 \phi)}}$$
$$(11.13)$$

$$= K_a \gamma z \quad (11.14)$$

where K_a is the coefficient of active pressure given by:

$$K_a = \cos \beta \frac{\cos \beta - \sqrt{(\cos^2 \beta - \cos^2 \phi)}}{\cos \beta + \sqrt{(\cos^2 \beta - \cos^2 \phi)}} \quad (11.15)$$

The total active thrust on a vertical wall of height H is obtained by integration of Eq. 11.14:

$$P_a = \tfrac{1}{2} K_a \gamma H^2$$

The pressure distribution is shown in Fig. 11.5. One set of failure surfaces will be the planes inclined at $(45 + \phi/2)$ to the direction of the major principal plane defined by angle θ in Fig. 11.4.

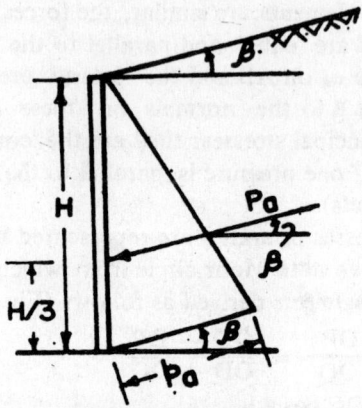

Fig. 11.5 Rankine pressure distribution, inclined backfill

(e) *Inclined wall*. Rankine considered only a vertical wall. If the back of the wall is inclined, an imaginary vertical back is assumed passing through the heel and extending upto the surface of the backfill or its extension, as shown in Fig. 11.6. Total active pressure P_1 is computed for this imaginary vertical back BC and the resultant pressure P is obtained as the vector sum (resultant) of P_1 and W, where W is the weight of the material in the $\triangle$ ABC.

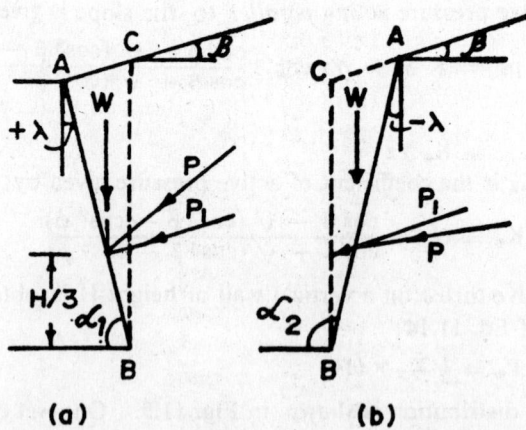

Fig. 11.6 Pressure on inclined wall (a) inclination away from backfill, positive batter (b) inclination towards backfill, negative batter

2. Active pressure—cohesive soil. Rankine considered a cohesionless ($c = 0$) backfill. Resal (1910) and Bell (1915) extended the Rankine type analysis to cohesive soils, i.e. soils having a 'cohesion intercept' in the Mohr-Coulomb failure criterion. Such are the overconsolidated and compacted clays. The failure envelope for a normally consolidated clay does not have a cohesion intercept and should be treated as a cohesionless soil, but such a state is rarely encountered in the field as clay soils nearly always have a cohesion intercept (See Section 9.11).

Considering a vertical wall with a $c - \phi$ backfill having a horizontal surface, the principal stresses on a soil element at plastic equilibrium are related by Eq. 9.12:

$$\sigma'_1 = \sigma_3' \tan^2 \alpha_f + 2 c' \tan \alpha_f$$

where σ_1' is the (effective) vertical pressure γz and σ_3' is the lateral, active pressure p_a; hence:

$$p_a = \gamma z \cot^2 \alpha_f - 2 c' \cot \alpha_f \qquad (11.16)$$

or

$$p_a = K_a \gamma z - 2 c' \sqrt{K_a} \qquad (11.17)$$

where

$$K_a = \cot^2 \alpha_f = \frac{1 - \sin \phi'}{1 + \sin \phi'}$$

and the resultant active pressure P_a is obtained by integrating Eq. 11.17:

$$P_a = \tfrac{1}{2} K_a \gamma H^2 - 2 c' H \sqrt{K_a} \qquad (11.18)$$

in which H is the depth of interest.

It will be seen from Eq. 11.16 or 11.18 that a cohesive soil is partially self supporting and, therefore, exerts a smaller pressure than a cohesionless soil. (Eq. 11.16 is known as *Bell's equation*.) Also, p_a is zero at a particular depth z_0 given by:

$$z_0 = \frac{2 c'}{\gamma \sqrt{K_a}} \qquad (11.19)$$

and the active pressure is negative above this depth; at surface ($z = 0$), the negative pressure is $- 2c' \sqrt{K_a}$ The pressure distribution diagram is shown in Fig. 11.7.

For a height $H_c = 2 z_0$ given by Eq. 11.20:

$$H_c = \frac{4 c'}{\gamma \sqrt{K_a}} \qquad (11.20)$$

the 'resultant' active pressure is zero. H_c represents the "critical height" to which a *vertical slope* (cut) in a cohesive soil could stand

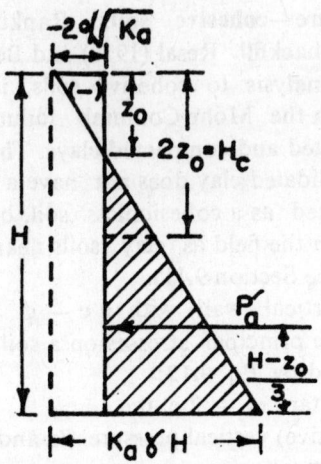

Fig. 11.7 Rankine active pressure distribution for a cohesive soil

without any lateral support, at least temporarily. In practice, however, a negative pressure (or tension in soil) cannot be relied upon to act on the wall, because cracks are likely to develop within the tension zone. It is usual to ignore the negative pressure upto the likely depth of tension cracks (z_0) and the total force P_a' for design purposes is taken as the full shaded area in Fig. 11.7 which is equivalent to:

$$P_a' = \tfrac{1}{2}K_a\,\gamma H^2 - 2\,c'\,H\sqrt{K_a} + \frac{2(c')^2}{\gamma} \qquad (11.21)$$

The above analysis is given in terms of effective stresses, which is applicable to a fully drained condition or where the pore pressures are known. For a submerged backfill below the water table, the water pressure should be considered separately. If tension cracks are also filled up, water pressure is taken as continuous below the ground surface. It is often convenient to use the total stress parameters c_T and ϕ_T because pore pressures in cohesive backfills are usually unknown. For short term conditions in case of saturated, or nearly saturated clays, ϕ_T is zero and c_T can be taken as half the average value of the unconfined compressive strength ($c_T = q_u/2$). For a $\phi_T = 0$ soil, $K_a = 1$ and the pressure distribution and the depth of tension crack are respectively given by:

$$p_a = \gamma_{sat}\, z - 2\, c_T \qquad (11.22)$$

$$z_0 = \frac{2\, c_T}{\gamma_{sat}} \qquad (11.23)$$

If the backfill carries a uniform surcharge of intensity q, the active pressure at every elevation increases by $K_a\, q$ or by q for $\phi_T = 0$ soil and z_0 will be reduced by q/γ.

Clay backfills tend to creep and, therefore, the active and passive states (limiting plastic states) will tend to seek the at-rest state under long term loading conditions.

3. Passive pressure. In the passive state, the vertical pressure γz is the minor principal stress σ_3 and the lateral pressure p_p is the major principal stress σ_1. Hence from the relation of σ_1 and σ_3 at failure (Eq. 9.12), the pressure intensities for cohesionless and cohesive soils can be expressed as:

$$p_p = \gamma z \tan^2 \alpha_f \quad (c = 0) \qquad (11.24)$$

$$p_p = \gamma z \tan^2 \alpha_f + 2\, c' \tan \alpha_f \quad (c > 0) \qquad (11.25)$$

or
$$p_p = K_p\, \gamma z \quad (c = 0) \qquad (11.26)$$

and
$$p_p = K_p\, \gamma z + 2\, c' \sqrt{K_p} \quad (c > 0) \qquad (11.27)$$

where
$$K_p = \tan^2 \alpha_f = \frac{1 + \sin \phi'}{1 - \sin \phi'} = 1/K_a \qquad (11.28)$$

The pressure distribution diagrams are shown in Fig. 11.8. The effects of water table, uniform surcharge or a stratified deposit are taken into account as for the active case.

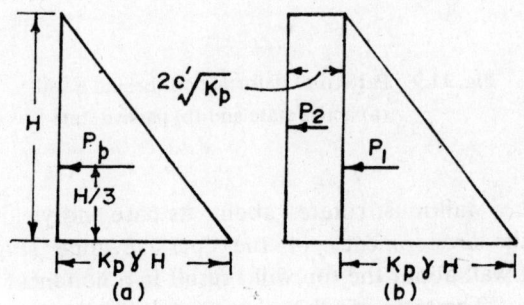

Fig. 11.8 Rankine passive pressure distribution
(a) cohesionless soil (b) cohesive soil

4. Deformation behind wall. The movement of a retaining wall of finite height cannot produce the active or passive state in the entire soil mass, as assumed in the Rankine theory. The active zone, for example, would be in the form of a wedge as shown in Fig. 11.9a. The soil mass beyond the outermost failure plane passing through the base will remain in a state of 'elastic' equilibrium. The width of the active zone reduces to zero at the base. A uniform strain within the wedge can be produced if the wall yields away from the fill by rotating about its base. Yielding of wall by rotation about its top does not satisfy the conditions for the development of the active state because there will not be adequate strain in soil near the surface. Pressure distribution behind a wall yielding by rotation about the top will be nonlinear.

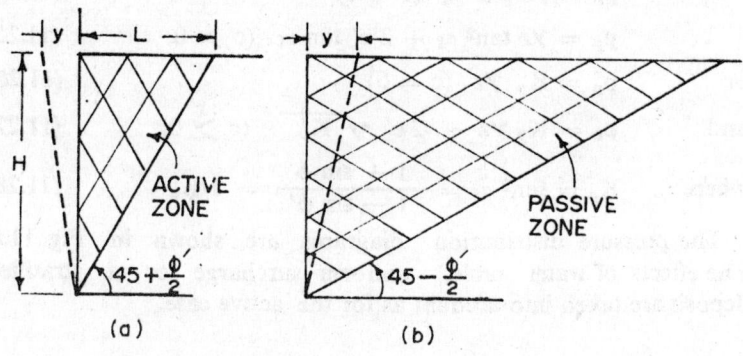

Fig. 11.9   Pattern of deformation behind a wall
(a) active state and (b) passive state

Similarly, a wall must rotate about its base and yield towards the fill for the development of the passive zone (Fig. 11.9b). Rotation of wall about the top will result in a nonlinear pressure distribution. The ratio of the horizontal displacement 'y' at top to the wall height H for a wall to achieve failure conditions is of the order of magnitude as given in Table 11.1. (CGS 1985)

Table 11.1
MAGNITUDE OF WALL ROTATION (y/H) TO REACH FAILURE
(CGS 1985)

Soil type	Active	Passive
Dense cohesionless	0.0005	0.002
Loose cohesionless	0.002	0.006
Stiff cohesive	0.01	0.02
Soft cohesive	0.02	0.04

Examples 11.1 — 11.6

11.1 A retaining wall, with a smooth, vertical back of height 8 m, supports a cohesionless backfill of unit weight 19 kN/m³ and angle of shearing resistance of 30°. The surface of the soil is horizontal. Find the total active pressure P_a per lineal metre of the wall by the Rankine theory.

What is the increase in horizontal pressure if the soil slopes up from the top of wall at an angle of 30° to the horizontal?

(a) *Horizontal soil surface.*

$$K_a = \frac{1 - \sin 30°}{1 + \sin 30°} = 0.33$$

$$P_{a1} = \tfrac{1}{2} K_a \gamma H^2 = 200.6 \text{ kN/lineal metre}$$

(b) *Sloping soil surface.*

$$K_a = \cos \beta \; \frac{\cos \beta - \sqrt{(\cos^2 \beta - \cos^2 \phi)}}{\cos \beta + \sqrt{(\cos^2 \beta - \cos^2 \phi)}}$$

$$\beta = \phi = 30°$$

$$K_a = \cos \beta = \cos \phi = 0.87$$

$$P_{a2} = \tfrac{1}{2} K_a \gamma H^2 = 526.5 \text{ kN/ lineal metre}$$

P_{a2} is assumed parallel to the surface, hence horizontal component = $P_{a2} \cos \beta$ = 456 kN and increase in pressure = 456 — 200.6 = 255.4 kN.

11.2 A smooth backed vertical wall, 10 m height, retains a soil having $\gamma = 20$ kN/m³, $\phi = 35°$. The soil surface is horizontal in level with the top of wall and carries a uniformly distributed load of 50 kPa. Find the total thrust per linear metre of wall and its point of application.

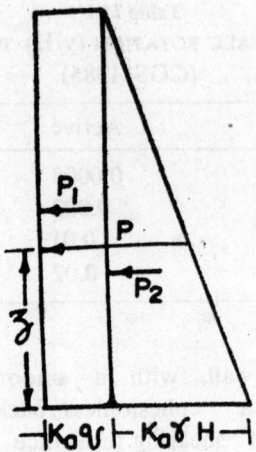

Fig. 11.10   Example 11.2

$\phi = 35°$, $K_a = 0.27$

$K_a q = 13.5$ kN/m²

$P_1 = K_a q H = 135$ kN at $H/2 = 5$ m above base

$P_2 = \frac{1}{2} K_a \gamma H^2 = 270$ kN at $H/3 = 3.33$ m above base

Total $P = P_1 + P_2 = 405$ kN/m

Taking moments about base,

$$P \times z = P_1 \frac{H}{2} + P_2 \frac{H}{3}$$

$$z = 3.87 \text{ m}$$

11.3 The top 4.5 m of fill behind a retaining wall with vertical back of 7.5 m height has $\gamma_1 = 20$ kN/m³, $\phi_1 = 35^0$ and c = 0; for the lower 3 m the values are $\gamma_2 = 18$ kN/m³, $\phi_2 = 30°$ and c = 0. The surface of the fill is horizontal in level with the top of wall. Find the magnitude and point of application of the thrust on the wall per lineal metre.

$\phi_1 = 35°$, $K_{a1} = 0.27$, $\phi_2 = 30°$, $K_{a2} = 0.33$

$p_1 = K_{a1} \gamma_1 H_1 = 24.3$ kN/m²

$p_2 = K_{a2} \gamma_1 H_1 = 29.7$ kN/m²

$p_3 = K_{a2} \gamma_2 H_2 = 17.82$ kN/m²

$P_1 = 54.67$ kN @ 4.5 m above base

$P_2 = 89.1$ kN @ 1.5 m above base

$P_3 = 26.73$ kN @ 1 m above base

$P = P_1 + P_2 + P_3 = 170.5$ kN/m

Taking moments about base, P acts at 2.38 m above base.

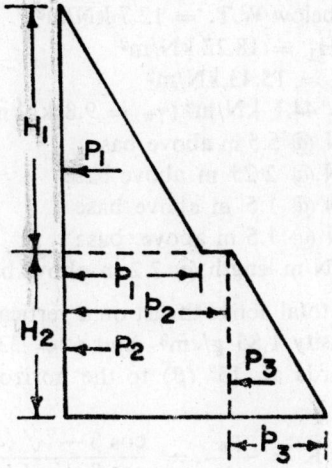

Fig. 11.11 Example 11.3

11.4 Vertical back of wall, H = 7.5 m. Water table behind the wall only at 3 m below top. Cohesionless fill in level with top of wall having $\gamma_{sat} = 22.5$ kN/m³, $\phi = 35°$. The top 3 m of fill above water table is also saturated by capillary moisture. Find the total thrust and its point of application.

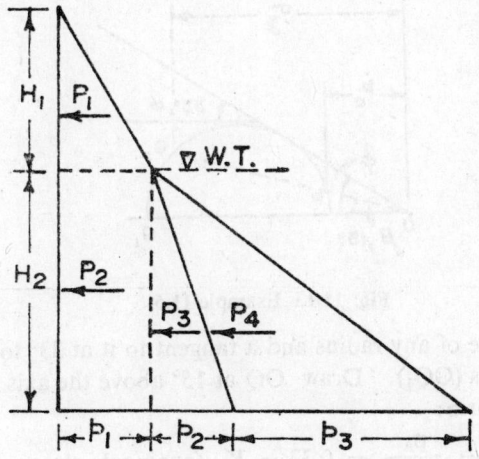

Fig. 11.12 Example 11.4

$\phi = 35°$, assumed the same above and below the water table

$K_a = 0.27$, γ' below W.T. $= 12.7$ kN/m³

$p_1 = K_a\, \gamma_{sat}\, H_1 = 18.22$ kN/m²

$p_2 = K_a\, \gamma' H_2 = 15.43$ kN/m²

$p_3 = \gamma_w\, H_2 = 44.1$ kN/m² ($\gamma_w = 9.8$ kN/m³)

$P_1 = 27.33$ kN @ 5.5 m above base

$P_2 = 81.99$ kN @ 2 25 m above base

$P_3 = 34.72$ kN @ 1.5 m above base

$P_4 = 99.23$ kN @ 1.5 m above base

$P = 243.27$ kN/m length @ 2.2 m above base

11.5 Calculate the total active thrust on a vertical wall, 9 m high, retaining sand of density 1.85 g/cm³ and $\phi = 33°$. The surface of sand slopes upwards at 15° (β) to the horizontal. Use the Rankine method.

Conjugate stress ratio $K = \dfrac{p_a}{\sigma_z} = \dfrac{\cos \beta - \sqrt{(\cos^2 \beta - \cos^2 \phi)}}{\cos \beta + \sqrt{(\cos^2 \beta - \cos^2 \phi)}}$

$\qquad = 0.33$

$K_a = \cos \beta \times K = 0.96 \times 0.33 = 0.32$

p_a @ base $= K_a\, \gamma\, H = 52.2$ kN/m², ($\gamma = 1.85 \times 9.8$ k N/m³)

$P_a = \frac{1}{2}\, p_a\, H = 234.9$ kN/m, acting parallel to surface

Graphically.

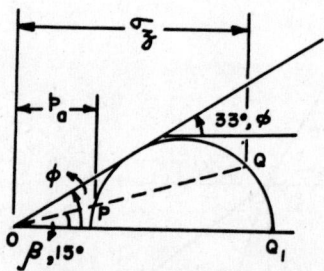

Fig. 11.13 Example 11.5

Draw a circle of any radius and a tangent to it at 33° to get point O on the x-axis (OQ_1). Draw OQ at 15° above the axis. Measure and find the ratio:

$$\frac{OP}{OQ} = \frac{p_a}{\sigma_z} = 0.33 = K \quad \text{(conjugate stress ratio)}$$

σ_z @ base $= \gamma H \cos \beta = 157.6$ kN/m²

p_a @ base $= 157.6 \times 0.33 = 52$ kN/m²

$P_a = 234$ kN/m

11.6 A 6 m high vertical wall retains a sandy clay of bulk unit weight 18 kN/m³ and $c' = 15$ kPa, $\phi' = 20°$. The surface of the fill is horizontal. Neglecting any effect of wall friction, determine the pressure intensities at the top and base of the wall and the likely depth of tension cracks in the fill.

Determine the pressure intensities and depths of tension cracks if the soil supports a uniform surcharge load of (a) 20 kPa or (b) 50 kPa. What is the increase in total pressure if water collects in the tension cracks in case of $q = 20$ kPa?

$\phi = 20°$, $K_a = 0.49$, $\sqrt{K_a} = 0.7$

$p_a = K_a \gamma z - 2 c' \sqrt{K_a}$

At top, $p_a = -2 c' \sqrt{K_a} = -21$ kPa

At base, $p_a = 31.92$ kPa

Depth of crack $z_0 = \dfrac{2c'}{\gamma \sqrt{K_a}}$ 2.38 m

Surcharge $q = 20$ kPa

At top, $p_a = -2 c' \sqrt{K_a} + K_a q = -11.2$ kPa

$z_0 = \dfrac{2 c'}{\gamma \sqrt{K_a}} - \dfrac{q}{\gamma} = 1.27$ m

At base, $p_a = K_a \gamma H - 2 c' \sqrt{K_a} + K_a q$

or $p_a = K_a \gamma (H - z_0) = 41.72$ kPa

Allowing for water in the tension cracks the total thrust is increased by:

$P_w = \frac{1}{2} \gamma_w z_0^2 = 7.9$ kN/m

Surcharge $q = 50$ kPa

At top, $p_a = 14$ kPa (+ ve)

At base, $p_a = p_a$ @ top $+ K_a \gamma H = 66.92$ kPa

There is no negative pressure and no cracking.

11.5 COULOMB ANALYSIS

Instead of considering the equilibrium of an element in a stressed mass like Rankine, Coulomb (1776) considered the stability, as a whole, of the *wedge* of soil between the wall and a trial, *single* failure surface. The basic assumptions of the analysis are as follows :
(1) The soil is dry, homogeneous, isotropic and elastically undeformable but breakable.
(2) When the wall yields a soil wedge is torn off from the rest of the soil mass. In the active state due to its own weight, the soil wedge slides downwards over a failure surface. In the passive state due to forcing the wall against the backfill, the soil wedge slides upwards on the failure surface. The wedge itself is considered as a rigid body. In either case, the wedge is in a condition of 'limiting equilibrium'.
(3) The failure surface, for the sake of convenience in analysis, is assumed to be a *plane* passing through the heel of the wall.
(4) Friction between the wall and the soil is taken into account.
(5) Failure is a two-dimensional problem (plane strain condition), i.e. a unit length of an infinite long wall is considered.
(6) The position and direction of the resultant earth pressure is assumed to be known. The resultant pressure is inclined at an angle δ with the normal of the back of the wall, where δ denotes the angle of friction between the wall and the soil and is termed as "angle of wall friction". For a backfill with plane surface, the resultant pressure acts on the back of the wall at a point one-third the height of the wall above the base.

Unless a wall is settling, friction on its back acts upwards on the active wedge (angle δ is positive), reducing active pressure. Wall friction acts downwards on a passive wedge (angle δ is negative) resisting its upward movement and increasing passive pressure. The effect of wall friction on active pressure is, in general, small and may be disregarded. The effect on passive pressure is large but definite movement is necessary for the mobilization of wall friction.

The main deficiency of the Coulomb theory is the assumption of a plane failure surface. Due to wall friction the failure surface is actually curved near the bottom of the wall. The curvature is slight in the active case and the error (under-estimation) involved in assuming a plane surface is relatively small (not likely to exceed 5% in the majority of practical problems, Scott 1974). In the passive case, however, the curvature is very much greater and increases with increasing value of δ. For values of δ less than $\phi'/3$, the error is usually small to be acceptable, but for $\delta > \phi'/3$ (or $\phi' >$ about 20°) the passive pressure may be over-estimated by 50% or more (Scott 1974). When $\delta = 0$, the result of the Coulomb theory is identical to that of the Rankine for the case of a vertical wall and a horizontal soil surface. Regardless of the above objection, the Coulomb theory usually gives quite useful results in practice. The theory is effective in proportion to the reliability of the soil parameters to be used in the analysis. The theory is not recommended for passive pressure determination in cohesive soils since the actual failure surface in these soils is markedly curved.

1. **Active pressure — cohesionless soil.** Consider a soil wedge ABC in Fig. 11.14a. BC is the trial failure plane inclined at angle θ to the horizontal. For the failure condition the soil wedge is in equilibrium under three forces: weight W of the wedge, *reaction* P to the force between the soil and the wall (which is equal to the active force) and the reaction R on the failure plane. The reaction P (active force) is inclined at the angle of wall friction δ to the normal to the back of the wall. The angle δ is commonly found to lie between $\phi'/3$ and $2\phi'/3$. For most

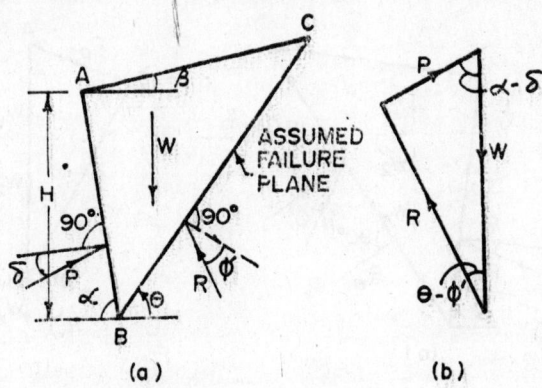

(a) (b)

Fig. 11.14 Coulomb solution for active pressure—trial wedge method

practical conditions, the backfill moves down relative to the wall and the inclination of P is below the normal. (Resultant pressure P_a will act at δ above the normal). However, if the wall were to settle more than the backfill, P would be inclined at the angle δ above the normal. The reaction R is the resultant of the normal and frictional forces on the failure surface and is inclined at ϕ' below the normal to the failure plane. The direction of all the three forces and the magnitude of W are known. The magnitude of P for the trial wedge in question can be determined by drawing a force triangle, as shown in Fig. 11.14b.

A number of trial failure planes are selected and the one which gives the maximum value of P is the 'critical failure plane' for the 'critical wedge'. The maximum value of P then represents the total active force (P_a) on the wall.

The above described graphical method of obtaining P_a by drawing force polygons for trial wedges, is known as the "trial wedge method". An analytical approach, as adopted by Coulomb, is to express P in terms of W and the angles in the triangle of forces. Then the maximum value of P corresponding to a particular value of θ is given by $\partial P/\partial\theta = 0$. This leads to the solution for P_a which can be expressed as:

$$P_a = \tfrac{1}{2} K_a \, \gamma \, H^2 \qquad (11.29)$$

where $K_a = \dfrac{\sin^2(\alpha + \phi')}{\sin^2\alpha \sin(\alpha-\delta)\left(1 + \sqrt{\dfrac{\sin(\phi'+\delta)\sin(\phi'-\beta)}{\sin(\alpha-\delta)\sin(\alpha+\beta)}}\right)^2}$

$$(11.30)$$

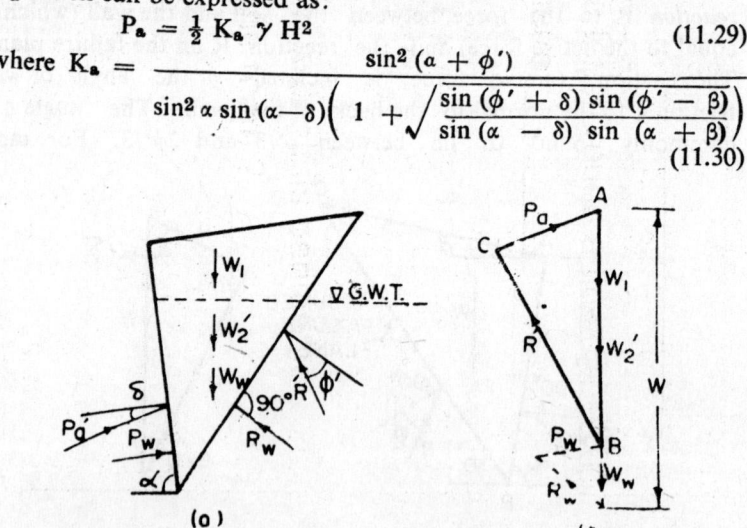

Fig. 11.15 Trial wedge method for submerged backfill, active case

(a) *Effect of submergence*. As considered in the Rankine analysis, partial or total submergence of a backfill results in an increase in the total pressure on the wall in the active case (or a decrease in the total pressure in the passive case). In the case where the ground-water table is horizontal and no seepage forces exist, the force system on the trial wedge is shown in Fig. 11.15.

R_w and P_w are the resultant water pressures acting normally on the failure plane and the back of wall respectively. The total weight W of the wedge consists of (i) total weight W_1 of soil above the water table and (ii) submerged weight W_2' of soil below the water table, given by:

$$W_2' = \frac{h^2}{2}(\cot \alpha + \cot \theta)\gamma' \qquad (11.31)$$

where h = height of water level above the base. Similarly, the weight of water W_w in the trial wedge below the water table is:

$$W_w = \frac{h^2}{2}(\cot \alpha + \cot \theta)\gamma_w \qquad (11.32)$$

The force polygon is shown in Fig. 11.15b. The vector AB represents $(W - W_w)$. The total thrust on the wall will be the vector sum of P_a and P_w. The vector sum of the two resultant water pressures P_w and R_w is equal to W_w. Since P_w may be easily calculated, lower triangle of the force polygon need not be drawn. P_w is the same for each wedge and hence, the maximum thrust results from the same wedge which gives the maximum value of P_a'.

2. Culmann construction. Culmann (1866) developed a convenient graphical procedure for the Coulomb solution. It is essentially the trial wedge method in which the force polygon is rotated so that it can be constructed directly on a scale drawing of the back of the wall and the backfill. The procedure, illustrated in Fig. 11.16, is described below:

(1) Draw to scale the cross-section of the backfill and the back of wall.

(2) From point B draw line BC at ϕ' to the horizontal. BC is known as the 'natural slope' line (ϕ – line), or the 'weight line'; (weight vector is plotted on this line).

(3) Lay off the line BD at an angle $(\alpha - \delta)$ with the line BC. BD is called the 'position line'. It is parallel to the direction which the P vector assumes in the rotated force polygon.

(4) Draw a trial failure plane BF_1. Determine the weight of the wedge and plot it on the weight line BC, starting at the base

point B. BF_1 also serves to define the direction of the R vector.

(5) Through the end of the weight vector, point b_1, draw a line b_1a_1 parallel to the position line BD and locate the point of intersection a_1 with BF_1. The length of the line a_1b_1 represents the magnitude of P vector for the particular assumed failure plane.

(6) Repeat steps 4 and 5 for several additional trial failure planes.

(7) Through the locus of end points of vector, $a_1, a_2 \ldots$, draw a smooth curve, called the "Culmann line".

(8) Draw a line parallel to BC and tangent to the Culmann line at the position of maximum force which is the active earth pressure P_a.

(9) From the point of tangency 'a' on the Culmann line (step 8), draw ab parallel to BD to intersect BC. The length of line ab gives the magnitude of P_a. The point of tangency 'a' is the point through which the 'critical' failure plane passes.

The Culmann construction is a universal method in the sense that it can be applied to any irregular backfill or loading condition on the backfill.

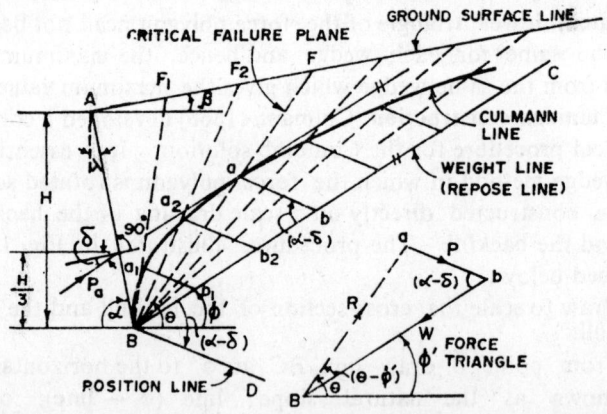

Fig. 11.16 Culmann graphical construction—active case, cohesionless soil

3. Poncelet construction. Poncelet (1840) gave a graphical solution of the Coulomb analysis, which is applicable to a wall with an unbroken, straight back and for a plane ground surface. The ground surface may have a uniformly distributed surcharge.

The method is also known as the *Rebhann method* (1871), and is illustrated in Fig. 11.17.

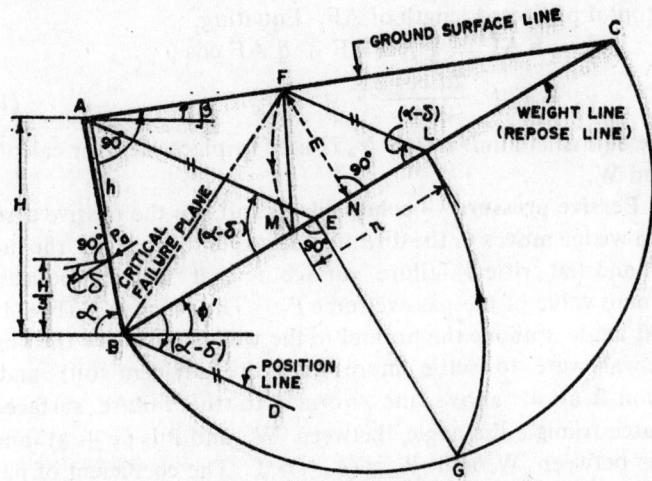

Fig. 11.17 Poncelet (Rebhann) graphical construction, active case, cohesionless soil

The ground surface line AC, the weight line BC and the position line BD are drawn as for the Culmann method. (AC and BC should meet for further construction). A semi-circle is drawn on BC. AE is drawn parallel to BD. From point E, EG is drawn perpendicular to BC to cut the semi-circle at G. With B as centre and BG as radius, point L is obtained on BC. LF is drawn parallel to AE, cutting AC at point F. The line BF now represents the Coulomb most dangerous failure plane and defines the size of the cirtical sliding wedge ABF.

With L as centre, LF as radious, point M is obtained on BC. FN is drawn perpendicular to BC. Let FN = m and LM = n. The area of Δ FLM ($= 0.5$ m n) multiplied by the unit weight of soil γ gives the total active pressure P_a acting at height H/3 and inclined at an angle δ to the normal of the back of the wall.

$$P_a = \tfrac{1}{2}\,\gamma\ m\,n \qquad\qquad (11.33)$$

and the weight of the sliding wedge W equals $0.5\,\gamma$ m BL

(a) *Surcharged ground surface.* If the ground surface is surcharged with a uniformly distributed load of intensity q*, an equivalent unit weight γ_e of the wedge is calculated as follows.

---

* Per unit area in plan

Let h be height of Δ ABF. The weight of wedge ABF with surcharge will be $(0.5 \, \gamma h \, AF + q \, AF \cos \beta)$, where AF $\cos \beta$ is the horizontal projected length of AF. Equating,

$$\tfrac{1}{2} \, \gamma_e \, h \, AF = \tfrac{1}{2} \, \gamma \, h \, AF + q \, AF \cos \beta$$

or $$\gamma_e = \gamma + \frac{2q \cos \beta}{h} \tag{11.34}$$

The equivalent unit weight γ_e is used in place of γ for calculating P_a and W.

4. **Passive pressure — cohesionlesss soil.** In the passive case, the failure wedge moves in the direction opposite to that of the active case, and the critical failure surface is that which produces the minimum value of the passive force P_p. The *reaction* P (Fig. 11.16) acts at angle δ above the normal to the wall (or δ below the normal, if the wall were to settle more than the adjacent soil) and the reaction R at ϕ' above the normal to the failure surface. In the force triangle the angle between W and P is $(\alpha + \delta)$ and the angle between W and R is $(\theta + \phi')$. The coefficient of passive pressure K_p is given by:

$$K_p = \frac{\sin^2 (\alpha - \phi')}{\sin^2 \alpha \sin (\alpha + \delta) \left(1 - \sqrt{\dfrac{\sin (\phi' + \delta) \sin (\phi' + \beta)}{\sin (\alpha + \delta) \sin (\alpha + \beta)}} \right)^2} \tag{11.35}$$

Example 11.7

The back of a 6 m high retaining wall is inclined *away* from the fill at 10° to the vertical. The surface of backfill slopes upwards at 8° from the horizontal. The backfill is dry, cohesionless and has the following properties: $\gamma = 17 \, kN/m^3$ and $\phi' = 32°$. Assuming the angle of wall friction as 25°, determine analytically and graphically the total active thrust and its point of application.

Referring to Fig. 11.16, $\alpha = 90 - \lambda = 90 - 10 = 80°$, $\beta = 8°$, $\phi' = 32°$, $\delta = 25°$, $\gamma = 17 \, kN/m^3$, H = 6 m

Eq. 11.30: $K_a = 0.4$

and $P_a = \tfrac{1}{2} \, K_a \, \gamma H^2 = 122.4 \, kN/m$ acting at 2 m above the base and inclined at $\delta = 25°$ above the normal to the back of wall.

(a) *Culmann construction.* (Fig. 11.16). In case of plane soil surface, it is convenient to have trial failure planes BF_1, BF_2,... such that $AF_1 = F_1F_2 = F_2F_3$... In Fig. 11.16, AF_1 is 2 m. The perpendicular distance h of point B above AF_1 is measured or can be calculated as:

$$h = H \ \frac{\cos(\beta \sim \lambda)}{\cos \lambda} = 6.089 \ m$$

Weight W of trial wedge $ABF_1 = \dfrac{h}{2} \times 2 \times 17 = 103.5$ kN,

which is marked to scale as Bb_1 on the weight line. The weight of the second trial wedge ABF_2 will be 2 W (as $AF_2 = 2 \ AF_1$) and will be represented by Bb_2 ($Bb_2 = 2 \ Bb_1$). The construction is completed to give 'ab' as the pressure vector, which is measured and on multiplication by the scale factor yields $P_a \approx 124$ kN.

(b) *Rebhann construction.* It is given in Fig. 11.17. Same value of P_a is obtained subject to the precision of graphical construction.

11.6 STABILITY OF RETAINING WALL

For stability a retaining wall should be safe against (i) overturning, (ii) sliding and (iii) bearing pressure failure.

1   Safety against overturning. There are two considerations for 'overturning'; The wall should not actually tip forward by rotating about its toe, and the tendency for rotation about the toe should not be strong enough to lead to a zone of zero contact pressure between the base of the wall and the soil near the heel. It is, of course, desirable to maintain a positive contact pressure (compressive) over the entire base. To ensure positive pressures over the entire base width B, the resultant R of all the forces on the wall must act within the middle third of the base, i. e. the eccentricity 'e' of the base resultant must not exceed B/6. This will fulfil both the above mentioned considerations of stability against overturning.

The first step in the design is to determine all the forces acting on the wall and find the position of the base resultant R. Considering a typical force system on a rigid retaining wall, as shown in Fig. 11.18, the position of force R is found by taking moments above the toe (or the heel). The contribution to stability of the soil in front of the wall is usually ignored, i.e. P_p is assumed as zero. Moments about the toe give:

$$y = \frac{W.a + P_v.d - P_h.b}{W + P_v} \tag{11.36}$$

and the safety factor F againt overturning is given by (Carter 1983, Lambe and Whitman 1973):

$$F = \frac{W.a}{P_h.o - P_v.d} \geq 1.5 \qquad (11.37)$$

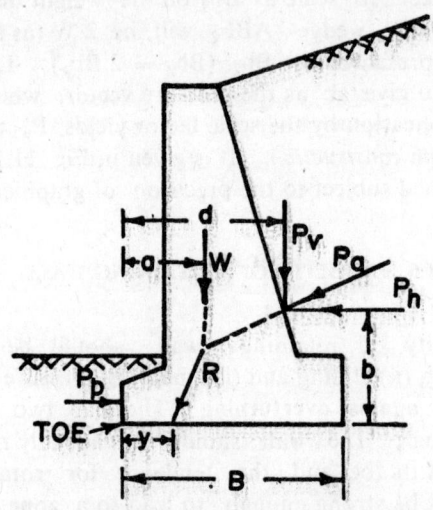

Fig. 11.18 Typical force system on a rigid retaining wall

If R is within the middle third (for soil) or the middle half (for rock), overturning may be neglected. The weight W includes the weight of wall and soil in front of the wall above the wall footing. In counterfort and cantilever walls, the weight of soil above the footing behing the wall is also included in W.

2. Safety against sliding. Sliding is resisted by the base friction and adhesion which may be developed. The safety factor is expressed as:

(a) Ignoring the soil in front of the wall,

$$F = \frac{(W + P_v) \tan \delta + c_a B}{P_h} \geq 1.5 \qquad (11.38)$$

(b) Allowing for passive pressure in front of wall,

$$F = \frac{(W + P_v) \tan \delta + c_a B + P_p}{P_h} \geq 2 \qquad (11.39)$$

where c_a and δ are the adhesion and angle of friction along the base. The values of tan δ (coefficient of friction between poured concrete base of retaining wall and soil) may be adopted as follows (Perloff and Baron 1976): coarse grained sands and gravels without silt 0.55, silty sands and gravels 0.45, cohesionless silts 0.35, sound rock, roughed surface 0.60. Adhesion c_a may be taken as the undrained shearing resistance, i.e. half the unconfined compression strength, provided $c_a \leqslant 48$ kPa.

3. **Safety against bearing pressure failure.** If R_v is the vertical component of the resultant force R, i.e. the total downward thrust, acting at an eccentricity 'e' not greater than B/6, the base pressures are positive, as shown in Fig. 11.19a. As the section modulus of the base is $B^2/6$, the maximum pressure p_{max} at toe is the sum of the direct pressure and that due to bending:

$$p_{max} = \frac{R_v}{B} + \frac{6 R_v e}{B^2} = \frac{R_v}{B} (1 + \frac{6 e}{B})$$

(11.40)

and the minimum pressure p_{min} at heel is:

$$p_{min} = \frac{R_v}{B} (1 - \frac{6 e}{B})$$

(11.41)

When R_v is at the middle third, i.e. e = B/6,

$$P_{max} = 2 R_v/B \text{ and } p_{min} = 0$$

(11.42)

If R_v lies outside the middle third, pressure at the heel will be negative, indicating tension, which is usually assumed not to develop and the pressure distribution is assumed as shown in Fig. 11.19b.

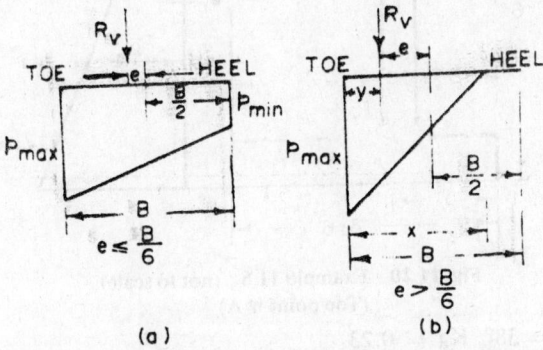

Fig. 11.19 Bearing pressures due to eccentric load

Equating forces per unit length vertically,
$$R_v = \tfrac{1}{2} \, p_{max} \, x$$
Taking moment about A,
$$R_v \, y = (\tfrac{1}{2} \, p_{max} \, x) \times \tfrac{1}{3} \, x = \tfrac{1}{6} \, p_{max} \, x^2$$
Hence, $p_{max} = \dfrac{2 \, R_v}{3 \, y}$ and $x = 3 \, y$ \hfill (11.43)

The maximum base pressures must not exceed the allowable bearing pressure for the soil. If the resultant acts within the middle third of the base the maximum pressure is less than twice the average pressure. The base pressures and the bearing capacity under eccentric loading can also be determined as described in Chapter 15.

Example 11.8 — 11.10

11.8 Examine the stability of a R.C.C. cantilever retaining wall as detailed in Fig. 11.20 with respect to bearing failure, sliding and overturning. The relevant properties are: γ_c (concrete) = 23.5 kN/m³, γ (soil) = 17 kN/m³, φ' = 38°, c' = 0, δ (at base) = 30°, safe bearing capacity = 200 kPa.

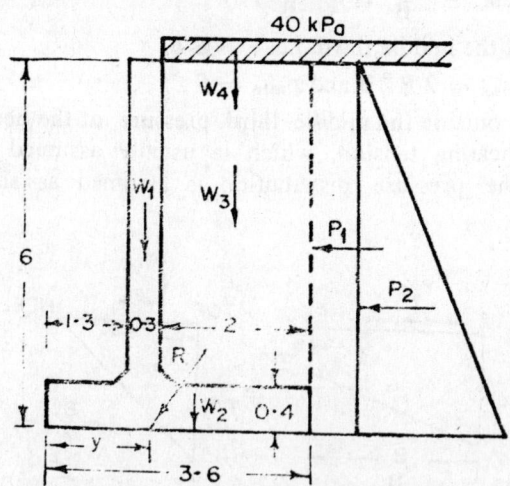

Fig. 11.20 Example 11 8   (not to scale)
(Toe point is A)

ϕ = 38°, K_a = 0.23
P_1 = $K_a \, q \, H$ = 55.2 kN @ 3 m from A

$P_2 = \frac{1}{2} K_a \gamma H^2 = 70.38$ kN @ 2 m from A

$R_h = P_1 + P_2 = 125.58$ kN

W_1 (stem) $= 0.3 \times 5.6 \times 23.5 = 39.48$ kN

@ 1.45 m from A

W_2 (base) $= 0.4 \times 3.6 \times 23.5 = 33.84$ kN

@ 1.8 m from A

W_3 (soil) $= 2 \times 5.6 \times 17 = 190.4$ kN @ 2.6 m from A

W_4 (load) $= 2 \times 40 = 80$ kN @ 2.6 m from A

$R_v = W_1 + W_2 + W_3 + W_4 = 343.72$ kN

Moment per metre about point A (M) = moment of vertical forces minus moment of horizontal forces,

$$= 821.19 - 306.36 = 514.63 \text{ kNm}$$

Lever arm of base resultant $y = \dfrac{M}{R_v} = 1.5$ m,

i.e. the resultant acts within the middle third of the base.

$e = 1.8 - y = 0.3$ m

$p_{max} = \dfrac{R_v}{B} (1 + \dfrac{6e}{B}) = 143.22$ kPa (less than the bearing capacity).

$p_{min} = \dfrac{R_v}{B} (1 - \dfrac{6e}{B}) = 47.74$ kPa

F against sliding $= \dfrac{R_v \tan \delta}{R_h} = 1.56$

F against overturning $= \dfrac{M \text{ (vertical forces)}}{M \text{ (horizontal forces)}}$

$$= \dfrac{821.19}{306.36} = 2.68$$

11.9 Gravity retaining wall shown in Fig. 11.21. γ (soil) = 18 kN/m³, $\phi = 35°$, c = 0, γ_m (masonry) = 24 kN/m³, coefficient of base friction $\mu = 0.65$. Use Rankine method. Find p_{max}, p_{min} and safety factors against sliding and overturning.

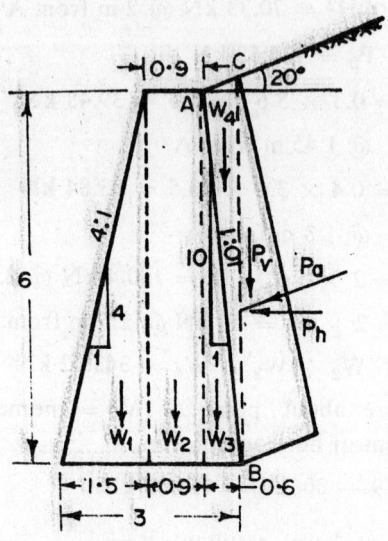

Fig. 11.21 Example 11.9
(Toe point is D)

$$K_a = \cos \beta \, \frac{\cos \beta - \sqrt{(\cos^2 \beta - \cos^2 \phi)}}{\cos \beta + \sqrt{(\cos^2 \beta - \cos^2 \phi)}} = 0.34$$

$BC = 6 + 0.6 \tan 20 = 6.22$ m

P_a on $BC = \frac{1}{2} K_a \gamma (BC)^2 = 118.39$ kN acting

parallel to soil surface

$P_h = P_a \cos 20° = 111.25$ kN @ 2.07 m from D

$P_v = P_a \sin 20° = 40.49$ kN @ 3 m from D

$W_1 = 108$ kN @ 1.0 m from D

$W_2 = 129.6$ kN @ 1.95 m from D

$W_3 = 43.2$ kN @ 2.6 m from D

W_4 (soil ABC) $= \frac{1}{2} \times 0.6 \times 6.22 \times 18 = 33.59$ kN

@ 2.8 m from D

Summation of forces and moments about D are given in Table 11.2.

Table 11.2 Example 11.9

Force	Vertical force (kN)	Horizontal force (kN)	Lever arm from D (m)	Moment about D (kNm)
W_1	108.00	—	1.0 (x_1)	+ 108.00
W_2	129.60	—	1.95 (x_2)	+ 252.72
W_3	43.20	—	2.6 (x_3)	+ 112.32
W_4	33.59	—	2.8 (x_4)	+ 94.05
$P_1 \sin 20°$	40.49	—	3.0	+ 121.47
$P_2 \cos 20°$	—	111.25	2.07	— 230.29
	354.88 (R_v)	111.25 (R_h)		458.27 (M)

$$y = \frac{M}{R_v} = 1.29 \text{ m (within middle third of base)}$$

$$e = 1.5 - 1.29 = 0.21 \text{ m}$$

$$p_{max} = \frac{R_v}{B}\left(1 + \frac{6 e}{B}\right) = 167.97 \text{ kPa}$$

$$p_{min} = 68.61 \text{ kPa}$$

$$F \text{ (sliding)} = \frac{\mu R_v}{R_h} = 2.07$$

$$F \text{ (overturning)} = \frac{W_1 x_1 + W_2 x_2 + W_3 x_3 + W_4 x_4}{(P_h \times 2.07) - (P_v \times 3)}$$

$$= 5.2$$

11.10 Gravity retaining wall shown in Fig. 11.22. Given: γ (soil) = 17 kN/m³, $\phi = 38°$, δ (for back and base) = 25°, γ_m (wall) = 23.5 kN/m³. Find p_{max}, p_{min} and safety factors against sliding and overturning.

K_a (Coulomb Eq. 11.30) = 0.39

$P_a = 119.34$ kN @ 2 m above base inclined at 25° above normal, i.e. at 35° above horizontal (*Note*. P_a by Coulomb analysis takes also into account the weight of soil between the back of wall and the vertical through B.)

P_2 (horizontal) = $P_a \cos 35° = 97.76$ kN @ 2 m from A.

Moment $M_6 = 195.5$ kN m (— ve)

P_1 (vertical) = $P_a \sin 35° = 68.45$ kN @ 2.65 m from A,

Moment $M_5 = 181.39$ kN m

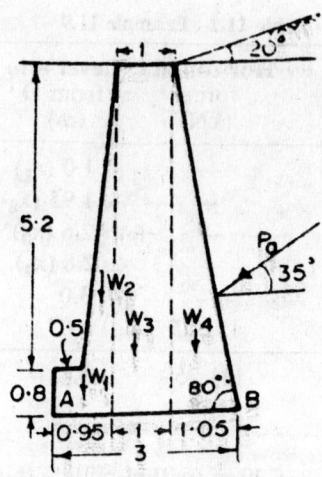

Fig. 11.22 Example 11.10

W_1 = 17.86 kN @ 0.475 m from A, M_1 = 8.48 kNm

W_2 = 27.49 kN @ 0.80 m, M_2 = 22.0 kNm

W_3 = 141.0 kN @ 1.45 m, M_3 = 204.45 kNm

W_4 = 74.03 kN @ 2.30 m, M_4 = 170.27 kNm

M = $M_1 + M_2 + M_3 + M_4 + M_5 - M_6$ = 391.09 kNm

R_v = $W_1 + W_2 + W_3 + W_4 + P_1$ = 328.8 kN

$y = \dfrac{M}{R_v}$ = 1.19 m (within middle third)

$e = 1.5 - y = 0.31$ m

$p_{max} = \dfrac{R_v}{B}\left(1 + \dfrac{6}{B}\right)$ = 177.55 kPa

p_{min} = 41.65 kPa

F (sliding) = $\dfrac{R_v \tan \delta}{P_2}$ = 1.57

F (overturning) = $\dfrac{\Sigma M \text{ (of wall only)}}{P_2 \times 2 - P_1 \times 2.65}$

$= \dfrac{405.2}{14.11} = 28.7$

11.7 SHEET PILE WALLS

Sheet piles are closely set piles of timber, steel, reinforced or prestressed concrete, which are driven vertically into the ground in a line to hold back soil and water. A wall of sheet piles is known as a "sheet pile wall" or "sheet piling", which may be a cantilever wall or anchored back at one or two levels. An anchored sheet pile wall is usually known as a "bulkhead".

A sheet pile wall, unlike a rigid gravity retaining wall is to some extent flexible. Its weight is negligible compared with the remaining forces involved. It depends for its stability on the passive resistance of the soil in front of, and behind, the lower part of the wall and the additional support given by anchors or struts.

1. Cantilever sheet pile wall. A cantilever sheet pile wall is used only when the retained height of soil is relatively small (upto about 5 m). The wall is generally considered to act as a rigid structure and it fails by rotation about some point above the base which develops passive resistance of soil in front of the wall above the point of rotation, as well as behind the wall below this point. The exact pressure distribution near the base of wall is not easily determined. An approximate pressure distribution is assumed as shown in Fig. 11.23. The force R is assumed to represent the passive resistance below the point of rotation.

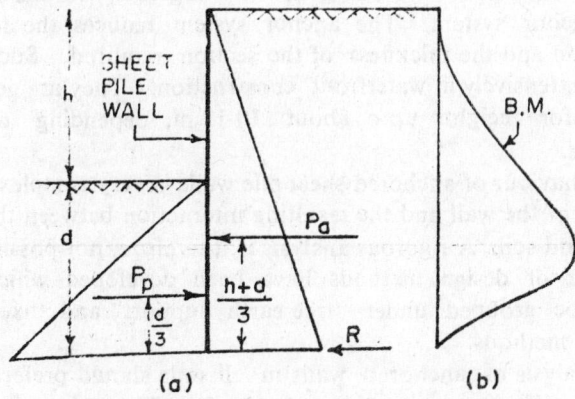

Fig. 11.23 Approximate pressure distribution and typical bending moment diagram for cantilever sheet pile wall, cohesionless soil

The minimum driving depth 'd' can be obtained by taking moments about the base. Thus,

$$P_p \frac{d}{3} = P_a \frac{h+d}{3} \qquad (11.44)$$

The calculated depth d is then increased arbitrarily by 20% to give a factor of safety and allow the passive resistance R to be developed. Alternatively, a factor of safety is probably better introduced by dividing the passive resistance of soil in front of the wall by a suitable factor (usually about 2). Thus,

$$\frac{P_p}{F} \cdot \frac{d}{3} = P_a \frac{h+d}{3} \qquad (11.45)$$

The active and passive pressure diagrams must include water pressure, if present. Surcharge pressures must also be added. if present on the active side. These walls are not suitable for permanent support in clays except those having low compressibility. Where used for permanent support, the analysis is based on effective angle φ', neglecting cohesion. For temporary support, analysis is based on undrained strength s_u ($= c_T$) and the computed pressures may be negative, but still a minimum earth pressure of $0.25 \, \gamma z$ at any depth z is assumed on the active side (CGS 1985).

2. Anchored sheet pile wall. Anchored sheet pile walls are supported near their top ends by rods or cables anchored in the soil some distance behind the wall. Their stability is due to the passive resistance developed in front of the wall together with the support of the anchor system. The anchor system reduces the depth of penetration and the thickness of the section required. Such walls are used extensively in waterfront construction. They are generally suitable for heights upto about 10-12 m, depending on soil conditions.

The behaviour of anchored sheet pile walls is very complex due to flexibility of the wall and the resulting interaction between the wall, anchors and soil. A rigorous analysis is, therefore, not possible and a number of design methods have been developed which may broadly be grouped under "free earth support" and "fixed earth support" methods.

The analysis of anchored walls in all soils should preferably be based on effective stress (ϕ', and $c' = 0$). The active pressure is calculated on the following basis (CGS 1985) : (i) using coefficient of active earth pressure K_a, where moderate movement is possible, (ii) using a coefficient $K = 0.5 (K_a + K_o)$, where foundations of buildings or services exist at shallow depth at a distance of 0.5 h to h (height of wall) behind the top of wall, and (iii) using coefficient of earth pressure at rest K_o, where foundations of buildings or services exist at shallow depth at a distance less than 0.5 h behind the top of wall. Water pressures and surcharge effects should be considered where present.

The design for short term works of stiff cohesive soils is based on c_T with $\phi_T = 0$. Where computed active pressure p_a is zero, a minimum pressure of $0.25\, \gamma z$ is assumed. Below the water table, water pressure is added where p_a is negative.

(a) *Free earth support method*. It is assumed that the depth of penetration below the dredge line (ground level in front of the wall) is insufficient to produce fixity at the bottom end of the wall; the wall is free to rotate about its base. The wall itself is assumed to be inflexible so that it also rotates about the anchor point. The assumed pressure distribution, according to the Rankine theory, and the bending moment are shown in Fig. 11.24.

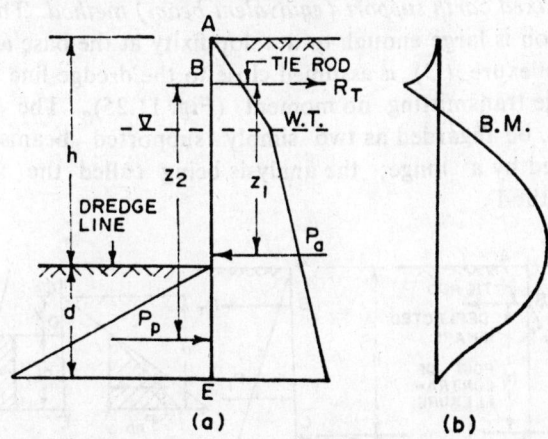

Fig. 11.24 Anchored sheet pile wall, free earth support

The required penetration depth d for stability is obtained by taking moments about the anchor point B,

$$P_p \times z_2 = P_a \times z_1 \qquad (11.46)$$

Out-of-balance water pressures must also be included in the calculations, if present. Eq. 11.46 is a cubic relation in d, which can be solved by giving trial values. The safety factor is introduced either by increasing the calculated depth d by 20 to 40% or, preferably, by dividing P_p by a factor (usually 2).

Once d has been determined, the force R_T on the rods is obtained by equating horizontal forces:

$$R_T = P_a - P_p/F \qquad (11.47)$$

Eq. 11.47 gives the force R_T per unit length of the wall; the *actual* force R_T' in each tie rod depends on the rod spacing. Finally, the bending moment diagram can be drawn; the maximum bending moment occurs at the point of zero shear and governs the pile section.

For a wall of known penetration depth, the safety factor with respect to P_p is obtained by taking moments about the anchor point which will give the value of mobilized passive resistance P_{pm} for equilibrium. The safety factor is the ratio of the ultimate, available P_p to P_{pm}. The force in the anchor R_T per metre of the wall is $P_p - P_{pm}$.

(b) *Fixed earth support (equivalent beam) method.* The depth of penetration is large enough to develop fixity at the base and a point of contraflexure (D) is assumed close to the dredge line which acts as a hinge transmitting no moment (Fig. 11.25). The wall can, therefore, be regarded as two simply supported beams, AD and DE, joined by a hinge; the analysis being called the 'equivalent beam method'.

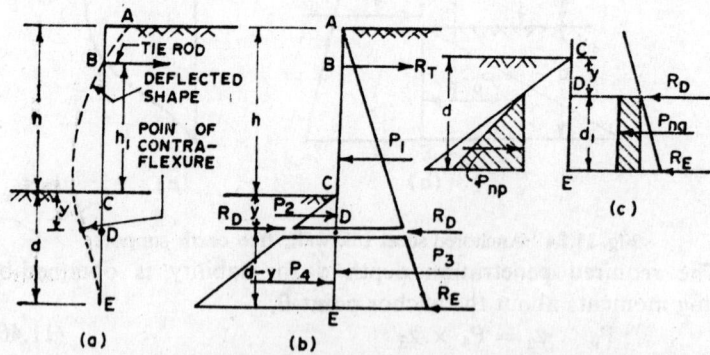

Fig. 11.25 Anchored sheet pile wall, fixed earth support (equivalent beam) method, (a) deflected shape, (b) forces on upper and lower parts, (c) net active and passive pressures on lower part

The value of depth 'y' to the point of contraflexure D is first assumed from values suggested by Terzaghai (1943), which are dependent on the values of ϕ' as given below:

ϕ'	20°	25°	30°	35°	40°
y	0.25 h	0.15h	0.08h	0.035h	— 0.007h

For most of the backfills the average ϕ' is 30° and y may be taken as 0.01 h involving only a little error.

Considering the upper beam AD:

$$P_1 = \tfrac{1}{2} K_a \gamma (h+y)^2 \qquad (11.48)$$

$$P_2 = \tfrac{1}{2} K_p \gamma y^2 \qquad (11.49)$$

Reaction R_D at the hinge can be found by taking moments about the anchor point B:

$$R_D (h_1 + y) = P_1 \left[\frac{2}{3} (h + y) - (h - h_1) \right]$$

$$- P_2 \left(h_1 + \frac{2}{3} y \right) \qquad (11.50)$$

The force R_T on the tie rods (per unit length of wall) can be obtained either by equating horizontal forces or by taking moments about D.

The forces on the lower beam DE consist of reaction R_D, active and passive soil pressures P_3 and P_4 and the unknown reaction R_E at E. Pressures P_3 and P_4 are calculated in terms of depth d. Moments of P_3, P_4 and R_D about E are equated to zero to solve for d. Any out-of-balance water pressures are also added in all the above calculations. For design purposes, the calculated value of d is increased by 20-40%.

The analysis for the lower beam DE can also be carried out conveniently in terms of *net* pressures. The total net pressure acting on the right hand side is given by:

$$P_{na} = p_a d_1 \text{ acting at } d_1/2 \text{ above E} \qquad (11.51)$$

where p_n = net pressure intensity at D

$$= \gamma [K_a (h + y) - K_p y] \qquad (11.52)$$

and the total net pressure on the left hand side is:

$$P_{np} = \tfrac{1}{2} (K_p - K_a) \gamma d_1^2 \text{ acting at } \frac{d_1}{3} \text{ above E}$$

$$(11.53)$$

Taking moments about E:

$$R_D \, d_1 = \tfrac{1}{6} (K_p - K_a) \, \gamma \, d_1^3 - \tfrac{1}{2} \, p_a \, d_1^2$$

and
$$d_1 = d - y = \frac{3 \, p_n^2 + [9 \, p_n^2 + 24 \, (K_p - K_a) \, \gamma \, R_D]^{0.5}}{2 \, (K_p - K_a) \, \gamma}$$

(11.54)

The value of p_n is often small, which may be taken as zero for simplicity and then:

$$d - y = \left[\frac{6 \, R_D}{(K_p - K_a)\gamma} \right]^{0.5}$$

(11.55)

and
$$d = y + \left[\frac{6 \, R_D}{(K_p - K_a) \, \gamma} \right]^{0.5}$$

(11.56)

Tie rod anchor. Tie rods are normally anchored in plates, beams or concrete blocks some distance behind the wall, as shown in Fig. 11.26. The anchor must be located beyond the plane FG to ensure that the passive wedge of the anchor does not encroach on the active wedge of the bulkhead. The lower end E of the active wedge is taken at the bottom of the wall for the free earth support and at two-third the embedded depth for fixed earth support (Cornfield 1975).

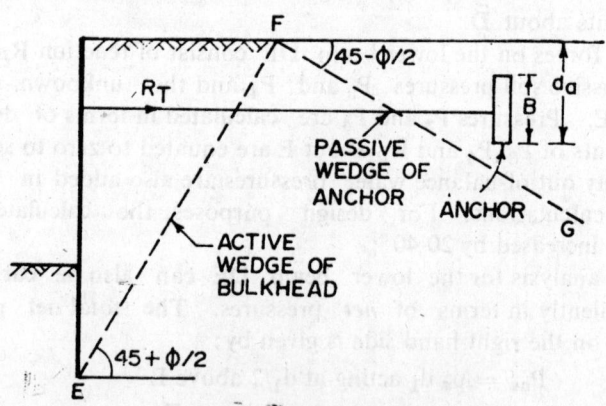

Fig. 11.26 Design of plate anchor

If an anchor of height B is buried to a depth d_a, where B is greater than $0.5 \, d_a$, the anchor is assumed to develop passive resistance over the depth d_a. Thus:

$$P_p = \tfrac{1}{2} K_p \gamma d_a^2, \quad p_a = \tfrac{1}{2} K_a \gamma d_a^2$$

and $\qquad R_T = P_p - P_a = \tfrac{1}{2} \gamma d_a^2 (K_p - K_a) \qquad\qquad (11.57)$

where $\qquad R_T = $ tie rod force per unit length of wall

If $\quad s = $ spacing of tie rods, $L = $ length of anchor per tie rod, and $F = $ safety factor,

$$R_T \, s = \frac{\gamma d_a^2 \, L}{2 \, F} (K_p - K_a) \qquad\qquad (11.58)$$

For a concrete anchor, δ/ϕ is taken as 0.5 for calculating K_p (Carter 1983). If the anchor is a continuous plate (wall), $s = L$ in Eq. 11.58.

Examples 11.11—11.15

11.11 A cantilever sheet pile wall is to support the sides of an excavation 2 m deep, in dry cohesionless soil having $\phi = 30°$ and $\gamma = 19 \text{ kN/m}^3$. Using a safety factor of 2.0 on the passive resistance of soil, determine the depth of penetration of the wall. Also determine the maximum bending moment in the sheet piling.

$$\phi = 30°, \, K_a = \tfrac{1}{3}, \, K_p = 3, \, \frac{K_p}{F} = 1.5$$

Referring to Fig. 11.23, h = 2 m

$$\frac{P_p}{F} \cdot \frac{d}{3} = P_a \, \frac{h+d}{3} \, ; \quad d = \text{ depth of penetration}$$

$$\tfrac{1}{2} \left(\frac{K_p}{F}\right) \gamma \, d^2 \cdot \frac{d}{3} = \tfrac{1}{2} K_a \gamma (h+d)^2 \frac{h+d}{3}$$

Substituting values and solving for d,
d = 3.08 m, say d = 3.1 m
The B.M. is maximum where shear force (S.F.) is zero.
Let S.F. be zero at d_0 below dredge line,
$\tfrac{1}{2} \times 1.5 \times 19 \times d_0^2 = \tfrac{1}{2} \times \tfrac{1}{3} \times 19 \times (d_0 + 2)^2$
$d_0 = 1.78$ m

B.M. at $d_0 = \tfrac{1}{2} K_a \gamma (h + d_0)^2 \frac{h + d_0}{3}$

$$- \tfrac{1}{2} \left(\frac{K_p}{F}\right) \gamma \, d_0^2 \, \frac{d_0}{3} = 30.2 \text{ kNm}$$

per metre run of wall

11.12 A sheet pile wall, anchored at a point 1 m below the top, is to support the sides of an excavation, 6 m deep, in dry sandy soil having $\phi = 35°$ and $\gamma = 19 \text{ kN/m}^3$. Using F = 2 on the passive resistance and assuming free earth support, calculate the depth of embedment and force in tie rods spaced at 2 metres.

$\phi = 35°$, $K_a = 0.27$, $K_p = 3.7$, $F = 2$

Referring to Fig. 11.24,

$P_a = \frac{1}{2} K_a \gamma (h + d)^2 = 2.565 (d + 6)^2$

$P_p = \frac{1}{2} K_p \gamma d^2 = 35.15 d^2$

Equating moments about anchor,

$$\left[-\frac{2}{3}(d + 6) - 1\right] P_a = \left[\frac{2}{3}d + 5\right]\frac{P_p}{F}$$

$d = 2.85$ m

$R_T = P_a - \dfrac{P_p}{F} = 58.144$ kN per metre run

Since spacing is 2 m, force in tie rod $= 2 R_T$

11.13 A sheet pile wall, anchored at 1 m below top, supports the sides of an excavation, 6 m deep in sandy soil having $\phi = 30°$ and $\rho = 1.9$ g cm^3. Find the factor of safety with respect to passive resistance if the sheet piling is embedded 4.2 m below dredge line. Find tension in tie rods spaced at 2 m centre to centre. Assume free earth support.

$K_a = 1/3$, $K_p = 3$, $\gamma = \rho \times 9.8 = 18.62$ kN/m^3

$P_a = 322.87$ kN @ 5.8 m from anchor point. Let P_{pm} be the mobilized passive resistance for equilibrium at $4.2/3 = 1.4$ m from bottom or 7.8 m from anchor point.

$P_{pm} \times 7.8 = P_a \times 5.8$; $P_{pm} = 240.082$ kN

P_p (available) $= \frac{1}{2} K_p \gamma d^2 = 492.685$ kN

$F = P_p/P_{pm} = 2.05$

$R_T = P_a - P_{pm} = 82.788$ kN/m

Force in one anchor $= 2 R_T = 165.6$ kN

11.14 Determine the minimum penetration depth of an anchored bulkhead to achieve fixed earth support conditions for the following data: dry cohesionless soil, $\rho = 1.9$ g/cm^3, $\phi = 30°$. depth of excavation to be supported $h = 6$ m, depth of anchor below top $= 1$ m (See Fig. 11.25).

$K_a = \frac{1}{3}$, $K_p = 3$. $\gamma = 18.62$ kN/m^3

For $\phi = 30°$, $y = 0.08 h = 0.48$ m, $h_1 = 5$ m

Considering upper part AD,

Eq. 11.48: $P_1 = 130.31$ kN

Eq. 11.49: $P_2 = 6.435$ kN

Eq. 11.50: $R_D = 72.699$ kN

Considering lower part DE,

Eq. 11.52: $p_a = 13.406$ kN

Eq. 11.54: $d_1 = d - y = 3.40$ m

$d = 3.88$ m

Neglecting p_n,

Eq. 11.56: $d = 3.44$ m

For design purposes, d is increased by 20-40%.

$R_T = P_1 - P_2 - R_D = 51.18$ kN

11.15 Design a continuous anchor to support the tie rods in Ex. 11.12. Use a safety factor of 2.

Eq. 11.58: $R_T s = \dfrac{\gamma d_a^2 L}{2F} (K_p - K_a)$

For a continuous anchor, $s = L$

$d_a^2 = \dfrac{2 F R_T}{\gamma (K_p - K_a)} = 3.569$

$d_a = 1.89$ m

$B > 0.5 d_a = 0.945$ m

Adopt $B = 1.2$ m for an anchor centred

1.0 m below surface

11.8 BRACED EXCAVATION

The sides of deep, narrow, excavations are normally supported by "bracing" which generally consists of vertical sheet piling or timbering supported by a series of struts and walings. A "strut" is a horizontal member in compression across the trench supporting the waling. A "waling is a horizontal member along the trench supporting piling boards or sheet piles. The construction sequence usually consists of first driving the piling, excavating the ground from inside the area enclosed by the piling and then installing the walings and struts as excavation proceeds. Such walls tend to rotate about a point in the upper portion. Hence, the active pressures do not vary linearly with depth. Field measurements have yielded a variety of curves for the pressure diagram.

The usual design procedure for braced walls is semi-empirical based on actual measurements of strut loads in a number of locations. The design pressure distribution is a hypothetical diagram which represents the envelope covering all the random distribution obtained from field measurements. For cohesionless soils, the design pressure distribution is shown in Fig. 11.27a, which is a uniform distribution of 0.65 times the Rankine value (Terzaghi and Peck 1967, CGS 1985).

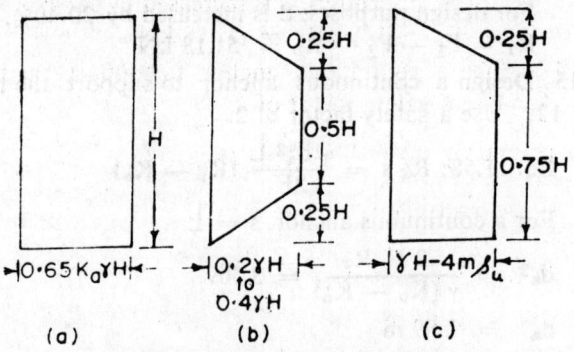

Fig. 11.27 Pressure distribution for braced excavations, (a) sands (b) stiff fissured clays (c) soft to firm clays

The behaviour of braced excavations in clay is dependent on the value of the "stability factor $\gamma H/s_u$" (See Chapter 12), where s_u is the average undrained shear strength of clay ($s_u = c_T$) adjacent to the excavation (Peck 1969). (*Note.* The stability factor increases with depth of excavation in a particular clay.) For stiff to very hard clays ($s_u > 50$ kPa), the pressure diagram is shown in Fig. 11.27b (CGS 1985). The maximum pressure intensity range (0.2 γH to 0.4 γH) depends on the character of clay, the degree of fissuring and the reduction in strength of clay with time. Appropriate value is adopted on the basis of experience. For soft to firm clays ($s_u = 12$ to 50 kPa) the pressure diagram is shown in Fig. 11.27c (CGS 1985). The value of 'm' is taken as 1.0 if a much more resistant layer exists at or near the base of excavation. If there is a great depth of soft clay below the base of excavation, the value of m is taken as 0.4.

If free water level is above the base of excavation, the water pressure must be added to the pressure distribution in sands. submerged density being used below the water table. Lateral pressures due to surcharge loading, if present at the ground surface adjacent to the excavation, should also be taken into account for both sands and clays (Terzaghi and Peck 1967).

Basal instability. Deep excavations in soft to firm clays (s_u = 12 to 50 kPa) are subject to base heave failures due to over-stressing the soil in shear. The safety factor F_b with respect to base heave is given by (CGS 1985):

$$F_b = \frac{N_b \, s_u}{\gamma H + q} \qquad (11.59)$$

where s_u = undrained shear strength of soil (c_T) below base level

   N_b = Janbu stability factor (See Chapter 12 also), depending on the geometry of excavation (See Fig. 11.28.)

   q = surcharge pressure acting on ground surface adjacent to the excavation, on both the sides of excavation

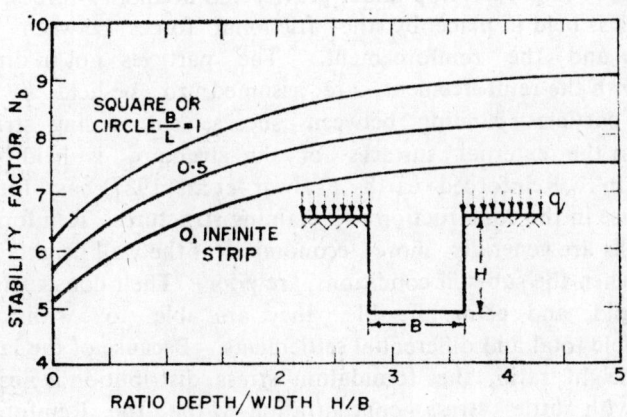

Fig. 11.28 Janbu stability factor for various geometries of excavation (Janbu 1954b CGS 1985)

Example 11.16

An excavation, 5 m wide, 10 m long and 5 m deep is to be made in a soft clay having $\gamma = 1_a$ kN/m³ and $s_u = 17$ kPa. The ground surrounding the excavation carries a uniform surcharge load of 10 kPa. Find the safety factor with respect to basal failure.

$$\frac{H}{B} = 1, \quad \frac{B}{L} = 0.5, \quad N_b = 7 \qquad \text{(Fig. 11.28)}$$

$$F_b = \frac{N_b \, s_u}{\gamma H + q} = 1.13$$

11.9 REINFORCED EARTH RETAINING WALL

The term "reinforced earth" is used to describe a soil mass reinforced with tension resistant elements. Reinforced earth (LCPC 1976) is primarily a composite material, invented by the French engineer Henri Vidal in 1966. It is formed by the association of soil, usually cohesionless, and tension resistant elements in the form of strips, sheets, nets or mats of metal, synthetic fabrics or fibre reinforced plastics. The reinforcement is so arranged in the soil mass as to reduce or suppress the tensile strains which might develop under gravity and boundary forces.

The soil is held in place by the frictional forces between the particles and the reinforcement. The particles not in direct contact with the reinforcement are assumed to be held by the arches of particles spanning between successive reinfocing strips. The soil at the external surfaces of the structure is held by a 'facing skin'. Reinforced earth (Talwar et al 1987) has found greatest use in the construction of retaining structures. Reinforced earth walls are generally more economical if the wall heights are large or when the sub-soil conditions are poor. Their construction is very rapid, and being flexible, they are able to withstand considerable total and differential settlements. Because of the large base to height ratio, the foundation stress distribution is nearly uniform with little stress concentration at the toe. Reinforced earth can also be used in embankments and bridge abutments. A typical reinforced earth system is shown in Fig 11.29.

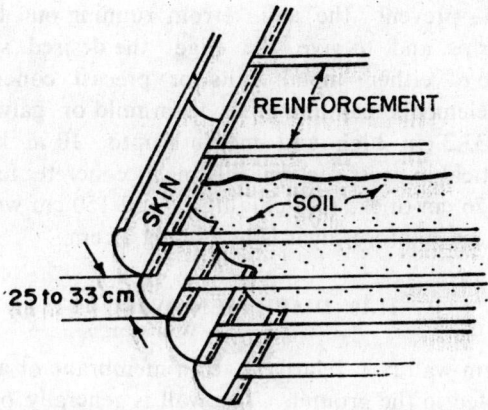

Fig. 11.29 Typical reinforced earth system

1. Soil and reinforcement. The essential feature of reinforced earth is the friction between the earth and the reinforcement. By means of friction, the soil stresses are transferred to the reinforcement which is thereby placed in tension. The soil used as a fill material should, therefore, be predominantly coarse grained. Typical specifications are as follows: percent passing 250 mm sieve = 100, 100 mm sieve = 75—100, and 80 micron sieve = 0—15. It is also essential to keep the fill adequately drained to prevent it from becoming saturated. The most commonly used material for the reinforcement is galvanized steel; the other materials used for reinforcement include stainless steel, aluminium alloys, plastics and synthetic fabrics. The length of the reinforcing strips and their density (cross section and spacing) are determined by calculations for the internal stability of the reinforced earth structure. The reinforcement, generally in the form of strips, range in width from 40 to 120 mm, and in thickness from 1.5 to 3.0 mm (LCPC 1976). Holes are drilled at one end in order to attach the strips to the facing. The width of the wall is determined by the length of reinforcement used and is usually constant, with the reinforcement varying in length from 0.8 to 1.2 times the height of the wall (Talwar et al 1987).

2. Facing elements. In a reinforced structure, it is essential to provide on the external face a *facing,* called the "skin" or "cladding" to prevent the soil from running out between the reinforcing strips, and to give the edge the desired shape. The facing is made of either metal units or precast concrete panels. Metal facing elements manufactured from mild or galvanized steel are generally 33.3 cm high and measure upto 10 m long. They are semi-elliptical in cross section. Typical concrete facing panels are 18, 22 or 26 cm thick, 150 cm high and 150 cm wide; vertical distance between reinforcement layers being 75 cm.

11.10 DIAPHRAGM WALL

A diaphragm wall is a relatively thin membrane or a separating wall constructed in the ground. The wall is generally of reinforced or mass concrete. It is constructed in a trench, the sides of which are supported prior to casting the wall by the hydrostatic pressure of a slurry of bentonite (a montmorillonite clay) in water.

Construction is started by first digging of pilot (guide) trench about one metre deep along the proposed alignment of the wall. This trench is faced with concrete (to serve as guide walls) and should be about 25 mm wider than the excavating device. The mechanical devices used for excavation are ordinary grab, grab operated with kelly, and rotary drilling with reverse circulation of bentonite.

Bentonite slurry is filled in the guide trench which acts as a reservoir for the slurry. The excavator is setup over the guide trench and the soil excavated to the full required depth over the length of one grab bite. Slurry is pumped into the hole, as excavation proceeds, to replace the removed soil. For practical purposes, a diaphragm wall is constructed in sections or panels varying in lengths from 1.5 m to 6 m (and widths between 0.5 m and 1.0 m). First one end, then the opposite end, and lastly the the cenral portion of the panel, are excavated. On completion of excavation the reinfocement cage is lowered and the length of trench is filled with concrete using, a tremie pipe in a continuous pour, displacing the slurry, which is pumped away to a storage for re-use or to waste. Wall panels are joined by means of a steel tube which is inserted as an end shutter or stop end in each bay. After setting of concrete the tube is withdrawn, leaving a vertical construction joint. The adjacent wall panel is cast against this joint. Once the full wall has been constructed and the concrete has attained adequate strength, the soil on one side of the wall can be excavated. To tie the wall into the retained soil, it is usual to install ground anchors at appropriate depths with the progress of excavation.

The bentonite slurry, not only accounts for the impermeability, but also stabilizes the faces of the trench. The suspension penetrates pervious soils, gels, and forms a skin of very low permeability, known as the 'filter cake' on the excavated surface. Consequently, full hydrostatic pressure of the slurry acts against the sides of the trench supporting the earth pressure. Typical slurry specifications are as follows. (Hutchison et al 1975, Bell 1978): concentration in excess of 4.5%, density 1.034 to 1.25 g/cm^3, plastic viscosity greater than 20 cP, 10 min gel strength 50 to 200 dyne/cm^2 and pH value less than 11.7. In soils of high permeability, the sealing mechanism of the slurry can be improved by adding a small quantity of fine sand, around 1 percent (Craig 1983).

Reinforced diaphragm walls are used as part of structural load bearing elements, as retaining walls or foundations. They need not be straight in plan and their depths range upto 60 m (Bell 1978). The technique is convenient for the construction of deep basements and underpasses and construction can be carried out close to adjoining structures, provided the soil is moderately compact. They are preferred to sheet pile walls because of their relative rigidity and their ability to ultimately form the part of the structure. High strength, precast wall elements can also be used. Mass concrete walls are simply used to form cut-offs, e.g. below dams. Use of a diaphragm wall for the protection of a bridge pier is also reported (Bhati 1982).

12
Landslides and slope stability

12.1 LANDSLIDE AND CREEP

The term "landslide" refers to a rather rapid displacement of a mass of rock or soil on sloping ground or down the hillside in which the centre of gravity of the moving mass advances in a downward and outward direction. "Creep", on the other hand, represents a slow, long term movement of soil or rock in steep slopes. Creep can be divided into two types: continuous and seasonal (Terzaghi 1950). The 'continuous creep' refers to a relatively deep movement produced by the force of gravity, often caused by the presence of various strata with different elastic properties which creep beyond their yield strength. Seasonal creep, which is caused by temperature and moisture variations in the soil or rock within the surface layer, takes place primarily in clayey or silty soils.

12.2 CAUSES OF LANDSLIDES

There are a large number of factors such as geological and hydrological conditions, topography, climate and weathering which affect the stability of a slope and can initiate a landslide. There is seldom a single factor responsible for a landslide. Terzaghi (1950) divides them into 'external' and 'internal' factors. External factors are those which cause an increase of the average shear stress along potential surface or planes in rock or soil, while the internal factors are those which cause a decrease of the average shear strength. Causes of increased stress may be: (i) external loads such as buildings, increased weight due to saturation,

(ii) removal of a part of mass by excavations, cuts, quarries; drainage of lakes and drawdown of reservoirs, (iii) undermining caused by tunnelling, seepage erosion, (iv) shock caused by earthquakes or blasting, machinery and traffic, (v) water pressures in cracks, etc. Causes of decreased strength may be: (i) swelling of clays,(ii) increase of pore water pressure, (iii) breakdown of soil struture as in loess and sensitive clays, (iv) thawing of frozen soil, (v) deterioration of cementing material, (vi) discontinuities such as faults, bedding planes, joints, etc.

12.3 TYPES OF LANDSLIDES

Four general types of landslides can be recognized: 'falls', flows', 'translational slides', and 'rotational slides'. There may also be a combination of the various types into a 'complex slide'.

1. **Falls.** Falls refer to the fall of soil or rock masses and where a sliding surface does not form. The moving mass travels mostly through air by free fall, leaping, bounding and rolling with little or no interaction between one moving unit and the other. Movements are very rapid.

2. **Flows.** Flows may be considered to be progressive mass movement of soil of very low shear strength and the mass takes on the physical appearance of a viscous material.

3. **Translational or plane slides.** Translational slides, limited to shallow depths only, take place along essentially planar surfaces. Such slides occur without any significant change in geometry of the unstable region and the force system which causes the failure remains constant. The failure surface is approximately parallel to the ground surface. This is in contrast to the rotational slides in which the geometry and, as a consequence, the force system are progressively modified. Translational slides can be divided into 'block slides', 'slab slides', 'multiple translational slides', and 'spreading failures'. "Block slide" is a translatory movement of blocks of soil or rock moving down a plane of weakness. Block slides can occur in overconsolidated jointed clays, and in rock along faults, bedding planes or joints. "Slab slides" occur primarily in weathered clay or shallow slope debris on bedrock. A sheet of soil moves as a mass down the slope. Typical depth-length ratios are less than 0.1 (Skempton

1953). "Multiple translational slides" are the progressive slides of the slab type which gradually spread upslope as the soil at the rear scarp is gradually weakened. "Spreading failure" describes a type of retrogressive translational slide which may occur where the slope of ground surface is relatively small. It is common in varved clay deposits where high pore pressures occur seasonally in sand or silt seams.

4. Rotational slides. When failure occurs by a mass of soil sliding along a curved surface of rupture, it is called a rotational slide. Rotational slides have been observed in slopes of both normally and overconsolidated clays, in shales and in man-made earthen embankments. These slides are relatively deep seated with the ratio of the depth to the total length of the slide being of the order of 0.15 to 0.33 (Skempton and Hutchinson 1969). This ratio increases as a rule with increasing inclination of the slope. Three types of rotational slides are recognized: rotational slip', 'multiple rotational slide' and 'successive slip'.

(a) *Rotational slip.* It is the most common form of failure in uniform cohesive soils, where the failure surface is approximately circular and the failing soil mass moves as a unit along a a relatively thin failure or slip surface. The critical slip surface is deep seated and is usually controlled by an underlying strong stratum. In a rotational slip, the modes of failures are classified as 'base failure', 'toe failure' and 'slope failure'. When the slip surface passes below the toe of slope and intersects the ground away from the toe, it is called a "base failure". Slip surface intersecting the toe representst "toe failure". If the slip surface intersects the slope above the toe, it is said to be a "slope failure."

(b) *Multiple rotational slides.* They are the progressive rotational types of slope failures. They develop gradually and spread backward along a common basal failure surface. These may occur in slopes of actively eroding overconsolidated clays, fissured clays and clay shales.

(c) *Successive slip.* It refers to a series of shallow rotational slides occurring typically in fissured, overconsolidated clays.

(d) *Wedge failure.* In addition to the above three types, wedge failure is also a deep seated slide, which may occur in stratified soil profile having bands of strong and weak material.

The failure surface is a combination of active and passive wedges with central sliding block depending on stratification.

(e) *Failure along an irregular surface.* Failure of slopes with complex soil and ground water conditions, and perhaps with an irregular face, may not follow any of the standard patterns described above. Such slopes fail along irregular surfaces.

12.4 ANALYSIS OF SOIL SLOPES

Soil slopes may be man-made, as in earth dams, cuts and fills for highways and railway lines, canal banks, and temporary excavations. They may also be naturally formed as stream banks and hillsides. A slope may be 'finite' or 'infinite' in extent. For practical purposes, any slope of great extent with soil conditions essentially the same along every vertical line can be considered an infinite slope.

1. **Basic assumptions of stability analysis.** The following basic assumptions are made for slope stability analysis:
   (1) Failure of an earth slope occurs along a slip surface, i.e. the failure can be represented as a two-dimensional plane problem. The failure surface may be plane or more often curved.
   (2) The sliding mass moves as an essentially rigid body, the deformations of which do not affect the problem.
   (3) The strength properties are isotropic, i.e. the shear resistance along the failure surface remains independent of the orientation of the failure surface.
   (4) The safety factor is determined by the 'limit equilibrium method' and is defined as the ratio of the average shear strength of soil(s) to the average shear stress (τ) required for equilibrium along a potential failure surface. In other words, safety factor is the ratio of the resisting forces to the driving forces.

2. **Effective stress and total stress methods.** In the effective stress method, the pore pressures along the assumed failure surface are estimated and the analysis utilizes effective strength parameters c' and φ' obtained from a CD shear test or a CU test with pore

pressure measurements. The effective stress analysis is used in the following situations (NAVFAC):

(1) Long term stability and drawdown in pervious, incompressible, coarse grained soils. Use φ' usually neglecting c'. Apply pore pressures from groundwater or seepage only.

(2) Dense, moderately compressible soil such as an earth dam. Use c' and φ'. Apply only seepage or drawdown or consolidation pore pressures if piezometers are installed to confirm pore pressures assumed in the design.

(3) Compressible soils where some drainage occurs during load application. Use c and φ from CU tests. Apply groundwater plus consolidation pore pressures, including an allowance for dissipation of hydrostatic excess pressures.

In the total stress method, the shear strength is determined from UU tests or from vane shear tests. φ_T is taken equal to zero. The total stress analysis is applied to the following situations.

(1) Failure in slopes of normally consolidated or slightly preconsolidated clays, where little dissipation of excess pore pressures occurs prior to critical stability conditions.

(2) Analysis of embankment or structure load applied rapidly on a clay stratum where no provision is made to drain pore water.

12.5 PLANE SLIDE ANALYSIS

Homogenous, infinite slopes, where the soil stresses and properties on a given plane parallel to the slope are more or less identical, tend to fail along a plane surface parallel to the slope at some depth. Finite slopes in homogeneous, cohesionless soils in a dry, fully saturated to the surface or submerged conditions may also have plane, surface slides. Upstream slopes of earth dams can also fail by translational instability as a result of rapid drawdown (Lee et al 1983). Presence of cohesion inhibits surface sliding and tends towards deep seated slips. However, a subsurface slide (a shallow depth plane slide) may occur at the base of a thin, weak cohesive material overlying a sound material.

1. Dry cohesionless soil. Assuming the failure plane parallel to the slope at a depth of of z below the surface, consider the equilibrium of a typical slice of one unit thickness, as shown in Fig. 12.1.

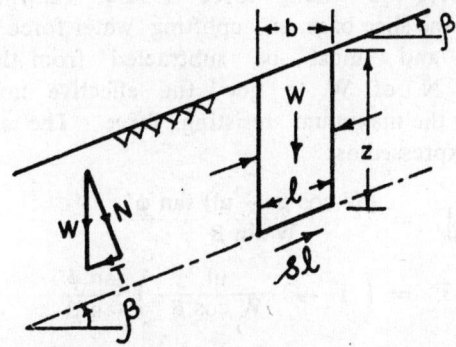

Fig. 12.1 Forces on a slice for plane slip analysis

In an infinite slope, the side forces will be the same on every vertical section and, therefore, will cancel out. The normal force N on the slice base due to its weight is:

$$N = W \cos \beta \qquad (12.1)$$

and if φ' is the effective angle of shearing resistance, the maximum resisting force developed is given by:

$$\text{Resisting force} = N \tan \varphi' = W \cos \beta \tan \varphi' \qquad (12.2)$$

The sliding force T due to the weight of slice is:

$$\text{Sliding force} = W \sin \beta \qquad (12.3)$$

and the safety factor F against sliding is defined as:

$$F = \frac{\text{max. resisting force}}{\text{sliding force}} = \frac{W \cos \beta \tan \phi'}{W \sin \beta}$$

$$= \frac{\tan \phi'}{\tan \beta} \qquad (12.4)$$

With $F = 1$, the maximum slope angle β is equal to φ'.

2. Submerged cohesionless soil. When a slope is under water (submerged) with free water standing against its face, the analysis is essentially the same as for a dry state except that the normal and shearing stresses are calculated on the basis of submerged unit weight rather than total unit weight. However, since the weights cancel, the safety factor for a submerged slope can also be expressed by Eq. 12.4.

3. Cohesionless slope with seepage. Consider a slope with seepage parallel to and at the surface for which a portion of the flow net is shown in Fig. 12.2. The weight W of the slice is calculated using the saturated unit weight and its component parallel to the slip plane gives the sliding force T (Eq. 12.3) If 'u' is the pore pressure at the slice base, an uplifting water force 'ul' acts normal to the base and must be subtracted from the total normal component N of W to get the effective normal force for determining the maximum resisting force. The safety factor is, therefore, expressed as:

$$F = \frac{(W \cos \beta - ul) \tan \phi'}{W \sin \beta} \qquad (12.5)$$

or

$$F = \left(1 - \frac{ul}{W \cos \beta} \right) \frac{\tan \phi'}{\tan \beta} \qquad (12.6)$$

The pore pressure can also be expressed by the parameter "pore pressure ratio r_u" which is defined as the ratio of pore pressure u at a given point in a soil mass to the total overburden pressure γz above it.

$$r_u = u/\gamma z \qquad (12.7)$$

Since $W = \gamma bz$ and $l = b/\cos\beta$, the safety factor (Eq. 12.6) may also be expressed in terms of r_u as follows:

$$F = (1 - r_u \sec^2 \beta) \frac{\tan \phi'}{\tan \beta} \qquad (12.8)$$

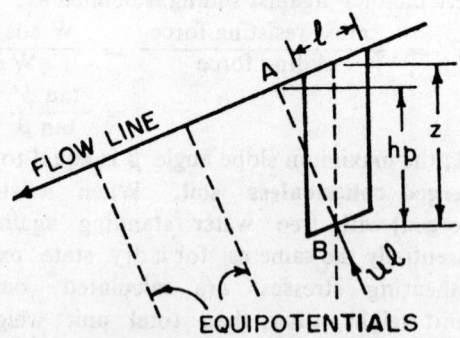

Fig. 12.2 Flownet when flow is parallel and at the surface

The pore pressure u at the slice base is represented by the piezometric head h_p. In Fig. 12.2, $h_p = AB \cos \beta = z \cos^2 \beta$. Hence,

$$u = \gamma_w z \cos^2 \beta \tag{12.9}$$

and

$$r_u = \frac{\gamma_w}{\gamma_{sat}} \cos^2 \beta \tag{12.10}$$

Substituting in Eq. 12.8,

$$F = \frac{\gamma'}{\gamma_{sat}} \frac{\tan \phi'}{\tan \beta} \tag{12.11}$$

By comparing Eqs. 12.4 and 12.11, it may be noted that seepage reduces the safety factor of a dry (or submerged) slope of cohesionless soil in the ratio of γ'/γ_{sat} ($\simeq 0.5$). In other words, the stable slope angle is also reduced by the same ratio. Further, the safety factor of an infinite slope is independent of the depth z to the potential failure surface. In dry loose sand, the stable angle ($\beta = \phi'$) is called the "angle of repose".

4. **Cohesive frictional soil with seepage parallel to slope.** The stability analysis of an infinite slope in 'c—ϕ' soil is similar to that for a 'c = 0' soil except that there is an additional resisting force on the slice base due to the effective cohesion c' acting over it. Thus,

$$F = \frac{c' 1 + (W \cos \beta - u 1) \tan \phi'}{W \sin \beta} \tag{12.12}$$

Eq. 12.12 may also be expressed as:

$$F = \frac{c' 1}{W \sin \beta} + (1 - r_u \sec^2 \beta) \frac{\tan \phi'}{\tan \beta} \tag{12.13}$$

or

$$F = \frac{c'}{\gamma z} \cdot \frac{\sec^2 \beta}{\tan \beta} + (1 - r_u \sec^2 \beta) \frac{\tan \phi'}{\tan \beta} \tag{12.14a}$$

or

$$F = \frac{c'}{\gamma z} \cdot \frac{\sec^2 \beta}{\tan \beta} + \frac{\gamma'}{\gamma_{sat}} \frac{\tan \phi'}{\tan \beta} \tag{12.14b}$$

Eq. 12.14 shows that F depends on the depth z of the failure plane. F will always be very high close to the surface but decreases as depth increases. This is why deep-seated slides, rather than surface slides, occur in cohesive soils unless there is a thin cover of weak cohesive material underlain by a strong material.

Example 12.1

A slope is to be made of a granular soil having a saturated density of 1.98 g/cm^3 and an angle of internal friction of $30°$. If the safety factor is to be 1.25, determine the safe angle of the slope (a) when the slope is dry or submerged, and (b) if seepage occurs at and parallel to the surface of the slope.

(a)   Dry or submerged slope.

$$F = \frac{\tan \phi'}{\tan \beta}; \quad \tan \beta = \frac{0.5774}{1.25}; \quad \beta = 25°$$

(b)   Slope with seepage.

$$F = \frac{\gamma'}{\gamma_{sat}} \frac{\tan \phi'}{\tan \beta}; \quad \beta = 13°$$

12.6 ROTATIONAL SLIP ANALYSIS

1. **Swedish circle method.** The concept of rotational slip along a circular failure surface originated in Sweden where failure surfaces in soft clays were observed to closely resemble arcs of circles. Fellenius presented in 1918 an analysis of slope failure using a circular slip surface, where the soil was regarded as a purely cohesive material ($\varphi_T = 0$). He further extended it to 'c-φ' soils. This analysis based on a circular slip circle has been known since about 1930 as the "Swedish Slip Circle Method".

The assumption of a circular failure surface gives the simplest solution, although it may lead to larger values of the safety factor (2 to 7%) because in most cases the critical shear failure surface for the minimum safety factor in non-circular.

2. **Fellenius method of slices and factor of safety.** The basic approach in the slip circle method is to consider the 'overall moment equilibrium' of a sliding mass of soil above a trial failure arc. The sliding mass is divided into a reasonal number of slices (6 to 10); a typical slice is shown in Fig. 12.3. Taking moment about the centre of rotation, the 'overall moment equilibrium' can be expressed as:

$$\Sigma \ W.x = \Sigma \ \tau \ l \ R \tag{12.15}$$

where $W =$ weight of slice

x = distance between C.G. of slice and centre
    of slip circle

l = length of slice base

R = radius of slip circle

τ = shear stress required for equilibrium

The side forces on the slices are not included in the moment equations, because when all the slices are considered, the net moment of side forces will be zero.

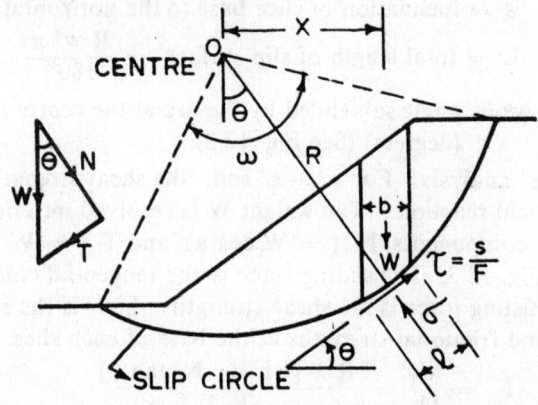

Fig. 12.3 Forces on a typical slice in the slip circle method

As stated earlier, the safety factor F is adopted with respect to shear strength 's', i.e.

$$F = \frac{s}{\tau}, \text{ or } \tau = \frac{s}{F} \qquad (12.16)$$

Assuming F is constant along the entire circular arc and substituting the value of τ from Eq. 12.16 in Eq. 12.15,

$$F = \frac{\Sigma\, s\, l\, R}{\Sigma\, W\, x} = \frac{M_r}{M_s} \qquad (12.17)$$

where M_r = resisting moment for fully developed
                shear strength

         M_s = resultant sliding (overturning or driving) moment
                produced by forces causing failure

Thus, when a circular slip surface is assumed the ratio between the resisting and sliding moments gives the safety factor with respect to the shear strength.

The analyses for '$\varphi_T = 0$' soils and 'c - φ' soils are described in the following sub-sections.

3. $\varphi_T = 0$ analysis. For slopes in clays where '$\varphi_T = 0$' condition applies, $s_u = c_T$ (undrained shear strength) and it is independent of the normal stress. Eq. 12.17 can be written as:

$$F = \frac{M_r}{M_s} = \frac{R \Sigma c_T l}{\Sigma W x} = \frac{\Sigma c_T l}{\Sigma W \sin \theta} = \frac{c_T. L}{\Sigma W \sin \theta}$$

$$(12.18)$$

where θ = inclination of slice base to the horizontal

$$L = \text{total length of slip surface} = \frac{R w° \pi}{180°}$$

w° = angle subtended by the arc at the centre of rotation (degrees) (See Fig. 12.3)

4. 'c - φ' analysis. For a 'c - φ' soil, the shear strength depends on the normal reaction. The weight W is resolved into normal and tangential components, N (= W cos θ) and T (= W sin θ). as shown in Fig. 12.3. The sliding force is the tangential component T and the resisting force is the shear strength which is the sum of the cohesive and frictional strengths at the base of each slice. Thus,

$$F = \frac{M_r}{M_s} = \frac{R \Sigma [c.l + N. \tan \phi]}{R \Sigma T} \qquad (12.19)$$

or $$F = \frac{c L + \tan \phi \Sigma W \cos \theta}{\Sigma W \sin \theta} \qquad (19.20)$$

If effective stress parameters are used, the normal force is reduced by the water force U = u.l, where u is the average pore pressure on the slice bottom. The safety factor is expressed as:

$$F = \frac{c' L + \tan \phi' \Sigma (N - U)}{\Sigma T} \qquad (12.21)$$

or $$F = \frac{c' L + \tan \phi' \Sigma (W \cos \theta - u l)}{\Sigma W \sin \theta} \qquad (12.22)$$

Assuming the base of each slice to be a straight line, its inclination to the horizontal gives θ. The volume of a slice of unit thickness (1 m normal to the section) is approximately the height of the mid-ordinate multiplied by the width 'b'. This quantity multiplied by the density (t/m³) and by 9.8 gives the weight W. The weights are assumed to act at the centre of each slice, except for the end ones which are nearly triangular in shape. N and T components can be determined either by drawing a force triangle or from $N = W \cos\theta$ and $T = W \sin \theta$. The calculations are best set out in a tabular form.

It may be noted that the slices to the left of the centre of rotation have a negative moment ($\sin \theta$ is negative). Although the side forces cancel out of the overall moment equations they do affect the magnitude of the normal reaction N on the slice base and thus the frictional shear component. By comparing the shear strength with the tangential force T for each slice, it is apparent that F for each slice varies; being lower for upper slices and higher for lower slices. F given by Eq. 12.21 thus represents an average safety factor of the individual slices. Eq. 12.21 is exact but approximations are induced in determining the normal force N and which, consequently, affects the value of F. In the 'ordinary method of slices' proposed by Fellenius, N equals $W \cos \theta$. The resulting safety factor is conservative (underestimated) for soils where $\varphi > 0$ (Dunn et al 1980); the error, compared with more accurate methods of analysis, is usually within the range 5-20% (Craig 1978). In the '$\varphi_T = 0$' analysis, as N does not appear in the equation of safety factor (Eq. 12.18), an exact value of F is obtained.

5. **Centre of critical circle.** The centre of the most critical circle giving the minimum safety factor can only be found by trial and error and the analyses of various slip circles. In case of homogeneous slopes, when $\tan \varphi' > 0.1$ ($\varphi' > 5°$) or the strength increases with depth such that $\triangle c/\triangle z > 0.1 \gamma$, the critical arc will pass through the toe of the slope (Terzaghi 1943, Gibson and Morgenstern 1962). Also when the slope angle $\beta > 53°$ (whatever the value of φ), the critical circle passes through the toe. However, if there is a layer of relatively stiff material at the base of the slope in level with the toe, the circle will be tangential to this stiff layer, and will intersect the slope. For cohesive soils with little angle of friction, the slip circle tends to be deeper and usually extends in front of the toe (base failure); the circle may essentially be limited in depth by the underlying stiff layer, if any. For slope angle $\beta < 53°$ in cohesive soils, the critical centre should be attempted on a vertical line bisecting the slope. It may be noted that the value of F is more sensitive to horizontal movements of the centre of rotation than to the vertical movements. Fellenius (1936) gave directional angles to determine the critical centre of rotation in case

of slopes of homogeneous cohesive soils. Jumikis (1962) further extended the construction for locating the critical centre in case of c - φ soils. The author suggests to use the Janbu graphs (Figs. 12.11 and 12.14) for locating the critical centres.

6. **Effect of seepage and submergence.** The forces acting on a soil element under steady seepage can be obtained by either of the two concepts (Taylor 1948):

(1) Vector sum of submerged weight of soil and seepage force.

(2) Vector sum of total weight of soil and the resultant water pressure on the whole boundary of the element

Fig. 12.4 shows a partially submerged slope with steady seepage through it. The free water contained in the boundary FCDE exerts an external pressure on the submerged face of the slope CDE giving an additional moment resisting sliding. If the whole body of water bounded by the segment BCFB is considered, it is symmetrical about a vertical line through O as BF is horizontal and it will, therefore, have no moment about O; its effect can be ignored in the calculation. In other words, the resisting effect due to the external body of water FCDE can be ignored provided the sliding effect due to a body of water bounded by BCDEB is also ignored, since the two are equal and opposite (Bishop)

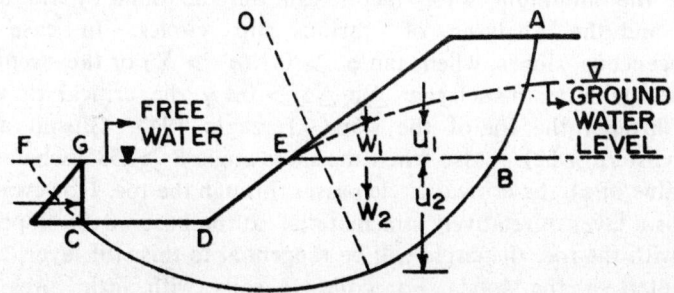

Fig. 12.4 Partially submerged slope

Ignoring the weight of water in the area BCDEB is equivalent to consider only the submerged weight of soil. It follows that the effect of external water surface can be included in the stability analysis by using the submerged unit weight, instead of the saturated unit weight, below the level of the external free water surface. Similarly, the pore pressure head u_2 upto the external water level is ignored, and only the portion u_1, measured above this level is considered (See Fig. 12.4). Thus in Eq. 12.22, W is replaced by $(W_1 + W_2)$, where W_1 is the *total* weight of slice above external water level and W_2 is the submerged weight below; and u is replaced by u_1, the pore pressure measured from the external water level.

For total stress analysis, the slip circle is extended through the water, which is treated as a material with weight but no shear strength. Instead of extension as an arc CF, an alternative is to take a vertical line CG through the water and take into consideration also the lateral hydrostatic thrust acting on CG. For total stress analysis, the weights of slices will be the total weights throughout the sliding sector inclusive of the body of water above the slip surface.

7. **Effect of tension crack.** Tension crack tends to develop in a cohesive soil near the top of the slope at an incipient failure. The depth of tension crack for an ideal soil is given by:

$$z_0 = 2.c_T/\gamma \tag{12.23}$$

In practice, the depth of tension crack is generally less than that given by Eq. 12.23. The effect of tension crack is to reduce the effective length of the arc contributing to resisting moment. To compensate for any water pressures that may be exerted if the crack gets filled with rain water, weight of whole of the sliding sector is used for calculating sliding moment. Where the bottom of the tension crack is below the phreatic line (seepage), the horizontal hydrostatic pressure set-up in the crack should be taken into consideration for calculating sliding moment.

12.7 INFLUENCE CHARTS FOR MOMENTS*

Influence charts have been devised by Sawaguchi and Takahashi (1976) for handy calculations of both sliding and resisting moments based on the Swedish method assuming a circular failure surface. The results obtained by these charts have been proved quite satisfactory on comparison with the more accurate values of safety factor obtained using an electronic computer.

1. **Influence chart for sliding moment.** The basic concept involved in preparing the influence chart is to divide a sliding mass by horizontal

*The author is grateful to Prof. M. Sawaguchi for permission of reproduction.

lines as well as vertical lines in such a way that each block has the same magnitude of sliding moment.

Consider two blocks adjoining horizontally, as shown in Fig. 12.5. The unit weight of the mass is assumed constant through the whole area. Each block is divided into equal vertical length, called the 'unit length' 'l_1'. For the two blocks to have equal moments about the centre of rotation (or the vertical line xx passing through the centre of the circle), the following relation (Eq. 12.24) has to be satisfied;

$$(l_n - l_{n-1}) \left(l_{n-1} + \frac{l_n - l_{n-1}}{2} \right) = (l_{n+1} - l_n)$$

$$\times \left(l_n + \frac{l_{n+1} - l_n}{2} \right) \qquad (12.24)$$

where l_{n-1}, l_n, l_{n+1} are the horizontal distances between the centre of the circle and the vertical boundary lines of respective blocks. $l_0 (= 0)$ means the vertical line passing through the centre of the circle, i.e. the original line of the vertical boundary line.

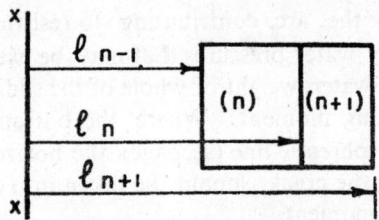

Fig. 12.5 Horizontally adjoining blocks

If l_1 is assumed to be the 'unit length', the solution of Eq. 12.24 is as follows:

$$l_n = \sqrt{n} \qquad (12.25)$$

If the sliding mass is divided by boundary lines whose length is one unit in vertical direction as well as by boundary lines of $l_n = \sqrt{n}$ in horizontal direction, all blocks have the same value of sliding moment. Assuming the unit length to be 1.0 metre, the same sliding moment of each block is:

$$\Delta M = \tfrac{1}{2} \rho \quad \text{(t-m/m)} \qquad (12.26)$$

where ρ = density in t/m^3 or g/cm^3

(In SI units use γ = kN/m^3)

Fig. 12.6 shows the influence chart drawn in this way. The 'standard scale' shown in the figure is the 'unit length', and the cross section of sliding mass should be reduced using this standard scale. The vertical lines in Fig. 12.6 are divided into equal scale (unit length) and the horizontal lines are divided into the scale of square root.

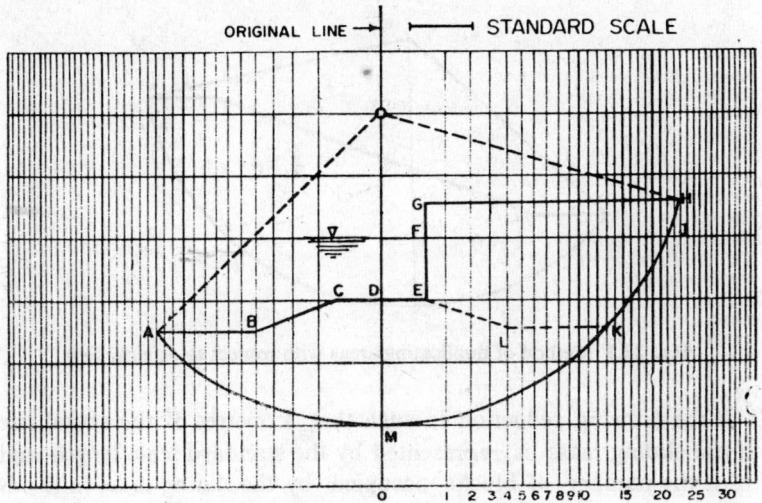

Fig. 12.6 Influence chart for sliding moment
(Courtesy: Prof. M. Sawaguchi)
(Fig. 12.9 is also superimposed)

To find the sliding moment by the influence chart, the cross section of sliding mass is first drawn on a tracing paper to such a reduction that the cross section occupies as wide part of the influence chart as possible in order to enlarge the accuracy by increasing the number of influence blocks occupied by the sliding mass. The number of influence blocks occupied by the area of the sliding mass are then counted. The moment given in one side with respect to the 'original line' has the contrary sign to that in other side, so that the (net) total sliding moment is obtained by subtracting the moment in one side from the moment in other side.

The influence chart for sliding moment can be used in quite the same way both for sands and clays.

The calculations can be simplified by duplicating the cross sectional area of the sliding mass in one side on to the other side

symmetrically with   respect to the original line   and counting
the number of blocks not doubled.   For example in Fig. 12.7, area
'B' is made symmetrical to area 'A' with respect to the original line,
so that the area A and B have   the same   number of blocks.
Accordingly, the total sliding moment of the sliding mass is given
by the balance area C only.

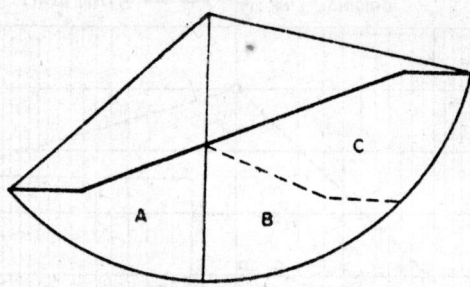

Fig. 12.7 Method of duplicating areas with respect to original line

If the drawing reduction is such that k metres of the actual size
of the sliding mass is represented by the standard scale length and
'n' is the number of blocks occupied by the sliding mass (n is not
necessarily an integer), the sliding moment of the mass is given by:

$$M_s = \tfrac{1}{2} \rho \, k^3 . \, n \quad (t\text{-}m/m) \tag{12.27}$$

where ρ = density (t/m³)

When a part of the soil mass is submerged with free water
standing against the slope, bulk density of soil is used for blocks
above the water level and submerged density for blocks below the
water table.

2. Influence chart for resisting moment.   An influence chart for
resisting moment has also been developed (Sawaguchi and
Takahashi 1976) for cohesionless soils only where the resisting
moment is attributed only to the frictional resistance due to effective
weight.   The calculation of the resisting moment for clay (or due
to the cohesion component of the shear strength) can be easily made
as described earlier.   The influence chart is shown in Fig. 12.8.   It
is so constructed that each block possesses the same influence value
of resisting moment equal to $0.01 \, \rho \, R^3 \tan \phi$ (t-m/m).

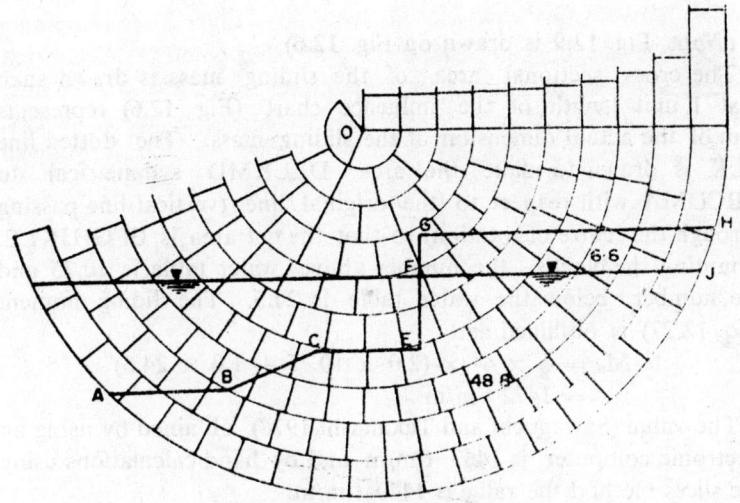

Fig. 12.8 Influence chart for resisting moment
(Courtesy: Prof. M. Sawaguchi)
(Fig. 12.9 is also superimposed)

To find the resisting moment of a cohesionless sliding mass, the area of the cross section is drawn with such a reduction scale as to make the slip circle coincide with the most outside boundry line and the horizontal line of the actual ground surface is parallel to the horizontal original line. If n is the number of blocks occupied by the area of the cross section, the total resisting moment M_r is given by:

$$M_r = n \times 0.01 \, \rho \, R^3 \tan \phi \quad (t\text{-}m/m) \qquad (12.28)$$

Where there are some strata with different densities or different friction angle ϕ, the number of blocks for areas of different properties should be counted separately and the total resisting moment is obtained as the sum of the resisting moments of different areas.

Example 12.2

Find the sliding moment, resisting moment and the safety factor of a slope having a sheet pile bulkhead, as shown in Fig. 12.9. Examine the slip circle of radius 20 m as drawn in the figure. The ground consists of homogeneous sand having $\phi = 40°$, $\rho = 2.0$ t/m³ above the water level, and $\rho' = 1.0$ t/m³ below the water level.

(*Note.* Fig. 12.9 is drawn on Fig. 12.6)

The cross sectional area of the sliding mass is drawn such that 1 unit length of the influence chart (Fig. 12.6) represents 4 m of the actual dimension of the sliding mass. The dotted line ELK is drawn to make the area DELKMD symmetrical to ABCDMA with respect to the original line (veitical line passing through the centre of rotation) so that the net area is EFGHJKLE. Counting the blocks, the number above water table is 10.75 and the number below the water table is 24.5. The sliding moment (Eq. 12.27) is obtained as:

$$M_s = \tfrac{1}{2} \times 4^3 \times (2.0 \times 10.75 + 1.0 \times 24.5)$$
$$= 1472 \text{ t-m/m}$$

The value (Sawaguchi and Takahashi 1976) obtained by using an electronic computer is 1467 t-m/m and by hand calculations using the slices method the value is 1440 t-m/m.

To find the resisting moment, the cross sectional area is drawn on the influence chart of Fig. 12.8. With centre of rotation at O, the scale is so chosen that the slip circle coincides with one of the circles of the chart (possible outer most). Counting the number as accurately as to 0.1 of one influence block, there are 6.6 blocks above the groundwater level and 48.8 below it (See Fig. 12.8). Accordingly, the resisting moment (Eq. 12.28) is calculated as:

$$M_r = 0.01 \, (n\rho + n'\rho') \, R^3 \tan \phi$$
$$= 0.01 \, (6.6 \times 2.0 + 48.8 \times 1.0) \times 20^3 \times \tan 40°$$
$$= 4162 \text{ t-m/m}$$

The value obtained by an electronic computer is 4155 t-m/m and by hand calculations 4147 t-m/m

On the basis of sliding and resisting moments as obtained by the influence charts, the electronic computer and hand calculations, the respective safety factors by the three methods are 2.83, 2.83 and 2.88. Thus the values obtained by these charts and the computer are quite the same, but that by calculations with hand is 1.8 percent larger than these two values (Sawaguchi and Takahashi 1976).

12.8 JANBU STABILITY CHARTS

The determination of the minimum safety factor requires, in general, a lengthy trial and error procedure for different locations of the slip surfaces.   For   circular   slip surfaces and simple regular slopes in fairly   homogeneous   soils, it is possible through analytical procedures to   determine   the location of the   critical circle and the corresponding   minimum F,   once and   for all. The final results can   be presented in   diagrams,   called the "stability charts", which include a   dimensionless   parameter   called   the "stability factor" or "stability number".   Slope   stability charts have been developed by Taylor   (1948), Janbu   (1954, 1968) and others (Hunter and   Schuster 1968).   They   provide very rapid analyses of slope stability   and are   perfectly   accurate for '$\phi_T = 0$' conditions (Wright   1969).   One of the   most   useful charts for clays is the   one   developed   by Janbu.   It   leads to numerical results in good   agreement   with the   Fellenius and Taylor solutions.   The   application   of   charts   proposed   by Janbu is substantially broad   as   they   including   the effects of an external overburden on   the   slope,   submergence and water-filled tension cracks. Janbu's   charts   are   also   available   for effective stress analysis.

For all practical purposes the accuracy of obtaining the minimum factor of safety by   the   Janbu   charts in about the   same as that of the Bishop analysis (See Section 12.10) (Janbu 1954); at the same time these   charts   are   simpler to   use.   These   charts are described below.

1. $\phi_T = 0$ **analysis.** In cohesive soils ($\phi_T = 0$) the critical height H_c or the maximum   possible   height of a slope with   a given slope angle β is directly   proportional   to the   undrained shear strength s_u and inversely proportional to the unit weight γ.   The critical height H_c may thus be expressed as:

$$H_c = N_0 \ \frac{s_u}{\gamma} \tag{12.29a}$$

or $$N_0 = \frac{\gamma H_c}{s_u} \tag{12.29b}$$

where N_0 is a pure   number,   called the "critical stability factor". Its value depends   only   on   the   slope angle β and on the factor 'd' (Janbu 'depth factor')   by   which   the   vertical height of a slope has to be multiplied to get the   depth   of the lowest point of the slip circle *below the toe* of the slope.   The   minimum safety factor F (with respect to s_u) of a stable slope of height H ($H < H_c$) can be expressed from Eq. 12.29 as:

$$F = \frac{N_o\, s_u}{\gamma H} = \frac{H_o}{H}; \quad (s_u = c_T) \tag{12.30}$$

The stability factor N_o is plotted versus slope angle β (or slope ratio $b = \cot \beta$) in Fig. 12.10. The critical circle may either pass through the toe ($\beta > 53°$), above the toe (slope circle, for firm base at shallow depths) or below the toe (base circle, for flat slopes, and deep seated firmer layers). The location of the critical circle may be determined by means of the centre coordinates as measured from the toe point,

$$X_o = x_o\, H \text{ and } Y_o = y_o\, H \tag{12.31}$$

where the critical unit coordinates x_o and y_o are obtained from Fig. 12.11.

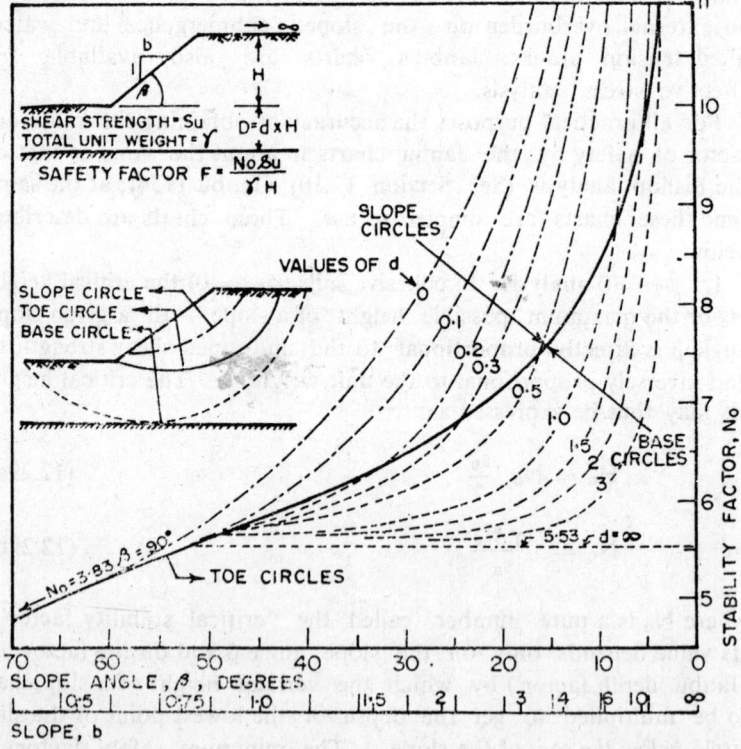

Fig. 12.10 Stability factors for slopes in cohesive soils ($\phi_T = 0$) (Janbu 1969)

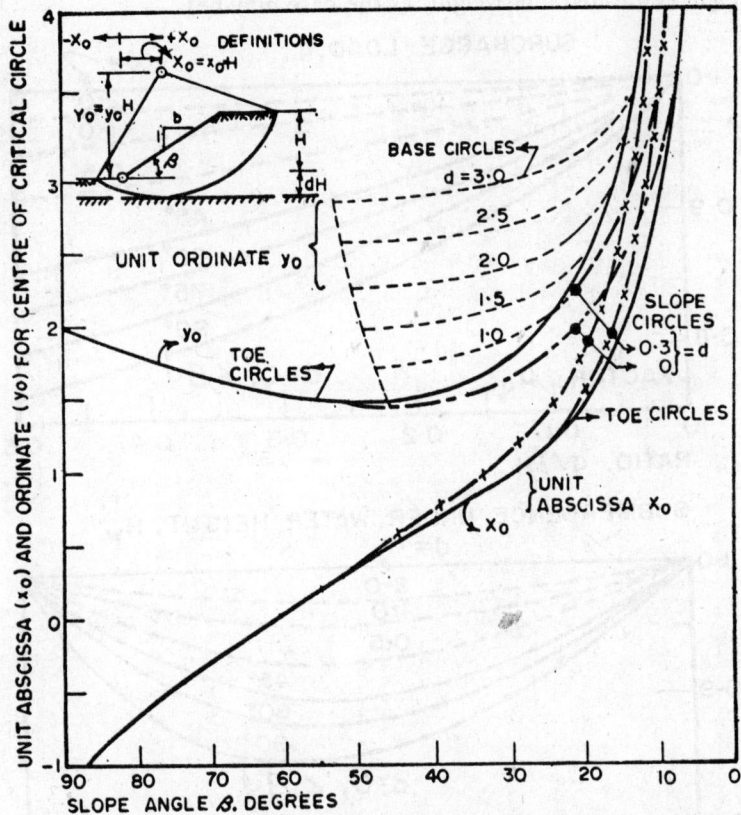

Fig. 12.11 Centre for critical circle for slopes in cohesive soil (Janbu)

To take into account the influence of surcharge q, submergence (external water level of depth H_w) and water filled tension cracks of depth H_t (See Fig. 12.12), the minimum safety factor is expressed as follows:

$$F = \frac{N_o \, s_u}{p_d} \qquad (12.32)$$

in which p_d is the total reference stress defined as:

$$p_d = \frac{\gamma H + q - \gamma_w H_w}{\mu_q \, \mu_w \, \mu_t} \qquad (12.33)$$

where the factors μ_q, μ_w and μ_t are obtained directly from

Fig. 12.12 on the basis of dimensionless ratios. γ is the total unit weight (saturated unit weight, as the case may be).

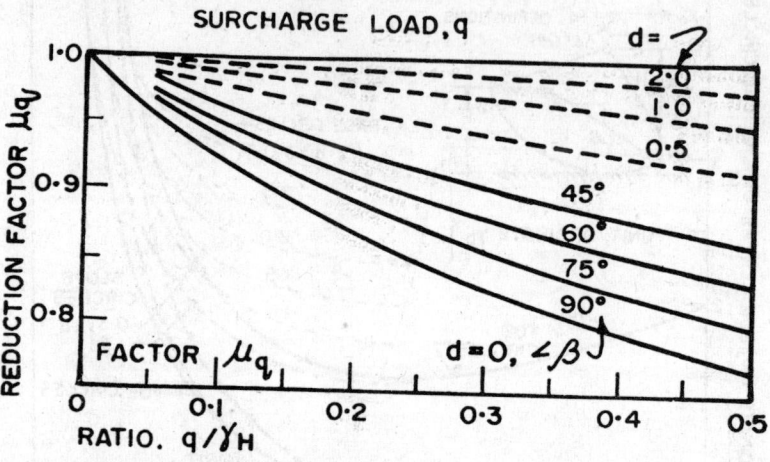

Fig. 12.12a Factors for influence of surcharge and submergence on stability (Janbu 1969, NAVFAC 1982)

2. **c—ϕ analysis.** The minimum safety factor is expressed as:

$$F = \frac{N_{of} \cdot c}{p_d} \qquad (12.34)$$

where N_{of} is the minimum stability factor (combined factor for cohesion and friction), and p_d is the total reference stress as given by Eq. 12.33. N_{cf} depends on the slope ratio b (= cot β) and the dimensionless parameter

$$\lambda c\phi = \frac{p_e \tan \phi}{c} \tag{12.35}$$

where p_e is defined as:

$$p_e = \frac{\gamma H + q}{\mu_q} \tag{12.36}$$

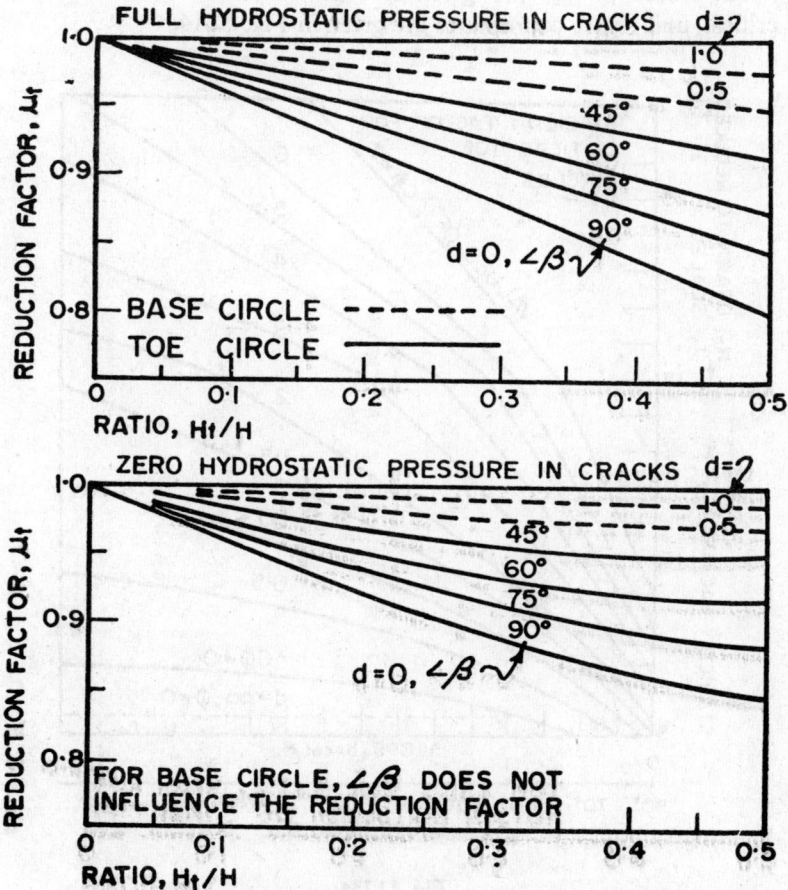

Fig. 12.12b   Factors for influence of tension cracks on stability
(Janbu 1969, NAVFAC 1982)

The μ-values are obtained from Fig. 12.12. It may be observed that if q is a short term live load on saturated clays (unconsolidated condition), one should use $q = 0$ and hence $\mu_q = 1$ in the p_e formula (Eq. 12.36) (but not in p_d of course). In case there is neither surcharge nor tension cracks and the groundwater level is below the failure surface,

$$p_d = p_e = \gamma H \qquad (12.37)$$

The values of N_{cf} are given in Figs. 12.13a and 12.13b. The critical unit centre coordinates are given in Fig. 12.14.

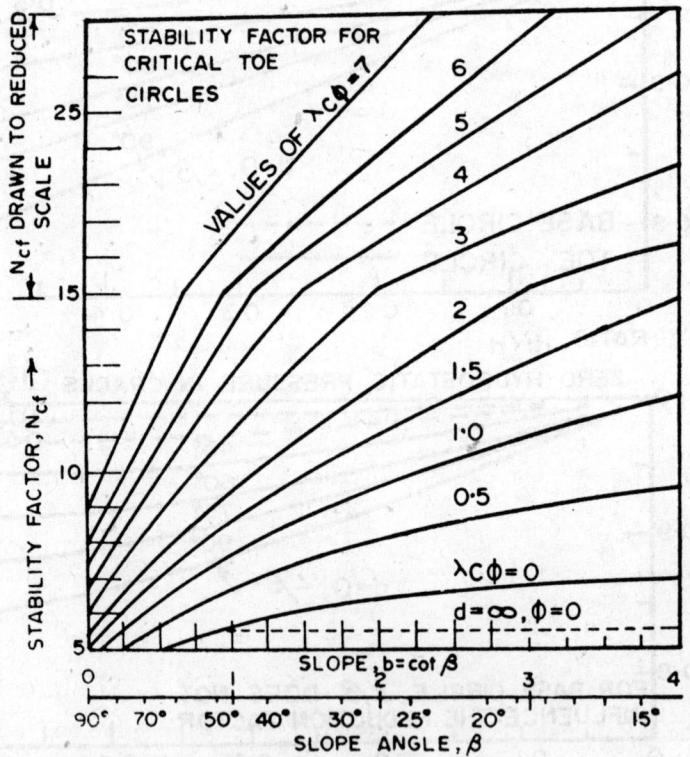

Fig. 12.13a

Stability factors for c—ϕ soils
(Critical toe circles) (Janbu)

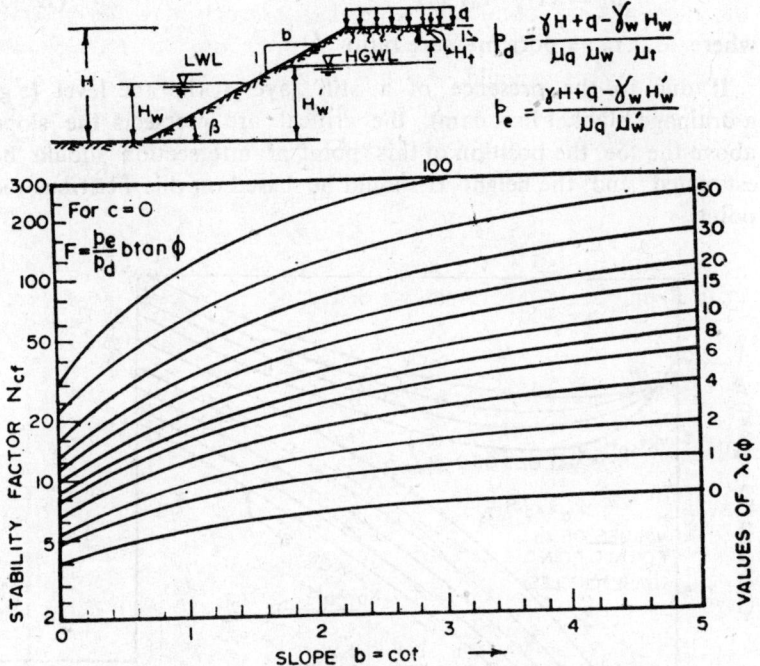

Fig. 12.13b
Stability factors for c—ϕ soils
(Critical toe circles) (Janbu)

The above mentioned stability charts (Fig. 12.13) can be used both for the total stress analysis (c_T, ϕ_T) and the effective stress analysis (c', ϕ'). p_d is defined by Eq. 12.33 for both the analyses. Eq. 12.36 is used for the total stress analysis; the effective reference stress p_o for c'—ϕ' analysis is, however, defined as:

$$p_o = \frac{\gamma H + q - \gamma_w H_w'}{\mu_q \mu'_w} \qquad (12.38)$$

where H_w' is the height of groundwater level *inside* the body of the slope, above the toe. In the absence of a surcharge and water filled tension crack and no free water standing against the slope, p_d and p_o may also be expressed as:

$$p_d = \gamma H \qquad (12.39)$$

$$p_e = (1 - r_u)\,\gamma H \tag{12.40}$$

where r_u = pore pressure ratio

If due to the presence of a stiff layer at the toe level (e.g. a drainage blanket in a dam), the critical arc intersects the slope above the toe, the position of this point of intersection should be estimated and the height H should be based on this fictitious toe point.

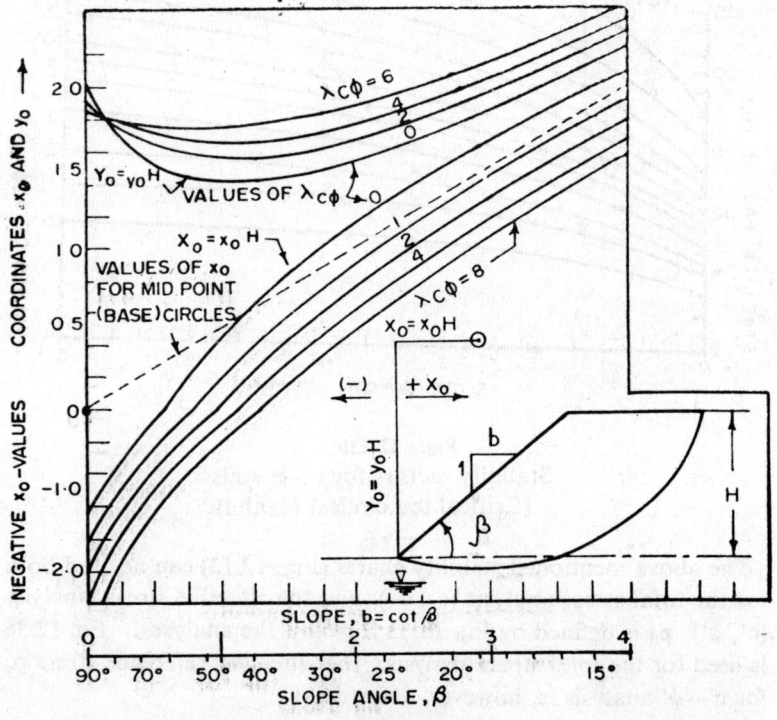

Fig. 12.14a

Critical unit centre coordinates for toe circles in $c-\phi$ soils. If $\lambda c\phi > 0$, critical slip circle intersects toe. Groundwater level and top of hard stratum are below critical circle.

$$F = \frac{N_{cf}\,c}{\gamma H}, \quad \lambda c\phi = \frac{\gamma H \tan \phi}{c} \quad \text{(Janbu)}$$

(See Fig. 12.14b also)

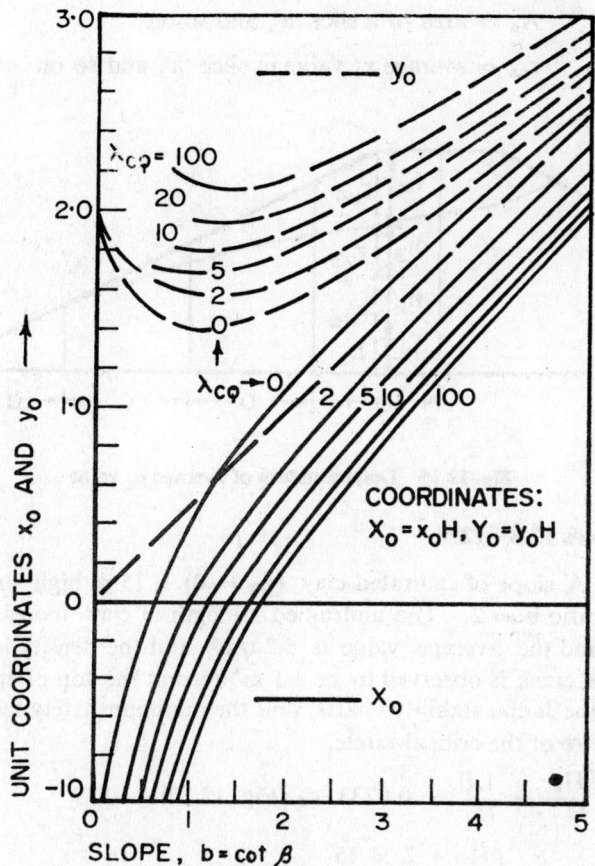

Fig. 12.14b

Unit coordinates for toe circles in c—ϕ soils

When r_u is variable, as is generally the case over the cross section of an embankment, a *weighted average value* is obtained by dividing the section into a suitable number of vertical slices (Fig. 12.15). r_u is determined for a series of points on the centre line of each slice and the average value for a particular slice is given by:

$$r_u = \frac{h_1 r_{u1} + h_2 r_{u2} + h_3 r_{u3} + ...}{h_1 + h_2 + h_3 + ...} \qquad (12.42)$$

The overall average for the whole section is:

$$r_u = \frac{A_a r_{ua} + A_b r_{ub} + A_c r_{uc} + ...}{A_a + A_b + A_c + ...} \qquad (12.43)$$

where A_a = area of a slice 'a', and so on

r_{ua} = average r_u value in slice 'a', and so on

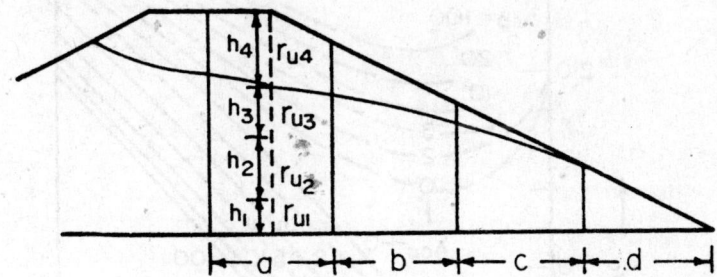

Fig. 12.15 Determination of average r_u value

Examples 12.3—12.9

12.3 A slope of saturated clay ($\phi_T = 0$) is 15 m high and has a slope ratio $b = 2$. The undrained strength of clay increases with depth and the average value is 5.2 t/m² and the density is 2 t/m³. Tension crack is observed to be 1.1 m deep at the top of the slope. Using the Janbu stability charts, find the minimum safety factor and the centre of the critical circle.

$$\frac{H_t}{H} = \frac{1.1}{15} = 0.0733, \ \mu_t \text{ (Fig. 12.12)} = 0.99$$

$$p_d = \frac{\rho H}{\mu_t} = \frac{2 \times 15}{0.99} = 30.3 \text{ t/m}^2$$

Increase of s_u with depth suggests toe failure circle, and for $b = 2$, N_o (Fig. 12.10) = 6.7.

$$F = \frac{N_o \ s_u}{p_d} = 1.15$$

For $b = 2$ and toe circle, the critical unit coordinates (Fig. 12.11) are: $x_o = 1.0$ and $y_o = 1.8$.

Using $H - H_t = 15 - 1.1 = 13.9$ m as effective height of the slope, the critical centre coordinates are:

$$X_o = x_o H = 13.9 \text{ m and } Y_o = y_o H = 25 \text{ m}$$

12.4 Cohesive soil, $\phi_T = 0$, density $\rho = 1.85$ t/m³, unit cohesion $c_T = 29$ kPa. Height of slope H = 8 m, slope angle $\beta = 35°$. Firm base at D = 6.4 m below toe. Assuming base circle failure, find F_{min}.

$$d = D/H = 0.8$$

For $\beta = 35°$, d = 0.8, $N_o = 5.8$ (Fig. 12.10)

$$F = \frac{N_o \, s_u}{\gamma \, H} = 1.16$$

12.5 Examine the stability of a slope, 8 m high at a slope angle 35°, of cohesive soil having $\gamma = 18.13$ kN/m³, $c_T = 29$ kPa, $\phi_T = 0$. A firm stratum exists at a depth of 6.4 m below the toe. The slope carries a uniform surcharge of 10.1 kPa and is submerged by free standing water against its face upto a height of 4.8 m. Tension cracks are observed upto a depth of 1.28 m at the top. Use the Janbu stability charts.

$q/\gamma H = 0.0696$, $\mu_q = 0.940$ (Fig. 12.12)

$H_w/H = 0.6$, $\mu_w = 0.945$

$H_t/H = 0.16$, $\mu_t = 0.985$ (zero hydrostatic pressure)

$$p_d = \frac{\gamma H + q - \gamma_w \, H_w}{\mu_q \, \mu_w \, \mu_t} = 123.55 \text{ kPa}$$

$D = 6.4$ m, d = D/H = 0.8

$\beta = 35°$, $N_o = 5.8$

$$F = \frac{N_o \, s_u}{p_d} = \frac{5.8 \times 29}{123.55} = 1.36$$

12.6 Cohesive-frictional soil, $c_T = 16$ kPa, $\phi_T = 17°$, $\gamma = 18.8$ kN/m³. Slope height H = 8.5 m, $\beta = 30°$. No surcharge or tension crack. Groundwater level is below the failure surface. Determine the safety factor and the coordinates of the critical centre.

$$p_d = p_e = \gamma \, H = 159.8 \text{ kPa}$$

$$\lambda c\phi = \frac{p_e \tan \phi}{c} = 3.06$$

$\beta = 30°$, $\lambda c\phi = 3.06$, $N_{cf} = 14.1$ (Fig. 12.13)

$$F = \frac{N_{cf} \cdot c}{p_d} = 1.41$$

Fig. 12.14: $x_o = 0.52$, $X_o = x_o H = 4.42$ m

$y_o = 1.65$, $Y_o = y_o H = 14.03$ m

12.7 Slope as in Ex. 12.6. $H = 8.5$ m, $\beta = 30°$, $\gamma = 18.8$ kN/m³, $\phi_T = 17°$, $c_T = 16$ kPa. There is also a surcharge $q = 2.88$ kPa. Depth of tension crack 3 m with full hydrostatic pressure. Partially submerged, $H_w = 1.8$ m.

$$q/\gamma H = 0.018, \ \mu_q = 0.995$$
$$H_w/H = 0.21, \ \mu_w = 0.97$$
$$H_t/H = 0.352, \ \mu_t = 0.97$$
$$p_d = \frac{\gamma H + q - \gamma_w H_w}{\mu_q \ \mu_w \ \mu_t} \simeq 155 \text{ kPa}$$
$$p_e = \frac{\gamma H + q}{\mu_q} = 163.5 \text{ kPa}$$
$$\lambda c\phi = \frac{p_e \tan \phi}{c} = 3.13, \ N_{ef} = 14.2$$
$$F = \frac{N_{ef} \cdot c}{p_d} = 1.47$$

12.8 Determine the minimum safety factor of a slope 17.4 m high which is submerged upto a height of 13.0 m by seepage but with no water standing on the exposed face of the slope. The slope angle $\beta = 26.5°$ ($b = 2$). The soil properties are: $c' = 1.0$ t/m², $\tan \phi' = 0.512$ and $\rho = 2$ t/m³. (Janbu 1969).

$$\frac{H_w'}{H} = \frac{13.0}{17.4} = 0.75, \ \mu_w' \text{ (Fig. 12.12)} = 0.97$$
$$p_e = \frac{\rho H - \rho_w H_w'}{\mu_w'} = \frac{2 \times 17.4 - 1 \times 13}{0.97}$$
$$= 22.5 \text{ t/m}^2$$
$$\lambda c\phi = \frac{p_e \tan \phi'}{c'} = 11.52$$

N_{ef} (Fig. 12.13) $= 35.0$

$H_t/H = 1.1/17.4 = 0.063$

$\mu_t \simeq 0.99$ (Fig. 12.12)

$$p_d = \frac{\rho H}{\mu_t} = 35.15$$
$$F = \frac{N_{ef} \ c'}{p_d} \simeq 1.0$$

12.9 Downstream slope of an earth dam, $H = 18.2$ m, slope ratio $b = 3.45$. Effective stress parameters $c' = 19.6$ kPa, $\phi' = 22°$.

Unit weight $\gamma = 19.6$ kN/m^3. Average pore pressure ratio $\gamma_{ua} = 0.4$. Determine F and the coordinates of the critical centre of the slip circle. (Janbu 1959).

$$p_o = (1 - r_u) \gamma H = 214.03 \text{ kPa}$$

$$p_d = \gamma H = 356.72 \text{ kPa}$$

$$\lambda c\phi = \frac{p_o \tan \phi'}{c'} = 4.37 \simeq 4.4$$

$$N_{of} = 26 \quad (\text{Fig. } 12.13)$$

$$F = \frac{N_{cf} c'}{p_d} = 1.43$$

$$x_o = 1.4, X_o = 25.5 \text{ m}$$

$$y_o = 2.15, y_o = 39.1 \text{ m}$$

12.9 TAYLOR STABILITY CHARTS

Taylor (1948) analysed the rotational, circular slip by introducing the device of a friction circle of radius R sin ϕ (where R = radius of slip surface) with the same centre O as that of the slip surface (Fig. 12.16). The forces acting on the sliding sector ABCD are: (i) self weight W acting through C.G. of the sector, (ii) cohesive force due to cohesion component of shear strength acting along the arc AD, (iii) frictional force due to frictional component (tan ϕ) acting along the arc, and the (iv) normal reaction acting on the arc. The resultant of the normal reaction and the frictional force, called 'the resultant frictional force R$_f$' is assumed to be tangent to the friction circle (ϕ-circle). Actually, R$_f$ is tangent to a circle of radius slightly larger than R sin ϕ. From the consideration of statics, the cohesive force acting *along the arc* AD can be replaced by its resultant C$_r$ acting *parallel to the chord* AD. The magnitude of C$_r$ will be (c $\times$ chord AD), where c is unit cohesion. Taking moment about O, the distance X$_o$ of C$_r$ from O is obtained as folloows:

$$(c \times \text{chord AD}) X_o = (c \times \text{arc AD}) R$$

or

$$X_c = R \frac{\text{arc AD}}{\text{chord AD}} \tag{12.44}$$

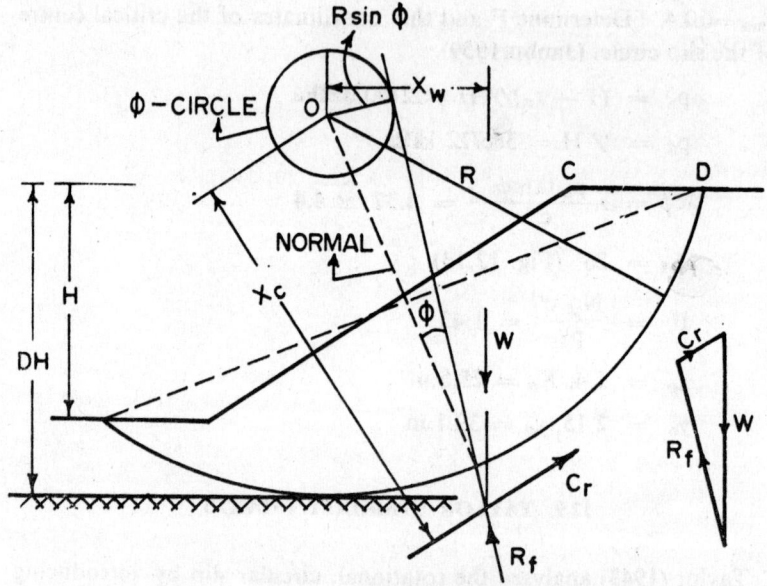

Fig. 12.16 Taylor analysis using friction circle
(Read lower arc point as A and toe of slope as B)

For equilibrium, the three forces W, C_r and R_f must be concurrent. The force W is first drawn vertically through the C.G. of the sliding sector and then C_r is drawn parallel to the chord AD at a distance X_o from O. The force R_f is then drawn passing through the point of intersection of W and C_r and tangentially to the ϕ-circle. As the magnitude of W is also known, the force triangle can be completed and the magnitude of C_r required for equilibrium can be determined; here it is assumed that the frictional force has been fully mobilised. If c_r is the unit cohesion required for equilibrium, $C_r = c_r \times$ chord AD and knowing C_r from the force triangle, c_r is obtained. If c_T is the ultimate unit cohesion of soil, the safety factor F_e with respect to cohesion is given by:

$$F_e = \frac{c_T}{c_r} \tag{12.45}$$

In the special case of a $\phi_T = 0$ analysis, moment equilibrium about O gives the expression for safety factor F as:

$$F = \frac{R (c_T \times arc\,AD)}{W.\,X_w} \qquad (12.46)$$

noting that R_f passes through O and does not contribute to the moment expression.

Additional centres and arcs must be analysed to get the minimum safety factor.

For all geometrically similar slip surfaces, the force diagrams obtained by the friction circle method are similar, irrespective of the height of the slope. (It is necessary to ignore the possibility of tension cracks to maintain geometriaclly similar slip surfaces.) Then C_r/W is constant. But $W \propto H^2\gamma$ and $C_r \propto (c_T/F_o) H$. Therefore $c_T/F_o\,\gamma H$ is constant. This coefficient, which depends only on ϕ and the geometry of the slope (slope angle β), is the Taylor "Stability number S_n". For $\phi_T = 0$, the value of S_n also depends on the 'depth factor D', where DH is depth of the lowest point of the slip circle *below the top* of the slope. For a slope of height H the stability number S_n for the failure surface along which safety factor is a minimum, is expressed as:

$$S_n = \frac{c_T}{F_o\,\gamma\,H} \qquad (12.47)$$

For specific values of β and ϕ (and hence for S_n), the safety factor F_o is directly proportional to c_T and inversely proportional to H. In a soil of constant c_T, F_o thus represents also the safety factor with respect to height, i.e. the ratio of the ciitical height H_0 to the actual height H:

$$F_o = \frac{c_T}{S_n\,\gamma\,H} = \frac{H_o}{H} = F_H \qquad (12.48)$$

It would be seen from Eqs. 12.30 and 12.48 that the Taylor "Stability number S_n" is the *reciprocal* of the Janbu "Stability factor N_o"

Based on friction circle analyses and the principle of geometric similarity, Taylor (1937) published stability numbers for the analysis of homogeneous slopes *in terms of total stresses*. To use them for effective stress analysis, may give rise to errors. The stability numbers for $\phi_T = 0$ analysis and c-ϕ analysis are plotted in Figs. 12.17 and 12.18.

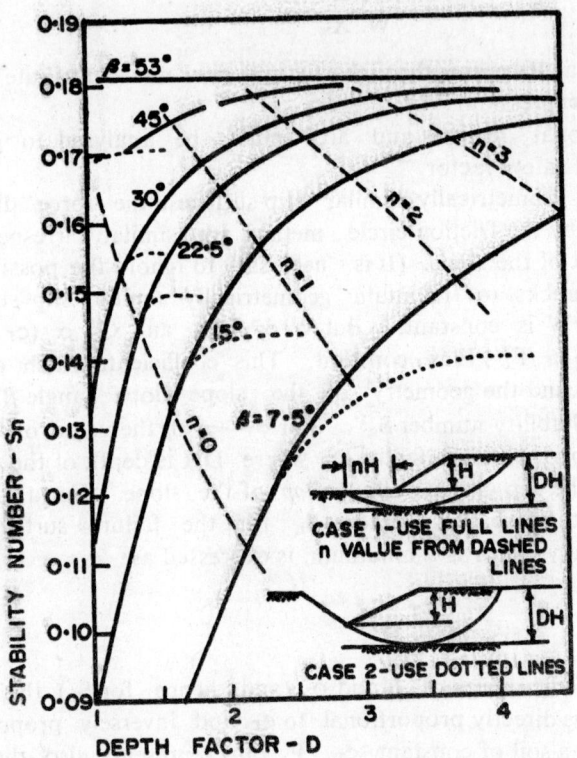

Fig. 12.17 Taylor stability chart for $\phi_T = 0$ analysis

The Taylor stability number gives safety factor with respect to cohesion, assuming full mobilization of frictional force. A safety factor $F\phi$ with respect to friction may also be defined as:

$$F\phi = \tan \phi / \tan \phi_r \qquad (12.49)$$

in which it is imagined that the cohesive component of the strength is fully developed and $\tan \phi$ is the required frictional component at a safety factor $F\phi$. To obtain the same value of the safety factor with respect to both cohesion and friction, $F\phi$ is initially assumed and ϕ_r determined from Eq. 12.49. Stability number is read corresponding to ϕ_r (instead of ϕ) and the resulting safety factor F_o is computed. The assumed $F\phi$ is

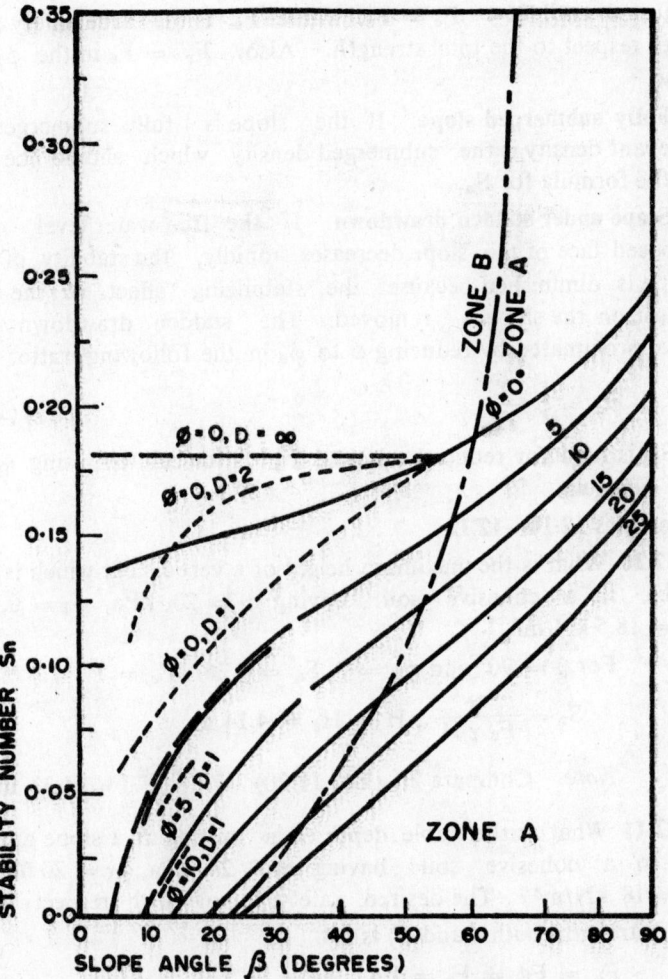

Fig. 12.18 Taylor stability chart for c—φ anlaysis

Zone A—Critical circle passes through the toe.

Zone B—*Case* 1. Critical circle passes through the toe (full lines).
Where full lines do not appear, this case is not
appreciably different from Case 3.

Case 2. Critical circle passes below toe (long dashed
lines). Where long dashed lines do not appear, the
circle passes through the toe.

Case 3. Strong stratum at toe level (D = 1) (dotted lines), adjusted until $F\phi = F_o = F_s$, where F_s is the true safety factor with respect to the total strength. Also, $F_c = F_s$ in the $\phi_T = 0$ case.

Fully submerged slope. If the slope is fully submerged, the relevant density is the submerged density which should be used in the formula for S_n.

Slope under sudden drawdown. If the free water level on the exposed face of the slope decreases rapidly, the stability of the slope is diminished because the stabilizing effect of the water thrust on the slope is removed. The sudden drawdown effect is approximated by reducing ϕ to ϕ_d in the following ratio:

$$\phi_d = \phi \ \frac{\gamma'}{\gamma_{sat}} \tag{12.50}$$

S_n is read for reduced ϕ_d and F_o is calculated using γ_{sat} in the formula.

Examples 12.10—12.12

12.10 What is the maximum height of a vertical cut which is just stable in a cohesive soil having $c_T = 20$ kPa, $\phi_T = 0$, and $\gamma = 18.5$ kN/m³.

For $\beta = 90°$ and $\phi_T = 0$, $S_n = 0.261$, $F_o = 1$

$$S_n = \frac{c_T}{F_o \gamma H} \ , \ H = H_c = 4.14 \text{ m}$$

Note. Compare H_c (Eq. 11.20) $= \dfrac{4 \ c_T}{\gamma} = 4.32$ m

12.11 What is the stable depth of a cutting at a slope angle of 30° in a cohesive soil having $c_T = 24$ kPa, $\phi_T = 20.6°$ and $\gamma = 18$ kN/m³? The desired safety factor with respect to the total strength (both c and ϕ) is 1.4.

$$F_o = F\phi = F_s = 1.4$$

$$F_\phi = \frac{\tan \phi_T}{\tan \phi_r} = \frac{\tan 20.6°}{\tan \phi_r} \ ; \ \phi_r = 15°$$

For $\phi_r = 15°$ and $\beta = 30°$, $S_n = 0.046$

$$S_n = \frac{c_T}{F_o \gamma H}; \ H = 20.7 \text{ m}$$

12.12 A canal is to be excavated to a depth of 3 m in a uniform silty clay layer, 7.5 m thick, underlain by a stiffer soil. The canal banks will be at a slope of 20°. The density of silty clay is

1.85 t/m³ and its total stress and effective stress strength parameters are respectively: $c_T = 25$ kPa and $\phi_T = 0$; $c' = 6$ kPa and $\phi' = 15°$. Examine the stability of banks under the following conditions: (a) immediate stability of free slope after construction, (b) long term stability of free slope, (c) immediate stability on full submergence, (d) long term stability after full submergence and (e) stability on sudden drawdown. What is the maximum depth upto which the canal could be excavated before failure?

(a) Immediate stability — free slope.

$H = 3$ m, $D\ H = 7.5$ m, $D = 2.5$ (depth factor)

For $\phi_T = 0$ and $\beta = 20°$, $S_n = 0.17$

$$S_n = \frac{c_T}{F_o\ \gamma\ H} = \frac{25}{F_o \times 18.13 \times 3} = 0.17$$

$F_o = 2.7$

(b) Long term stability — free slope.

Use c', ϕ' and γ.

For $\beta = 20°$ and $\phi' = 15°$, $S_n = 0.018$

$$S_n = \frac{c'}{F_o\ \gamma\ H}, F_o = 6.13$$

To obtain true safety factor F_s with respect to both c' and ϕ', $F\phi$ is assumed and S_n corresponding to a reduced ϕ_r is used to get F_o. The procedure is repeated until $F_o = F\phi = F_s$.

(c) Immediate stability on submergence.

Use $\phi_T = 0$, $c_T = 25$ kPa, $\gamma' = \gamma - \gamma_w = 18.3 - 9.8$
$$= 8.33 \text{ kN/m}^3$$

$$S_n = \frac{c_T}{F_o\ \gamma'\ H} = 0.17$$

$F_o = 5.88$

In the submerged state, F_o increases due to stabilizing effect of free water against the face of slope. F_s is found by trial and error.

(d) Long term stability — full submergence.

Use c', ϕ', γ'.

For $\beta = 20°$, $\phi' = 15°$, $S_n = 0.018$

$F_o = 13.3$

F_s is found by trial.

(e) Sudden drawdown.

$$\phi_d = \frac{\gamma'}{\gamma_{sat}}\ \phi' = \frac{8.33}{18.13} \times 15 = 6.9°$$

For $\beta = 20°$, $\phi_d = 6.9°$, $S_n = 0.08$

Use c′ and γ_{sat}.

$$F_o = \frac{6}{0.08 \times 18.13 \times 3} = 1.38$$

Thus F_o is considerably reduced on sudden drawdown.

(f) Maximum depth of excavation before failure.

F_o of free slope of height $H = 3$ m immediately on excavation is 2.7.

Since $F_e = F_H = H_o/H$

$$H_e = 2.7 \times 3 = 8.1 \text{ m}$$

12.10 BISHOP ANALYSIS

The Fellenius method of slices satisfies only the overall moment equilibrium, neglecting the moment equilibrium of the individual slice and only approximating the force equilibrium of each slice. The force system on a slice considered by Bishop (1955) is shown in Fig. 12.19, where X and V represent respectively the horizontal and vertical components of the side forces acting on the slice. Bishop found that by including horizontal side forces

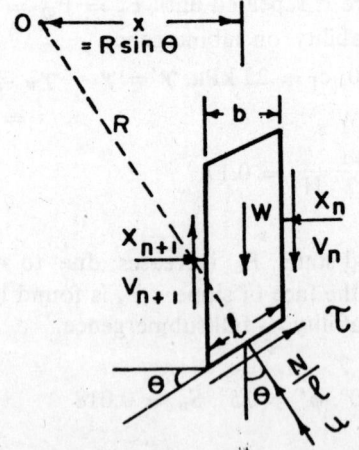

Fig. 12.19 Force system for a slice in the Bishop analysis

to compute the normal force on the slice base and also satisfying the overall moment equilibrium, the resulting safety factor was only slightly less than that found by more rigorous methods (0 — 6%; Wright et al 1973).

Each slice is assumed to have the same safety factor F and the shear stress τ on the base of any slice required for equilibrium is defined as:

$$\tau = \frac{c'}{F} + \left(\frac{N}{l} - u \right) \frac{\tan \phi'}{F} \qquad (12.51)$$

and the safety factor as before:

$$F = \frac{\Sigma \, s \, l \, R}{\Sigma \, W. \, x} = \frac{\Sigma \, c' \, l + (N - u \, l) \tan \phi'}{\Sigma \, W \sin \theta} \qquad (12.52)$$

since $\quad x = R \sin \theta$

Resolving forces in the vertical direction:

$$N \cos \theta = W - \tau \, l \sin \theta - (V_{n+1} - V_n) \qquad (12.53)$$

As a simplification, Bishop assumed $(V_{n+1} - V_n)$ as negligible in comparison with the other force, i.e. $(V_{n+1} - V_n) = 0$. Substituting the value of τ from Eq. 12.51 into Eq. 12 53 and rearranging:

$$N - ul = \frac{W - \dfrac{c'l \sin \theta}{F} - u \, l \cos \theta}{\cos \theta + \sin \theta \tan \phi'/F}$$

Putting $l \cos \theta = b$ and substituting the above expression into the equation for F (Eq. 12.52) and after some rearrangement:

$$F = \frac{1}{\Sigma \, W \sin \theta} \, \Sigma \left[\frac{\{c' \, b + (W - u \, b) \tan \phi'\} \sec \theta}{1 + (\tan \theta \tan \phi')/F} \right] \qquad (12.54)$$

Using $r_u = u/\gamma z$ and $W = \gamma b z$, Eq. 12.54 may be rewritten as:

$$F = \frac{1}{\Sigma \, W \sin \theta} \, \Sigma \left[\frac{\{c' \, b + W (1 - r_u) \tan \phi'\} \sec \theta}{1 + (\tan \theta \tan \phi')/F} \right] \qquad (12.55)$$

Eq. 12.54 or 12.55 is known as the 'Bishop simplified solution'. It has been found to give reliable results in a large number of case records. Although it is expressed in terms of effective stress parameters, it can be used for total stress parameters also.

If the soil possesses a constant strength ($\phi' = 0$), the Bishop simplified equation reduces to the Fellenius solution Eq. 12.18:

$$F = \frac{\Sigma c' b \sec \theta}{\Sigma W \sin \theta} = \frac{\Sigma c' l}{\Sigma W. x/R} = \frac{c' L R}{\Sigma W. x}$$

$$= N_s \frac{c'}{\gamma H} \tag{12.56}$$

where N_s is the 'stability factor' (Mitchell 1983). It is found that deeper failure surface becomes critical, the extent of which is generally limited by a stiff base.

For a cohesionless soil ($c'=0$), the equation reduces to:

$$F = \tan \phi' \Sigma \left[\frac{1 - r_u \sec^2 \theta}{\tan \theta} \right] \tag{12.57}$$

It has been shown by computer analyses that very shallow failure surfaces are critical in cohesionless soils (Mitchell 1983), the physical reason being the increased frictional resistance with increase in the normal stresses on deeper circles. A very shallow failure surface approaches a plane failure (Section 12.2), where θ is constant ($\theta \rightarrow \beta$) and Eq. 12.57 then becomes:

$$F = \frac{\tan \phi'}{\tan \theta} (1 - r_u \sec^2 \theta) \tag{12.58}$$

where $\theta =$ inclination of the failure plane which approaches the slope angle β for long slopes (Eq. 12.8)

As the safety factor F occurs on both sides of the Bishop equation, it is calculated from successive trials starting with an initial estimate for F. If the calculated value of F is used in succeeding trials, convergence is quite rapid. The method is very suitable for solution on a computer.

For a partially submerged slope with seepage, W in Eq. 12.54 is replaced by (W_1+W_2), where W_1 is total weight of a slice above external water level and W_2 is the submerged weight below this level; and 'u' is replaced by 'u_1', the pore pressure measured from the external water level (See Section 12.6.) For a fully submerged slope with external water against the slope, $r_u = 0$ and submerged density is used in the stability analysis.

1. Average pore pressure ratio and critical centre. The average pore pressure ratio r_u for a homogeneous groundwater condition can be estimated from the position of the top flow line (phreatic surface) or from a flow net. A method is given in Section 12.8

(Eq. 12.43). As an alternative method according to Mitchell (1983), once the critical slip arc has been approximately located, the potential sliding sector is divided into slices and individual values of $r_u = u/\gamma z$ are obtained for about 10 equal-width slices comprising the lower two-thirds of the sliding sector. This is done using a flow net. For relatively flat slopes ($\beta = 30°$ or less) without artesian or drawdown pressures, u may be assumed equal to $h\gamma_w$, where h is the vertical distance between the slice base and the phreatic line; this assumption results only in a marginal overestimate of u in a homogeneous soil. The average γ_{ua} is then calculated as:

$$r_{ua} = \sum_1^n (u/\gamma z)/n \tag{12.59}$$

where   n = number of slices comprising the lower two-thirds of the slope

Worst seasonal and site conditions such as rain and infiltration, etc., should be taken into account for evaluating r_u. Mitchell (1983) further recommends the following values of r_u for three common flow conditions:

(1) Flow parallel to the slope without any slope seepage (dipping strata):

$$r_u = \frac{\gamma_w}{\gamma_{sat}} \cos^2 \beta \text{ for } \frac{h}{H} > 0.8, \text{ or}$$
$$(H - h) < 3 \text{ m} \tag{12.60}$$

(2) Horizontal flow with full slope seepage ($k_h > k_v$, artesian condition, drawdown):

$$r_u = \frac{\gamma_w}{\gamma_{sat}} \text{ for } \frac{h}{H} > 0.8, \text{ or } (H - h) < 3 \text{ m} \tag{12.61}$$

(3) Parabolic top flow line with seepage face near the toe (homogeneous condition):

$$r_u = \frac{\gamma_w}{\gamma_{sat}} \cos \beta \text{ for } \frac{h}{H} > 0.8, \text{ or } (H - h) < 3 \text{ m} \tag{12.62}$$

where   H = height of slope
        h = maximum height of water table or top flow line measured above the toe of the slope in the region just beyond the slope

The coordinates (X_0, Y_0) of the centre of critical circle in homogeneous soil, defined with respect to the toe as origin (X_0 = horizontal distance from toe, Y_0 = vertical distance from toe, See Fig. 12.14a), may be estimated from Fig. 12.20 (Mitchell 1983). When tan ϕ' is greater than 0.1, the critical arc will pass through the toe of the slope. For low values of r_u (<0.3), the critical centre is well approximated by:

$$X_0 = H \cot \beta \, (0.6 - \tan \phi') \qquad (12.63)$$

$$Y_0 = H \cot \beta \, (0.6 + 2 \tan \phi') \qquad (12.64)$$

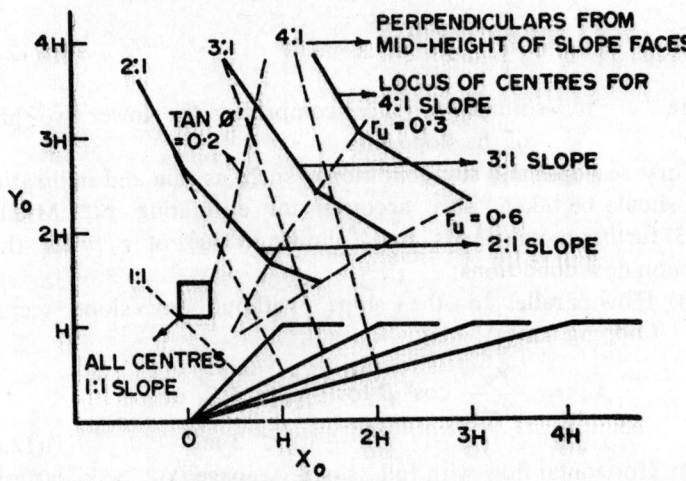

Fig. 12.20 Approximate locations of critical centres for $F = 1$. © Mitchell 1983. Reproduced with permission of George Allen and Unwin, Publishers, Ltd., London.

2. Design charts. For a given slope angle and given soil properties the safety factor varies linearly with r_u and can thus be expressed as:

$$F = m - n \, r_u \qquad (12.65)$$

where m and n are dimensionless "stability coefficients" which are functions of β, ϕ, the dimensionless number $c'/\gamma H$ and the depth factor D. The Bishop simplified equation (Eq. 12.54) is of the form of Eq. 12.65. Based on computer solution of the Bishop equation,

Bishop and Morgenstern (1960) obtained the values of m and n and prepared design charts for homogeneous slopes. The charts were further extended by O'Connor and Mitchell (1977). Using the format of Hoek and Bray Charts (1974) for rock slope design, Mitchell (1983) has given design charts for soil slopes

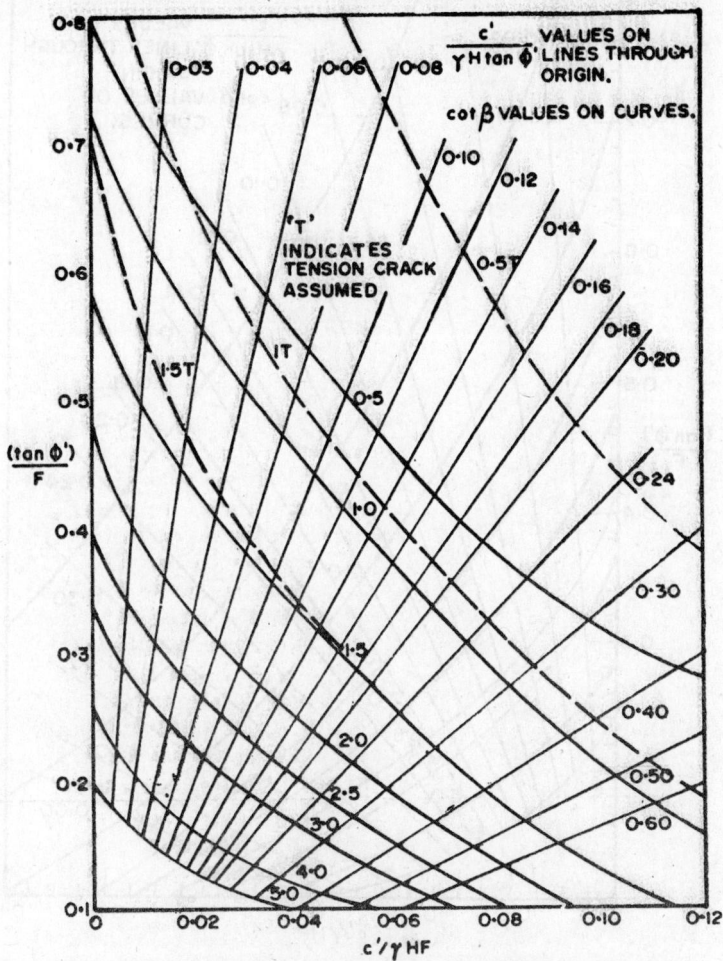

Fig. 12.21.   Design chart for rotational slip, $r_u = 0$
(© Mitchell 1983. Reproduced with permission of
George Allen & Unwin, Publishers. Ltd.; London

(Figs. 12.21-12.23) which have greater advantage of operational simplicity. When a slope is fully submerged, use $c'/\gamma'H$ in place of $c'/\gamma H$ and equate r_u to zero. For a full rapid drawdown condition, use $c'/(\gamma_{sat}H)$ and $r_u = \gamma_w/\,$.

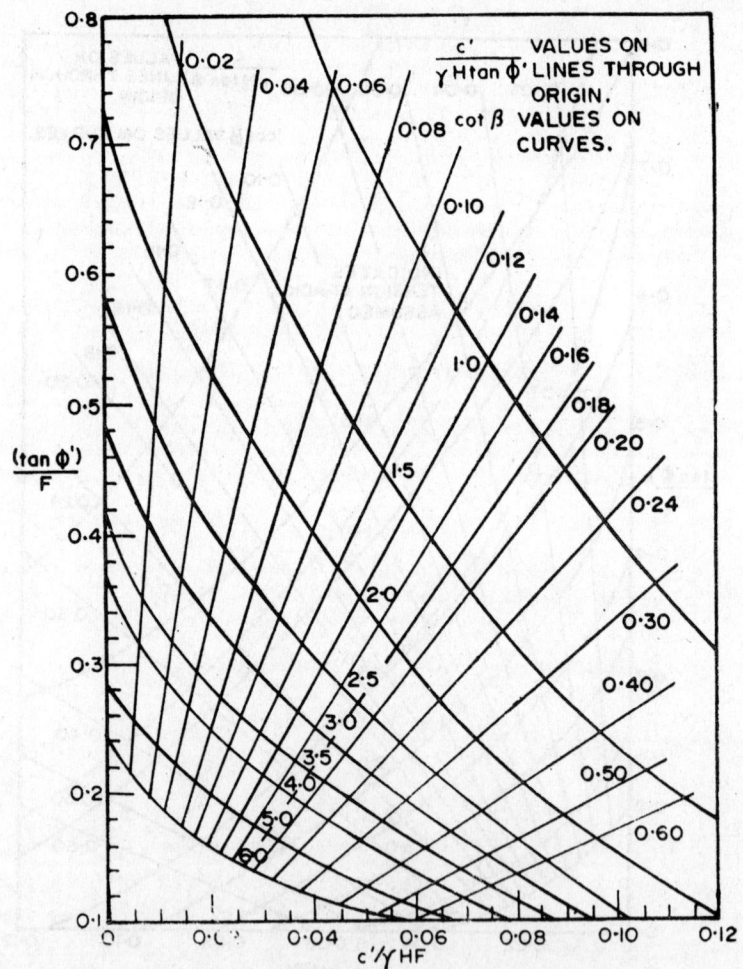

Fig. 12.22. Design chart for rotational slip. $r_u = 0.3$
© Mitchell 1983. Reproduced with permission of George Allen & Unwin, Publishers, Ltd., London.

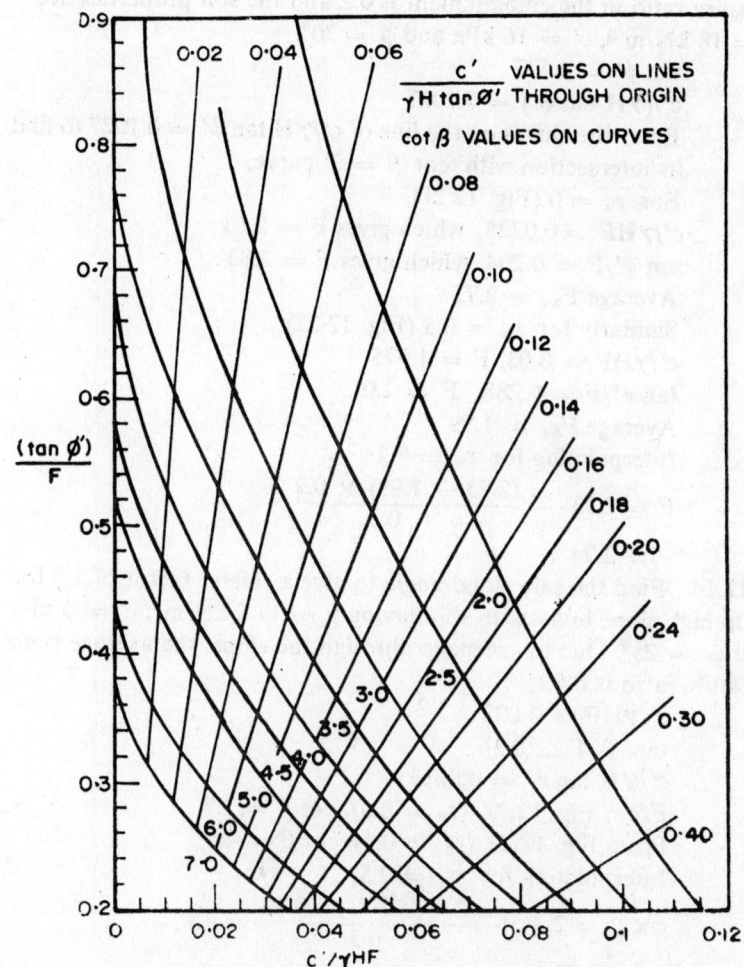

Fig. 12.23. Design chart for rotational slip, $r_u = 0.6$
ⓒ Mitchell 1983. Reproduced with permission of
George Allen & Unwin Publishers, Ltd., London.

Examples 12.13—12.14

12.13. Using the stability coefficient design charts (Figs. 12.21-12.23), find the safety factor of an embankant slope, 15 m high, constructed at a slope ratio of 3 (cot $\beta = 3$). The average pore pressure ratio in the embankment is 0.2 and the soil properties are: $\gamma = 18$ kN/m^3, $c' = 16$ kPa and $\phi = 30°$.

$\tan \phi' = 0.577$

$c'/(\gamma H \tan \phi') = 0.1027$

Enter the charts on the line of $c'/\gamma H \tan \phi' = 0.1027$ to find its intersection with 'cot $\beta = 3$' curve.

For $r_u = 0$ (Fig. 12.21),

$c'/\gamma HF = 0.0225$, which gives F = 2.63

$\tan \phi'/F = 0.204$, which gives F = 2.83

Average $F_{av} = 2.73$

Similarly for $r_u = 0.3$ (Fig. 12.22),

$c'/\gamma HF = 0.03$, F = 1.975

$\tan \phi'/F = 0.288$, F = 2.0

Average $F_{av} = 1.99$

Interpolating for $r_u = 0.2$

$$F = 2.73 - \frac{(2.73 - 1.99) \times 0.2}{0.3}$$

$$= 2.24$$

12.14 Find the safe slope angle to give a safety factor of 1.3 for 10 m high slope built with soil having $\gamma = 17.5$ kN/m^3, $c' = 5$ kPa and $\phi' = 25°$. Due to seepage through the slope, the average pore pressure ratio is 0.5.

$c'/\gamma HF = 0.022$

$\tan \phi'/F \simeq 0.36$

$c'/\gamma H \tan \phi' = 0.0613$

From Fig. 12.22 ($r_u = 0.3$), cot $\beta = 2.7$

From Fig. 12.23 ($r_u = 0.6$). cot $\beta = 4.8$

Interpolating for $r_u = 0.5$,

$$\cot \beta = 2.7 + \frac{(4.8 - 2.7) \times 0.2}{0.3} = 4.1$$

$$\beta = 13.7°$$

12.11 GREENWOOD SIMPLE SOLUTION

An effective stress, simple equation for slope stability has been derived by Greenwood (1983) which is easier to be solved in comparison to the Bishop equation (Eq. 12.55) that has F on both sides. The equation can be used for both circular and non-circular slips.

The safety factor F is defined as before by the expression:

$$F = \frac{\Sigma (c'l + N' \tan \phi')}{\Sigma W \sin \theta} \qquad (12.66)$$

where N' is the effective normal stress on the slice base which is determined by considering a soil element on the potential slip surface, as shown in Fig. 12.24.

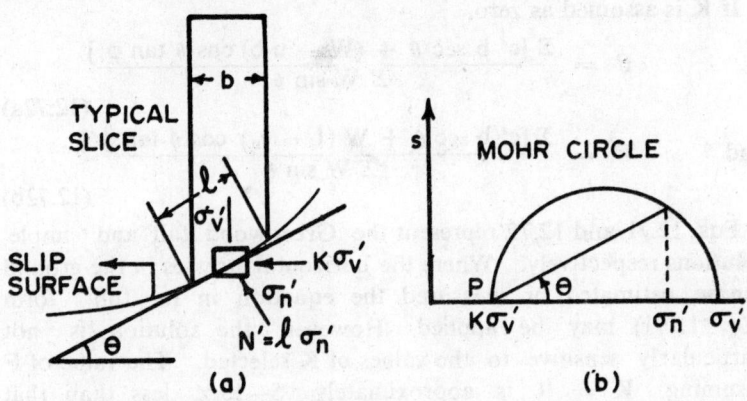

Fig. 12.24. Consideration of soil element on slice base in Greenwood solution (Greenwood 1983)

If σ_v' is the effective vertical stress on the base of slice and K is the ratio of horizontal to vertical effective stresses, the horizontal effective soil pressure is given by $K \sigma_v'$ (the effect of ignoring possible rotation of principal stresses is insignificant). From the Mohr circle (Fig. 12.24):

$$\sigma_n' - K\sigma_v' = (\sigma_v - K\sigma_v') \cos^2\theta \qquad (12.67)$$

or $\quad \sigma_n' = \sigma_v' [\cos^2\theta + K (1 - \cos^2\theta)]$

$$= \sigma_v' (\cos^2\theta + K \sin^2\theta) \qquad (12.68)$$

But $\quad N' = l\sigma_n'$

and $\quad \sigma_v' = (W - ub)/b$ (neglecting inter-slice forces)

Hence $N' = \dfrac{l\,(W - ub)}{b}(\cos^2\theta + K \sin^2\theta)$

Substituting $l = b/\cos\theta \qquad (12.69)$

$$N' = \frac{W - ub}{\cos\theta} (\cos^2\theta + K \sin^2\theta)$$

or $\quad N' = (W - ub) \cos \theta (1 + K \tan^2\theta) \qquad (12.70)$

Substituting for N' and l (Eqs. 12.69 and 12.70) in Eq. 12.66:

$$F = \frac{1}{\Sigma\,W\,\sin\,\theta}\,\Sigma\,[c'\,b\,\sec\,\theta + (W - u\,b)\,(1 + k\,\tan^2\,\theta)$$
$$\times\,\cos\,\theta\,\tan\,\phi'] \qquad (12.71a)$$

Eq. 12.71 can be expressed in term of the pore pressure ratio $r_u = u/\gamma z = u\,b/W$,

$$F = \frac{1}{\Sigma\,W\,\sin\,\theta}\,\Sigma\,[c'\,b\,\sec\,\theta + W\,(1 - r_u)\,(1 + k\,\tan^2\,\theta)$$
$$\times\,\cos\,\theta\,\tan\,\phi'] \qquad (12.71b)$$

If K is assumed as zero,

$$F = \frac{\Sigma\,[c'\,b\,\sec\,\theta + (W - u\,b)\,\cos\,\theta\,\tan\,\phi']}{\Sigma\,W\,\sin\,\theta}$$
$$(12.72a)$$

and

$$F = \frac{\Sigma\,[c'\,b\,\sec\,\theta + W\,(1 - r_u)\,\cos\,\theta\,\tan\,\phi']}{\Sigma\,W\,\sin\,\theta}$$
$$(12.72b)$$

Eqs. 12.71 and 12.72 represent the Greenwood 'full' and 'simple' solutions respectively. Where the horizontal stresses in the ground can be estimated or measured the equation in its 'full' form (Eq. 12.71) may be applied. However, the solution is not particularly sensitive to the values of K selected. The value of F assuming K = 0 is approximately 5—15% less than that calculated for K = 0.5 and for routine calculations, it may be conservatively assumed that K = 0. The difference between the Greenwood simple solution (Eq. 12.72) and the Bishop simplified solution (Eq. 12.54) is generally less than 15% with Eq. 12.72 giving higher values for shallow slips and the Bishop equation giving higher values for deeper slip circles (Greenwood 1983)

12.12 SLOPES IN CLAYS OF INCREASING STRENGTH

The methods of slope analysis so far described assume that the strength is uniform with depth. In general, however, the undrained strength of clays ($\phi_T = 0$) is not uniform but increases with depth. In most cases of normally consolidated clays, the undrained strength c_T may be expressed as:

$$c_T = c_0 + k\,z \qquad (12.73)$$

where $c_0 = c_T$ at the ground surface

k = rate of increase in c_T with depth

$= \triangle c_T / \triangle z$

z = depth below the ground surface

The value of k for particular soil can be expressed as $(c_T/\sigma_o')\,\gamma'$, where c_T/σ_o' is a constant for particular soil and γ' is the effective unit weight of soil (Nakase 1966); k has therefore the same dimension as density.

1. Nakase method for resisting moment. For slopes in clays of increasing strength, Nakase (1966) proposed a method for calculating the resisting moment M_r to be used in Eq. 12.18 for determining the safety factor. Considering the simple case, where c_T is uniform with depth, the resisting moment $\triangle M_r$ produced by the shear resistance along a part of the slip surface BC (Fig. 12.25) can be expressed as:

$$\triangle M_r = R^2. \triangle\alpha. c_T \tag{12.74}$$

and along the whole slip surface as:

$$M_r = R^2 (\beta_1 + \beta_2)\, c_T \tag{12.75}$$

where $\triangle\alpha$, β_1 and β_2 are measured in radians.

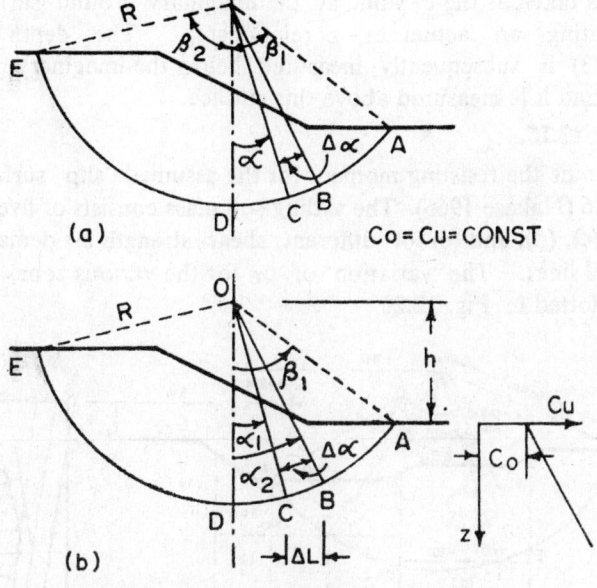

Fig. 12.25 Key sketch for obtaining resisting
moment (Nakase 1966)

If the undrained shear strength is expressed in the form of Eq. 12.73, $\triangle M_r$ along the part are BC (Fig. 12.25) is expressed as:

$$\triangle M_r = R^2 \left[(c_0 - kh)(\alpha_2 - \alpha_1) + k. \triangle L \right] \qquad (12.76a)$$
$$= R^2 \left[c_0 (\alpha_2 - \alpha_1) + k \{ \triangle L - h (\alpha_2 - \alpha_1) \} \right]$$
$$\qquad (12.76b)$$

In general, the resisting moment M_r is expressed in the form:

$$M_r = R^2 \, \Sigma \left[(c_0 - k\,h) \triangle \alpha + k \, \triangle L \right] \qquad (12.77a)$$
$$= R^2 \, \Sigma \left[c_0 \triangle \alpha + k \, (\triangle L - h \, \triangle \alpha) \right] \qquad (12.77b)$$

where   h = height of the centre of slip circle
             above the ground surface

$\triangle L$ = horizontal distance between points B and C

In the above equations, the angle α should be measured from the vertical line passing through the centre of slip circle   The value of c_0 is equal to the c_T value at point 'A'. When a similar computation is to be made for the part ED, the c_T value at point E should be taken for c_0.   In practice, however, it is convenient to assume that the datum level for construction is an 'imaginary ground surface' and c_0 is taken as the c_T value at the imaginary ground surface by extrapolating an actual c_T - z relationship.   The depth of z (Eq. 12.73) is subsequently measured below the imaginary ground surface, and h is measured above this surface.

Example 12.15

Determine the resisting moment for the assumed slip surface in Fig. 12.26 (Nakase 1966). The sliding soil mass consists of five zones (a), (b), (c), (d), and (e) of different shear strength as demarcated by dotted lines. The variation of c_T for the various zones is also shown plotted in Fig. 12.26.

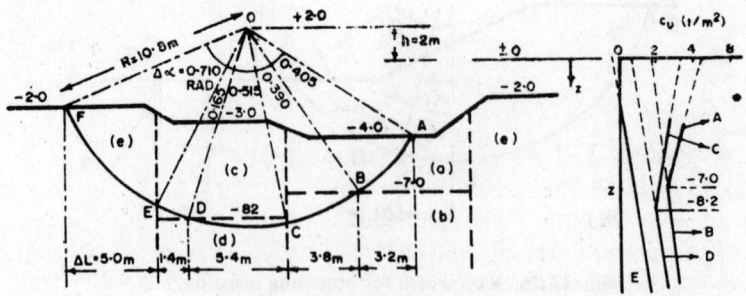

Fig. 12.26   Example 12.15

An imaginary ground surface is assumed at EL $\pm$ 0 m. The c_T -z plot in Fig. 12.26 is extended upto the imaginary ground surface to get c_0 values for different zones.

Expressions for c_T (z = depth below EL $\pm$ 0 m):

(a) $c_T = 4.6 - 0.30 z$
(b) $c_T = 1.8 + 0.10 z$
(c) $c_T = 3.36 - 0.18 z$
(d) $c_T = 0.72 - 0.14 z$
(e) $c_T = - 0.28 + 0.14 z$

Table 12.1   Example 12.15

Zone	c_0 (t/m^2)	k (t/m^3)	$c_0 - kh$ (t/m^2)	
(a)	4.6	-0.30	$+5.2$	R $= 10.8$ m
(b)	1.8	$+0.10$	$+1.6$	R$^2 = 116.64$ m^2
(c)	3.36	-0.18	$+3.72$	h $= 2$ m
(d)	0.72	$+0.14$	$+0.44$	
(e)	-0.28	$+0.14$	-0.56	

Table 12.2   Example 12.15

Arc	Zone	$\Delta\alpha$ (rad)	$\Delta\alpha (c_0 - kh)$	ΔL (m)	$k.\Delta L$	(4) + (6)
(1)	(2)	(3)	(4)	(5)	(6)	(7)
AB	(a)	0.405	$+2.106$	3.2	-0.960	1.146
BC	(b)	0.390	$+0.624$	3.8	$+0.380$	1.004
CD	(c)	0.515	$+0.227$	5.4	$+0.756$	0.983
DE	(d)	0.165	$+0.614$	1.4	-0.252	0.362
EF	(e)	0.710	-0.398	5.0	$+0.700$	0.302
					sum $=$	3.797

Eq. 12.77a: $M_r = 3.797 R^2 = 3.797 \times 116.64 = 442.9$ t-m/m

2. Gibson and Morgenstern method. Gibson and Morgenstern (1962) analysed the stability for a $\phi_T = 0$ soil with increasing undrained strength. The minimum safety factor F_0 for slopes in a deep deposit of normally consolidated clay (D→∞) with increasing strength can be expressed as:

$$F_0 = \frac{c_1}{s_n \gamma z} \qquad (12.78)$$

where c_1 = undrained strength at depth z
            S_n = stability number, values of which
                   are given in Table 12.3

Since $c_1/\gamma z$ is a constant for normally consolidated clay, the stability is independent of the height of the slope. Table 12.3 also gives the values of S_n for deep layer of clay having strength constant with depth (Taylor analysis).

Table 12.3
VALUES OF STABILITY NUMBER FOR A DEEP LAYER OF CLAY,
$\phi_T = 0$ (GIBSON AND MORGENSTERN 1962, LEE ET AL 1983)

Slope angle	Stability number S_n	
$\beta°$	Increasing strength	Constant strength
90	0.500	0.261
80	0.416	0.230
70	0.361	0.209
60	0.303	0.190
50	0.264	0.182
40	0.220	0.182
30	0.182	0.182
20	0.141	0.182
10	0.091	0.182

Example 12 16

Find the safety factor of a 40° cutting made in a normally consolidated clay. The bulk density of clay is 17.5 kN/m³ and the undrained shear strength which increases with depth is measured as 45 kPa at a depth of 4 m.

$\beta = 40°$, $S_n = 0.22$ (Table 12.3)
$F_o = c_1/S_n \gamma z = 2.92$

3. Mitchell method. For short term stability analysis and design of temporary slopes in $\phi_T = 0$ soils, an undrained strength profile (c_T versus depth z) is plotted. The best-fit line through the lower points is extended upwards to get the strength value c_o at the surface (Fig. 12.27). The strength profile may be represented by Eq. 12.73.

$c_T = c_o + k z$

where $k = \Delta c_T / \Delta z$

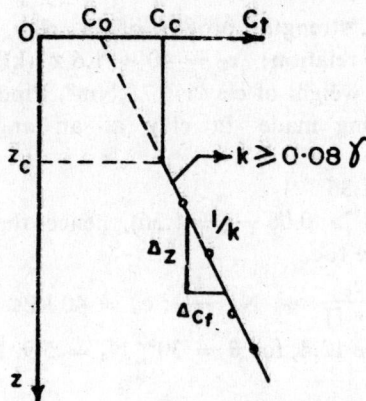

Fig. 12.27 Typical undrained strength profile

If the strength increases with depth sufficiently such that $k > 0.08 \, \gamma$, the critical circle will pass through the toe and the safety factor is given by (Mitchell 1983):

$$F = N_e \frac{c_o}{\gamma H} + N_o \frac{k}{\gamma} \qquad (12.79)$$

where H = height of slope (depth of excavation)

N_s, N_o = stability factors (as given in Table 12.4.)

If the strength increases only marginally with depth ($k < 0.08 \, \gamma$), the soil is considered to be of constant strength and an average value of c_T is used. The location of the *critical* circle is usually governed by the underlying stiffer layer.

Table 1 .4

STABILITY FACTORS FOR TEMPORARY SLOPES IN
$\phi_T = 0$ SOIL WITH INCREASING STRENGTH (MITCHELL 1983)

$\beta°$	18.5	25	30	40	50	60	70
N_s	6.7	6.3	5.9	5.3	4.8	4.3	3.8
N_o	7.7	6.4	5.5	4.6	3.8	3.3	2.8

Referring to Fig. 12.27, Mitchell recommends:

(1) For a slope of height $H \leqslant 0.7 \, z_o$, consider z_o as the depth of stiff base ($DH = z_o$) and take $c_T = c_o$ = constant;

(2) For $0.7 z_o < H < 1.7 \, z_c$, use toe circle and $F = N_s (c_o/\gamma H)$

(3) For $H > 1.7 \, z_o$, use toe circle and F is given by Eq. 12.79.

Example 12.7

The undrained strength profile of a clay deposit can be represented by the relation: $c_T = 40 + 1.6 z$ (kPa), where z is in metres. The unit weight of clay is 17 kNm³. Find the safety factor of 3 m deep cutting made in clay at an angle of 30° to the horizontal.

$0.08\ \gamma = 1.36$

$k\ (= 1.6) > 0.08\ \gamma\ (= 1.36)$, hence the circle will pass through the toe.

$$F = N_s \frac{c_0}{\gamma H} + N_0 \frac{k}{\gamma}; c_u = 40 \text{ kPa}$$

From Table 12.4, for $\beta = 30°$, $N_s = 5.9$, $N_0 = 5.5$

$F = 5.15$

12.13 WEDGE FAILURE ANALYSIS

When there are weak strata within or beneath the slope and also when the slope rests upon a very strong stratum, the surface of sliding is likely to consist of three or more sections which do not merge smoothly into one another and cannot be replaced by a continuous curve without introducing an error on the unsafe side.

Fig. 12.28 shows a typical slope underlain by a thin layer of very soft clay. On failure, the slip occurs along some 'composite surface' 'a b c d'. In this method, it is usual to divide the sliding mass into two or three large sections or 'wedges'. The upper and

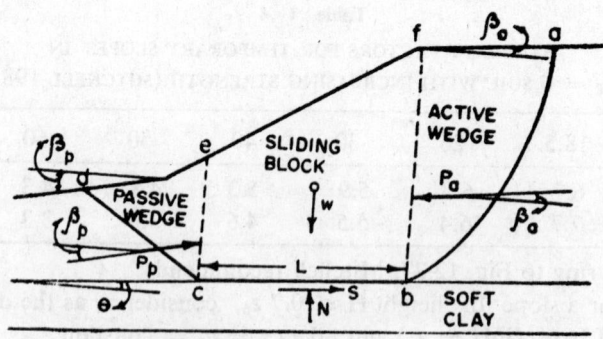

Fig. 12.28 Wedge failure analysis

lower wedges are respectively called the 'active' and 'passive' wedges; and the middle wedge, in a three wedge system, is generally called the 'sliding block'.

Assuming that sufficient deformation occurs to generate active and passive failure wedges, the stability of the sliding block 'b c e f' is analysed. The maximum available shear resistance along the base 'b c' is compared with the magnitude of the shear force required for equilibrium S_{eq} to define the safety factor:

$$F = \frac{S_{max}}{S_{eq}} \qquad (12.80)$$

It is necessary to repeat the analysis for different positions of the points 'b' and 'c' to yield the least factor of safety.

Assuming P_a and P_p to act parallel to the slope which creates them, the equilibrium shear force S_{eq} is given by:

$$S_{eq} = P_a \cos (\beta_a - \theta) - P_p (\beta_p - \theta)$$
$$+ W \sin \theta \qquad (12.81)$$

where θ, β_a, β_p are as shown in Fig. 12.28.

The maximum shear resistance along the surface bc, in term of effective stresses, is:

$$S_{max} = c' L + [W \cos \theta + P_a \sin (\beta_a - \theta) + P_p \sin$$
$$(\beta_p - \theta)] \tan \phi' \qquad (12.82)$$

The above values of S_{eq} and S_{max} are substituted in Eq. 12.80 to get the safety factor.

If $\theta = \beta_a = \beta_p = 0$, the above equations reduce to the following expressions:

$$S_{eq} = P_a - P_p \qquad (12.83)$$

$$S = c' L + W \tan \phi' \qquad (12.84)$$

and

$$F = \frac{c' L + W \tan \phi'}{P_a - P_p} \qquad (12.85a)$$

or

$$F = \frac{c' L + (W - U) \tan \phi'}{P_a - P_p} \qquad (12.85b)$$

where $U = ul$, pore pressure on the sliding block

In terms of total stress analysis,

$$F = \frac{c_T \cdot L}{P_a - P_p} \qquad (12.86)$$

The active and passive pressures can be calculated either by the Rankine method assuming planar failure or by other sophisticated methods.

12.14 DESIGN FACTOR OF SAFETY

The design (or safe value) which is selected for F depends on the risk involved with a failure as well as the method of analysis, reliability of the strength test results and the estimated pore pressures. It also depends on whether the work is temporary or permanent. For temporary slopes in fine grained soils, assuming that long term conditions will be achieved in about three months time for silty clays (or varved clays) and in about six months time for intact clays, Mitchell (1983) recommends the safety factors given by:

$$F = 1.3 + 0.2 \, M \quad \text{for silty clays} \tag{12.87a}$$

$$F = 1.3 + 0.1 \, M \quad \text{for intact clays} \tag{12.87b}$$

where M is the number of months for which the excavation will remain open.

While considering the stability of earth dams, there are three generally recognised critical stages based on pore pressure conditions: (i) end of construction, (ii) steady state seepage, and (iii) rapid drawdown. Construction pore pressures usually attain their maximum value when the full height of the embankment has been constructed. The steady state seepage condition is attained after the reservoir remains filled for a long time and the pore pressure can be estimated from a flow net. Rapid lowering of the reservoir water level produces the third critical condition, particularly for slow draining soils. The pore pressures beneath the upstream slope following a rapid drawdown may be assumed to be hydrostatic with surface of upstream slope, i.e. $u = \gamma_w \, h_p$, where h_p is the vertical depth of the point under consideration below the slope. The pore pressures in the central portion of the embankment do not exceed those for steady state seepage. The stability of the upstream slope should be examined for the end of construction and rapid drawdown conditions and that of the downstream slope for the construction and steady state conditions. Using the Bishop analysis. the minimum safety factors should be: (i) end of construction 1.3, (ii) steady state seepage 1.5, and (iii) rapid drawdown 1.3. For transient loads, such as earthquakes safety factors as low as 1.2 or 1.15 may be tolerated.

12.15 SLOPE STABILIZATION

The methods commonly used to increase the stability of slopes and preventing slides can be divided into three main groups: (1) geometrical improvemet or profile control, (2) slope drainage, and (3) slope strengthening and use of retaining devices.

1. Geometrical improvements. To increase stability, the slope profile is regraded by flattening, removing material producing sliding forces, or adding material to increase resisting forces. Generally flattening of the slope is applicable to deep masses of cohesive soils susceptible to rotational slip and to relatively small slide areas where toe has been oversteepened. Excavation at top of slope is also helpful in deep cohesive soils. Benching is appropriate on steep slopes difficult to be flattened. Benching can increase stability if it produces a reduced overall r_u factor and the material possesses frictional strength. Benches should be sloped to collect runoff and convey it off the slide. Complete removal of unstable mass is feasible for relatively small sliding mass or shallow creep movement. Stabilizing berms in the form of rock, gravel or sand fills placed at the toe of the slope (where upward movement of the sliding mass may occur) increase resisting forces and at the same time prevent internal erosion in situations where the latter is likely to occur. Graded filters or synthetic filter fabrics can be used to prevent internal erosion or piping.

2. Slope drainage. Slope drainage is one of the most important methods to inciease slope stability. Surface drains are sometimes effective but in most cases internal drains are required. Open ditches around a possible slide area can be used to divert any surface water (rain, streams, etc.). Surface treatment by seeding, paving or drainage blanket is applied in critical locations for erosion control. Horizontal or inclined drain holes which are drilled by helical hollow-stem augers or by rotary drilling methods are frequently the most economical method of diainage. The drill holes are generally lined with slotted or perforated metal or plastic pipes approximately 5 cm in diameter. The length of the drains may be upto 60 to 70 m long and they are generally spaced 5 to 15 m apart (Broms 1975). Wick drains (sand drains) are useful where there is a pervious drainage layer at depth. Well points and electro-osmos can be effective in temporary stabilization during the planning and construction stages. Interception of seepage by tunnels is useful for draining large slide masses where structures of great value are threatened.

3. **Slope strengthening and retaining devices.** Shear strength of moderately permeable soils can be increased by injections or grouts of cement, asphalt emulsions or silicates. Proper compaction is the conventional procedure for soil placement. Freezing and electro-osmosis are the temporary methods of strengthening. Retaining walls, sheet pile walls, toe walls and piles may be required to retain the sliding mass and prevent failure. Rock bolting is appropriate for steep slopes of weathered rock, fractured hard rock or stiff soils where shallow failure occurs on joint system.

13

Vertical pressure under surface loads

Any load placed on a soil mass induces stress changes within the soil. The changes are greatest at shallow depths, close to the point of load application, and they become smaller as the vertical distance below the load or the horizontal distance from the load increases. The stress disribution depends on (i) the nature of loading, i.e. shape of the loaded area, load distribution, and the manner of load placement, and (ii) the physical properties of soil, i.e. modulus of elasticity, Poisson's ratio, compressibility and stratification. Because of great variations in these factors, an exact solution is not possible and simplifying assumptions are made to arrive at reasonably accurate results. Most of the methods currently used for studying stress distributions within soil masses are based on elastic theory or empirical modifications to precise analytical solutions of elasticity. The commonly used assumptions are that the soil mass is (ı) semi-infinite in extent, (ii) homogeneous, (iii) isotropic, and (iv) elastic, and obeys Hooke's law. Natural soils seldom comply with any of these assumptions but the lack of acceptable alternative approaches makes their use a practical necessity.

13.1 ELASTIC SOLUTIONS

1. **Point load.** The analytical solutions for stresses due to a concentrated load on the surface is generally attributed to Boussinesq (1885) who made the following assumptions:

(1) The soil mass is elastic, homogeneous, isotropic and semi-infinite which extends infinitely in all directions from a level surface and which obeys the Hooke's law.

(2) The soil is weightless.

(3) Load acts vertically at a point on the horizontal boundary surface (ground surface).

Referring to Fig. 13.1, the vertical normal pressure σ_z at a depth 'z' and a horizontal, radial distance 'r' from the point of application of a concentrated load Q is given by:

$$\sigma_z = \frac{3 \, Q}{2 \, \pi} \frac{z^3}{(r^2 + z^2)^{5/2}} \tag{13.1a}$$

or
$$\sigma_z = \frac{Q}{z^2} \cdot \frac{3}{2\pi} \left(\frac{z}{R} \right)^5 = \frac{Q}{z^2} \cdot \frac{3}{2\pi} \cos^5 \psi \tag{13.1b}$$

where $R = \sqrt{r^2 + z^2}$

It would be seen that neither the modulus of elasticity E nor the Poisson's ratio μ appears in the equation, which means that the pressure is independent of the elastic properties, provided they are the same at all points and in all directions.

Eq. 13.1 may also be expressed as:

$$\sigma_z = \frac{Q}{z^2} \cdot \frac{3}{2\pi \, [\, 1 + (r/z)^2]^{5/2}} = \frac{Q}{z^2} \cdot I_p \tag{13.2}$$

where $I_p = \dfrac{3}{2\pi \, [\, 1 + (r/z)^2 \,]^{5/2}}$, an influence factor (13.3)

The vertical pressure is thus a function of the factor I_p depending on the dimensionless ratio r/z. Typical values of I_p in terms of r/z are given in Table 13.1.

Table 13.1
INFLUENCE FACTOR I_p FOR VERTICAL PRESSURE DUE TO POINT LOAD

r/z	I_p	r/z	I_p	r/z	I_p
0.00	0.4775				
0.10	0.4657	1.1	0.0658	2.1	0.0070
0.20	0.4329	1.2	0.0513	2.2	0.0058
0.30	0.3849	1.3	0.0402	2.3	0.0048
0.40	0.3294	1.4	0.0317	2.4	0.0040
0.50	0.2733	1.5	0.0251	2.5	0.0034
0.60	0.2214	1.6	0.0200	3.0	0.0015
0.70	0.1762	1.7	0.0160	4.0	0.0004
0.80	0.1386	1.8	0.0129	5.0	0.0001
0.90	0.1083	1.9	0.0105	10.0	0.0000
1.00	0.0844	2.0	0.0085		

The form of variation of σ_z with z and r is illustrated in Fig. 13.1b. σ_z thus decreases with increasing z. It also decreases on any horizontal plane with increasing radial distance r. The vertical pressure directly under the load (r = 0) is given by:

$$\sigma_z = \frac{0.478\ Q}{z^2}$$ (13.4)

The vertical pressure directly under the load thus decreases with the square of the depth.

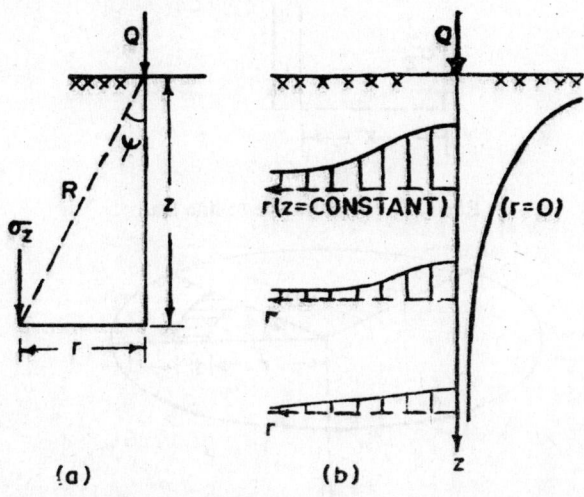

Fig. 13.1 Vertical pressure due to point load (a) definition of z, r and R, (b) variation of σ_z with z and r

2. **Line load.** The vertical pressure due to a line load Q_1 per unit length on the surface (Fig.13.2) at a point located at depth z and distance x laterally away is given by:

$$\sigma_z = \frac{2\ Q_1}{\pi} \cdot \frac{z^3}{(x^2 + z^2)^2}$$ (13.5)

3. **Uniformly loaded circular area.** Where the load is uniformly distributed over a circular area (radius = a) of the surface (Fig.13.3), the resultant vertical pressure σ_z at depth z below the *centre of the area* may be determined by integrating the Boussinesq expression (Eq. 13.1) over the area.

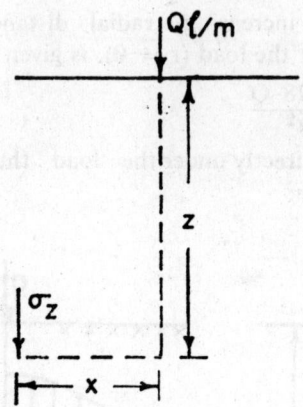

Fig. 13.2   Pressure due to line load

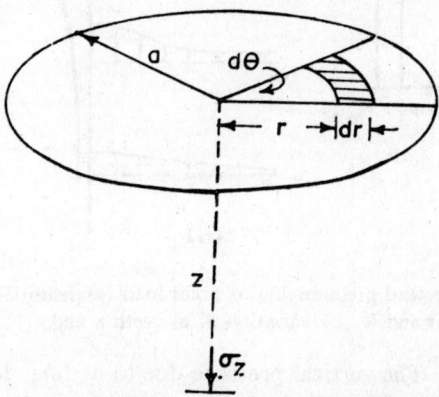

Fig. 13.3   Pressure calculation below the centre of a uniformly loaded circular area

The load on an element of the area is $q \, r \, d\theta \, dr$, where q is the pressure intensity. It may be considered as a point load at distance 'r' from the centre. At depth z below the centre of the area the vertical pressure σ_z is obtained by integration:

$$\sigma_z = \int_0^{2\pi} \int_0^a \frac{q . r . d\theta \, dr}{z^2} . \frac{3}{2\pi} . \frac{z^5}{R^5}$$

$$= \frac{3q\,z^3}{2\pi} \int_{0}^{2\pi} \int_{0}^{a} \frac{r}{R^5}\, d\theta\, dr$$

$$= \frac{3q\,z^3}{2\pi} \int_{0}^{2\pi} \int_{0}^{a} \frac{r\,d\theta\,\, dr}{(r^2 + z^2)^{5/2}}$$

or $\qquad \sigma_z = q \left[1 - \frac{1}{\{1 + (a/z)^2\}^{3/2}} \right] = q\,I_o \qquad (13.6)$

where $\qquad I_o = 1 - \frac{1}{[\,1 + (a/z)^2]^{3/2}} \qquad (13.7)$

Values of the influence factor I_o for this case are given in Table 13.2.

TABLE 13.2

INFLUENCE FACTOR I_o FOR VERTICAL PRESSURE BELOW THE
CENTRE OF A UNIFORMLY LOADED CIRCULAR AREA
OF RADIUS 'a'

z/a	I_o	z/a	I_o	z/a	I_o
0.0	1.000				
0.1	0.999	1.1	0.595	2.1	0.264
0.2	0.992	1.2	0.547	2.2	0.245
0.3	0.970	1.3	0.502	2.3	0.229
0.4	0.949	1.4	0.461	2,4	0.214
0.5	0.911	1.5	0.424	2.5	0.200
0.6	0.864	1.6	0.390	3.0	0.146
0.7	0.818	1.7	0.360	4.0	0.087
0.8	0.756	1.8	0.332	5.0	0.057
0.9	0.701	1.9	0.307	10.0	0.015
1.0	0.646	2.0	0.284		

4. Uniformly loaded strip. The principal stresses and the vertical pressure due to a uniform pressure q on a strip area of width B and infinite length are given in terms of α and β defined in Fig. 13.4:

$$\sigma_1 = \frac{q}{\pi} (\alpha + \sin \alpha) \qquad (13.8)$$

$$\sigma_3 = \frac{q}{\pi} (\alpha - \sin \alpha) \qquad (13.9)$$

and $\qquad \sigma_z = \frac{q}{\pi} (\alpha + \sin \alpha \cos 2\beta) \qquad (13.10)$

where $\qquad \beta = \alpha/2 + \beta'$ (angle between vertical and σ_1)

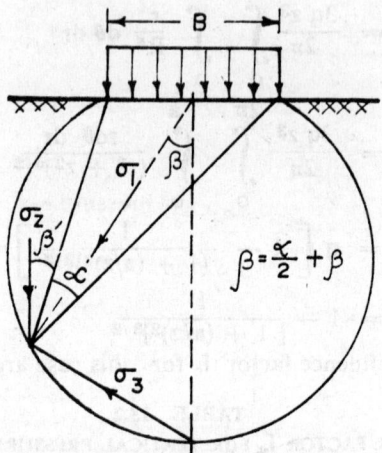

Fig. 13.4 Stresses below strip load

Below the centre of the strip, σ_1 is of course the vertical pressure σ_z at depth z:

$$\sigma_z = \frac{q}{\pi} (\alpha + \sin \alpha) = q \, I_s \qquad (13.11)$$

The values of the influence factor I_s are given in Table 13.3.

Table 13.3

INFLUENCE FACTOR I_s FOR VERTICAL PRESSURE BELOW THE CENTRE
OF A UNIFORMLY LOADED STRIP OF WIDTH B

z/B	I_s	z/B	I_s	z/B	I_s
0.2	0.978	1.4	0.420	2.6	0.240
0.4	0.880	1.6	0.374	2.8	0.223
0.6	0.755	1.8	0.336	3.0	0.209
0.8	0.642	2.0	0.306	4.0	0.16
1.0	0.551	2.2	0.280	5.0	0.13
1.2	0.477	2.4	0.270	6.0	0.11

5. Uniformly loaded rectangular area. According to the solution given by Love (1935, See Tsytovich 1976), the vertical pressure σ_{zr} for any point lying on the vertical below the *corner* of a loaded rectangle with sides L and B is given by:

$$\sigma_{zr} = \frac{q}{2\pi} \left[\frac{LBz}{D} \cdot \frac{L^2 + B^2 + 2z^2}{D^2 z^2 + L^2 B^2} \right.$$

$$\left. + \arcsin \left(\frac{LB}{\sqrt{(L^2 + z^2)} \; \sqrt{(B^2 + z^2)}} \right) \right] \quad (13.12)$$

where $D = 2 \sqrt{L^2 + B^2 + z^2}$ (13.13)

The above equation for corner point pressure σ_{zr} may be expressed as:

$$\sigma_{zr} = q. I_r \quad (13.14)$$

and the maximum pressure under the *centre* of the loaded area, σ_z may be expressed as :

$$\sigma_z = q I_0 \quad (13.15)$$

The influence factors I_r and I_0 are the functions of the ratio depths m and n, where

$$m = \frac{z}{B} \text{ (for } I_r\text{), or } \frac{2z}{B} \text{ (for } I_0\text{)} \quad (13.16)$$

and $n = L/B$ (13.17)

The values of I_0 and $4I_r$ are given in Table 13.4.

The corner point influence factors can be used to find the vertical pressure at any point under a rectangular area by suitable division of the area into smaller rectangles such that the point considered is a *corner* point. Even if the point lies outside the loaded area, the rectangles are extended upto it. The resultant vertical presssure will then be equal to the algebraic sum of the pressures from the various rectangles for which this point is a corner point. For a fictitions area outside the loaded area, the pressure is considered negative.

6. Equivalent point load method. The point load equation 13.2 can be used to find the vertical pressure on a point under a uniformly loaded square or rectangular area by dividing the area into smaller area units with loads acting as point loads at the centroids of the individual unit areas. The Boussinesq solution is applied to each unit area in turn and the results summed. If the pressures are to be determined at a depth z, the larger side 'l' of the unit area should be chosen such that :

$$l \leqslant z/3 \quad (13.18)$$

TABLE 13.4

VALUES OF INFLUENCE FACTORS 4 I_F (CORNER POINT) AND I_0 (CENTRE POINT) FOR UNIFORMLY LOADED RECTANGLE AND I_0 (CENTRE POINT) FOR STRIP AND CIRCULAR AREAS (TSYTOVICH 1976)

m	Rectangular area with n = L/B											Strip n>10 (I_0)	circular area (I_0) 13.4
	1	1.2	1.4	1.6	1.8	2	2.4	2.8	3.2	4	5		
0.0	1.000	1.000	1.000	1.000	1.000	1.000	1.000	1.000	1.000	1.000	1.000	1.000	1.000
0.4	0.960	0.968	0.972	0.974	0.975	0.976	0.976	0.977	0.977	0.977	0.977	0.977	0.949
0.8	0.800	0.830	0.848	0.859	0.866	0.870	0.875	0.878	0.879	0.880	0.881	0.881	0.756
1.2	0.606	0.652	0.682	0.703	0.717	0.727	0.740	0.746	0.749	0.753	0.754	0.755	0.547
1.6	0.449	0.496	0.532	0.558	0.578	0.593	0.612	0.623	0.630	0.639	0.639	0.642	0.390
2.0	0.336	0.379	0.414	0.441	0.463	0.481	0.505	0.520	0.529	0.540	0.545	0.550	0.285
2.4	0.257	0.294	0.325	0.352	0.374	0.392	0.419	0.437	0.449	0.462	0.470	0.477	0.214
2.8	0.201	0.232	0.260	0.284	0.304	0.321	0.350	0.369	0.383	0.400	0.410	0.420	0.165
3.2	0.160	0.187	0.210	0.232	0.251	0.267	0.294	0.314	0.329	0.348	0.360	0.374	0.130
3.6	0.130	0.153	0.173	0.192	0.209	0.224	0.250	0.270	0.285	0.305	0.320	0.337	0.106
4.0	0.108	0.127	0.145	0.161	0.176	0.190	0.214	0.233	0.248	0.270	0.285	0.306	0.087
4.4	0.091	0.107	0.122	0.137	0.150	0.163	0.185	0.203	0.218	0.239	0.256	0.280	0.073

4.8	0.077	0.092	0.105	0.118	0.130	0.141	0.161	0.178	0.192	0.213	0.230	0.258	0.062
5.2	0.066	0.079	0.091	0.102	0.112	0.123	0.141	0.157	0.170	0.191	0.208	0.239	0.053
5.6	0.058	0.069	0.079	0.089	0.099	0.108	0.124	0.139	0.152	0.172	0.189	0.223	0.046
6.0	0.051	0.060	0.070	0.078	0.087	0.095	0.110	0.124	0.136	0.155	0.172	0.208	0.040
6.4	0.045	0.053	0.062	0.070	0.077	0.085	0.098	0.111	0.122	0.141	0.158	0.196	0.036
6.8	0.040	0.048	0.055	0.062	0.069	0.076	0.088	0.100	0.110	0.128	0.144	0.184	0.032
7.2	0.036	0.042	0.049	0.056	0.062	0.068	0.080	0.090	0.100	0.117	0.133	0.175	0.028
7.6	0.032	0.038	0.044	0.050	0.056	0.062	0.072	0.082	0.091	0.107	0.123	0.166	0.024
8.0	0.029	0.035	0.040	0.046	0.051	0.056	0.066	0.075	0.084	0.098	0.113	0.158	0.022
8.4	0.026	0.032	0.037	0.042	0.046	0.051	0.060	0.069	0.077	0.091	0.105	0.150	0.021
8.8	0.024	0.029	0.034	0.038	0.042	0.047	0.055	0.063	0.070	0.084	0.098	0.144	0.019
9.2	0.022	0.026	0.031	0.035	0.039	0.043	0.051	0.058	0.065	0.078	0.091	0.137	0.018
9.6	0.020	0.024	0.028	0.032	0.036	0.040	0.047	0.054	0.060	0.072	0.085	0.132	0.016
10	0.019	0.022	0.026	0.030	0.033	0.037	0.044	0.050	0.056	0.067	0.079	0.126	0.015
11	0.017	0.020	0.023	0.027	0.029	0.033	0.040	0.044	0.050	0.060	0.071	0.114	0.011
12	0.015	0.018	0.020	0.024	0.026	0.028	0.034	0.038	0.044	0.051	0.060	0.104	0.009

Note. For intermediate values of n and m, the influence factors are found by interpolation.

This is because when z/B for a square unit area (width B) is 3 the σ_z/q ratio by the equivalent point load method is 0.053 and by theoretical analysis it is 0.05.   If z/B is less than 3, the difference between the two values increases (Rosenak 1963).

Examples 13.1

13.1   Compare the vertical pressures at depths of 2 m and 4 m directly below a load of 600 kN when (a) the load acts as a point load, and (b) the load is spread over a circular area of radius 1.0 m on the surface.

(a) $\sigma_z = \dfrac{0.478 \ Q}{z^2}$ $= 71.7$ kPa when $z = 2$ m

$= 17.92$ kPa when $z = 4$ m

(b) $q = \dfrac{Q}{\pi a^2} = \dfrac{600}{\pi (1)^2} = 191.08$ kPa

$\sigma_z = q \ I_o$

Table 13.2: $I_o = 0.284$ for $z/a = 2$

$= 0.087$ for $z/a = 4$

$\sigma_z = 54.27$ kPa for $z = 2$ m

$= 16.62$ kPa for $z = 4$ m

Alternatively, use Eq. 13.7.

13.2   A load of 1600 kN is distributed uniformly on a 2 m square area on the surface.   Compute the vertical pressures 2 m and 6 m below the centre of the area.   Also determine the vertical pressures at the same points if the load is assumed to act as a point load at the centre of the area.

(a) $q = 1600/4 = 400$ kPa

$L = B = 2$ m, $n = L/B = 1$

$z = 2$ m, $2z/B = 2$, $I_o = 0.336$ (Table 13.4)

$z = 6$ m, $2z/B = 6$, $I_o = 0.051$ (Table 13.4)

$\sigma_z = q \ I_o = 134.4$ kPa at $z = 2$ m

$= 20.4$ kPa at $z = 6$ m

(b) The same results are obtained if the area is divided into 4 smaller unit areas and the corner point method is used, e.g.

$L = B = 1$, $n = 1$, $m = z/B = 2/1 = 2$

Table 13.4 : $I_r = 0.336/4 = 0.084$

$\sigma_{zr} = q \ I_r = 33.6$ kPa (due to one unit area at corner)

$\sigma_z = 4 \ \sigma_{zr} = 134.4$ kPa at $z = 2$ m

(c) Equivalent point load method:
$\sigma_z = 0.478 \ Q/z^2 = 191.2$ kPa at z = 2 m
$= 21.2$ kPa at z = 6 m

13.3 A rectangular footing, 6 m × 3 m distributes a uniform pressure of 250 kPa near the surface of a soil deposit. Determine the vertical pressure at 6 m (a) below the centre of footing, (b) below one corner of footing, (c) below a point on the middle of the long edge, and (d) below a point located along the short axis of the footing at a distance of 3 m from the long edge.

(a) Centre of footing: L = 6 m, B = 3 m, L/B = 2 = n
m = 2 z/B = 4, I_o = 0.190 (Table 13.4)
σ_s = q I_o = 47.5 kPa

(b) Corner of footing: n = 2, m = z/B = 2
4 I_r = 0.481, I_r = 0.120
σ_{zr} = 30 kPa

(c) Centre of long edge (Fig. 13.5):
L/B = 3/3 = 1 = n, m = z/B = 6/3 = 2
4 I_r = 0.336, σ_s = 2 q I_r = 42 kPa

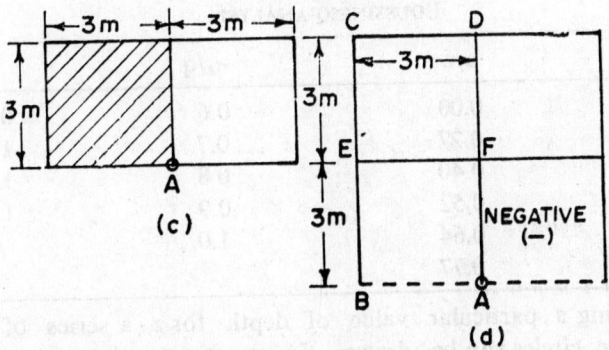

Fig, 13.5 Example 13.3

(d) Point 'A' outside footing (Fig. 13.5):
Area ABCD: L = 6 m, B = 3 m (+ve)
n =: 2, m = 2 I_r = 0.12
σ_{s1} = 250 × 0.12 = 30 kPa (+ve)
Area ABEF: L = 3 m, B = 3 m (— ve)
n = 1, m = 2, 4 I_r = 0.336

$\sigma_{z2} = 21$ kPa ($-$ ve)

Net $\sigma_z = \sigma_{z1} - \sigma_{z2} = 9$ kPa (due to half areas)

Total pressure intensity $= 2 \times 9 = 18$ kPa

13.2 NEWMARK CHART

Newmark (1942) evolved an influence chart on the basis of the Boussinesq solution which can be used for finding the vertical pressure below any irregularly shaped area carrying a uniform load.

Rewriting Eq. 13.6 as follows:

$$\frac{a}{z} = \left\{ \left(1 - \frac{\sigma_z}{q} \right)^{-2/3} - 1 \right\}^{0.5} \qquad (13.19)$$

if a series of values are assigned to the ratio σ_z/q (say, 0, 0.1, 0.2, etc), the corresponding values of the ratio a/z are obtained as given in Table 13.5.

Table 13.5
VALUES OF THE RATIOS σ_z/q AND a/z

BOUSSINESQ ANALYSIS

σ_z/q	a/z	σ_z/q	a/z
0.0	0.00	0.6	0.92
0.1	0.27	0.7	1.11
0.2	0.40	0.8	1.39
0.3	0.52	0.9	1.91
0.4	0.64	1.0	∞
0.5	0.77		

Choosing a particular value of depth for z a series of ten concentric circles can be drawn. In practice, only nine circles will be drawn as the tenth circle has an infinite radius. If the diagram of concentric circles is further divided into smaller areas by drawing a set of equally spaced radial rays from the centre (say m divisions), there will be $10 \times m$ such area units (influence units). Each influence unit will exert $\sigma_z/10$ m pressure, where σ_z is the total pressure. If, for example, m $= 20$ (i.e. total units $= 200$), each influence unit contributes $\sigma_z/200 = 0.005 \, \sigma_z$. Thus the influence factor is 0.005. Adopting a convenient dimension for z (say 3 cm), the radii of the circles are 0.81 cm, 1·20 cm, and 1.56 cm, etc. The resulting diagram is shown in Fig. 13.6. The scale line AB represents 3 cm.

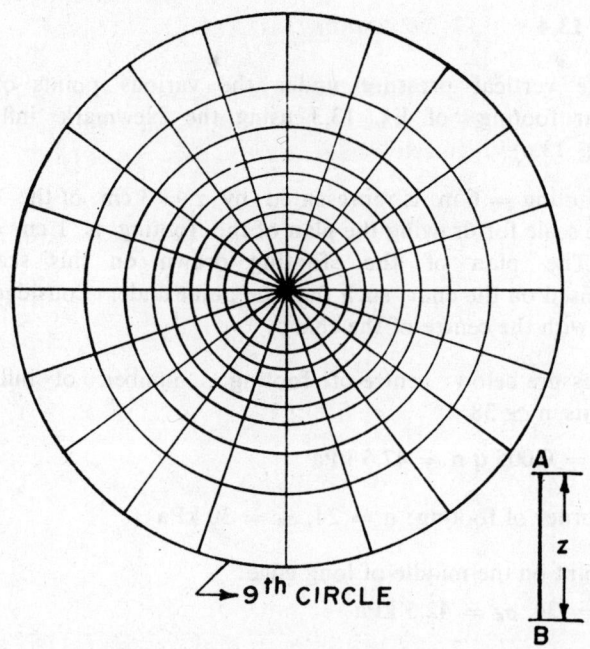

Fig. 13.6   Newmark influence chart based on Boussinesq analysis
Influence factor 0.005.   A B $=$ z $=$ 3 cm

The chart can be used for a uniformly loaded area of any shape. The plan of the loaded area is drawn on a tracing paper to a scale that the actual depth z at which the pressure is required is represented by the scale line AB (= 3 cm). The plan is then placed over the chart, and the point below which pressure is required, is coincided with the centre of the chart. The point below which pressure is required may lie within or outside the area. The number of influence units (including fractions of units) contained within the boundaries of the plan are added to get the total number of units. The vertical pressure σ_z is then given by:

$$\sigma_z = 0.005 \ q \times \text{number of influence units} \qquad (13.20)$$

If the plan of the loaded area extends beyond the 9th circle, it may be assumed to approach the 10th circle for the purpose of counting the influence units.

Example 13.4

Find the vertical pressure under the various points of the rectangular footing of Ex. 13.3 using the Newmark influence chart (Fig. 13.6).

z for footing = 6 m is represented by z = 3 cm of the chart. Hence the scale for drawing the plan of the footing is 1 cm = 6/3 = 2 m. The plan of the footing drawn on this scale is superimposed on the chart such that the point under consideration coincides with the centre of the chart.

(a) Pressure below centre of footing: number of influence units $n \simeq 38$

$\sigma_z = 0.005$ q n = 47.5 kPa

(b) Corner of footing: n = 24, σ_z = 30 kPa

(c) Point on the middle of long edge:
n = 34, σ_z = 42.5 kPa

(d) Point 3 m away from long edge:
n_1 = 24 for area ABCD, σ_{z1} = 30 kPa

n_2 = 17 for area ABEF, σ_{z2} = 21.25 kPa

$\sigma_z = 2 (\sigma_{z1} - \sigma_{z2}) = 17.5$ kPa

13.3 JANBU CHART

A very useful chart for estimating the increase in vertical pressure σ_z below the *centre* of a uniformly loaded flexible area of strip, rectangular or circular shape, carrying a net pressure q, is shown in Fig. 13.7. The chart was given by Janbu, Bjerrum and Kjaensli (1956).

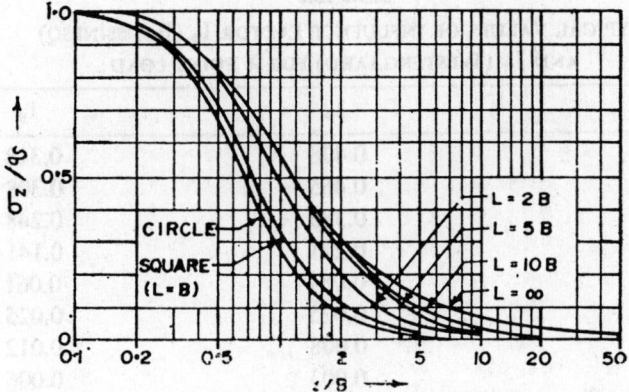

Fig. 13.7 Chart for determining increase in vertical pressure below the centre of uniformly loaded flexible footings (Janbu et al 1956)

13.4 WESTERGAARD ANALYSIS

Westergaard (1938) analysed the stresses in the elastic material in which horizontal deformation is prevented by the inclusion of a large number of thin inextensible sheets. This model is thought to give a better approximation to the behaviour of certain stratified or 'varved' silts and clays having thin layers of granular material which partially prevent lateral deformations. The Westergaard solutions are for a Poisson's ratio of zero, because this assumption gives the highest stresses. The vertical pressure below a point load for a Poisson's ratio of zero may be expressed as:

$$\sigma_z = \frac{Q}{z^2} \cdot \frac{1}{\pi \left[1 + 2\,(r/z)^2\right]^{3/2}} \tag{13.21}$$

or

$$\sigma_z = \frac{Q}{z^2}\, I_w \tag{13.22}$$

The Westergaard solution gives smaller values of vertical stresses than the Boussinesq solution, specially below the centre of loading. Typical values of the Boussinesq influence factor I_D (Eq. 13.2) and the Westergaard influence factor I_w (Eq. 13.22) for a point load are given in Table 13.6.

Table 13.6

TYPICAL VALUES OF INFLUENCE FACTOR I_p (BOUSSINESQ)
AND I_w (WESTERGAARD) FOR A POINT LOAD

r/z	I_p	I_w
0.0	0.478	0.318
0.1	0.465	0.308
0.3	0.385	0.248
0.6	0.221	0.141
1.0	0.084	0.061
1.5	0.025	0.025
2.0	0.008	0.012
2.5	0.003	0.006

An influence chart similar to the Boussinesq chart (Fig. 13.6) can also be prepared. For pressure σ_z at points below the centre of a circular loaded area of radius 'a', the Westergaard equation (Poisson's ratio $\mu = 0$) may be written as:

$$\frac{a}{z} = + \left[\frac{1}{2} \left(1 - \frac{\sigma_z}{q} \right)^{-2} - \frac{1}{2} \right]^{0.5} \tag{13.23}$$

Typical values of the ratios σ_z/q and a/z are given in Table 13.7.

Table 13.7

VALUES OF THE RATIOS σ_z/q AND a/z, WESTERGAARD ANALYSIS, $\mu = 0$

σ_z/q	a/z	σ_z/q	a/z
0.0	0.00	0.6	1.620
0.1	0.343	0.7	2.249
0.2	0.530	0.8	3.464
0.3	0.721	0.9	7.036
0.4	0.943	1.0	∞
0.5	1.227		

Using 200 influence units for the chart, as for the Boussinesq chart, the influence factor is 0.005. For $z = 2$ cm, the radii of the concentric circles are 0.686 cm, 1.06 cm, and 1.442 cm, etc. The resulting influence chart is shown in Fig. 13.8. The chart is drawn to include only $\sigma_z/q = 0.8$, using the radius of 3.464 AB ($= 6.93$ cm) and thus the chart includes only 160 of the 200 influence units.

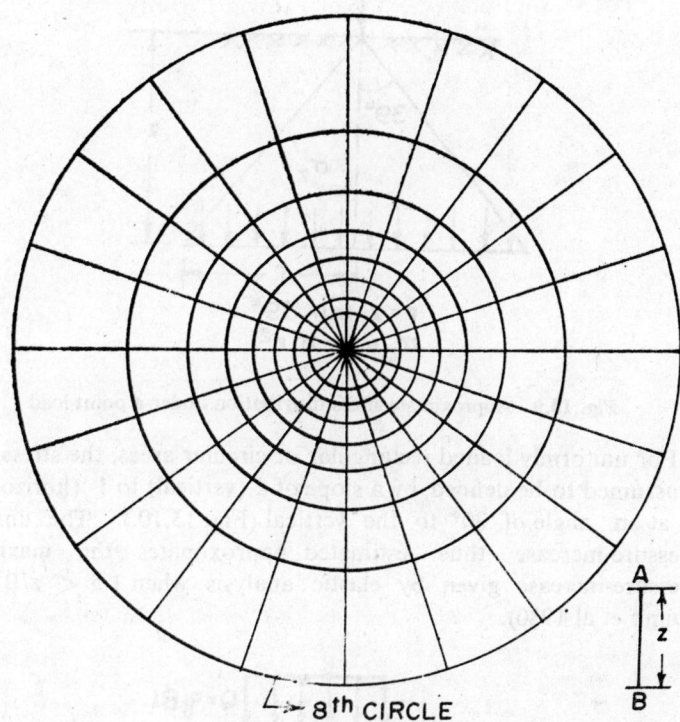

Fig. 13.8 Influence chart for vertical pressure at depth
z = AB (= 2 cm) based on Westergaard analysis
($\mu = 0$). Influence factor = 0.005

13.5 APPROXIMATE SOLUTIONS

An approximate stress distribution under a point load may be
found by assuming the cone of influence formed by the surface
sloping at angle of 39° to the vertical (Fig. 13.9). The average
vertical pressure by this assumption is reasonably close to the
maximum pressure given by a more exact elastic solution for the
point load (Dunn et al 1980).

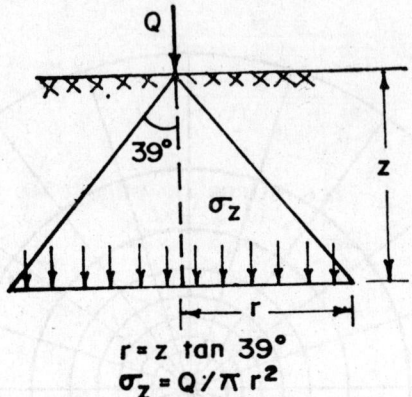

$$r = z \tan 39°$$
$$\sigma_z = Q / \pi\, r^2$$

Fig. 13.9 Approximate stress distribution under a point load

For uniformly loaded rectangular or circular areas, the stress zone is assumed to be defined by a slope of 2 (vertical) to 1 (horizontal), or at an angle of 30° to the vertical (Fig. 13.10.). The uniform pressure-increase thus estimated approximates the maximum pressure-increase given by elastic analysis when $1.5 < z/B < 5$ (Dunn et al 1980).

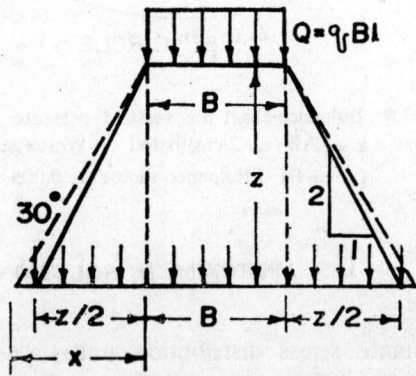

Fig. 13.10 Approximate stress distribution under a uniformly loaded rectangle B × L

The increase in vertical pressure σ_z for a rectangular area (B × L) at a depth z is thus given by:

$$\sigma_s = \frac{q\,B\,L}{(B + z)\,(L + z)} \text{ for 2:1 slope} \qquad (13.24)$$

or $$\sigma_s = \frac{q\,B\,L}{(B + 2\,x)\,(L + 2\,x)} \text{ for 30° spread} \qquad (13.25)$$

where $\quad x = z \tan 30°$

13.6 BULBS OF PRESSURE

If points of equal vertical pressure are plotted on a cross section through the foundation, a diagram of the form shown in Fig. 13.11

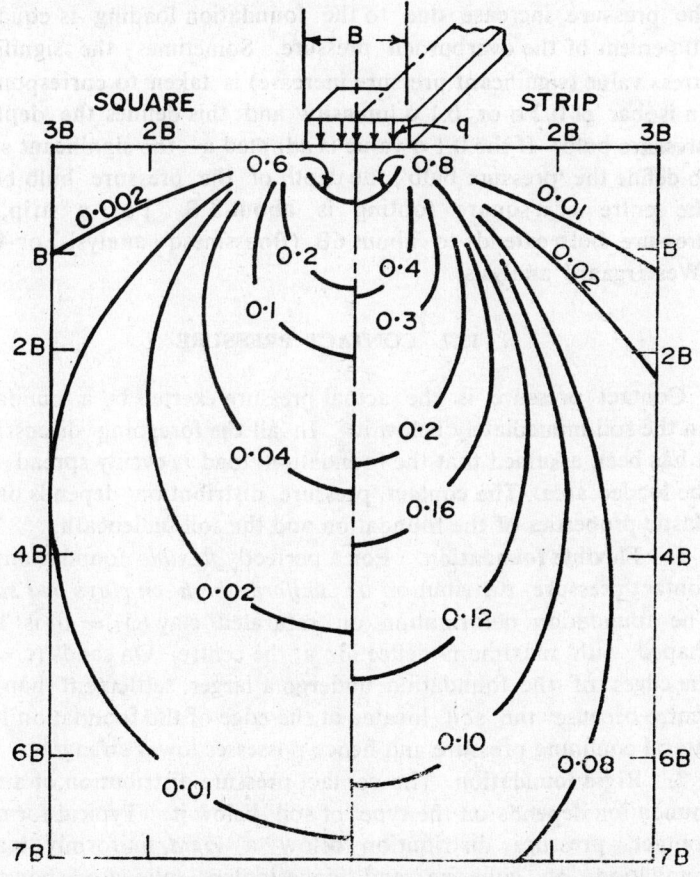

Fig. 13.11 Bulbs of pressure under uniformly loaded square and strip areas, Boussinesq solution

is obtained. These diagrams are known as "bulbs of pressure" or "pressure bulbs". A contour of equal vertical pressure is called an "isobar". Fig. 13.11 shows the bulbs of pressure for uniformly loaded square and strip footings. Stresses below a circular footing of the same area would approximate the stresses below the square footing.

"Significantly stressed zone" or "significant depth" represents the depth below a foundation upto which the pressure increase may be assumed to cause significant deformation of the soil. For practical purposes, this depth may be taken to be the level at which the pressure increase due to the foundation loading is equal to 20 percent of the overburden pressure. Sometimes the significant stress value (significant pressure increase) is taken to correspond to an isobar of 0.2 q or 0.1 q intensity and this defines the depth of pressure bulb. If the 0.1 q value is adopted as the significant stress to define the 'pressure bulb', the depth of the pressure bulb below the centre of a square footing is about 2 B. For a strip, the pressure bulb extends to about 6B (Boussinesq analysis) or 4.5 B (Westergaard analysis).

13.7 CONTACT PRESSURE

Contact pressure is the actual pressure exerted by a foundation on the soil immediately below it. In all the foregoing discussions, it has been assumed that the foundation load is evenly spread over the loaded area. The contact pressure distribution depends on the elastic properties of the foundation and the soil underneath.

1. Flexible foundation. For a perfectly *flexible* foundation, the contact pressure distribution *is uniform both on clays and sands*. The foundation deformation on saturated clay ($\phi_T = 0$) is bowl shaped with maximum deflection at the centre. On sands (c = 0), the edges of the foundation undergo a larger settlement than the centre because the soil located at the edge of the foundation lacks lateral confining pressure and hence possesses lower strength.

2. Rigid foundation. The contact pressure distribution of a rigid foundation depends on the type of soil below it. Typical forms of contact pressure distribution below a *rigid*, uniformly loaded foundation on cohesive and cohesionless soils are shown in Fig. 13.12.

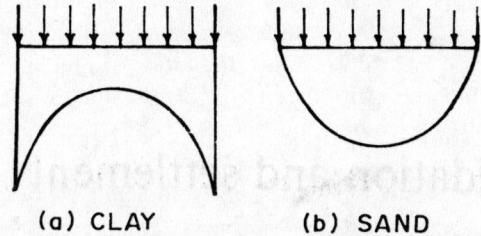

(a) CLAY   (b) SAND

Fig. 13.12   Contact pressure distribution below a rigid, uniformly loaded
foundation on (a) cohesive soil and (b) cohesionless soil

On the assumption of a uniform settlement of a rigid foundation,
it is found from the elastic theory that the pressure intensity at the
edge of a foundation on cohesive soils is infinite. Obviously, the
soil will yield locally under the edges until the resultant distribution
is of the form shown in Fig. 13.12a.   Thus on clays, a portion of
the load is transferred from the centre to the edges.   With increase
of pressure when whole of the soil reaches failure (yield stress), the
contact pressure distribution again tends to uniformity.

In case of a rigid foundation on cohesionless soil, the stress at
the edge is zero as there is no overburden to give the soil shear
strength, whilst the pressure distribution is roughly parabolic as
shown in Fig. 13.12b; the effective modulus of deformation is
greater below the centre than near the edges. As the depth of
foundation increases, greater shear strength develops at the edges
of the foundation and the pressure distribution tends to be more
uniform.

For computing settlement of rigid foundations, Sowers (1962)
recommends the correction factors of Table 13.8 to be applied to
the maximum pressures computed at the foundation centre line so
that the average pressure increase may be obtained.

Table 13.8

CORRECTION FACTOR TO CONVERT CENTRE LINE PRESSURE TO
AVERAGE PRESSURE BELOW RIGID FOUNDATION (SOWERS 1962)

Depth	Factor
0 — 0.5 B	0.85
B	0.90
1.5 B	0.95
2 B	1.00

14

Consolidation and settlement

14.1 SOIL COMPRESSION

When a soil mass is stressed, it deforms. Deformation in general can be either a change of shape, i.e. "distortion" or a change of volume, i.e. 'compression', or both. The property of a soil pertaining to its susceptibility to volume change due to changes in stress is called "compressibility". Compressibility of soil means the reduction in its volume by the application of an external pressure. On loading, a soil may compress due to (i) deformation of soil particles, (ii) compression of pore fluid (air and water), and or (iii) expulsion of water and air from the voids. Under typical engineering loads, the compression of soil mineral particles themselves and of pore water may be neglected. However, for organic soils, especially peat, the compressibility of solid matter may be considerable. Therefore, the compression of air and the expulsion of air and water from the voids contribute the most to the volume change of loaded soil deposits. These volume changes may be brought out by two distinct processes: compaction and consolidation. "Compaction" is the process by which the solid soil particles are packed more closely together by mechanical means (dynamic loading) such as rolling, tamping and vibration, etc. It is achieved through the reduction of air voids, with little or no reduction in the water content. "Consolidation" is the process whereby soil particles are packed more closely together over a period of time under the application of continued pressure (static loading). It is accompanied mainly by the gradual drainage of water from the soil pores. Consolidation is strictly applicable only to saturated or nearly saturated clays or other soils of low permeability. In a free draining soil such as saturated sand, the escape of water can take place rapidly, but in clays the movement of water occurs very much more slowly and, therefore, considerable time is required for compression.

14.2 CONSOLIDATION PROCESS

When an external pressure (termed the *total* pressure σ) is applied to a saturated cohesive soil, the entire pressure is at first carried by the pore water, because the soil structure must compress to take any additional load. The pore water pressure which is induced, is called the "excess pore water pressure" or simply the "excess pore pressure" since it is in excess of the original hydrostatic pressure. The *initial* excess pore pressure u_0 is thus equal to the total external pressure σ. If the soil is bounded by surfaces from which water can escape (such as sand or gravel layer), the excess pore pressure will cause the water to drain out into the adjoining layers. This will, however, occur slowly because of the low permeability of the soil. As water drains out, the soil begins to compress and a portion of additional pressure is transferred to the soil grains forming the soil 'skeleton', and the pore pressure correspondingly falls. The difference between the total external pressure σ and the pore pressure 'u' at any instant is the effective pressure σ' carried by the soil particles. At equilibrium when the drainage of water ceases, whole of the external pressure is carried by the soil particles as the effective pressure which is known as the "consolidation pressure σ_0' and the excess pore pressure reduces to zero. This process of expulsion of water from soil due to excess pore pressure with gradual reduction in soil volume accompanied by transfer of pressure from pore water to soil skeleton is called the 'process of consolidation'. Opposite to it is the 'process of swelling' which involves an increase of water content due to an increase in the volume of voids. The change in the volume of voids and the decrease of pore water may also be brought out by the self weight of soil. The consolidation process is thus the adjustment of the volume of pore water with subsequent compression of soil under an applied load.

The compression of soil due to the dissipation of the excess pore pressure under loading is termed "primary consolidation". It is a time-dependent phenomenon and the time lag in compression is known as the "hydrodynamic lag". Some compression continues ever after the excess pore pressure has virtually dissipated and no more water drains out. This phase of compression is known as "secondary compression" or "secondary consolidation". This process is complex, but it is thought to be due to continued movement of particles as the soil structure adjusts itself to the increasing effective stress. The saturated double layer water surrounding the particles may also get deformed and expelled.

14.3 CONSOLIDATION TEST

The consolidation (or swelling) characteristics of a soil can be studied by means of one-dimensional (vertical) consolidation test, also known as the oedometer test. The test apparatus is called the consolidometer.

A test specimen of virtually saturated soil, usually 75 mm diameter and 20 mm high for routine testing, is held in a metal ring (the consolidation ring). The consolidation ring designed by the author (Singh 1981) holds a specimen 50 cm^2 in sectional area (79.8 mm diameter) and 24 mm high. An undisturbed specimen is prepared from the sample extruded from a sampling tube. A remoulded specimen can be prepared from the soil which has been compacted in a standard compaction mould (Chapter 21). The ring fits into a circular metal cell assembly (Fig. 14.1) which locates the ring in position with porous stones covering the bottom and top faces of the specimen and allows the specimen to be flooded. The ring confining the specimen may be 'fixed type', i.e. clamped to the body of the cell with only the top porous stone capable of moving inside the ring, or 'floating type', i.e. free to move vertically with both the top and bottom porous stones capable of moving inside the ring and compressing the specimen. The consolidation cell is mounted on a loading frame which allows a load to be applied, usually through a lever system. The compression of the specimen under pressure is measured by means of a dial gauge operating on the loading cap which rests on the top porous stone.

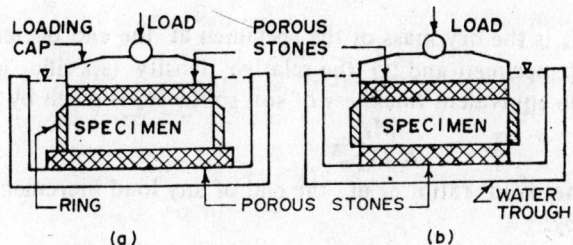

Fig. 14.1 P.inciples of consolidation cell
(a) fixed ring (b) floating ring

A load increment ratio (LIR) of unity is used in conventional testing. The LIR is defined as the pressure increment $\Delta\sigma$ divided by the initial (previous) pressure σ_i. An LIR of unity means that the load is doubled each time.

The loading pressures are usually selected from the sequence 25, 50, 100, 200, 400, 800, 1600 and 3200 kPa, the choice of loads depending mainly on the expected site pressures, including overburden pressures. Each pressure is normally maintained for a period of 24 hours. The specimen consolidates with free drainage occurring from both the top and bottom faces. Compression readings are taken usually at 30 seconds, 1, 2, 4, 8, 15 and 30 minutes, and 1, 2, 4, 8 and 24 hours. When consolidation under the final pressure is complete, the specimen is unloaded and expansion is measured. The swelling characteristics of the soil can also be measured at the end of any intermediate stage of loading by successive decreases in applied pressure.

The results are presented in the form of a graph between the void ratio at the end of each pressure increment plotted as ordinate on a linear scale and the corresponding applied pressure (effective pressure) as abscissa on a logarithmic scale (or also on a linear scale).

The equilibrim or final void ratio corresponding to each applied pressure can be calculated from the dial gauge readings and either the dry weight or the moisture content of the specimen at the end of the test, as described below.

1. Void ratio by dry weight method. This method is preferable as it is applicable to both saturated and partly saturated specimens.

If M_s is the dry mass of the specimen at the end of test, 'A' the area of specimen and 'G' the relative density (specific gravity) of soil, the equivalent thickness of soil solids H_s is given by:

$$H_s = \frac{M_s}{A \, G \, \rho_w} \qquad (14.1)$$

and the void ratio e_1 at the end of any load increment period is given by:

$$e_1 = \frac{H_1 - H_s}{H_s} = \frac{H_1}{H_s} - 1 \qquad (14.2)$$

where H_1 = specimen height at equilibrium under any pressure. The specimen height H_1 at equilibruim under a pressure is given by:

$$H_1 = H_o \pm \Sigma \Delta H \qquad (14.3)$$

where H_o = specimen height at the beginning of test

          $\Sigma \Delta H$ = cumulative change in specimen height upto the end of the particular pressure increment period

2. Void ratio by final moisture content method. The method is applicable where full saturation ($S_r = 100\%$) of the specimen may be assumed. If w_f' is the moisture content at the end of test, the void ratio at the end of test e_f' is given by:

$$e_f' = w_f' \, G \qquad (14.4)$$

In a laterally confined specimen where the height H corresponds to $(1 + e)$ in the phase diagram, the void ratio change Δe is proportional to the height change ΔH and, therefore,

$$\frac{\Delta e}{1 + e} = \frac{\Delta H}{H} \qquad (14.5)$$

Using Eq. 14.5 for the known values of e_f' and H_f' at the end of test, the void ratio changs Δe is given by:

$$\Delta e = \frac{1 + e_f'}{H_f'} \, \Delta H \qquad (14.6)$$

By working backwards from the known value of e_f' the void ratio under any pressure can be calculated.

14.4 COMPRESSIBILITY PARAMETERS

The compressibility of a soil is indicated by the following three parameters which are derived from a consolidation test: (i) coefficient of compressibility a_v, (ii) coefficient of volume compressibility m_v, and (iii) compression index C_c.

Typical compression curves which can be obtained from a consolidation test performed on a clay sample initially dispersed in water are shown in Fig. 14.2, where the equilibrium void ratio is plotted against the corresponding consolidation effective pressure. After consolidating the specimen upto pressure point B, the specimen is allowed to expand by pressure decrements. During expansion, the specimen, never returns to the original volume due to some irreversible or permanent compression. On reloading, the recompression curve CD is obtained which lies somewhat above the expansion curve. When the previous pressure corresponding to point B is reached, the recompression curve has a slightly lower void ratio. As the pressure is further increased and the test continued, the resulting curve is more or less the extension of the initial portion AB.

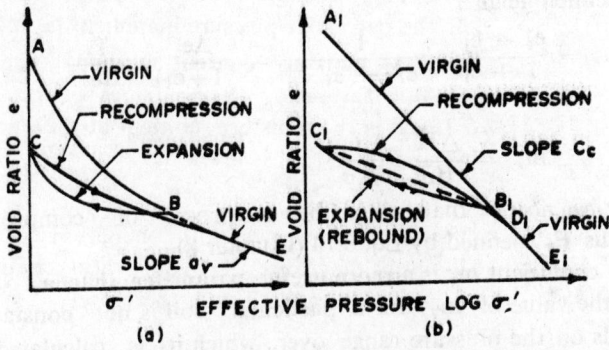

Fig. 14.2 Typical compression curves for clay initially dispersed in water
(End point of recompression curve is D)

The portions $A_1 B_1$ and $D_1 E_1$ of the curve on either side of point B_1 on the semi-log plot are approximately straight in the range of pressure ordinarily encountered in geotechnical engineering problems. These straight line portions of the curve are referred to as the *virgin compression* curves. A virgin curve represents the void ratio versus effective pressure relationship for pressure levels exceeding the maximum past pressure experienced by the soil (See Sections 14.5 and 14.6).

1. **Coefficient of compressibility.** The coefficient of compressibility a_v for a particular pressure increment is equal to

the change in void ratio Δe for that increment, divided by the incremental pressure $\Delta\sigma'$:

$$a_v = \frac{e_f - e_i}{\Delta\sigma'} = -\frac{\Delta e}{\Delta\sigma'} \tag{14.7}$$

where e_i and e_f are the void ratios at the beginning and end of consolidation under the pressure increment. The negative sign indicates that e decreases as σ' increases. The units of a_v are the reciprocal of the pressure units, i.e. m²/kN in SI units. The value of a_v depends on the pressure range over which it is calculated. The coefficient a_v is rarely used in practice.

2. Coefficient of volume compressibility. The coefficient of volume compressibiltiy m_v, also known as "coefficient of volume change (decrease)", or "modulus of volume change", is defined as the volume change per unit volume per unit increase in pressure. The volume change may be expressed in terms of either void ratio or specimen height.

$$m_v = \frac{e_f - e_i}{1 + e_i} \cdot \frac{1}{\sigma_f' - \sigma_i'} = -\frac{\Delta e}{1 + e_i} \cdot \frac{1}{\Delta\sigma'} = \frac{a_v}{1 + e_i} \tag{14.8 a}$$

or

$$m_v = \frac{\Delta H}{H_i} \cdot \frac{1}{\Delta\sigma'} \tag{14.8b}$$

It may be noted that m_v is the reciprocal of 'compressibility modulus E_v' defined by Eq. 8.14 (Chapter 8).

The coefficient m_v is a more useful parameter than a_v. Similar to a_v, the value of m_v for a particular soil is not constant but depends on the pressure range over which it is calculated. The BS 1377 specifies the use of m_v calculated for a pressure increment of 100 kPa in excess of the insitu effective overburden pressure. The units of m_v are the same as of a_v, but are usually multiplied by 1000 to express m_v in m²/MN. Thus,

$$m_v = 1000 \frac{\Delta H}{H_i} \cdot \frac{1}{\Delta\sigma'} \quad (m^2/MN) \tag{14.9}$$

if $\Delta\sigma$ is measured in kN/m².

Typical values of m_v (m²/MN) are as follows (Smith 1974) : hard clay (boulder clays): 0.125—0.0625; stiff clay : 0.25 — 0.125; plastic clay (NC alluvial clays) : 2.0 — 0.25; peat : 10.0 — 2.0

3. Compression index. The compression index C_c is the slope of the *linear portion* of the 'e' versus log σ' curve and is dimensionless. Thus for any two points on the linear portion of the curve:

$$C_c = \frac{\Delta e}{\Delta \log \sigma'} = \frac{\Delta e}{\log \sigma_t'/\sigma_i'} = \frac{e_i - e_f}{\log \sigma_t'/\sigma_i'}$$

$$(14.10)$$

or

$$e_t = e_i - C_c \log_{10} \frac{\sigma_i' + \Delta \sigma'}{\sigma_i'} \qquad (14.11)$$

Numerically, C_c is equal to the change in void ratio for one log cycle of pressure change.

Similar to C_c, the 'swell index C_s' is equal to the slope of the expansion (unloading) curve of 'e' plotted against log σ'. The swell index or expansion index (or the recompression index C_r) is usually taken as the average slope of the expansion — recompression hysteresis loop.

Although e versus log σ' has been the most common way of plotting a compression curve, strain is more direct than void ratio for presenting the results of one dimensional consolidation test. The strain is based on the original specimen height H_0 (at the beginning of of the test) and the change in height, i.e. $\epsilon = \Delta H/H_0$.

On a strain versus log σ' curve the slope of the virgin curve (straight portion) is called the modified compression index $C_{o\epsilon}$ (Holtz and Kovacs 1981), and is expressed as:

$$C_c\epsilon = \frac{\Delta \epsilon}{\log \sigma_t'/\sigma_i'} = \frac{\Delta H/H_0}{\log \sigma_t'/\sigma_i'} \qquad (14.12)$$

Sometimes $C_{o\epsilon}$ is called the "compression ratio". $C_{o\epsilon}$ and C_o are related as:

$$C_c\epsilon = \frac{C_o}{1 + e_o} \qquad (14.13)$$

The vertical strain versus log pressure curve may be preferable because it can be plotted while the test is being performed. A e/log σ' curve requires the knowledge of the initial and final values of 'e' for which the dry mass of soil solids can be determined only at the *end* of the test. Estimating field settlements is also simpler from a ϵ/log σ' curve.

It may be noted that the slope of the compression curve obtained by plotting void ratio versus pressure arithmetically (both on linear scales) is the coefficient of compressibility a_v ($= \Delta e/\Delta \sigma'$). When the results are plotted in terms of strain versus pressure arithmetically, the slope of the compression curve is the coefficient of volume compressibility m_v ($= \Delta\epsilon/\Delta\sigma'$, $\Delta\epsilon = \Delta H/H_i$). In one-dimensional compression ϵ is equal to $\Delta e/(1 + e_0)$.

14.5 NORMALLY CONSOLIDATED CLAY

A clay which has been subjected *only to an increase in pressure* is a normally consolidated (NC) clay. The existing pressure is thus the greatest pressure so far applied to the clay. Compression characteristics of an element of normally consolidated clay in the field are similar to the straight portions of the curve in Fig. 14.2b (virgin curves). NC clays are generally compressible and are usually soft for a considerable depth.

A typical laboratory consolidation curve for an undisturbed sample of NC clay is shown in Fig. 14.3a. Due to disturbance of the soil sample during sampling and test preparation, the straight portion (virgin portion) of the laboratory curve will be slightly different from the compression of soil in its natural state in the field, viz. the slope of the field virgin compression curve will be slightly greater than that of the laboratory curve. The field curve (Fig. 14.3) may be obtained by the procedure suggested by Schmertmann (1953, 1955).

(1) Plot point B on the $e/\log \sigma'$ graph of an undisturbed laboratory specimen. This point represents the field equilibrium condition, i.e. the *insitu* effective pressure σ_0' and the void ratio e_0 at the sample location. The insitu void ratio e_0 may be calculated from the insitu moisture content and the relative density G of soil on the assumption of full saturation ($e_0 = w_n G$). No appreciable error will be involved if e_0 is taken as being equal to the void ratio at the start of the laboratory test. For NC clay, $\sigma_0' = \sigma_p'$, the preconsolidation pressure (See Section 14.6) and which can also be determined by the Casagrande method described in Section 14.6.

(2) From a point on the e-axis equal to $0.42 e_0$ draw a horizontal line to meet the extension of the laboratory virgin curve at point F. Schmertmann (1955) found that laboratory curves for varying degrees of disturbance intersected the field virgin compression curve at approximately this point.

(3) Connect the two points B and F by a straight line. This line BF is the estimated field virgin compression curve with a slope C_c.

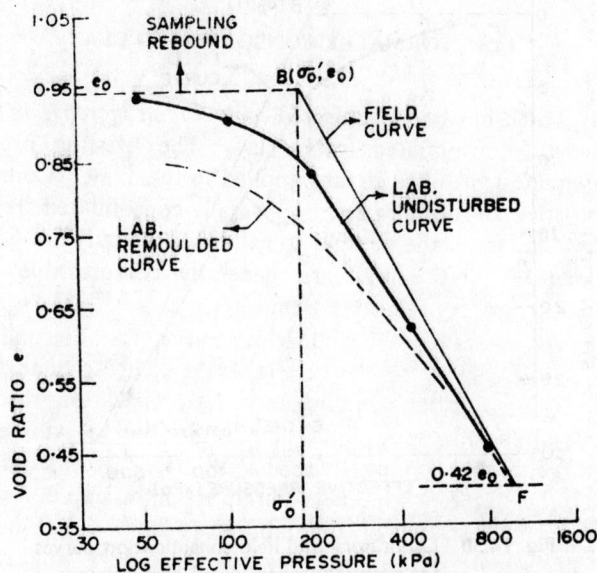

Fig. 14.3a Laboratory and field consolidation curves for normally consolidated clay (void ratio versus log pressure)

A field compression curve can be plotted also in terms of strain versus log σ' (Fig. 14.3b). In that case point B in Fig. 14.3b representing field equilibrium condition (i.e. zero strain) will have the coordinates $(\sigma_0', 0)$. The second point F (Fig. 14.3b) for the field compression curve is found by extending the linear portion of the laboratory curve $(\epsilon/\log \sigma')$ to intersect the horizontal line through a strain of approximately (Dunne et al 1980):

$$\frac{0.6 \, (H_o - H_s)}{H_o} \tag{14.14}$$

where H_o is the intial height of specimen and H_s the height of soil solids (Eq. 14.1). This strain corresponds to a void ratio of $0.4 \, e_o$, where e_o is the void ratio at zero strain. The field curve is obtained by drawing a straight line between B and F and it has the equation:

$$\epsilon = C_c \epsilon \log (\sigma'/\sigma_o') \tag{14.15}$$

where $C_{c\epsilon}$ = modified compression index

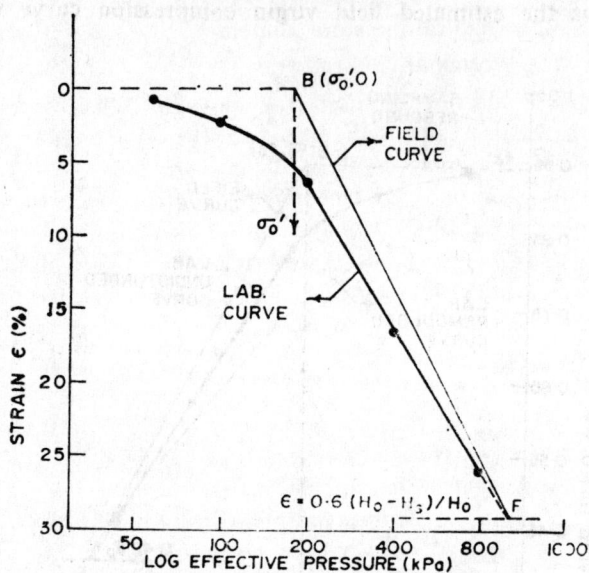

Fig. 14.3b   Laboratory and field consolidation curves
for normally consolidated clay (strain
versus log pressure)

For a correct estimate of field compression, the values of C_c or $C_{c\epsilon}$ and m_v should be based on the field curve. If this correction is not made, the calculated estimates would be under-estimated.

On the basis of tests performed on many different remoulded soils and adjusted to reflect the difference in behaviour of remoulded and undisturbed samples, the compression index C_c for NC clays of low or medium sensitivity has been empirically related to the liquid limit w_L (%) of clay to a reasonable degree of approximation by the following equation (Terzaghi and Peck 1967):

$$C_c = 0.009 \, (w_L - 10\%) \tag{14.16}$$

The other approximate relations are (NAVFAC DM-7.1):

$$C_c = 0.0115 \, w_n \quad \text{(organic soils, peat)} \tag{14.17}$$

$$C_c = (1 + e_o) \, [0.1 + (w_n - 25) \, 0.006]$$
$$\text{(varved clays)} \tag{14.18}$$

and according to Nishida (1956):

$$C_c = 0.54 \ (e_o - 0.35) \quad \text{(all clays)} \tag{14.19}$$

where w_a = natural moisture content

   e_o = original void ratio

Eq. 14.16 should not be used for clays of sensitivity greater than 4, if the w_L is greater than 100, or if the clay contains a high percentage of organic matter. Typical values of C_c are: (i) Normally consolidated medium sensitive clay: 0.2 — 0.5, (ii) Organic clays (OH): 4 and up; (iii) Organic silt and clayey silts (ML-MH): 1.5—4.0, and (iv) Mexico City clay: 7—10 (Holtz and Kovacs 1981). Approximate values of C_c for uniform sands in the pressure range of 100 to 400 kPa may vary from 0.05 to 0.06 (loose state) and from 0.02 to 0.03 (dense state) (NAVFAC DM-7.1).

A more general equation for estimating the compression index of fine grained soils of any stress history has been proposed by Herrero (1980), which empirically relates the compression index C_c' of virgin laboratory curves with the insitu dry density or void ratio e_o and the relative density G of soil particles:

$$C_c' = 0.141 \ G^{6/5} \ (\gamma_w/\gamma_d)^{12/5} \tag{14.20a}$$

or $$C_c' = 0.141 \ G^{6/5} \left[\frac{1 + e_o}{G} \right]^{2 \cdot 382} \tag{14.20b}$$

For normally consolidated clays, it is observed from available laboratory tests data that on an average the virgin void ratio (start of virgin curve) is approximately 95 % of the initial (insitu) void ratio e_o. Therefore γ_d may be calculated on the basis of a value of 0.95 e_o. In case of overconsolidated clays (further discussion given ahead), Eq. 14.20 will approximately correspond to the recompression portion of the laboratory curve and γ_d is calculated from the insitu soil index properties.

14.6 OVERCONSOLIDATED CLAY

An overconsolidated or precompressed clay is a clay which has been consolidated in the past under an effective pressure which is greater than the existing pressure. The maximum pressure to which a clay was subjected in the past is called the "preconsolidation pressure σ_p'". This pressure value, σ_p', forms the boundary between recompression and virgin compression ranges. The ratio of the preconsolidation pressure to the present effective overburden pressure is called the "overconsolidation ratio OCR". Thus 'OCR $>$ 1' indicates an overconsolidated clay and 'OCR $=$ 1' indicates a normally consolidated clay. Over consolidation of clay due to changes in stress may be caused by a variety of factors such as: erosion of pre-existing overburden material; removal of prior load of glaciers, wind-blown dune sands or old structures; drying of surface exposed to air with a resultant decrease in void ratio by capillary forces (this process is referred to as overconsolidation by desiccation); and rise in groundwater table resulting in reduction of the effective overburden pressure, etc.

The compression characteristics of an element of overconsolidated clay in the field are similar to the recompression curve $C_1 B_1$ in Fig. 14. 2b. Comparing the virgin curve and the recompression curve it can be seen that an overconsolidated clay is much less compressible than a normally consolidated clay. The stress history of a soil has thus a great influence on its stress-strain behaviour. Whether a clay is overconsolidated can be judged from the following considerations.

Geological history of the clay, if known, can indicate whether previous overburden has been removed or the groundwater table is now higher than in the past. As a rough guide, the natural moisture content of overconsolidated clays is near or at the plastic limits; the liquidity index (Eq. 2.70) of overconsolidated clays varies from 0 to about 0.6 and that of normally consolidated clays from about 0.6 to 1.0 (Simons and Menzies 1977). The maximum past effective pressure σ_p' can be estimated from consolidation of undisturbed samples by a procedure such as suggested by Casagrande (1936) and compared with the existing pressure. The Casagrande method is described below.

1. Determination of preconsolidation pressure. According to Casagrande's empirical procedure based on the results of many laboratory tests, a point on the e versus log σ' curve of an undisturbed specimen is marked by observation which has the smallest radius of curvature; point A in Fig. 14.4. A horizontal line and a tangent are drawn at this point and the angle batween them is bisected. The straight line portion of the virgin compression curve is extended upwards to intersect the bisector of the angle. The point of intersection B defines approximately the magnitude of the most probable preconsolidation pressure σ_p'.

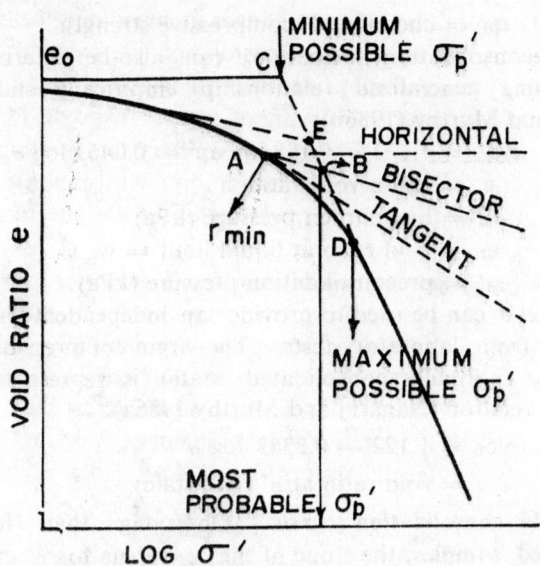

Fig. 14.4 Casagrande graphical procedure for determination of the most probable preconsolidation pressure σ_p'

For overconsolidated clays, $\sigma_p' > \sigma_o'$ (present effective overburden pressure) and for normally consolidated clays $\sigma_p' = \sigma_o'$. Points of minimum possible pressure (point C) and maximum possible pressure (point D) are also shown in Fig. 14.4. An even simpler method used by some engineers to estimate σ_p' is to extend the two straight line portions of the laboratory curve to get the intersection point E, which defines another most probable σ_p' (Holtz and Kovacs 1981).

For clays with natural moisture content w_n at w_L (liquidity index of 1), σ_p' ranges between about 10 to 80 kPa depending on soil sensitivity; σ_p' increasing with increasing sensitivity. For w_n at w_p ($I_L = 0$), σ_p' ranges from about 1150 to 2400 kPa.

The σ_p' can also be estimated approximately from the following relation (NAVFC DM-7.1):

$$\sigma_p' = \frac{q_u/2}{0.11 + 0.0037\, I_p} \qquad (14.21a)$$

where q_u = unconfined compressive strength

The preconsolidation pressure σ_p' can also be determined from the following generalized relationship empirically established by Nagaraj and Murthy (1986b):

$$e/e_L = 1.122 - 0.188 \log \sigma_p' - 0.0463 \log \sigma_0' \qquad (14.21b)$$

where e = insitu void ratio

σ_0' = overburden pressure (kPa)

e_L = void ratio at liquid limit = $w_L\, G$

σ_p' = preconsolidation pressure (kPa)

Eq. 14.21b can be used to provide an independent check on σ_p' computed from laboratory tests. The virgin compression response beyond σ_p' (normally consolidated state) is represented by the following relation (Nagaraj and Murthy 1986a):

$$e/e_L = 1.122 - 0.2343 \log \sigma' \qquad (14.21c)$$

where e = void ratio at σ' (NC state)

2. Field consolidation curve. Observing that for good undisturbed samples, the slope of the e versus $\log \sigma'$ curve in the overconsolidation range is approximately equal to the slope of the rebound (expansion) curve and that the slope of the rebound curve is not very sensitive to sample disturbance, Schmertmann (1953, 1955) proposed the following graphical method for constructing the field consolidation curve for overconsolidation clays (Fig. 14.5).

(1) Determine σ_p', σ_0' and e_0.

(2) From e_0 point on the void ratio axis, draw a horizontal line to get point 'A' on the σ_0' line.

(3) From point A, draw a line parallel to the mean slope of the rebound curve to get point B on the σ_p' line. The straight line AB is assumed to be the field *recompression* curve with slope C_r. Although the rebound curves from any effective stress are nearly parallel, the average slope of the rebound-recompression curve depends on the stress at which the cycle starts, the slope increasing with increasing stress. It is therefore, advisable to conduct the rebound recompression cycle at a stress slightly less

than σ_p' (Perloff and Baron 1976, Leonards 1976) or at $\sigma_o' + \triangle\sigma$, where $\triangle\sigma$ will be the stress increase in the field (Holtz and Kovacs 1981).

(4) Through point B draw a straight line to intersect the straight portion of the laboratory curve at F where $e = 0.42\ e_o$ (as was done for the NC clay). The line BF represents the field virgin compression curve with slope C_o.

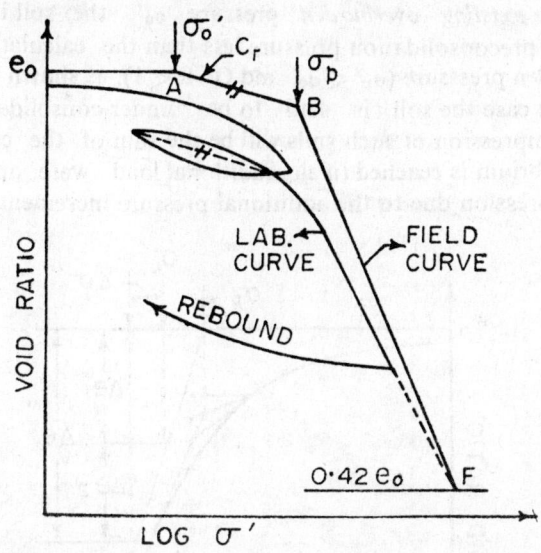

Fig. 14.5 Field consolidation curve for overconsolidated clay

The recompression index C_r (average slope) is often assumed to be 5% to 10% of C_o. Typical values of C_r range from 0.015 to 0.035 (Leonards 1976): the lower values being for clays of lower plasticity and low OCR.

A field consolidation curve can be obtained on the $\epsilon/\log\sigma'$ plot as well. In that case, point A (Fig. 14.5) will have the coordinates ($\sigma_o', 0$) which is the equilibrium condition in the field. The straight line portion of the laboratory curve and the field virgin curve (for pressures greater than σ_p') are assumed to meet at a strain of $[0.6\ (H_o - H_s)/H_o]$ which corresponds to a void ratio of $0.4\ e_o$, where e_o is the void ratio at zero strain. The slope of the recompression portion AB on the $\epsilon/\log \sigma'$ plot is called the the "modified recompression index $C_{r\epsilon}$".

14.7 UNDER-CONSOLIDATED CLAY

In areas of recent landfill or where the groundwater table has recently been lowered, a deposit may not have reached equilibrium under the *existing overburden* pressure σ_0'; the soil is found to exhibit a preconsolidation pressure less than the calculated existing overburden presssure ($\sigma_p' < \sigma_0'$ and OCR < 1), as shown in Fig. 14.6. In such a case the soil is said to be "under-consolidated". The total compression of such soils will be the sum of the compression till equilibrium is reached (if no additional load were applied) and the compression due to the additional pressure increment.

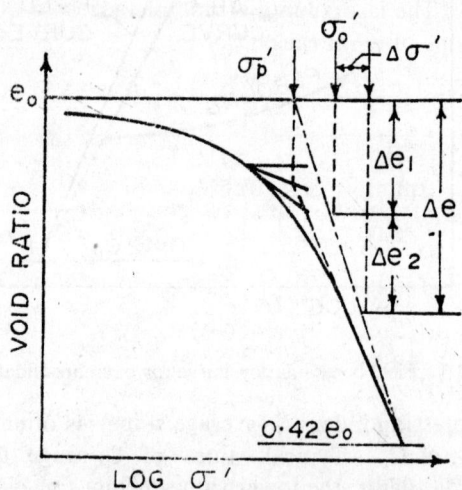

Fig. 14.6 Compression of under-consolidated clay. Δe_1 due to overburden, Δe_2 due to structure, Δe = total compression due to net pressure increment $(\sigma_0' - \sigma_p' + \Delta\sigma')$

Examples 14.1 − 14.3

14.1 A specimen of normally consolidated undisturbed clay, 75 mm in diameter and 20 mm high, is tested in a consolidometer under pressures as given in Table 14.1. The equilibrium height at each stage of loading is recorded in column 2 of the table. Plot the void ratio versus pressure and the strain versus pressure relationships using the following available data: natural moisture content $w_n = 35.2\%$, relative density $G = 2.68$, and final moisture content at end of test $w_t' = 21\%$, and dry mass of specimen at end of test $M_s = 120.89$ g.

(a) Moisture content method.

Assuming full saturation,

$e_o = \bar{w}_n G = 0.352 \times 2.68 = 0.943$

e'_f (end of test) $= w'_t G = 0.563$

$$\Delta e = \frac{1 + e_f}{H_f} \Delta H = \frac{1.563}{16.02} \Delta H = 0.0976 \Delta H$$

The computations are completed in a tabular form as given in Table 14.1. The last column is completed, by working upwards from the value of e_f' at the end of test.

Table 14.1

EXAMPLE 14.1

σ (kPa)	H (mm)	ΔH (mm)	Δe	e
0	20.00	...	...	0.950
		— 0.20	— 0.019	
50	19.80	...		0·931
		— 0.25	— 0.024	
100	19.55	...	...	0.907
		— 0.87	— 0.085	
200	18.68	...	...	0.822
		— 1.99	— 0.194	
400	16.69	...	...	0.628
		— 1.89	— 0.184	
800	14.80	...	...	0.444
		+ 1.22	+ 0.119	
0	16.02	...	...	0.563

(b) Dry weight method.

Sectional area of specimen $A = 4418$ mm^2

$$H_s = \frac{M_s}{A.G. \rho_w} = \frac{120.89 \times 1000}{4418 \times 2.68 \times 1} = 10.21 \text{ mm}$$

Table 14.2 gives the computed values of void ratio and also of the strain. The void ratios at the beginning of the test $e_o = 0.959$ and at the end of test $e_f' = 0.569$ can be used to check the degrees of saturation.

$$S_r \text{ (beginning of test)} = \frac{w_n \, G}{e_o} = 98.4\%$$

$$S_r \text{ (end of test)} = \frac{w_f' \, G}{e_f'} = 98.9\%$$

As S_r is not 100 percent the first method based on moisture content does not yield identical results. However, the differences are insignificant from practical considerations.

Table 14.2

EXAMPLE 14.1

σ (kPa)	H (mm)	$\dfrac{H}{H_s}$	$e = \left(\dfrac{H}{H_s} - 1 \right)$	$\Sigma\Delta H$ (mm)	Strain $\Sigma\Delta H/H_o$ (%)
0	20.00	1.959	0.959	0	0
50	19.80	1.939	0.939	0.20	1.0
100	19.55	1.915	0.915	0.45	2.3
200	18.68	1.830	0.830	1.32	6.6
400	16.69	1.635	0.635	3.31	16.6
800	14.80	1.449	0.449	5.20	26.0
0	16.02	1.569	0.569	3.98	19.9

The relationships are plotted in Figs. 14.3a and 14.3b. These figures also show the extrapolated field curves, as discussed in Section 14.5.

14.2 Calculate the values of the coefficient of volume change m_v for the pressure ranges 200 to 100 kPa, 200 to 400 kPa, and 400 to 800 kPa for the specimen of Ex. 14.1, using data of Table 14.2.

m_v can be calculated from either of the following relations:

$$m_v = \frac{\Delta H}{H_i} \cdot \frac{1}{\Delta \sigma'}, \text{ or } m_v = \frac{\Delta e}{1 + e_i} \cdot \frac{1}{\Delta \sigma'}$$

where H_i and e_i correspond to the initial values of the pressure for a particular pressure range. The results are given in Table 14.3. It may be noted that m_v varies with the pressure range. It decreases with increasing pressure on the linear portion of the curve.

Table 14.3

EXAMPLE 14.2

σ (kPa)	H (mm)	ΔH (mm)	m_v (m²/kN)	e	Δe	m_v (m²/kN)
100	19.55	—	—	0.915		
		0.87	4.45×10^{-4}		0.085	4.44×10^{-4}
200	18.68	—	—	0.830		
		1.99	5.33×10^{-4}		0.195	5.33×10^{-4}
400	16.69	—	—	0.635		
		1.89	2.83×10^{-4}		0.186	2.83×10^{-4}
800	14.80	—	—	0.449		

14.3 Determine the values of the compression index C_c and the modified compression index $C_{o\epsilon}$ for the laboratory and field curves in Fig. 14.3, considering the linear portion of the curves between the pressure range 200 to 800 kPa.

Table 14.4

EXAMPLE 14.3

Pressure	Lab. curve		Field curve	
σ' (kPa)	Void ratio	Strain %	Void ratio	Strain %
200	0.830	6.6	0.920	1.75
800	0.449	26.0	0.465	25.50
	Δe	$\Delta \epsilon$	Δe	$\Delta \epsilon$
	0.381	0.194	0.455	0.2375

$$C_c = \frac{\Delta e}{\log \sigma_f'/\sigma_1'} \; ; \; C_{o\epsilon} = \frac{\Delta \epsilon}{\log \sigma_f'/\sigma_1'}$$

C_o (lab) $= 0.633$ $C_{o\epsilon}$ (lab) $= 0.322$

C_o (field) $= 0.756$ $C_{o\epsilon}$ (field) $= 0.395$

Check by formulae $C_o = C_{o\epsilon} (1 + e_o)$ and

$C_{c\epsilon} = C_c / (1 + e_o)$,   where $e_o = 0.943$.

14.8 RATE OF CONSOLIDATION

Terzaghi (1925) presented a theory to describe the time rate of consolidation which is based on the following main assumptions:

(1) The soil layer under consolidation is horizontal, homogenous, fully saturated, of uniform thickness and is laterally confined.

(2) Soil particles and water are incompressible.

(3) Compression and flow are one-dimensional (vertical).

(4) Darcy's law is valid at all hydraulic gradients.

(5) The time taken by the clay to consolidate depends entirely on the permeability k.

(6) Strains are small; k, a_v and m_v remain constant over the increment of applied pressure.

(7) As a_v is assumed constant then there is *unique* relationship, independent of time, between void ratio and effective pressure (i.e. $\partial e/\partial \sigma'$ = constant).

(8) The applied pressure is uniform along a horizontal plane. The intial excess pore pressure due to the applied pressure is also uniform throughout the depth of the clay layer.

(9) One or both of the strata adjacent to the clay layer are perfectly free draining in comparison with clay.

In practice, the drainage of water and displacements which take place during consolidation are nearly always three dimensional. But for cases in which a wide foundation rests on a relatively thin layer of clay between pervious layers, the drainage pattern can be assumed one dimensional.

The first three assumptions may thus be closely met with in practice. There is evidence of deviation from the Darcy law at low hydraulic gradients. The soil properties k, a_v and m_v also decrease during consolidation since e/σ' relationship is non-linear. The main limitation of the theory arises from assumptions of unique relationship between e and σ' and its being independent of time. This implies that there is no secondary compression (Section 14.13).

Consider a soil element Δx, Δy and Δz (height) within a clay layer at depth 'z' below the top of the layer. (It is assumed that the top and bottom surfaces of clay are in contact with coarse materials and are thus drainage faces). At any instant during the process of consolidation:

$$\Delta \sigma = \Delta \sigma' + u \tag{14.22}$$

also $h = u/\gamma_w$ \hfill (14.23)

and according to the Darcy law the flow velocity v_s is given by:

$$v_s = k\, i_s = -k\, \frac{\partial h}{\partial z} = -\frac{k}{\gamma_w} \frac{\partial u}{\partial z} \tag{14.24}$$

The rate of change of velocity with depth is given by:

$$\frac{\partial v_s}{\partial z} = -\frac{k}{\gamma_w} \cdot \frac{\partial^2 u}{\partial z^2} \tag{14.25}$$

Quantity of water entering the element per unit time is $v_s\,\Delta x\,\Delta y$ and the quantity flowing out is $(v_s + \dfrac{\partial v_s}{\partial z}\,\Delta z)\,\Delta x\,\Delta y$.   The net quantity of water flowing out per unit time is given by:

$$\frac{\Delta V}{\Delta t} = -\frac{k}{\gamma_w} \cdot \frac{\partial^2 u}{\partial z^2} \cdot \Delta x\,\Delta y\,\Delta z \tag{14.26}$$

Eq. 14.26 represents the continuity condition. The rate of volume change may also be expressed in terms of m_v:

$$\frac{\Delta V}{\Delta t} = m_v\, V_i \frac{\partial \sigma'}{\partial t} = m_v\,\Delta x\,\Delta y\,\Delta z\, \frac{\partial \sigma'}{\partial t}$$

$$\tag{14.27}$$

The total pressure increment $\Delta \sigma$ is gradually transferred to the soil skeleton as u decreases. Differentiating Eq. 14.22 (where $\Delta \sigma = $ constant):

$$\frac{\partial \sigma'}{\partial t} = -\frac{\partial u}{\partial t} \tag{14.28}$$

Therefore $\dfrac{\Delta V}{\Delta t} = -m_v \dfrac{\partial u}{\partial t}\,\Delta x\,\Delta y\,\Delta z$ \hfill (14.29)

Combining Eqs. 14.26 and 14.29,

$$m_v \frac{\partial u}{\partial t} = \frac{k}{\gamma_w} \frac{\partial^2 u}{\partial z^2}$$

or $\dfrac{\partial u}{\partial t} = c_v \dfrac{\partial^2 u}{\partial z^2}$ \hfill (14.30)

where $c_v = \dfrac{k}{m_v\,\gamma_w} = \dfrac{k}{m_v\,\rho_w\,g}$ \hfill (14.31)

c_v is called the *coefficient of consolidation.*

The "coefficient of consolidation" (Eq. 14.31) is a parameter which relates the change in 'u' with respect to time 't', to the amount of water draining out of the voids of a clay prism during the same time, due to consolidation. It indicates the combined effects of permeability and compressibility of a soil on the rate of volume change. As compression proceeds both k and m_v decrease, but c_v which depends on the ratio k/m_v remains fairly constant within a considerable range in pressure. The coefficient c_v has dimensions of L^2T^{-1}. The practical unit of c_v is m²/y_r (1 m²/y_r = 31.7 $\times$ 10⁻⁹ m²/s; 1 m²/s = 31.56 $\times$ 10⁶ m²/yr).

Eq. 14. 30 is the basic , one dimensional consolidation equation. The solution of this differential equation describes the excess pore pressure distribution as a function of depth and time. The solution is expressed in terms of "degree of consolidation U", which is defined as the ratio of the excess pore pressure lost after a certain time due to drainage, to the initial excess pore pressure u_o at any instant during the consolidation process:

$$U = \frac{u_o - u}{u_o} \times 100 \tag{14.32}$$

The value of U is sometimes referred to as the "percentage pore pressure dissipation." The pore pressure dissipates more rapidly near the drainage surfaces than at points remote from them. If the percentage U is related to the *average* pore pressure at any time t, it may be assumed that average U is proportional to the amount of settlement (compression) which has taken place by time t. Thus average U can also be defined as the ratio of settlement S_t upto time t to the final or ultimate settlement S_f (i.e. when U = 100%).

$$U = \frac{S_t}{S_f} \times 100 \tag{14.33}$$

Eq. 14.30 is solved for the following boundary conditions:

1. At t = 0 at any depth z, $\triangle u = u_o = \triangle \sigma$ (constant)
2. At t = ∞ at any depth z, $\triangle u = 0$
3. At t > 0, $\triangle u = 0$ both at z = 0 and z = H (i.e. complete drainage both at top and bottom of clay layer of thickness H)

The solution can be expressed as:

$$U = f \left(\frac{c_v\, t}{d^2} \right) \tag{14.34}$$

where 'd' is the 'drainage path' defined as the maximum distance which pore water under pressure, in a particular soil layer, has to travel. A layer for which both the top and bottom boundaries are free draining is called an 'open layer' and d = H/2. A layer for which only one boundary is free draining is a 'half-closed' layer and d = H (layer thickness).

The expression $(c_v t/d^2)$ is an independent dimensionless variable known as the *time factor* T_v:

$$T_v = \frac{c_v.\, t}{d^2} \tag{14.35}$$

Eq. 14.34 may thus be written as:

$$U = f\, (T_v) \tag{14.36}$$

The dimensionless parameter T_v which is related to c_v and d is used for defining the theoretical rate of consolidation curve.

For conditions of drainage both at top and bottom, the solution of Eq. 14.36 relating the average value of U over the layer thickness can be expressed as follows (Taylor 1948):

or $$U = 1 - \frac{8}{\pi^2}\, \exp \left(- \frac{\pi^2\, T_v}{4} \right) \tag{14.37}$$

The values of U derived from Eq. 14.37 are given in Tables 14.5 and 14.6.

Table 14.5

VARIATION OF T_v WITH U FOR CONDITIONS OF DRAINAGE BOTH AT TOP
AND BOTTOM (APPLICABLE TO ALL LINEAR PRESSURE DISTRIBUTIONS)

U (%)	T_v	U (%)	T_v
10	0.008	55	0.238
15	0.018	60	0.287
20	0.031	65	0.342
25	0.049	70	0.403
30	0.071	75	0.477
35	0.096	80	0.567
40	0.126	85	0.684
45	0.160	90	0.848
50	0.197	95	1.129
		100	∞

Note. The above values of T_v may also be used for single drainage at only top
or bottom If there is uniform distribution of consolidation pressure.

Table 14.6

Variation U with T_v for conditions of drainage both at top
and bottom (applicable to all linear pressure distributions). Also
applicable to single drainage but having uniform distribution of
consolidation pressure across the layer.

T_v	U (%)	T_v	U (%)
0.004	7.14	0.200	50.41
0.008	10.09	0.250	56.22
0.012	12.36	0.300	61.32
0.020	15.96	0.350	65.82
0.028	18.88	0.400	69.79
0.036	21.40	0.500	76.40
0.048	24.72	0.600	81.56
0.060	27.64	0.700	85.59
0.072	30.28	0.800	88.79
0.083	32.51	0.900	91.20
0.100	35.68	1.000	93.13
0.125	39.89	1.500	98.00
0.150	43.70	2.000	99.42
0.175	47.18		

Eq. 14.37 may be very closely represented by the following empirical relations:

$$\text{For } U < 0.6: T_v = \frac{\pi}{4} U^2 \tag{14.38a}$$

$$U = 1.13 \sqrt{T_v} \tag{14.38b}$$

For $U > 0.6$: $T_v = 1.781 - 0.933 \log_{10} (100 - U\%)$ (14.39)

Sivaram and Swamee (1977) gave the following relation valid for all values of average U:

$$T_v = \frac{0.25 \pi U^2}{(1 - U^{5.6})^{0.357}} \tag{14.40}$$

The relationship between U and T_v is shown plotted in Fig. 14.7.

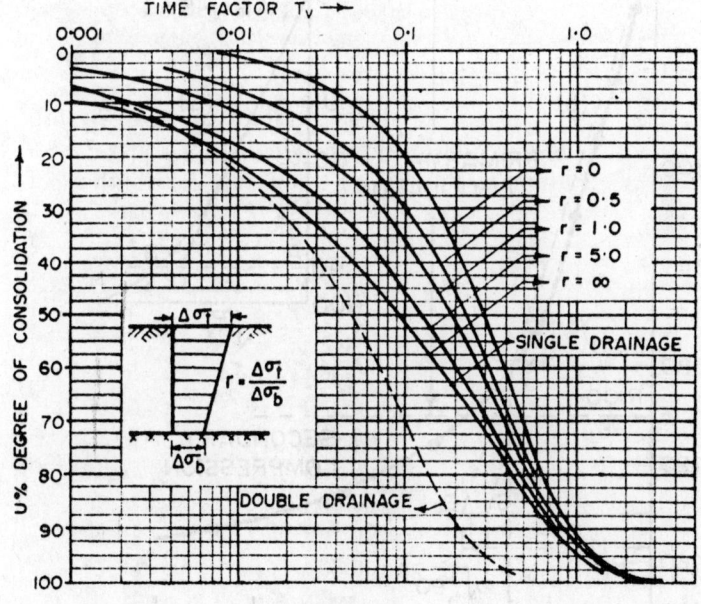

Fig. 14.7

Relationship between U and T_v (Janbu et al 1956). The double drainage curve is plotted using *total* thickness H of the layer; multiply T_v - values by 4 for $d = H/2$.

14.9 DETERMINATION OF COEFFICIENT OF CONSOLIDATION

The value of the coefficient of consolidation c_v for a *particular* pressure increment in the consolidation test is determined by comparing the characteristics of the laboratory and theoretical consolidation curves, the process being known as "curve fitting". There are two curve fitting methods, one using the square-root-time versus settlement curve (the square root time method) and other using the log-time versus settlement curve (the log time method). Curve fitting relates to the primary consolidation phase only.

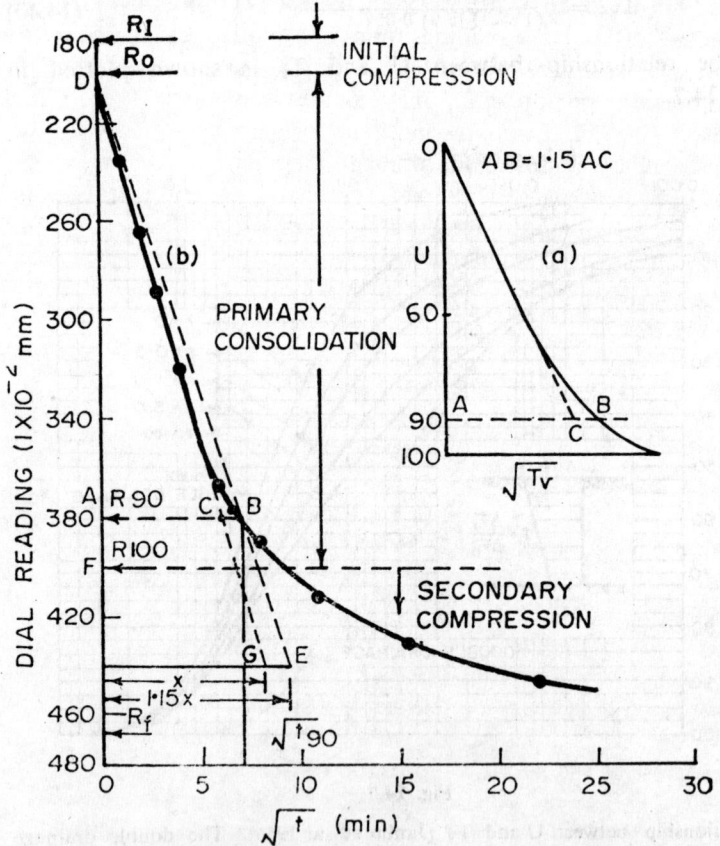

Fig. 14.8   Square root time method (a) theoretical curve
(b) laboratory curve

1. Square root time method. This method is known as the Taylor's method (Taylor 1942). The theoretical curve of U versus $\sqrt{T_v}$ is shown in Fig. 14.8a. The curve is linear upto about U 60% (Eq. 14.38) and at 90% consolidation of abscissa (AB) is 1.15 times the abscissa (AC) of the extension of linear part of the curve. This characteristic is used to determine the 90% consolidation point on the laboratory curve.

For a laboratory curve, the degree of consolidation is represented by the amount of compression (i.e. settlement in terms of dial gauge readings) of the specimen at a particular time after load application. For convenience of obtaining $\sqrt{t}$ values, the dial readings may be recorded at the following elapsed times t: 0.25, 1.0, 2.25, 4.0, 6.25, 9.0, 12.25, 16.0, 20.25, 25, 36, 49, 64, 121, 240, 484 and 1444 minutes. Dial gauge readings (mm) are plotted as shown in Fig. 14.8b against square root of time in minutes $\sqrt{t}$. For analytical purposes, the compression may be divided into three distinct phases: (i) initial compression, (ii) primary consolidation, and (iii) secondary compression, although these phases overlap and the time dependent phases (ii) and (iii) probably occur simultaneously.

(a) *Initial compression.* It is the compression of a test specimen which takes almost simultaneously with the application of a pressure increment and before commencement of drainage. It is partly due to compression of air within the pore spaces (degree of saturation being marginally below 100%) and partly to bedding down of contact surfaces in the cell and in the load frame.

(b) *Primary consolidation.* It is the time dependent compression due to the dissipation of the excess pore pressure, and is accounted for by the Terzaghi theory.

(c) *Secondary compression.* It is the compression which continues at a very slow rate for an indefinite period of time after the excess pore pressure of the primary phase has virtually dissipated.

The early straight time portion of the laboratory curve is produced back to the ordinate at zero time. The point of intersection D is the corrected zero reading R_0 corresponding to U = 0%. The difference between the initial dial reading R_i and R_0 represents the initial compression of the specimen. A straight time DE is then drawn having abscissae 1.15 times the corresponding abscissae on the linear part of the laboratory curve, extended to any orbitrary point G. The intersection of DE with the curve locates the point B having coordinates ($\sqrt{t_{90}}$, R_{90}) corresponding to U = 90%. The value of t_{90} is thus obtained. The theoretical value of T_v for U = 90% is 0.848. The coefficient of consolidation c_v is calculated from Eq. 14.35 as:

$$c_v = \frac{0.848 \, d^2}{t_{90}} \qquad (14.41)$$

where   d = half the *average* thickness of specimen for the particular pressure increment

The point F on the settlement axis (R_{100}) corresponding to U = 100% is found by setting off DF equal to DA/0.9. All settlement below this level is assumed to be secondry compression.

2. Log time method. This method was derived by Casagrande and hence, is known as the Casagrande method.

The laboratory curve obtained by plotting the dial gauge reading against the logarithm of time in minutes (Fig. 14.9) is compared with the theoretical curve of average U versus log T_v. The initial portion of the theoretical curve (Fig. 14.9a) approximates

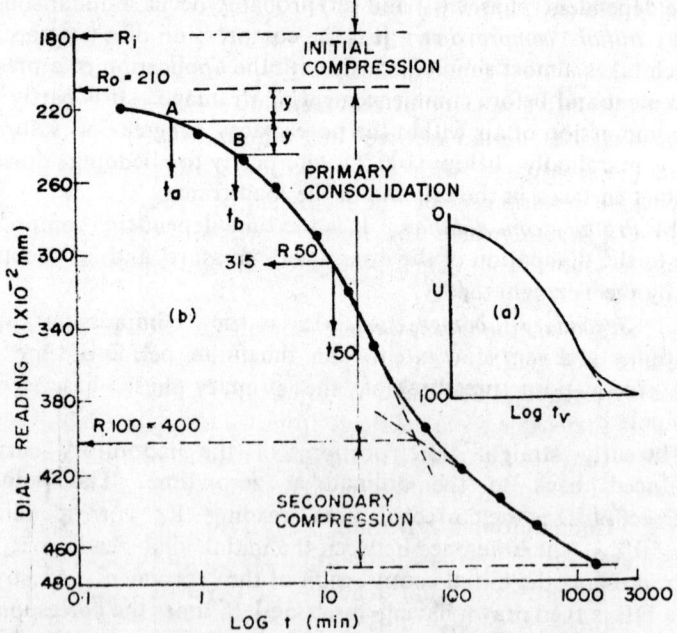

Fig 14.9 Log time method (a) theoretical curve (b) laboratory curve

closely to a parabola which is followed by linear portion and finally the curve becomes asymptotic to the horizontal axis at $U = 100\%$.

In Fig. 14.9b, R_i is the initial dial reading ($t = 0$). The point corresponding to $U = 0$ is found from the consideration that the initial portion of the laboratory curve also should approximate at a parabola. Two points A and B are selected on the early part of the curve for which the ratio of times $t_b/t_a = 4:1$. The corrected dial reading R_o corresponding to $U = 0$ is then located a distance above point A equal to the difference in dial readings between points A and B. The difference between R_i and R_o is the initial compression. The dial reading R_{100} representing 100% primary consolidation is taken as the point of intersection of a tangent drawn to the curve at its point of inflexion (which occurs at about 75% U) with the backward extension of the secondary compression portion of the curve. The point corresponding to $U = 50\%$ is then located midway between the R_o and R_{100} points and the corresponding time t_{50} is obtained. The value of c_v can, therefore, be calculated from Eq. 14.35 and knowing that T_v for $U = 50\%$ is 0.197:

$$c_v = \frac{0.197 \ d^2}{t_{50}} \qquad (14.42)$$

3. **Relative merits of the methods.** Dial readings covering a much shorter period of time are required in the square root time method compared with the log time method, which requires an accurate development of the secondary compression portion of the curve. In many cases, the linear portion of curve is not obtained on the square root time plot and the log time method should be used.

The two methods may yield somewhat different results. The square root method usually gives a larger value of c_v than does the log method (Lambe and Whitman 1973).

It is preferable to calculate c_v from t_{50} (log time method) rather than from t_{90} because the middle portion of a laboratory curve agrees most closely with the theoretical curve.

4. **Computational method.** Sivaram and Swamee (1977) proposed the following computational method for determining c_v:

(1) Record two dial readings, R_1 and R_2, corresponding to times, t_1 and t_2, during the early phase of consolidation ($< 50\%$ compression).

(2) Also record a dial reading R_3 at time t_3 after considerable compression of the specimen.

(3) Determine R_o as:

$$R_o = \frac{R_1 - R_2 \sqrt{t_1 / t_2}}{1 - \sqrt{t_1 / t_2}} \qquad (14.43)$$

(4) Determine R_{100} as:

$$R_{100} = R_o - \frac{R_o - R_3}{[1-\{(R_o-R_3)(\sqrt{t_2}-\sqrt{t_1})/(R_1-R_2)\sqrt{t_3}\}^{5 \cdot 6}]^{0 \cdot 179}}$$

$$(14.44)$$

(5) Determine c_v as:

$$c_v = \frac{\pi}{4} \left(\frac{R_1 - R_2}{R_0 - R_{100}} \cdot \frac{d}{\sqrt{t_2} - \sqrt{t_1}} \right)^2 \qquad (14.45)$$

where   d = drainage path = $H_{av}/2$

(6) Repeat the above computation for 3 different sets of readings and take the average c_v.

5. **Values of c_v.** The value of c_v varies from increment to increment and is different for loading and unloading. It may even vary considerably among specimens of the same soil. It may thus be difficult to adopt an accurate value of c_v for a particular field problem. Actual rate of settlement has been observed to be 2 to 4 times faster than the rate predicted on the basis of c_v. Such differences arise partly due to difficulties in measuring c_v and partly due to the limitations of the one dimensional consolidation theory.

The values of c_v obtained in a consolidation test for various pressure increments are plotted on a semilog graph of c_v versus log of average pressure for each load increment. The value of c_v for the field increment is then read for the average field pressure (Dunn et al 1980).

Typical values of c_v (m^2/yr) for undisturbed clays (montmorillonite) varying with plasticity index are as follows (Lambe and Whitman 1979, Head 1982):
(i) $I_p > 25$, $c_v = 0.1-1$,   (ii) $I_p = 25-5$, $c_v = 1-10$,
(iii) $I_p \leqslant 5$, $c_v = 10-100$. For silts, $c_v > 100$. Disturbed values are about $25-50\%$ of undisturbed values.

6. Compression ratios. The relative magnitudes of the initial compression, primary consolidation and the secondary compression are expressed by the following ratios:

(a) Initial compression ratio $r_o = \dfrac{R_i - R_o}{R_i - R_f}$ (14.46)

(b) Primary compression ratio (root time):

$$r_p = \frac{10 (R_o - R_{90})}{9 (R_i - R_f)}$$ (14.47)

(c) Primary compression ratio (log time):

$$r_p = \frac{R_o - R_{100}}{R_i - R_f}$$ (14.48)

(d) Secondary compression ratio:

$$r_s = \frac{R_{100} - R_f}{R_i - R_f}$$ (14.49)

or $r_s = 1 - (r_o + r_p)$ (14.50)

7. Coefficient of permeability. The coefficient of permeability k can be calculated from Eq. 14.31 when the values of c_v and m_v have been obtained:

$$k = c_v\, m_v\, \rho_w\, g = c_v\, m_v \times 0.31 \times 10^{-9} \text{ m/s}   (14.51)$$

where c_v is expressed in m²/yr and m_v in m²/MN.

Example 14.4

The following compression reading were obtained during a consolidation test on a saturated clay sample when the applied pressure was increased from 100 kPa to 200 kPa. Time (min): 0, 0.25, 1, 2.25, 4, 9, 16, 25, 36, 49, 64, 121, 240, 484 and 1444; corresponding dial readings $(1 \times 10^{-2}$ mn); 180, 220, 235, 247, 265, 289, 320, 351, 368, 380, 391, 415, 433, 448 and 468. At the beginning of the pressure increment (t = 0), the height of the sample was 21.11 mm. Determine c_v from both the square root and log time methods and the values of the three compression ratios. Find also the value of k. If the moisture content of the sample after the 1444 minutes reading was measured as 20.2% and the relative density of solids is 2.7, determine the void ratios after consolidation under 100 kPa and 200 kPa. Determine also the value of c_v from Sivaram computational method.

$H_i = 21.11$ mm, $\triangle H = R_f - R_i = 4.68 - 1.80 = 2.88$ mm

$H_f = H_i - \triangle H = 18.23$ mm; $(H_i + H_f)/2 = 19.67$ mm

Average drainage path $d = 19.67/2 = 9.835$ mm

The test data are plotted in Figs. 14.8 and 14.9. The $\sqrt{t}$ values are as follows: 0, 0.5, 1, 1.5, 2, 3, 4, 5, 6, 7, 8, 11, 15.5, 22, and 38 (min) which are plotted in the root time graphs (Fig. 14.8).

(a) Root time method.

$$\sqrt{t_{90}} = 7.25 \text{ min}, t_{90} = 52.56 \text{ min}$$

$$c_v = \frac{0.848 \, d^2}{t_{90}} = 1.56 \text{ mm}^2/\text{min}$$

$$= \frac{1.56 \times 60 \times 24 \times 365}{10^6} = 0.82 \text{ m}^2/\text{yr}$$

$$r_o = \frac{R_i \sim R_o}{R_i \sim R_f} = \frac{200 - 180}{468 - 180} = \frac{20}{288} = 0.0694$$

$$r_p = \frac{10 \, (R_o \sim R_{90})}{9 \, (R_i - R_f)} = \frac{10 \, (380 - 200)}{9 \, (288)} = 0.694$$

$$r_s = 1 - (r_o + r_p) = 0.2366$$

(b) Log time method.

$$t_{50} = 12.5 \text{ min}$$

$$c_v = \frac{0.197 \, d^2}{t_{50}} = 1.52 \text{ mm}^2/\text{min} = 0.80 \text{ m}^2/\text{yr}$$

$$r_o = \frac{R_i \sim R_o}{R_i \sim R_f} = \frac{210 - 180}{288} = 0.104$$

$$r_p = \frac{R_o \sim R_{100}}{R_i \sim R_f} = \frac{400 - 210}{288} = 0.66$$

$$r_o = 1 - (r_o + r_p) = 0.236$$

(c) $m_v = \dfrac{\Delta H}{H_i} \dfrac{1}{\Delta \sigma'} = \dfrac{2.88}{21.11} \cdot \dfrac{1}{100} = 1.36 \times 10^{-3} \text{ m}^2/\text{kN}$

$$= 1.36 \text{ m}^2/\text{MN}$$

$k = c_v \, m_v \, \rho_w \, g$

$$= c_v \cdot m_v \times 0.31 \times 10^{-9} \text{ m/s}$$

$$= 0.82 \times 1.36 \times 0.31 \times 10^{-9} = 3.46 \times 10^{-10} \text{ m/s}$$

(d) $w_i' = 0.36$, $e_f' = 0.36 \times 2.7 = 0.972$ @ 200 kPa

$$\Delta e = \Delta H \frac{1 + e_f}{H_f} = 0.311$$

$e_i = e_f' + \Delta e = 1.283$ @ 100 kPa

$$\text{Check, } m_v = \frac{\Delta e}{1 + e_i} \cdot \frac{1}{\Delta \sigma} = 1.36 \text{ m}^2/\text{MN}$$

(e) Sivaram method.

t_1 = 4 min R_1 = 2.65 mm   d = 9.835 mm

t_2 = 9 min R_2 = 2.89 mm

t_3 = 36 min R_3 = 3.68 mm

R_0 = 2.17 mm R_{100} = 4.04 mm

c_v = 1.25 mm²/min

It may be noted that the computed values of R_0 and R_{100} are very close to those obtained by graphical methods.

14.10 CONSOLIDATION SETTLEMENT

The final (ultimate) consolidation settlement S_f may be calculated using any of the following methods:

(a) Based on change in void ratio $\triangle e$, or strain ϵ:

$$S_f = \frac{\triangle e}{1 + e_0} H_o = \frac{e_0 - e_f}{1 + e_0} H_o \qquad (14.52)$$

$$S_f = \epsilon H_o \qquad (14.53)$$

(b) Based on coefficient of volume compressibility m_v:

$$S_t = m_v H_o \triangle \sigma' \qquad (14.54)$$

(c) Based on compression index C_c (or recompression index C_r) or on modified indices $C_c\epsilon$, and $C_r\epsilon$:

$$S_t = \frac{C_c}{1 + e_0} H_o \log \frac{\sigma_0' + \triangle \sigma}{\sigma_0'} \quad \text{(NC clay)} \quad (14.55)$$

$$S_f = C_o\epsilon H_o \log \frac{\sigma_0' + \triangle \sigma}{\sigma_0'} \quad \text{(NC clay)} \qquad (14.56)$$

$$S_t = \frac{C_r}{1 + e_0} H_o \log \frac{\sigma_0' + \triangle \sigma}{\sigma_0'} \quad \text{(OC clay)} \quad (14.57)$$

$$S_t = C_r\epsilon H_o \log \frac{\sigma_0' + \triangle \sigma}{\sigma_0'} \quad \text{(OC clay)} \qquad (14.58)$$

where H_o is the thickness of the compressible layer and $\triangle \sigma$ is the average increase in pressure on the layer. The settlement at a point on the ground surface or at the base of a foundation is assumed to equal the settlement (compression) of the clay layer directly below the point in question.

Although any of the above methods can be used for NC clays and OC clays, it is convenient to use the void ratio change and compression index when dealing with NC clays and m_v for OC clays.

The insitu void ratio e_0 can be calculated from the relation $e_0 = w_n G$ on the assumption of full saturation. The pressure increase $\Delta \sigma$ will normally vary with depth. Similarly m_v varies with the pressure range (depth), even for ranges on the linear part of the e/log σ' curve. To take these variations into account, a layer may be divided into a suitable number of sublayers. σ_0' and $\Delta \sigma$ are determined for the mid-depth of a sublayer and m_v is calculated for the particular pressure range applicable to the sublayer. The settlement of the whole layer is equal to the sum of the sublayer settlements.

The compression indices are the same for any pressure range on the linear part of their respective consolidation curves. C_0 and $C_{c\epsilon}$ are used for NC clay, and C_r and $C_{r\epsilon}$ for OC clay. The compression indices should be based on field consolidation curves. For NC clay, C_0 can be determined from empirical relationships also such as Eqs. 14.16, 14.20. For OC clay, if $(\sigma_0' + \Delta \sigma) > \sigma_p'$, settlement will be calculated in two parts, first along the field recompression line (AB in Fig. 14.5) using C_r and then along the field virgin compression line (BF in Fig. 14.5) using C_0. The total settlement then becomes:

$$S_f = C_r \; \frac{H_0}{1 + e_0} \; \log \frac{\sigma_0' + (\sigma_p' - \sigma_0')}{\sigma_0'}$$

$$+ \; C_0 \; \frac{H_0}{1 + e_0} \; \log \frac{\sigma_p' + (\sigma_0' + \Delta \sigma' - \sigma_p')}{\sigma_p'}$$

$$(14.59a)$$

$$= C_r \; \frac{H_0}{1 + e_0} \; \log \frac{\sigma_p'}{\sigma_0'} + C_0 \; \frac{H_0}{1 + e_0}$$

$$\times \log \frac{\sigma_0' + \Delta \sigma'}{\sigma_p'} \qquad (14.59b)$$

In terms of the modified indices,

$$S_f = C_{r\epsilon} \, H_0 \log \frac{\sigma_p'}{\sigma_0} + C_{0\epsilon} \, H_0 \log \frac{\sigma_0' + \Delta \sigma'}{\sigma_p'}$$

$$(14.60)$$

Although technically, the void ratio corresponding to σ_p' on virgin compression curve should be used in the right hand term of

Eq. 14.59, the use of e_0 does not make any significant difference in the result.

The consolidation settlement of an under-consolidated soil is the sum of (i) settlement which occurs till equilibrium is reached (consolidation is complete) under existing overburden pressure σ_0' and (ii) the settlement in response to additional pressure increase $\Delta\sigma$ (See Fig. 14.6, Section 14.7). The calculations are made in the same manner as for normally consolidated deposits.

$$S_t = C_c \frac{H_0}{1 + e_0} \log \left(\frac{\sigma_p' + (\sigma_0' - \sigma_p') + \Delta\sigma'}{\sigma_p} \right)$$

$$(14.61 \text{ a})$$

$$= C_c \frac{H_0}{1 + e_0} \log \left(\frac{\sigma_0' + \Delta\sigma'}{\sigma_p'} \right) \qquad (14.61 \text{ b})$$

where σ_p' = current effective pressure in the soil

14.11 RATE OF SETTLEMENT

The rate of settlement of a clay layer due to additional loading can be predicted using the results of consolidation tests on undisturbed specimens of clay. The final settlement S_t is first calculated as given in Section 14.10. An appropriate value of c_v for the average pressure during consolidation is then selected. The time rate of settlement is calculated using Eqs. 14.62 and 14.63:

$$t = \frac{T_v d^2}{c_v} \qquad (14.62)$$

$$U = \frac{S_t}{S_f} \qquad (14.63)$$

where d = drainage path

 S_t = settlement in time t

 T_v = Time factor corresponding to U

The time values are calculated from Eq. 14.62 and the settlement values from Eq. 14.63. The results can be reported in a tabular form with vertical columns of U, T_v , t and S_t.

1. Model law of consolidation. If two layers of the same clay with different drainage paths d_1 and d_2 are subjected to the same pressure increase and reach the same degree of consolidation in times t_1 and t_2 respectively, then theoretically their coefficients of consolidation must be equal as must their time factors. From Eq. 14.35,

$$T_{v\,1} = \frac{c_{v\,1} \cdot t_1}{d_1{}^2} \,, \; T_{v\,2} = \frac{c_{v\,2} \cdot t_2}{d_2{}^2}$$

Equating $c_{v\,1} = c_{v\,2}$ and $T_{v\,1} = T_{v\,2}$,

$$\frac{t_1}{d_1{}^2} = \frac{t_2}{d_2{}^2} \tag{14.64}$$

Eq. 14.64 gives a simple method for predicting the time for a clay layer in the field to reach a certain degree of consolidation by testing an undisturbed sample of clay and measuring the time it takes to reach the same degree of consolidation; the time is directly proportional to the square of the drainage path.

2. Multi-layered deposit. A clay deposit consisting of a system of layers with different coefficients of consolidation can be converted to an equivalent layer with equivalent properties. Consider, for example, three layers of thicknesses H_1, H_2 and H_3 with coefficients of consolidation as $c_{v\,1}$, $c_{v\,2}$ and $c_{v\,3}$ respectively. The thicknesses of the second and third layers, H_2 and H_3, can be transformed to an equivalent thickness of a layer possessing the properties of the first layer as follows (NAVFAC DM 7.1):

$$H_2{}' = H_2 \, (c_{v\,1}/c_{v\,2})^{0 \cdot 5} \tag{14.65}$$
$$H_3{}' = H_3 \, (c_{v\,1}/c_{v\,3})^{0 \cdot 5} \tag{14.66}$$

where $H_2{}'$, $H_3{}'$ = equivalent thicknesses of second and third layer in terms of first layer

The total thickness of the equivalent layer $H_t{}'$ possessing the properties of the first layer will be given by:

$$H_t{}' = H_1 + H_2{}' + H_3{}' \tag{14.67}$$

where $H_t{}'$ is assumed to possess a coefficient of consolidation $c_{v\,1}$. The degree of consolidation at various times t will now be determined for a layer thickness $H_t{}'$ with $c_{v\,1}$.

3. Effect of internal drainage. Where a compressible layer contains one or more seams of pervious material which act as drainage layer, the average degree of consolidation U_a at any time t can be found by equating the settlements of the individual sublayers to the settlement of entire layer calculated on the basis of average U_a. Considering the case of a layer of thickness H which is divided into two sublayers, H_1 and H_2, due to the presence of an internal drainage layer (of negligible thickness):

$$U_a \, S_t = U_1 \, S_{t1} + U_2 \, S_{t2} \qquad (14.68)$$

where $\quad S_t =$ final settlement of the entire layer

$\quad S_{t1}, \ S_{t2} =$ final settlements of the sublayers

Since final settlement is proportional to the layer thickness, Eq. 14.68 may also be written as:

$$U_a H \ = U_1 H_1 + U_2 H_2 \qquad (14.69)$$

Examples 14.5—14.13

14.5 During a pressure increment a test specimen, 20 mm thick, attained 50% primary consolidation in 45 minutes. How long would it take a 10 m thick layer of the same soil to reach the same degree of consolidation if (a) the clay layer was drained on both surfaces, and (b) it was drained on the top surface only?

(a) For specimen with double drainage, $d_1 = 20/2 = 10$ mm
Clay layer with double drainage, $d_2 = 10/2 = 5$ m

$$t_2 \ = \ \frac{t_1}{d_1{}^2} \cdot d_2{}^2 = \frac{45 \, (5 \times 10^3)^2}{(10)^2} \ \times \ \frac{1}{60} \times \frac{1}{24}$$

$$\times \ \frac{1}{365} \ = \ 21 \ 4 \ \text{yr}$$

(b) Clay layer with single drainage, $d_2 = 10$ m

$\quad t_2 \ = 4 \times 21.4 = 85.6$ yr

14.6 An 18 mm thick laboratory specimen, drained top and bottom, reached 25% consolidation in 10 minutes. How long would it take the same specimen to reach 50% consolidation?

Upto $U = 0.6$, $U = 1.13 \ \sqrt{T_v}$, $d = 9$ mm

$(T_v)_{25} \ = (0.25/1.13)^2 = 0.049$

$c_v \qquad = T_v \ \dfrac{d^2}{t} = 0.3969 \ \text{mm}^2/\text{min}$

$(T_v)_{50} \ = 0.197$

$t_{50} \qquad = \dfrac{(T_v)_{50} \, d^2}{c_v} = 40.2$ min

14.7 The present, average effective pressure on a normally consolidated clay layer, 5 m thick, is 180 kPa. Estimate the ultimate consolidation settlement if the structural loads on the surface increase the average pressure in the layer by 60 kPa. The natural moisture content of clay is 35.2%, and G = 2.68. The results of a consolidation test performed on an undisturbed clay specimen, 75 mm in diameter and 20 mm thick, taken from the mid-point of the layer are as given in Example 14.1. Determine settlement using (a) e_o, e_f values on laboratory and field curves, (b) C_c and $C_{c\epsilon}$ on laboratory and field curves, and (c) m_v on field curve (Fig. 14.3).

$$\sigma_o' = 180 \text{ kPa}, \quad \triangle\sigma' = 60 \text{ kPa}, \quad \sigma_f' = 240 \text{ kPa}$$

(a)   Using e values.

Lab curve: e_f at σ_f' (240 kPa) = 0.775

$$S_t = \frac{e_o - e_f}{1 + e_o} H_o = 0.432 \text{ m}$$

Field curve: e_f at $\sigma_f' = 0.86$

$$S_f = 0.213 \text{ m}$$

(b)   Using C_c.

Lab curve: $C_c = 0.633$

$$S_t = \frac{C_c}{1 + e_o} H_o \log \frac{\sigma_f'}{\sigma_o'} = 0.204 \text{ m}$$

Field curve: $C_c = 0.756$

$$S_t = 0.243 \text{ m}$$

(c)   Using $C_{c\epsilon}$.

Lab curve: $C_{c\epsilon} = 0.322$

$$S_t = \epsilon H_o = H_o \, C_{c\epsilon} \log \frac{\sigma_f'}{\sigma_o} = 0.201 \text{ m}$$

Field curve: $C_{c\epsilon} = 0.395$

$$S_f = 0.247 \text{ m}$$

(d)   Using m_v.

$$\sigma_o' = 180 \text{ kPa}, \quad e_o = 0.943$$
$$\sigma_f' = 240 \text{ kPa}, \quad e_f = 0.860$$
$$m_v = \frac{\Delta e}{1 + e_o} \frac{1}{\Delta\sigma'} = 7.12 \times 10^{-4} \text{ m}^2/\text{kN}$$
$$S_t = m_v \, H_o \, \Delta\sigma' = 0.2!4 \text{ m}$$

The result using C_c or $C_{c\epsilon}$ on the basis of field consolidation curve should be considered more reliable for a normally consolidated clay.

14.8 A building constructed on a compressible layer with double drainage settles by 80 mm in 4 years. What will be the settlement in 9 years, if the final settlement is expected to be about 300 mm. What time will be required to settle by 210 mm? What will be the settlement in 25 years.

(a) Assume that at $t = 9$ years, $U \leqslant 0.6$

$$U = 1.13 \sqrt{T_v}, U \propto \sqrt{T_v}, T_v = c_v \ t/d^2, T_v \propto t$$

$$\frac{S_{t1}}{S_{t2}} = \frac{U_1}{U_2} = (T_{v1}/T_{v2})^{1/2} = (t_1/t_2)^{1/2}$$

$$\frac{80}{S_{t2}} = (4/9)^{0.5}, S_{t2} = 120 \text{ mm}$$

Check $U_2 = S_{t2}/S_t = 120/300 = 0.4 \ (< 0.6)$

(b) When $S_{t3} = 210$ mm, $U_3 = 210/300 = 0.7 \ (> 0.6)$.

T_{v3} as read from Table 14.5 or graph (Fig. 14.7) (double drainage)
$$= 0.403$$

When $t_1 = 4$ yr., $U_1 = 80/300 = 0.267 \ (< 0.6)$

$$T_{v1} = \frac{\pi}{4} U^2 = 0.056$$

$$\frac{t_3}{t_1} = \frac{T_{v3}}{T_{v1}}, t_3 = 28.78 \text{ yr}$$

(c) When $t_1 = 4$ yr, $T_{v1} = 0.056$

$$t_4 = 25 \text{ yr}, T_{v4} = T_{v1} \times \frac{t_4}{t_1} = 0.35$$

From Table 14.6, $U_4 = 0.66$
$$S_{t4} = U_4 \times S_t = 198 \text{ mm}$$

14.9 A 6 m thick layer of saturated clay is sandwiched between pervious sand at top and rock at bottom. For the clay, $m_v = 2.5 \times 10^{-4}$ m²/kN and $c_v = 1.24$ m²/yr. Find the final settlement if the clay is subjected to an increase of uniform pressure of intensity 80 kPa. Find settlement at the end of 3 years.

$$S_t = m_v \ H_o \ \Delta\sigma' = 0.12 \text{ m} = 120 \text{ mm}$$
$$T_v = c_v \ t/d^2 = 0.1033$$
$$U = 1.13\sqrt{T_v} = 0.363$$
$$S_t = U \ S_t = 44 \text{ mm}$$

14.10 If a very thin layer of free draining sand existed at 1.6 m above the bottom of clay layer in Ex. 14.9, what would be the final settlement and the 3 year settlement? Determine also the individual settlement of the sublayers of clay above and below the thin sand layer.

Ignoring the thickness of sand, S_f will still be 120 mm (Ex. 14.9), but for drainage conditions, there is now a 4.4 m thick sub-layer with double drainage and a 1.6 m thick sub-layer with single drainage.

$$d_1 \ (H_1 = 4.4 \ m) = 2.2 \ m$$
$$d_2 \ (H_2 = 1.6 \ m) = 1.6 \ m$$

$$T_{v1} = \frac{c_v t}{d_1{}^2} = 0.768, \quad U_1 \simeq 88\%, \ t = 3 \ yr$$

$$T_{v2} = \frac{c_v t}{d_2{}^2} = 1.453, \quad U_2 \simeq 97\%$$

$$U_a \ H = U_1 \ H_1 + U_2 \ H_2, \quad H = 6 \ m$$
$$U_a = 90.4\% \ (\text{average U})$$
$$S_t \ (\text{at } t = 3 \ yr) = 0.904 \ S_f = 108 \ mm$$

Since $S_f \propto H$

$$S_{f1} \ (\text{top sublayer}) = 120 \ \frac{4.4}{6} = 88 \ mm$$

$$S_{f2} \ (\text{bottom sublayer}) = 32 \ mm$$
$$S_{t1} \ (\text{top sublayer @ 3 yr}) = U_1 \ S_{f1} = 77 \ mm$$
$$S_{t2} \ (\text{bottom sublayer @ 3 yr}) = U_2 \ S_{f2} = 31 \ mm$$

14.11 A 6 m thick clay deposit, sandwiched between coarse sand, has $c_{v1} = 1.5 \ m^2/yr$ in its upper 3 m thickness and $c_{v2} = 24 \ m^2/yr$ in the lower portion. What will be the percentage consolidation at the end of 6 months if it is subjected to structural loading?

As c_v is different for the upper and lower portions, each 3 m thick, find the equivalent layer thickness in terms of c_{v1}.

$$H_2' = H_2 \ (c_{v1}/c_{v2})^{1/2} = 3 \ (1.5/24)^{1/2} = 0.75 \ m$$
$$H_t' = H_1 + H_2' = 3.75 \ m, \text{ having } c_{v1} = 1.5 \ m^2/yr$$
$$d = 3.75/2 \ m$$
$$T_v = c_v t/d^2 = 0.213, \ t = 0.5 \ yr$$
$$U = 1.13 \ \sqrt{T_v} = 0.52$$

14.12 A square raft, 10 m × 10 m, placed at a depth of 2 m in a sandy soil carries a load of intensity 110 kPa. The unit weight of sand is 20 kN/m³. A 6 m thick saturated clay layer exists at a depth of 3.5 m below the raft, for which $m_v = 0.2 \ m^2/MN$ and

$c_v = 1.1 \ m^2/yr$. The clay layer is under-lain by sand. Assuming the raft pressure to be uniformly distributed, determine the final settlement under the centre of the raft due to consolidation of clay. Also determine the times required for 50% and 90% of the ultimate settlement. Do the computations considering (a) total thickness of clay as one layer, and (b) total thickness divided into 2 sublayers, each 3 m thick.

Net raft pressure $q_n = 110 - 2 \times 20 = 70$ kPa

(a) Single layer.

Determine $\triangle \sigma'$ at middle of layer, i.e. at $z = 3.5 + 3$
$= 6.5$ m below raft.

$z/B = 0.65$

From Fig. 13.7, $\dfrac{\triangle \sigma'}{q_n} = 0.57, \triangle \sigma' = 39.9$ kPa

$S_t = m_v \ H_o \ \triangle \sigma' = 47.88 \simeq 48$ mm

(b) Two sublayers.

Upper layer: $z_1/B = 5/10 = 0.5$

$\quad\quad\quad \triangle \sigma_1'/q_n = 0.7, \ \triangle \sigma_1' = 49$ kPa

Lower layer: $z_2/B = 0.8$

$\quad\quad\quad \triangle \sigma_2'/q_n = 0.46, \ \triangle \sigma_2' = 32.2$ kPa

$S_t = m_v \ H_1 \ \triangle \sigma_1' + m_v \ H_2 \ \triangle \sigma_2'$

$\quad = 29.4 + 19.32 \simeq 49$ mm

Rate of consolidation depends upon the drainage path ($d = 3$ m) and is independent of the sublayers in which the total thickness is divided.

$(T_v)_{50} = 0.197, (T_v)_{90} = 0.848, \ T_v = c_v t/d^2$

$t_{50} = 1.61$ yr, $t_{90} = 6.94$ yr

14.13 A clay layer, 5 m thick, is subjected to a pressure of 50 kPa. What is the coefficient of consolidation, if the layer undergoes 50% consolidation with double drainage in one year? If k is 0.025 m/yr, how much settlement has taken place at the end of one year? Determine the rate of flow of water per unit area of the surface of clay at this time.

(d) $d = 2.5$ m, $(T_v)_{50} = 0.197$

$\quad c_v = T_v \ d^2/t = 1.23 \ m^2/yr$

(b) $m_v = k/c_v \ \gamma_w = 2.074 \times 10^{-3} \ m^2/kN$

$S_f = m_v \ H_o \ \triangle\sigma' = 0.518 \ m$

$S_t \ (\text{at } U = 50\%) = 0.5 \ S_f = 0.259 \ m$

(c) For $U < 60\%$, the time-settlement relation is parabolic of the form:

$$t = AS^2$$

When $t = 1$ yr, $S = 0.259$ m

Hence $A = 1/(0.259)^2 = 14.91$

Equation of parabola is:

$$t = 14.91 \ S^2$$

Settlement rate $\dfrac{dS}{dt} = \dfrac{1}{2 \times 14.91 \ S}$

Settlement rate @ 1 yr is:

$$\frac{dS}{dt} = \frac{1}{2 \times 14.91 \times 0.259} = 0.1295 \ m/yr$$

As the settlement of clay is due to flow of water out of it, q per unit surface area of clay having double drainage is given by:

$$q = 0.1295/2 = 0.0647 \ m^3/yr/m^2$$

14.12 CONSTRUCTION PERIOD SETTLEMENT

In practice, structural loads are applied to the soil over a period of time and not instantaneously, as is the load on the laboratory specimen. For short duration, routine construction, it is sufficiently accurate to assume that the entire foundation load is applied halfway through the construction period. For large constructions spread over several years, it is sometimes useful to know the settlement that takes place by the end of construction, the problem being of consolidation due to increasing load. An empirical method of correction for the estimated settlement due to gradually increasing load is shown in Fig. 14.10.

The effective construction period is measured fromt he time when the net load is zero, i.e. when additional loading equals the weight of soil excavated. It is assumed that the net load increases uniformly to its full value P' over the time t_o and the settlement at time t_o is the same as if the load P' had been acting as a constant load for the period $t_o/2$. Thus the settlement at any time t during

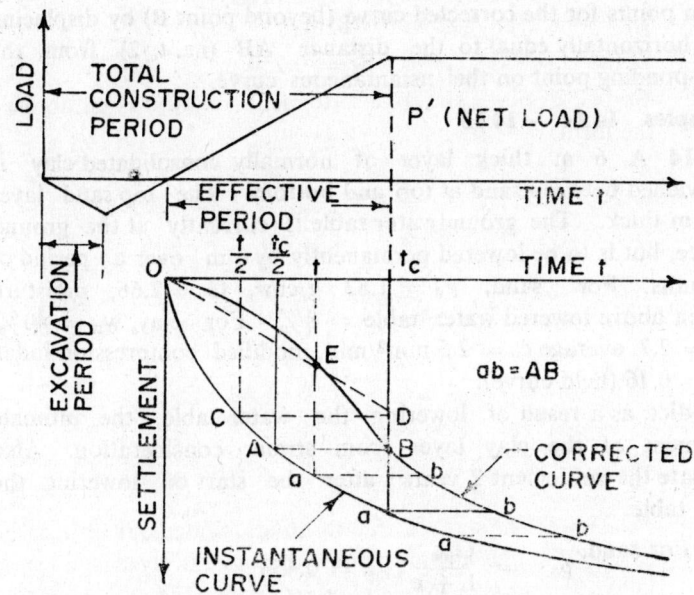

Fig. 14.10 Correction for gradual loading during construction

construction is equal to that occurring for instantaneous loading at $t/2$. However, since total load will not be acting at $t < t_e$, the value of settlement so obtained must be reduced in the proportion of that load to the total load. The graphical construction (Fig. 14.10) will be as follows:

(1) Plot a settlement/time curve showing the settlement that would occur if the load P' had been applied instantaneously (instantaneous curve).

(2) Obtain point A on the instantaneous curve corresponding to $t_e/2$ and project it horizontally to get point B on a vertical line through t_e. Point B represents the corrected settlement at t_e.

(3) To obtain other points on the corrected curve, select a time t. Determine the settlement on the instantaneous curve for $t/2$ (point C). Draw a horizontal from C to meet the vertical line from t_e at D. Join OD. Where OD cuts the vertical line from t gives the point E on the corrected curve. Repeat the procedure for different values of t to establish the corrected curve.

(4) For the period subsequent to the completion of construction, obtain points for the corrected curve (beyond point B) by displacing them horizontally equal to the distance AB (i.e. $t_o/2$) from the corresponding point on thei nstantaneous curve.

Examples 14.14 — 14.15

14.14 A 6 m thick layer of normally consolidated clay is sandwiched between sand at top and bottom. The top sand layer is 12 m thick. The groundwater table is currently at the ground surface, but is to be lowered permanently by 4 m over a period of 6 months. For sand, $\rho_d = 1.83$ g/cm³, $G = 2.66$, moisture content above lowered water table = 8%. For clay, $w_n = 40\%$, $G = 2.7$, average $c_v = 2.5$ mm²/min, modified compression index $C_c\epsilon = 0.16$ (field curve).

Predict, as a result of lowering the water table, the ultimate settlement of the clay layer from strain consideration. Also calculate the settlement 2 years after the start of lowering the water table.

(a) For sand, $\qquad \rho_d = \dfrac{G\rho_w}{1+e}$, $e = 0.453$

$$\rho_{sat} = \frac{G+e}{1+e} \rho_w = 2.14 \text{ g/cm}^3$$

ρ (above W.T.) $= \rho_d (1 + w) = 1.976$ g/cm³

For clay, $\qquad e_o = w_a G = 1.08$, $\rho_{sat} = 1.82$ g/cm³

σ_o' before lowering of W.T. at middle of clay = 158.17 kPa

σ_t' after lowering of W.T. = 190.94 kPa

Strain ϵ at middle of clay $= C_c\epsilon \log \dfrac{\sigma_f'}{\sigma_0'} = 0.01308$

$S_t = \epsilon H_o = 78$ mm

Alternatively, $C_c = C_c\epsilon \times (1 + e_0) = 0.3328$

$$S_t = \frac{C_o}{1 + e_o} H_o \log \frac{\sigma_t'}{\sigma_o'} = 78.5 \text{ mm}$$

(b) Assume start of consolidation at half the lowering time.

$t = 2 - \dfrac{1}{4} = 1.75$ yr

$c_v = 2.5 \dfrac{60 \times 24 \times 365}{10^6} = 1.314$ m²/yr

$d = 3$ m, $T_v = 0.2555$, $U \simeq 56.8\%$

$S_t = U S_t = 44.5$ mm

14.15 A 5 m thick clay layer has 6 m of sand deposit on top and rock at bottom. The depth of water table in sand is 2 m; ρ (sand) above W.T. $= 1.75$ g/cm^3 and ρ_{sat} (sand) $= 2.03$ g/cm^3. Clay is normally consolidated; $c_v = 1.4$ m^2/yr, $\rho_{sat} = 1.94$ g/cm^3, $w_L = 45\%$, $w_a = 30\%$, $G = 2.7$. Over a period of one year a fill of density 1.9 g/cm^3 will be dumped on the surface of sand over an extensive area upto a height of 1.0 m. Calculate final consolidation settlement and settlements after 2 and 4 years from the start of dumping.

(a) σ_o' at middle of clay $= 97.7$ kPa

$\triangle \sigma = 3 \times 1.9 \times 9.8 = 55.86$ kPa

$e_o = w_a G = 0.81$

$C_c = 0.009 (w_L - 10) = 0.315$

$S_f = \dfrac{C_o}{1 + e_o} H_o \log \dfrac{\sigma_f'}{\sigma_o'} = 171$ mm

(b) Assume that settlement commences at half the dumping period.

$t_1 = 1.5$ yr, $t_2 = 3.5$ yr, d $= 5$ m

$T_{v1} = 0.084, T_{v2} = 0.196, T_v = \dfrac{c_v t}{d^2}$

$U_1 = 0.327, U_2 = 0.5, U = 1.13 \sqrt{T_v}$
$S_1 = 56$ mm, $S_2 = 85$ mm

14.13 SECONDARY COMPRESSION

Secondary compression is different from primary consolidation that it occurs at a constant effective pressure, i.e. after essentially all the excess pore pressure has dissipated. To describe the magnitude and rate of secondary compression the following terms are introduced (Raymond and Wahls 1976, Mesri and Godlewski 1977, Holtz and Kovacs 1981).

(a) Secondary compression index $C\alpha$:

$$C\alpha = \frac{\triangle e}{\triangle \log t} = \frac{\triangle e}{\log (t_2/t_1)} \qquad (14.70)$$

where $\triangle e =$ void ratio change between times t_1 and t_2 along a part of e/log t curve (after completion of primary consolidation)

$\triangle t =$ time between t_2 and t_1

Eq 14.70 is analogous to Eq. 14.10 ($C_c = \triangle e/\triangle \log \sigma'$) which defines the primary compression index.

(b) Modified secondary compression index $C\alpha\epsilon$:

$$C\alpha\epsilon = \frac{C\alpha}{1 + e_p} \qquad (14.71)$$

where e_p = void ratio at the start of the *linear part* of the e/log t curve. (The insitu void ratio e_0 may also be used with no appreciable loss of accuracy).

The index $C\alpha\epsilon$ is also called "rate of secondary compression" and defined as (Ladd et al 1977):

$$C\alpha\epsilon = \frac{\triangle \epsilon}{\triangle \log t} = \frac{\triangle H/H}{\triangle \log t} \qquad (14.72)$$

The indices $C\alpha$ and $C\alpha\epsilon$ may be determined from the straight line portion of dial reading R versus log t curve *which occurs after the completion of primary consolidation*. The dial change $\triangle R$ is usually measured over one log cycle of time. The corresponding void ratio change $\triangle e$ (or strain change $\triangle \epsilon$) is then calculated.

The magnitude of secondary compression in a given time is generally greater in NC clays than in OC clays. Secondary compression effects are often neglected for inorganic clays; but they are of much significance in the settlement of certain highly plastic clays and organic soils and they increase as the applied load increases. The average value of $C\alpha/C_c$ for a wide variety of natural soils is about 0.05 (not exceeding 0.1). Typical range of $C\alpha/C_c$ for inorganic soils is 0.025 to 0.06 and for organic clays and silts 0.035 to 0.06 (Mesri and Godlewski 1977). Typical values of the modified index $C\alpha\epsilon$ are as follows (Ladd et al 1977, Lambe and Whitman 1979): NC clays 0.005 to 0.02, OC clays (OCR > 2) less than 0.001, very plastic clays and organic clays 0.03 or higher.

The secondary settlement S_s can be calculated by substituting $\triangle e$ from Eq. 14.70 into Eq. 14.52 and using e_p for e_0.

$$S_s = \frac{C\alpha}{1 + e_p} H_o \log (t_1/t_p) \qquad (14.73)$$

The secondary settlement may also be computed in terms of modified secondary compression index $C\alpha\epsilon$:

$$S_s = C\alpha\epsilon \, H_o \log (t_1/t_p) \qquad (14.74)$$

where t_p is the time for the completion of primary consolidation (years) which may be taken corresponding to 90% U and t_1 is the service life of the proposed structure (years) or any desired period (t_1 is measured after the application of structural load). For preliminary estimates, $C\alpha$ may be determined from an average $C\alpha/C_c$ value of 0.05; the modified index $C\alpha\epsilon$ may be adopted as 0.02 or may be based on the empirical relationship established by Mesri (1973) between $C\alpha\epsilon$ and the natural moisture content, as shown in Fig. 14.11

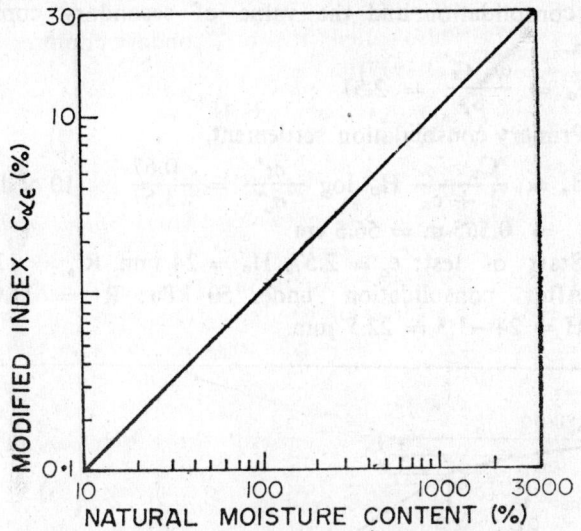

Fig. 14.11   Modified secondary compression index versus natural moisture content (Mesri 1973)

Example 14.16

A 10 m thick layer of saturated marine clay has the following properties: $w_L = 85\%$, $w_p = 25\%$, $w_n = 95\%$, $S_r = 99.8\%$, $G = 2.7$, $C_c = 0.67$. The existing overburden pressure is 50 kPa which will be increased to 100 kPa due to structural loading. A consolidation test was performed on an undisturbed specimen of clay, 24 mm high and 50 cm² in sectional area, by setting the dial gauge to read initially 12.50 mm at zero load. After consolidation under 50 kPa the dial reading was 11.00 mm. The pressure was then increased to 100 kPa and the time versus dial readings are given in Table 14.7. Compute for the pressure increment of 50 to 100 kPa the following parameters :(a) final consolidation (primary) settlement of the clay, (b) coefficient of consolidation, (c) time for 90% primary consolidation of the clay layer in the field, (d) secondary settlement of the clay layer during the service life t_1 of the structure assumed as 100 years or 200 years after completion of construction. Assume that the time rate of deformation for the pressure range in the test approximates that occurring in the field. Determine also e_p the void ratio of specimen after completion of primary consolidation, and the value of secondary compression index $C\alpha$.

(a) $e_0 = \dfrac{w_n\,G}{S_r} = 2.57$

Primary consolidation settlement,

$$S_c = \frac{C_c}{1 + e_0}\,H_0\,\log\frac{\sigma_1'}{\sigma_0'} = \frac{0.67}{3.57} \times 10 \times \log\frac{100}{50}$$
$$= 0.565 \text{ m} = 56.5 \text{ cm}$$

(b)   Start of test: $e_0 = 2.57$, $H_0 = 24$ mm. $R'_0 = 12.50$ mm
        After consolidation under 50 kPa: $R = 11.00$ mm, $H = 24 - 1.5 = 22.5$ mm

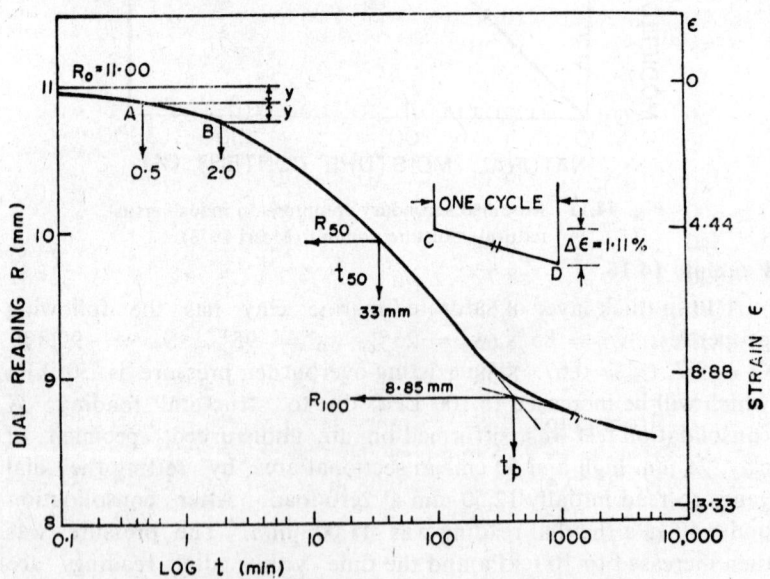

Fig. 14.12   Compression curve for example 14.16

Time, dial reading, dial change ΔH and strain ϵ for the pressure range 50 kPa to 100 kPa on the specimen are given in Table 14.7. The test date are plotted in Fig. 14.12 to give a compression curve of dial reading versus log t (or strain versus log t).

Table 14.7
EXAMPLE 14.16

Time	R	$\triangle H$	$\epsilon = \dfrac{\Delta H}{22.5} \times 100$
(min)	(mm)	(mm)	(%)
0	11.00	0.00	0
0.1	10.95	0.05	0.22
0.25	10.92	0.08	0.35
0.5	10.88	0.12	0.53
1.0	10.82	0.18	0.80
2.0	10.74	0.26	1.15
3.0	10.66	0.34	1.51
5.0	10.51	0.49	2.17
10	10.35	0.65	2.88
15	10.20	0.80	3.55
30	9.98	1.02	4.53
60	9.72	1.28	5.68
100	9.52	1.48	6.57
180	9.30	1.70	7.55
300	9.05	1.95	8.66
540	8.88	2.12	9.42
1320	8.72	2.28	10.13
1800	8.68	2.32	10.31
2850	8.65	2.35	10.44
4320	8.58	2.42	10.75

From graph, $t_{50} = 33$ min, $(T_v)_{50} = 0.197$

H @ $t_o = 22.5$ mm (specimen height)

$H_t = 22.5 - 2.42 = 20.08$ m

$d = \dfrac{22.5 + 20.08}{4} = 10.65$ mm

$c_v = \dfrac{(T_v)_{50} \, d^2}{t_{50}} = 0.677$ mm^2/min $= 0.356$ m^2/yr

(c)   d in the field $= 10/2 = 5$ m

For 90% U, $T_v = 0.848$

$$t_{90} = \frac{0.848 \times 5^2}{0.356} = 59.5, \text{ say 60 years}$$

(d)   Dial change $\triangle R$ for one log cycle after the completion of primary consolidation $= 0.25$ mm

$$\triangle \epsilon \text{ for 1 log cycle} = \frac{0.25}{22.5} = 1.11\%$$

$\triangle \epsilon$ for 1 log cycle can also be read directly from the strain versus long t curve

Hence $C\alpha\epsilon = 0.0111$ (Eq. 14.72)

Check from Mesri relationship (Fig. 14.11),

For $w_n = 90\%$, $C\alpha\epsilon = 0.95 \% = 0.0095$

Assume that the time for completion of primary consolidation corresponds to 90% U. Thus t_p in the field $= 60$ years. Secondary settlements are to be found from $t_p = 60$ yr to $t_1 = 100$ yr and from $t_p = 60$ yr to $t_1 = 200$ yr.

$$S_{s1} = C\alpha\epsilon \, H_o \log \frac{t_1}{t_p} = 0.0111 \times 10 \times \log \frac{100}{60}$$

$$= 0.0246 \text{ m} = 2.5 \text{ cm (for 100 yr)}$$

$$S_{s2} = 0.0111 \times 10 \times \log \frac{200}{60} = 5.8 \text{ cm (for 200 yr)}$$

(e)   In a phase diagram of soil element, the void ratio e, the total height H and height of solids H_s are related as:

$$H_s/H = 1/(1 + e)$$

$$H_s = \frac{H_o}{1 + e_o} = \frac{24}{1 + 2.57} = 6.723 \text{ mm}$$

$$e = \frac{H_v}{H_s} = \frac{H_o - H_s}{H_s} = \frac{H_o - (R_o' - R) - H_s}{H_s}$$

$$= \frac{(H_o - H_s) - (R_o' - R)}{H_s} \; ; \; \begin{array}{l} \text{R after consolidation} \\ \text{under 50 kPa} = 11.00 \text{ mm} \end{array}$$

$$= \frac{(24 - 6.723) - (12.50 - 11.00)}{6.723} = 2.347$$

Void ratio e when 100 kPa is applied $= 2.347$

e_p when $R_{100} = 8.85$ mm (complection of primary consolidation)

$$= \frac{(22.5 - 6.723) - (11.00 - 8.85)}{6.723} = 2.027$$

$C\alpha = C\alpha\epsilon (1 + e_p) = 0.0111 (1 + 2.027) = 0.0336$

If e_o is used in place of e_p,

$C\alpha \simeq 0.0111 (1 + 2.57) = 0.0396$

Check the value of $C\alpha$ from the approximate relation:

$C\alpha/C_o = 0.05$

$C\alpha = 0.05 \times 0.67 = 0.0335$

Check C_o from the approximate relation:

$C_o = 0.009 (w_L - 10) = 0.009 (85 - 10) = 0.756$

14.14 VERTICAL DRAINS AND THREE DIMENSIONAL CONSOLIDATION

1. **Purpose of vertical drains.** The slow rate of consolidation of a loaded cohesive soil can be accelerated by providing horizontal drainage in addition to the normal vertical drainage by means of installing vertical drains, commonly known as "sand drains". Sand drains are vertical columns of sand or other pervious material inserted through a compressible stratum. They hasten the process of consolidation by decreasing the drainage path and quite often taking advantage of a higher coefficient of consolidation in the horizontal direction. Sand drains, providing supplementary outlets for the expelled pore water, also accelerate the increase in shear strength of the consolidating soil. A typical application of the sand drains is in the construction of highway and airport embankments over compressible soils to ensure that most of the settlements will occur during, and not after, construction. Sand drains are also used under some circumstances to drain and to control land slides (Jumikis 1965). They have also been used in the subsoil supporting foundations of buildings and other civil engineering structures (Dastidar 1985).

2. **Description and installation.** The diameter of sand drains varies in practice from 300 to 600 mm and the spacings vary from 1.5 to 4.5 m. To be effective, the spacing of sand drains should be less than the thickness of the consolidating layer. The depth of sand drains is governed by the subsoil conditions, i.e. by the depth of the firm layer below the ground surface. Sand drains have been installed to depths of upto 45 m.

A sand drain is usually installed by driving boreholes through the clay layer and backfilling with a suitably graded sand. The sand should permit efficient drainage of water without permitting fine soil particles to be washed in. A typical arrangement of sand drains under an embankment is shown in Fig. 14.13. Sometimes the sand drains are allowed to puncture an impervious layer if it is underlain by pervious layer. This creates two-way vertical drainage as well as lateral, which further hastens consolidation. After installation, a blanket of gravel and sand, 0.3 to 1.0 m thick, is spread over the entire sand-drain area to provide lateral drainage at the base of the fill.

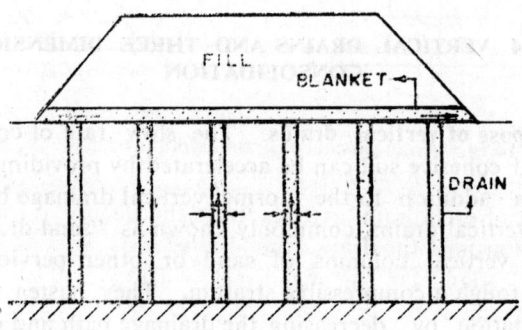

Fig. 14.13 Vertical drains

The permanent fill is placed on top of the drainage blanket. Often an overfill or overload fill is placed as an extra fill over the permanent fill to accelerate consolidation, which is later on removed. Fills as high as 15 to 20 m have been stabilized by vertical drains (Jumikis 1965). When used for supporting foundations of buildings, a preload is used for acceleration of consolidation.

Apart from backfilled sand drains, pre-fabricated drains are also used and they are generally more economical. The "fabric drains" are made from various types of artificial fabric and plastic materials. They are usually 100 mm wide and 3 to 5 mm thick and are inserted in the ground by special mandrel and equipment. A quite efficient, 'preformed' drain is the "drainage wick" (trade name *sandwick*) developed by Dastidar (1985). It consists of a cylindrical bag, standard diameter 65 mm, made of either jute fabric or any pervious fabric or mat and filled with sand. Another type is the 'rope drain' developed by the CBRI, Roorkee, which is made by rolling coir mat in the shape of a pipe.

Vertical drains are laid out in either (a) square or (b) triangular patterns, as shown in Fig. 14.14. A triangular pattern is more economical (Barron 1947).

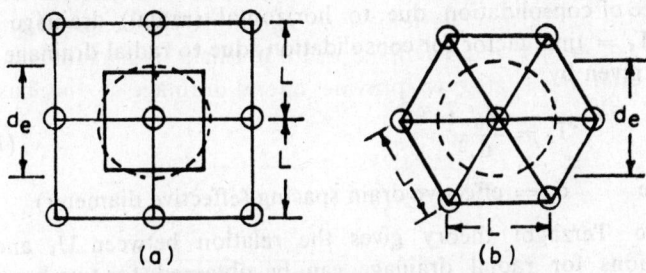

Fig. 14.14. Arrangements of drain layout (a) square pattern (b) triangular pattern

The effect of each drain extends upto the surface of a cylindrical block of soil around the drain, the diameter d_e of which differs for square and triangular arrangements. The diameter d_e is the diameter of the zone of influence of a drain and is called the "effective (or equivalent) drain spacing (or diameter)".

For a square system, the area of square enclosed by the grid is L^2, where L is the actual drain spacing. Equating it to the area of an equivalent circle of diameter d_e (i.e. $\pi d_e^2/4 = L^2$), $d_e = 1.13$ L. In a triangular system, a hexagon is formed by bisecting the various grid lines joining adjacent drains. The total area of the hexagon is 0.865 L^2 from which the diameter of the equivalent circle is $d_e = 1.05$ L.

3. Theory of vertical drains. The theory of functioning of a vertical drain is based on the general theory of 'three-dimensional consolidation' of soil, which is described by the following partial differential equation in the cylindrical (polar) coordinate system as:

$$\frac{du}{dt} = c_h \left(\frac{d^2u}{dr^2} + \frac{1}{r} \frac{du}{dr} \right) + c_v \frac{d^2u}{du^2} \qquad (14.75)$$

where r = radial cylindrical coordinate, z = axial cylindrical coordinate, c_h = coefficient of consolidation for horizontal drainage, and c_v = coefficient of consolidation for vertical drainage. The solution to Eq. 14.75 can be written in two parts:

$$U_v = f(T_v) \text{ and } U_r = f(T_r)$$

where U_v = average degree of consolidation due to vertical drainage only, T_v = vertical time factor (Eq. 14.35), U_r = average degree of consolidation due to horizontal (radial) drainage only, and T_r = time factor for consolidation due to radial drainage only. T_r is given by:

$$T_r = \frac{c_h \cdot t}{d_e^2} \tag{14.76}$$

where d_e = effective drain spacing (effective diameter)

The Terzaghi theory gives the relation between U_v and T_v. Solutions for radial drainage can be obtained for two boundary conditions: (i) uniform vertical surcharge on the ground surface ('free strain' condition), and (ii) uniform vertical deformation of the surface ('equal strain' condition). The solutions for these two conditions, in fact, give similar results, but that for uniform vertical strain is simpler. The U_r for this uniform strain condition is expressed as:

$$U_r = 1 - \exp\left[-\frac{8T_r}{F(n)}\right] \tag{14.77}$$

where $F_a = \left(\frac{n^2}{n^2 - 1}\right)\log_e n - \left(\frac{3n^2 - 1}{4n^2}\right) \tag{14.78}$

$$n = \frac{d_e}{d_w} = \text{ratio of effective drain spacing to drain}$$
$$\text{diameter} \tag{14.79}$$

Superimposing the solutions for vertical drainage and radial drainage, the resultant average degree of three dimensional consolidation U_T can be expressed as:

$$U_T = 1 - (1 - U_v)(1 - U_r) \tag{14.80}$$

The solution for radial drainage, due to Barron (1947), giving U_r/T_r relationship is shown in Fig. 14.15. The dotted curve gives the relation between U_v and T_v. Typical values of T_r are given in Table 14.8.

Table 14.8

VALUES OF T_r FOR RADIAL DRAINAGE—EQUAL VERTICAL STRAIN CONDITION (LEONARD 1962)

Average degree of consolidation U_r (%)	$n = d_e/d_w$										
	5	10	15	20	25	30	40	50	60	80	100
0	0	0	0	0	0	0	0	0	0	0	0
5	0.006	0.010	0.013	0.014	0.016	0.017	0.019	0.020	0.021	0.023	0.025
10	0.012	0.021	0.026	0.030	0.032	0.035	0.039	0.042	0.044	0.048	0.051
15	0.019	0.032	0.040	0.046	0.050	0.054	0.060	0.064	0.068	0.074	0.079
20	0.026	0.044	0.055	0.063	0.069	0.074	0.082	0.088	0.092	0.101	0.107
25	0.034	0.057	0.071	0.081	0.089	0.096	0.106	0.114	0.120	0.131	0.139
30	0.042	0.070	0.088	0.101	0.110	0.118	0.131	0.141	0.149	0.162	0.172
35	0.050	0.085	0.106	0.121	0.133	0.143	0.158	0.170	0.180	0.196	0.208
40	0.060	0.101	0.125	0.144	0.158	0.170	0.188	0.202	0.214	0.232	0.246
45	0.070	0.118	0.147	0.169	0.185	0.198	0.220	0.236	0.250	0.270	0.288
50	0.081	0.137	0.170	0.195	0.214	0.230	0.255	0.274	0.290	0.315	0.334
55	0.094	0.157	0.197	0.225	0.247	0.265	0.294	0.316	0.334	0.363	0.385
60	0.107	0.180	0.226	0.258	0.283	0.304	0.337	0.362	0.383	0.416	0.441
65	0.123	0.207	0.259	0.296	0.325	0.348	0.386	0.415	0.439	0.477	0.506
70	0.137	0.231	0.289	0.330	0.362	0.389	0.431	0.463	0.490	0.532	0.564
75	0.162	0.273	0.342	0.391	0.429	0.460	0.510	0.548	0.579	0.629	0.668
80	0.188	0.317	0.397	0.453	0.498	0.534	0.592	0.636	0.673	0.730	0.775
85	0.222	0.373	0.467	0.534	0.587	0.629	0.697	0.750	0.793	0.861	0.914
90	0.270	0.455	0.567	0.649	0.712	0.764	0.847	0.911	0.963	1.046	1.110
95	0.351	0.590	0.738	0.844	0.926	0.994	1.102	1.185	1.253	1.360	1.444
99	0.539	0.907	1.135	1.298	1.423	1.528	1.693	1.821	1.925	2.091	2.219
100	∞	∞	∞	∞	∞	∞	∞	∞	∞	∞	∞

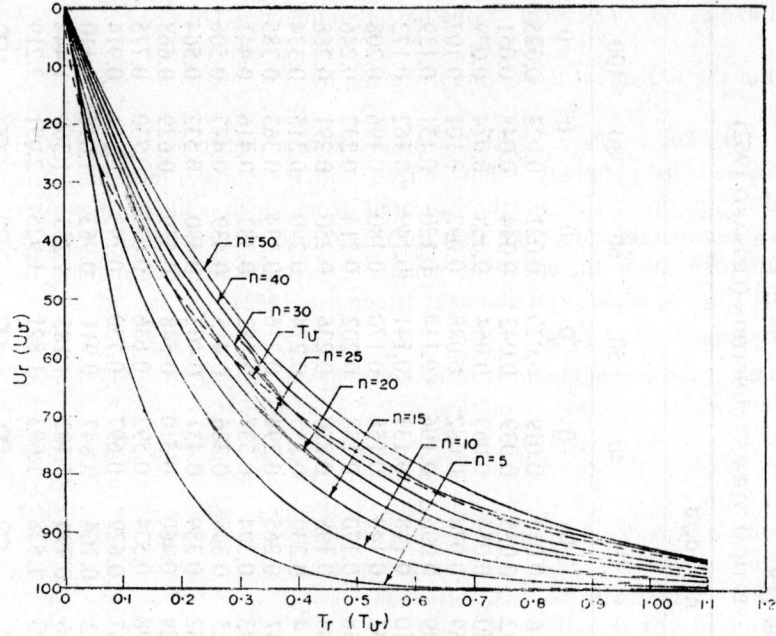

Fig. 14.15 Barron's solution for radial consolidation
(plotted by the author)

(a) *Coefficient of consolidation for horizontal drainage* c_h. The coefficient c_h is given by:

$$c_h = \frac{k_h}{m_v \gamma_w} \qquad (14.81)$$

From Eqs. 14.31 and 14.81,

$$\frac{c_h}{c_v} = \frac{k_h}{k_v} \qquad (14.82)$$

where k_h and k_v are the coefficients of permeability in the horizontal and vertical directions respectively.

To determine c_h, the ratio k_h/k_v is generally assumed and c_h is estimated from c_v (Eq. 14.82) which is determined from one-dimensional consolidation test.

(b) *Settlement calculations*. The final settlement due to consolidation of the compressible soil layer is calculated using the normal procedures described earlier (Section 14.13), assuming that the presence of vertical drains has no influence. Thus,

$$s_t = m_v \, H_o \, \triangle \sigma' \qquad (14.54)$$

But the settlement S_t at time t is given by:

$$S_t = S_f \, U_T \qquad (14.83)$$

(c) *Smear effect.* Installation of a drain disturbs a zone of soil around its periphery. This effect, called *smear*, modifies the drainage properties of the smeared zone which is likely to exhibit k_h considerably less than that for the undisturbed material. As an approximation, the effect of smear may be taken as equivalent to halving the actual well diameter (Leonards 1962).

Examples 14.17-14.20

14.17· An embankment is to be constructed over a 10 m thick clay layer overlying rock. Sand drains, 0.4 m in diameter, are installed in the clay layer in a square pattern at a centre to centre (c/c) spacing of 3.54 m. For the clay, $c_v = 4$ m²/yr, $c_h = 8$ m²/yr $m_v = 0.25$ m²/mN. Determine the settlement of the clay layer in 6 months time commencing from the middle of the construction period, if the embankment increases the mean effective vertical pressure by 80 kPa. What will be the settlement if (a) the drain spacing is doubled, (b) the drain diameter is doubled, keeping the spacing unchanged.

(a) $d_w = 0.4$ m, L = 3.54 m, $d_e = 1.13$ L = 4 m

   n $= d_e/d_w = 10$

   For vertical drainage: d = H = 10 m

   $T_v = \dfrac{c_v \, t}{d^2} = 0.02$, $U_v = 0.1596$

   For radial drainage: $T_r = \dfrac{c_h t}{d_e{}^2} = 0.25$

   For n = 10 and $T_r = 0.25$, $U_r = 0.725\%$

   (The value of U_r can be calcula'ed from Eq. 14.77 or read from Fig. 14 15 or Table 14.8)

   $U_T = 1 - (1 - U_v) (1 - U_r) = 0.769$

   Final settlement $S_f = m_v H_o \wedge \sigma' = 200$ m

   $S_t = S_f \, U_T = 154$ mm

(b)   Drain spacing doubled.

   L = 7.08 m, $d_e = 8$ m, n = 20, $T_r = 0.0625$

   $U_r \simeq 20\%$, $U_v = 15.96\%$ (unchanged)

$$U_T = 1 - (1 - U_v)(1 - U_r) = 0.328$$
$$S_t = S_f U_T = 66 \text{ mm}$$

(c) Drain diameter doubled.

$$L = 3.54 \text{ m}, d_e = 4 \text{ m}, d_w = 0.8 \text{ m}, n = 5$$
$$T_r = 0.25, U_r \simeq 88\%$$
$$U_T = 0.899$$
$$S_t = 180 \text{ mm}$$

14.18 Drainage wicks, 6.5 cm in diameter, are provided at 1.9 m c/c in a triangular pattern in a clay layer for which $c_v = 2.4 \times 10^{-3}$ cm^2/s and $k_h = 5 k_v$. Find the time required for 90% of the radial consolidation to occur.

$$d_w = 0.065 \text{ m}, L = 1.9 \text{ m}, d_e = 1.05 L = 1.995 \text{ m}$$
$$n = d_e/d_w = 30.7$$
$$U_r = 90\%, T_r = 0.77$$
$$\frac{c_h}{c_v} = \frac{k_h}{k_v}, \quad c_h = 1.2 \times 10^{-6} \text{ m}^2/\text{s}$$
$$T_r = \frac{c_h t}{d_e{}^2}, \quad t = 29.5 \text{ days, say 30 days}$$

14 19 A compressible clay layer overlying rock is 8 m thick and has the following properties: $m_v = 3.8 \times 10^{-4}$ m^2/kN, $c_v = c_h = 4.2$ m^2/yr. An embankment which is to carry a road, will be constructed over the clay in 4 months time and this will result in an increase of average pressure by 75 kPa. The road pavement will be placed on top of the embankment one year after the start of construction and the maximum allowable settlement after this is to be limited to 25 mm. Design a suitable sand drain system.

Final settlement $S_f = m_v H_o \Delta \sigma' = 228$ mm

Minimum settlement that must occur before the construction of road pavement $= 228 - 25 = 203$ mm $= S_t$

U_T (required) $S_t/S_f = 0.89$

Assuming that settlement commences at the middle of the construction period for the embankment, time to reach $U_T = 0.89$ is 10 months.

$$T_v = \frac{c_v t}{d^2} = \frac{4.2 \times 10}{(8)^2 \times 12} = 0.055$$
$$U_v = 1.13 \sqrt{T_v} = 0.26$$
$$(1 - U_T) = (1 - U_v)(1 - U_r)$$
$$U_r \text{ (required)} \geq 0.85$$

Try 300 m drain at L = 2.84 m in a triangular pattern

$$d_e = 1.05\ L = 2.99\ m \simeq 3\ m$$

$$\frac{d_e}{d_w} = n = 10$$

$$T_r = \frac{c_h t}{d_e{}^2} = 0.389$$

For $T_r = 0.389$ and $n = 10$, $U_T = 86\%$ which is greater than required $U_T\ (85\%)$.

In practice, a number of possible layouts with different diameters and spacings are investigated and the solution adopted would be based both on technical and economical grounds. An alternative method is given in the next example.

14.20 Determine the spacing, in triangular pattern, of 65 mm diameter sand wicks for installation in a 10 m thick clay layer underlain by an impervious hard stratum such that the clay layer may undergo 91% consolidation in 6 months after being loaded. The clay has the following properties: $c_v = 2.5\ m^2/yr$ and $k_h/k_v = 2$.

$$c_h = 2\ c_v = 5\ m^2/yr$$

$$d_e/d_w = n,\ d_e = 0.065\ m$$

$$T_v = \frac{c_v\ t}{d^2} = 0.0125$$

$$U_v = 1.13\ \sqrt{T_v} = 0.126$$

$$(1 - U_T) = (1 - U_v)\ (1 - U_r)$$

$$(1 - 0.91) = (1 - 0.126)\ (1 - U_r)$$

$$U_r\ (\text{required}) = 0.897,\ \text{say}\ 0.9$$

$$T_r = \frac{c_h\ t}{d_e{}^2} = \frac{591.7}{n^2}$$

$$n = \sqrt{(591.7/T_r)}$$

The suitable value of n is found by trial and error. Starting with a certain value of n, T_r is read for $U_r = 0.9$ (from graph or table). Using this value of T_r, $\sqrt{(591.7/T_r)}$ is calculated and a graph is plotted between $\sqrt{(591.7/T_r)}$ and n as shown in Fig. 14.16.

n	T_r	$\sqrt{(591.7/T_r)}$
30	0.764	27.8
25	0.712	28.8
20	0.649	30.2

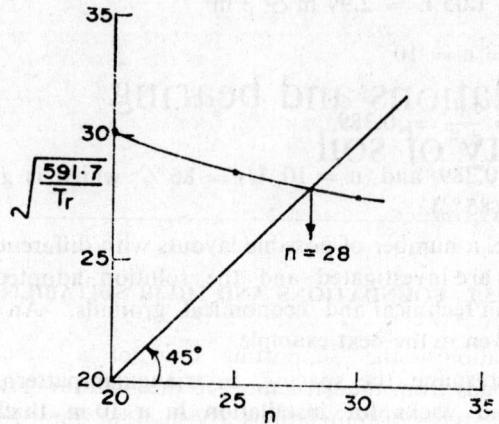

Fig. 14 16

EXAMPLE 14.20

From Fig. 14 16 it is seen that n = 28, ($T_r = 0.755$)

$\quad d_e = 0.065$, n = 1.82 m

Triangular pattern, $d_e = 1.05$ L

Drain spacing L $= \dfrac{1.82}{1.05} = 1.73$ m, say 1.7 m

15
Foundations and bearing capacity of soil

15.1 FOUNDATIONS AND THEIR SUITABILITY

A 'foundation' is the supporting base of a structure which transmits loads from the structure to the natural ground. There are two basic types of foundations: (i) shallow and (ii) deep. A 'shallow' foundation generally derives its support from the soil or rock close to the lowest part of the structure which it supports. The depth of the bearing area below the adjacent ground is usually about equal to or less than the width of the bearing area. Shallow foundations in turn may be classified into (i) footings and (ii) rafts or mats. A 'footing' is a unit foundation constructed as an enlargement or spread of the base of a column or wall for the purpose of transferring the load to the subsoil. When used to support and distribute an individual load (column load) it is called a 'spread footing' or a 'pad footing'. A 'strip' or 'continuous' footing is normally provided for load bearing walls and for rows of columns which are spaced so closely that pad footings would nearly touch one another. Thus a strip footing is one that has its length considerably greater than its width. A 'raft' or 'mat' foundation is a single slab, laid directly on the soil, usually reinforced, cast of uniform thickness, or with ribs. A raft supports many columnar and/or wall loads. A raft is required on soils of low bearing capacity, or where the loads are so large that the individual or continuous footings will occupy close to 50 percent of the projected area of the building; it becomes more economical to use a raft covering the entire area. Rafts are useful in reducing differential settlements on variable soils or where there is a wide variation in loading between adjacent columns. 'Combined footing', is inter-mediate between single spread footings and mats. It is a form of a little mat supporting two or more columns. A 'strap footing' consists of two footings joined by a tie beam If a raft, usually in the form of a monolithic box, is placed at such a depth that the intensity of imposed loading due to structure and foundation does not exceed or becomes less than the initially existing pressure at the depth prior to excavation, it is called a 'floating foundation' or a 'compensated foundation'. It is indicated in deep soil deposits of medium, high and very highly compressibility and low bearing capacity. A compensated foundation may also have additionally friction piles underneath.

A 'deep' foundation transmits all, or a portion of, the load to some significant depth below the base of the structure. The deep foundation is ordinarily intended to bypass weak or undesirable materials and carry the loads to some stratum below the unsuitable soils. Part or whole of the structural load may also be carried by side friction or adhesion of a deep foundation. Piles are the most common type of deep foundation for buildings. Wells (caissons) and piers are also classified under the deep foundations.

The suitability of different types of foundations for various site conditions can be preliminarily ascertained from Tables 15.1 and 15.2

Table 15.1

GUIDE TO TENTATIVE SELECTION OF FOUNDATION
TYPES BASED ON SITE CONDITION (JUMIKIS 1971)

Site condition	Type of structure	
	Light flexible	Heavy rigid
1. Thick, firm stratum	Footings (individual, strip or combined)	Footings Rafts
2. Firm stratum over soft stratum	Footings, Light surface rafts	Rafts Friction piles
3. Thick soft stratum	Friction piles Rafts	Rafts Friction piles
4. Soft stratum overlying easily attainable firm stratum	End bearing piles Piers Caissons	End bearing piles. Piers Caissons

Table 15.2

SUITABILITY OF FOUNDATIONS FOR BUILDINGS (NOT MORE THAN
4 STOREYS HIGH) BASED ON SOIL TYPE AND SITE
CONDITION (BRE 1977)

Soil type and site condition	Foundation
1. Rock, solid chalk, sands and gravels or sands and gravels with only small proportions of clays, dense silty sands	Shallow strip or pad footings
2. Uniform, firm and stiff clays	
(a) Where vegetation is insignificant	Bored piles and ground beams, or strip foundations at least 1 m deep
(b) Where trees and shrubs are growing or to be planted close to site	Bored piles and ground beams
(c) Where trees are felled to clear the site and construction is due to start soon afterwards	Reinforced bored piles of sufficient length with the top 3 m sleeved from the surrounding ground and with suspended floors, or thin reinforced rafts supporting flexible buildings, or basement rafts
	Note. Downhill creep may occur on slopes greater than 1 in 10. Unreinforced piles have been broken by slowly moving slopes.
3. Soft clays, soft silty clays	Strip footings upto 1 m wide if bearing capacity is sufficient, or rafts
	Note. Settlement of strips or rafts must be expected. In soft soils of variable thickness it is better to pile to firmer strata.

4. Peat, fill	Bored piles with temporary steel lining or precast or insitu piles driven to firm strata below. *Note.* Design with large safety factor on end resistance of piles only as peat or fill consolidating may cause downward drag on pile. If fill is sound, carefully placed and compacted in thin layers, strip footings are adequate. Fills containing combustible or chemical wastes should be avoided.
5. Mining and other subsidence areas	Thin reinforced rafts for individual houses with load-bearing walls and for flexible buildings *Note.* A layer of granular material should be placed between the ground surface and the raft to permit relative horizontal movement.
6. Deep deposit of loose sand (Mc Carthy 1977)	Rafts, augered cast-in-situ piles. Spread footings may settle excessively or require use of very low bearing pressures. Compact sand by vibroflotation or other method, then use spread footings. Driven piles could be used and would densify sand.

15.2 FOUNDATION REQUIREMENTS AND PLACEMENT

The two basic requirements of a foundation are:
(1) It must be safe against failure; there should be neither the bearing capacity failure of the supporting soil nor the structural failure of the foundation.
(2) It must not exceed tolerable settlements; the probable maximum and differential settlements of the soil, viz. various parts of the foundation must be limited to safe and tolerable magnitudes.

In addition, a foundation must have an adequate depth from considerations of adverse environmental influences. It must also be economically feasible in terms of the overall structure.

The safety criteria against shear failure and excessive settlement are discussed in the subsequent sections. To be safe against adverse environmental influences, the foundation must be carried below: (i) zones of high volume change due to moisture fluctuations, (ii) depth of frost penetration, (iii) topsoil or organic matter, peat and muck, (iv) unconsolidated material such as abandoned garbage dumps and similar filled-in areas and (v) zone significantly weakened by root holes or cavities by burrowing animals or worms. The depth should also be enough to prevent the rain water scouring below the footing. Ordinarily, all foundations should extend to a depth of atleast 50 cm below natural ground level (IS: 1904).

The foundations should be adequately placed with respect to adjacent structures, existing or anticipated, to minimize the possibility of damage due to mutual interference. The minimum edge-to-edge clear distance between the spread footings placed at different level should be greater than the difference in elevations of the bases of the footings; this is an approximation for reducing the pressure overlap (Bowles 1983). For footings at different levels located adjacent to sloping ground, the IS Code (1904) specificies: (i) The sloping ground surface should not encroach upon a frustrum of bearing material under the footing having sides which make an angle with the horizontal of 60° for rock and 30° for soil and the horizontal distance from the lower edge of the footing to the sloping surface should be at least 60 cm for rock and 90 cm for soil. (ii) In case of footings in granular soils, a line drawn between the lower adjacent edges of the two footings should not have a steeper slope than 2 (horizontal) to 1 (vertical). (iii) In clayey soils, a line drawn between the lower adjacent edge of the upper footing and the upper adjacent edge of the lower footing should not have a steeper slope than 2 (horizontal) to 1 (vertical).

15.3 TERMINOLOGY

It may be useful to define at this stage the various terms relating to bearing capacity and bearing pressure.

(1) *Total overburden pressure* σ. It is the intensity of total pressure, due to the weight of both soil and water, on any horizontal plane. The total overburden pressure at the base level of the foundation before commencement of construction operation will be represented by σ_0. This may also be called the 'surcharge pressure'.

(2) *Effective overburden pressure* σ_0'. It is the total overburden pressure minus the pore water pressure at the foundation level.

(3) *Total foundation pressure q* (or gross load intensity). It is the total pressure at the foundation level after the structure has been constructed and fully loaded. It may also be termed the 'total bearing pressure'.

(4) *Net foundation pressure or net bearing pressure* q_n (or net loading intensity). It is the net increase in pressure at the foundation level due to the dead and live loads applied by the structure.

$$q_n = q - \sigma_0 \qquad (15.1)$$

The increase in stress distribution at any depth below the foundation level is determined by using q_n.

(5) *Ultimate bearing capacity* q_f. It is the gross loading intensity (gross bearing pressure) at which the soil fails in shear.

(6) *Net ultimate bearing capacity* q_{nf}. It is the net loading intensity at which the soil fails in shear, i.e.

$$q_{nf} = q_f - \sigma_0 \qquad (15.2)$$

(7) *Net safe bearing capacity* q_{ns} (or presumed bearing value). It is the net ultimate bearing capacity divided by a safety factor or a load factor F (usually 3):

$$q_{ns} = \frac{q_{nf}}{F} \qquad (15.3)$$

(8) *Safe bearing capacity* q_s. It is the maximum intensity of loading which the soil will carry without risk of shear failure irrespective of settlement considerations.

$$q_s = q_{ns} + \sigma_0 = \frac{q_{nf}}{F} + \gamma D \qquad (15.4)$$

where γ = unit weight of soil

D = depth of foundation

The term 'safe bearing capacity' is also used in literature as the gross ultimate bearing capacity divided by a safety factor, i.e. $q_s = q_f/F$.

(9) *Bearing pressure for allowable settlement* q_{sct}. It is the net bearing pressure which will not allow the settlement of a foundation to exceed the given maximum allowable value S.

(10) *Allowable bearing pressure* q_a. It is the maximum allowable *net* bearing pressure which gives safety against both the shear failure and the excessive settlement. It is thus the lower of the values of q_{ns} and q_{sot}. The gross allowable bearing pressure will be $q_a + \gamma D$.

15.4 MODES OF SHEAR FAILURE

The three principal modes of shear failure under foundations have been described as 'general shear failure' (Caquot 1934, Terzaghi

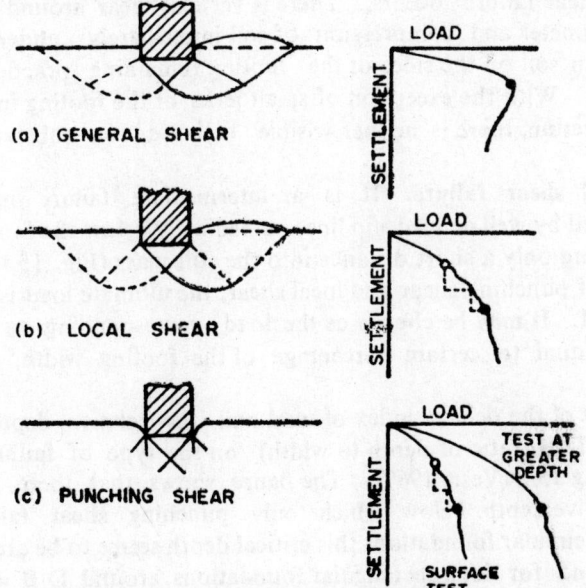

Fig. 15.1 Modes of bearing capacity failure

1943), 'local shear failure' (Terzaghi 1943, De Beer and Vesic 1958, Vesic 1967, 1973) and 'punching shear failure' (De Beer and Vesic 1958, Vesic 1967, 1973). Considering a strip footing (L $>$ 5 B) on the ground surface, the failure modes are illustrated in Fig. 15.1.

1. General shear failure. Well defined slip lines are assumed to extend from the edge of the footing to the adjacent ground surface. In stress-controlled conditions, under which most foundations operate, failure is sudden and catastrophic (Vesic 1973) with well defined ultimate load. Adjacent soil tends to bulge during loading, although the final soil collapse occurs only on one side. In strain-controlled conditions (e.g. load applied by jacking), a visible decrease of load necessary to produce footing movement after failure may be observed (Fig. 15.1a). This type of failure is characteristic of narrow, surface footing or of shallow depth resting on stronger, denser soils which are relatively incompressible.

On weaker and more compressible soils, and under wider and deeper footings two other failure modes—'punching shear' and 'local shear' are observed.

2. Punching shear failure. On soils of high compressibility punching shear failure occurs. There is vertical shear around the footing perimeter and compression of soil immediately under the footing, with soil on the sides of the footing remaining practically uninvolved. With the exception of small jerks of the footing in the vertical direction, there is neither visible collapse nor substantial tilting.

3. Local shear failure. It is an intermediate failure mode characterized by well defined slip lines immediately below the footing but extending only a short distance into the soil mass (Fig, 15.Ic)

In case of punching shear and local shear, the ultimate load is not well defined. It may be chosen as the load corresponding to the settlement equal to certain percentage of the footing width, say at 10%B.

The effect of the density index of sand and the relative depth of foundation D/B (ratio of depth to width) on the type of failure is shown in Fig 15.2 (Vesic 1967). The figure shows that there is a critical relative depth below which only punching shear failure occurs. For circular foundations this critical depth seems to be around D/B = 4 and for long rectangular foundations around D/B = 8. Also, more compressible soil generally has lower critical depths.

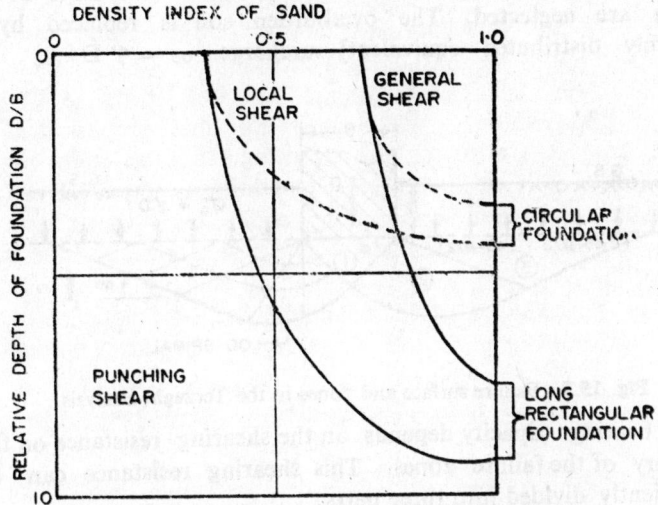

Fig. 15 2   Effect of density index of sand and relative depth
of foundation on type of failure (Vesic 1967)

15 5 BEARING CAPACITY OF SHALLOW FOOTINGS
KARL TERZAGHI ANALYSIS

Theoretical methods for predicting ultimate bearing capacity are
generally based only on the general shear failure case. For the
other failure modes, a reduction in the ultimate bearing capacity
due to compressibility effects is applied to the value obtained for
the general shear case.

The Terzaghi analysis makes the following assumptions to arrive
at an approximate value of the ultimate bearing capacity.

(1) The footing is a strip at shallow depth and has a rough
base; (L > 5B, D ≯ B).

(2) The soil is homogeneous, isotropic and relatively incompressible. There is two-dimensional general shear failure with well
defined failure surfaces and zones as indicated in Fig. 15.3. The
wedge of soil directly beneath the footing, elastic zone 1, acts as
part of the footing and pushes side ways the plastic zones — the
radial shear zone 2 and the Rankine passive zone 3.

(3) The failure zones do not extend above the horizontal plane
through the base of the footing. The shearing resistance of soil

above the base level as well as friction between soil and sides of the footing are neglected. The overburden soil is replaced by a uniformly distributed equivalent] surcharge $\sigma_0 = \gamma D$.

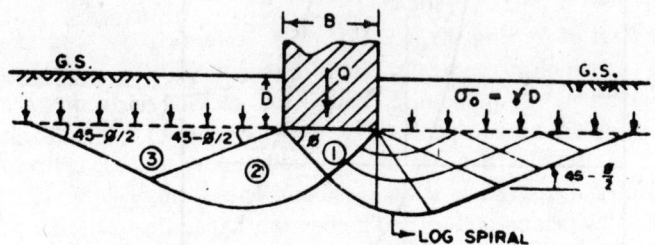

Fig. 15.3 Failure surface and zones in the Terzaghi analysis

The bearing capacity depends on the shearing resistance on the boundary of the failure zones. This shearing resistance can be conveniently divided into three parts:

(1) Cohesive resistance.

(2) Frictional resistance resulting from the surcharge σ_0 at the footing level.

(3) Frictional resistance resulting from the weight of soil within the failure zones.

Although these quantities are not entirely independent, they may be considered separately and on superposition the ultimate bearing capacity may be expressed by Eq. 15.5 given below (Tomlinson 1980). This superposition is believed to lead to errors on the safe side, (not exceeding 17% to 20% for $\phi = 30°$ to 40°, while being equal to zero for $\phi = 0$, Vesic 1973).

$$q_t = c_T N_c + \sigma_0' N_q + 0.5 \gamma BN\gamma \qquad (15.5a)$$

where
$\quad c_T$ = undrained cohesion of soil

$\quad \sigma_0'$ = effective overburden pressure at foundation level ($= \gamma'D$, if submerged)

$\quad \gamma$ = unit weight of soil below foundation level (use γ', if submerged)

N_c, N_q and $N\gamma$ are bearing capacity factors (coefficients) depending only on the value of ϕ.

Eq. 15.5a may also be written as:

$$q_t = [c_T N_e + \sigma_o' (N_q - 1) + 0.5\gamma BN\gamma] + \sigma_o \quad (15.5b)$$

where σ_0 = total overburden pressure $= \gamma D$

The terms within the brackets represent the *net* ultimate bearing capacity q_{nt}

1. Effect of footing shape. Eq. 15.5 refers to a continuous footing for which plane strain conditions (two-dimensional failure) can be assumed. For square, round or rectangular footings, the shape of the failure surface is three dimensional. Thus the bearing capacity is altered. Because of the complexity of the problem, a semi-empirical approach is suggested. The terms of Eq. 15.5 are multiplied by 'shape factors'. The bearing capacities of square and circular footings of width/diameter B are expressed as:

Square footing: $q_t = 1.2 \, c_T \, N_c + \gamma DN_q + 0.4\gamma BN\gamma$ (15.6)

Circular footing: $q_t = 1.2 \, c_T \, N_e + \gamma DN_q + 0.3\gamma BN\gamma$ (15.7)

The original Terzaghi value of shape factor for the first terms on the right hand sides of Eqs. 15.6 and 15.7 was 1.3 which has been subsequently changed to 1.2 (Terzaghi and Peck 1967). On analogy, the bearing capacity of a rectangular footing (L × B) can be expressed as follows (author):

Rectangular footing:

$$q_t = c_T \, N_c \left(1 + 0.2 \frac{B}{L} \right) + \gamma \, DN_q + 0.5 \, \gamma \, BN\gamma$$

$$\times \left(1 - 0.4 \frac{B}{L} \right) \quad (15.8)$$

2. Effect of compressibility. For footings on soft clay or sand of low density index where local shear or punching shear failure may occur, reduced values of cohesion c_r and friction angle ϕ_r as given by Eqs. 15.9 and 15.10, have been suggested (Terzaghi and Peck 1967).

$$c_r = 0.67 \, c_T \quad (15.9)$$

$$\phi_s = \tan^{-1} (0.67 \tan \phi) \quad (15.10)$$

Vesic (1973) suggests that for sands the factor 0.67 in Eq. 15.10 should be replaced by a correction factor as given below which varies with the density index (applicable in the range $0 \leqslant I_D \leqslant 0.67$).

Correction factor $= (0.67 + I_D - 0.75 \, I_D^2)$ (15.11a)

Thus for local shear failure, the bearing capacity factors are determined with respect to ϕ_r and c_T is replaced by c_r in the bearing capacity equations.

According to Zeevaert (1983), the simplest way to take compressibility into consideration is to assume that q_f is approximately proportional to the density index and the correction factor can be expressed as:

$$\text{Correction factor} = (I_D + 0.1) \tag{15.11b}$$

where the above factor is assumed equal to unity for dense state, equal to 0.6 for semi-dense state, and equal to 0.3 for a very loose state. The bearing capacity calculated on the assumption of general shear failure by Eqs. 15.5 to 15.7 is to be multiplied by the above factor (Eq. 15.11b).

3. Bearing capacity of clays. For $\phi_T = 0$, the Terzaghi bearing capacity factors are $N_c = 5.7$, $N_q = 1$ and $N\gamma = 0$. In place of $N_o = 5.7$, the Prandtl value (Terzaghi and Peck 1967) of $N_o = 2 + \pi = 5.14$ is currently adopted. The bearing capacity equations for clays are thus written as:

Strip: $q_f = 5.14\,c_T + \gamma D$ (15.12)

Square and round: $q_f = 6.2\,c_T + \gamma D$ (15.13)

In the above equations, unit cohesion may be taken as one-half the unconfined compressive strength ($c_T = 0.5\,q_u$).

If $\phi_T = 0$ and $c_T > 0$, the increase in bearing capacity per unit area produced by the surcharge γD is exactly compensated by the weight of soil excavated for construction. It is, therefore, convenient to deal with the *net* bearing capacity:

$$q_{nf} = q_f - \gamma D = 5.14\,c_T \quad \text{(strip)} \tag{15.14}$$

$$q_{nf} = 6.2\,c_T \quad \text{(square and round)} \tag{15.15}$$

4. Skempton's values of N_o. Rounding conservatively N_c equal to 5 for a strip, Skempton (1951) proposed the following equation for the net bearing capacity of a rectangular footing on saturated clay ($\phi_T = 0$) for values of D/B not exceeding 2.5.

$$q_{nf} = 5\,c_T \left(1 + 0.2\,\frac{B}{L}\right)\left(1 + 0.2\,\frac{D}{B}\right) \text{ for } \frac{D}{B} \not> 2.5 \tag{15.16}$$

The increase in N_o with increase in D/B ratio is conditional upon there being no soft or loose soil above foundation level (Tomlinson 1980). For D/B > 2.5, Skempton suggested that $N_c = 1.5\,N_c$

at ground surface.   Thus the maximum value of N_o for a deep strip footing is 7.5 and for a deep square or circular footing maximum $N_o = 9$.

In many practical cases, the   undrained shear strength c_T varies with depth   below   a footing.   In this   case, the   average value of c_T should be taken over a depth below the footing equal to 2/3 B, provided c_T of any layer does not depart from the average strength by more than $\pm$ 50 percent (Skempton 1951).   This average value of c_T is used in all the bearing capacity equations for clays.

5.   Choice of shear parameters.   In the form   presented by Terzaghi, the bearing capacity solution can be applied strictly only to cases where the groundwater table is deep (the total stresses equal the effective stresses) and the shear parameters should be expressed in terms of effective stresses.   The solution   could   also be applied to an   undrained, total   stress condition;   using in this case   total stress shear strength parameters c_T (c_u) and ϕ_T (ϕ_u).

The choice of using c', ϕ' or c_T, ϕ_T depends on the type of soil and the construction time.   For free draining granular soils (sand and gravels)   where   the   excess   pore pressure is likely to dissipate during or immediately after the construction period and application of load, a drained analysis in terms of effective stresses is made. For these soils c will also be zero.   For clays or clayey soils (c — ϕ soils) where the permeability is so low that very little pore pressure dissipation occurs   during   construction, the   condition should be considered   as essentially undrained and the total stress parameters should be used.   Of course with time, a clay will consolidate and gain in strength resulting in an   increase of the long-term bearing capacity.   *To be on the safe side,* the undrained cohesion c_T is used in all bearing capacity calculations which are valid for immediate end-of-construction stability of foundation on clays and clayey soils.

6.   Effect of water table.   The position of water table has a significant effect on the bearing capacity because submergence of soil reduces the cohesion (the apparent cohesion due to capillary stresses or from weak cementation bonds is all lost), and reduces the effective unit weight to about half of the value above the water table,   although the angle of shearing resistance is not appreciably changed.   Thus through submergence, all the three terms of the bearing capacity equation may become considerably smaller.   For

this reason, it is essential that the bearing capacity is determined assuming the likely worst moisture conditions. For clays and clayey soils (c — ϕ soils) where there is no possibility of immediate pore pressure dissipation, it is advisable that undrained cohesion c_T obtained on submergence is used in the first term of the bearing capacity equation. For different positions of the water table, the bearing capacity is expressed as follows:

(a) Water table at ground surface:

$$q_t = c_T N_o + \gamma' D (N_q - 1) + 0.5 \gamma' B N_\gamma + \gamma_{sat} D \quad (15.17)$$

(b) Water table between ground surface and base of footing:

$$q_t = c_T N_o + \sigma_o' (N_q - 1) + 0.5 \gamma' B N_\gamma + \sigma_o \quad (15.18)$$

where σ_o' = effective overburden pressure at base level, taking into account the bulk unit weight above the water table and submerged unit weight below

σ_o = total overburden pressure

(c) Water table at base of the footing:

$$q_t = c_T N_o + \gamma D (N_q - 1) + 0.5 \gamma' B N_\gamma + \gamma D \quad (15.19)$$

(d) Water table between the base of footing and a distance B below the footing:

$$q_f = c_T N_c + \gamma D (N_q - 1) + 0.5 \gamma_a' B N_\gamma + \gamma D \quad (15.20)$$

where γ_a' = weighted (average) unit weight (Dunn et al 1980)

$$= \frac{\gamma x + \gamma' (B - x)}{B} \quad (15.21)$$

x = depth of water table below the footing

(e) If the water table is at a distance $\geqslant$ B below the footing, it is assumed to have no effect.

Since the submerged unit weight of granular soils is about one half of the moist, dry or saturated unit weight, it will be seen that the net bearing capacity q_{nf} of granular soils will be reduced to about one half of its value when the water table rises to the ground surface. Assuming a 50 percent reduction in bearing capacity due to rise of water table upto ground surface (or due to its location at the ground surface) and no reduction when the water table is at a depth B below the footing, Peck et al (1974) proposed the following correction factor:

$$C_w = 0.5 + 0.5 \frac{D_w}{D + B} \quad (15.22)$$

where D_w = depth of water table below the ground surface
$C_w = 0.5$ for $D_w = 0$, and $C_w = 1$ for $D_w = D + B$

The bearing capacity of granular soils for any depth D_w of water table can be approximately obtained as:

$$q_t = [\gamma_1 D (N_q - 1) + 0.5 \gamma_2 B N\gamma] C_w + \gamma_1 D \quad (15.23)$$

where γ_1 = average bulk unit weight above the footing

γ_2 = average bulk unit weight below the footing within depth B

The assumption of 50% reduction in the bearing capacity due to rise of water table upto ground surface is valid for normal soils. For metastable soils this reduction may be still greater. Limited model footing tests on surface of dune sand have indicated a reduction of bearing capacity to almost one-fourth the value when the water table rose upto the surface (Mathur 1982). Moreover, for the bearing capacity to remain entirely unaffected, the depth of water table below the footing should be greater than 1.5B (Meyerhof 1955).

7. Other influencing factors. It has been contended often in literature that the roughness of footing base increases the bearing capacity. However, Vesic (1973) concludes that base roughness has little effect on the bearing capacity as long as applied external loads remain vertical.

When footings are placed close enough that their zones of action overlap, the bearing capacity is affected. The effects of adjacent strip footing vary considerably with the angle of shearing resistance ϕ. For low ϕ values, they are negligible; however, for high ϕ values they appear to be significant, particularly if a footing is surrounded by others on both sides (Vesic 1973). In general, interference between footings on dense sand is observed to cause an increase in bearing capacity and decrease in settlement with reduction in spacing (Singh et al 1973. 1976, Ohri et al 1981).

These effects are considerably reduced as L/B→1. For design purposes, it is not recommended to consider interference effects in bearing capacity computations (Vesic 1973).

Studies indicate that in case of shallow footings, the average shear strengh mobilized along a slip line under the footing decreases with size. The actual bearing capacity of larger footings on sand is probably less than the value conventionally computed. This is due to the decrease in the $N\gamma$ value with increase in size of footing (Vesic 1973, NGI 1972). For large footings on sand, a reduction of $2°—4°$ in the value of ϕ is tentatively suggested (NGI 1972) for calculating the value of $N\gamma$.

The effects of eccentricity and inclination of load and depth factors are described in Section 15.6.

Some of the important points regarding bearing capacity can now be summarized as follows:

(1) The net ultimate bearing capacity q_{nf} depends on the (effective) over-burden pressure at the footing level and thus increases with the depth of footing.

(2) For cohesive soils ($\phi_T = 0$), q_t is independent of the footing size.

(3) For frictional soils ($\phi > 0$), q_t is directly dependent on footing size but the depth of footing is more important than size.

(4) Position of water table influences the bearing capacity by reducing the unit weight of soil to the submerged value γ' which is nearly one-half the bulk unit weight. q_{nf} depends on the submerged unit weight.

(5) q_t is very sensitive to ϕ, particularly at large values of ϕ (See Table 15.3). Small increase in ϕ leads to large increases in N_q and $N\gamma$.

(6) It is only in case of narrow (less than about 1 m) and shallow footings on submerged loose sand that the allowable bearing pressure may be governed by shear failure consideration. In all cases of footings on dry cohesionless soils, and most cases of wide or deep footings on submerged cohesionless soils, the allowable bearing pressures are governed by considerations of tolerable settlement.

Examples 15.1 — 15.6

15.1 Determine the diameter of a circular footing to carry a concentric column load of 800 kN ($\simeq$ 80 t)· The depth of footing is 1.0 m. The soil is partly saturated and has $\phi_T = 10°$, $c_T = 50$ kPa ($\simeq$ 5 t/m²), and $\gamma = 18$ kN/m³. Use a safety factor (load factor) of 3 and ignore the weight of the footing.

Table 15.3 : $\phi_T = 10°$, $N_o = 8.3$, $N_q = 2.5$, $N_\gamma = 1.4$

$q_{af} = 1.2\,c_T\,N_o + \gamma D\,(N_q - 1) + 0.3\gamma BN_\gamma$
$\quad = 525 + 2.16\,B$

$q_{ns} = \frac{1}{3}\,q_{af} = 175 + 0.72\,B$

$q_s = q_{ns} + \gamma D = 193 + 0.72\,B$

$\dfrac{800}{\pi\,B^2/4} = q_s = 193 + 0.72\,B$

$B \simeq 2.29 = 2.3\ \text{m}$

15.2 Determine the width of a strip footing for the following data. $D = 1$ m, $c_T = 1.4$ t/m², $\phi_T = 28°$, $\rho = 1.78$ t/m³, F = 3, load per metre run = 90 t. $N_o = 25.8$. $N_q = 14.7$, $N_\gamma = 10.6$.

$$q_s = \frac{1}{F}\left[\,c_T\,N_o + \rho D\,(N_q - 1) + 0.5\rho BN_\gamma\,\right] + \rho D$$
$$= 90/B$$
$$B = 2.9\ \text{m}$$

15.3 A square foundation is to be designed to carry a load of 50 t (inclusive of foundation) at a depth of 1.5 m in a loose granular soil having $\phi' = 28°$. The water table is at the base of the foundation. The soil has bulk density ρ of 1.7 t/m³ above the water table and a saturated density ρ_{sat} of 1.95 t/m³. If the excavation will be backfilled to its original level, determine a suitable size of the foundation which will have a safety factor of 3 against shear failure.

$\phi = 28°$, $N_q = 14.7$, $N_\gamma = 10.6$, $c_T = 0$, $\rho' = 0.95$ t/m²

$q_{ns} = \rho D\,(N_q - 1) + 0.4\,\rho'\,BN_\gamma$

$q_s = \frac{1}{3}\,q_{nf} + \rho D = 14.18 + 1.34\,B = 50/B^2$

$B = 1.75\ \text{m}$

15.4 A 2.5 m square footing is located in a dense sand at a depth of 1.5 m, the shear strength parameters being $c' = 0$, $\phi' = 38°$. Determine the ultimate bearing capacity for the following water table positions: (a) at ground surface, (b) at 1 m below ground surface, (c) at footing level, (d) at 0.5 m below the footing, and (e) at a depth greater than B below the footing. The moist unit weight of sand above the water table is 18 kN/m³ and the saturated unit weight is 20 kN/m³. For $\phi = 38°$, $N_q = 48.9$ and $N_\gamma = 58.9$.

(a) Water table at ground surface.

$q_f = \gamma'\,D\,(N_q - 1) + 0.4\,\gamma'\,B\,N_\gamma + \gamma_{sat}\,D$

$\gamma' = 20 - 9.8 = 10.2$ kN/m³

$q_f = 1364$ kPa

(b)   Water table at 1 m below ground surface.

$q_t = \sigma_0' (N_q - 1) + 0.4\, \gamma'\, B\, N\gamma + \sigma_0$

$\sigma_0' = (1 \times 18) + (0.5 \times 10.2) = 23.1 \text{ kPa}$

$\sigma_0 = (1 \times 18) + (0.5 \times 20) = 28 \text{ kPa}$

$q_f = 1735 \text{ kPa}$

(c)   Water table at footing level.

$q_t = \gamma\, D\, (N_q - 1) + 0.4\, \gamma'\, BN\gamma + \gamma\, D$

$ = 1921 \text{ kPa}$

(d)   Water table at 0.5 m below the footing.

Calculate weighted average unit weight γ_a' below the footing to a depth equal to $B = 2.5$ m.

$\gamma_a' = \dfrac{(0.5 \times 18) + (2 \times 10.2)}{2.5} = 11.76 \text{ kN/m}^3$

$q_t = \gamma DN_q + 0.4\, \gamma_a'\, BN\gamma = 2013 \text{ kPa}$

(e)   Water table at great depth ($\geqslant$ B below footing).

$q_t = \gamma\, DN_q + 0.4\, \gamma\, BN\gamma = 2380 \text{ kPa}$

15.5 Determine the net safe bearing capacity of a square footing, 3 m × 3 m, placed at a depth of 2 m in a sandy soil of average bulk density ρ of 1.7 t/m³ and having $\phi = 35°$. Use a safety factor of 3. What will be the net safe bearing capacity if the water table rises to an elevation 1 m below the ground surface?

$\phi = 35°,\ c_T = 0,\ N_q = 33.3,\ N\gamma = 33.9$

$q_s = \dfrac{1}{F} [\rho\, D\, (N_q - 1) + 0.4\, \rho\, B\, N\gamma] = 59.7 \text{ t/m}^2$

$C_w = 0.5 + 0.5\, \dfrac{D_w}{D + B} = 0.6$

$q_s = 59.7 \times 0.6 = 35.8 \text{ t/m}^2$

15.6 A raft, 20 m × 12 m, is placed at a depth of 2 m below the ground surface on a site where the soil profile consists of 1.5 m of fill material overlying a deep bed of saturated clay. The bulk unit weights of the fill and clay are respectively 18 kN/m³ and 20 kN/m³. The undrained unit cohesion of clay increases from 40 kPa ($\simeq$ 4 t/m²) at foundation level to 80 kPa ($\simeq$ 8 t/m²) at 14 m below the ground surface. Determine the safe bearing capacity using a load factor of 2.5.

For the rectangular raft,

$$q_{nf} = c_T N_c \left(1 + 0.2 \frac{B}{L} \right) + \gamma D (N_q - 1)$$
$$+ 0.5 \gamma B N_\gamma \left(1 - 0.4 \frac{B}{L} \right)$$

For $\phi_T = 0$, $N_c = 5.14$, $N_q = 1$, $N_\gamma = 0$

$$q_{nf} = c_T N_c \left(1 + 0.2 \frac{B}{L} \right)$$

c_T at 2/3 B below raft $= 40 + (80 - 40) \dfrac{8}{12}$

$\qquad = 66.66$ kPa

Average c_T from base upto 2/3 B below it

$$= \frac{40 + 66.66}{2} = 53.33 \text{ kPa}$$

Using $c_T = 53.33$ kPa, $q_{nf} = 307$ kPa

$$q_s = \frac{q_{nf}}{F} + \sigma_o = \frac{307}{2.5} + (1.5 \times 18 + 0.5 \times 20)$$
$$= 159.8 \text{ kPa}$$

15.6 GENERAL BEARING CAPACITY EQUATION BRINCH HANSEN ANALYSIS

Apart from the Terzaghi solution there have been recently several proposals for the computation of the ultimate bearing capacity. The use of the Terzaghi equations is gradually decreasing, even though the Terzaghi bearing capacity factors are not substantially different numerically from factors proposed by others. The principal reason is that these equations are based on obviously incorrect failure patterns (Vesic 1973, Bowles 1983). Also these equations do not have provisions for including other boundary conditions.

The most comprehensive solutions which take into account the shape and depth of the foundations, the eccentricity and inclination of loading and inclination of the foundation have been derived by Hansen (1970) and Meyerhof (1963). Both expressed the general bearing capacity equation in the same form (Eq. 15.24), but the shape, depth, inclination and N_γ factors are computed in a different way.

The Hansen analysis gives more conservative values (Tomlinson 1980) and are described here. The Hansen analysis seems to provide better computed bearing capacities than the Terzaghi analysis. According to Hansen (1970, DGI 1985) the general bearing capacity equation is expressed as:

$$q_f = c_T N_o s_o i_o d_o + \gamma D N_q s_q i_q d_q$$
$$+ 0.5 \, \gamma \, B \, N_\gamma \, s_\gamma \, i_\gamma \, d_\gamma \qquad (15.24)$$

where $s_o \, s_q, \, s_\gamma$ = shape factors
$i_o, i_q, \, i_\gamma$ = inclination factors
d_o, d_q, d_γ = depth factors

The equation for short term bearing capacity of clay ($\phi_T = 0$) is written as:

$$q_f = c_T \, N_o \, s_o \, i_o \, d_o + \gamma \, D \qquad (15.25)$$

and $$q_{nf} = c_T \, N_o \, s_o \, i_o \, d_o \qquad (15.26)$$

where $N_o = 5.14$

The bearing capacity factors N_q and N_o are the same as obtained by Prandtl and Reissner (Vesic 1973).

$$N_q = \tan^2 (45 + \frac{\phi}{2}) \exp (\pi \tan \phi) \qquad (15.27)$$

$$N_o = (N_q - 1) \cot \phi \qquad (15.28)$$

The factor N_γ is given by the following empirical formula corresponding closely to the solution given by Christensen (Hansen 1970).

$$N_\gamma = 1.5 \, (N_q - 1) \tan \phi = 1.5 \, N_o \tan^2 \phi \qquad (15.29)$$

The Danish Code gives the following formula for N_γ (Eq. 15.30):

$$N_\gamma = (0.08705 + 0.3231 \sin 2 \, \phi - 0.04836 \sin^2 2 \, \phi)$$
$$\times (N_q \, \exp \frac{\pi}{2} \tan \phi - 1) \qquad (15.30)$$

The values of N_o, N_q and N_γ according to Hansen (1970) and the Danish Code (DGI 1985) are given in Table 15.3 and are shown plotted in Fig. 15.4. It is recommended that plane strain values of ϕ (ϕ_p) should be used for the general bearing capacity equation (Eq. 15.24) which may be estimated from the following relation (Meyerhof 1963):

$$\phi_p = (1.1 - 0.1 \frac{B}{L}) \phi \qquad (15.31)$$

where $\phi =$ frictional angle from triaxial test. Eq. 15.31 is recommended for dense granular soils only; for loose soils, $\phi_p = \phi$.

Table 15.3

BEARING CAPACITY FACTORS (HANSEN 1970, DGI 1985)

ϕ	N_c	N_q	N_γ
0	5.14	1.0	0.0
5	6.5	1.6	0.1
10	8.3	2.5	0.4
14	10.4	3.6	1.0
15	11.0	3.9	1.2
20	14.8	6.4	2.8
22	16.9	7.8	4.0
24	19.3	9.6	5.5
25	20.7	10.7	6.6
26	22.3	11.9	7.6
28	25.8	14.7	10.6
30	30.1	18.4	14.8
32	35.5	23.2	20.6
34	42.2	29.4	29.0
35	46.1	33.3	33.9
36	50.6	37.8	41.1
38	61.4	48.9	58.9
40	75.3	64.2	85.6
42	93.7	85.4	126.0
44	118.0	115.0	190.0
45	134.0	135.0	200.0
46	152.0	159.0	291.0
50	267.0	319.0	563.0

1. **Shape factors.** The shape of the footing is considered by using the following values of the shape factors (DGI 1985).

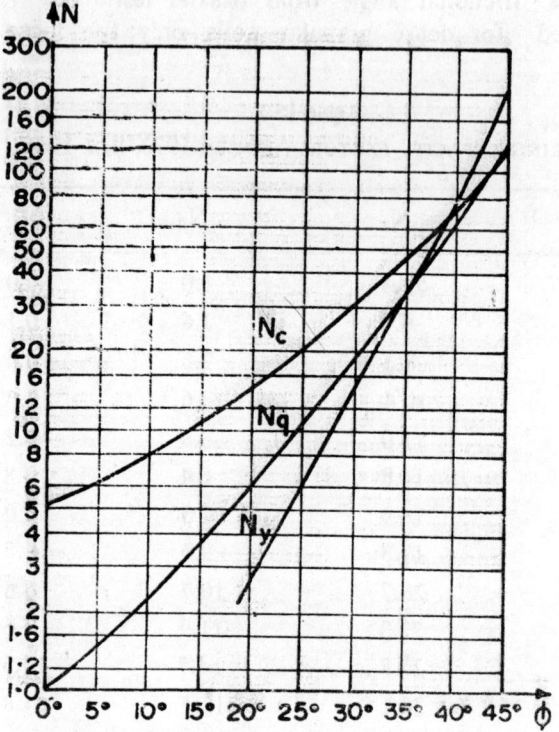

Fig. 15.4 Bearing capacity factors (DGI 1985)

(a) Strip: $s_c = s_q = s\gamma = 1$ (15.32)

(b) Square, circular or rectangular:

$$s_c = s_q = 1 + 0.2\frac{B}{L}$$ (15.33)

$$s\gamma = 1 - 0.4\frac{B}{L}$$ (15.34)

2. Inclination factors. For vertical loading, the inclination factors, i_c, i_q and $i\gamma$ are all equal to unity. If the resultant load R is inclined at an angle θ to the vertical and its horizontal and vertical components are H and V respectively, the inclination factors are given by (DGI 1985):

$$i_c = i_q = \left[1 - \frac{H}{V + A\, c_T \cot \phi} \right]^2$$ (15.35)

and $i\gamma = i_q^2$ (15.36)

For $\phi_T = 0, i_e = 0.5 + 0.5 \sqrt{1 - \dfrac{H}{A c_T}}$ (15.37)

where H and V = total horizontal and vertical loads respectively on area A

Foundations influenced by horizontal forces should also be examined for sliding:

$$H < A c_T + V \tan \delta \qquad (15.38)$$
$$H < A c_T \quad \text{for } \phi_T = 0 \qquad (15.39)$$

and it should furthermore be checked that

$$H/V < 0.4 \qquad (15.40)$$

3. Depth factors. The depth factors account for the shear resistance of soil above the base of the footing. The depth factors according to Hansen (Vesic 1973) are given by Eqs. 15.41 to 15.44. The factor $d\gamma$ may be taken as unity for all cases. Also when ϕ is zero, d_q is equal to unity.

$$\frac{D}{B} \leqslant 1, d_q = 1 + 2 \tan \phi (1 - \sin \phi)^2 \frac{D}{B} \qquad (15.41)$$

$$d_e = 1 + 0.4 \frac{D}{B} \quad (\phi = 0) \qquad (15.42)$$

$$\frac{D}{B} > 1, d_q = 1 + 2 \tan \phi (1 - \sin \phi)^2$$

$$\times \tan^{-1}\left(\frac{D}{B}\right) \qquad (15.43)$$

$$d_e = 1 + 0.4 \tan^{-1}\left(\frac{D}{B}\right) \quad (\phi = 0) \qquad (15.44)$$

A simplified value of d_e and d_q when ϕ is less than $25°$ is $(1 + 0.35$ D/B) (Tomlinson 1980).

4. Meyerhof's factors. The values of the various factors suggested by Meyerhof (1963) are as follows:

(a) *Shape factors.*

    (i) Strip: $s_e = s_q = s\gamma = 1$ (15.32)

    (ii) Square, circular or rectangular:

$$s_e = 1 + 0.2 \frac{B}{L} \tan^2\left(45 + \frac{\phi}{2}\right) \qquad (15.45a)$$

$$s_q = s\gamma = 1, \text{ for } \phi = 0° \qquad (15.46)$$

$$S_q = S\gamma = 1 + 0.1 \frac{B}{L} \tan^2\left(45 + \frac{\phi}{2}\right) \text{ for } \phi > 10° \quad (15.47)$$

The Canadian FE Manual (CGS 1985) recommends the

Meyerhof factors in the simplified form as:

$$s_0 = s_q = 1 + (B/L)(N_q/N_o) \qquad (15.45b)$$

$$s\gamma = 1 - 0.4(B/L) \text{ (same as Hansen)} \qquad (15.34)$$

(b) *Inclination factors.*

$$i_0 = i_q = (1 - \theta/90°)^2 \qquad (15.48)$$

$$i\gamma = (1 - \theta/\phi)^2 \qquad (15.49)$$

(c) *Depth factors.*

$$d_c = 1 + 0.2 \frac{D}{B} \tan(45 + \frac{\phi}{2}) \qquad (15.50)$$

$$d_q = d\gamma = 1, \text{ for } \phi = 0 \qquad (15.51)$$

$$d_q = d\gamma = 1 + 0.1 \frac{D}{B} \tan(45 + \frac{\phi}{2}) \text{ for } \phi > 10° \qquad (15.52)$$

5. Vesic's recommendations: According to Vesic (1973), the depth factors are effective in increasing the bearing capacity only when the method of placement of foundation (driving) causes significant lateral compression. This effect is practically negligible if the foundation is drilled in or buried and back-filled, or if the overburden strata are relatively compressible. For this reason, *it is advisable not to introduce depth factors in the design of shallow foundations*, i.e. take $d_0 = d_q = d\gamma = 1$.

6. Eccentricity, If 'e' is the eccentricity of a vertical load on the base of a footing of width B, the *effective* width B over which the pressure is assumed to be uniformly distributed is taken as follows (Meyerhof 1963):

$$B' = B - 2e \qquad (15.53)$$

For eccentricity in two directions, the *effective* width and length are taken as:

$$B' = B - 2e_B \qquad (15.54a)$$

$$L' = L - 2e_L \qquad (15.54b)$$

and effective area $A' = L' B'$ \qquad (15.55)

If the resultant load R is inclined and there is also eccentricity (Fig. 15.5), both the inclination factors and the eccentricity corrections should be applied.

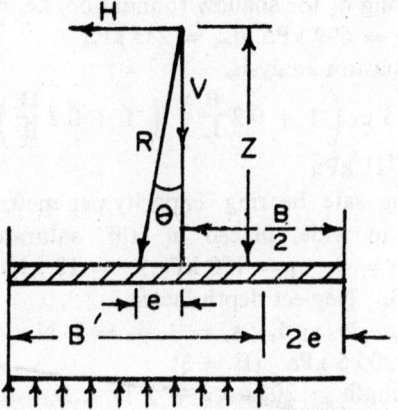

Fig. 15.5 Inclined and eccentric load

$$(e/Z = \tan \theta = H/V)$$

Where there is eccentricity, B, L and A should be replaced by B', L' and A' in all bearing capacity calculations and factors.

7. **Large eccentricity.** For large eccentricities (e > 0.3 B), it is necessary to examine the bearing capacity for a failure pattern extending partly below the unloaded part of the foundation. This can be done by disregarding the bearing capacity component from surcharge σ_0 and multiplying the components from c_T and γ in the general bearing capacity equation by (1.05 + tan³ φ) and 2 respectively (DGI 1985).

Examples 15.7—15.12

15.7 Find the total net load which can be supported by a rectangular footing 3 m × 2 m, at a safety factor of 3. The depth of footing is 0.6 m. The sub-soil is saturated clay having c_T = 120 kPa ($\simeq$12 t/m²).

$$\phi_T = 0, N_e = 5.14$$

(a) $q_{nf} = c_T N_e s_c d_c$

$$s_c = 1 + 0.2 \frac{B}{L} = 1.133$$

$$d_c = 1 + 0.4 \frac{D}{B} = 1.12$$

$$q_{nf} = 783 \text{ kPa}$$

$$q_{ns} = 261 \text{ kPa}$$

Total net load = 1566 kN

(b)   Neglecting d_c for shallow foundation, i.e. $d_c = 1$

$q_{nf} = 699$ kPa, $q_{ns} = 233$ kPa

(c)   By Skempton analysis.

$$q_{nf} = 5\, c_T \left(1 + 0.2 \frac{B}{L} \right) \left(1 + 0.2 \frac{D}{B} \right)$$

$$= 721 \text{ kPa}$$

15.8 Find the safe bearing capacity per metre length of a long wall footing, 2 m wide, placed in stiff, saturated clay at depth $D = 0.6$ m. $\phi_T = 0$, $c_T = 120$ kPa, $\gamma = 18$ kN/m³. $F = 3$. Neglect depth factor.

$q_{nf} = c_T\, N_c\, s_c\, d_o$; $s_c = 1$, $d_c = 1$, $N_c = 5.14$

$q_{ns} = 205.6$ kPa $(F = 3)$

q_s/m length $= 205.6 \times 2 + \gamma\, D \times 2 = 433$ kN/m

15.9 Find the net total load for a rectangular footing, 3 m × 2 m, if the resultant load is vertical but has an eccentricity $e = 0.15$ m in each direction. $D = 0.6$ m, $c_T = 120$ kPa, $\phi_T = 0$, $N_c = 5.14$. Neglect depth factor.

$L' = L - 2e = 2.7$ m

$B' = B - 2e = 1.7$ m

$$s_c = 1 + 0.2 \frac{B'}{L'} = 1.126$$

$q_{nf} = c_T\, N_c\, s_c = 694\,5$ kPa

Total net load $= \frac{1}{3}\, q_{nf} \times B'\, L' = 1062$ kN

15.10 Square footing, 3 m × 3 m, $D = 2$ m, medium dense sand, $c_T = 0$, $\phi = 36°$, $\rho = 1.75$ t/m³, vertical load without eccentricity, $F = 3$. Find net safe capacity.

$\phi = 36°$, $N_q = 37.8$. $N\gamma = 41.1$

$q_{nf} = \rho\, D\, (N_q\, s_q - 1) + 0.5\, \rho\, B\, N\gamma\, s\gamma$

$$s_q = 1 + 0.2 \frac{B}{L} = 1.2$$

$$s\gamma = 1 - 0.4 \frac{B}{L} = 0.6$$

$q_{ns} = 73$ t/m²

15.11 A 24 m long retaining wall has a 6 m wide foundation at a depth of 1.5 m in a silty sand having $c_T = 1.5$ t/m², $\phi = 25°$, $\rho = 2$ t/m³. The wall carries a horizontal load of 25 t/m run at a point 2 m above the base and a centrally applied load of 100 t/m run. Determine the safety factor against general shear failure of the base of the wall.

$\phi = 25°$, $N_c = 20.7$, $N_q = 10.7$; $N\gamma = 6.6$

$$\frac{e}{Z} = \tan \theta = \frac{H}{V} = \frac{25}{100} = 0.25 \ ; Z = 2 \text{ m}$$

$e = 0.5$ m, $\theta = 14°$

$B' = B - 2 e = 5$ m, $L = 24$ m, $A' = 120$ m^2

$$s_c = s_q = 1 + 0.2 \frac{B'}{L} = 1.04$$

$$s\gamma = 1 - 0.4 \frac{B'}{L} = 0.92$$

$$i_c = i_q = \left[1 - \frac{H}{V + A' c_T \cot \phi} \right]^2$$

$H = 25 \times 24$ t, $V = 100 \times 24$ t

$i_c = i_q = 0.616$

$i\gamma = i_q^2 = 0.379$

$q_t = c_T N_c s_c i_c + \rho D N_q s_q i_q + 0.5 \rho B' N\gamma s\gamma i\gamma$
$\quad = 52$ t/m^2

Actual bearing pressure for $B' = \dfrac{100}{5} = 20$ t/m^2

$$F = \frac{52}{20} = 2.6$$

15.12 The ground surface of a 2.5 m wide strip footing is only 0.75 m above the base on one side and higher on the other side. The soil is saturated clay whose unconfined compressive strength q_u is 250 kPa ($\simeq$ 25 t/m^2), $\gamma = 20$ kN/m^3 and $\phi = 0$. It carries a horizontal load of 80 kN/m run and a centrally applied vertical load of 400 kN/m run, such that the resultant load has an eccentricity of 0.2 m. Check the safety factor against shear failure of base.

Minimum depth of footing $= 0.75$ m

$c_T = \frac{1}{2} q_u = 125$ kPa

$B' = B - 2 e = 2.1$ m

$\phi_T = 0$, $N_c = 5.14$, $s_c = 1$

$$i_c = 0.5 + 0.5 \sqrt{1 - \frac{H}{A' c_T}}$$

For 1 m length, $H = 80$ kN, $A' = 2.1$ m^2

$i_c = 0.917$

$q_t = c_T N_c s_c i_c + \gamma D = 604$ kPa

q for $B' = \dfrac{400}{2.1} = 190.5$ kPa

$F = q_t/q = 3.2$

15.7 PLATE LOAD TEST

The insitu tests (Singh 1981) which may be used for estimating ultimate bearing capacity and/or allowable bearing pressure are:
(1) Plate load test
(2) Standard penetration test
(3) Dynamic cone penetration test
(4) Static cone penetration test
(5) Pressuremeter test

The plate load test is performed in a test pit dug upto the proposed base level of the footing. A rigid, steel plate is seated in the centre of the pit. The author recommends the minimum size of the plate to be 0.32 m square or 0. 36 m in diameter giving an area of 0.1 m². The test pit should be 5 times the size of the plate (at least 1.5 m square) to eliminate the surcharge effects; the plate is seated at the level bottom of the pit.

Reaction loading is found convenient. For this purpose a, steel truss is anchored to the ground across the pit. A hydraulic jack with an attached pressure gauge is interposed between the underside of the truss and the test plate. The settlement of the plate is measured by means of atleast two dial gauges (accurate to 0.02 mm) resting on the plate and fixed to an independent datum bar.

To commence the test, a seating pressure of about 0 5 t/m² (5kPa) is first applied to the plate. It is then removed and dial gauges are set to read zero. Load is then applied in cumulative equal increments of not more than 10 t/m² (100 kPa) or of not more than one-fifth of the estimated allowable bearing pressure. For weaker soils, the incremental pressure may be 5 t/m² or 2.5 t/m². Each load is maintained constant until, according to the author, the rate of settlement is less than 0.05 mm in 10 minutes (0.3 mm/hour), or until 2 hours have elapsed, whichever occurs first, but not less than 1 hour in any case. On clayey soils, the load is maintained constant until about 75 percent of the probable ultimate settlement at that stage is reached, or for 24 hours. Recording settlement at some convenient intervals and the final settlement at the end of that loading period, the load is increased to the next higher value, and the process is repeated. Testing is continued until one of the following stages is attained: (i) the settlement becomes definitely progressive indicating shear failure, (ii) the applied pressure exceeds

3 times the allowable pressure, or (iii) the total settlement exceeds 10 percent the width or diameter of the plate. The load is then released. If desired, rebound observations may be taken.

1. **Interpretation.** The test is interpreted by plotting 'net load intensity' versus 'final settlement' curves on linear scales and on log-log scales, as shown in Fig. 15.6.

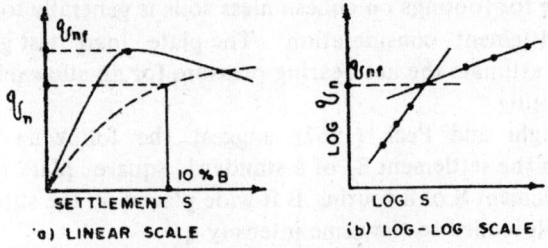

SETTLEMENT S   LOG S

'a) LINEAR SCALE   (b) LOG-LOG SCALE

Fig. 15.6 Plate load test curves

The net ultimate bearing capacity of the plate can be read as marked on the curves. The log-log plot often indicates more clearly the failure point. In undefinable cases, failure may be assumed at a settlement of 10% the plate width.

The net ultimate bearing capacity of actual footings on clays may be taken to be the same as that of the plate. On dense sands, the bearing capacity increases with width. The values obtained by the 0.1 m² or 0.2 m² plate may be conservatively increased upto 50 percent. The size effect on loose to medium sands is not appreciable, and no increase in bearing capacity is recommended. It may, however, be noted that the design of footings on sands is generally governed by settlement considerations rather than by shear failure. A safety factor of 3 should be used to arrive at the net safe bearing capacity.

2. **Limitations.** The pressure bulb of 0.2 q intensity of a test plate of width B_p extends upto about 1.5 B_p below it, whereas the pressure bulb of the same intensity for a footing of width B ($> B_p$) will extend upto 1.5 B. The presence of a weak stratum beyond a depth of 1.5 B_p but within 1.5 B will only affect the footing and not the plate. Under such situations, the results of a plate load rest are misleading and unsafe. Similarly, the proximity of a water table may be within the influence of the footing and not that of the

small test plate. To verify the validity and safe applicatiou of the results of a plate load test, it is necessary to explore the sub soil conditions to atleast 1.5 B below the footing. A plate test is essentially a short duration test and it gives no information whereby the magnitude and rate of long term consolidation settlement in clays may be calculated.

3. **Allowable bearing pressure.** As stated earlier, the bearing pressure for footings on cohesionless soils is generally to be obtained from settlement consideration. The plate load test graph can be used to estimate the net bearing pressure for an allowable settlement of a footing.

Terzaghi and Peck (1967) suggest the following relationship between the settlement S_1 of a standard square plate 1 ft wide and the settlement S of a footing B ft wide placed on the surface of sand and both loaded to the same intensity q :

$$\frac{S}{S_1} = \left(\frac{2 B}{1 + B} \right)^2 \tag{15.56}$$

Expressing B in metres, Eq. 15.56 is written as:

$$\frac{S}{S_1} = \left(\frac{6.56 B}{1 + 3.28 B} \right)^2 \tag{15.57}$$

where S_1 = settlement of standard test plate 0.305 m square

Eq. 15.57 may be applied to circular, square, rectangular or continuous footings of width B metres. If B_p is the width of test plate in metres, Eq. 15.57 is written as:

$$\frac{S}{S_p} = \left(\frac{B}{B_p} \right)^2 \left(\frac{3.28 B_p + 1}{3.28 B + 1} \right)^2 \tag{15.58}$$

Investigations of Bjerrum and Eggestad (1963) produced evidence to show that the settlement ratio S/S_1 was not a unique function of the width ratio as in Eq. 15.56, but varied with the state of compaction of sand. They presented curves for sands varying from a very loose state to a very dense state. Their investigations were based on a 0.1 m^2 plate. The Terzaghi-Peck relation (Eq. 15.56) was found valid for a dense sand only.

Assigning a density index of 85 percent to the Terzaghi-Peck relation and 35 percent to the average curve of Bjerrum and Eggestad, Arnold (1980) proposed the following modified relationship:

$$\frac{S}{S_1} = \left(\frac{2B}{1 + B^m}\right)^2 \qquad (15.59)$$

where B = width of footing (ft)

m = function of density index I_D

$$= 0.788 + 0.0025\, I_D \qquad (15.60)$$

Expressing B in metres, Eq. 15.59 is written as:

$$\frac{S}{S_1} = \frac{43.06\, B^2}{[1 + (3.28\, B)^m]^2} \qquad (15.61)$$

When $m = 1$, i.e. at $I_D = 85\%$, Eqs. 15.61 and 15.59 reduce to the corresponding Terzaghi Eqs. 15.57 and 15.56.

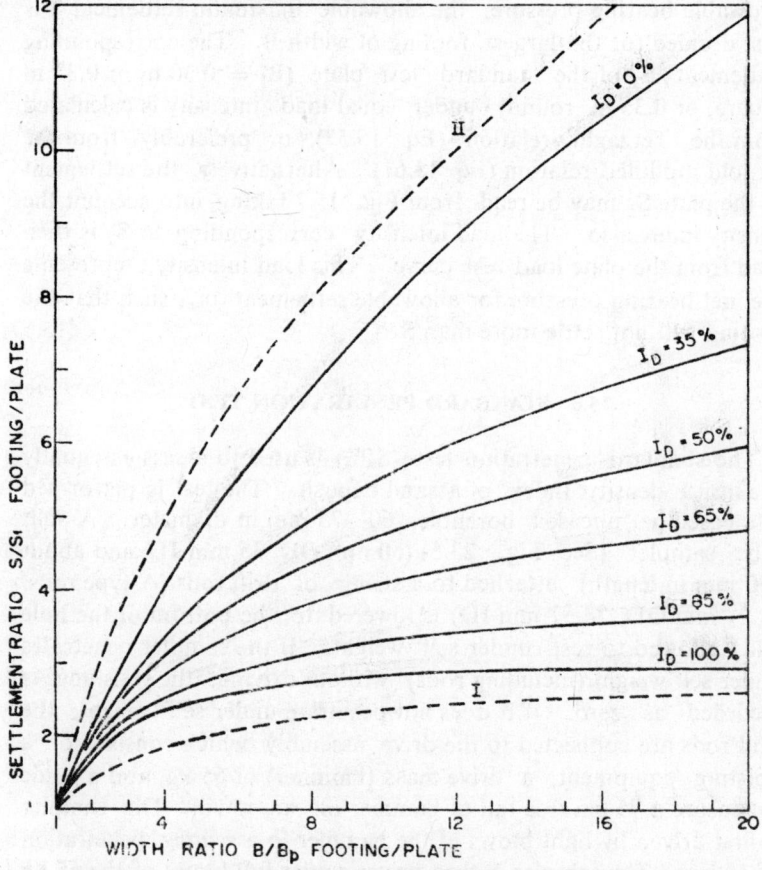

Fig. 15.7   Relationships between settlement ratio and width ratio

Arnold's modified equations are shown plotted by the author in Fig. 15.7 Because of the approximations involved, the width ratio B/B_p and the settlement ratio S/S_1 may be assumed to be applicable to both the 1.0 ft^2 and 0.1 m^2 plates. Fig. 15.7 also shows the limiting curves (dotted curves) obtained by Bjerrum and Eggestad with respect to 0.1 m^2 plate.

For ordinary buildings, if a differential settlement of ΔS can be tolerated, the maximum settlement of the largest footing is not likely to exceed about $4 \, \Delta S/3$ (Terzaghi and Peck 1967); thus for a tolerable differential settlement of 20 mm, the maximum settlement will be of the order of 25 mm, and vice versa. To determine the allowable bearing pressure, the allowable maximum settlement S is first decided for the largest footing of width B. The corresponding settlement S_1 of the standard test plate ($B_1 = 0.30$ m or 0.32 m square, or 0.36 m round) under equal load intensity is calculated from the Terzaghi relation (Eq. 15.57), or preferably from the Arnold modified relation (Eq. 15.61). Alternatively, the settlement of the plate S_1 may be read from Fig. 15.7 taking into account the density index also. The load intensity corresponding to S_1 is then read from the plate load test curve. This load intensity represents the 'net bearing pressure for allowable settlement' q_{set} such that the footing will not settle more than S.

15.8 STANDARD PENETRATION TEST

The standard penetration test (SPT) is used to assess essentially the insitu density index of a sand deposit. The test is performed in a cased or uncased borehole, 60—75 mm in diameter. A split tube sampler (See Fig. 23.5) (50 mm OD, 35 mm ID and about 650 mm in length) attached to a string of drill rods (A-type rods, 41.27 mm OD, 28.57 mm ID) is lowered to the bottom of the hole and allowed to rest under self weight. If the sampler penetrates under self weight (including rods) without driving, the resistance is recorded as zero. If it does not penetrate under self weight, the drill rods are connected to the drive assembly which consists of a hoisting equipment, a drive mass (hammer) of 65 kg, and a guide to ensure a 75 cm free fall of hammer on an anvil. The sampler is first driven by light blows of the hammer to a *seating* penetration of 15 cm. The sampler is then driven under full blows of the 65 kg hammer falling from a height of 75 cm to an additional penetration of 30 cm and the number of blows is recorded as the 'standard penetration resistance' N. If 50 blows are reached before a penetration of 30 cm, no further blows should be applied but the actual penetration should be recorded. At the end of the test, the sampler is withdrawn and the sand extracted. Tests are normally carried out at intervals between 0.75 m and 1.50 m to a depth at least equal to the width B of the foundation. In gravelly soils, the driving shoe of the sampler is replaced by a solid 60° cone.

The SPT is subject to many errors in practice, including the skill of the operator. Boring process or upward flow of water frequently disturbs the material at the bottom of the borehole leading to an under-estimation of N. The free fall of the hammer is freqnently too small and the friction of the rope on the shieves and winch drum is frequently too high, leading to an over-estimation of N.

1. Overburden pressure correction. Terzaghi and Peck (1967) and Peck et al (1974) have proposed empirical correlations between the penetration resistance N, allowable bearing pressure and other soil properties such as density index and angle of shearing resistance, etc. Studies indicate that the penetration resistance reflects both the insitu density index and the effective stress to the depth of testing; therefore, a number of combinations of stress level and density index may result in the same measured N value. To arrive at a standardized value of N for use in the empirical correlations, an overburden pressure correction is necessary. For a constant density index, the N value increases with increasing effective overburden pressure for which corrections have been proposed by Gibbs and Holtz (1957), Thornburn (1963), and others. In laboratory controlled tests, the Terzaghi and Peck relation between N and I_D is observed to correspond to an overburden pressure of about 210 kPa or 280 kPa ($\simeq$ 21 or 28 t/m²) (Gibbs and Holtz 1957, Sanglerat 1972). There is a wide spread feeling that the Gibbs and Holtz corrections provide over-estimates of the density index (Thornburn 1978, Arnold 1980). The author has also proposed a correction chart which is shown in Fig. 15.8. It can be used to estimate the field density index of dry or moist sands and silty sands from the measured N values and the corresponding overburden pressure. The chart was developed from laboratory tests and has been verified to some extent in the field (Singh 1972).

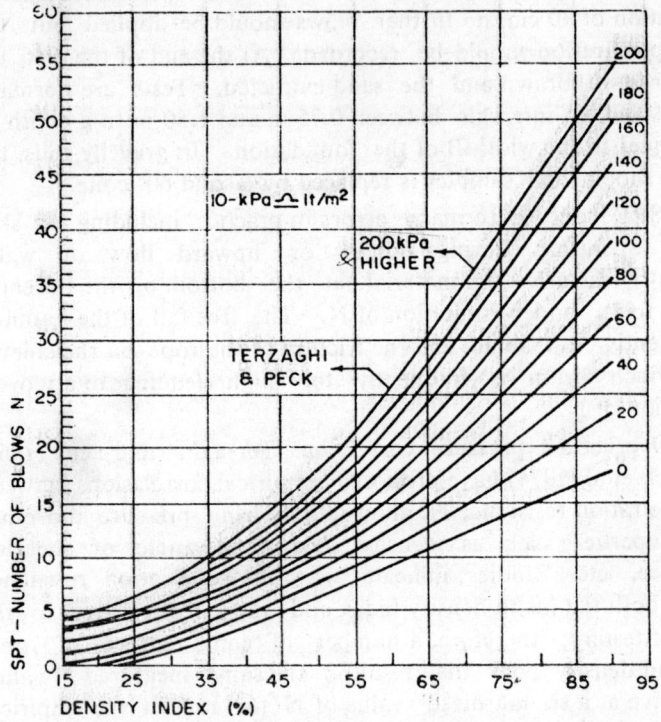

Fig. 15.8 Author's correction chart for overburden pressure (1982)

Peck et al (1974) proposed a conservative correction. Adopting about 100 kPa ($\simeq$ 10 t/m²) as the reference overburden pressure at which the SPT values should be reported or converted, the corrected blows (called the normalized blows or penetration resistance N_n) are obtained from the measured blows N by using the following correction:

$$N_n = C_n N \tag{15.62}$$

where C_n = overburden pressure correction (normalizing factor)

$$\simeq 0.77 \, \log \, \frac{200}{\sigma_0}, \, \sigma_0 \text{ in t/m}^2 \tag{15.63}$$

or $\qquad \simeq 0.77 \log \dfrac{2000}{\sigma_0}$, σ_0 in kPa $\qquad$ (15.64 a)

where $\qquad \sigma_0 \simeq$ *effective* overburden pressure at the test level

The above correction is valid for $\sigma_0 \geqslant 2.5$ t/m^2 (25 kPa). At the reference pressure $\sigma_0 = 10$ t/m^2, $C_n = 1$.   The maximum value of C_n at surface ($\sigma_0 = 0$) is 2. The other values are: At $\sigma = 1$ t/m^2 (10 kPa), $C_n = 1.8$; at $\sigma_0 = 2$ t/m^2, $C_n = 1.6$; at $\sigma_0 = 2.5$ t/m^2, $C_n = 1.5$.   The correction C_n becomes less than unity as σ_0 increases beyond 10 t/m^2.   At $\sigma_0 = 45$ t/m^2, $C_a \simeq 0.5$.

On the basis of studies conducted at the U.S. Waterways Experiment Station (Marcuson and Bieganousky 1976) more

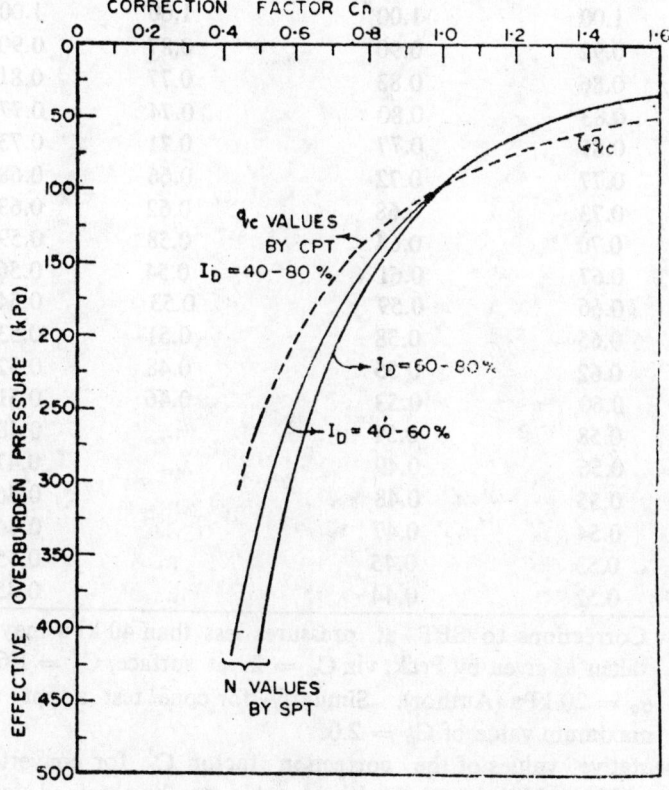

Fig. 15.9   Recommended curves to determine C_a for normalizing N and q_c (Marcuson et al 1976, Seed et al 1983) (converted to SI units by the author)

Table 15.4

VALUES OF NORMALIZING FACTOR C_n

(CONVERTED TO SI UNITS AND TABULATED BY THE AUTHOR)

Effective Pressure (kPa)	SPT $I_D=60—80\%$	SPT $I_D=40—60\%$	CPT $I_D=40—80\%$	CPT Robertson (1985)
0	...	...	...	...
20	...	...	...	2.80
40	1.46	1.46	...	1.76
50	1.33	1.33	1.58	1.58
60	1.23	1.23	1.39	1.40
80	1.09	1.09	1.15	1.18
100	1.00	1.00	1.00	1.00
120	0.92	0.90	0.87	0.90
140	0.86	0.83	0.77	0.81
150	0.83	0.80	0.74	0.77
160	0.81	0.77	0.71	0.73
180	0.77	0.72	0.66	0.68
200	0.73	0.68	0.62	0.63
220	0.70	0.64	0.58	0.59
240	0.67	0.61	0.54	0.56
250	0.66	0.59	0.53	0.54
260	0.65	0.58	0.51	0.53
280	0.62	0 55	0.48	0.52
300	0.60	0.53	0.46	0.51
320	0.58	0.51	...	0.48
340	0.56	0.49	...	0.47
350	0.55	0.48	...	0.46
360	0.54	0.47	...	0.46
380	0.53	0.45	...	0.45
400	0.52	0.44	...	0.45

Note. Corrections to SPT at pressures less than 40 kPa may be taken as given by Peck, viz $C_n = 2$ at surface, $C_n = 1.6$ at $\sigma_0 = 20$ kPa (Author). Similarly, for cone test adopt the maximum value of $C_n = 2.0$.

representative values of the correction factor C_n for converting measured blows N to normalized belows N_a can be obtained from Fig. 15.9, (Seed et al 1983). The reference effective overburden pressure for the correction is 2000 lb/sq. ft. (98 kPa $\simeq$ 100 kPa).

The author has tabulated the correction factors at specific pressures in Table 15.4. The figure and the table also give the correction for normalizing the measured values of the static cone resistance q_o (See Section 15.10.)

A simple relation (Eqs. 15.64b and 15.64c) for the correction factor C_n, comparable to the values of Fig. 15.9, has been proposed by Lio and Whitmen (1986)and is recommended by the author for adoption.

$$C_n = \sqrt{\frac{10}{\sigma_o}}, \quad \sigma_o \text{ in } t/m^2 \tag{15.64b}$$

or
$$C_n = \sqrt{\frac{100}{\sigma_o}}, \quad \sigma_o \text{ in kPa} \tag{15.64 c}$$

2. **Submergence correction.** Where the soil consists of *very fine or silty sand below the water table*, the measured N value, if greater than 15, should be corrected for the increased resistance due to excess pore water set up during driving and unable to dissipate immediately. The corrected value N' is given by:

$$N' = 15 + \tfrac{1}{2}(N - 15) \tag{15.65}$$

Wherever both the overburden and submergence corrections are necessary, *the overburden correction is applied first.*

3. **Correlation of soil properties.** Terzaghi and Peck (1967) have described the density index of sand on the basis of N values (normalized values). An approximate correlation with angle of shearing resistance ϕ is given in Table 15.5 (Peck et al 1974).

Table 15.5

STANDARD PENETRATION TEST AND ANGLE OF SHEARING
RESISTANCE (PECK ET AL 1974)

Corrected N_n	5	10	15	20	25	30	35	40	45	50
ϕ' (degrees)		28.5	30	32	33	35	36	37.5	39	40	41

Mitchell and Katti (1981) presented the currently accepted correlation of soil properties with SPT and CPT, which is given in Table 2.1 (Chapter 2).

4. Ultimate bearing capacity. Estimating the approximate value of ϕ from Table 15.5 or 2.1, and the corresponding values of N_q and $N\gamma$ from Table 15.3, the ultimate bearing capacity q_{at} can be calculated.

5. Bearing pressure for allowable settlement. Peck et al (1974) have proposed empirical design charts to estimate net bearing pressure on the basis of N value such that the maximum settlement of shallow footings on sand does not exceed 25 mm. For footings about 1.0 m or greater in width, the charts can be represented by the following relation (author).

$$q_{25} = 1.025\ N_a\ (t/m^2) = 10.25\ N_a\ (kPa) \qquad (15.66)$$

where q_{25} = net pressure for settlement not exceeding 25 mm

In general, for a settlement S mm and allowing for water table correction, the net bearing pressure q_{set} can be expressed as:

$$q_{set} = 0.041\ N_a\ C_w\ S\ (t/m^2) \qquad (15.67)$$
$$= 0.41\ N_a\ C_w\ S\ (kPa) \qquad (15.68)$$

where C_w = water table correction (Eq. 15.22)

N_a = *average* corrected N value for overburden (and submergence, if necessary)

The proximity of ground water is likely to reduce the bearing capacity by 50 percent or increase the settlement by 100 percent. Hence corrections are recommended when the bearing capacity and settlement are evaluated on the basis of N values. Some engineers, however, are of the opinion that the effect of groundwater gets already reflected in the measured values of N and no correction is necessary where the water table already exists at the time of tests. Corrections are justified only when the water table rises after the tests.

For design of footings, the general practice is to determine N at 0.75 m intervals as the test boring is advanced. The average corrected values of N over a distance from the base of footing to a depth B below the footing is calculated. When several borings are made, the lowest average value should be used.

6. Bearing pressure for rafts and piers. On account of the large size of rafts, the safety factor against a bearing capacity failure of the underlying sand is always large. Moreover, differential settlements are less and higher maximum settlements may be permitted. Experience has shown that for a differential settlement of 20 mm, the maximum settlement of raft may be about 50 mm instead of 25 mm as for a footing and, therefore, a pressure approximately twice as great as that allowed for individual footings may be used (Peck et al 1974):

$$q_{50} = 2.05 \; N_n \; C_w(t/m^2) = 20.5 \; N_n \; C_w \; (kPa) \qquad (15.69)$$

where q_{50} = net pressure for S = 50 mm, (or differential settlement ΔS = 20 mm)

SPT should be performed in a number of boreholes and N_n to be used in Eq. 15.69 should be the *minimum average* for depth B below the base of the raft. If N_a is less than 5, sand is too loose and should be compacted or else, alternative foundations on piles or piers should be considered. If the depth of raft D is less than 2·5 m, the edges of the raft settle appreciably more than the interior because of lack of confinement of sand. If bedrock is encountered at depth less than B/2, q_{50} (Eq. 15.69) may be increased somewhat. (Peck et al 1974). If allowable settlement S other than 50 mm is adopted, q_{set} is known from Eq. 15.67 or 15.68.

The settlement of a pier under a given net pressure is less than that of a shallow footing on sand of comparable I_D, because of the confining overburden pressure. At comparatively larger depths, the confining pressure also correspondingly increases the N values. Hence, unless the final ground level differs greatly from that at the time the SPT was made, the q_{25mm} for piers may be obtained from Eq. 15.66 with measured N values *uncorrected for the confining pressure* (Peck et al 1974).

15.9 DYNAMIC CONE TEST

The test consists in determining the resistance of soil to dynamic penetration of a 60° cone attached to a string of drill rods (A-type) and driven directly into the ground without pre-boring or casing under the blows of a 65 kg hammer falling from a heigh of 75 cm. Two sizes of cones are recommended by the ISI, cones of base diameters 50 mm and 62.5 mm (IS: 4968-I & II.) The 62.5 mm cone has the provision of injecting a bentonite slurry through the drill rods, which gets circulated upwards past the drill rods, thereby reducing the friction on rods. The number of blows for 30 cm penetration is termed the penetration resistance N_{o50} (for 50 mm cone) or $N_{o62\cdot5}$ (for 62.mm cone).

Like the SPT, dynamic cones can only be used to have a qualitative evaluation of insitu compactness of cohesionless soils or to have a qualitative comparison of subsoil stratification. They may be correlated with SPT or static cones. Desai(1972) developed an empirical chart for estimating the allowable bearing pressure from the results of a 50 mm cone (Singh 1982).

Examples 15.13 − 15.17

15.13 Determine the net bearing pressure for a 3 m $\times$ 3 m footing at a depth of 2 m in a medium dense sand so that the total settlement does not exceed 25 mm. The average SPT blows below the footing are 28 per 30 cm. The average moist density is 1.75 t/m³. The water table is more than 3 m below the footing.

σ_0' at B/2 below footing $= 1.75 (2 + 1.5) = 6.125$ t/m²

$$C_n = 0.77 \log \frac{200}{\sigma_0'} = 1.165$$

$$N_a = C_n N = 33$$

$$q_{25} = 0.041 N_n S = 33.8 \text{ t/m}^2$$

15.14 Determine the net bearing pressure for the footing in Ex. 15.13, if water table rises to an elevation 1 m below the footing ($D_w = 3$ m).

$$C_w = 0.5 + \frac{D_w}{D + B} = 0.8$$

$$q_{25} = 1.025 N_n C_w = 27 \text{ t/m}^2$$

15.15 A water tank has a concrete foundation slab, 5 m wide by 20 m long, constructed at a depth of 1 m in medium sand where the water table is 2 m below the ground surtace. The depth of water in the tank is 8 m. The density of sand above the water table is 1.7 t/m³ and the submerged density is 0.9 t/m³. The SPT blows in a borehole are given in Table 15.6. Examine whether the total settlement will not exceed 25 mm.

$$C_w = 0.5 + 0.5 \frac{D_w}{D + B} = 0.67$$

$$q_{25} = 0.041 N_n C_w S = 8.9 \text{ t/m}^2 \text{ for } S = 25 \text{ mm}$$

Actual net pressure $= q - \sigma_0 \simeq (8 \times 1) - (1 \times 1.7) = 6.3$ t/m²
Hence the settlement will remain less than 25 mm.

Table 15.6
EXAMPLE 15.15

Depth below G.S. (m)	N	σ_0' (t/m²)	C_a	N_n	Average N_n
1	6	1.7	1.59	10	
2 (W.T.)	8	3.4	1.36	11	
3	10	4.3	1.28	13	13
4	12	5.2	1.22	15	
5	12	6.1	1.17	14	
6	14	7.0	1.12	16	

15 16 Determine the net and gross bearing pressures for a raft, 8 m × 15 m, plac:d at a depth of 3 m in coarse sand underlain by fine silty sand below a depth of 7.5 m. The water table is 3 m below ground surface. SPT was performed in 5 bore-holes. The SPT blows for one of the boreholes representing the *weakest* spot are given in Table 15.7. ρ (above W.T.) = 1.9 t/m³, ρ' (coarse sand, below W.T.) = 1.12 t/m³, ρ' (silty sand) = 1.05 t/m³. The maximum settlement is limited to 50 mm. Use $c_n = 0.77 \log \dfrac{200}{\sigma_0'}$.

Table 15.7
EXAMPLE 15.16

Depth below G.S. (m)	N	σ_0' (t/m²)	c_a	N_u	N'
3 (W.T.)	8	5.70	1.19	10	10
3.75	10	6.54	1.14	11	11
4.50	12	7.38	1.10	13	13
5.25	10	8.22	1.07	11	11
6.00	14	9.06	1.03	14	14
6.75	15	9.90	1.00	15	15
7.50 (silty sand)	18	10.74	0.98	18	!7
8.25	20	11.53	0.95	19	17
9.00	18	12.32	0.93	17	16
9.75	22	13.10	0.91	20	18
10.50	20	13.90	0.89	18	17
11.25	24	14.68	0.87	21	18
				Total	177

Average blows $N' = \dfrac{177}{12} = 15$

$C_w = 0.5 + 0.5 \dfrac{3}{3+8} = 0.64$

$q_{50} = 2.05 \ N' \ C_w = 19.68 \ t/m^2$

$q \ (gross) = 19.68 + \rho \ D = 25.38 \ t/m^2$

15.17 Find the net total load which can be supported by a 3.5 m × 3.5 m footing at a depth of 1.15 m in a sandy soil. The bulk density of sand is 1.74 t/m³. The SPT blows at the footing level are 13. The maximum settlement should not exceed 25 mm and the factor of safety against shear failure should not be less than 3. A plate load test on a 0.1 m² plate ($B_1 = 0.36$ m dia) is also performed in an open pit at the footing level. The plate settlement is observed as 1.2, 3, 4.6, 7, 10 and 14 mm under 5, 10, 15, 20, 25 and 30 t/m² respectively.

$\sigma_0' = \rho D = 1.74 \times 1.15 = 2 \ t/m^2$

From author's chart (Fig. 15.8), I_D (for N = 13 at $\sigma_0' = 2 \ t/m^2$) is 65%.

$\dfrac{B}{B_1} = \dfrac{3.5}{0.36} \simeq 10$

From Fig. 15.7, using the curve for $I_D = 65$,

$\dfrac{S}{S_1} = 4$ for $\dfrac{B}{B_1} = 10$

Hence S_1 (plate) $= S/4 = 25/4 \simeq 6$ mm

From the plate load test graph (not shown here),

q_a (for $S_1 = 6$ mm) $= 18 \ t/m^2$

Hence from settlement consideration q_{25} for the footing is 18 t/m² (as obtained from the plate test).

At $\sigma_0' = 2 \ t/m^2$, Peck correction $C_a \simeq 1.6$

$N_a = 1.6 \times 13 \simeq 20$, and ϕ from Table 15.5 $\simeq 33°$.

From Table 15.3, $N_q \simeq 25$, $N\gamma \simeq 24$

$q_{ns} = \dfrac{1}{F} [\ \rho \ D \ (N_q - 1) + 0.4 \ \rho \ B \ N\gamma \ ]$

$= 35.5 \ t/m^2 > q_{25} \ (18 \ t/m^2)$

q_{25} will govern the design.

Net total load $= 18 \times 3.5 \times 3.5 = 221$ t

Check from Peck formula:

$q_{25} = 1.025 \ N_n = 1.025 \times 20 = 20.5 \ t/m^2$

Net total load $= 20.5 \times 3.5 \times 3.5 = 251$ t

15.10   DUTCH STATIC CONE PENETRATION TEST

A continuous   evaluation   of density and   strength   of sands and undrained shear   strength of clays can be obtained by means of a Dutch static cone penetration test (CPT).   The Dutch cone has an apex angle of 60° and a base   area of 10   cm^2 (35.7 mm   diameter). It is attached to a string of solid rods   (called the   sounding rods) running inside hollow outer tubes (called the mantle or jacket tubes). The external diameter of mantle tube is equal to the cone diameter. The point is shielded by a mantle of reduced diameter   to   prevent intrusion of soil particles.   A manually or hydraulically operated driving mechanism is used to push   the cone into the ground. No boring is required for the test.

To obtain the cone   resistance q_0, the cone is pushed vertically into the soil at a uniform speed of 1 cm/sec or 2 cm/sec through a distance of 5 cm (or 4 cm) by   means of the sounding   rods, the mantle (jacket) tubes remaining stationary. The pressure   required for pushing is recorded on pressure gauges. The outer mantle tubes are then pushed downwards to collapse the   extended cone tip and advance the cone to the next depth.   The cycle is repeated to obtain cone resistance   at   depth   increments   not ordinarily exceeding 20 cm.

An alternative version is the 'friction jacket (mantle) cone' in which a friction jacket is located   above the   cone. The cone and the friction jacket can be advanced separately by means of the sounding rods.   Initially the cone is pushed through a distance of 5 cm to record the penetration resistance.   With further advance-ment of the cone, a flange engages the friction jacket.   Now both the cone and the friction jacket advance together to give the total resistance. Subtracting cone   resistance from   total resistance, jacket friction f_c can be obtained.

A further development is the 'electric cone' (Holden 1974). The cone resistance and the side resistance (jacket friction) can be measured independently and continuously with penetration by means of load cells fixed inside the body of the instrument. Typical cones are shown in Fig. 15.10.

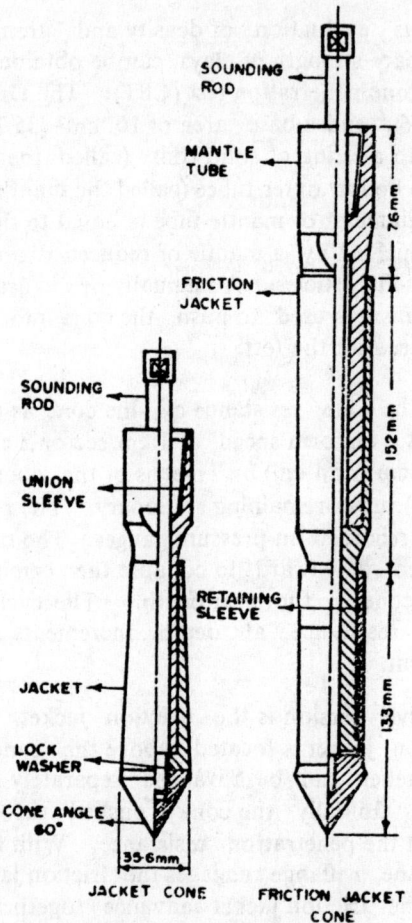

Fig. 15.10   Dutch static cones

The CPT is free of many of the objections of the SPT (Section 15.8). The two tests are correlated roughly as given in Table 15.8, the cone resistance has the units of kg/cm² or 100 kPa.

Table 15.8

CORRELATION BETWEEN DUTCH CONE TEST AND STANDARD
PENETRATION TEST (SCHMERTMANN 1970)

Soil type	q_c/N
Silts, sandy silts, slightly cohesive silt-sand mixtures	2
Clean, fine to medium sands, and slightly silty sands	3.5
Coarse sands and sands with little gravel	5
Sandy gravel and gravel	6

Note. q_c in $kg/cm^2 \simeq 100$ kPa

Based on extensive field data, a more rational correlation is given in Fig. 15.11 in which the effect of grain size is also reflected.

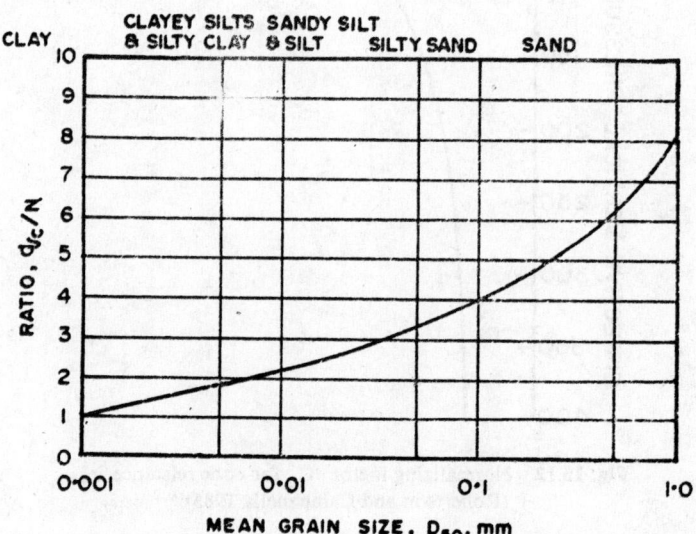

Fig. 15.11 Variation of q_c/N ratio with mean grain size (D_{50} mm) (q_c in 100 kPa) (Robertson et al 1983)

For the correlation of q_o values with other soil properties, the q_o values should be normalized at the standard overburden pressure of 100 kPa (as is done for the SPT resistance N).

$$q_{on} = C_n \, q_o \tag{15.70}$$

where q_{on} = normalized value of cone resistance
C_n = normalizing factor (over-burden correction)
q_o = measured value of cone resistance

The normalizing factor C_n can be read from Fig. 15.9 or Table 15.4. Similar values of C_n are shown in Fig. 15.12 which are also based on comprehensive data (Robertson and Campanella 1985). These values are tabulated in the last column of Table 15.4.

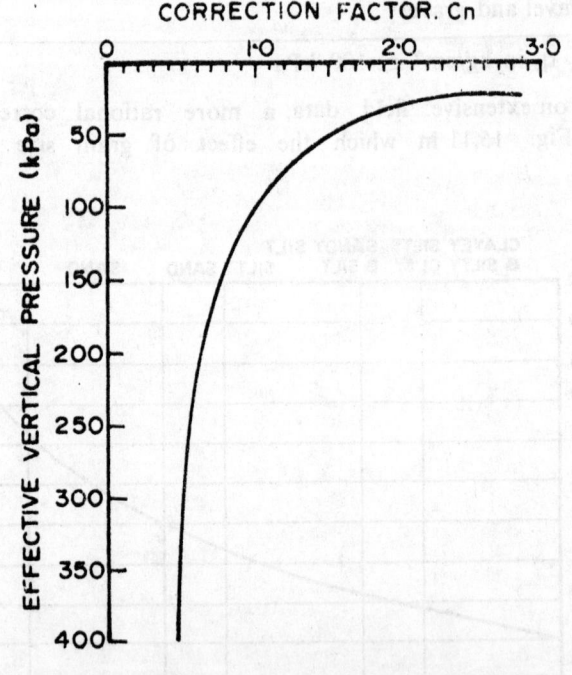

Fig. 15.12   Normalizing factor C_n for cone reistance q_o
(Robertson and Campanella 1985)

The cone resistance q_o can be used to estimate ϕ'.   Fig. 15.13 gives the correlation between q_o and ϕ' as suggested by Meyerhof. Table 2.1 gives the correlation with normalized value of q_o.

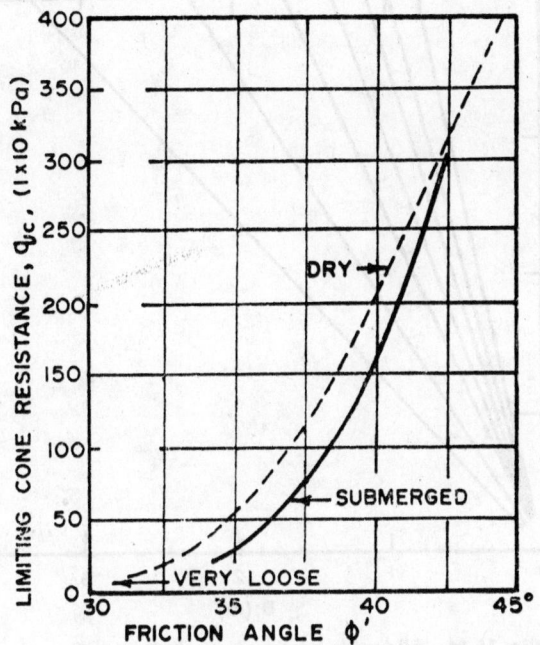

Fig. 15.13   Correlation between q_o and ϕ'(Meyerhof 1976)

For shallow footings of commonly used dimensions on dry sand with an embedment of about 1 m, the bearing pressure for 25 mm settlement can be approximately estimated from the following relation (Meyerhof 1956, CGS 1985):

$$q_{25} = \frac{1}{10}\, q_c \tag{15.71}$$

Eq. 15.71 should be used with caution for simple cases only. For other cases, the bearing pressure may be estimated from Fig. 15.14 (Meyerhof 1956, CGS 1985). q_{25} obtained from Eq. 15.71 or Fig. 15.14 should be halved if the sand within the stressed zone is submerged (Tomlinson 1980). Also q_{25} may be doubled for rafts, in a way similar to Terzaghi and Peck (1967).

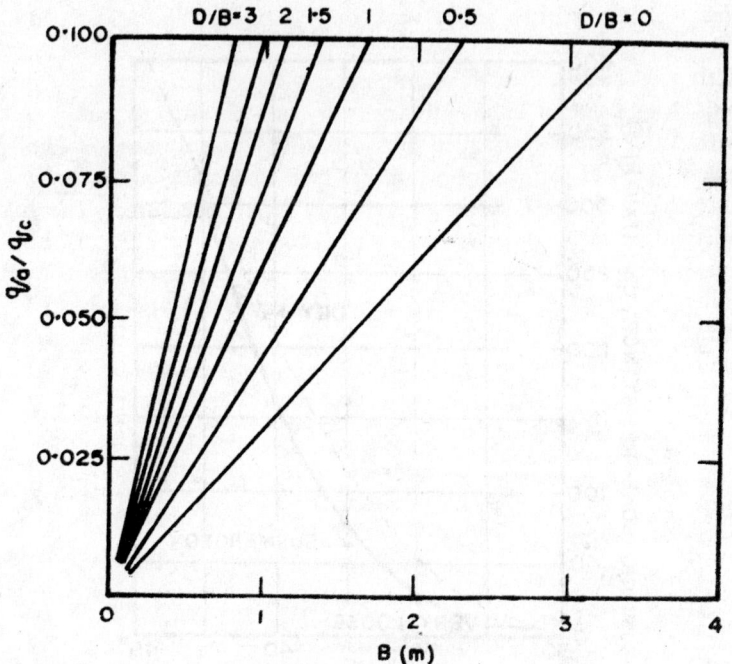

Fig. 15.14 Allowable bearing pressure from static cone penetration test (CGS 1985)

The use of CPT for estimating the settlement of foundations on sand in described in Chapter 18. The test is best suited for the design of piles in sand (Chapter 19). Tests in clay are reliable only when used in conjunction with vane tests.

15.11 MENARD PRESSUREMETER TEST

The pressuremeter test (MPT) developed by Menard (1965) is an insitu loading test carried out in a borehole by means of a cylindrical probe for determining the strength and deformation modulus of soils and weak rocks. The two parameters determined in particular are: (i) pressuremeter deformation modulus E_p representative of the elasticity of soil, which permits the estimation of the settlement, and (ii) limit pressure p_l related to shear strength of soil, which is used for computing the bearing capacity.

The apparatus consists of three components as shown in Fig. 15.15: a probe, a pressure and volume control unit, and connecting tubes. The probe consists of a metal cylinder covered with an inflatable rubber membrane under which three independent cells are located. The central cell or the measuring cell is filled with water under controlled pressure from the volumeter. The two guard cells at top and bottom are inflated with a gas or water automatically maintained at a slightly lower pressure. The guard cells ensure uniformity of stress and deformation conditions around the central cell which is used for the test measurements. The volume changes of the central cell during the test are known from the volume of water expelled from the volumeter. The details of the test procedure and the determination of p_l and E_p are described by LCPC (1971), Baguelin et al (1978) and Singh (1981).

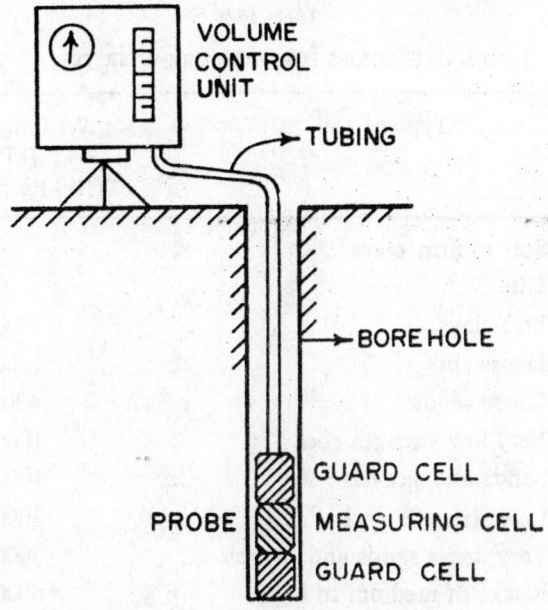

Fig. 15.15 Menard pressuremeter

The ultimate bearing capacity q_t is related to the limit pressure p_l (which corresponds to the failure of soil) by the following linear function (Menard 1975, CGS 1985):

$$q_{nf} = q_f - \gamma D = K_g (p_1 - p_0) = K_g \, p_{1n} \qquad (15.72)$$

where K_g = bearing capacity factor varying from 0.8 to 9 according to the embedment, the shape of foundation and the nature of soil

p_0 = 'at rest' horizontal earth pressure at the foundation (at the time of the test)

p_1 = limit pressure (within a zone extending 1.5 times the width of foundation above and below the foundation level)

$$p_{1n} = \text{net limit pressure} = p_1 - p_0 \qquad (15.73)$$

and $q_{ns} = \frac{1}{3} K_g (p_1 - p_0)$

To know the value of K_g, soils have been divided into 4 categories according to Table 15.9.

Table 15.9

SOIL CATEGORIES FOR PRESSUREMETER TEST

Category	Type of soil		Net limit pressure (kPa) (10 kPa $\simeq$ 1 t/m²)
I	Soft to firm clays	...	0 — 1200
	Silts	...	0 — 700
II	Stiff clays	...	1800 — 4000
	Dense silts	...	1200 — 3000
	Loose sands	...	400 — 800
	Very low strength rock	...	1000 — 3000
III	Sands and gravels	...	1000 — 2000
	Low strength rock	...	3000 — 6000
IV	Very dense sands and gravels	...	3000 — 6000
	Rocks of medium to high strength	...	6000 — 10000

The bearing capacity factor K_g for shallow foundations is given in Fig. 15.16. K_g for rectangular footing is assumed to vary linearly with the values of B/L.

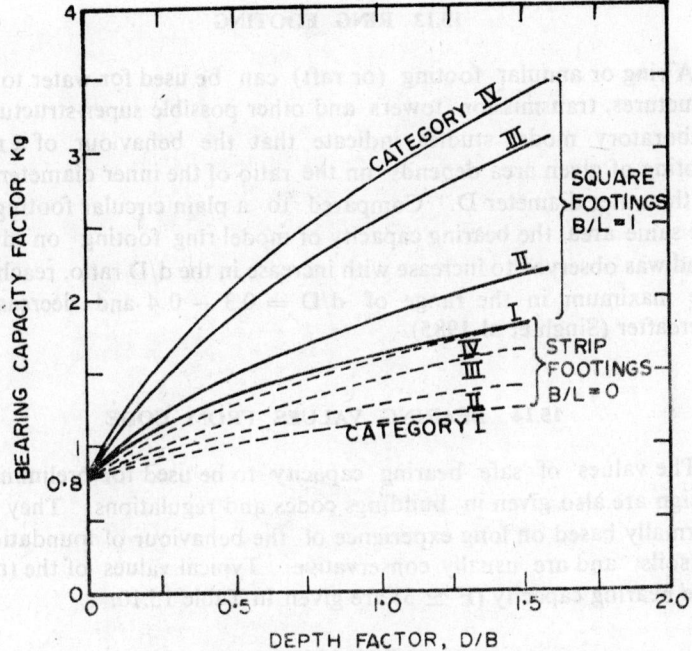

Fig. 15.16 Bearing capacity factor K_g for shallow
foundations (Menard 1975)

15.12 SKIRTED FOOTING

A 'skirt' is a wall-like projection or enclosure around a footing
on its underside. It may be constructed as an integral part of the
footing along its periphery or independently adjacent to the footing.
The object of the skirt is to provide lateral confinement of the
subsoil below the footing and thus contribute to bearing capacity.
Laboratory studies on model square and strip footings on dune sand
indicate that an integral skirt increases the ultimate bearing and
decrease the settlement (Singh et al 1982, Singh et al 1985). The
depth of skirt was half the footing width and the most efficient
range of angle with the vertical was observed to be 30° to 45°.
Beneficial effects of vertical skirt around a footing in increasing
bearing capacity and reducing settlement are reported by Rao and
Bhandari (1977).

15.13 RING FOOTING

A ring or annular footing (or raft) can be used for water tower structures, transmission towers and other possible super-structures. Laboratory model studies indicate that the behaviour of a ring footing of given area depends on the ratio of the inner diameter 'd' to the outer diameter D. Compared to a plain circular footing of the same area, the bearing capacity of model ring footing on dune sand was observed to increase with increase in the d/D ratio, reaching the maximum in the range of d/D = 0.3 − 0.4 and decreasing thereafter (Singh et al 1985).

15.14 BEARING VALUES FROM CODE

The values of safe bearing capacity to be used for preliminary design are also given in buildings codes and regulations. They are normally based on long experience of the behaviour of foundations on soils and are usually conservative. Typical values of the (net) safe bearing capacity (F $\simeq$ 3) are given in Table 15.10.

Table 15.10
CANADIAN SPECIFICATIONS FOR SAFE BEARING
CAPACITY OF SOILS (CGS 1985)

Type and conditions of soils	Safe bearing capacity (kPa)	Remarks
Cohesionless soils		
1. Dense gravel, dense sand and gravel	> 600	Foundation width B not less than 1 m.
2. Compact (medium dense) gravel, compact sand and gravel	200-600	Ground water level not less than B below base.
3. Loose gravel, loose sand and gravel	< 200	(100 kPa $\simeq$ 10 t/m²)
4. Dense sand	> 300	
5. Compact sand	100-300	
6. Loose sand	< 100	

Cohesive soils
1. Very stiff to hard clays or   300-600   Cohesive soils are
   heterogeneous mixtures   susceptible to long
   such as till   term consolidation,
                                               and swelling or
2. Stiff clays   150-300   shrinking
3. Firm clays   75-150   If $I_p > 30$ and
                                               clay content $>$
4. Soft clays and silts < 75 25%, examine
                                               swelling properties.
5. Very soft clays and silts   Not
                                 applicable

15.15 BEARING CAPACITY OF DEEP FOOTINGS

In case of a deep strip foundation the potential surface of sliding for a fairly isotropic soil mass may be assumed to be a logarithmic spiral starting under the foundation and ending with a vertical tangent, as shown in Fig. 15.17.

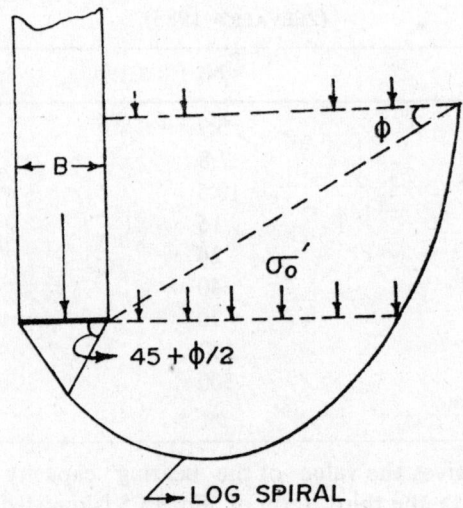

Fig. 15.17 Failure surface for a deep strip foundation
(Zeevaert 1983)

The overburden pressure σ_0' is assumed to act at the base level of the footing. With the conservative assumption that the mass of soil within the logarithmic spiral is weightless, Zeevaert (1983) analysed the limiting plastic equilibrium conditions and obtained the values of the bearing capacity factors N_c and N_q which are given in Table 15.11. Further, neglecting for all practical purposes the contribution of the third term of the bearing capacity equation (Eq. 15.5), the ultimate bearing capacity for a deep footing may be written as:

$$q_t \geqslant s \, [c_T \, N_c + \sigma_0' \, N_q] \, (I_D + 0.1) \qquad (15.74)$$

where   s = shape factor

= 1.0 for strip and 1.2 for square and circular footings

and $I_D + 0.1 = 1.0$ for dense state

The ultimate bearing capacity on a 'long-term basis' for footings on clays may be estimated by replacing the value of I_D by the consistency index I_c.

Table 15.11

BEARING CAPACITY FACTORS FOR DEEP STRIP FOUNDATIONS
(ZEEVAERT 1983)

ϕ (degrees)	N_c	N_q
0	5.7	1.0
5	7.8	1.7
10	10.5	2.7
15	15	4.8
20	24	8.1
25	40	15
30	70	30
35	140	65
40	300	150
45	~	420

Eq. 15.74 gives the value of the bearing capacity on the safer side because (i) the third term of Eq. 15.5 is omitted, and (ii) the side friction on the lateral surfaces of the footing is neglected; however, this effect is accounted for to some extent approximately

by the shape factor.   The safety factor for getting the safe bearing capacity should not be less than 2.0.

Example 15.8

Find the unit ultimate bearing capacity of a deep circular footing placed in a soil having $c_T = 30$ kPa, $\phi = 35°$ and $I_D = 0.4$. Assume $\sigma_0' = 120$ kPa.

For $\phi = 35°$, $N_e = 138$, $N_q = 66$

$I_D + 0.1 = 0.5$

$$q_t = 1.2 [(30 \times 138) + (120 \times 66)] \times 0.5$$

$$= 7236 \text{ kPa}$$

16

Bearing pressure on rocks

16.1. METHODS OF DETERMINATION

Rocks are usually considered as the best foundation material, however, unfavourable rock conditions may be dangerous and require at least the same care as a foundation on any type of soil. Depending on the rock quality, the following methods may be used to determine the allowable bearing pressure. (CGS 1985).

(1) Preliminary estimates of the allowable bearing pressure on sound rock and broken rock with joint spacing exceeding 0.6 m can be made fairly accurately from values given in codes, which are based on rock description.

(2) The allowable bearing pressure for sound rock (defined here as the rock with joint spacing exceeding 0.3 m) can be determined from their core strength.

(3) The pressuremeter test is suited best for determining the bearing pressure for rock of very weak to weak strength (1 to 25 MPa) and for rock mass with discontinuities at very close (0.02 — 0.06 m) to close spacing (0.06 — 0.20 m).

(4) Poor rocks which are very weak (q_u < 5 MPa) and rock mass with very close joint spacing (< 0.06 m) or are weathered and fragmented should be treated as a granular mass for determining their bearing pressure on the basis of conventional soil mechanics approach. However, it is difficult to determine the strength parameters necessary for design. Here also the pressuremeter allows for a direct determination of the strength parameters.

16.2 CORE STRENGTH METHOD

In case of a sound rock mass (joint spacing exceeding 0.3 m but including very weak rock material) with favourable conditions (i.e. horizontal foundations with vertical loading, and no open discontinuities), the (net) allowable bearing pressure may be estimated from the following expression (CGS 1985):

$$q_a = K_{sp} \, q_{u\text{-}core} \qquad\qquad (16.1)$$

where $q_{u\ core}$ = average unconfined compressive strength of rock
cores

K_{sp} = empirical coefficient depending on joint spacing and including a factor of safety of 3

= 0.4 (joint spacing > 3 m), 0.25 (joint spacing 1 — 3 m), 0.1 (joint spacing 0.3—1 m)

16.3 VALUES FROM CODE

1. Canadian recommendations. Typical values of the (net) allowable bearing pressure for vertically loaded foundations and carried down to unweathered rock are given in Table 16.1.

Table 16.1

CANADIAN SPECIFICATIONS FOR ALLOWABLE BEARING PRESSURE
ON ROCKS (CGS 1985)

Type and conditions of rocks	Rock material strength (MPa)	Allowable bearing pressure (MPa)
1. Massive igneous and meta-morphic rocks (granite, diorite, basalt, gneiss) in sound condition.	High to very high (> 50)	10
2. Foliated metamorphic rocks (slate, schist) in sound condition, (and where the strata or foliation are level or nearly so and the area has ample lateral support).	Medium to high (15—200)	3

3. Sedimentary rocks: cemented   Medium to high   1—4
shale, siltstone, sand-stone,   (15—200)
limestone without cavities,
thoroughly cemented con-
glomerates, all in sound condition,
(and level strata or foliation).

4. Compact shale and other   Low to medium   0.5
argillaceous rocks in sound   (4—50)
condition. (These rocks are
apt to swell on release of
stress, and on exposure to
water they are apt to soften
and swell appreciably).

5. Broken rocks of any kind   1
with joint spacing >0.3 m,
except argillaceous rocks (shale).

Note. The allowable bearing pressure for thinly bedded limestone, sand-stone and shale, and for heavily shattered or weathered rocks should be assessed by examination in situ, including testing if necessary. *Sound* rock conditions for the above table allow joint spacing not closer than 1 m.

2. IRC recommendations. The values of allowable bearing pressure on rock recommended by IRC: 78 are given in Table 16.2.

Table 16.2

IRC SPECIFICATIONS FOR ALLOWABLE BEARING PRESSURE ON ROCKS
(IRC: 78)

Types of rocks	Allowable bearing pressure for average conditions (MPa)
1. Hard igneous and gneissic rocks in sound conditions	10
2. Hard lime stones and hard sand stones	4
3. Schists and slates	3
4. Hard shales, hard mud stones and soft sand stones	2
5. Soft shales and soft mud stones	0.6—1
6. Soft lime stone	0.6
7. Heavily shattered rocks, con-glomerates and laterites	To be assessed by insitu tests

Note. These values are based on the assumption that the foundations are carried down below the weathered rock.

17

Foundations on expansive and collapsible soils

17.1 CONSTRUCTION ON EXPANSIVE SOIL

Expansive soils undergo volume changes upon wetting and drying which may result in structural damage to buildings. This damage is caused mainly by differential heave. Differential heave has been reported to be approximately $\frac{1}{4}$ to $\frac{1}{2}$ of the total heavy (Jennings and Kerrich 1962). In general, three basic approaches may be adopted for foundations on expansive clays (O'Neill and Poormoayed 1980): (i) altering the condition of the expansive clay, (ii) bypassing the expansive clay by isolating the foundation from its effects, and (iii) providing a shallow foundation capable of withstanding differential movements and mitigating their effects in the superstructure.

Referring to the USAEWES classification of expansive soils (Table 3.9, Chapter 3), volume changes in soils of 'low' potential expansion are minimal and normal construction procedures may be used; for marginal and high potential expansive soils, it is desirable to quantify the swell and use a rational foundation design. Table 17.1 gives recommendations for foundation types based on total computed heave and experience with various foundations for various wall spans (O' Neill and Poormoayed 1980).

These methods of construction for a certain range of heave are primarily ways for: (i) strengthening structures to withstand movement, (ii) making the structure sufficiently flexible to absorb movement without failure, or (iii) construction of the structure to be independent of soil heave (Gromko 1974). A common method of isolating the foundations from the effects of expansive soils is to construct a system of drilled piers and grade beams. In India, under-reamed piles and grade beams have proved quite successful in 'black cotton' expansive soils; these are described in Section 17.3.

Table 17.1

TOTAL HEAVE AND PREFERRED FOUNDATION CONSTRUCTION ON EXPANSIVE SOILS

Heave (mm) Length-to-height ratio of wall panel		Preferred foundation construction
1.25	2.5	
0—6	0—12	No special precautions
6—12	12—50	Stiffened mat; strip footings with high bearing pressure
12—50	50—100	Stiffened mat; cellular mat (several independent sections); drilled piers (piles) with suspended grade beams but floor slabs supported on fill with flexible joints; three-point support where feasible
50	100	Drilled piers with suspended floors

17.2 ALTERATION OF SOIL CONDITION

Based on the economics and practicality of the operation, methods which can be used to minimize heave of a particular soil may include: lime stabilization, moisture control, prewetting, compaction control, and replacement.

1. Lime stabilization. Addition of lime to soil supplies an excess of multivalent calcium cations which tend to replace monovalent cations such as sodium and potassium, and improve swell, shrinkage and workability characteristics. Lime soil pozzolanic reaction forms cementing agents providing a major increase in strength which is useful in retarding swell forces. Normally 5—7% hydrated lime by weight provides adequate alteration (O' Neill and Poormoayed 1980). Intimate mixing of soil with lime can be carried out practically only to a depth of about 1 m, but it is usually confined to the top 15 cm. Pressure injected lime slurry (PIL) has also been tried to con'rol swell (Thompson and Robnett 1976).

2. Moisture control. Most moisture control methods are applied around the perimeter of structures in order to minimize edge wetting or drying of foundations and to maintain uniform water conditions beneath the structure. Both horizontal and vertical moisture barriers are used; the former appear to be less effective. These barriers may take the form of either surface side walks, paved areas or buried impervious membrane. Vertical trenches, about 15 cm wide by 1.5 m deep and filled with gravel (capillary barrier) lean concrete, or mixtures of granulated rubber, lime and fly ash have been observed to be quite effective moisture barrier (Thompson and Robnett 1976, Johson 1979). All moisture barriers should be supplemented with adequate drainage systems. Vegetation and planting of trees and shrubs adjacent to the buildings should be avoided.

3. Prewetting. The purpose of prewetting is to raise the moisture content (i.e. to lower the suction) of the near-surface clays prior to placement of the structures. One of the most common wetting methods is ponding or submerging of an area in water. Ponding has been shown to be effective in minimizing subgrade heaving under highways, but it may not be effective for shallow foundation system.

4. Compaction control. Expansive clays expand very little when compacted at low densities and high moisture. Gromko (1974) recommends compaction at $2\% - 5\%$ above optimum moisture content. An excellent approach is to recompact swelling clays at moisture contents slightly above their natural moisture content and at a low density (Chen 1975). Compaction is also used in conjunction with replacement.

5. Replacement. A simple and easy solution for slabs and footings on expansive soils is to replace the foundation soil with non-swelling soils. Experience indicates that there is no danger of foundation movement if the subsoil consists of more than about 1.5 m of non-swelling soil underlain by highly expansive soils (Chen 1975). The main requirement for the replacement soils is that it should be non-expansive. All granular soils ranging from GW to SC may fulfill this requirement. However, granular soils with fines are preferable. The following criteria are found satisfactory (Chen 1975): (I) $w_L >$ 50%, fraction minus 75 μm = 15-30%, (ii) w_L = 30-50%, fraction minus 75 μm = 10-40%, (iii) $w_L < $ 30%, fraction minus 75 μm = 5-50%. In case it is difficult to locate materials to meet the above requirements, any selected fill may be used provided it is non-expansive. The degree of compaction should be 90 percent of standard Proctor density for supporting slabs and 95 to 100 percent for supporting footings. A minimum thickness of the fill beneath the bottom of footings and floor slabs should be 1 m, although 1.5 m is preferable, The replacement should also extend laterally for more than about 2 m beyond the building lines. (Chen 1975).

The use of a cohesive non-swelling soil (CNS) layer in replacing expansive, black cotton soil has been tried in India (Katti 1979) for canal linings, foundations of cross drainage structures and buildings. For building foundations, the normal thickness of the CNS layer below the footing level was kept 1 m, extending 1 m beyond the footing dimensions. It was compacted to modified Proctor compaction (ISI heavy compaction). A CNS layer 1 m wide is also recommended to be compacted on sides of the foundation course upto ground level. Tentative specifications for CNS material are given in Table 17.2

Table 17.2

TENTATIVE SPECIFICATIONS FOR CNS MATERIAL FOR REPLACEMENT OF
EXPANSIVE SOILS (KATTI 1979)

Properties		Specifications range
1. Grain size analysis		
Clay ($<$ 0.002 mm)	(%)	15—25
Silt (0.06 — 0.002 mm)	(%)	30—45
Sand (2 — 0.06 mm)	(%)	30—40
Gravel ($>$ 2 mm)	(%)	10
2. Consistency limits		
Liquid limit	(%)	30—50
Plastic limit	(%)	20—25
Plasticity index		10—25
Shrinkage limit	(%)	15 and above
3. Swelling pressure when compacted to standard Proctor optimum conditions and at no volume change (kg/cm²)		Less than 0.1
4. Approximate thickness of CNS layer:		
For swelling pressure (kg/cm²)		Thickness (cm)
1—1.5		75—85
2—3		90—100
3.5—5.0		105—115
		(Field trial necessary to arrive at optimum thickness)

6. High bearing pressure. The amount of heave decreases with an increase in the bearing pressure of the foundation. The bearing pressure to control soil expansion to a tolerable value (not precisely defined) may be estimated from the following equation (Bowles 1983) proposed by Komornik and David (1969) based on satistical analysis of some 200 soils:

$$\log \sigma_s = \bar{2}.132 + 2.08 \, (w_L) + 0.665 \, (\rho_d) - 2.69 \, (w_o) \qquad (17.1)$$

where σ_s = required bearing pressure (kg/cm²)

   w_L = liquid limit (decimal)

   ρ_d = dry density (g/cm³)

   w_o = natural moisture content (decimal)

The probable differential settlement may be estimated by knowing the percent free swell S_p given by the following two statistically obtained equations. Eq. 17.2 is given by Johson and Snethen (1979) and Eq. 17.3 by O'Neill and Ghazzaly (1977).

$$\log S_p = 0.0367 \, w_L - 0.0833 \, w_o + 0.458 \quad (\%) \quad (17.2)$$

and $S_p = 2.27 + 0.131 \, w_L - 0.27 \, w_o \quad (\%) \qquad (17.3)$

where w_L and w_o are in percents

The confining pressure σ_s (kPa) required to reduce the 'free swell' obtained from the above equations may be estimated from the following equation proposed by Bowles (1983):

$$S_p' = S_p \, (1 - A \, \sqrt{\sigma_s}) \qquad (17.4)$$

where S_p = percent swell from Eq. 17.2 or 17.3

   S_p' = reduced swell under σ_s

   A = 0.0735 for σ_s in kPa

The above equations for estimation of swell are expected to be valid within a ± 50 percent error, and the results may not differ greatly from those based on consolidation tests (Bowles 1983).

Loading the soil to sufficient pressure to balance swell pressure is used in many fills where the fill mass balances the swell pressure. This method can also be used beneath buildings by making footing of high bearing pressure. It can be used in combination with replacement layer (surcharge) also. The method may not be practical for ordinary buildlngs developing small soil pressures.

Example 17.1

Estimate the bearing pressure to limit the swell of a footing to an acceptable value. The soil has $w_L = 65\%$, natural moisture content $= 25\%$, $\rho_d = 1.4$ g/cm³ and depth of potential swell $= 2.5$ m. What will be the swell under the adopted bearing pressure?

From Eq. 17.1: σ_s = 0.552 kg/cm^2 = 54.1 kPa
From Eq. 17.2: S_{p1} = 5.77%
From Eq. 17.3: S_{p2} = 4.03%

These two values compare very well. Use the average value=4.9 %. Free swell S_p' under σ_s = 54.1 kPa is given by Eq. 17.4:

$$S_p' = 4.9 \,(1 - 0.0735 \sqrt{54.1}) = 2.25\%$$

$$\text{Total swell} = \frac{2.5 \times 2.25 \times 1000}{100} = 56 \text{ mm}$$

17.3   UNDER-REAMED PILES

The underreamed pile foundation developed by the CBRI, Roorkee (CBRI 1978, De 1978) is found to be economical and suitable for expansive clays. The principle of this type of foundation is to anchor the structure at a depth where ground movement due to seasonal moisture changes is negligible; in India this depth is about 3 to 4 m.

1. Dimensions. The underreamed pile is a short bored pile with an enlargement or bulb formed at or near the base. The following dimensions have been standardized (De 1978, IS 2911).

(1) Length 3 to 8 m. Minimum length in deep deposits of expansive clay 3.5 m, neglect top 1.2 m for skin friction. In shallow deposits, length may be reduced but take at least 50 cm in the stable zone.

(2) Diameter of underreamed bulb 2 to 3 times the shaft diameter, normally 2.5 times the shaft. For heavier loads, double or multiple bulbs may be used.

(3) Diameter of shaft of manually bored piles 200 to 500 mm.

(4) In case of multiple-bulb design (Mohan et al 1966), vertical spacing between two bulbs 1.5 times bulb diameter for piles upto 30 cm diameter. For larger diameter piles, bulb spacing 1.25 times diameter. The top most bulb is kept at a minimum depth of 1.5 m or twice bulb diameter, whichever is greater, from the ground surface.

(5) Minimum horizontal spacing of piles 2 times the bulb diameter under normal loading. If adjacent piles are of different diameters, take the average value. If the spacing is reduced to 1.5 times, reduce bearing capacity by 10 percent. Maximum spacing generally not exceeding 2.5 m to avoid heavy capping beams.

Typical underreamed piles with one and two bulbs and a capping
beam are shown in Fig. 17.1.

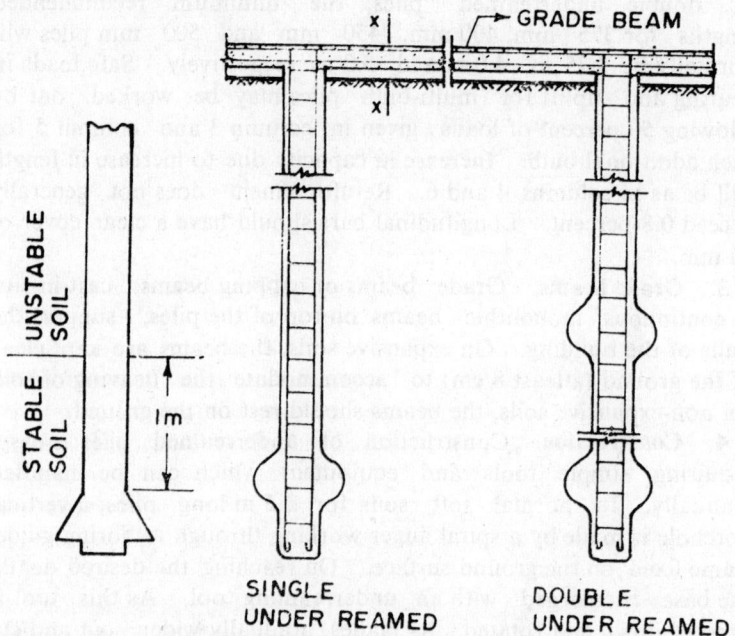

SINGLE
UNDER REAMED

DOUBLE
UNDER REAMED

Fig. 17.1 Underreamed pile foundation

2. Load capacity. The ultimate load carrying capacity of an
underreamed pile can be determined by static analysis as described
in Chapter 19.   A pile load test (IS: 2911-8) is more reliable than
static analysis.

Because of uncertainties in sampling and laboratory testing, safe
loads for piles have been worked out from a large number of full
scale, short and long term tests carried out in different part of
India.   These values (CBRI 1978) are given in Table 17.3 and are
applicable for clays of medium consistency $(4 < N < 8)$ and for
medium compact sands $(10 < N < 30)$.   For stiff clays $(N \geqslant 8)$ and
dense sands $(N \geqslant 30)$, the safe loads may be increased by 25 percent.
Here N is the penetration resistance in terms of number of blows in
the standard penetration test (Chapter 15). A reduction of 25 percent
should be made in case of loose sands and soft clays. Only 75 percent
of the safe loads should be used where the boreholes are full of subsoil water during concreting.   For a group of piles, the safe load of

individual piles is multiplied by the number of piles in the group,
The values given in the table are for a minimum pile length of 3.5 m.
In double underreamed piles, the minimum recommended
lengths for 375 mm, 400 mm, 450 mm and 500 mm piles will
normally be 3.75 m, 4.0 m and 5.0 m respectively. Safe loads in
bearing and uplift for multi-bulb piles may be worked out by
allowing 50 percent of loads given in column 3 and column 5 for
each additional bulb. Increase in capacity due to increase in length
will be as in columns 4 and 6. Reinforcement does not generally
exceed 0.8 percent. Longitudinal bars should have a clear cover of
40 mm.

   3.   Grade beams. Grade beams or capping beams, cast-in-situ
as continuous monolithic beams on top of the piles, support the
walls of the building. On expansive soils, the beams are kept clear
of the ground (atleast 8 cm) to accommodate the heaving of soil.
On non-expansive soils, the beams should rest on the ground.

   4.   Construction. Construction of underreamed piles is easy,
requiring simple tools and equipment which can be handled
manually. In normal soft soils for 3.5 m long piles, a vertical
borehole is made by a spiral auger working through a 'boring guide'
frame fixed on the ground surface. On reaching the desired depth,
the base is enlarged with an underreaming tool. As this tool is
pressed down and rotated, its blades gradually widen out and cut
the soil from the sides which is collected in a bucket provided with
the underreamer. When the bucket is full, the underreamer is
taken out. The process is repeated until full underreaming is
complete. For multi-under-reamed piles, after the first bulb is
complete, the underreamer is taken out and the borehole is
advanced to the next depth where the other bulb is to be formed.
For boring deeper than 3.5 m, and also for boreholes larger than
30 cm diameter, a portable tripod hoist with a manually operated
winch is used to handle the equipment.

   Concreting is done soon after the borehole is ready. The
reinforcement cage is lowered and concrete is powered in. Slump
of concreting is kept 70 mm to 150 mm (120 mm is desirable).
Compaction of concrete by rodding is sufficient. In clayey soils,
boring and underreaming under water do not normally pose any
problem. In sandy soils where the sides tend to cave in, the
borehole is kept filled with a drilling mud made by mixing about 5
percent bentonite in water (Jain et al 1969). Concrete is deposited

Table 17.3

SAFE LOAD FOR SINGLE BULB, UNDERREAMED PILES IN CLAYEY AND SANDY SOILS, INCLUDING BLACK COTTON SOILS

Shaft dia	Bulb dia	Bearing-resistance		Uplift-resistance		Lateral thrust	Reinforcement	
		Single bulb	Increase per 30 cm length	Single bulb	Increase per 30 cm length		Longitudinal bars No./dia	Rings 6 mm φ spacing
(mm)	(mm)	(t)	(t)	(t)	(t)	(t)	(mm)	(mm)
(1)	(2)	(3)	(4)	(5)	(6)	(7)	(8)	(9)
200	500	8	0.9	4	0.65	1.0	3/10	180
250	625	12	1.15	6	0.85	1.5	4/10	220
300	750	16	1.4	8	1.05	2.0	4/12	250
375	940	24	1.8	12	1.35	3.0	5/12	300
400	1000	28	1.9	14	1.45	3.4	6/12	300
450	1125	35	2.15	17.5	1.60	4.0	7/12	300
500	1250	42	2.4	21	1.80	4.5	9/12	300

by means of a tremie pipe with a suitable valve assembly at its bottom.

17.4 CONSTRUCTION ON COLLAPSIBLE SOIL

If a structure is to be supported on shallow foundation on or above collapsible soils and the isolated footing occupy upto 30 percent of the foundation area, or when allowable settlements cannot be satisfied, it is more economical and safe to use continuous strip footings (Zeevaert 1972). In a typical design (Fig. 17.2) suggested by Zeevaert, the foundation system consists reaction beams formed by footing beams and load balancing beams in the longitudinal direction; the load balancing beams are reinforced to make the system sufficiently stiff. The ultimate bearing capacity is determined from the expression (Zeevaert 1983).

$$q_t = (\sigma_0' N_q + 0.5 \gamma B N_\gamma)(I_D + 0.1) \tag{17.5}$$

where I_D = density index. Assume $(I_D + 0.1)$ = 1 for dense soil;
= 0.6 for semicompact or medium compressible soil;
= 0.3 for loose or highly compressible soil.

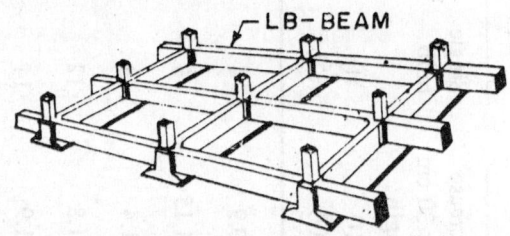

Fig. 17.2 Continuous footings on compressible soils
(Adapted from Zeevaert 1972)

If the footing area becomes greater than 50 pescent of the entire area of the building, a mat or raft foundation should be considered. The performance may be increased by making it a compensated foundation. In many cases, deep foundations (piles, caissons, etc.) may be required to transmit load to suitable bearing strata below the collapsible deposit.

Soil treatment. On some sites, it may be feasible to either stabilize or cause collapse of the soil deposit prior to construction on it (Clemence 1983). The various treatment methods can be summarized as follows (Bora 1976, Clemence 1981): (a) *For depth* *of treatment   0—1.5 m*: moistening and compaction (conventional, extra-heavy, impact, or vibratory rollers). (b)   *For depth of treatment 1.5—10 m*: (i)   Over-excavation and recompaction (earth pads with or without stabilization by additives such as lime or cement). (ii)   Control of potential water sources through drainage and/or impervious membranes.   (iii)   Vibroflotation (free draining soils). (iv)   Rock columns (Vibroreplacement).   (v)   Displacement piles. (vi)   Injection of silt or lime.   (vii)   Ponding or flooding (if no impervious layers exist).   (c)   *For depth of treatment 10 m*: (i) Ponding and infiltration wells.   (ii)   Ponding and infiltration wells with the use of explosives.   (iii)   Any of the above or combinations of the above methods where applicable. The possible future methods may be: (i) Heat treatment to solidify the soils in place. (ii)   Ultrasonics to produce vibrations that will destroy the bonding mechanisms.   (iii) Chemical additives to strengthen the bonding mechanism of the metastable soil structure (possibly electro-chemical methods).   (iv) Use of grout-like additives to fill the pore spaces before solidification.

The Russians have conducted extensive studies with chemical stabilization techniques. The methods currently employed are (Clemence 1983): (i) gaseous silicatization of sandy and loessial soils; (ii) strengthening of carbonate cements by polymers; and (iii) chemical strengthening of alluvial soils by clay-silicate solutions.

18
Foundation settlement

18.1 COMPONENTS OF SETTLEMENT

In general, the total settlement S of a foundation may be regarded as consiting of three components:

$$S = S_i + S_o + S_s \tag{18.1}$$

where, S_i is the 'immediate' settlement, S_o is the 'primary consolidatiion,' settlement (time-dependent) and S_s is the 'secondary settlement', or *creep* which is also time dependent. The immediate settlement is the settlement which takes place as the initial construction load is applied. It is a short duration settlement and is assumed to take place as soon as the load is applied, (or during a short period, say within 7 days of the application of the load). Immediate settlement keeps pace with construction. Immediate settlement is also defined as the 'distortion' settlement which takes place without change in volume, or moisture content, and is sometimes referred to as the 'elastic' settlement.

The constant-volume, distortion settlement under undrained conditions may be possible only in case of saturated clays where significant lateral strain is possible; immediate settlement is zero if the lateral strain is zero. For example, immediate settlement may be neglected for thin clay layer sandwiched between cohesionless soil layers or between a cohesionless soil layer at top and rock at bottom. In caseof rapid draining coarse grained soils and in case of partly saturated fine giained soils (degree of saturation less than about 90 percent), immediate settlement is accompanied by volume changes.

18.2 IMMEDIATE SETTLEMENT OF SATURATED CLAYS

The immediate, distortion settlement, although not actually elastic, is usually estimated by the elastic theory. The immediate settlement S_i below a uniformly loaded foundation is given by:

$$S_i = q_n \ B \ \frac{1 - \mu^2}{E_u} \ I \qquad (18.2)$$

where q_a is the net applied pressure, B is the width or diameter of loaded area, I is the combined influence factor for shape and rigidity (Table 18.1), E_u is the 'undrained' modulus of deformation and μ is the Poisson's ratio (ranging from 0.3 to 0.5) which is taken as 0.5 for saturated clay with no volume change.

Table 18.1
INFLUENCE FACTORS I FOR SETTLEMENT OF UNIFORMLY LOADED
FOUNDATION ON THE SURFACE OF AN ELASTIC
MASS (TSYTOVICH 1976)

Side ratio L/B	For half space			For finite thickness layers of H/B (flexible, average)				
	Flexible (centre) (max.)	Flexible (average)	Rigid	0.25	0.5	1	2	5
1 (circle)	1.00	0.85	0.79 (π/4)	0.22	0.38	0.58	0.70	0.78
1 (square) (rectangle)	1.12	0.95	0.88	0.22	0.39	0.62	0.77	0.87
2	1.53	1.30	1.22	0.24	0.43	0.70	0.96	1.16
3	1.78	1.53	1.44	0.24	0.44	0.73	1.04	1.31
5	2.10	1.83	1.72	—	—	—	—	—
10	2.53	2.25	2.12	0.25	0.46	0.77	1.15	1.62

The influence factor for settlement at a corner of square and rectangular, flexible foundations is half of the value for the centre. I for edge of a circle is 0.64. H in Table 18.1 is the finite thickness of the compressible layer.

The ratio of the settlement of a rigid foundation to the settlement of a flexible foundation loaded to the same, uniform intensity is called the 'rigidity factor'. As an approximation,

$$S_i \ (\text{rigid}) \approx 0.8 \ (S_i)_{max} \ (\text{flexible})$$
$$\approx 0.9 \ (S_i)_{average} \ (\text{flexible}) \qquad (18.3)$$

The undrained modulus may be determined from the undrained shear strength s_u (van $\geq$ strength) with the following empirical relations (CGS 1978):

$$E_u = 500\, s_u \text{ (soft sensitive clay, NC clay)} \tag{18.4}$$

$$E_u = 1000\, S_u \text{ (firm to stiff clay, OCR} < 2) \tag{18.5}$$

$$E_u = 1500\, s_u \text{ very stiff clay, OCR} > 2) \tag{18.6}$$

1. Depth factor. The calculated settlement for surface loading requires correction if the foundation is placed below the ground surface. The correction factor — depth factor f_D — is a function of D/B, L/B and μ (Fox 1948). Solving the equation for depth factor proposed by Fox, Bowles (1983) plotted a chart for depth factor, from which the author has tabulated the values as given in Table 18.2. The corrected settlement becomes:

$$S_i \text{ (corrected} = S_i \text{ (surface)} \times f_D \tag{18.7}$$

Table 18.2

VALUES OF FOX DEPTH FACTOR f_D (TABULATED BY THE AUTHOR FROM BOWLES — 1983)

$\dfrac{D}{B}$	L/B = 1		L/B = 5	
	$\mu = 0.3$	$\mu = 0.5$	$\mu = 0.3$	$\mu = 0.5$
0.5	0.775	0.85	0.87	0.925
0.6	0.74	0.82	0.85	0.91
0.7	0.715	0.785	0.83	0.89
0.8	0.685	0.76	0.815	0.875
0.9	0.67	0.74	0.80	0.86
1.0	0.65	0.72	0.78	0.85
2.0	0.565	0.625	0.68	0.76
3.0	0.53	0.585	0.62	0.69
4.0	0.515	0.57	0.58	0.65
5.0	0.50	0.55	0.56	0.625
6.0		0.54	0.545	0.615
7.0		0.54	0.54	0.595
8.0		0.535	0.53	0.58
9.0		0.525	0.52	0.57
10.0		0.520	0.515	0.565

2. Janbu analysis. Janbu et al (1956) proposed the following generalized equation (Eq. 18.8) which is more convenient to use for finding the average immediate settlement of a uniformly loaded flexible foundation placed on saturated, clay having a constant undrained modulus E_u:

$$S_i \text{ (average)} = f_1 \ f_o \ \frac{q_n B}{E_u} \quad \text{(for } \mu = 0.5) \qquad (18.8)$$

where f_1 = correction factor for finite thickness H of compressible layer

f_o = correction factor for depth D of foundation

B = width or diameter of foundation

The factors f_1 and f_o have been evaluated by Christian and Carrier (1978). Values of factor f_1 are given in Fig. 18.1. Values of factor f_o on the basis of Christian and Carrier chart are given by the author in Table 18.3. Rigidity factors are not applied to the settlements calculated from Eq. 18.8 (Tomlinson 1980).

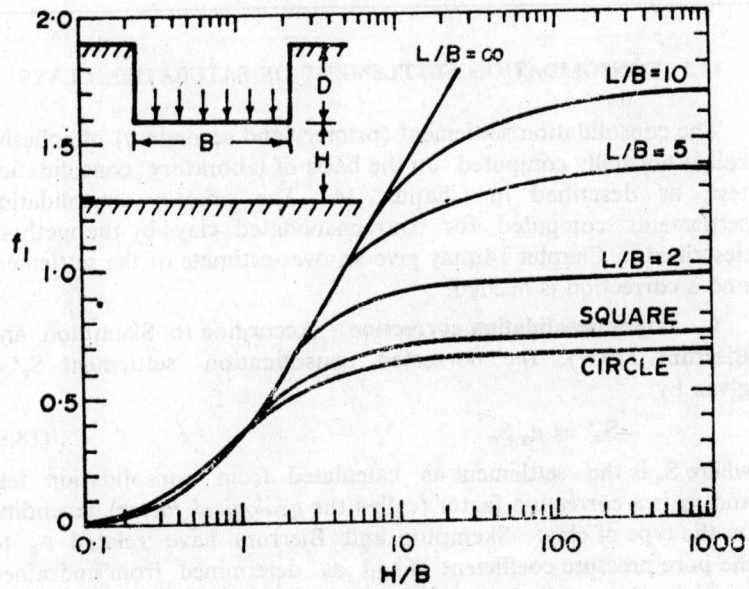

Fig. 18.1 Improved correction factor f_1 for use in Janbu Eq. 18.8 (Christian and Carrier 1978)

Table 18.3

CORRECTION FACTOR f_0 FOR USE IN JANBU EQ. 18.8

(CHRISTIAN AND CARRIER 1978)

D/B	f_0	D/B	f_0
0.0	1.000	3.0	0.890
0.2	0.964	4.0	0.884
0.4	0.950	5.0	0.878
0.6	0.934	6.0	0.874
0.8	0.924	7.0	0.872
1.0	0.920	8.0	0.870
1.2	0.914	9.0	0.869
1.4	0.908	10.0	0.868
1.6	0.906	12.0	0.864
1.8	0.904	15.0	0.860
2.0	0.900	20.0	0.858

18.3 CONSOLIDATION SETTLEMENT OF SATURATED CLAYS

The consolidation settlement (primary and secondary) of cohesive soil is normally computed on the basis of laboratory consolidation test, as described in Chapter 14. The primary consolidation settlements computed for overconsolidated clays by the methods described in Chapter 14 may give an over-estimate of the settlement and a correction is needed.

1. **Over-consolidation correction.** According to Skempton and Bjerrum (1957), the corrected consolidation settlement S_o' is given by:

$$S_o' = \mu_g S_o \tag{18.9}$$

where S_o is the settlement as calculated from consolidation test and μ_g is a correction factor (called the *geological factor*) depending on the type of clay. Skempton and Bjerrum have related μ_g to the pore pressure coefficient of soil as determined from undrained triaxial tests and also to the dimensions of the foundation. For practical purposes, the values of μ_g may be adopted as given in Table 18.4 (Tomlinson 1980).

Table 18.4

GEOLOGICAL FACTOR μ_g FOR CORRECTION OF CONSOLIDATION
SETTLEMENT (TOMLINSON 1980)

Type of clay	Factor
Soft, sensitive clays	1.0—1.2
Normally consolidated clays	0.7—1.0
Lightly overconsolidated clays	0.5—0.7
Heavily overconsolidated clays	0.2—0.5

To account for small departure from one dimensional consolidation, Leonards (1976) (NAVFAC DM-7.1) has related the correction factor μ_g to the overconsolidation ratio OCR, the width B of the foundation and the thickness H of the compressible layer, as given in Fig. 18.2. The correction is suggested to be applied to the settlement of overconsolidated clays. If B $>$ 4H, $\mu_g = 1$ may be used. Also, if the depth to the top of clay layer exceeds 2 B, use $\mu_g = 1$ (Das 1983).

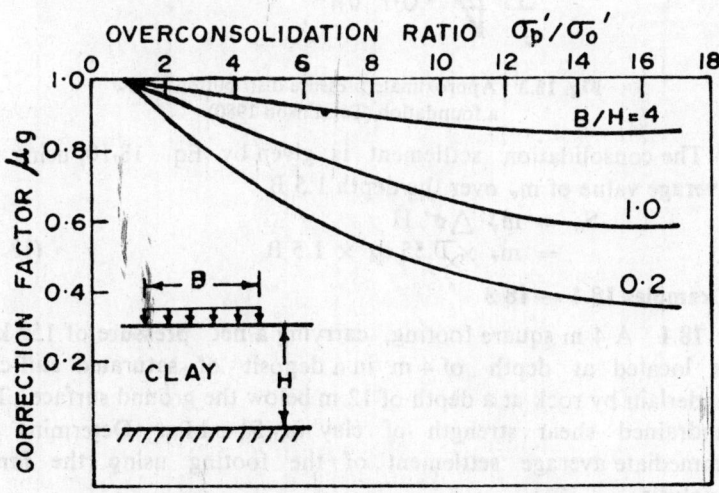

Fig. 18.2 Geological correction factor μ_g for settlement of over-consolidated clays (Leonards 1976)

2. Depth correction. Theoretically, Fox correction (Table 18.2) applies only to immediate settlement, but it is considered logical to apply the correction to consolidation settlement S_o also to allow for the depth of the foundation (Tomlinson 1980, IS: 8009-1).

3. **Approximate pressure distribution method.** For fairly small size foundations on uniform stiff and relatively incompressible clays, Tomlinson (1980) has suggested an approximate pressure distribution, as shown in Fig. 18.3.

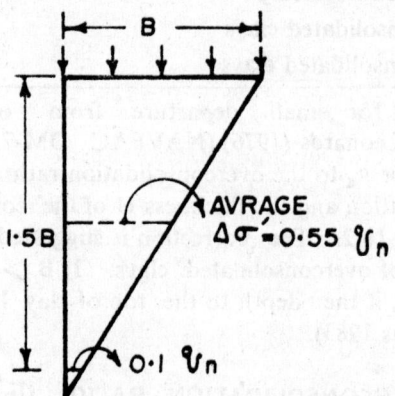

Fig. 18.3 Approximate pressure distribution below a foundation (Tomlinson 1980)

The consolidation settlement is given by Eq. 18.10, using an average value of m_v over the depth 1.5 B.

$$S_o = m_v \, \Delta \sigma' \, H$$
$$= m_v \times 0.55 \, q_n \times 1.5 \, B \tag{18.10}$$

Examples 18.1 — 18.3

18.1 A 4 m square footing, carrying a net pressure of 150 kPa is located at depth of 4 m in a deposit of saturated stiff clay underlain by rock at a depth of 12 m below the ground surface. The undrained shear strength of clay is 50 kPa. Determine the immediate average settlement of the footing using the Janbu analysis.

$$D = B = 4 \text{ m}, D/B = 1, \quad f_o = 0.92 \quad \text{(Table 18.3)}$$
$$H = 8 \text{ m}, H/B = 2, \qquad f_1 = 0.5 \quad \text{(Fig. 18.1)}$$

Assume $E_u = 1000 \, s_u = 50,000 \text{ kPa}$

$$S_i = f_1 \, f_o \frac{q_n \, B}{E_u} = 5.5 \text{ mm}$$

18.2. A R.C.C. foundation, 20 m $\times$ 10 m, exerts a net pressure of 200 kPa on the surface of a saturated clay extending to a great depth. The undrained modulus of deformation is estimated to be 45 MN/m². Determine the immediate settlement under the footing. What would be the settlement if rock existed at a depth 10 m below the surface. What will be the settlement by the Janbu analysis, when rock exists?

L/B $=$ 2, μ $=$ 0.5, I (rigid) $=$ 1.22 (Table 18.1)

$$S_i = q_a B \frac{1 - \mu^2}{E_u} I = 41 \text{ mm}$$

Rock at 10 m.

Assume the foundation flexible,

H/B $=$ 1, I $=$ 0.7 (Table 18.1)

S_i $=$ 23.3 mm (average)

S_i rigid) $\simeq$ 0.9 S_i (average, flexible) $=$ 21 mm

$$S_i = f_1 f_0 \frac{q_a B}{E_u}$$

f_0 $=$ 1, f_1 $=$ 0.35 for H/B $=$ 1 and L/B $=$ 2

S_i $=$ 16 mm

18.3 A concrete footing, 5 m $\times$ 5 m, is placed at a depth of 3 m in a bed of stiff clay underlain by rock at 5 m below the ground surface. The footing carries a net pressure of 180 kPa. The clay is overconsolidated, OCR $=$ 3 and has m_v $=$ 0.09 m²/MN and E_u $=$ 55 MN/m². Estimate the total immediate and consolidation settlement.

(a)   Immediate settlement.

$$S_i = q_n B \frac{1 - \mu^2}{E_u} I$$

I (Table 18.1) $=$ 0.62, for L/B $=$ 1 and H/B $=$ 1

Assume μ $=$ 0.25

S_i $=$ 7.61 mm

S_i (rigid) $\approx$ 0.9 S_i (average, flexible)

$=$ 0.9 $\times$ 7.61 $=$ 6.85 mm

Depth factor f_D $=$ 0.82, for D/B $=$ 0.6 (Table 18.2)

Corrected S_i $=$ 6.85 $\times$ f_D $=$ 5.6 mm

(b)   Consolidation settlement.

$$S_c = m_v \Delta\sigma' H$$

$\Delta\sigma'$ at mid-depth of layer by assuming 2 : 1 pressure distribution $=$ 80 kPa

$S_o = 36$ mm

Geological factor $\mu_g = 0.9$ (Fig. 18.2)

Depth factor $f_D = 0.82$ (Table 18.2)

Corrected $S_c = 36 \times 0.9 \times 0.82 = 26.57$ mm

Total $S = S_i + S_o = 32$ mm

If the Janbu method is used for immediate settlement:

$$S_i = f_i \, f_o \quad \frac{q_n \, B}{u}$$

$f_1 = 0.3$ (Fig.18.1), $f_o = 0.934$ (Table 18.3)

$S_1 = 4.6$ mm, no further correction for rigidity or depth is required.

18.4 SETTLEMENT OF COHESIONLESS SOILS

The settlement of foundations on coarse grained soils occurs almost exclusively by volume change at a rate essentially equivalent to the rate of stress change. Some coarse grained soils exhibit a delayed volume change phenomenon known as 'friction lag'. This is analogous to the secondary compression of fine grained soil. Deformation moduli of coarse grained soils are markedly nonlinear with respect to stress change and depend on the initial state of stress. Because of the difficulty in obtaining reasonably undisturbed samples for laboratory testing, the settlements are normally estimated by semi-empirical methods based on insitu testing.

1· **Buisman De Beer method.** Using Buisman empirical formula, De Beer and Marten (1957) proposed a 'constant of compressibility C' given by:

$$C = \frac{\beta_o \, q_c}{\sigma_o{}'} \tag{18.11}$$

where $q_c =$ static cone resistance (CPT) (kPa)

$\sigma_o{}' =$ effective overburden pressure at point of measurement (kPa)

$\beta_c = 1.5$

The total settlement S (immediate + long term, if any) under a shallow foundation is calculated by using C in the classic Terzaghi formula for consolidation settlement and is given by:

$$S = \frac{H}{C} \, 2.3 \log \frac{\sigma_o{}' + \angle \sigma'}{\sigma_o{}'} \tag{18.12}$$

where $\sigma_0' =$ average, intial effective overburden pressure on
the layer of thickness H

$\Delta\sigma' =$ pressure increase at middle of layer due to net
foundation pressure q_n

A cone resistance versus depth of layer (thickness H of the
compressible layer or twice the width of foundation, whichever is
less) is plotted. The total depth being considered for compression
is divided into separate layers, each having approximately equal q_c,
and an *average* q_c is assigned to each layer. The settlement of each
layer is separately calculated and the results are added together to
get the total settlement.

By using the value of $\beta = 1.5$ in Eq. 18.11, the calculated settlements are generally overestimated. A less conservative value of
$\beta = 1.9$ suggested by Meyerhof of (1965) has been more widely
used. As a recommended modification (Merritt and Gardner 1983)
which considers the influence of the density index I_D and increased
deformation modulus, β_0 for use in Eq. 18.11 is given by:

$$\beta_0 = 2 (1 + I_D{}^2) \qquad (18.13)$$

The method strictly only applies to normally loaded sands. The
settlement will be small in a preloaded sand. But it is, of course,
difficult to determine the degree of overconsolidation of a granular
deposit.

2. Schmertmann method. The Schmertmann method (1970)
which is also based on CPT, employs a simplified distribution of
vertical strain below foundation and may give predictions which are
closer to reality though somewhat on the conservative side.

The distribution of vertical strain ϵ under a foundation transmitting a net, uniform pressure q_n is expressed as:

$$\epsilon = \frac{q_n}{E_V} I_s \qquad (18.14)$$

where E_V is the compressibility modulus and I_s is the 'strain
influence factor' which is a function of Poisson's ratio μ and depth
z. The compression of an elementary thickness Δz is thus $\epsilon \Delta z$.
According to Schmertmann et al (1978), I_s immediately below
a rigid foundation is assumed as:

$$I_s = 0.1 \text{ for square foundation (L = B)} \qquad (18.15)$$

$$I_s = 0.2 \text{ for long foundation (L} \geqslant \text{10 B)} \qquad (18.16)$$

and the peak value I_{sp} is assumed as:

$$I_{sp} = 0.5 + 0.1 (q_n/\sigma_{op}')^{0.5} \qquad (18.17)$$

where σ_{op} is the initial, effective overburden pressure at the depth where the peak occurs. The depth of peak strain is assumed as 0.5 B for a square foundation and B for a long foundation. The thickness of layer contributing to settlement is taken as 2 B for square and 4 B for long foundation, i.e. I_z is zero at these two depths respectively. Based on these assumptions. the simplified I_z distribution is shown in Fig. 18.4.

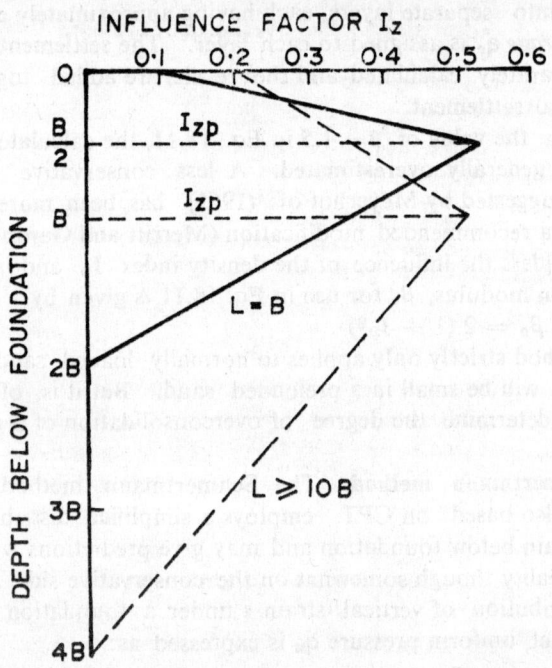

Fig. 18.4   Distribution of strain influence factor for settlement analysis of rigid footings on sand (Schertmann at al 1978)

The values of E_V for clean, fine to medium sands and silty sands are empirically estimated from the cone resistance q_o:

$$E_V = 2.5\ q_o \quad (L/B \leqslant 2) \tag{18.18}$$

$$E_v = 3.5\ q_o \quad (L/B \geqslant 10) \tag{18.19}$$

The settlement of a long foundation induces plane strain. Because of additional confinement, E_V (plane strain) exceeds E_V for a square area (axisymmetric case); i.e. settlement of the former is less.

According to Lee's experiments (1970), the modulus for plane strain is 1.4 times the modulus for axisymmetrical case.

Introducing the correction factors for foundation embedment (depth factor C_1) and time-dependent, creep settlement (frictional lag) (creep factor C_2) the total settlement S is given by:

$$S = C_1 \, C_2 \, q_a \, \Sigma(I_z/E_v) \, \Delta z \qquad (18.20)$$

where $\qquad C_1 = 1 - 0.5 \, (\sigma_o'/q_u) \qquad\qquad (18.21)$

$$C_2 = 1 + 0.2 \log_{10} (t/0.1) \qquad\qquad 18.22)$$

and $\qquad \sigma_o' =$ initial, effective overburden pressure at foundation level

$\qquad\qquad t =$ time in years after construction

For intermediate cases of foundation, $B < L < 10 \, B$, interpolation is made between the solutions obtained for $L=B$ and $L=10 \, B$. No correction is indicated for water table. The cone resistance is obtained with approximately the same water table as when the foundation is loaded. The method is not applicable to large footings and foundation rafts (Merritt 1983).

Computations are made in a tabular form following the steps given below:

(1) Plot the strain influence factor diagram.

(2) Based on q_o profile, divide the compressible depth (2 B or 4 B, as the case may be) into convenient number of layers.

(3) Assign average q_c to each layer, and calcualte E_v.

(4) Locate mid-depth of each layer, below foundation and find I_z for this depth.

(5) Calculate $(I_z/E_v) \, \Delta z$ for each layer and sum the results.

(6) Find C_1 and C_2.

(7) Calculate S from Eq. 18.20.

If SPT resistance (N values) is measured, E_v is estimated as given in Table 18.5 (NAVFAC DM-7, 1982).

Table 18.5

VALUES OF COMPRESSIBILITY MODULUS (E_v) ESTIMATED FROM SPT RESISTANCE (N) FOR USE IN SCHMERTMANN METHOD

Soil type	E_v (kPa)
1. Silts, sandy silts, slightly cohesive silt-sand mixtures	400 N
2. Clean, fine to medium sands, and slightly silty sands	700 N
3. Coarse sands, and sands with little gravel	1000 N
4. Sandy gravels, and gravel	1200 N

3. Plate load test method. The use of plate load test for estimating the foundation settlement and obtaining bearing pressure for desired settlement has been described in Chapter 15. The following equation (Eq. 18.23) proposed by Bond (1961) is also found to give a reasonably good estimate of the foundation settlement S from the plate settlement S_p, loaded to the same intensity (Bowles 1983) (CGS 1985):

$$S = S_p (B/B_p)^{n+1} \qquad\qquad ((18.23)$$

where B is the foundation width, B_p is the plate width and the exponent n, depending on soil type, may be assumed as (Bowles 1983): (i) clay: 0.03—0.95, (ii) sandy clay: 0.8—0.10, (iii) dense sand: 0.40—0.50; (iv) medium dense sand: 0.25—0.35, and (v) loose sand: 0.20—0.25. The exponent n can also be determined by performing two or more plate tests with different plate sizes and solving Eq. 18.23.

The plate load test method is suitable only for cohesionles soils where time-dependent settlement relationships are negligible. It tests only a shallow depth of soil and extrapolation to large footings should be carried out with caution.

4. Meyerhof method. The total settlement S (mm) of shallow foundations (width B m) on saturated sands and gravels, according to Meyerhof empirical correlation (1976) is given by:

$$S = 0.96 \frac{q_a \sqrt{B}}{N} \simeq \frac{q_n \sqrt{B}}{N} \text{ (mm)} \qquad (18.24)$$

where q_n is the net foundation pressure (kPa) and N is the average corrected SPT blows within the seat of settlement ($\simeq$ B). For silty sands, twice the right hand side of Eq. 18.24 should be used. For deep foundation (D > 4B), Eq. 18.24 can be used with a 50% reduction for settlement.

Examples 18.4 — 18.7

18.4 A footing, 3 m square, carries a net pressure of 220 kPa at a depth of 0.5 m in a deep deposit of sand of bulk unit weight 19 kN/m³. The average cone resistance with depth is given in Table 18.6. Estimate the settlement of the footing, using the Schmertmann method, at the end of construction and after 5 years of construction. What will be the settlement by the Buisman-De Beer method?

q_n = 220 kPa = 0.22 MPa

σ_{op}' at B/2 = $(0.5 + 1.5) \times 19 = 38$ kPa

I_{sp} = $0.5 + 0.1 (q_n/\sigma_{op}')^{0.5} = 0.74$

σ_0' at footing level = 9.5 kPa

C_1 = $1 - 0.5 (\sigma_0'/q_n) = 0.98$

C_2 = $1 + 0.2 \log (t/0.1) = 1.34$

A strain influence diagram (not shown) is plotted with $I_s = 0.1$ at footing level and $I_{sp} = 0.74$ at 1.5 m below the footing. $I_z = 0$ at 2 B = 6 m. Calculations are shown in Table 18.6. Take $E_V = 2.5\ q_o$.

Table 18.6

EXAMPLE 18.4   (SCHMERTMANN METHOD)

Depth below footing (m)	Average q_o (MPa)	E_v (MPa)	Average depth below footing (m)	I_s (1×10^{-2})	$\dfrac{I_s}{E_v}$	Δz (m)	$\dfrac{I_s}{E_v} \Delta z$ (1×10^{-2})
0—1	2.0	5.0	0.5	0.313	6.26	1	6.26
1—2	3.0	7.5	1.5	0.740	9.86	1	9.86
2—4	3.5	8.75	3.0	0.493	5.63	2	11.26
4—6	6.0	15.0	5.0	0.164	1.09	2	2.18
							29.56

S = $C_1 C_2 q_n \Sigma (I_s/E_V) \Delta z$

At t = 0, $C_2 = 1$

S_o = $0.98 \times 1 \times 0.22 \times 29.56 \times 10^{-2}$ mm = 6.37 cm

At t = 5 yr

S_5 = $C_2 S_o = 1.34 \times 6.37 = 8.54$ cm

Buisman De-Beer Method

$$C = \frac{1.5\ q_o}{\sigma_0'}$$

$$S = \frac{H}{C}\ 2.3 \log \frac{\sigma_t'}{\sigma_o}.$$

The initial over-burden pressure σ_0' is calculated at mid-depth of each layer. Total load on footing is 1980 kN. The pressure increase $\Delta\sigma'$ at mid-depth is calculated assuming 2 vertical to 1 horizontal spread of pressure under the footing. Other more accurate methods of pressure distribution can also be used, as given in Chapter 13. Calculations are given in Table 18.7.

Table 18.7

EXAMPLE 18.4   (BUISMAN DE-BEER METHOD)

Layer depth below footing (m)	σ_o' (kPa)	q_c (kPa)	C	Area at mid-depth (m)^2	$\Delta\sigma'$ (kPa)	σ_t' (kPa)	S (cm)
0—1	19	2000	157.89	12.25	161.60	180.60	1.42
1—2	38	3000	118.42	20.25	97.78	135.78	1.07
2—4	66 5	3500	78.94	36.00	55.00	121.50	1.52
4—6	104.5	6000	86.12	64.00	30.90	135.40	0.60
						Total	4.61

18.5 A square footing, 3.6 m $\times$ 3.6 m, carrying a gross pressure of 268 kPa is placed at a depth of 1 m below ground surface in a sandy deposit of unit weight 18 kN/m^3. The sand layer is underlain by rock at a depth of 4 m. The static cone resistance is recorded in Table 18.8. Find the total settlement by the Schmertmann method.

$$q_a = q - \gamma D = 250 \text{ kPa} = 0.25 \text{ MPa}$$
$$\sigma_{op}' \text{ at } B/2 = (1 + 1.8) \times 18 = 50.4 \text{ kPa}$$
$$I_{zp} = 0.5 + 0.1 (q_n/\sigma_{op}')^{0.5} = 0.723 \text{ at } B/2$$
$$I_s \text{ at footing level} = 0.1$$
$$C_1 = 1 - 0.5 (\sigma_o'/q_a) = 0.964$$
$$C_2 = 1 \text{ (end of construction)}$$

A strain influence diagram is plotted (not shown) with $I_s = 0.1$ at footing level and $I_{zp} = 0.723$ at 1.8 m below the footing. $I_s = 0$ at 2 B = 7.2 m. As rock exists at a depth of 3 m below the footing, compression can occur upto 3 m only. Calculations are shown in Table 18.8.

Take $E_v = 2.5 \, q_c$

$$S = C_1 C_2 q_n \Sigma (I_s/E_v) \Delta z$$
$$= 0.964 + 1 \times 0.25 \times 16.3 \times 10^{-2} \text{ m} = 3.93 \text{ cm}$$

Table 18.8

EXAMPLE 18.5

Depth below footing	Average q_c	E_v	Average depth below footing	I_s	I_s/E_v	Δz	$\dfrac{I_s \cdot \Delta z}{E_v}$
(m)	(MPa)	(MPa)	(m)	(1×10^{-2})		(m)	(1×10^{-2})
0—1	3.0	7.50	0.5	0.273	3.64	1	3.64
1—2	3.50	8.75	1.5	0.619	7.07	1	7.07
2—3	4.50	11.25	2.5	0.629	5.59	1	5.59
							16.30

18.6 Footing of a wall, width $= 2$ m, depth in sand deposit $= 1$ m, $\gamma = 18$ kPa, gross pressure $= 258$ kPa, cone resistance given in Table 18.9. Find settlement by the Schmertmann method at the end of construction.

$$q_n = q - \gamma D = 240 \text{ kPa} = 0.24 \text{ MPa}$$
$$\sigma_{op}' \text{ at } B = (1+2) \times 18 = 54 \text{ kPa}$$
$$I_{sp} = 0.5 \times 0.1 \ (q_a/\sigma_{op}')^{0.5} = 0.711$$

I_s at footing level $= 0.2$, $I_s = 0$ at $4B = 8$ m below footing. A diagram I_s versus depth below footing is plotted (not shown).

$$E_v = 3.5 \ q_o \quad \text{(for strip footing)}$$
$$C_1 = 1 - 0.5 \ (\sigma_o'/q_n) = 0.96$$
$$C_2 = 1$$

The calculations are given in Table 18.9.

$$S = C_1 C_2 q_a \Sigma \ (I_s/E_v) \ \Delta z$$
$$= 0.96 \times 1 \times 0.24 \times 25.61 \times 10^{-2} \text{ m} = 6 \text{ cm}$$

Table 18.9

EXAMPLE 18.6

Depth below footing	Average q_c	E_v	Average depth below footing	I_z	I_s/E_v	Δz	$\dfrac{I_s \ \Delta z}{E_v}$
(m)	(MPa)	(MPa)	(m)	(1×10^{-2})		(m)	(1×10^{-2})
0—1	2.5	8.75	0.5	0.328	3.748	1	3.748
1—3	3.5	12.25	2.0	0.711	5.804	2	11.608
3—6	4.0	14.00	4.5	0.415	2.964	3	8.892
6—8	5.0	17.50	7.0	0.119	0.680	2	1.360
							25.608

18.7 Estimate the maximum settlement of a 4.5 m square footing placed at a depth of 10 m in saturated sand. The corrected SPT blows within the depth 4.5 m below the footing are 30. The net pressure on the footing is 200 kPa.

Meyerhof Eq. 18.24: $S \simeq \dfrac{q_n\sqrt{B}}{N}$ mm

$$S = \frac{200 \sqrt{4.5}}{30} = 14 \text{ mm}$$

Comparing with Peck formula   Eq. (15.66)

q_n (for 25 mm settlement) $= 10.25 \ N$ (kPa)

$$= 10.25 \times 30 = 307.5 \text{ kPa}$$

$$S \text{ under } 200 \text{ kPa} = \frac{25 \times 200}{307.5} = 16 \text{ mm}$$

It is assumed that the effect of water table is already reflected in measured values of N.

18.5 TOTAL DEFORMATION MODULUS METHOD

1. Determination of total deformation modulus*. The total deformation modulus E_o can be determined in the field by a plate load test or in the laboratory from the data of a consolidation test. It can also be empirically estimated from the results of insitu penetration tests.

The equation for total settlement S_p (elastic + residual) of a rigid, circular plate of diameter B_p on the surface of *soil* half space can be written from the theory of elasticity by using E_o in place of the Young's modulus E of the seminfinite body:

$$S_p = q_a \ B_p \ \frac{1 - \mu^2}{E_o} \ I \ ; \ I = \frac{\pi}{4} \tag{18.25}$$

For a square plate of width B, the equivalent diameter $B_p = 2B/\sqrt{\pi}$ is used. In terms of total load Q, Eq. 18.25 may be written as:

$$E_o = (1 - \mu^2) \ \frac{Q}{S_p \ B_p} \tag{18.26}$$

where $Q = \pi B_p^2 \ q_a/4$

---

* For a detailed description, please refer to the author's book Geotechnical Testing & Instrumentation, Vol. 2 of Soil Engineering in Theory and Practice (Singh 1981).

The test is performed in a test pit using a 79.8 cm diameter plate (area 0.5 m^2) and the E_0 can be calculated from Eq. 18.26. In case of boreholes, a rigid circular plate, 0.04 to 0.05 m^2 in area, is used. The calculated value of E_0 is multliplied by a correction factor for depth, which may be taken equal to 0.5 for $D/B_p \geqslant 10$, where D is the depth of borehole.

In the consolidation test, the compressibility modulus E_v is first calculated from Eq. 8.14:

$$E_v = \frac{\Delta\sigma_z}{\Delta H/H_1} = \frac{1}{m_v}$$

and E_0 is then calculated as:

$$E_0 = \beta E_v = \frac{\beta}{m_v} \qquad (18.27)$$

where m_v is the coefficient of volume change and β is a coefficient characterizing lateral expansion of soil, which is a function of Poisson's ratio μ.

$$\beta = \frac{(1 + \mu)(1 - 2\mu)}{(1 - \mu)} \qquad (18.28)$$

Average values of μ and β are given in Table 18.10.

Table 18.10
AVERAGE VALUES OF μ AND β

Soil type	μ	β
1. Dense sand	0.25	0.84
2. Loose sand, and clayey sands and silts	0.30	0.74
3. Firm and stiff sandy clays and silty clays	0.35	0.63
4. Firm and stiff clays	0.40	0.47
5. Very stiff clays, and gravelly soils	0.20	0.90

Approximate estimates of compressibility modulus E_v (kPa) can also be made from empirical correlations with SPT (Bowles 1983): (i) gravelly sand: $1200 (N + 6)$, (ii) sand: $500 (N + 15)$ and also $(18000 + 750 N)$, (iii) clayey sand: $320 (N + 15)$, (iv) silty sand: $300 (N + 6)$. The empirical correlations with CPT are as follow, where E_v will have the units of q_0 (Bowles 1983): (i) soft clay: $6-8 q_0$, (ii) clayey sand: $3-6 q_0$, (iii) silty sand: $1-2 q_0$.

For sands: $E_0 = 1.5—3.5 \, q_o$, also $2 (1 + I_D^2) \, q_c$ (Vesic 1970). The higher the q_o, the lower is the multiplying factor. (See also Singh 1981).

The standard values of the total modulus E_0 recommended in the Russian Code (Dalmatav 1962) are given in Table 18.11, which can be adopted for design purposes.

Table 18.11

STANDARD VALUES OF TOTAL DEFORMATION MODULUS E_0 (MPa)

(DALMATAV 1962)

Soil type	Plastic limit (%)	Void ratio (%)					
		41—50	51—60	61—70	71—80	81—95	95—100
Gravelly and coarse sand	—	46	40	33			
Medium sand	—	46	40	33			
Fine sand	—	37	28	24			
Silty sand	—	14	12	10			
Clayey soils	< 9.4	18	14	11			
	9.4—12.4	23	16	13			
	12.5—15.4	35	21	15	12		
	15.5—18.4	—	30	19	13	10	8
	18.5—22.4	—	—	30	18	13	9
	22.5—26.4	—	—	—	26	16	11
	26.5—30.4	—	—	—	—	22	14

2. Calculation of settlement. In general, the total settlement (elastic + residual) of foundations on soil can be dectermined according to the theory of elasticity by using total modus E_0 in place of elastic modulus E and rewriting Eq. 18.2 as:

$$S = q_n \, B \, \frac{1 - \mu^2}{E_0} \, I \qquad (18.29)$$

In practice, it is more convenint to find settlement according to the method using coefficient of volume compressibility m_v adopted for cohesive soils. The long term settlement which is the total settlement for cohesionless soils and consolidation settlement for cohesive soils is thus expressed as a sum of the settlements of individual compressible layers:

$$S = \Sigma H \, m_v \, \Delta\sigma' = \Sigma H \frac{\beta}{E_o} \, \Delta\sigma' \qquad (18.30)$$

where $\Delta\sigma'$ = average increase in pressure on layer

Tsytovich (1976) proposed an alternative equivalent stratum method in which the settlement of a certain *equivalent stratum* of limited thickness H_e subjected to a uniform increase of pressure q_n is equated to that calculated by Eq. 18.29. Thus,

$$H_e\, m_v\, q_a = H_e\, \frac{\beta}{E_o}\, q_n = \frac{1 - \mu^2}{E_o}\, I\, B\, q_n \qquad (18.31)$$

or
$$H_e = \frac{(1 - \mu)^2}{(1 - 2\mu)}\, IB = I_e\, B \qquad (18.32)$$

where I_e = coefficieat of equivalent stratum

Tsytovich has tabulated values of I_e as function μ and L/B ratio. The equivalent stratum H_e for a foundation of width B can be easily computed and the settlement is then given by:

$$S = H_e\, m_v\, q_n \qquad (18.33)$$

Reference may be made for further details to Tsytovich (1976) or to the author's other book (Singh 1981).

Example 18.8

18.8 A 0.6 m square test plate is placed on the surface of sand and loaded to 200 kPa. Determine the total deformation modulus, if the plate settles by 4 mm. What wiil be the settlement of a rigid footing, 2.5 m × 2.5 m, if it is loaded to the same intensity of pressure.

$E_o = (1 - \mu^2)\, Q/S_p\, B_p$

$B_p = 2B/\sqrt{\pi} = 2 \times 0.6/\sqrt{\pi} = 0.677$ m

Assume $\mu = 0.3$

$E_o = 24.2 \times 10^3$ kPa

$S = q_n\, B\, \dfrac{1 - \mu^2}{E_r}\, I,\quad B = 2.5$ m, $I = 0.88$

$= 16.5$ mm

Compare with the result obtained by the Bond equation (Eq. 18.23), assuming n=0.2. S= 24 mm.

18.6 PRESSUREMETER MODULUS METHOD

Foundation settlements can be estimated to an acceptable degree of accuracy from the pressuremeter test results. In fact, these estimates are considered presently the most reliable for granular soils (CGS 1985). The pressurementer deformation moduls E_p is given by (Singh 1981):

$$E_p = K \frac{\Delta p}{\Delta V} \qquad (18.34)$$

where Δp = pressure increment

ΔV = volume change of cell

K   = probe coefficient of compression

The compressibility modulus E_v can be calculated from E_p as:

$$E_v = \beta_m E_p \qquad (18.35)$$

where β_m is the structure factor depending on the type of soil and on E_p/p_1, as given in Table 18.12

Table 18.12

VALUES OF FACTOR β_m (CGS 1985) (WILUN AND STARZEWSKI 1975)

Type	Peat		Clay		Silt		Sand		Sand & gravel	
		E_p/p_1	β_m	$\dfrac{E_p}{p_1}$	β_m	$\dfrac{E_p}{p_1}$	β_m	$\dfrac{E_p}{p_1}$	β_m	β_m
OC or very dense	—	> 16	1.0	> 14	1.5	> 12	2.0	> 10	3.0
NC or dense	1	9—16	1.5	8—14	2.0	7—12	3.0	6—10	4.0
Underconsoli- dated or loose	—	< 9	2.0	< 8	2.0	< 7	3.0	< 6	4.0

The total settlement S (cm) of a footing of width B (cm) is given by Eq. 18.36 (CGS 1985) (Wilun and Starzewski 1975). The first term of the equation represents the settlement due to volumetric deformation,

$$S = \frac{4}{9 E_p} \, q \, R_0 \left[\lambda_1 \, \frac{B}{2 R_0} \right]^{1/\beta_m} + \frac{q}{9\beta_m E_p} \lambda_2 B$$

$$(18.36)$$

where   q   = applied pressure in units of E_p (kPa)

R_0 = reference width = 30 cm

$\lambda_1 \lambda_2$ = shaper factors as given in Table 18.13, for a circle, $\lambda_1 = \lambda_2 = 1.0$

Table 18.13

SHAPE FACTORS λ_1 AND λ_2

L/B	1.0	2.0	3.0	5.0	10.0	20.0
λ_1	1.12	1.53	1.70	2.14	2.40	2.65
λ_2	1.10	1.20	1.30	1.40	1.45	1.50

18.7 ALLOWABLE SETTLEMENT

The allowable settlement is the settlement for a particular structure which is limited by (i) allowable maximum settlement and (ii) allowable differential settlement, both for the structure as a whole and between parts of the structure. A uniform settlement causes theoretically no damage to the structure, e.g. Palace of Arts and National Theatre in Mexico City have withstood settlements of 3 m and 1.5 m respectively. However, settlements exceeding certain limits may interfere with the function of the structure, or may create an unacceptable maintenance or aesthetic problem. Differential settlements may damage the structure by overstressing. The foundation must be so designed that both the criteria of allowable maximum and differential settlements are satisfied. Based on observations of settlement of buildings, the maximum settlement of a part of a building has been roughly correlated to the differential settlement between two parts. According to Terzaghi and Peck, the differential settlements generally do not exceed 75% of the maximum settlement. Grant et al (1974) observed the differential settlement to be roughly 54% of the maximum settlement on clay and 58% of the maximum settlement on sand.

As maximum settlements can be predicted with some accuracy (but not differential settlements), it is usual to relate allowable settlement to maximum settlement. Allowable maximum settlements suggested by Skempton and MacDonald (1955) and also recommended by IS: 1904 are as follows:

(1) Isolated foundations on sand   = 40 mm
(2) Isolated foundations on clay   = 65 mm
(3) Rafts on sand   = 40 to 65 mm
(4) Rafts on clay   = 65 to 100 mm

Greater settlement limits are allowed on clays because progressive settlements on clays permit strain adjustments in the building and also due to the fact that clays tend to be more homogeneous than granular soils. Bjerrum (1963) related structural damage to angular distortion, defined as the settlement difference (ΔS) between two points divided by the horizontal distance (L) apart.

These limits ($\Delta S/L$) are given below and may be considered in design.

(1) Limit where difficulties with machinery sensitive to settlements are to be feared $= 1/750$

(2) Limit of danger for frames with diagonals $= 1/600$

(3) Safe limit for buildings where cracking is not permissible $= 1/500$

(4) Limit where first cracking in panel walls is to be expected, and limit where difficulties with overhead cranes are to be expected $= 1/300$

(5) Limit where tilting of high, rigid buildings might become visible $= 1/250$

(6) Considerable cracking in panel walls and brick walls; safe limit for flexible brick walls ($h/L < \frac{1}{4}$); and limit where structural damage of buildings is to be feared $= 1/150$

19

Pile foundation

19.1 TYPES OF PILE

A 'pile' is a columnar element in a foundation which performs the function of transferring load from the superstructure to underlying stronger and less compressible strata through weak and compressible strata or through water. A pile may be required to carry uplift and lateral loads when used to support structures subjected to overturning forces. A pile may also be used to mainly densify a loose, cohesionless soil to increase its bearing capacity.

Piles may be classified in a number of ways, e g. by the material of which they are made, by their method of installation, or by the manner they transmit their load.

According to the method of installation, the main types of piles are as follows:

(a) *Driven piles.* Preformed piles, usually of concrete, steel or timber, which are driven into the soil by the blows of a hammer.

(b) *Driven and cast-in-place piles.* Piles formed by driving a tube fitted with a driving shoe into the soil and filling the tube with concrete. The tube may or may not be withdrawn.

(c) *Jacked and screwed piles.* Steel or concrete piles jacked or screwed into the soil.

(d) *Bored and cast-in-place piles.* Piles formed by making a borehole into the soil and filling it with concrete.

(e) *Composite piles.* Combinations of two or more above types or combinations of different materials in the same type of pile.

The first three of the above types are called "displacement" piles since their installation causes displacement and disturbance of soil. But in the case of steel H piles and tubes without a driving shoe, soil displacement is small. The bored piles and some forms of composite piles are called "non-displacement" piles.

Any pile sustaining and transmitting a superimposed load is called a "bearing pile". Depending on the manner in which bearing piles transmit loads to the soil, they are classified as:

(1) end-bearing or point bearing piles (free standing columns)
(2) friction piles, and
(3) combined resistance piles.

If a pile supports the load mainly by the resistance developed at its point or base, it is called an "end bearing pile". The base (tip) of an end bearing pile bears on a hard stratum such as rock, dense gravel or sand. A pile installed in a homogeneous weak soil transmits its load to soil mainly by shearing resistance (friction or adhesion) along its surface (shaft resistance) and is called a "friction pile," or "floating pile" A combined resistance pile supports the load partly by base resistance and partly by shaft resistance.

19.2 LOAD CAPACITY

The ultimate load capacity (ultimate bearing capacity) of a single pile is generally estimated by:

(1) Static analysis
(2) Dynamic analysis (for driven piles)
(3) Test loading

The static analysis considers the pile-soil ineraction *after* installation and equates the load on the pile to the shaft resistance and the base resistance. The dynamic analysis is based on equating the driving energy to the work done by the pile and energy losses during driving.

19.3 STATIC ANALYSIS

The basis equation for the net, ultimate pile load Q_t, i.e. the load which can be supported by a pile top at failure, is to express Q_t as the sum of the net load carried by the pile base Q_b (net, total base resistance) and the load carried by the pile shaft Q_s (total shaft resistance) at failure.

$$Q_t = Q_b + Q_s \tag{19.1}$$

or
$$Q_t = f_b A_b + f_s A_s \tag{19.2}$$

where f_b = net, ultimate bearing capacity of pile base (unit base resistance, or point resistance)

f_s = unit shear resistance of soil adjacent to pile shaft (also called frictional resistance or skin friction)

A_b = area of pile base (point)

A_s = surface area of pile shaft

Since the mechanism of resistance offered by cohesionless and cohesive soils are different, their cases are discussed separately.

19.4 ANALYSIS OF PILE CAPACITY IN SAND

The base resistance and skin friction of a pile of width or diameter B and length L_p embedded in cohesionless soil can be evaluated on the basis of (i) plastic equilibrium of soil, (ii) standard penetration test, and (iii) static cone test.

1. Method based on plastic equilibrium. The ultimate, net unit base resistance f_b in sand according to the theory of bearing failure of soil can be expressed as:

$$f_b = \frac{Q_b}{A_b} = \sigma_0' N_{qp} + 0.4\,\gamma' B N_{\gamma p} - \frac{W_p}{A_b} \qquad (19.3)$$

where N_{qp} and $N_{\gamma p}$ are the pile bearing capacity factors and σ_0' is the effective overburden pressure at the base level. The term $0.4\gamma' B N_{\gamma p}$ is neglected to compensate for the weight of pile itself (W_p/A_b) and because of the usually small value of B, the pile width. (Here the author uses the *net* resistance to mean the gross resistance less the pressure due to pile weight and not in the usual sense where original overburden pressure is subtracted from gross resistance. In many cases the pressure due to weight of pile may be assumed to be approximately equal to the pressure release due to the soil displacement. Moreover, the pile weight W_p itself is small in relation to Q_t and may be ignored to obtain the load at the pile top.) Thus Eq. 19.3 can be re-written as:

$$f_b = \sigma_0' \, N_{qp} = \gamma' \, L_p \, N_{qp} \qquad (19.4)$$

It has been observed from tests on full scale and model pile tests that the base resistance as well as the skin friction of piles in sand increase with depth only upto a certain value of depth, known as the 'critical depth D_c' which is roughly 7B to 20B. Both f_b and f_s below the critical depth remain approximately constant to limiting values in uniform soil conditions. This is thought to be due to arching of soil around the lower part of the pile. The overburden pressure in Eq. 19.4 is, therefore, multiplied by a reduction factor β_p (Berezantzev et al 1961) whose values are given in Table 19.1, and Eq. 19.4 is written as:

$$f_p = \beta_p \, \sigma_o' \, N_{qp} \tag{19.5}$$

Table 19.1

REDUCTION FACTOR β_p FOR OVERBURDEN PRESSURE

(BEREZENTZEV ET AL 1961)

L_p/B	ϕ' (degrees)				
	26	30	34	37	40
5	0.75	0.77	0.81	0.83	0.85
10	0.62	0.67	0.73	0.76	0.79
15	0.55	0.61	0.68	0.73	0.77
20	0.49	0.57	0.65	0.71	0.75
$\geqslant 25$	0.44	0.53	0.63	0.70	0.74

Instead of applying β_p, an alternative method is to calculate σ_o' taking into account the likely critical depth D_c. Thus,

$$\sigma_o' = \gamma' L_p \text{ for } L_p < D_o \tag{19.6}$$

and

$$\sigma_o' = \gamma' D_o \text{ for } L_p \geqslant D_o \tag{19.7}$$

Suggested values of D_o (as a ratio D_o/B) are given in Table 19.4 (SAA 1978). Koerne (1984) recommends a limit of approximately 20B for calculating overburden pressure for piles longer than 15 m. *The maximum value of unit base resistance as indicated by published pile tests is* 11000 kPa (Tomlinson 1980).

The pile bearing capacity factor N_{qp} depends on ϕ' and the depth-diameter ratio L_p/B. For a given value of ϕ', the value of N_{qp} decreases with increasing L_p/B ratio. Suggested values of N_{qp} (Berezantzev et al 1961) are given in Table 19.2. Values of N_{qp} for $L_p/B = 50$ are the extrapolated values (Craig 1983).

Table 19.2

VALUES OF PILE BEARING CAPACITY FACTOR N_{qp}
(BEREZANTZEV ET AL 1961, CRAIG 1983)

ϕ'	Ratio of embedded length to shaft diameter				
(degrees)	5	20	25	50	70
25	15	8	—	—	—
28	—	—	12	9	5
30	28	20	17	14	12
32	—	—	25	22	—
34	—	—	40	37	—
35	67	52	—	—	42
36	—	—	58	56	—
38	—	—	89	88	—
40	180	145	137	136	125

The values of N_{qp} recommended by NAVFAC DM-7-2 (1982) are given in Table 19.3, and those by the Australian Pile Code (SAA 1978) in Table 19.4. The author suggests to adopt the values of Table 19.3 for design purposes. For smaller values of ϕ', N_{qp} may be adopted the same as for footing.

For $\phi \leqslant 20°$, N_{qp} may also be adopted as twice the values of N_q given in Table 15.3 for shallow foundations (DGI 1978).

Table 19.3

RECOMMENDED VALUES OF N_{qp} FOR DRIVEN DISPLACEMENT PILES
(NAVFAC DM-7-2(1982) (LIMIT ϕ' TO 28°, IF JETTING IS USED)

ϕ' (degrees)	N_{qp}	ϕ' (degrees)	N_{qp}
26	10	35	50
28	15	36	62
30	21	37	77
31	24	38	86
32	29	39	120
33	35	40	145
34	42		

The average, ultimate unit skin friction f_s in homogeneous sand may be expressed as:

$$f_s = K_s \sigma_s' \tan \delta \leqslant f_{sm} \qquad (19.8)$$

where K_s = average coefficient of earth pressure on pile shaft

σ_s' = average effective overburden pressure along pile shaft

δ = angle of friction between sand and pile

f_{sm} = maximum value of f_s at and below D_c

For design purposes K_s may be taken as 1.0 for loose sand and 2.0 for dense sand. These values should be halved for steel H piles. Suggested values of δ are $0.75\phi'$ for concrete pile and $20°$ for steel piles. The values of $K_s \tan \delta$ are influenced by various factors such as angle of shearing resistance, method of installation, compressibility and original horizontal stress is soil, and pile size and shape. Values of $K_s \tan \delta$ suggested by SAA (1978) are given in Table 19.4. Reliable values of $K_s \tan \delta$ can only be obtained from load tests.

Table 19.4

SUGGESTED VALUES OF D_c/B, $K_s \tan \delta$ AND N_{qp}
FOR FRICTION PILES IN SANDS (SAA 1978)

Density index		D_c/B	$K_s \tan \delta$		N_{qp}	
State	I_D		Driven piles	Bored or cast in place piles	Driven piles	Bored piles
Loose	0.2—0.4	6	0.8	0.3	60	25
Medium	0.4—0.75	8	1.0	0.5	100	60
Dense	0.75—0.9	15	1.5	0.8	180	100

The skin friction f_s is assumed to increase linearly with depth upto D_c, and a constant value is taken beyond D_c. *Values of f_s higher than 100 kPa have not been recorded and this should be considered as the upper limit for design* unless justified by a full scale load test (Wawryk 1972). For straight sided piles driven deeper then 20B, average values of skin friction, as suggested by Tomlinson (1980) and given in Table 19.5 may be used for design purposes.

Table 19.5
AVERAGE SKIN FRICTION IN SANDS (TOMLINSON 1980)

Density index		f_s (kPa)
State	I_D	
Loose	< 0.35	10
Medium-dense	$0.35-0.65$	$10-25$
Dense	$0.65-0.85$	$25-70$
Very dense	> 0.85	$70-110$

Allowable load. The allowable load is obtained by applying to f_b and f_s a minimum safety factor of 3. The allowable load Q_a is thus given by:

$$Q_a = \frac{1}{3} \left[\left(f_b \frac{\pi B^2}{4} \right) + \left(\frac{f_s}{2} \pi BL_p \right) \right] \text{ for } L_p < D_c$$

(19.9)

where f_b and f_s are calculated at depth L_p.

and $$Q_a = \frac{1}{3} \left[\left(f_b \frac{\pi B^2}{4} \right) + \left(\frac{f_s}{2} \pi BD_c \right) + f_s \pi B \ (L_p\text{-}D_c) \right]$$
$$\text{for } L_p > D_c$$ (19.10)

where f_b and f_s are calculated at depth D_c.

2. Method based on SPT. The values of f_b and f_s can be ob'ained from the empirical correlations proposed be Meyerhof (1976, 1983) for driven piles.

$$f_b = 40 N \frac{L_p}{B} \leqslant 400 N \text{ (kPa) (sands)} \quad (19.11a)$$

$$\leqslant 300 N \text{ (kPa) (non-plastic silts) (19.11}_b)$$

where $N =$ average N value of SPT in the vicinity of pile base (corrected for overburden pressure)

$L_p =$ pile length *actually* embedded within sand (deducting the portion in weak soil, if any)

and $f_s = 2 N_{av}$ (kPa) 19.12)

where $N_{av} =$ average N value over embedded length of pile

also $f_s = N_{av}$ (kPa) (small displacement H piles) (19.13)

For bored piles, f_b and f_s are approximately one-third and one-half, respectively, of the corresponding values for driven piles.

The ultimate load Q_f is calculated from Eq. 19.2 and the allowable load is obtained by applying to Q_f a minimum safety factor of 4 (CGS 1985).

3. Method based on CPT. The static cone can also be used to estimate empirically the values of f_b and f_s (Meyerhof 1970) for driven piles.

$$f_b = q_c \tag{19.14}$$

$$f_s = (f_c)_{av} \tag{19.15}$$

or $\quad f_s \simeq (q_0)_{av}/200 \text{ (sand)} \tag{19.16a}$

$$\simeq (q_0)_{av} /150 \text{ (non-plastic silt)} \tag{19.16b}$$

where $\quad q_0 =$ cone penetration resistance at base level

$(f_s)_{av} =$ average skin friction measured by a cone fitted with friction sleeve

$(q_c)_{av} =$ average cone resistance over embedded pile length

The units of f_b and f_s are the same as those of q_0 and f_0. For H piles, f_s is reduced by half. The cone resistance q_c for equating to the pile base resistance is popularly taken as the average between 3B above and B below the base. This may lead to overestimation in some cases. To reach the full base resistance, the pile must be driven to atleast 5B into the bearing stratum. To correct for pile driving effects in sand, a correction factor is also suggested (Wawryk 1972) as follows: (i) for $q_0 \leqslant 6$ MPa , correction factor $= 1.2$, (ii) for $q_0 = 6-12$ MPa, factor $= 1.0$, (iii) for $q_c = 12-15$ MPa, factor $= 0.8$ (loosening effect) (iv) for $q_c > 15$ MPa, factor $= 0.5$.

It is recommended to use f_b equal to the minimum measured value of q_c for piles with B > 0.5 m (CGS 1985). (See *Scale effect*). Knowing f_b and f_s, Q_f is calculated from Eq. 19.2. In all the above methods, frictional resistance of H sections is calculated for all surfaces and not for the gross area, and the base resistance is calculated on the net cross sectional area of steel only. *The normal upper limit for f_s is 100 kPa (or 50 kPa for H piles).* The limiting value of q_c for point bearing pressure is 15 MPa (Lee et al 1983); it is doubtful whether large diameter piles can be driven 8 pile diameters into sand when $q_0 > 15$ MPa. (Wawryk 1972). *It is advisable not to exceed the upper limit of $f_b = 11000$ kPa in design* as this is the maximum value indicated by published pile tests (Tomlinson 1980).

Since CPT results are more reproducible then SPT results giving a greater confidence in the design method based on CPT, a safety factor between 2.5 and 3 is applied to Q_f for getting Q_a (CGS 1985).

When piles are driven to refusal in very dense sands or gravels or to rock, the maximum allowable load is usually limited by the strength of pile rather than the support of soil.

Scale effect. The above mentioned methods of estimating the pile capacity from SPT and CPT results are valid for piles less than about 0.5 m diameter. For piles of larger diameter, the ultimate base resistance is recommended to be reduced by the following empirical reduction factor R_b (Meyerhof 1983):

$$R_b = \left(\frac{B + 0.5}{2B} \right)^n \leqslant 1 \text{ for } B > 0.5 \text{ m} \qquad (19.17)$$

where $n = 1$ for loose sand, $n = 2$ for medium dense sand, and $n = 3$ for dense sand. The ultimate unit skin friction of piles in sand of a given density is, however, practically independent of the pile diameter.

Examples 19.1–19.3

19.1 A cast-in-place concrete pile in medium sand is 0.5 m in diameter and 10 m long. The water table is 2 m below the top of pile. The bulk density of sand above the water table is 1.8 t/m³ and below the water table the density is 1.95 t/m³. Angle of shearing resistance is 38°. Assume $N_{qp} = 60$ and $K_s \tan \delta = 0.5$ (Table 19.4). Determine the ultimate load and the safe load for the pile with a safety factor of 3.

(a) *Base resistance.*

Table 19.4: $D_o/B = 8$, $D_o = 8 \times 0.5 = 4$ m

$L_p > D_o$

$f_b = \sigma_o' N_{qp}$, where $\sigma_o' =$ effective vertical pressure at D_o

$\sigma_c' = (1.8 \times 9.8 \times 2) + (0.95 \times 9.8 \times 2) = 53.9$ kPa

$f_b = 3234$ kPa

$A_b = \pi B^2/4 = 0.196$ m²

$Q_b = f_b . A_b = 635$ kN

(b) *Frictional resistance.*

The frictional resistance is calculated by considering the pile in 3 parts: 0 to 2 m, 2 m to 4 m (D_c) and 4 m to 10 m. The frictional resistance will increase upto $D_o = 4$ m only.

σ' at 2 m = $2 \times 1.8 \times 9.8 = 35.28$ kPa

σ' at 4 m = $35.28 + (0.95 \times 9.8 \times 2) = 53.9$ kPa

Average vertical stress for $0-2$ m = 17.64 kPa

$\text{Av. } f_{s1} = 17.64 \text{ (}K_s \tan \delta\text{)} = 8.82 \text{ kPa}$

Av. stress for 2 m to 4 m = 44.59 kPa

$\text{Av. } f_{s2} = 44.59 \times 0.5 = 22.3 \text{ kPa}$

Vertical stress below D_c is assumed constant and equal to 53.9 kPa

$f_{s3} = 53.9 \times 0.5 = 26.95 \text{ kPa}$

$Q_s = f_{s1} \pi B \times 2 + f_{s2} \cdot \pi B (D_c - 2) + f_{s3} \cdot \pi B (L_p - D_c)$

$\quad = 352 \text{ kN}$

(c) *Load capacity.*

$Q_f = Q_b + Q_s = 987 \text{ kN}$

$Q_a = \frac{1}{3} Q_f = 329 \text{ kN}$

19.2 A cased concrete pile, 0.42 m outside diameter, is driven through 11 m of loose sand for a depth of 3 m into underlying medium dense sand. The skin friction is measured with a static cone fitted with a friction sleeve; $f_s = (f_c)_{av} = 20 \text{ kPa}$ in loose sand and $f_s = (f_c)_{av} = 55 \text{ kPa}$ in medium-dense sand. The cone resistance at base level is 12000 kPa. Determine the safe load for the pile: $F = 2.5$.

$f_b = q_c = 12000 \text{ kPa}$

$Q_b = \frac{\pi}{4} B^2 f_b = 1662.53 \text{ kN}$

$Q_{s1} = \pi B \times 11 \times 20 = 290.28 \text{ kN}$

$Q_{s2} = \pi B \times 3 \times 55 = 217.71 \text{ kN}$

$Q_f = Q_b + Q_{s1} + Q_{s2} = 2170.5 \text{ kN}$

$Q_a = \frac{1}{2.5} Q_f = 868 \text{ kN}$

19.3 Determine the penetration depth for a 40 cm diameter concrete pile which is required to carry load of 900 kN at a safety factor of 2.5. The soil consists of silty sand upto a depth of 10 m in which the average cone resistance $(q_c)_{av}$ is 4000 kPa. Silty sand is underlain by dense sand. The average cone resistance increases to 13000 kPa within the depth 10 m to 10.5 m and to 18000 kPa below 10.5 m.

To reach full base resistance the pile must be driven to atleast 5B, i.e. 2 m, in bearing stratum of dense sand. Examine the allowable load for the pile with a total penetration depth of 12 m.

q_c at base = 18 MPa, but it should be limited to 15 MPa only for design (upper limit, Lee et al 1983).

$f_b = q_c = 15000 \text{ kPa}$

$Q_b = \frac{\pi B^2}{4} \times f_b = 1884.95 \text{ kN}$

$$f_s \ (0.10 \ \text{m}) \quad = \frac{(q_c)_{av}}{200} = \frac{4000}{200} = 20 \ \text{kPa}$$

$$(10\text{-}10.5 \ \text{m}) \ = \frac{13000}{200} = 65 \ \text{kPa}$$

$$(10.5\text{-}12 \ \text{m}) \ = \frac{18000}{200} = 90 \ \text{kPa} \ (< 100 \ \text{kPa})$$

$$Q_s = \pi \ (0.4)^2 \ [(10 \times 20) + (0.5 \times 65) + 1.5 \times 90)]$$
$$= 461.81 \ \text{kN}$$
$$Q_t = 2346.76 \ \text{kN}$$
$$Q_a = \frac{1}{2.5} \ Q_t = 938.7 \ \text{kN} > 900 \ \text{kN (applied load)}$$

Hence safe.

19.5 ANALYSIS OF PILE CAPACITY IN CLAY

In saturated clays, the undrained load capacity is generally considered to be the more critical value and a total stress analysis can be made by assuming $\phi_T = 0$ and $\delta = 0$.

1. **Base resistance.** The net, unit base resistance f_b in saturated clay is expressed as:

$$f_b = c_T \ N_{cp} + \gamma \ L_p - \frac{W_p}{A_b}, \ (N_{qp} = 1) \tag{19.18}$$

where $\quad c_T =$ total stress, undrained cohesion below base level

Assuming that the pile weight $W_p = \gamma \ L_p \ A_b$ (providing the pile base is not significantly enlarged), Eq. 19.18 can be re-written as:

$$f_b = c_T \ N_{cp} \tag{19.19}$$

where the pile bearing capacity factor N_{cp} is generally taken as 9.0, but where the ratio L_p/B is less than 4, progressive reduction in the value of N_{cp} to 5.6 at the surface is made (Poulos 1981). For larger diameter (> 0.5 m) piles, adopt $N_{cp} = 7$ ($B = 0.5$ to 1 m), $N_{cp} = 6$ ($B > 1$ m) (CGS 1985).

The base resistance can also be estimated from the CPT (De Ruiter and Beringen 1979, Briaud and Meyer 1983):

$$f_b = N_{cp} \ \frac{q_c}{N_k} \tag{19.20}$$

with $\quad N_{cp} = 9, \ N_k \simeq 20$

2. **Skin friction.** The skin friction, in terms of total stress, is the ultimate *adhesion* c_a between the clay and the pile shaft which is generally related to the undrained shear strength c_T as follows:

$$f_s = c_a = \alpha \ (c_T)_{av} \tag{19.21}$$

where α = adhesion factor

$(c_T)_{av}$ = average undrained cohesion over pile shaft

The adhesion factor depends on the type of clay, the method of installation and the pile material. The appropriate value of α is obtained from the results of load tests. For driven piles, the values of α range on the average roughly from unity for soft clay to 0.5 or less for stiff clay, while for bored piles in stiff clay, α is roughly from 0.3 to 0.5. In the absence of other data, the following values of α may be adopted for preliminary design purposes:

α = 0.3 for $c_T \geqslant$ 250 kPa, α = 0.4 for c_T = 100 kPa, α = 0.5 for $c_T \leqslant$ 50 kPa, *with a maximum value of f_s = 100 kPa.* A value of α = 0.45 for the skin friction of firm to stiff clays may be adopted (with f_s not more than 100 kPa), when no previous experience is available (Skempton 1959, Tomlinson 1980).

The skin friction may be estimated from CPT (Ruiter and Beringen 1979, Briaud and Meyer 1983) as follows:

$$f_s = a_o \frac{(q_c)_{av}}{N_k} \tag{19.22}$$

where N_k = 20, a_o = 1 for NC clay, α_c = 0.5 for OC clay, $(q_c)_{av}$ = average cone resistance.

An alternative approach is to estimate skin friction in terms of effective stress because it may be assumed that the excess pore pressure set up during pile installation in the relatively thin zone of disturbance around the pile should virtually dissipate by the time the structural load is applied. Using drained shear strength parameters of saturated clay (c' = 0, and ϕ') and taking the friction angle between pile and soil $\delta' \doteq \phi'$, the f_s is expressed as:

$$f_s = K_s \, \sigma_s' \tan \phi' \leqslant c_T \tag{19.23}$$

where K_s = average coefficient of earth pressure
($\simeq$ coefficient at rest K_o)

σ_s' = average effective over-burden pressure along pile shaft

Expressing the product $K_s \tan \phi'$ by β, the skin friction factor, Eq. 19.23 is written as:

$$f_s = \beta \, \sigma_s' = (K_o \tan \phi') \, \sigma_s' \tag{19.24}$$

The coefficient of earth pressure at rest K_o can be approximately represented as follows (Meyerhof 1976):

$$K_o = (1 - \sin \phi') \quad \text{(NC clay)} \tag{19.25}$$

$$K_o = (1 - \sin \phi') \, (OCR) \quad \text{(OC clay)} \tag{19.26}$$

Using the above values of K_o, β can be evaluated. The value of β for NC clays is usually within the range 0.25 to 0.40 and a typical value of 0.3 may be used for design purposes. For OC clays, the values are significantly higher and vary within relatively wide limits.

3. Allowable load. The allowable load is normally obtained by using an overall safety factor of 2.5, but which should be increased to 3 for large diameter piles (> 0.5 m) if the soil parameters show a big scatter. The base resistance requires larger deformations for full mobilization than the shaft resistance. To reflect this effect, a safety factor of 3 is applied to base resistance and a factor 1.0 or 1.5 to shaft resistance. Thus Q_a is calculated as:

$$Q_a \leqslant \frac{1}{2.5} (Q_b + Q_s) \qquad (19.27)$$

or $\qquad Q_a \leqslant \left(\frac{Q_b}{3} + \frac{Q_s}{1.5} \right) \qquad (19.28)$

4. Scale effect. For piles larger than 0.5 m diameter, the following reduction factors are recommended to be applied to Q_b: no reduction is needed in Q_s (Meyerhof 1983):

Driven piles in stiff fissured clay:

$$R_b = \left(\frac{B + 0.5}{2 \, B} \right) \leqslant 1 \qquad (19.29)$$

Bored piles in stiff fissured clay:

$$R_b = \left(\frac{B + 1}{2 \, B + 1} \right) \leqslant 1 \qquad (19.30)$$

where $\qquad$ B = pile base diameter (m)

2. Negative skin friction. The negative skin friction f_n is the downward force (drag) which is exerted on the pile shaft by the soil around it. Such a condition arises when the clay deposit, in which or through which a pile has been installed, is subject to consolidation. This downward drag tends to reduce the useable pile capacity. Negative skin friction develops in the following cases.

(1) Pile is driven through a clay layer (e.g. recent cohesive fill) which is still undergoing consolidation under its own weight.

(2) Pile is driven through a soft, cohesive layer to a firm bearing stratum below, but a cohesionless fill is placed on the surface of soft layer and the soft layer starts consolidating.

(3) A cohesive fill is placed on the surface of compressible layer only a short time before driving the piles. The fill starts consolidating under its own weight and the compressible layer starts consolidating under the weight of the fill. Both will exert negative friction (dragdown).

(4) The clay deposit in which a pile has been placed is subject to general subsidence resulting from lowering of groundwater table or other causes.

The principal effect of skin friction is to increase the axial load in a pile. This tends to increase additional settlement. Due to negative skin friction a part of soil weight around the pile is carried by the pile resulting in reduction of overburden pressure at base and which reduces the ultimate base resistance.

The negative skin friction increases gradually as consolidation of clay proceeds and the effective overburden pressure σ_s' gradually increases. The effective stress analysis is appropriate and the negative skin friction f_n may be determined from Eq. 19.24 using $\beta = 0.25$ as a reasonable upper limit.

$$f_n = \beta\sigma_s' = 0.25 \, \sigma s' \tag{19.31}$$

The empirical factor β is reported by Garlanger (1973) from pile tests as follows: clay, $\beta = 0.20 - 0.25$; silt, $\beta = 0.25 - 0.35$; sand, $\beta = 0.35 - 0.50$. Negative skin friction develops along that portion of the pile shaft where the settlement of adjoining soil exceeds the downward displacement of the shaft. Observations indicate that a relative downward moment of 15 mm may be sufficient to mobilize full negative skin friction (Vesis 1977)

In the *total stress* analysis, f_n is calculated from Ex. 19.21. Thus, $f_n = \alpha \, (c_T)_{av}$

Since negative skin friction is usually estimated on the safe side, the safety factor of usually unity is applied to the total negative frictional force, and the allowable load for a pile is reduced by this amount.

Negative friction on driven piles may be reduced by applying bituminous or other viscous coatings to the pile surfaces, or in the case of steel piles by using the electro-osmosis technique (See Chapter 22). Floating sleeves have been used successfully for cast-in-place piles.

Example 19.4

A bored concrete pile, 30 cm in diameter and 6.5 m long, passes through stiff fissured clay subjected to seasonal shrinkage and swelling upto a depth of 1.5 m. The average undrained strength of clay varies linearly from 50 kPa at 1.5 m to 186 kPa at 10 m. Find the ultimate load capacity.   Assume $\alpha = 0.3$.

Since soil shrinkage may cause all adhesion to be lost, the top 1.5 m is ignored.

c_T at 1.5 m $= 50$ kPa

c_T at base $= 50 + \dfrac{186 - 50}{8.5} \times 5 = 130$ kPa

average $(c_T)_{av} = \dfrac{50 + 130}{2} = 90$ kPa

$f_s = \alpha\,(c_T)_{av} = 27$ kPa

$Q_s = 27 \times \pi\,B \times 5 = 127.23$ kN

$f_b = 9\,c_T = 9 \times 130 = 1170$ kPa

$Q_b = 82.7$ kN

$Q_t = 210$ KN

19.6 LOAD CAPACITY OF UNDER-REAMED PILES

1. Piles in clays. For underreamed piles installed in clays, the unit base resistance f_b can be calculated from Eq. 19.19 taking the values of N_{cp} as 9.0. The unit skin friction along the shaft can be calculated from Eq. 19.21 taking the value of adhesion factor α as 0.4. The ultimate load is given by Eq. 19.2:

$$Q_f = f_b\,A_b + f_s\,A_s$$

The base area of an underreamed pile is enlarged by under-reaming and providing a bulb or giving a bell shape. For a bell shaped base A_b is the area of enlarged base. When a bulb is provided a little above the tip, it is practically accurate enough to calculate A_b from the diameter of the bulb, ignoring the projected stem below and using the average value of c_T below the bulb for the determination of f_b (Eq. 19.19). If the bulb is higher enough and there is considerable difference in the value of c_T at the bulb level and the tip level, the base resistance is calculated as:

$$Q_b = (0.25 \pi B^2) \times 9 c_T + 0.25 \pi (B_b{}^2 - B^2)$$
$$\times 9 c_T' \qquad (19.32)$$

where   B   = diameter of pile shaft

            B_b = diameter of bulb

            c_T = unit cohesion at the tip (base) of pile (total stress)

            c_T' = unit cohesion at bulb level (total stress)

As a result of settlement of the pile, there is a possibility of a small gap developing between the top of the under-ream and the overlying soil, leading to a downward drag of soil on the shaft. It is, therefore, advisable to ignore 2B length of the shaft above the under-ream in the calculation of skin friction. Similarly, the little portion of the shaft projecting below the bulb should also be neglected. When two or more bulbs are provided, frictional resistance is calculated as follows:

$$Q_s = \alpha \, (c_T)_{av} \, A_s + c_{Ta} \, A_{sb} \qquad (19.33)$$

where A_s = surface area of shaft above the top bulb (ignoring 2 B length)

            A_{sb} = surface of the cylinder (between top and bottom bulbs) circumscribing the bulbs

            $(c_T)_{av}$ = average cohesion on A_s

            c_{Ta} = average cohesion on A_{sb}

The allowable loads are calculated on the basis of Eqs. 19.27 and 19.28, using the lower one. If a certain depth of top soil is subject to seasonal shrinkage and swelling, it should be neglected for calculating frictional resistance (adhesion) on the pile shaft.

2. Piles in sand. The load capacity of bored, underreamed piles installed in cohesionless soil, (although their use is restricted because of practical difficulty in under-reaming), can be calculated similarly by using the formulae for base resistance and skin friction applicable to sands (Section 19.4) making due allowance for under-reaming as for clays.

Example 19.5

Find the ultimate and safe loads for an underreamed pile installed in clay with the following data: diameter of shaft = 40 cm, diameter of underream (bulb) provided near the base = 100 cm, length of shaft above bulb = 5 m, average undrained strength along shaft = 115 kPa, and at base = 150 kPa, adhension factor = 0.4.

$f_b = N_{\phi p}\, c_T = 9 \times 150 = 1350$ kPa

$f_s = \alpha\, (c_T)_{av} = 46$ kPa

A_b (considering bulb diameter for bearing) $= 0.785$ m²

Neglect $2\, B\, (= 0.8$ m) of shaft above bulb for adhesion. (If a certain top portion is subjected to shrinkage and swelling, it should also be neglected.)

$A_s = \pi\, B \times 4.2 = 5.28$ m²

$Q_b = A_b.\, f_b \quad = 1059.75$ kN

$Q_s = A_s.\, f_s \quad = 272.78$ kN

$Q_t = 1302.5$ kN

$Q_a = \dfrac{1}{2.5}\, Q_t = 521$ kN

or $\quad Q_a = \dfrac{1}{3}\, Q_b + \dfrac{1}{1.5}\, Q_s = 515$ kN

19.7 DYNAMIC ANALYSIS

The dynamic analysis is based on the assumption that the dynamic resistance to driving is equal to the ultimate load capacity of the pile under static loading. Upon striking the pile, the kinetic energy of the driving hammer is equated to the work done by the pile in penetrating the soil plus the energy losses (due to friction, heat, hammer rebound, vibration and elastic compressions of pile, packing assembly and soil). A number of pile driving formulae have been proposed. Some of the most commonly used formulae are given below.

1. Engineering news formula. The Engineering News formula was published by the editor, A.M. Wellington, of *Engineering News* (New York) in 1888. It takes into account the energy loss due to temporary, elastic compression C_1 of the pile and this energy loss is calculated assuming the resistance to increase linearly from 0 to Q_t while the compression develops. Equating the energy applied by the hammer to the work done by pile and energy loss:

$$WH = Q_t\, S + \tfrac{1}{2}\, Q_t\, C_1 = Q_t\, (S + C_1/2) \qquad (19.34)$$

where W = weight of hammer

H = equivalent free fall

S = 'set' or penetration of pile per blow

C_1 = elastic compression of pile

Q_t = average resistance of soil to penetration, assumed as ultimate load capacity of pile

In practice, empirical values are given to the term $C_1/2$ and the formula is generally written as:

$$Q_f = \frac{WH}{S + C} \qquad (19.35)$$

where   C = empirical constant

S and C should be in the units of H. In 'mm' units,

C = 25 mm for drop hammers

C = 2.5 mm for single and double acting steam hammers

The set S is taken as the average penetration of pile for the last 5 strokes of a drop hammer, or 20 strokes of a steam hammer.

Eq. 19.35 is the form of the formula when used for drop hammer and single acting steam hammer. To account for the additional energy of steam, the formula in case of double acting steam hammer is written as:

$$Q_t = \frac{(W + ap) H}{S + C} \qquad (19.36)$$

where   a = effective area of piston

p = mean effective steam pressure

(For double acting hammers the rated energy per blow is usually given.)

A safety factor of 6 is recommended to get the allowable load.

$$Q_a = \frac{1}{6} \left(\frac{WH}{S + C} \right) \qquad (19.37)$$

The Engineering News formula has been modified to take into account the energy losses in the hammer system and due to impact. The modified Engineering News formula is written as:

$$Q_f = \frac{\eta WH}{S + C} \cdot \frac{W + e^2 W_p}{W + W_p} \qquad (19.38)$$

where η = hammer efficiency (< 1)

e = coefficient of restitution

W_p = weight of pile

The hammer efficiency η depends on the pile driving equipment, driving procedure, type of pile and the ground conditions. In the absence of known values, η may be adopted as follows (Bowles

1983) in reasonably good operating conditions: drop hammer, $\eta =$ 0.75—1.0; single acting hammer, $\eta = 0.75$—0.85; double acting or differential hammer, $\eta = 0.85$; diesel hammer, $\eta = 0.85$—1.0. The representative values of the coefficient of restitution 'e' are as follows: broomed timber pile, $e = 0$; good timber pile, $e = 0.25$; driving cap with timber dolly on steel pile, $e = 0.3$; driving cap with plastic dolly on steel pile, $e = 0.5$; helmet with composite plastic dolly and packing on reinforced concrete pile, $e = 0.4$. The safety factor used on the modified formula also is six.

2. Danish formula. The Danish formula (Sorensen and Hansen 1957) is derived with the following assumptions:

(1) Due to friction loss in the hammer system, the actual energy of impact is $\eta\,WH$, where the constant η called the 'hammer efficiency', is less than 1.0.

(2) Under a blow the pile is compressed elastically as if it were a strut under a static load Q_t:

$$\text{Elastic compression } C_1 = \frac{Q_t\,L}{A\,E} \tag{19.39}$$

where

$L = $ length of pile

$A = $ cross sectional area of pile

$E = $ Young's modulus of pile

(3) Elastic compression is that which would occur if all the available hammer energy were used in causing compression.

$$\eta WH = \tfrac{1}{2}\,Q_t\,C_1 \tag{19.40}$$

Substituting the value of Q_t from Eq. 19.40 into Eq 19.39,

$$C_1 = \left(\frac{2\eta\,W\,H\,L}{A\,E}\right)^{0.5} \tag{19.41}$$

$$\text{Elastic energy } = \frac{Q_t}{2}\,C_1 = \frac{Q_t}{2}\left(\frac{2\,\eta\,W\,H\,L}{A\,E}\right)^{0.5} \tag{19.42}$$

Equating the hammer energy to the work done by pile and the elastic energy lost,

$$\eta\,W\,H = Q_t.\,S + \tfrac{1}{2}\,Q_f\left(\frac{2\eta\,W\,H\,L}{A\,E}\right)^{0.5}$$

or

$$Q_t = \frac{\eta\,W\,H}{S + \tfrac{1}{2}\left(\dfrac{2\,\eta\,W\,H\,L}{A\,E}\right)^{0.5}} \tag{19.43}$$

Eq. 19.43 is known as the Danish formula. A safety factor of 3 is used to get the allowable load Q_a. The hammer efficiency η

varies from 0.7 to 0.95. A value of $\eta = 0.7$ may be used in good driving conditions when no direct measurements are available.

3. **Janbu formula.** The energy losses considered in the Janbu formula (1953) are:

(i) in the hammer system: actual imparted energy $= \eta\, W\, H$

(ii) due to elastic compression of pile:

$$\text{energy loss} = \tfrac{1}{2}\, Q_t\, C_1 = \tfrac{1}{2}\, Q_t \left(\frac{2\eta\, W\, H\, L}{A\, E} \right)^{0.5}$$

(iii) due to impact, and this may be stated empirically.

Introducing empirically derived constants, the Janbu formula is written as:

$$Q_f = \frac{1}{K_u} \cdot \frac{\eta\, W\, H}{S} \tag{19.44}$$

where

$$K_u = C_d \left[1 + \left(1 + \frac{\lambda}{C_d} \right)^{0.5} \right] \tag{19.45}$$

$$\lambda = \frac{\eta\, W\, H\, L}{A\, E\, S^2} \tag{19.46}$$

$$C_d \text{ (driving coeff.)} = 0.75 + 0.15 \frac{W_p}{W} \tag{19.47}$$

$W_p =$ weight of pile

The driving coefficient may be assumed as unity, which simplifies the value of K_u without any sacrifice of accuracy (Olson and Flaate 1967):

$$K_u = 1 + \sqrt{(1 + \lambda)} \tag{19.48}$$

values of η may be adopted as follows (Simons and Menzies 1977): good driving conditions, $\eta = 0.70$; average driving condition, $\eta = 0.55$; difficult or bad conditions, $\eta = 0.40$.

4. **Hiley formula.** The Hiley formula (Hiley 1925, 1930), said to be a 'complete' formula, takes into account the energy loss in the hammer system, the energy losses due to elastic compression of pile (C_1), the soil (C_2) and the packing assembly on top of the pile (C_3), all represented by a term C, and the energy losses due to impact, represented by an efficiency factor n_i. Thus,

$$Q_t = \frac{\eta\, W\, H\, n_i}{(S + C/2)} \tag{19.49}$$

where
η = hammer efficiency factor
C = total elastic compression $= C_1 + C_2 + C_3$

$$n_l \quad = \text{efficiency factor of impact energy loss (efficiency of blow)}$$

$$= \frac{W + e^2 W_p}{W + W_p} \tag{19.50}$$

$$e \quad = \text{coefficient of restitution}$$

H, S and C should be in the same units. For double acting hammers, the rated energy per below is substituted for WH. The values of hammer efficiency η and the coefficient of restitution 'e' may be adopted as given for the modified Engineering News formula (Eq. 19.38). The elastic compression of the pile and the soil can be obtained from the driving trace of the pile (measured elastic recovery). The temporary compression C_3 of driving cap and dolly increases with increasing driving resistance; typical values are: 1.2 to 5.0 mm for direct driven timber piles; 0.6 to 2.5 mm for 12 to 25 mm thick composite plastic dolly on concrete pile; and $C_3 = 0$ for direct driven steel piles (Wilun and Starzewski 1975). The allowable load is obtained by applying a safety factor of 2.5 or 3.0.

   5. **Reliability of formulae.** The estimates of load capacity of piles by different dynamic formulae vary through a wide range and the formulae cannot be used with a degree of certainty. Even then they are recommended to be retained as a method of rapid determination of capacity and controlling it under job conditions.

   The Danish formula, and the slightly more refined Janbu formula, not only possess the merit of simplicity but are also considered reliable over a wide range of conditions. If safety factor of 3 is used with the Janbu formula and a safety factor of 2.7 with the Hiley formula, the results are not much different (Flaate 1964). It has been reported that the true safety factors when using the Engineering News formula may range from less than 1 to 17 (Michigan SHC 1965), but the modified Engineering News formula is found reasonably valid. The dynamic formulae should be used only for piles in sands and gravels and if possible should be calibrated against the results of field loading tests which are always more reliable.

Examples 19.6 — 19.7

   19.6 A steel H-pile of sectional area (A) 14×10^{-3} m^2 is driven into sand by a double acting steam hammer supplying a rated energy (WH) of 35 kN-m per blow. Weight of pile + cap (W_p) = 19.5 kN, weight of hammer including casing (W) = 35.5 kN, length of pile (L) = 12 m, Young's modulus (E) = 210×10^6 kPa, average set per blow (S) = 16 mm, hammer efficiency (η) = 0.85, coefficientof restitution (e) = 0.5. Find ultimate capacity of the pile by (a) Engineering News formula, (b) Modified Engineering News formula, (c) Danish formula, and (d) Janbu formula.

(a)   Engineering News formula.

$$Q_t = \frac{W\,H}{S + C} = \frac{35 \times 1000}{16 + 2.5} = 1892 \text{ kN}$$

If η is applied, $Q_t = \frac{\eta\,W\,H}{S + C} = 1605 \text{ kN}$

(b)   Modified Engineering News formula.

$$Q_t = \frac{\eta\,W\,H}{S + C} \cdot \frac{W + e^2\,W_p}{W + W_p} = 1180 \text{ kN}$$

(c)   Danish formula.

$$\tfrac{1}{2}\left(\frac{2\,\eta\,W\,H\,L}{A\,E}\right)^{0.5} = \tfrac{1}{2}\left(\frac{2 \times 0.85 \times 35 \times 12}{14 \times 10^{-3} \times 210 \times 10^6}\right)^{0.5}$$
$$\times 1000$$
$$= 7.79 \text{ mm}$$

Eq. 19.43 : $Q_t = 1250 \text{ kN}$

(d)   Janbu formula.

$$C_d = 0.75 + 0.15\left(\frac{W_p}{W}\right) = 0.832$$

$$\lambda = \frac{\eta\,(W\,H)\,L}{A\,E\,S^2} = \frac{0.85 \times 35 \times 12}{14 \times 10^{-3} \times 210 \times 10^6 \times (0.016)^2} = 0.474$$

$$K_u = C_d\left[1 + \left(1 + \frac{\lambda}{C_d}\right)^{0.5}\right] = 1.874$$

$$Q_t = \frac{1}{K_u}\,\frac{\eta\,(WH)}{S} = 992 \text{ kN}$$

If $K_v = 1 + \sqrt{(1 + \lambda)} = 2.214$ is used,

$$Q_t = 840 \text{ kN}$$

19.7   Find the ultimate load and the allowable load for a R.C.C. pile, 40 cm square in section and 20 m long, driven by a 30 kN single acting steam hammer into medium dense sand to a final set of 4 mm per blow. Stroke of hammer = 1.5 m, weight of pile and dolly = 80 kN, coefficient of restitution = 0.4, hammer efficiency = 0.85, total elastic compression of soil, pile and dolly = 20 mm, $E = 20 \times 10^6$ kPa.

(a) Engineering News formula.

$$Q_f = 6923 \text{ kN}; \quad Q_a = \frac{1}{6} Q_f = 1154 \text{ kN}$$

(b) Modified Engineering News formula.

$$Q_t = 2290 \text{ kN}; \quad Q_a = \frac{1}{6} Q_t = 382 \text{ kN}$$

(c) Hiley formula.

$$Q_t = 1063 \text{ kN}; \quad Q_a = \frac{1}{2.5} Q_t = 425 \text{ kN}$$

(d) Danish formula.

$$Q_f = 2562 \text{ kN}; \quad Q_a = \frac{1}{3} Q_t = 854 \text{ kN}$$

(e) Janbu formula.

$$\lambda = 14.94, C_d = 1.15, K_u = 5.45$$

$$Q_f = 1754; \quad Q_a = \frac{1}{3} Q_t = 585 \text{ kN}$$

If $C_d = 1, K_u = 4.99$
$Q_t = 1915$ kN.

19.8 PILE LOAD TESTS

Load testing of piles is the most positive method of determining load capacity and assessing the accuracy of predicted values. The purpose of pile load tests performed during the *initial* stages of design and construction, called 'initial tests' (IS: 2911-1), is to ascertain that the allowable loads obtained by design are appropriate and that the installation procedure is satisfactory. Such tests are usually performed on 'test piles' installed exclusively for testing. 'Routine tests' performed after construction on selected working piles, called 'routine testing or proof testing', are meant to check the completed work and verify the allowable load or find the settlement upto a certain multiple of the allowable load. The two commonly used methods of pile testing are the 'maintained load test' and the 'constant rate of penetration test'.

1. **Maintained load test.** The pile is loaded usually by jacking against a kentledge (tension piles or anchors) but sometimes by

direct loading of the kentledge. Equal load increments of about 0.2 Q_a are usually applied. E ch incremental load is maintained constant until the rate of settlement is less than 0.1 mm/5 min; the next load increment is then applied. A friction pile in cohesive soils should be tested at least 24 hours after installation.

For initial testing on a 'test pile', loading is continued upto 2.5 or 3 times Q_a or to a settlement of B/10 (or 7.5 % of bulb diameter of an underreamed pile) or to failure, whichever is earlier. In a routine test on a 'working pile', load is increased upto 1.5 Q_a. Some-times the pile is unloaded at the proposed Q_a to know the rebound, and then reloaded back to the same load after which the loading is continued. When the maximum load has been applied and final settlement recorded. the pile is unloaded (in stages) and the final rebound is recorded after 24 hours of removal of the entire load. A load settlement curve is plotted.

The ultimate load can be read from the load settlement curve if the test has been carried to failure and a well defined failure point is indicated by the curve plotted on linear scales. A double-log plot may sometimes define the failure point better. When no definite failure point is indicated, it may be assumed to be at a settlement equal to 10% pile diameter.

The ultimate load divided by a safety factor of 2 or 2.5 gives the allowable load. According to IS: 2911-1 & 3 the allowable load is taken as the least of the following:

(i) 2/3 of final load causing a total settlement of 12 mm, (ii) 2/3 of load causing a net settlement of 6 mm (total settlement minus rebound), (iii) 1/2 of load causing a total settlement of 10% pile diameter (or 7.5 % of bulb diameter of underreamed pile).

2. **Constant rate of penetration (CRP) test.** In the CRP test, developed by Whitaker (1963) (Whitaker and Cooke 1961), the pile is made to penetrate the soil at a constant speed from its position as installed; the readings of the pressure in the jack and the settlement of the pile head are taken at regular time intervals not greater than 3 minutes. Suitable rates of penetration for tests in sands and clays are 1.5 mm/min and 0.75 mm/min respectively. Provided the rates are steady, a variation of the rate of penetration between 0.5 to 3.5 mm/min does not affect the results significantly. The test is continued until the load starts decreasing after reaching a maximum value or till the penetration is equal to atleast 10% of the diameter of the

pile base plus the elastic deformation of the pile; the elastic deformation being obtained either by direct measurement or by assuming that the test load acts on the full length of the pile. (The elastic deformation $= QL_p/AE$.) It is recommended to load the pile upto at least 2.5 Q_a, if possible.

The results are plotted in the form of a load penetration curve. The ultimate load for the pile is taken equal to the load (i) at failure point if well defined by decreasing load after attaining a maximum, (ii) when the load reaches a maximum value which substantially remains constant for 50 mm or more penetration, (iii) when penetration is 10% B, if (i) and (ii) criteria are not satisfied. The allowable load is obtained by applying a safety factor of 2 or 2.5.

19.9 CAPACITY OF PILE GROUP

1. Types of groups. Piles are seldom used singly. They are generally used in a group with a foundation slab or 'cap' cast on their heads to distribute the load. When the cap is in contact with the ground surface (as in the most common cases), the system is called a 'piled foundation' (Whitaker 1976). When the cap is above the ground surface (as for offshore structures), the system is called 'free standing group'.

2. Pile spacing. If piles are driven very closely into dense or incompressible soil, there may be upheaval of the ground surface. Also, deviation of a few piles from their intended line may lead to concentration of toe loads in a small area, which will cause different settlement between adjacent piles. If the spacing is too large, on the other hand, uneconomical pile caps may result. While driving piles in granular soils, it is advisable to start driving at the centre of a group and then work outwards, to avoid difficulty of 'tightening up' of the ground. The centre-to-centre pile spacing in clay, in terms of pile diameter or width B, should not be less than 3 B, with a minimum of 1.0 m (Tomlinson 1977). In sands, the spacing should not be less than 2.5 B, with a minimum of 0.8 m. For end bearing piles, the minimum spacing is 2 B.

3. Types of failures. A pile group may fail either by individual penetration of the piles or as an equivalent 'block' containing the piles and the soil between them. Model experiments on free standing groups in homogeneous clay indicate a block failure for pile spacings less than of the order of 2 to 2.5 B (Whitaker 1976); this spacing is known as the 'critical spacing.' Block failure will not occur with pile spacing of one pile perimeter or greater, but is a real possibility with pile spacing of 2.5 B if the length diameter ratio exceeds about 45 (Wawryk 1972). When the cap is in contact with ground surface turning the group into a piled foundation, failure is observed to occur like a block even at spacings larger than the critical value, which would not have been the case had the group been free standing. In practice, the soil immediately below a pile cap is often weaker than the average strength of the bed in which the group is installed, therefore the cap may not be as fully effective in producing block action as it is in the case of a homogeneous clay; and the resulting failure load will be lower, but not so low as for a free standing group. Block failure is also observed to occur in free standing groups at small pile spacings in sand. If a pile group is safe from block failure then settlement consideration usually governs the design.

4. Group action and efficiency. Group action refers to the dissimilarity in behaviour of a pile with respect to its load capacity and settlement when used singly and in a group and to the mechanism of interference between adjacent piles which causes this dissimilarity. Group action is usually interpreted in terms of two indices: *group efficiency* and *group settlement ratio*.

The 'group efficiency E_g', also termed group efficiency factor or ratio, is defined as the ratio of the average load on a pile of a group at failure to the failure load on the isolated pile under similar conditions:

$$E_g = \frac{Q_g}{n\,Q_f} \tag{19.51}$$

where Q_g = failure load for the group
Q_f = failure load for an isolated pile
n = number of piles in the group

The 'group settlement ratio r_g' is defined as the ratio of the settlement of the group to that of a comparable single pile when the average load per pile in the group is the same as that on the single pile. The settlement ratio is discussed in the next section.

The group efficiency E_g depends on the size and shape of the pile group, the size, spacing and length of piles, and the type of soil surrounding the group and underlying the group in the stressed zone. In loose and medium-dense sands, pile driving in group compacts the surrounding sand, provided the spacing is less than about 8 B and consequently E_g is greater than unity, (E_g may be as high as two). With bored piles, boring tends to reduce rather than increase compaction so that E_g is unlikely to exceed unity (due to overlapping of individual stress fields without the compensating effect of compaction). When piles are driven into dense sand, E_g is less than unity due to loosening of sand and the overlapping of zones of shear. No reduction due to grouping occurs with end bearing piles driven to refusal and the group capacity should be as great as the sum of the individual pile capacities.

In cohesive soils with closely spaced piles, E_g is less than unity because of overlapping zones of pressures. Several 'group efficiency formulae' are available to determine E_g but they are fundamentally deficient in that they do not take into account the nature of soil in which the piles are installed. They are not recommended for use. As an approximate estimate of E_g for pile groups in clay, the following values suggested by Kerisel (1965) may be used: pile spacing in terms of pile diameter or width (B) = 2.5 B, 3 B, 4 B, 5 B, 6 B, 8 B and 10 B; corresponding E_g = 0.55, 0.65, 0.75, 0.85, 0.90, 0.95 and 1.0.

The ultimate load capacity (Q_g) of a pile group thus may or may not be equal to the sum of the ultimate load capacities (nQ_t) of the individual piles in the group; in soft clays with closely spaced piles Q_g may be considerably less than nQ_t, whereas in loose sands, Q_g may be greater. The group capacity may be estimated as described below for different cases.

5. Free standing group in clay. The ultimate load is taken as the lesser of the following two values:

(1) $E_g.n. Q_f$ (See Eq. 19.51)

(2) Ultimate load for block failure (Q_B)

The group efficiency factor E_g may be adopted on the basis of Kerisel's suggested values. The ultimate load Q_B for *block failure* is given by:

$$Q_B = B_0 L_0 f_b + 2 L_p (B_0 + L_0) f_s \qquad (19.52)$$

where B_0 and L_0 are the overall width and overall length along the outer perimeter of the group beneath the cap and L_p is embedded length of the piles. The unit base resistance f_b is given by:

$$f_b = c_T N_{op} \qquad (19.53)$$

where c_T = undisturbed unit cohesion below base

N_{op} = bearing capacity factor, may be adopted as 9 (Skempton's value).

The frictional resistance f_s along the perimeter area is taken as the undrained strength of the remoulded clay unless loading is to be delayed for at least six months, in which cases c_T of the undisturbed clay can be used.

6. Piled foundation in clay. The ultimate load is taken as the lesser of the two values:

(1)   n Q_t

(2) Q_B (Eq. 19.52)

Here it is assumed that the soil immediately beneath the cap is strong enough to be relied on to take load. If $c_T < 100$ kPa, multiply n Q_t with E_g (70%) also as for free standing group. Recent tests and effective stress consideration suggest that such a reduction is overly conservative and that $E_g = 100\%$ may be used **(CGS 1985).** For $c_T > 100$ kPa, it is a common practice to neglect group effect, i.e. E_g is taken as unity (CGS 1985). When the pile spacing is greater than 7 B, the piles may be assumed to act as individual piles ($E_g = 1$) both for the free standing and piled groups in clay.

7. Pile groups in sand. If the individual pile has an adequate safety factor against failure there can be no risk of the block failure of a pile group terminated in and applying stress to sand.   For friction piles in loose sand, both for free standing and piled cases, the ultimate group load may be taken as nQ_t, E_g being assumed as unity. (For bored piles, E_g may be taken as 2/3; Lee et al 1983). Although block failure is unlikely to occur but it is necessary to investigate this mode of failure for close spacing; Q_B may be calculated from Eq. 19.52, using appropriate values of f_b and f_s for sand. The piles of a group in sand act as individual piles at spacing $> 7B$ (CGS 1985). For most engineering structures, the group load for piles in cohesionless soil is calculated from settlement considerations rather than by calculating the ultimate load and dividing it by an arbitrary safety factor of 2 or 3 (Tomlinson 1980).

Experience has shown that a pile cap in contact with the ground surface develops a bearing capacity which increases the apparent group efficiency. For piled foundations in clays and sands, Poulos (1981) recommends to take the ultimate group load as the lesser of:

(1) $n\,Q_f$ plus the ultimate bearing capacity of the net area of the cap.

(2) The ultimate load of the equivalent block Q_B plus the bearing capacity of the area of the cap outside the perimeter of the equivalent block.

8. End bearing piles. For end bearing piles on rock, dense sand and gravel, the ultimate group load may be taken as $n\,Q_f$.

Where a group is founded in fi m layer overlying a softer layer, the group load is calculated as for 'piled foundation in cl ly', except that a reduced base resistance is used for the single piles and the equivalent block. An approximate method of making such reduction is to decrease f_b linearly from the value for the pile or group resting directly on the stiff layer (if the softer layer is more than two pile or group widths below the base of the piles) to the value for the pile or group resting directly on the soft layer, when no stiff layer exists (Poulos 1981).

19.10 SETTLEMENT OF PILES IN SAND

Elastic solutions based on the elastic properties and the pile material are available for estimating the settlement of a single pile and pile group (Poulos and Davis 1974). The theoretical analysis, however, cannot include many factors which influence the actual settlement of piles and the results generally are not accurate enough to be of practical value. Instead, estimates of settlements are based on empirical relationships.

1. Settlement of single pile. The settlement of a single pile in cohesionless soils is observed to be a function of the ratio of applied load Q to Q_t and pile diameter B. The only satisfactory method is to measure the settlement directly in a load test. As a preliminary guide, for normal loads level (safety factor > 3) the settlement of displacement piles S_1 can be estimated from the empirical formula given by Vesic (1970,1977) (CGS 1985):

$$S_1 = \frac{B}{100} + S_e \qquad (19.54)$$

where S_1 = settlement of single pile (cm)

  B = pile diameter (cm)

  S_e = elastic compression of pile (cm)

The elastic compression for the above relation is commonly assumed as:

$$S_e = \frac{100\,Q\,L_p}{A\,E} \tag{19.55}$$

where Q = applied load (kN)
  A = average cross sectional area (m²)
  L_p = length of pile (m)
  E = modulus of elasticity of pile (kPa)

The settlement under Q_t will be approximately 3 times as great.

2. Settlement of pile group. The settlement of a pile group is always greater than the settlement of a corresponding single pile because of the overlapping of the individual pressure bulbs of the piles in a group. Typical pressure bulbs for a single pile and a closely spaced group are shown in Fig. 19.1. The resultant isobar for the group at an intensity equal to that of the single pile isobar extends to greater depths. Significant stresses are thus developed over a much wider area and much greater depth, causing greater settlement of the group.

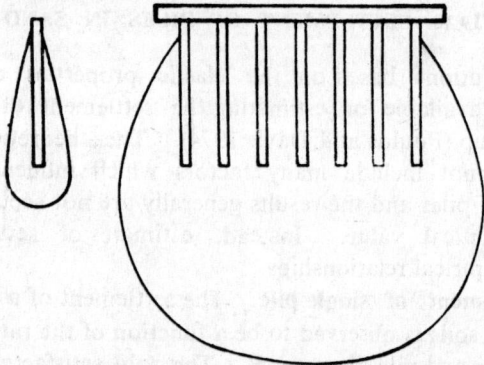

Fig. 19.1 Pressure bulbs for a single pile and a pile group

Skempton method. The group settlement can be estimated from the following empirical relation proposed by Skempton (Skempton et al 1953):

$$\frac{S_g}{S_1} = \left(\frac{4\,B_o + 3}{B_o + 4} \right)^2 \tag{19.56}$$

or $S_g = r_g S_1$ (19.57)

where S_g = settlement of group

S_1 = settlement of single pile at the same load intensity

B_o = overall width of pile group (m)

The 'group settlement ratio r_g' is a function of the ratio of width of pile group to the pile diameter (B_o/B). Its values are given in Table 19.6 (CGS 1985).

Table 19.6

VALUES OF SETTLEMENT RATIO

B_o/B	1	5	10	20	40	60
r_g	1	3.5	5	7.5	10	12

Vesic method. Vesic's empirical relation (Vesic 1977, NAVFAC DM-7-2 1982, CGS 1985) for settlement ratio is as follows:

$$S_g/S_1 = \sqrt{B_o/B}$$ (19.58)

3. Load transfer and settlement analysis. The action of piles is to transfer the load to some lower stratum. The total load carried by a group of friction piles can be assumed to be transferred unifomly on to an equivalent footing or raft $(B_o \times L_o)$ located at $L_p/3$ up from the pile tips. In case of end bearing piles, the equivalent footing is assumed at the base level of the piles. For a pile group driven through soft clay to combined skin friction and end bearing in a layer of granular soil, the equivalent footing is assumed at $L_p'/3$ above the base, where L_p' is the effective depth of embedment in granular layer. The load from the equivalent footing is then assumed to spread from its perimeter to the underlying soil at a slope of 2 vertical to 1 horizontal (or 30°) as shown in Fig. 19.2. The compression of the stressed layers, whether sand or clay below the equivalent footing, can be calculated as described in Chapter 18. The settlement of sand layers is determined by the SPT or CPT methods.

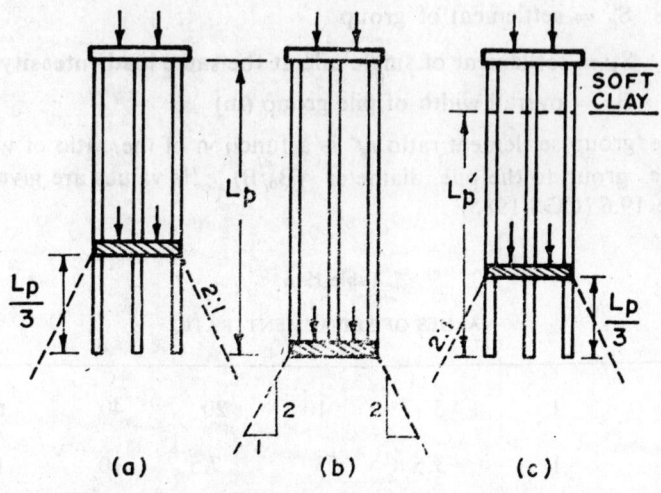

Fig. 19.2 Equivalent footing concept, (a) friction piles in sand (b) end bearing piles in sand, (c) piles passing through soft clay to a layer of granular soil offering combined resistance of skin friction and end bearing.

Meyerhof method. The settlement of pile groups in saturated sands can also be estimated by the following empirical relation given by Meyerhof (1976) :

$$S_g \text{ (mm)} = 0.96 \frac{q_n \sqrt{B_o}}{N} I \simeq \frac{q_n \sqrt{B_o}}{N} I \quad (19.59)$$

where q_n = net pressure of the equivalent footing (kPa)

B_o = width of equivalent footing (m)

N = average corrected SPT blows within the seat of settlement ($\simeq B_o$)

I = influence factor of *effective* group embedment L_p', (L_p — soft zone)

$$I = 1 - \frac{L_p'}{8B_o} > 0.5 \quad (19.60)$$

19.11 SETTLEMENT OF PILES IN CLAY

1. Settlement of single pile. The settlement of single pile in soft clays ($c_T < 100$ kPa) is due to local shear deformations along the pile shaft rather than due to consolidation settlement and is, therefore, very limited. It can be known from load tests. The settlement analysis of single piles in stiff clay is difficult at the present time (CGS 1985) because little data are available on actual behaviour of such piles. A load test can be used to evaluate the settlement.

2. **Settlement of pile group.** The settlement of a pile group in clay is due to consolidation settlements. As in case of friction pile group in sand, the load carried by the pile group in homogeneous clay is assumed to be transferred to the soil through an equivalent footing or raft ($B_o \times L_o$) located at $L_p/3$ up from the bottom of the piles. The total load is then assumed to spread uniformly at a slope of 2 vertical to 1 horizontal (or 30°). Where a top layer of poor soil of negligible friction or a layer of negative friction (L_n) exists, the equivalent footing is assumed at a level of $(L_p - L_n)/3$ above the pile base.

3. **Negative skin friction.** Negative skin friction will act on the piles of a group where the fill and any underlying compressible clay moves downwards relative to the shaft. The magnitude of this skin friction cannot increase as a result of decreasing the pile spacing in the group. The total negative friction acting on the group cannot exceed the total weight of soil enclosed by the piles. Thus the total friction drag on a pile of a group is taken as the smaller of the values given by:

$$F_n = A_s \times (c_T)_{av} \times \alpha \tag{19.61}$$

or
$$F_n = \frac{W}{n} = \frac{(B_o \times L_o)(L_n \times \gamma')}{n} \tag{19.62}$$

where A_s = circumferential area of pile for length L_n

             W = effective weight of soil contained in the group of overall dimensions B_o and L_o for Length L_n (ignoring the presence of piles)

             γ' = effective unit weight of soil

             n = number of piles in the group

             L_n = depth over which the soil is moving downwards relative to pile

If L_{n1} and L_{n2} are respective depths of the fill and underlying compressible clay and γ_1' and γ_2' are their respective unit weights, W is given by:

$$W = (B_o \times L_o)(L_{n1} \gamma_1' + L_{n2} \gamma_2') \tag{19.63}$$

At close spacings, the frictional drag on piles of a group will generally be limited by the weight of soil enclosed by the piles, and the total load Q_T on top of the group will be given by:

$$Q_T = Q_a + (B_0 \times L_0) (L_a \times \gamma') \qquad (19.64)$$

or

$$Q_T = Q_a + (B_0 \times L_0) (L_{n1} \gamma_1' + L_{n2} \gamma_2') \qquad (19.65)$$

where Q_a allowable structural load acting on the group

If Q_T is to be limited to the maximum allowable load for the group, the structural load will have to be reduced by the amount equivalent to the total frictional drag on the group. The total load Q_T is assumed to act on the equivalent footing or raft $(B_0 \times L_0)$ located at $(L_p - L_n)/3$ above the base for calculating the compression of the underlying strata.

The consolidation settlements are calculated as described in Chapters 14 and 18. As indicated earlier, the settlement is normally the limiting criterion for pile groups both in sands and clays.

19.12 PILES IN MULTI-LAYER DEPOSIT

A pile driven through a multi-layer deposit may derive its load capacity from both skin friction and base resistance. However, the evaluation of the relative importance of skin friction and base resistance is difficult and may need to be confirmed by load tests. Wherever possible, the piles should be driven to a stronger layer of sufficient thickness so as to assume that the load capacity is derived entirely from that layer.

The possibility of a pile group driven to a strong layer (say of sand) punching through into an underlying soft layer should also be examined. The total load Q on the pile group is assumed to spread from the equivalent footing located at the base of the piles at a slope of 2 : 1. The vertical pressure increment q' at the top of the soft layer should not exceed the presumed bearing value of the soft layer. For example, in the general case of cohesive soil, q' should not exceed $3 c_T$.

The settlement of a pile group in layered deposits can be estimated from the methods given in Sections 19.10 and 19.11 provided the layer in which the pile tips are located extends to a depth at least equal to 3 times the width of the pile group below the base. The equivalent footing is assumed at the base of the group. Where alternating compressible and non-compressible layers exist below the base of the group, the settlement is assumed to originate in the compressible layers only.

Examples 19.8 — 19.14

19.8 A free standing, square group of 16 piles extends to a depth of 12 m in stiff clay which is underlain by rock at 24 m depth. The piles are 0.5 m in diameter and are spaced at 1.5 m centres. c_T at base = 140 kPa, average $(c_T)_{av}$ along shaft = 90 kPa, adhesion $\alpha = 0.45$, coefficient of volume decrease m_v below base = 0.08 m²/MN, remoulded c_T along shaft = 50 kPa. Assume group efficiency $E_g = 0.65$. Find the allowable load for the group with F = 2.5. Also estimate the settlement under the allowable load.

(a)   Considering individual piles.

$$f_b = c_T N_{cp} = 140 \times 9 = 1260 \text{ kPa}$$
$$f_s = \alpha c_T = 0.45 \times 90 = 40.5 \text{ kPa}$$
$$Q_t = \frac{\pi}{4} B^2 f_b + \pi B L_p f_s = 1010.8 \text{ kN}$$
$$Q_g = E_g . n Q_t = 10512 \text{ kN}$$

(b)   Considering block failure.

$$Q_B = B_0 L_0 f_b + 2 L_p (B_0 \times L_0) f_s; B_0 = L_0 = 5 \text{ m}, L_p = 12 \text{ m}$$
$$f_b = c_T N_{cp} = 140 \times 9 = 1260 \text{ kPa}$$
$$f_s = c_T (\text{remoulded}) = 50 \text{ kPa}$$
$$Q_B = 43500 \text{ kN} > Q_g$$
$$Q_a = \frac{1}{2.5} Q_g = \frac{10512}{2.5} = 4205 \text{ kN}$$

For settlement the load is assumed to act at an equivalent raft (5 m x 5 m) located at 12/3 = 4 m above the base. The compressible zone of 16 m thickness is divided into 4 layers, each 4 m thick. Pressure increase at the middle of each layer is calculated by assuming 2 : 1 spread of load Q_a. The calculations are given in Table 19.7. The total settlement is 48 mm.

Table 19.7
EXAMPLE 19.8

Layer	Mid-depth (m)	Area (m²)	$\Delta\sigma$ (kPa)	$m_v \Delta\sigma H$ (mm)
1	2	49	85.80	27.45
2	6	121	34.75	11.12
3	10	225	18.68	5.97
4	14	361	11.65	3.73
				48.28

19.9 A group of 9 piles, arranged in a square pattern at 1 m centres, is driven through soft clay to bear on a bed of dense sand which is 2.7 m thick. The pile diameter is 30 cm. The total load on the group is 2250 kN. A 2 m thick layer of saturated soft clay exists below dense sand, and it is further underlain by another layer of dense sand. Find the settlement of the pile group due to consolidation of clay and also examine the possibility of bearing failure of clay. For clay, $c_T = 50$ kPa, $m_v = 1.2 \times 10^{-4}$ m^2/kN.

For load transfer the equivalent footing ($B_0 = L_0 = 2.3$ m) is assumed at the pile tips and load is assumed to spread at 2 vertical to 1 horizontal.

$$\Delta\sigma \text{ at middle of clay} = \frac{2250}{(2.3 + 3.7)^2} = 62.5 \text{ kPa}$$

$$S = m_v \Delta\sigma H = 15 \text{ mm}$$

With load spread at 2 : 1, loaded area at top of clay

$$= (2.3 + 2.7)(2.3 + 2.7) = 25 \text{ m}^2$$

$$\Delta\sigma = \frac{2250}{25} = 90 \text{ kPa} < 3 c_T \text{ hence safe against bearing failure}$$

19.10 A group of piles consists of 3 rows of 4 piles spaced at 1.0 m centres in both directions. The pile diameter is 0.35 m. Each pile carries an average working load of 250 kN. The piles are driven to a depth of 9 m below ground level into stiff clay which has a fairly uniform compressibility ($m_v = 0.8 \times 10^{-4}$ m^2/kN). Calculate the settlement of the group, assuming geological factor μ_g for correction to consolidation settlement as 0.7.

Total load on group = $12 \times 250 = 3000$ kN

Equivalent footing at $9/3 = 3$ m above base,

$$B_0 = 2.35 \text{ m and } L_0 = 3.35 \text{ m}$$

$$q_n = \frac{3000}{2.35 \times 3.35} = 381 \text{ kPa}$$

Significant depth for settlement can be taken upto the level at which $\Delta\sigma = q_n/10$, and with approximate pressure distribution this depth will be at 1.5 B_0 below the equivalent footing (See Fig. 18.3).

Eq. 18.10: $S = m_v \times 0.55 q_n \times 1.5 B_0 = 59$ mm

Corrected $S = 0.7 \times 59 = 41$ mm

19.11   A square group of 9 piles, spaced at 0.75 m centres both ways, consists of 0.3 m diameter piles driven through a 5 m thick recently placed cohesive fill to bear on a firm stratum. The average, safe load on each pile is not to exceed 250 kN. Determine the maximum working load for the group taking into consideration the negative friction. $(c_T)_{av} = 90$ kPa, $\alpha = 0.45$, $\gamma = 1.8$ t/m³.

$$F_{n1} = A_s (c_T)_{av} \alpha = (\pi BL_n) (c_T)_{av} \alpha = 191 \text{ kN}$$

$$F_{n2} = \frac{(B_o \times B_o) L_a \gamma}{n} = 31.75 \text{ kN}$$

Lower of F_{n1} and F_{n2} (=31.75 kN) is the frictional drag on each pile.

Total frictional drag on group = 31.75 × 9 = 285.77 kN

Net working load on top of group = (250 × 9) — (285.77)
= 1964 kN

19.12   A group consists of 4 rows of 5 driven piles spaced at 0.75 m centres in both directions. The soil is soft clay in the top 1.0 m depth underlain by loose sand of great depth. Groundwater table is at 3 m below the ground surface. Length of piles = 13 m, diameter of piles = 0.25 m, γ (clay) = 17.4 kN/m³, γ (sand) = 17.5 kN/m³, submerged $\gamma' = 8.6$ kN/m³. The average cone resistance q_c along the 12 m length of the shaft is 3000 kPa, but average q_c along the lower 4 m length is 3200 kPa. q_c at pile tips and upto 1.0 m below is 3500 kPa. Find the working load (F = 3) for the group and also the settlement under the working load using the Buisman De Beer method.

Effective depth of embedment = 12 m

$$f_s = \frac{3000}{200} = 15 \text{ kPa}$$

$$f_b = q_c \text{ at tips} = 3500 \text{ kPa}$$

$$Q_s = (\pi \times B \times 12) \times f_s = 141.4 \text{ kN}$$

$$Q_b = \frac{\pi B^2}{4} \times f_b = 171.8 \text{ kN}$$

Ultimate group load Q_g ($E_g = 1$) = 20 × 313.2 = 6264 kN

Working load = 6264/3 = 2088 kN

Overall dimensions, $B_o = 2.5$ m, $L_o = 3.25$ m

Equivalent raft is assumed at 12/3 = 4 m above base. The compressible zone is assumed to extend upto 2 B_o (= 5 m) below the equivalent raft, which is divided into two layers: first layer 4 m thick with $(q_0)_{av}$ = 3200 kPa and second layer 1 m thick with $(q_0)_{av}$ = 3500 kPa.

(a) First layer.

$$\sigma'_0 \text{ at middle} = (1 \times 17.4) + (2 \times 17.5) + (8 \times 8.6)$$
$$= 121.2 \text{ kPa}$$

$$\Delta\sigma' = \frac{2088}{(2.5 + 2)(3.25 + 2)} = 88.38 \text{ kPa}$$

$$C = \frac{1.5 \, q_0}{\sigma_0'} = \frac{1.5 \times 3200}{121.2} = 39.6$$

$$S = \frac{H}{C} \, 2.3 \log \frac{\sigma_0' + \Delta\sigma}{\sigma_0'}$$

$$= \frac{4}{39.6} \times 2.3 \log \frac{121.2 + 88.38}{121.2} = 0.0552 \text{ m}$$

(b) Second layer.

$$\sigma_0' = 121.2 + (2.5 \times 8.6) = 142.7 \text{ kPa}$$

$$\Delta\sigma = \frac{2088}{(2.5 + 4.5)(3.25 + 4.5)} = 38.49 \text{ kPa}$$

$$C = \frac{1.5 \times 3500}{142.7} = 36.8$$

$$S = \frac{1}{36.8} \, 2.3 \log \frac{142.7 + 38.49}{142.7} = 0.0064 \text{ m}$$

Total settlement 62 mm.

(Settlement consideration rather than bearing failure governs the allowable load for a pile group in sand.)

19.13 A piled foundation group has 16 piles arranged in a square pattern with centre-to-centre (c/c) spacing of 0.75 m between the piles. The piles are 0.25 m square in section and the total length is 16 m. The soil profile consists of a recently placed cohesionless fill from 0 to 1.0 m, soft clay from 1.0 m to 4.0 m, stiff clay from 4.0 m to 20.0 m, and dense gravel bed at 20.0 m below ground surface. Water table is at top of soft clay. γ (fill) = 17.6 kN/m³, γ' (soft clay) = 8.8 kN/m³, $(c_T)_{av}$ (soft clay) = 40 kPa, adhesion factor (soft clay) = 0.6, $(c_T)_{av}$ (stiff clay, undisturbed) = 150 kPa, $(c_T)_{av}$ (stiff clay, remoulded) = 80 kPa, adhesion (stiff clay) = 0.45, c_T at base of piles = 200 kPa. Find the allowable load for the group with an overall safety factor of 2.5. Determine the consolidation settlement taking the negative friction into consideration. Assume $m_v = 0.6 \times 10^{-4}$ m²/kN, $E_g = 0.65$, $\mu_g = 0.5$.

3 m thick layer of soft clay to a depth of 4 m into an underlying

There will be some drag in the fill zone but the principal down-drag is in the soft clay layer below. The cohesionless fill is assumed to act as surcharge.

(i) Considering individual piles.

$f_n = \alpha\,(c_T)_{av} = 0.6 \times 40 = 24$ kPa

F_a per pile $= 4 \times 0.25 \times 3 \times 24 = 72$ kN

Total frictional drag on 16 piles $= 1152$ kN

(ii) Considering weight of soil within the group ($B_o = 2.5$ m),

Total weight $= [(1 \times 17.6) + (3 \times 8.8)] \times (2.5 \times 2.5)$
$= 275$ kN

Hence total negative friction $= 275$ kN

The group capacity is determined considering the effective embedment of 12 m into stiff clay.

(a) Individual pile behaviour.

$f_s = \alpha(c_T)_{av} = 0.45 \times 150 = 67.5$ kPa

$f_b = c_T\,N_{cp} = 200 \times 9 = 1800$ kPa

$Q_f = (4 \times B \times 12)\,f_s + B^2 \times f_b = 922.5$ kN

$Q'_g = n\,Q_f\,E_g = 9594$ kN

(b) Block behaviour.

$f_s = 80$ kPa (remoulded, $\alpha = 1$)

$f_b = 200 \times 9 = 1800$ kPa

$Q_B = (4 \times 2.5 \times 12)\,f_s + (2.5)^2\,f_b = 20850$ kN $> Q_g$

$Q_a = \dfrac{1}{F}\,Q_g = 3837$ kN

Total load on group $= Q_a +$ frictional drag
$= 3837 + 275 = 4112$ kN

Assume equivalent footing (2.5 m $\times$ 2.5 m) at 12/3 $= 4$ m above base. Consider 3 layers, each 4 m thick. Calculations are given in Table 19.8. Corrected settlement $= 68.7 \times \mu_g = 34$ mm.

Table 19.8
(Example 19.12)

Layer	Av. depth (m)	Area (m²)	σΔ (kPa)	S (mm)
1	2	20.25	20.3	48.72
2	6	72.25	56.9	13.66
3	10	156.25	26.3	6.32
				68.70

19.14 A square group of 25 precast concrete piles, 0.25 m square in section and spaced at 0.75 m centres, is driven through a

3 m thick layer of soft clay to a depth of 4 m into an underlying deposit of medium sand. The SPT blows in sand are recorded in Table 19.9. Water table is at the surface of sand and clay is saturated by capillary water. Unit weight of clay $= 17.3$ kN/m^3, unit weight of sand $= 19.8$ kN/m^3. Find the working load for the group with a safety factor of 4. Also estimate the probable maximum settlement.

Average of SPT blows is determined for 4 m depth from which f_s is estimated. Base resistance f_b is calculated from the average N value in the vicinity of the base (corrected for overburden). For settlement estimation by the Meyerhof empirical formula, average of N blows corrected for overburden within the seat of settlement, ($\simeq$ overall width $B_o = 3.25$ m below the equivalent footing) is determined. Equivalent footing is assumed at $L'_p/3$ above base ($= 1.33$ m). Overburden correction factor is taken from Fig. 15.9 (Table 15.4); the Peck formula (Eq. 15.64a) or the Lio formula (Eq. 15.64c) could also be used. Corrections to N values are given in Table 19.9

Table 19.9

Example 19.14

Depth in sand (m)	N values	σ'_o (kPa)	correction C	Corrected N_n
0 25	15			
1.00	18			
1.75	21			
2.50	22			
3.25	21	84.4	1.06	22
4.00	24	91.9	1.03	25
4.75	26	99.4	1.00	26
5.50	32	106.9	0.95	30
6.25	34	114.4	0.92	31
7.00	38	121.9	0.89	34
7.75	38	129.4	0.86	33

Average N (uncorrected) upto 4 m depth $= 20$
Average N (corrected) in the vicinity of base $= 25$
Average N (corrected) (3.25 m to 7.75 m) $= 29$
Neglect friction of clay.

$f_s = 2 (N)_{av} = 2 \times 20 = 40$ kPa

f_b = 400 × N = 400 × 25 = 10,000 kPa (< 11000 kPa)
Q_f = $(0.25)^2$ f_b + (4 × 0.25 × 4) f_s = 785 kN
Q_a = $\frac{1}{4}$ Q_f = 196 kN
Allowable load for group $(E_g = 1)$ = 25 × 196 = 4900 kN

$$q_n = \frac{4900}{(B_o)^2} = \frac{4900}{(3.25)^2} = 464 \text{ kPa}$$

$$I = 1 - \frac{L_p'}{8 B_o} = 0.85$$

$$S = 0.96 \frac{q_n \sqrt{B_o}}{N} I \; ; \quad N = 99$$

$$= 24 \text{ mm}$$

19.13 UPLIFT RESISTANCE

1. Single pile. The ultimate uplift resistance of a pile with straight shaft is equal to the skin friction which can be mobilized along the surface of the shaft. The skin friction is commonly assumed equal to that for a pile under compressive load.

For a pile with enlarged base, e.g. underreamed pile, the uplift resistance Q_{up} is given by:

$$Q_{up} = 0.25 \; \pi \; (B_b{}^2 - B^2) \; f_b + A_s . \; f_s \qquad (19.66)$$

where B_b = diameter of enlarged base (bulb)

B = diameter of shaft

A_s = surface area of shaft above the bulb and below the active zone of soil, if any, which is subject to expansion.

The base resistance f_b and the skin friction are determined as described for piles under compression. A safety factor of 2.5 may be used to find the allowable uplift load.

2. Pile group. The uplift resistance of a pile group is taken as the lesser of the two following values.

(1) Sum of uplift resistance of piles in the group.

(2) Sum of shear resistance mobilized on the surface perimeter of the group plus total weight of soil and piles enclosed in the perimeter.

Example 19.15 — 19.16

19.15 Find the ultimate uplift resistance and the safe uplift resistance (F = 2.5) of the underreamed pile of Ex. 19.5, if the top 2 m of clay is active and subject to swelling.

Projected area of bulb $= \dfrac{\pi}{4}(B_b{}^2 - B^2) = 0.659 \text{ m}^2$

Neglecting top 2 m,

Surface area of shaft $A_s = \pi \, B \times 3 = 3.77 \text{ m}^2$

$Q_{uplift} = 0.659 \, f_b + 3.77 \, fs,$

where $f_b = 1350$ kPa and $f_s = 46$ kPa

$Q_{uplift} = 1063$ kN

Safe uplift load $= \dfrac{1063}{2.5} = 425$ KN

19.16 Find the penetration depth for a 35 cm square R.C. pile driven through sea bed for a jetty structure to carry a maximum compressive load of 500 kN and a net uplift load of 300 kN. The soil below sea bed consists of a 10 m thick layer of saturated, medium-dense sand overlying dense sand and gravel. The average SPT blows in the top 10 m sand are 12 and on the bottom sand-gravel layer 40. No erosion is expected. The minimum safety factors should be 2.5 on the uplift load and 4 on the compressive load.

f_s (medium-dense sand) $= 2 \, N_{av} = 24$ kPa

f_s (dense sand-gravel) $= 2 \times 40 = 80$ kPa

Ultimate uplift resistance required $= 300 \times 2.5 = 750$ kN

Q_s in top 10 m sand $= (4 \times 0.35 \times 10) \times 24 = 336$ kN

Balance $= 750 - 336 = 414$ kN

Frictional resistance per metre length in sand-gravel

$= (4 \times 0.35 \times 1) \times 80 = 112$ kN/m

Penetration required $= \dfrac{414}{112} = 3.69 \text{ m} \simeq 3.7 \text{ m}$

Total penetration $= 10 + 3.7 = 13.7$ m

f_b (sand gravel) $= 400 \, N = 400 \times 40 = 16000$ kPa

Adopt a maximum of $f_b = 11000$ kPa

$Q_b = A_b . f_b = 0.35 \times 0.35 \times 11000 = 1347.5$ kN

$Q_f = Q_b + Q_s = 1347 + 750 = 2097$ kN

$F = \dfrac{2097}{500} = 4.2 > 4 \text{ (safe)}$

which indicates that the required penetration is governed by uplift resistance.

19.14 LATERAL RESISTANCE

The lateral resistance of piles subjected to horizontal loads depends on various factors such as the size and type of pile, nature of the surrounding soil, pile spacing and the amount of fixity at the top. Amongst the various available theoretical solutions the Golubkov's method (1950) is of great practical utility (Wilun and Starzewski 1975) and is given below.

The allowable horizontal loads P_a (applied at ground surface) for a pile fully fixed to cap with no possibility of rotations and for a free standing pile are given by:

$$P_a = \frac{12\,EI}{\alpha L_f^3}\, y_a \quad \text{(fixed head, zero rotation)} \qquad (19.67)$$

$$P_a = \frac{3\,E\,I}{\alpha L_f^3}\, y_a \quad \text{(head free to rotate)} \qquad (19.68)$$

where E = Young's modulus of the pile; I = moment of inertia; EI = flexural stiffness of pile; y_a = allowable horizontal deflection of pile head at ground level; L_f = effective length of deflection (depth of virtual fixity from ground surface); for reinforced concrete piles of B = 35—45 cm:

$L_f = 1.25$ m for $y_a = 5$ mm and $L_f = 1.40$ m for $y_a = 10$ mm;
α = dimensionless coefficient depending on pile spacing and type of pile and soil.

The values of the coefficient α may be assumed as follows: for reinforced concrete pile with fixed head and spacing 3 B c/c, $\alpha = 16$ if embedded in weak soil (loose sand, soft silty clay) and $\alpha = 10$ if embedded in strong soil (stiff sandy silty clay, sand); for piles spaced at 6 B, $\alpha = 8$ and 5 respectively. For piles of different sizes and for major structures, values of αL_f^3 should be based on field tests with measurement of lateral deflections and using Eqs. 19.67 and 19.68 for interpretation of result (Wilun and Starzewski 1975).

A simple, approximate method of calculating the deflection (y) of a laterally loaded pile under working loads (P) is given by Tomlinson (1977) by assuming that the pile acts as a cantilever above the point of virtual fixity.

$$y = \frac{P\,(e + L_f)^3}{12\,E\,I} \quad \text{(fixed head)} \qquad (19.69)$$

$$y \simeq \frac{P (e + L_f)^3}{3 \, EI} \quad \text{(free head)} \qquad (19.70)$$

where e = height from ground surface to the point of application
. of load P

L_f = depth from ground surface to point of virtual fixity

$\simeq 1.5$ m for compact granular soil or stiff clay (below shrinkage zone) and 3 m for soft clay or silt

If the head carries only a moment M instead of a horizontal force, M can be replaced by a horizontal force P at 'e' above ground surface, where $M = P \times e$.

20
Well foundation

20.1 WELL AND ITS ELEMENTS

A 'well' or an 'open caisson' is a large hollow open ended structure which is generally built in parts and sunk through ground or water to the prescribed depth and which subsequently becomes an integral part of the permanent foundation. If a well is provided with a working chamber at its lower end in which air is maintained above atmospheric pressure to prevent the entry of water during sinking operations, it is called a 'pneumatic well or caisson'. Well foundations are used where it is necessary to transmit large vertical and horizontal loads to deeper and stronger strata because of the low bearing capacity of the overlying soils or for protection against scour. In India, wells are more commonly provided under bridge piers and abutments when a firm load supporting stratum is located deeper than approximately 5 m below a saturated soil and open excavation and placement of foundation become difficult and uneconomical. Masonry and concrete wells are the common types used in India. A typical section of well foundation for a bridge is shown in Fig. 20.1.

The various elements of a well foundation are as follows:

(1) *Steining.* It is the wall or shell of the well which transfers the loads down below. It acts as a cofferdam (enclosure) for excavating the soil so that the well may penetrate into the soil.

(2) *Curb.* It is a reinforced concrete (ring) beam with a steel cutting edge at bottom. The curb supports the well steining and facilitates sinking due to its wedge shape in cross section. The curb is kept slightly projected (say 25 mm) beyond the steining, which helps in reducing skin friction on the steining. The curb resists the hoop forces to which it is subjected during sinking and afterwards.

(3) *Cutting edge.* It is the lowermost portion of the well curb which cuts into the soil during sinking.

(4) *Bottom plug.* After completion of sinking, a concrete seal or plug is provided at the bottom of the well. Confined by the well curb the bottom plug acts as a raft against soil reaction from below. Thus the whole base area of the well becomes effective in transmitting loads to the soil.

(5) *Backfill.* After setting the bottom plug, the water is pumped out of the well and it is backfilled with sand or excavated material.

(6) *Top plug.* It is a concrete plug provided over the filling inside the well.

(7) *Well cap.* It is generally a R.C.C. slab laid at the top to transmit the load of the superstructure to the steining. Its top level is generally kept at the low water level or the river bed level. (The minimum thickness is about 0.75 m.)

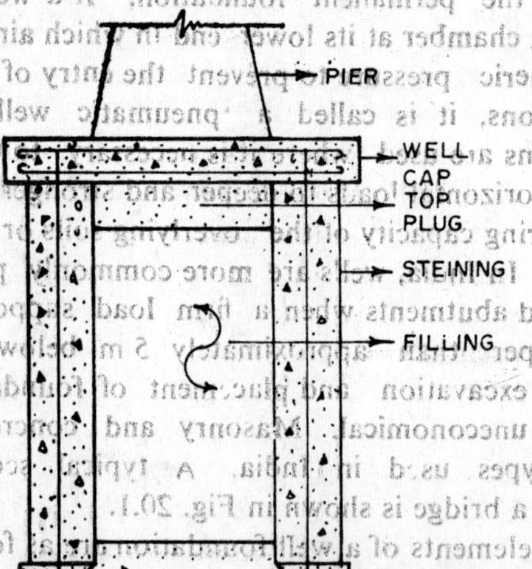

Fig. 20.1 Typical section of well foundation

20.2 SHAPE AND LATERAL SIZE

In horizontal cross section, the common types of well are: single wells of circular, square, rectangular or elliptical section, and twin wells of double-D or dumb-bell section. There may be a combination of more than two wells also, (a well with multiple dredge holes), e.g. combination of two square wells with D-shaped ends giving four dredge holes. The shape and combination are from the considerations of the size of the pier or abutment to be supported, the ease and cost of sinking and the likely tilts and shifts.

The most advantageous shape for a well is the circular one because (i) it has the least perimeter for a given area of the base, which means the least frictional resistance during sinking, (ii) the external lateral pressure of soil and water induces compressive stresses in the circular well steining, and (iii) excavation in a spiraled fashion (starting from the middle) making the soil fall uniformly from all sides towards the centre helps in a uniform, untilted sinking of the well. The disadvantage is the obstruction to flow which is caused when a large diameter well is used to accommodate a large oblong pier. Two or three independent wells or combined wells are used for large oblong piers.

The size of the base, i.e. contact area, is proportioned for the loads to be transmitted to the soil, taking into consideration the allowable bearing pressure; usually ignoring the skin friction between soil and steining (Jumikis 1971). The minimum inside diameter of the well must be such as to provide a convenient working space for excavation and easy hoisting of soil and materials. The IRC Code 78 recommends a diameter not less than 2 m for a dredge hole (IRC 78). The thickness of the steining (t in metres) is chosen empirically by experience as (Jumikis 1971):

Masonry or concrete well: $t \simeq 0.1 + D/10$ (m) (20.1)
Reinforced concrete well: $t \simeq 0.1 + D/12$ (m) (20.2)
where D = external diameter (m) of circular well

The minimum thicknesses from practical considerations for steinings of masonry, concrete and R.C.C. are also suggested as 0.6 m, 0.45 m, and 0.25 m respectively (Saxena 1971).

Generally adopted sizes of single circular wells for road bridges in India under a pier for different span and nature and depth of foundations are given in Table 20.1. Under an abutment or in a severe seismic zone, the size of a well is slightly larger than the values given in Table 20.1. As a rough guide, Table 20.2 gives the steining thickness for single circular wells sunk through sandy strata. The maximum external diameter for a circular well is recommended to be 6 m for brick masonry steining and 9 m for cement concrete steining. (Saxena 1971). The IRC 78 recommends that the external diameter of plain and reinforced concrete single well should not exceed 12 m and the brick masonry well should not be used for depths greater than 20 metres.

When a group of wells are sunk, the minimum spacing between them depends on the depth of well. For general guidance in design, a spacing of about 1.0 m may be used (IS: 3955). The IRC 78 recommends the minimum clearance between the wells not less than half the external diameter. As far as possible, the wells should be sunk plumb; however, tilts and shifts do occur. A tilt of 1 in 80 and a shift of 150 mm in a resultant direction should be considered in the design. (IRC: 78).

Table 20.1
SIZES OF SINGLE CIRCULAR WELLS GENERALLY ADOPTED FOR ROAD BRIDGES (SAXENA 1971)

Span arrangment	External diameter (m)	Depth and nature of foundation
(1) Simple supported spans 9 m to 12 m	4.3—4.5	Foundations resting on good strata and at reasonable depths, 9 m to 12 m
(2) Simply supported spans 12 m to 18 m	4.5—5.0	
(3) Simply supported spans 18 m to 25 m	4.5—5.5	
(4) Balanced cantilever spans 21 m to 24 m	4.3—5.0	Foundations resting on uniform strata and at depths 12 m to 23 m
(5) Balanced cantilever spans 24 m to 34 m	5.0—6.0	
(6) Balanced cantilever spans 34 m to 38 m	6.0—6.7	
(7) Box girder balanced cantilever spans 38 m to 50 m	6.7—7.3	Difficult and very deep foundations, more than 23 m
(8) Simply supported prestressed concrete spans 38 m to 53 m	6.4—7.3	
(9) Spans above 53 m	7.5 and above	

Table 20.2

RECOMMENDED STEINING THICKNESS AS A ROUGH GUIDE FOR
SINGLE CIRCULAR WELLS IN SANDY STRATA (SAXENA 1971)

External diameter of well (m)	Stening thickness for brick well (m)	Steining thickness for concrete well
3.7	0.6—0.7	0.55—0.6
4.3	0.7—0.8	0.6—0.7
4.9	0.8—0.9	0.7—0.75
5.5	0.9—1.1	0.75—0.85
6.1	1.1 and above	0.85—0.9
6.4—7.3	Not recommended	0.9—1.1
7.5—9.0	Not recommended	1.1—1.2

20.3 DEPTH OF WELL

A well should be sunk to a depth to satisfy the following requirements:

(1) It is safe against the maximum possible scour and has adequate embedded length, called the 'grip length', below the lowest scour level so as to resist the overturning moments due to horizontal forces.

(2) It rests on sound strata of adequate bearing capacity and is safe from settlement considerations also.

1. Depth of scour. Whenever possible the scour depth should be determined by taking soundings in the vicinity of the proposed bridge site during or immediately after the flood. Allowance should be made in the observed depth for increased scour due to (i) the design discharge being greater than the flood discharge, (ii) velocity increase due to obstruction in flow caused by the construction of the bridge, and (iii) increase in scour in the proximity of piers and abutments. The *mean depth of scour* (d in metres) below the highest flood level (HFL) for natural channels flowing in cohesionless alluvium can be estimated from the following equation (IRC: 78):

$$d = 1.34 \, (q_1{}^2/f)^{1/3} \quad (m) \tag{20.3}$$

where q_1 is the discharge in cumecs per metre width. The discharge q_1 is taken to be the maximum of the following: (i) the total design discharge divided by the effective linear waterway between abutments or guide bunds, as the case may be, (ii) the value obtained after considering concentration of flow though a portion of the waterway assessed from the study of the cross section of the river, and (iii) actual observations, if any. If it is not possible to obtain information as per (ii) and (iii), the discharge obtained as per (i) above should atleast be increased by 50 percent for calculating the scour depth (Sastry 1981). Special studies are to be undertaken to determine the maximum scour when (i) the bridge is located in a bend of river, (ii) deep channel hugs to one side, (iii) obliquity of flow is considerable, and (iv) the bridge has very thick piers inducing heavy local scour.

The factor f in Eq. 20.3 is called the "silt factor" and is given by:

$$f = 1.76 \sqrt{D_m} \tag{20.4}$$

where D_m = weighted mean diameter (mm) of a representative sample of bed material upto normal depth of scour.

Typical values of silt factor are given in Table 20.3.

Table 20.3

TYPICAL VALUES OF SILT FACTOR (IRC: 78)

Type of material	Mean diameter (mm)	Silt factor
Silt	0.081	0.50
	0.120	0.60
	0.158	0.70
	0.233	0.85
	0.323	1.00
Sand	0.505	1.25
	0.725	1.50
	0.988	1.75
	1.290	2.00

The *maximum depth of scour* for the design of well foundations below HFL for piers and abutments located in a straight reach and having no floor protection works is taken as follows (IRC: 78):

(1) 2.0 d, in the vicinity of piers

(2) 2.0 d, near abutments (scour all round)

(3) 1.27 d, near abutments (approach retained)

2. Grip length. The grip length D, or the depth of embedment below the anticipated maximum scour level should not be less than one-third the maximum depth of scour below HFL (IRC; 78), i.e.

$$D \nless \tfrac{1}{3} (2d) \quad \text{(for pier wells)} \tag{20.5}$$

The lateral resistance of the well is discussed in the next section.

Example 20.1

Determine the mean depth and the maximum depth of scour for a bridge pier in a river flowing in alluvium. The sieve analysis of a representative sample of bed material is given in Table 20.4. The total discharge is 36000 cumecs over an effective linear waterway of 500 m. What should be the grip length for a well foundation? (Data from bridge site across Godawari river near Bellur village in Nanded district; Bedse et al 1984.)

Table 20.4

Example 20.1

Sieve size	Weight retained	Percentage retained	Average size of sieve	(3) × (4)	Mean dia D_m (mm) (5)/100
(mm)	(g)	(%)	(mm)		
1	2	3	4	5	6
4.75	0	0			
2.00	49	4.90	3.3750	16.537	87.200
1.00	190	19.00	1.5000	28.500	100
0.600	295	29.50	0.8000	23.600	= 0.872
0.425	253	25.30	0.5125	12.966	
0.300	83	8.30	0.3625	3.009	
0.150	104	10.40	0.2250	2.340	
0.075	20	2.00	0.1125	0.225	
Pan	6	0.60	0.0375	0.023	
	1000	100		87.200	

Silt factor $f = 1.76\sqrt{D_m} = 1.76\sqrt{0.872} = 1.64$

$q_1 = 36,000/500 = 72$ cumecs/m

$d = 1.34 (q_1^2/f)^{1/3} = 19.61$ m

Maximum depth of scour near pier $= 39.22$ m

Grip length $D \nless \tfrac{1}{3} (2d) \nless 13.1$ m

20.4 LATERAL RESISTANCE — COHESIONLESS SOIL

The system of various forces to which a well may be subjected can be represented by a vertical force and two horizontal forces in the directions across and along the pier. The force system in the directions across the pier, i.e. ´ along the tranverse axis of the pier (along the longitudinal axis of the bridges) is often more critical for the stability analysis. The well should be able to resist the overturning moment induced by the horizontal force.

The IRC: 45 has recemmonded a procedure for estimating the resisting moment of well foundations of bridges resting on cohesionless soils like sand and surrounded by the same soil below maximum scour level. It is applicable only if the depth of embedment is not less than 0.5 times the width of foundation in the direction of lateral forces. The resisting moments for a well under a pier are calculated as follows:

1. Base resisting moment. The base resisting moment M_b is the moment of the frictional force mobilized along the surface of rupture which is assumed to be cylindrical passing through the corners of the base for a square well, as shown in Fig. 20.2. In case of cricular wells, the rupture surface becomes spherical with its centre at the point of rotation, and passes through the periphery or the base.

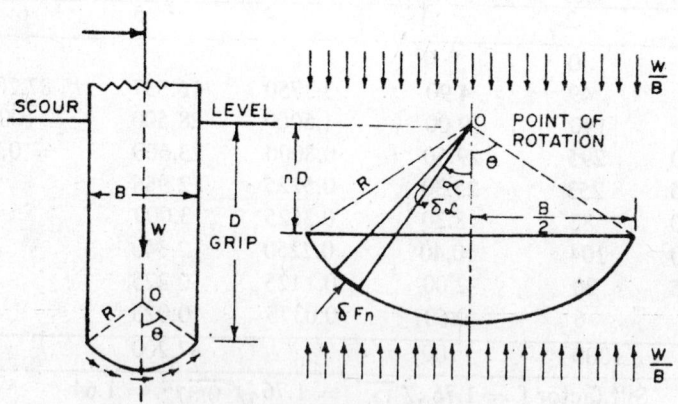

Fig. 20.22 Base resisting moment

If B is the width in case of square and rectangular wells *parallel* to the direction of lateral forces and diameter for circular wells, and W is the total vertical load, the applied load per unit width is W/B which causes an upward reaction of W/B per unit width. For a square or rectangular base, consider a small arc length $R\,\delta\alpha$ at an angle α from the vertical axis, where R is the radius of the rupture surface.

Horizontal component of $R\,\delta\alpha = R\,\delta\alpha\cos\alpha$

Vertical force on the element $= R.\,\delta\alpha\cos\alpha.\,W/B$

Normal component of the force $\delta F_n = R\,\delta\alpha\cos\alpha\,W/B.\,\cos\alpha$

$$F_a = 2\int_0^\theta \frac{WR}{B}\cos^2\alpha\,d\alpha = \frac{2WR}{B}\int_0^\theta \left(\frac{1+\cos 2\alpha}{2}\right)d\alpha$$

$$F_a = \frac{W\,R}{B}\left(\theta + \sin\theta\cos\theta\right) \tag{20.6}$$

$$\sin\theta = \frac{B/2}{R}, \quad \cos\theta = \frac{nD}{R}, \quad \tan\theta = \frac{B}{2\,nD}$$

$$R = \sqrt{(B/2)^2 + (nD)^2} = \frac{B}{2}\sqrt{1 + \frac{4n^2\,D^2}{B^2}}$$

and $$F_n = \frac{W}{2}\sqrt{1 + \frac{4n^2\,D^2}{B^2}}\left[\tan^{-1}\frac{B}{2nD} + \frac{2n\,B\,D}{B^2 + 4n^2D^2}\right]$$
$$\tag{207}$$

The moment of resistance of the base M_b about the point of rotation O is given by:

$$M_b = (F_n\tan\phi)\,R \tag{20.8}$$

In the case of a circular base where the rupture surface is spherical, a shape factor of 0.6 is used expressing the M_b as:

$$M_b = 0.6\,F_a\tan\phi\,R \tag{20.9}$$

Assuming the point of rotation at $nD = 0.2\,D$, the formula (Eq. 20.8) simplifies to:

$$M_b = K\,W\,B\tan\phi \tag{20.10}$$

where K = constant depending on the shape of well and D/B ratio

The values of K for square and rectangular wells are given in Table 20.5. A shape factor of 0.6 is to be multiplied for wells with circular base. The values of K for intermediate values of D/B may be linearly interpolated.

Table 20.5

VALUES OF CONSTANT K FOR BASE MOMENT (IRC: 45)

D/B	0.5	1.0	1.5	2.0	2.5
K	0.41	0.45	0.50	0.56	0.64

2. Side resisting moment. The side resisting moment M_s due to earth pressure is calculated by assuming the 'net' ultimate soil pressure distribution at the front and back faces of the well as shown in Fig. 20.3.

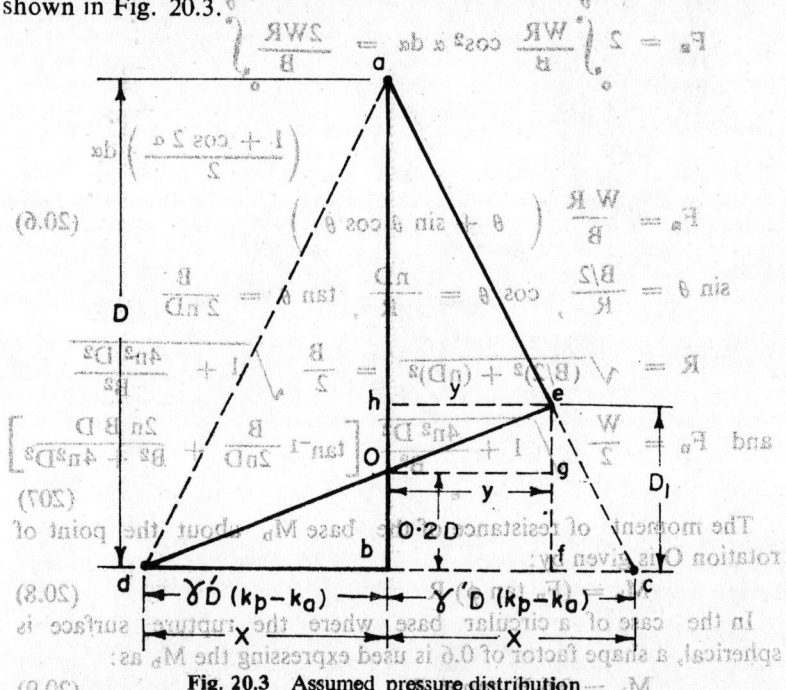

Fig. 20.3 Assumed pressure distribution

The point of rotation is assumed at 0.2D above the base. The resisting moment M_s per unit length of the well is calculated as follows.

From Δ dob and Δ oeg:

$$\frac{ob}{db} = \frac{0.2\,D}{x}, \quad \frac{eg}{og} = \frac{D_1 - 0.2D}{y}$$

or $\quad \dfrac{D}{x} = \dfrac{5D_1 - D}{y}$

From $\triangle$ abc and $\triangle$ ahe:

$$\frac{D}{x} = \frac{D - D_1}{y} = \frac{5D_1 - D}{y}$$

or $\quad D_1 = \dfrac{D}{3}$ $\hfill$ (20.11)

The moment of side resistance about o is the algebraic sum of moments of $\triangle$ abc and $\triangle$ dec.

$$M_s = \frac{1}{2} D \times \left(\frac{D}{3} - \frac{D}{5} \right) + \frac{1}{2} \left(\frac{D}{3} \right)$$
$$\times 2 \times \left(\frac{D}{5} - \frac{D}{9} \right)$$
$$= \frac{x D^2}{15} + \frac{4 \times D^2}{135} = 0.096 \, x D^2$$
$$\simeq 0.1 \times D^2$$

or $\quad M_s = 0.1 \, \gamma' D^3 (K_p - K_a)$ per unit length of well $\hfill$ (20.12)

If the base length of the well is L, (in case of circular well, take L = 0.9 B), the total side resisting moment is given by:

$$M_s = 0.1 \, \gamma' D^3 (K_p - K_a) L \hfill (20.13)$$

where K_p and K_a = coefficients of passive and active earth pressures

The IRC: 45 recommends to determine K_p and K_a by the Coulomb theory. However, the author would . suggest to use the Rankine theory. The Coulomb theory significantly over estimates the passive resistance when ϕ is greater than about 20° and δ is greater than $\phi/3$. Moreover, the lateral displacements of the well may not be sufficient to develop fully the earth pressures. The friction angle δ between the well and soil will, however, be considered for determining the resisting moment due to frictional forces.

3. Resisting moment due to friction on faces. A resisting moment M_f also develops due to the vertical frictional forces on the front and back faces of the well, (the latter being the vertical component of the net passive pressure), the resultant of which is assumed to be inclined to the horizontal at an angle δ, the angle of wall friction.

Total frictional force per unit length

$$= (\triangle \text{ aoe} + \triangle \text{ bod}) \sin \delta$$

As $D_1 = D/3$, pressure at $e = \dfrac{2}{3} \gamma' D (K_p - K_a)$

Area of Δ aoe $= \dfrac{2}{3} \gamma' D (K_p - K_p) \dfrac{0.8 D}{2}$

$= \dfrac{0.8}{3} \gamma' D^2 (K_p - K_a)$

Area of Δ bod $= \dfrac{0.2D}{2} \gamma' D (K_p - K_a)$

$= 0.1 \gamma' D^2 (K_p - K_a)$

Total frictional force per unit length

$$= \dfrac{1.1}{3} \gamma' D^2 (K_p - K_a) \sin \delta$$

In case of rectangular well of base length L, the moment about the centre of rotation is given by:

$$M_f = \left[\dfrac{1.1}{3} \gamma' D^2 (K_p - K_a) \sin \delta . L \right] \dfrac{B}{2}$$

$$= 0.183 \gamma' (K_p - K_a) LBD^2 \sin \delta$$

or $\quad M_f \simeq 0.18 \gamma' (K_p - K_a) LBD^2 \sin \delta \qquad (20.14)$

In case of circular wells, the lever arm is taken as B/π and

$$M_f = \dfrac{1.1}{3} \gamma' D^2 (K_p - K_a) \sin \delta . L . \dfrac{B}{\pi}$$

Putting $L = 0.9 B$ in case of circular wells,

$$M_f = \dfrac{0.33}{\pi} \gamma' (K_p - K_a) B^2 D^2 \sin \delta$$

or $\quad M_f \simeq 0.11 \gamma' (K_p - K_a) B^2 D^2 \sin \delta \qquad (20.15)$

4. Total resisting moment. The total resisting moment M is given by:

$$M_t = M_b + M_s + M_f \qquad (20.16)$$

As the passive resistance may not develop fully and also to take into account the special nature of the risk of foundation failure, the resisting moment is reduced by multiplying with a strength factor 0.7. Thus the effective resisting moment M_t' is taken as:

$$M_t' = 0.7 (M_b + M_s + M_f) \qquad (20.17)$$

For stability, the total, effective resting moment M_t given by Eq. 20.17 should not be less than the total applied moment M about the point of rotation (i.e. located at 0.2D above the base) taking appropriate load factors and combinations of applied loads.

20.5 BEARING CAPACITY OF WELL

The diameters of wells for bridges in India vary from a few metres to about 20—25 metres and the depth of embedment from a few metres to about 40—50 metres. The depth to diameter ratio is found to lie between 1 to 2.5. Although the D/B ratio is limited to 2.5 only, the absolute value of depth is considerable, and the behaviour of well foundations is much different from those of shallow foundations ordinarily used for buildings. In case of ordinary shallow foundations, moments, if any, are resisted by the base. In case of wells, both the sides and the base resist the moments and it is not appropriate to neglect the shear resistance of the overburden above the base level, as is done in other shallow foundations. Wells are also different from piles, the other deep foundations, as the moment taken by the pile base is negligible.

The ultimate bearing capacity of well foundations can be estimated from the Hansen or Meyerhof analysis (Chapter 15) taking into consideration the depth factors also. The bearing capacity factor N_c, N_q and $N\gamma$ can be taken from Table 15.3, which are on the conservative side. For clays ($\phi_T = 0$), the Skempton analysis can also be used. In sandy soils, the ϕ-values can be estimated from the standard penetration test. A safety factor of three should be applied to the ultimate value to get the safe bearing capacity.

The IS Code: 3955 recommends the following formula (Eq. 20.18) for determining the allowable bearing pressure (kg/m²) of well foundations in sandy strata:

$$q_a = 5.4 N^2 B + 16 (100 + N^2) D \quad (kg/m^2) \quad (20.18)$$

where   N = corrected SPT blows
        B = smaller dimension of well (m)
        D = grip depth (m)

The author suggests to ignore the contribution of skin friction to the bearing capacity.

20.6 SETTLEMENT OF WELL

The settlement of well foundations in cohesionless strata can be estimated by the methods described in Chapter 18 using the results of CPT and SPT. The Meyerhof formula (Eq. 19.59) is also worth trying:

$$S \text{ (mm)} = \frac{q_n \sqrt{B}}{N} I \qquad (20.19)$$

where $\qquad I = 1 - \frac{D}{8B} \geqslant 0.5 \qquad (20.20)$

N = average SPT blows within depth B (m) below the well

q_a = pressure at base (kPa)

If sand is overconsolidated, Eq. 20.19 will over-estimate the settlement.

The Peck formula (Eq. 15.66) may be used to find the pressure for a settlement not exceeding 25 mm:

$$q \text{ (kPa)} = 10.25 N \quad (S \not> 25 \text{ mm}) \qquad (20.21)$$

To take into account the effect of confining overburden pressure in reducing settlement (as compared to that of shallow foundation), it is recommended not to apply overburden correction to N values in Eqs. 21.19 and 20.21. The water table correction may also be ignored because its effect should be reflected in the measured values of N.

The consolidation settlement of wells on clays can be estimated as described in Chapters 14 and 18. The settlement should be reduced by using the depth factor (Table 18.2).

Example 20.2

A circular well of 6 m outside diameter is sunk to a depth of 12 m below the anticipated scour level. The resultant horizontal force on the well is 2000 kN and the applied moment at base level is 32000 kN-m. The total net load on the well, inclusive of self weight, is 9000 kN, the bed soil is sand for which $\phi' = 35°$, $c = 0$, $\gamma' = 7.8$ kN/m^3. The angle of friction δ between sand and well may be assumed as 23°. The average of SPT blows below the base level is 26. Examine the safety of the well against the applied moment. What are the ultimate bearing capacity and the probable settlement of the well ?

(a) *Resisting moment.*

ϕ = 35°, δ = 23°, K_a = 0.27, K_p = 3.7

D/B = 2.0, K = 0.56 (Table 20 5)

M_b = 0.6 KWB tan ϕ = 12700.8 kN-m

M_s = 0.1γ' D^3 ($K_p - K_a$) L ; L = 0.9 B for circular well

$\qquad$ = 24964.7 kN-m

$M_f = 0.11\ \gamma'\ (K_p - K_a)\ B^2\ D^2\ \sin \delta$

$\qquad = 5949.9\ \text{kN-m}$

$M_t = M_b + M_s + M_f = 43615\ \text{kN-m}$

Effective $M_t' = 0.7\ M_t = 30530\ \text{kN-m}$ (about point of rotation at 0.2 D above base)

Applied moment at base = 32000 kN-m

Distance of resultant horizontal force from base = 32000/2000

$\qquad = 16\ \text{m}$

Distance of horizontal force above point of rotation

$\qquad = 16 - 0.2\ D = 13.6\ \text{m}$

Applied moment M about point of rotation

$\qquad = 2000 \times 13.6 = 27200\ \text{kN-m}$

Effective $M_t' > M$ (applied moment), hence safe.

(b) *Ultimate bearing capacity.* The well base is subjected to both vertical load and moment resulting in eccentric loading. Assuming that the three moments M_b, M_s and M_f are mobilized to the same fraction α of their ultimate values at equilibrium and equating to the external moment about the point of rotation,

$M = \alpha\ (M_b + M_s + M_f) = \alpha\ M_t$

$27200 = \alpha\ (43615),\ \text{or}\ \alpha = 0.62$

Base moment at equilibrium M_b' about the point of rotation is given by:

$M_b' = M - \alpha\ (M_s + M_f) = 8032.9\ \text{kN m}$

Equivalent horizontal force H is given by:

$H \times 13.6 = 8032.9$

$\text{or}\ H = 590.66\ \text{kN} \simeq 590\ \text{kN}$

Height of H above base = 13.6 + 2.4 = 16 m

Thus the well base is subjected to a vertical load V = 9000 kN and a horizontal load H = 590 kN at Z = 16 m above base. If 'e' is the eccentricity and θ is the inclination of the resultant with the vertical,

$$\frac{e}{Z} = \tan \theta = \frac{H}{V} = \frac{590}{9000}$$

Hence e = 1.05 m, $\theta = 3.75°$

$B' = B - 2\ e = 6 - 2.1 = 3.9\ \text{m}$

Using Hansen's analysis,

$\phi = 35°,\ N_q = 33.3,\ N_\gamma = 34,\ \gamma = 7.8\ \text{kN/m}^3$

$$s_q = 1 + 0.2 \frac{B'}{L} = 1 + 0.2 \frac{3.9}{6} = 1.13$$

$$s_\gamma = 1 - 0.4 \frac{B'}{L} = 0.74$$

$$i_q = (1 - H/V)^2 = 0.87$$

$$i_\gamma = i_q^2 = 0.76$$

$$d_q = 1 + 2 \tan \phi (1 - \sin \phi)^2 \tan^{-1} (D/B) = 1.32$$

$$d_\gamma = 1$$

$$q_t = \gamma' D N_q s_q i_q d_q + 0.5 \gamma' B' N_\gamma s_\gamma i_\gamma d_\gamma$$
$$= 4044.75 + 290.84 = 4335.6 \text{ kPa}$$

$$q_s = q_t /3 = 1445 \text{ kPa}$$

Checking with IS code formula,

$$q_a = 5.4 N^2 B + 16 (100 + N^2) D, \quad N = 26$$
$$= 170894 \text{ kg/m}^2 = 1675 \text{ kPa}$$

(c) Settlement.

Meyerhof: $S = \dfrac{q_n \sqrt{B}}{N} I$

$$q_a = \frac{9000 \times 4}{\pi B^2} = 318.47 \text{ kPa}$$

$$I = 1 - \frac{D}{8B} = 0.75$$

$$S = 22.5 \text{ mm}$$

Checking with the Peck formula for 25 mm settlement and ignoring water table correction,

$$q = 10.25 N = 266 \text{ kPa (for S} = 25 \text{ mm)}$$

The Meyerhof and the Peck empirical formulae are comparable within the expected accuracy in this particular case, which may not be the case in general.

21
Compaction and pavement design

21.1 COMPACTION AND ITS OBJECTIVES

Compaction is the process of increasing the density of soil by the application of mechanical energy, such as by tamping, rolling and vibration. Compaction is achieved by forcing the particles closer together with a reduction in *air* voids; it may also involve a modification of the moisture content as well as the gradation of the soil.

The objective of compaction is the improvement of the engineering properties of soil. Several benefits are obtained through compaction, such as:

(1) Increase in shear strength and bearing capacity.
(2) Decrease in volume changes and settlement.
(3) Decrease in permeability and ingress of water.

21.2 THEORY OF COMPACTION

The principles of compaction were developed by R.R. Proctor (1933) when he was building dams for the old Bureau of Waterworks and Supply in Los Angeles (USA). Compaction is measured in terms of the dry density achieved. Proctor established that compaction is a function of three variables: (i) moisture content, (ii) compaction effort and (iii) type of soil. The effects of these factors are discussed by referring to the laboratory method of studying compaction (See Section 21.3) where a sample of soil is compacted, using a standard compactive effort, into a cylindrical mould of 1000 cm³ volume. The soil in the mould is weighed, the moisture content is measured and from these measurements the dry density is calcualted, using the following equation:

$$\rho_d = \frac{\rho}{1 + w} = \frac{M/V}{1 + w} \qquad (21.1)$$

where M = mass of moist soil, V = volume of mould and w = moisture content.

1. Effect of moisture content. Of all the factors affecting compaction, moisture has the most important effect. At low moisture contents the shearing resistance to relative movement of the soil particles is large, the soil tends to be stiff and is difficult to compact. As moisture content is increased, water 'lubricates' the soil making it more workable and resulting in higher dry density and lower void ratio. It is, however, observed that for a given compactive effort the dry density reaches a peak with increase of moisture content and then starts decreasing with further increase in moisture. The moisture content at which the dry density is maximum for a given compactive effort is called the "optimum moisture content OMC". Beyond the OMC the water tends to keep the soil particles apart without causing an appreciable decrease in the air voids, which results in low dry densities and high void

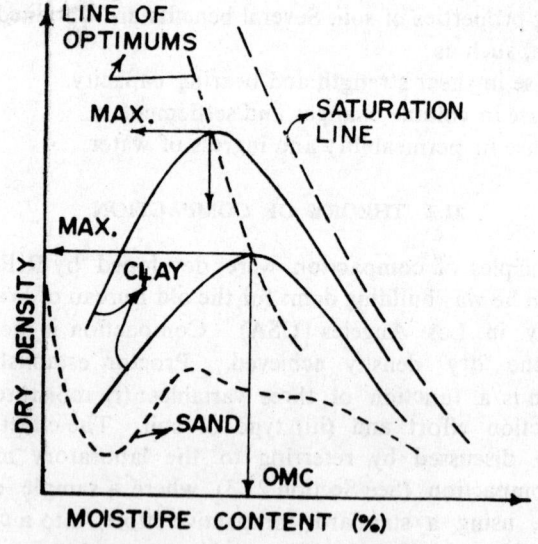

Fig. 21.1 Typical compaction curves obtained from laboratory compaction tests

ratio.   A plot of dry density against moisture content is known as 'a compaction curve'.   Typical compaction curves obtained from laboratory compaction tests are shown in Fig, 21.1.

2.   Effect of compactive effort.   Compactive effort is a measure of the mechanical energy applied to soil.   (In the   laboratory, compactive effort is usually reported in kg. m/m^3 or kJ/m^3, where J $=$ joules: 1 J $=$ 1 N.m.   In the field, compactive effort is the number of passes or 'coverages' of the roller of certain type and weight   on a given volume of soil.) Increasing the compactive effort tends to increase the maximum dry density, as expected, but also decreases the optimum moisture   content. The compaction curve is shifted upwards and leftwards as shown in Fig. 21.1.   This is true for all types of soils and all methods of compaction.   It may be noted that maximum dry density for a soil is only a maximum for a specific compactive effort and method of compaction. A line drawn through the peak points of several compaction curves at different compactive efforts for the same soil is called the "line of optimums".

3.   Effect of soil type.   Different soils attain different maximum dry densities at different optimum moisture contents for a specific compactive effort.   The   highest densities   at lower optimum moisture contents are produced in well graded coarse grained soils, with rounded particles.   Clay soils have much higher optimum moisure contents and consequently lower maximum dry densities. The effect of increasing the compactive effort is also much greater on clay soils. With sands, the resulting compaction curve is often much flatter (a low maximum dry density or even with no pronounced peak); there may be two peaks also, the higher one being in the dry state, as shown in Fig. 21.1.   Typical values of maximum dry unit weight and optimum moisture contents under light compaction test are as follows: gravel-sand-clay mixture (20 kN/m^3, 9%), sand (19 kN/m^3,   11%), sandy clay (18.1 kN/m^3, 14%) silty clay (16.3 kN/m^3, 21%), and clay (15.2 kN/m^3, 28%) (HMSO 1952),

4.   Saturation line.   The air voids of a soil go on decreasing as compaction proceeds with increasing moisture contents upto the optimum value.   With further increase in moisture the air voids do not decrease appreciably because the remaining air takes the form of small occluded bubbles, entirely surrounded by water and held in position by surface tension.   Even a higher compactive effort results only in a momentary increase in pore pressure so that there is less and less permanent volume change. It is not possible to expell all the air and make the sample fully saturated by compaction alone.

If all the air could be expelled by compaction, the soil would become fully saturated and the dry density would be the maximum possible value for that moisture content at saturation. This value of dry density is called the 'saturated dry density' or 'zero air voids dry density'. Since full saturation is unattainable in practice by compaction, the relationship between dry density and moisture content at saturation ($S_r = 1$) is known as the theoretical 'saturation line', or 'zero air voids' line which can be plotted from the expression.

$$\rho_d = \frac{G}{1 + w_{sat} G} \rho_w \tag{21.2}$$

In general, the dry density and moisture content relationship for any degree of saturation or air content n_a can be obtained from the following equations (See Chapter 2):

$$\rho_d = \frac{G}{1 + \dfrac{w G}{S_r}} \rho_w \tag{21.3}$$

$$\rho_d = \frac{G (1 - n_a)}{1 + w G} \rho_w \tag{21.4}$$

The experimental compaction curve for a particular campactive effort must lie completely to the left of the saturation line. If curves of different n_a or S_r are plotted, they enable to know the value of n_a or S_r at any point on the experimental curve.

Some other factors affecting compaction in the field are discussed in Section 21.6.

21.3 LABORATORY COMPACTION TESTS

The purpose of a laboratory compaction test is to establish a dry density/moisture content relationship for a soil under controlled conditions which can form a standard for comparison with field compaction and on the basis of which field specifications can be drawn. The first laboratory test was devised by Proctor (1933) to produce results comparable to field compaction generally

obtainable at that time. In his honour, the test is known as the
'standard Proctor test', adapted by ISI as the 'light compaction
test' (IS: 2720—7). As compaction plants increased in weight
giving higher compaction, the American Association of State
Highway Officials (AASHO) devised a modified test for comparison
The test is known as the 'modified Proctor test', adapted by the ISI
as the 'heavy compaction test' (IS: 2720—8).

   1. **Light compaction test.** The main apparatus consists of a
rammer (2.6 kg mass, 310 mm drop, with 50 mm diameter tamping
foot) and a cylindrical mould (100 mm internal diameter, 127.3 mm

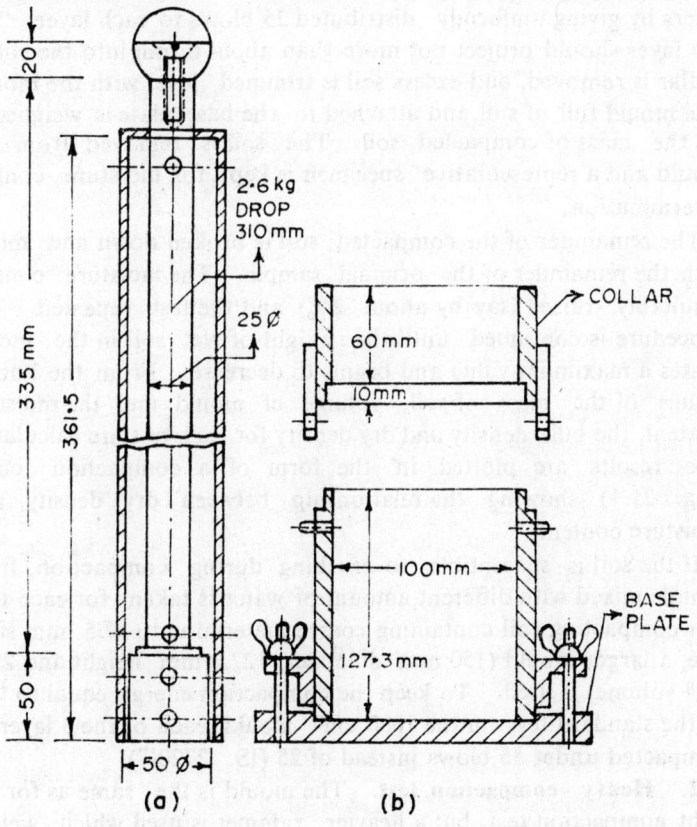

Fig. 21.2 Rammer and mould for compaction test

height, 1000 cm³ volume) (IS: 9198 and 10074). The mould can be fitted to a base plate and a 60 mm high collar. The rammer and the mould are shown in Fig. 21.2.

The test is normally performed on air dried sample passing a 19 mm sieve. About 4 kg of soil is thoroughly mixed with water to give a fairly low moisture content (say about 5% for coarse grained soils and 10% for fine grained soils). The wet sample is covered (sealed) and left for a suitable maturing time (15 minutes to 2 hours or so, depending on soil type) to permit proper absorption of moisture. The mould is attached to the base plate and weighed. The collar is then fitted. Soil is compacted in 3 equal layers by giving uniformly distributed 25 blows to each layer. The top layer should project not more than about 6 mm into the collar. Collar is removed, and excess soil is trimmed level with the mould. The mould full of soil and attached to the base plate is weighed to get the mass of compacted soil. The soil is removed from the mould and a representative specimen is kept for moisture content determination.

The remainder of the compacted soil is broken down and mixed with the remainder of the original sample. The moisture content is suitably raised (say by about 2%) and the test repeated. The procedure is continued until the weight of wet soil in the mould passes a maximum value and begins to decrease. From the known values of the mass of soil, volume of mould and the moisture content, the bulk density and dry density for each test are calculated. The results are plotted in the form of a compaction curve (Fig. 21.1) showing the relationship between dry density and moisture content.

If the soil is susceptible to crushing during compaction, fresh sample mixed with different amount of water is taken for each test. For compacting soil containing coarse material upto 37.5 mm sieve size, a larger mould (150 mm diamater, 127.3 mm height and 2250 cm³ volume) is used. To keep the compaction energy equal to that of the standard test in the 1000 cm³ mould, each of the 3 layers is compacted under 55 blows instead of 25 (IS: 2720-7).

2. Heavy compaction test. The mould is the same as for the light compaction test, but a heavier rammer is used which weighs 4.9 kg with a drop of 450 mm. Test procedure is similar to that of light compaction with the difference that soil is compacted in 5 layers instead of three, each layer being given 25 blows. For testing soil upto 37.5 mm size, the larger mould (2250 cm³) is used; soil is compacted in 5 layers, each layer receiving 55 blows. (IS: 2720-8).

The compactive effort in the heavy compaction test is 2700 kJ/m³ (275.6 × 10³ kg-m/m³) and that in the light compaction test 592.4 kJ/m³ (60.45 × 10³ kg-m/m³), the former being about 4.56 times greater.

3. Jodhpur Mini-Compactor test. This test devised by the author (Singh 1965, Singh and Punmia 1965) gives the values of maximum dry density and optimum moisture content very nearly the same as those obtained by the light compaction test.

The Jodhpur Mini-Compactor (Fig. 21.3) consists of a mould, 60 mm high, 50 cm² in internal cross sectional area (79.8 mm

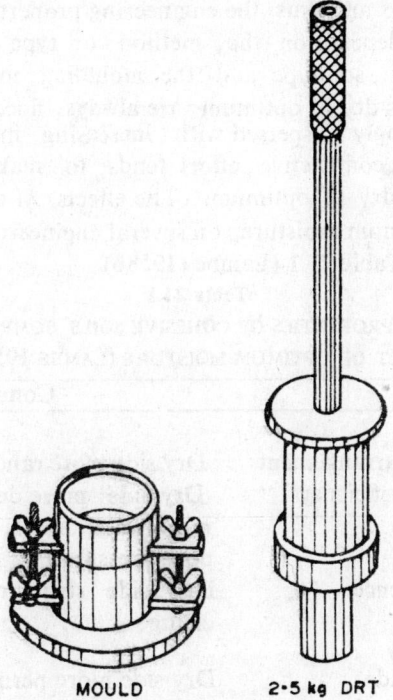

MOULD   2·5 kg DRT

Fig. 21.3 Jodhpur Mini-Compactor

diameter) and 300 cm³ volume, and a 2.5 kg Dynamic Ramming Tool (2.5 kg DRT). The 2.5 kg DRT has a 2.5 kg drop weight which falls freely through a height of 250 mm on a tamping foot, 40 mm in diameter and 75 mm high.

Air dried soil passing a 4.75 mm sieve is used for the test. Starting with a low moisture content, soil is compacted in the mould (attached to the base plate and the collar) in two layers, each layer being given 15 blows of the 2.5 kg DRT. The test is repeated with increasing moisture contents. Dry density and moisture content are determined for each test and a compaction curve is plotted to get the maximum dry density and optimum moisture (Singh 1981). The compactive effort used in the test is 612.5 kJ/m³ (62.5 × 10³ kg-m/m³).

21.4 PROPERTIES OF COMPACTED SOIL

The structure and thus the engineering properties of compacted *cohesive* soils depend on the method or type of compaction, compactive effort, soil type and the moulding moisture content. In general, soils dry of optimum are always flocculated and they become increasingly dispersed with increasing moisture content. Increasing the compactive effort tends to make the soil more dispersed even dry of optimum. The effects of compaction, dry and wet of optimum moisture, on several engineering properties are summarized in Table 21.1 (Lambe (1958b).

Table 21.1

COMPARISON OF PROPERTIES OF COHESIVE SOILS COMPACTED DRY AND WET OF OPTIMUM MOISTURE (LAMBE 1958b)

Property	Comparison
(1) Structure	
A. Particle arrangement	Dry side more random (flocculated)
B. Water deficiency	Dry side more deficient, imbibes more water, swells more, has lower pore pressure
C. Permanence	Dry side structure sensitive to change
(2) Permeability	
A. Magnitude	Dry side more permeable
B. Permanence	Dry side permeability reduced much more by permeation

(3) Compressibility

A. Magnitude	Wet side more compressible in low pressure range, dry side in high pressure range
B. Rate	Dry side consolidates more rapidly

(4) Strength

A. As moulded	
(a) Undrained	Dry side much higher
(b) Drained	Dry side somewhat higher
B. After saturation	
(a) Undrained	Dry side somewhat higher if swelling prevented, wet side can be higher if swelling permitted
(b) Drained	Dry side about the same or slightly greater
C. Pore pressure at failure	Wet side higher
D. Stress-strain modulus	Dry side much greater
D. Sensitivity	Dry side more apt to be sensitive

21.5 FIELD MEASUREMENT OF COMPACTION

Compaction in the field is measured by finding out the insitu density. The two commonly used methods are the 'core cutter method' and the 'sand replacement method.'

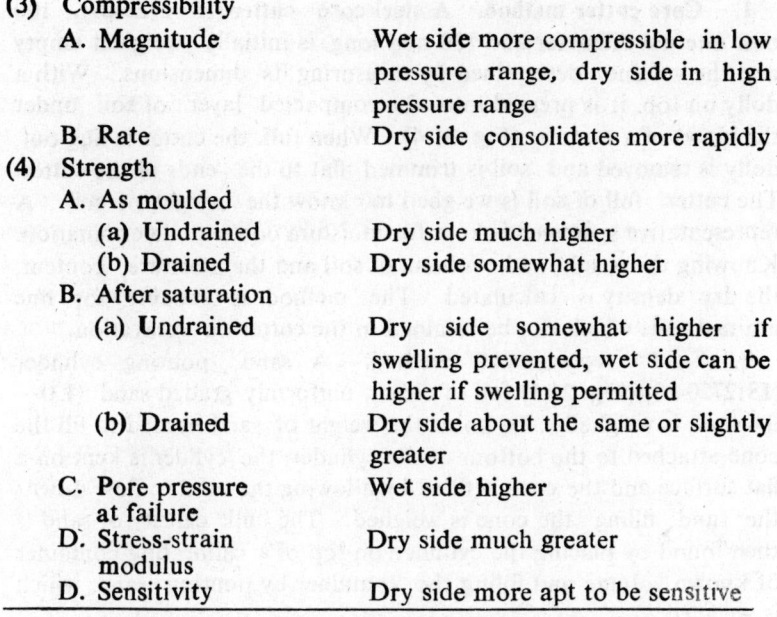

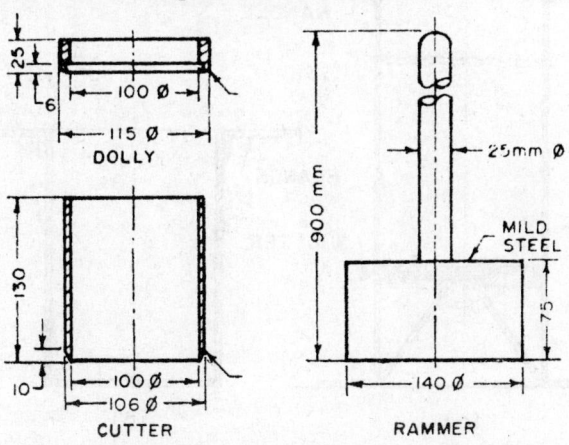

Fig. 21.4 Core cutter apparatus

1. Core cutter method. A steel core cutter (IS: 2720-29), 100 mm internal diameter and 130 mm long, is initially weighed empty and the volume determined by measuring its dimensions. With a dolly on top, it is pressed into the compacted layer of soil under the blows of a rammer (Fig. 21.4). When full, the cutter is dug out, dolly is removed and soil is trimmed flat to the ends of the cutter. The cutter full of soil is weighed to know the weight of soil. A representative specimen is kept for moisture content determination. Knowing the weight and volume of soil and the moisture content, the dry density is calculated. The method is suitable for fine grained soils which can be retained in the cutter on excavation.

2. Sand replacement method. A sand pouring cylinder (IS:2720-28) (Fig. 21.5) full of clean, uniformly graded sand (1.0 — 0.6 mm) is weighed. To know the weight of sand needed to fill the cone attached to the bottom of the cylinder, the cylider is kept on a flat surface and the cone is filled by allowing the sand to flow down; the sand filling the cone is weighed. The bulk density of sand is then found by placing the cylinder on top of a calibrating container of known volume and filling the container by pouring sand, which is weighed.

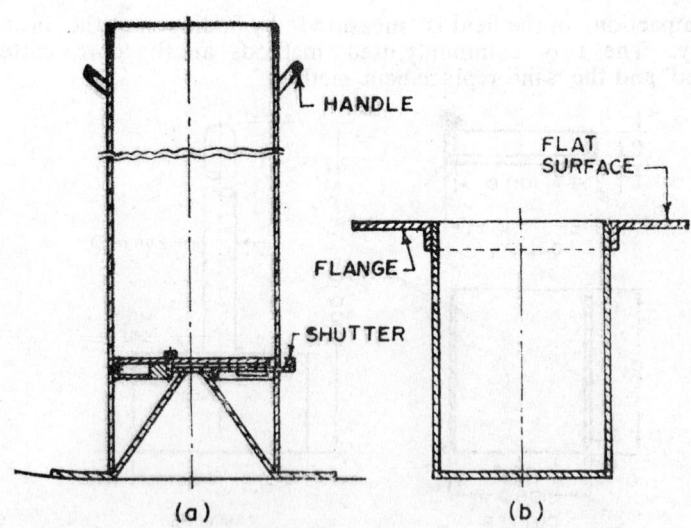

Fig. 21.5   Sand pouring cylinder (a) cylinder (b) caliprating container

A small hole is made at the test site and the soil taken from it is weighed. The hole is then filled by pouring sand from the cylinder. The cylinder with the remaining sand is weighed to get the weight of sand needed to fill the cone and the hole. Deducting the weight of sand in the cone as determined earlier, the weight of sand in the hole is known. The volume of hole (volume of excavated soil) is calculated from the known bulk density and weight of sand in the hole. Knowing the weight and volume of excavated soil, the bulk density is found. The moisture content of soil is determined and the dry density is calculated. The test is suitable for both coarse grained and fine grained soils.

Examples 21.1 — 21.3

21.1 The maximum dry density of a sample by the light compaction test is 1.84 g/cm³ at an optimum moisture content of 14 percent. Find the air content and degree of saturation if the relative density (G) of soil is 2.67. What would be the corresponding value of ρ_d on the zero air voids line at OMC ?

Eq. 21.4: n_a = 5.3 %

Eq. 21.3: S_r = 82.8 %

Eq. 21.2: ρ_d = 1.94 g/cm³

21.2 The following observations are obtained for determining the bulk density of a compacted embankment by the sand replacement method.

(1) Initial mass of cylinder and sand = 3.200 kg

(2) Final mass of cylinder and sand after filling hole = 1.476 kg

(3) Mass of soil taken from hole = 1.865 kg

(4) Mass of sand required to fill the cone of the cylinder = 0.330 kg

(5) Bulk unit weight of dry sand = 14.90 kN/m³

Find the bulk density and the dry density, if the moisture content is 15%.

Mass of sand in the hole = 3.200 — 1.476 — 0.330

= 1.394 kg

Bulk unit weight of soil = $\dfrac{1.865}{1.394}$ × 14.9 = 19.93 kN/m³

$$\text{Dry unit weight} = \frac{19.93}{1 + 0.15} = 17.3 \text{ kN/m}^3$$

21.3 A sample weighing 840 g is excavated from a compacted soil layer and is coated with wax. The mass of wax coated sample is 885 g. The volume of waxed sample is found by immersing it in water and measuring the volume of displaced water. The volume of displaced water is 479 cm³. What is the bulk density of soil, if the density of wax is 0.91 g/cm³ ?

$$\text{Mass of wax used} = 885 - 840 = 45 \text{ g}$$

$$\text{Volume of wax} = \frac{45}{0.91} = 49.45 \text{ cm}^3$$

$$\text{Volume of sample} = 479 - 49.45 = 429.55 \text{ cm}^3$$

$$\text{Bulk density of soil} = \frac{840}{429.55} \; 1\text{-}96 \text{ g/cm}^3$$

21.6 FIELD COMPACTION EQUIPMENT

The following common types of equipment are used in the field for compacting embankments, subgrades and road bases; amongst them the rollers are by far the most common.

1. Smooth wheel rollers. They consist of hollow steel drums, the weight of which can be increased by filling water or sand ballast. The usual type has a single drum at the front and the two rollers (cylindrical drums) of larger diameter at the rear. The other type is the *tandem roller* having two identical drums following in the same track.

2. Pneumatic tyred rollers. In the usual form they have a loaded open body or platform carried on two axles. Rubber tyred wheels are mounted close together on the axles, the rear set over-lapping the lines of the front set. The wheels compact soil by kneading action and pressure.

3 Sheepsfoot rollers. They consist of a hollow steel drum from which numerous tapered or club-shaped 'feet' project about 0.2 m. Dead weight is provided by placing water or wet sand inside the drum. The roller compacts soil by kneading action under pressure.

4. Vibrating rollers. These are smooth wheeled rollers fitted with a power driven vibration mechanism.

5. Power rammers. Manually controlled power rammers (generally petrol driven) are used for compacting small areas with restricted working space.

6. Factors affecting field compaction. In addition to the main factors, described in Section 21.2, the other factors which influence compaction in the field are: (a) lift thickness, (b) contact pressure and (c) speed of rolling of soil (IRC 1978). Greater the thickness of the layer, less is the compaction under each pass of the roller. There is also variation in the degree of compaction from the top to the bottom of the layer which increases with lift thickness. Contact pressure depends on the weight of the roller and the contact area. In case of pneumatic rollers, the tyre inflation pressure also determines the contract pressure in addition to the wheel load. A higher contact pressure increases the dry density and lowers the optimum moisture content. Speed of rolling mainly affects the output and the effect on compaction is not significant, except in case of vibratory rollers; slower the speed of a vibratory roller the greater will be the compaction.

7. Suitability of compaction equipment. The IRC recommendations (IRC: 36-1970) on the selection of compaction equipment are given in Table 21.2 and CRRI recommendations (IRC 1978a) are given in Table 21.3. It is always advantageous to supplement the information given in these tables with field trials.

Table 21.2

GENERAL GUIDE TO THE SELECTION OF COMPACTION PLANT FOR
DIFFERENT TYPES OF SOIL (IRC : 36)

Type of plant	Cohesive soil	Well graded granular and dry cohesive soil	Uniformly graded materials
(1) Smooth wheel roller	Suitable	Suitable	Suitable only if the roller is towed by tractor and the load per cm width of roller is less than 55 kg
(2) Pneumatic tyred	Suitable	Suitable when load on each	Suitable only if the roller is towed by

roller		wheel is more than 2 tonnes	tractor and the load on each wheel is less than 1.5 tonnes
(3) Sheepsfoot roller	Suitable	Unsuitable	Unsuitable.
(4) Virbratory roller	Suitable only when the static load per cm width of roller is more than 7 kg	Suitable	Suitable, but when the static load per cm width is more than 12 kg, the roller should be towed by tractor
(5) Power rammer	Suitable	Suitable	Unsuitable

Notes. For the purpose of the above table, soils are grouped as follows:

(1) 'Cohesive soils' include clays with upto 20% gravel and having a moisture content not less than the value of plastic limit minus 4.

(2) 'Well graded grannular and dry cohesive soils' include clays containing more than 20% of gravel and/or having a moisture content less than the value of the plastic limit minus 4, well graded sands and gravels with a uniformity coefficient exceeding 10 and all shales and clinker ash.

(3) 'Uniformly graded' materials include sands and gravel with uniformity coefficient of 10 or less and all silts and pulverized fuel ashes. Any soil containing 80% or more of material in the size range 0.06—0.002 mm will be regarded as silt for this purpose.

Table 21.3

RECOMMENDED PLANT FOR DIFFERENT SOILS AND NUMBER OF PASSES FOR SUFFICIENT COMPACTION (CRRI, IRC 1978a)

Plant or combination of plants	Clayey soil	Silty soil	Sandy soil	Sand
Smoothwheel roller (7 t)	—	8—16	8—16	8—16
Pneumatic tyred roller	—	—	8—16[1]	8—16[1]

Sheepsfort roller	—	—
Sheepsfoot roller[2] with	8—16	—
smoothwheel roller	4—8	—

1. If the material is fairly wet with higher moisture content than optimum.
2. In case of the sheepsfoot roller, another 4-8 passes of the smooth wheel roller will be required to finish off the mulch.

21.7 COMPACTION SPECIFICATIONS

The required standard of field compaction can be specified in terms of 'relative compaction' or 'percent compaction' defined as:

$$\text{Precent compaction} = \frac{\text{Field density}}{\text{Lab. max. dry density}}$$

The laboratory maximum dry density may be either the value obtained from the standard light compaction test or the heavy compaction test.

Typical percent compaction requirements (relative to maximum dry density obtained by light compaction test) specified by the IRC are glven in Table 21.4. For further requirements, reference may be made to IRC:36.

Table 21.4
PERCENT COMPACTION REQUIREMENTS FOR EMBANKEMENTE (IRC 1978b)

Type of work/material	Percent compaction
(1) Top 0.5 m of embankment below subgrade level and shoulders	Not less than 100
(2) Other portions of embankment	Not less than 95
(3) Highly expansive clays	85 to 90

The soil is compacted in layers not exceeding 25 cm in loose thickness. In case of a sheepsfoot roller, the loose layer thickness should not exceed the length of the projected feet by more than 5 cm. The moisture content of each layer for road woiks at the time of compaction, making due allowance for evaporation losses, should be in the range of 1% above to 2% below the OMC. Highly expansive soils should be compacted at 3 to 4% above the optimum value and to a density not exceeding 90% of the laboratory maximum value. (IRC:36).

21.8 COMPACTION CONTROL

Compaction control involves two basic operations:
(1) Determination of field dry density
(2) Determination of moisture content

To know the dry density, the bulk density is first determined by the methods described in Section 22.5 and the moisture content can be determined by any of the methods described in Chapter 2. Test time is, however, an important factor in field compaction so that work is not held up. Quick methods of moisture content determination should therefore be used. Proctor needle method can also be used for rapid moisture determination in addition to the methods given in Chapter 2.

1. Proctor needle method. The Proctor needle apparatus, devised by Proctor, is a small penetrometer used for rapid determination of moisture content of fine grained soils in the field. The apparatus (Fig. 21.6) consists of a set of interchangeable cylindrical

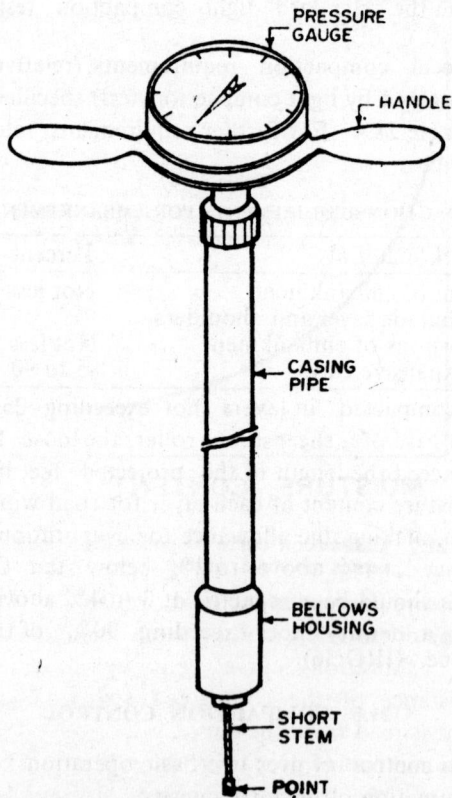

Fig. 21.6 Proctor needle penetrometer

needle points of face areas 6, 5, 3, 2, 1.5, 1.0, 0.5 and 0.25 cm², which are selected according to the type of soil.

A calibration curve is first prepared. Soil is compacted at varying moisture contents by the light compaction method in the 100 cm diameter mould. A suitable needle point is attached to the penetrometer and forced down vertically in the compacted soil at a rate of about 12.5 mm per second to a depth of not less than 75 mm. If the apparatus is hydraulically operated (Fig. 21.6), the penetration resistance is read from the pressure gauge. If it has a spring loaded plunger, a sliding ring on the plunger indicates the resistance. A calibration curve (Fig. 21.7) is plotted between penetration resistance and moisture content. This curve is usually superimposed on the compaction curve.

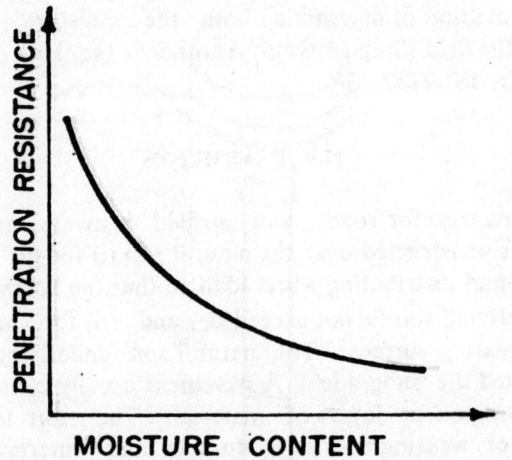

Fig. 21.7 Calibration curve of penetration resistance versus moisture content curve

To know moisture content in the field, wet soil is compacted in the 100 cm diameter mould in the same manner as used for preparing the calibration curve (e.g. by light compaction). Penetration resistance of the compacted soil is found and the moisture content is read from the curve.

An alternative method of preparing the calibration curve, *which is independent of the compactive effort*, has been suggested by Marcovici and Ciuraru (1963). In this method, a relationship is established between moisture content and the ratio p/γ, where p is the penetration *force* (kg) for a particular needle point (8 mm diameter in their studies) and γ is the bulk density. Samples are compacted in the laboratory to various densities and moisture contents (even using varying compactive effort) and the moisture content versus p/γ relationship is established. The curve is similar to that of Fig. 21.7 and is concave upward. In the field, a sample of compacted soil is taken in a core cutter and the bulk density is established. Attaching the needle point used in calibration, 2 or 3 penetration force readings are taken and an average value of p is obtained. The ratio p/γ is calculated and moisture content is read from the calibration curve. The method cannot be applied to soils with a gravel content of more than about 20 percent.

    2. Other methods. The 'neuclear method' is a very efficient and quick method of determinng both the moisture content and density in the field (Singh 1981). Another is the 'Hilf method' as described by IS: 2720—38.

21.9 PAVEMENTS

    Pavements used for roads and airfield runways are relatively stable crusts constructed over the natural soil (i) for the purpose of supporting and distributing wheel load so that the bearing capacity of the underlying soil is not exceeded, and (ii) for providing an adequate wearing surface. The natural soil underlying the pavement is called the 'subgrade'. A pavement usually consists of two or more component layers of material. The top layer, called 'surfacing' or 'wearing course' is durable and waterproof. Below the surfacing is the 'base' which is the main load spreading layer of the pavement. For economic reasons, the base material is sometimes split into two layers, a 'base' and a 'sub-base'. The sub-base is thus the second load spreading layer under the base. The sub-base is also called 'soling', and 'base' is also called 'metalling'. In a concrete pavement, the sub-base may or may not be used and the function of surfaciug may also be performed by the concrete slab itself.

    Pavements are mainly of two types: 'flexible' and 'rigid'. A flexible pavement is capable of resisting only very small tensile stresses and any change in the surface of subgrade results into a corresponding change in the pavement. Flexible pavements can have water, tar or bitumen bound macadam, lean concrete base, or cement bound granular base, all overlain with bitumen surfaces. Rigid pavements consists of reinforced concrete base and surface. They can take appreciable tensile stresses and are capable of bridging small weaknesses and depressions in the subgrade. Pavements with lean concrete base and cement or lime stabilized base are sometimes classified as 'semi-flexible', but for design purposes they are generally grouped under flexible pavements as indicated above.

21.10 CALIFORNIA BEARING RATIO

The strength of subgrades and base courses of pavements can be expressed in terms of California Bearing Ratio (CBR) value. The CBR value is measured by an empirical test devised by the California State Highway Association. The CBR is defined as the ratio of the load required to penetrate a soil mass with a circular plunger of 50 mm diameter at the rate of 1.25 mm/min to the standard load for corresponding penetration of a standard material (crushed stone). The CBR is usually determined for penetrations of 2.5 mm and 5 mm. The standard loads are 1370 kg (13.43 kN) for 2.5 mm penetration and 2055 kg (20.14 kN) for 5.0 mm penetration. The higher of the values for 2.5 mm and 5 mm is used for design. Usually the 2.5 mm value is higher.

The laboratory CBR test (IS:2720-16) is performed in a cylindrical mould, 150 mm diameter and 175 mm high, which can be fitted to a detachable perforated base plate and a collar. A spacer (displacer) disc is placed inside the mould during sample compaction. With the displacer disc inside, the effective height of the mould remains only 127.3 mm (volume 2250 cm^3) which is the same as that of the large size mould used for campaction tests.

1. **Sample preparation.** An undisturbed or field compacted sample can be obtained by attaching a steel cutting collar to the mould and pressing the mould into the soil insitu. Remoulded samples are prepared at dsired density and moisture content in the mould by either static or dynamic compaction. The test dry density is either the field density or the maximum dry density

obtained by compaction tests. The moisture content for compaction is the optimum moisture content or the field equilibrium moisture content. The material used for the test should pass a 19 mm sieve. Allowance for larger material is made by replacing it by an equal amount of material which passes a 19 mm sieve but is retained on 4.75 mm sieve.

For static compaction, the mass (M) of wet soil with moisture content (w) to give the desired dry density (ρ_d) is calculated from the expression:

$$M = \rho_d (1 + w) V \qquad (21.6)$$

where V = volume of compacted specimen = 2250 cm^3

The above mass of soil is filled in the mould, the displacer disc is placed on top of loose soil and pressed down in a compression machine. For dynamic compaction, either the light compaction or heavy compaction method is adopted according to the field requirements. To simulate worst moisture conditions in the field, the compacted sample is kept submerged in water and allowed to soak for about 3-4 days. During soaking period, surcharge weights are kept on the sample to represent pavement loading as described under 'penetration test' below.

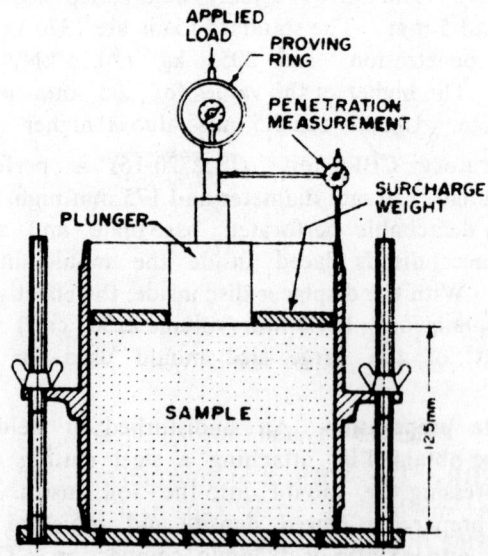

Fig. 21.8 Setup for CBR test

2. Penetration test. The mould containing the soil sample is kept in a loading machine fitted with the standard plunger (Fig. 21.8). Surcharge weights with a central hole or slot to permit penetration of plunger and each weighing 2.5 kg are placed on the sample to represent pavement loading. (Each 2.5 kg weight is approximately equivalent to 7 cm of pavement construction. The minimum surcharge is 5 kg,) The plunger is penetrated at 1.25 mm/min and load readings are taken at penetrations of 0.5, 1.0. 1.5, 2.0, 2.5, 3, 4, 5, 7.5, 10 and 12.5 mm.

3. CBR calculation. A graph is plotted between load on plunger (as ordinate) and penetration (as abscissa). The graph is normally convex upwards. If the initial portion of the curve is concave upwards, the corrected origin point is obtained by drawing a tangent to the curve at the points of greatest slope, as shown in Fig. 21,9. The test loads P_1 and P_2 for 2.5 mm and 5 mm penetrations are read from the graph. These loads divided by the standard loads give the the CBR. The higher of the values is used for design.

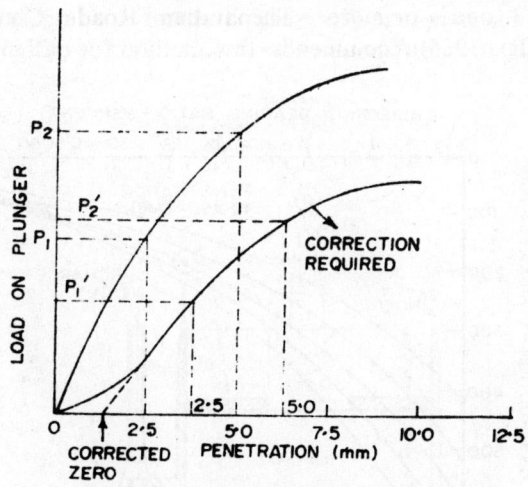

Fig. 21.9 Typical CBR test curves

4. Field CBR. The CBR of a subgrade can be determined insitu also. A mechanical screw type jack fitted with a load measuring proving ring and the standard plunger is attached to the underside of a truck chasis or a load truss. Surcharge weights are

250 mm in diameter through which the plunger penetrates the soil. Each 2.5 kg surcharge weight is approximately equivalent to 25 mm of construction. Minimum surcharge is 15 kg. The penetration test is performed as in the laboratory and the CBR determined.

21.11 FLEXIBLE PAVEMENT DESIGN

Flexible pavements are usually designed on the basis of the CBR of the subgrade for which empirical design curves have been developed. These curves relate the thickness of pavement to the expected 'traffic' on the pavement. Two methods of traffic evaluation and pavement design are used.

1.   Design based on traffic in term of number of vehicles. This is the earlier method developed by the Transport and Road Research Laboratory of U.K. (TRRL) and is generally followed in India. Volume of traffic likely to be carried by a pavement over its design life (10 to 20 years) is considered in units of 'heavy' vehicles per day in both the directions. Heavy vehicle means any vehicle with a laden weight of 3 tonnes or more. The Indian Roads Congress (IRC: 37-1984, IRC 1985) recommends this method for design traffic not

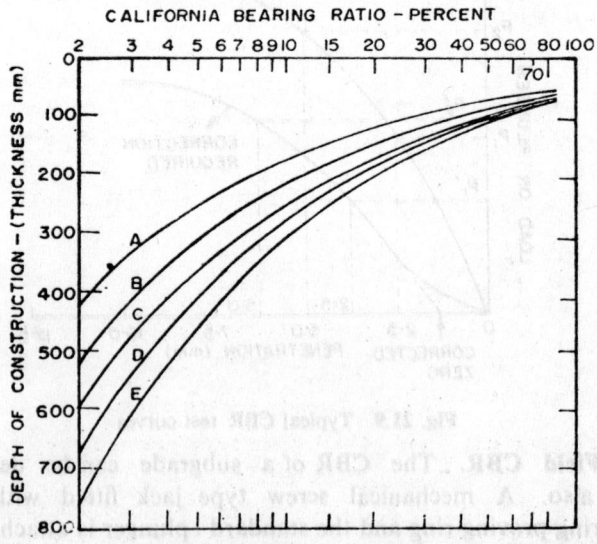

Fig. 21.10   CRB curves for flexible pavement design (IRC: 37-1984)

exceeding 1500 vehicles per day. Traffic is divided into 5 classes for which 5 design curves, A,B,C,D and E, are shown in Fig 21.10. Traffic classification and the corresponding curves are given in Table 21.5.

Table 21.5

TRAFFIC CLASSIFICATION AND CBR CURVES

Curve	Number of vehicles per day
A	0—15
B	15—45
C	45—150
D	150—450
E	450—1500

The curves give the total thickness of construction necessary over the underlying material for which the CBR is determined. Thus the CBR of the subgrade will give the total thickness of the pavement and the CBR of the sub-base will give the combined thickness of base and surfacing.

2. **Designs based on traffic in terms of standard axles.** These are the revised design procedures of the TRRL (U.K.) published in Road Notes 29 and 31. The Roade Note 29 method is more complex and relates specifically to U.K. conditions and practices. The Road Note 31 gives a simple design method suitable for roads carrying light to medium traffic in tropical and sub-tropical countries, and this method is briefly described here.

Traffic is measured in terms of 'equivalent 8200 kg (18000 lb) standard axles'. Vehicles with axle loads other than 8200 kg are converted into equivalent numbers of standard axle by the use of factors given in Table 21.6, obtained from the AASHO road test, 1962.

Table 21.6

FACTORS FOR CONVERTING AXLE LOADS TO EQUIVALENT
STANDARD AXLES

Axle load (kg)	Equivalence factor	Axle load (kg)	Equivalence factor
910(2000 lb)	0.0002	11790	5.2
1810	0.0025	12700	7.2
2720	0.01	13610	9.9

3630	0.04	14520	13.3
4540	0.08	15430	17.6
5440	0.2	16320	22.9
6350	0.3	17230	29.4
7260	0.6	18140	37.3
8160	1.0	19070	47.0
9070	1.6	19980	58.0
9980	2.4	20880	72.0
10890	3.6	21790	87.0
		(48000 lb)	

Total traffic flow is calculated as the cumulative number of standard axles in one direction over the design life of the pavement (say 10 to 20 years). Thus if the total number of vehicles (and their axle loadings) is known the total traffic can be estimated; e.g. design life = 20 years, standard axles per day = 500,

Cumulative axles = 500 × 365 × 20 = 3650000

= 3.65 million standard axles (msa)

Knowing the subgrade CBR and the cumulative number of axles, the thickness of the pavement is found from Fig. 21.11.

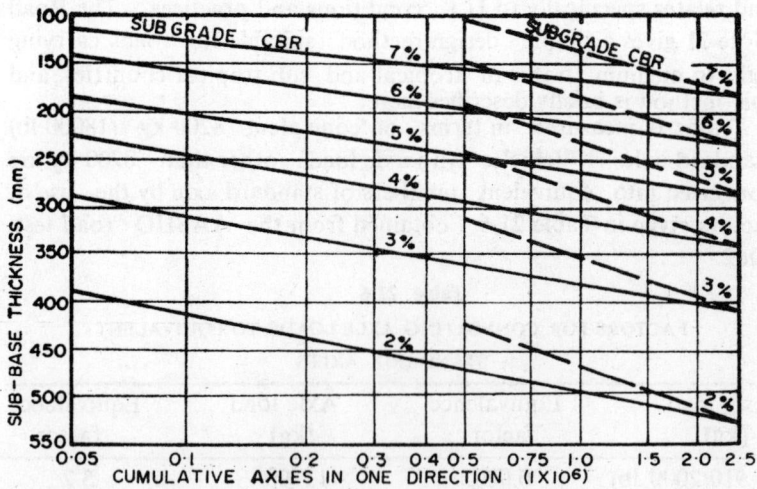

Fig. 21.11 Flexible pavement design chart by the Road Note 31 method (TRRL, U.K.).

The chart assumes a minimum thickness of sub-base 100 mm with subgrade CBR of 8 to 24% (sub-base CBR to be not less than 25%).   Upto 0.5 million standard axles (msa), the chart assumes a standard 150 mm thick base   with surface dressing and the additional thickness of sub-base is read from the chart.   For traffic flow more   than 0.5 msa, either a 150 mm base with 50 mm of bituminous surfacing or a 200 mm base with a double layer of surface dressing may be used.   As an initial economic measure, a 150 mm base with double surface dressing may be constructed in the beginning and an overlay of 50 mm thick bituminous surfacing may be provided later on.   It will be seen from the chart that no sub-base will be required where the subgrade CBR exceeds 24.   The sub-base should be compacted to atleast 100% of maximum dry

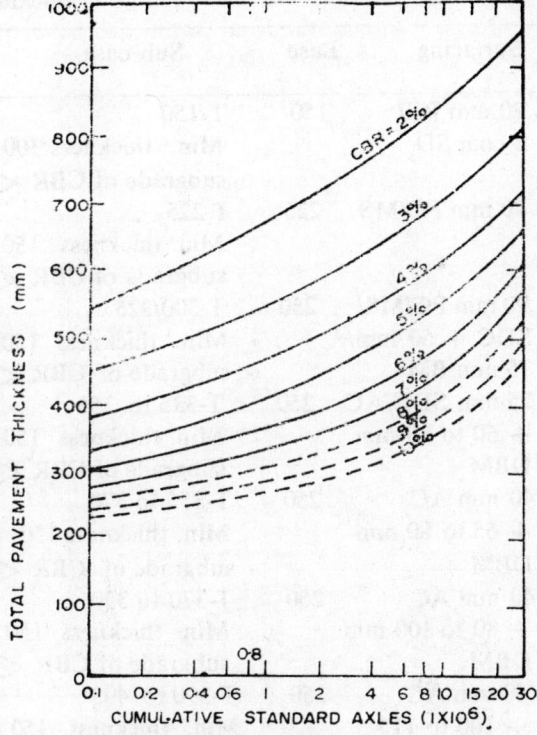

Fig 21.12 Pavement thickness design curves based on standard axles (IRC : 37-1984)

density obtained by the light compaction test. Crushed rock, natural gravel or crushed gravel, or soil stabilized by cement, lime or bitumen may be used base construction.

3. IRC recommendations. The design curves adopted and recommended by the IRC: 37-1984 (IRC 1985) are shown in Fig 21.12. The total pavement thickness is read from continuous curves and the dotted curves are used for proportioning sub-base thickness. Subgrade CBR is mentioned on the curves. The recommended minimum thickness and the compositions of component layers for new construction are given in Table 21.7.

Table 21.7

RECOMMENDED STRUCTURAL SECTION OF FLEXIBLE PAVEMENT DESIGN ON THE BASIS OF STANDARD AXLES (IRC : 37)

Cumulative standard axles (msa)	Minimum thickness of component layers, compacted thickness (mm)		
	Surfacing	Base	Sub-base
0.5	20 mm PC/ 2-coat SD	150	T-150 Min. thickness 100 mm on subgrade of CBR $< 20\%$
0.5—2	20 mm PC/MS	225	T-225 Min. thickness 150 mm on subgrade of CBR $< 20\%$
2—5	20 mm PC/MS/ SDC + 50 mm/ 75 mm BM	250	T-300/325 Min. thickness 150 mm on subgrade of CBR $< 30\%$
5—10	25 mm SDC/AC + 60 to 80 mm DBM	250	T-335 to 355 Min. thickness 150 mm on subgrade of CBR $< 30\%$
10—15	40 mm AC + 65 to 80 mm DBM	250	T-355 to 370 Min. thickness 150 mm on subgrade of CBR $< 30\%$
15—20	40 mm AC + 80 to 100 mm DBM	250	T-370 to 390 Min. thickness 150 mm on subgrade of CBR $< 30\%$
20—30	40 mm AC + 100 to 115 mm DBM	250	T-390 to 405 Min. thickness 150 mm on subgrade of CBR $< 30\%$

Note. T = total thickness, SD = surface dressing, PC = premix carpet, MS = mix seal surfacing, SDC = semi-dense carpet, AC = asphaltic concrete, BM = bituminous macadam binder course, DBM = dense bituminous macadam binder course. msa = million standard axles.

If CBR of subgrade is more than the requirement for the sub-base, no sub-base is required. Binder course of thickness more than 80 mm should be laid in two layers.

While using the CBR design curves of Fig. 21.10, the minimum thickness of different layers and composition are given in Table 21.8

Table 21.8

RECOMMENDED STRUCTURAL SECTION OF FLEXIBLE PAVEMENT DESIGNED ON THE BASIS OF CBR CURVES (IRC : 37)

Traffic range (Fig. 21.10)	Adopt composition and minimum layer thickness applicable to msa of Table 21.7
(1) Upto 150 vehicles/ day (curves A,B & C)	Upto 0.5 msa
(2) 150-450 vehicles/ day (curve D)	0.5 to 2 msa
(3) 450-1500 vehicles/ day (curve E)	2 to 4 msa

21.12 MODULUS OF SUBGRADE REACTION

The strength of subgrade is also expressed in terms of "modulus (or coefficient) of subgrade reaction K_v", which is defined as the ratio between the vertical pressure p per unit area and the corresponding settlement ΔS of the surface of subgrade,

$$K_v = \frac{p}{\Delta S} \tag{21.7}$$

The modulus of subgrade reaction is used in the design of rigid pavements and raft foundations.

K_v is determined in the field (Singh 1981) by loading a rigid circular plate (75 cm diameter, or preferably of 0.5 m^2 area). A seating pressure of 0.7 kg/cm^2 (or 7 kPa) is first applied. Without releasing the seating load, an additional pressure of 0.7 kg/cm^2 (70 kPa) is applied and held until practically complete settlement has taken place. The settlement is noted and K_v is calculated from:

$$K_v \text{ (kg/cm}^2/\text{cm)} = \frac{0.7}{\Delta S \text{ (cm)}} \qquad (21.8)$$

or

$$K_v \text{ (kN/m}^2/\text{m)} = \frac{70 \times 10^3}{\Delta S \text{ (mm)}} \qquad (21.9)$$

In an alternative method, pressure/settlement readings are taken at suitable intervals with pressure increments until a total settlement of not less than 1.75 mm occurs. A pressure/settlement graph is plotted and the pressure for 1.25 mm settlement is read from the graph. K_v is calculated from:

$$K_v \text{ (kg/cm}^2/\text{cm)} = \frac{10\ p}{1.25}, \quad (p \text{ in kg/cm}^2) \qquad (21.10)$$

or

$$K_v \text{ (kN/m}^2/\text{m)} = \frac{1000\ p}{1.25}, \quad (p \text{ in kPa}) \qquad (21.11)$$

For details of the methods, reference may be made to the author's book on Geotechnical Testing and Instrumentation (Singh 1981).

22

Soil improvement techniques

22.1 SOIL IMPROVEMENT

Soil improvement, in the broad sense, is the alteration of any property of soil and the treatment of ground such that the soil and ground may serve better their intended engineering purpose. 'Soil stabilization' is an alternative term in vogue for 'soil improvement', the latter getting greater usage recently. Soil property alteration and ground treatment are commonly undertaken to:

(i) improve shear strength, supporting capacity and stability;

(ii) reduce compressibility and frost susceptibility; and

(iii) reduce permeability.

In describing the various soil improvement techniques, it is convenient to consider them in two parts: those applicable to shallow or surface layers and those applicable to deep or thick layers.

22.2 SHALLOW COMPACTION

The process of densifying, i.e. compaction (Chapter 21), is the most obvious and simple way of increasing the stability and supporting capacity of soil. Compaction alone at proper moisture content will often solve a particular soil problem and has necessarily to be judiciously adopted in the construction of pavements, embankments and other fills. Compaction also constitutes an essential part of a number of other methods of soil improvements. Compaction improves the performance of both granular and cohesive soils.

22.3 MECHANICAL TREATMENT

Mechanical treatment or 'stabilization' is based on the principles of (i) controlled grading of soil and (ii) proper compaction. Where suitable soil materials are economically available, mechanical stabilization may prove to be the simplest and a cheap method of constructing bases for pavements and improving the subgrades.

For stabilization purposes, soils are conveniently sub-divided into (i) those which in the compacted state possess a granular bearing skeleton composed of particles larger than 0.075 mm (75 micron sieve) and (ii) those without such a bearing skeleton. The former possess frictional resistance properties and volume stability but they may require bonding or cementation and also decrease in permeability, for which a suitable proportion of 'soil fines' (silt and clay fraction finer than 75 micron) is necessary. By a suitable addition of soil fines (also called as 'binder') to a granular soil followed by proper compaction a dense mix can be obtained which meets the requirements of strength, volume stability and permeability. For constructing mechanically stabilized bases and surfacings for pavements, specifications for proper mix giving the gradation and plasticity requirements are available. These are mainly based on experience. Typical specifications are given in Table 22.1. Blending of materials or mix can be designed to meet the specifications by combining granular and cohesive soils.

Table 22.1

TYPICAL SPECIFICATIONS FOR GRANULAR PAVEMENT BASES AND SURFACINGS (HMSO 1952, BRITISH CROWN COPYRIGHT)

	Grading requirements (Percentage passing)			
	Base		Base or surfacing	
I.S. Sieve	Nominal max. size (mm)		Nominal max. size (mm)	
	37.5	19	9.5	4.75
37.5 mm	100	—	—	—
19.0 mm	80—100	100	—	—
9.5 mm	55—80	80—100	100	—
4.75 mm	40—60	50—75	80—100	100
2.00 mm	30—50	35—60	50—80	80—100
1.18 mm	—	—	40—65	50—80
600 μm	15—30	15—35	—	30—60
300 μm	—	—	20—40	20—45
75 μm	5—15	5—15	10—25	10—25

Note. The material passing a 425 μm sieve should have the following characteristics.

For bases   : $w_L \not> 25\%$, $I_p \not> 6$ (preferably NP)

For surfacing : $w_L \not> 35\%$, $I_p = 4$—9

22.4 USE OF ADMIXTURES

A number of admixtures or materials other than soils have been tried to improve the physical properties of soils. These include lime, cement, bitumen, industrial wastes such as molasses from sugar industry and lignin from paper industry, water proofers such as natural and artificial resins, aggregants and dispersants, etc. 'Chemical stabilisation' is sometimes used as a general term to include all the above mentioned admixtures. The more widely used admixtures are lime, cement and bitumen, and amongst them it is relatively more economical in India to use lime as soil stabilizer.

22.5 LIME STABILIZATION

1. Reactions and physical changes. When lime is added to a lime-reactive soil there is a reduction in plasticity and a gradual increase in, strength with time. after compaction, which may be attributed to the following reactions (Thompson 1964, Herrin and Mitchell 1961).

(a) *Cation exchange*: replacement of the exchangeable cations (sodium, hydrogen, potassium, etc.) of the soil by the calcium cations derived from the lime.

(b) *Flocculation and agglomeration*: an increase in grain size created by the suppression of the double water layer surrounding the clay particles due to an increased electrolyte concentration, which results in flocculation.

(c) *Carbonation (of minor influence)* ; reaction of lime and carbon dioxide from the atmosphere to form relatively weak cementing agents, calcium and/or magnesium carbonate.

(d) *Pozzolanic reactions*: reactions between the silica and alumina present in the soil minerals and the calcium from the lime to form new cementitious minerals.

The physical changes produced by lime treatment of clay soils can be summarised as follows: (i) The plasticity index decreases markedly by a factor of 3 or more in some cases; w_p generally increases and w_L decreases (some conflicting opinions exist regarding effects of lime on w_L). (ii) The linear shrinkage and swell decreases significantly. (iii) Soil becomes friable and can be pulverized and worked easily. (iv) Unconfined compressive strength, CBR and bearing capacity increase substantially. (v) Treated sub-bases or bases of pavements form a water resistant barrier. (vi) The optimum moisture content increases and the maximum dry density decreases.

2. Suitability and applications. Lime treatment is specially effective for heavy clays or for silty and clayey granular soils which are likely to loose strength because of high water affinity in their silt-clay fraction. Lime is less effective for silty soils and is not recommended for sandy soils except in combination with added clay or other pozzolanic materials. Organic soils are not suitable for lime stabilization. Lime may be used as a stabilizer for soils in the sub-base and base courses of pavements, under concrete foundations, on embankment slopes and canal lining. Lime treatment is also used as a preparatory measure for subsequent stabilization of clays with cement, bitumen and other water proofing agents. Quicklime treatment has been used for drying of soils by the heat of hydration which is generated. Lime piles and lime columns are used for treating deep layers. (See Section 22.16).

Several types of limes are available, notably quicklime (CaO), slaked or hydrated lime [Ca (OH)$_2$] and dolomitic lime (CaO + MgO). Quicklime is difficult to handle and is suitable specially for marshy lands where quick drying is also aimed at. Dolomitic lime containing significant proportions of magnesium oxide is not an effective stabilizer. Hydrated lime with a high percentage of calcium oxide is better for stabilization and also easy to handle.

3. Lime stabilized base courses. Based on U.S. experience (Winterkorn 1975) lime requirements for stabilizing base and sub base courses of pavements are as follows:

(1) Five to ten percent of lime Ca(OH)$_2$, by weight of soil, for heavy clay soils to serve as bases, or one to three percent in sub-bases.

(2) Two to four percent of lime for clay-gravel materials to serve as bases.

In India, the amount of lime needed for stabilization is recommended to be decided by the criteria of pH value, CBR and unconfined compressive strength (IRC 1983). The minimum percentage of lime is that which on mixing with soil in the form of a slurry gives a pH of 12.4. This quantity (percentage) is termed the 'modified optimum' or 'lime fixation limit', because at this point the plasticity of soil reduces to zero and the soil becomes saturated with calcium cations attaining maximum *modification* conditions, without significant gain in strength. Addition of lime in excess of 'modification optimum' contributes to increase in strength. It is observed that if lime is 4 percent above the modification optimum, the stabilized soil generally becomes durable and resistant to frost provided soil activity is 0.75 or less.

Starting with the modification optimum, soil-lime mixes are prepared with increasing lime contents. Their optimum moisture contents and maximum dry densities are determined using light compaction. Samples for CBR and unconfined compressive strength tests (sample size 50 mm diameter and 100 mm height, or 100 mm diameter and 200 mm height) are prepared at the above values. The CBR samples are first cured for 3 days and then soaked for 4 days. The quantity of lime is taken to be the value which gives a minimum soaked CBR of 20 to 25% for sub-base stabilization and a CBR of 80 to 100% for base stabilization, provided the base is covered with a bituminous crust of 5 cm thickness. For design purposes, the field CBR may be assumed as 60 to 70 percent of the laboratory value. The unconfined compressive strength of 28 days cured samples should be a minimum of 14 kg/cm^2 for bases and 7 kg/cm^2 for sub-bases.

For effective stabilization, a soil must have not less than 15% fraction passing a 425 micron sieve and its plasticity index should be atleast ten. The organic content should not be more than 2% and the sulphate content not more than 0.2%. For proper mixing, the soil should be pulverized to about 25 mm and smaller size, about 50 to 60% passing a 4.75 mm sieve.

Pulverization and mixing of soil and lime in the field can be done manually or mechanically. The mix is compacted at OMC making allowance for moisture losses. It is recommended that all compaction should be completed within four hours after mixing, of soil, lime and water.   The base should be cured for 7 to 28 days   under moist conditions.

22.6 LIME FLY ASH STABILIZATION

Sandy and silty soils lacking in clay fraction can be effectively stabilized by lime in combination with pozzolanic materials such as fly ash, pulverized blast furnance slag or expanded shale.   In many parts of India, fly ash is available as a waste product from the thermal power plants.   It is a finely divided residue resulting from the combustion of ground or pulverized coal. Where available, lime fly ash stabilization can be economically adopted for sub-base and also for bases of pavements.   Fly ash by itself has little cementitious value but in the presence of moisture it reacts chemically with lime to form cementitious compounds.

Fly ash, if wet, should be dried and pulverized to the gradation of 100% passing a 12.5 mm sieve, 95% (min) passing a 9.5 mm sieve and 75% (min) passing a 2.0 mm sieve (IRC 1981). Lime should be the hydrated high calcium lime. Granular sandy and silty soils with low clay content are suitable. A high clay content may require larger amount of lime. Clayey silts and low plasticity clays having I_p between 5 to 20 and w_L less than 25 are quite suitable.

The required amount of lime normally varies from 3 to 7 percent and that of fly ash from 10 to 20 percent of dry soil (Winterkorn 1975). Lime to fly ash ratios of 1:3 to 1:4 are found to give optimum strength for various soil types (IRC 1981). The lime, fly ash and soil mix may be designed on CBR and compressive strength criteria, which are recommended as follows for construction of sub-bases of pavements (IRC 1981). CBR (%): 3 days cured + 4 days soaking = 10, 28 days cured + 4 days soaking = 25, unconfined compressive strength: 7 days cured = 3 kg/em², 28 days cured = 7.5 kg/cm². Test samples should be prepared at optimum moisture content and maximum dry density and curing should be at a temperature of 38°C. Trial mixes of 'lime : fly ash: soil' may be as follows: (i) 2.5:5:92.5, (ii) 2.5:7.5:90, (iii) 4:12:84, (iv) 5:15:80, (v) 2:8:90, (vi) 3:12:85, (vii) 4:16:90. The minimum sub-base thickness should be 10 cm or 15 cm depending on the type of mixing and compaction equipment available on site. *Note.* If the compacted and soaked CBR of the subgrade soil is of the order of 25 percent or more, a sub-base may not be necessary.

22.7 CEMENT STABILIZATION

Portland cement and soil mixed at the proper moisture content produce 'soil cement', a structral material which is hard and durable. Soil cement has been used mainly as bases under concrete pavement for highways and air fields.   It is also used for wave protection on earth dams, as canal lining and as a cheap building material.

1. **Forms of soil cement.** Three forms of soil cement are recognized:

(a) *Compacted soil cement* contains sufficient cement (usually 5 to 15% by dry weight) and only enough moisture to facilitate compaction and satisfy the hydration requirements of cement. The resulting material is hard and durable and offers a well defined resistance to weathering and mechanical forces.

(b) *Plastic soil cement* has enough water at the time of placement to produce a wet consistency similar to that of plastering mortar.   It is used for canal linings and for erosion protection on steep slopes where road building equipment is difficult to be used. It also results in a hardened product.

(c) *Cement modified soil* is a less rigid or semihardened product containing relatively small quantity of cement with (generally less than 5%) which is sufficient to improve the engineering properties of a soil and reduce its water affinity, but not enough to impart the properties of a hardened soil cement.

2. **Soil type and cement content.** A wide range of soils can be treated and improved by mixing with cement the notable exception of highly organic soils. Well graded granular soils with 10-35 percent fines ($w_L < 40$, $I_p < 20$) give best results and require lesser amounts of cement. Soils with a high clay content are difficult to be pulverized.   The cement requirement increases with the amount and water affinity of clay.   For pavement construction, the cement content is usually decided on the basis of the CBR or the unconfined compressive strengh (UCS) of the stabilized samples. Typical requirement is a soaked CBR of 100% or a UCS of 1700 kPa ($17 kg/cm^2$), tested at 7 days. To allow for a lower mixing

efficiency in the filed and soil variations, the cement content should either be increased by 1.0% above that determined by the mix or should be sufficient to give CBR or UCS of the trial mix at least 20% above the specified minimum (Carter 1983). Typical cement requirements (percentage by weight) are as follows: well graded granular soils with or without small fraction of fines = 2 to 5 (ii) uniformly graded sands = 4—8, (iii) non-plastic or moderately plastic silty soils = 10, (iv) plastic clay soils = 13 or more.

Plastic clays which are difficult to be pulverized may be initially improved by adding a small quantity of lime (say 4%), Organic soils which are otherwise unsuitable (organic matter 2—4%) may be stabilized by adding about 2% calcium chloride along with cement.

3. **Properties of soil cement.** The properties of cement stabilized soil depend on the soil types, cement content and efficiency of mixing and compaction. Several properties are given in Table 22.2 (Mitchell 1976).

Table 22.2

PROPERTIES OF CEMENT STABILIZED SOIL (MITCHELL 1976)

Property	Granular soils	Fine grained soils	Remarks
Density (t/m³)	1.6—2.2	1.4—2.0	May be higher or lower than untreated soil. Density and strength reduces by delay between mixing and compaction.
Unconfined compressive strength (kPa)	(500 to 1000) C	(300 to 600)C	C = cement content (%)
Cohesion (kPa)	$60+0.225q_u$	$60+0.225q_u$	q_u = unconfined compressive strength
Friction angle	40°—45°	30°—40°	
Flexural strength	1/5 to 1/3 q_u	1/5 to 1/3 q_u	Need 1—3% cement to develop

CBR	$0.0038 (q_u)^{1.45}$	$0.0038 (q_u)^{1.45}$	q_u in kPa
Modulus (compression) (kPa)	$(7{-}35)10^3$	$7 \times 10^2 - 7 \times 10^3$	
Poission's Ratio	0.1—0.2	0.15—0.35	
Permeability	$<1 \times 10^{-6}$ cm/s	$<1 \times 10^{-6}$ cm/s	k parallel to compaction planes may be upto 20 times greater than normal to them.

22.8 BITUMINOUS STABILIZATION

Bituminous materials (asphalts and tars) are used in various consistencies to improve the engineering properties of soils. In cohesive soils which usually have satisfactory bearing capacity at low moisture contents, bitumen mainly serves the purpose of water-proofing the soil and reducing water absorption. In the case of sandy soils lacking cohesion, bitumen serves as a bonding or cementing agent. Bitumen requirements commonly range from 4 to 7% of the dry weight of the cohesive soils and from 4 to 10% for sandy soils. Bituminous soil stabilization has been mainly used for base courses.

22.9 OTHER CHEMICALS

Hygroscopic and water retentive chemicals (calcium chloride and sodium chloride) are used as supplementary to mechanical stabilization to retain a certain amount of moisture for stability and serve as dust pallatives. Natural and artificial resins (aniline-furfural) are used for the purposes of waterproofing and cementing soil; only small amounts of stabilizer (less than 2%) are required. Dispersing agents are used to modify the compaction and strength characteristics and to decrease permeability.

22.10. FREEZING AND HEATING

Artificial soil refrigeration (freezing the pore water) has been used to facilitate construction of open and under-ground excavations. Heating is adopted for clay soils. A cohesive soil may be burnt in kilns to produce artificial aggregate. Heating is also adoptedinsitu for road bases by the use of travelling road burners at temperatures 500° to 600°C. Collapsible loessial soils have been stabilized by burning liquid or gaseous fuel in sealed boreholes.

22.11 GEOTEXTILES

Geotextiles are permeable synthetic fabrics which can be used to improve soils. They are currently used in many civil engineering works including pavements, embankments, retaining structures, reservoirs, canals, dams, bank protection and coastal engineering. Geotextiles are used to perform four basic functions and in many situations several of them simultaneously (Leflaive 1985).

(1) Separation -- between the clean granular base of a road and a soft cohesive formation soil.

(2) Filtration — e.g. in the wrapping of coarse aggregate fill placed in a drainage trench.

(3) Drainage — beneath railway ballast over a clay formation where the geotextile can conduct water in its plane.

(4) Reinforcement -- at the base of an embankment on soft foundation soil where the additional mechanical resistance introduced by the geotextile is sufficient to prevent failure.

Geotexiles for use in civil engineering have many qualities which include high strength, ability to elongate without rupture under loads and repetitive stresses, flexibility, high abrasion resistance, stable in contaminated groundwater, and resistance to fungal attack (Cannon 1976). Geotextiles are complemented by "geomembranes" which are impermeable membranes used widely as cut-offs and as canal and reservoir liners. Another important use of geomembranes is for containment of hazardous wastes and their leachates.

22.12 IMPROVEMENT OF DEEP LAYERS

The various techniques adopted for treating and improving the engineering properties of thick or deep layers are: dynamic compaction and consolidation, precompression and drainage by sand drains, electro-osmosis, lime piles and columns, stone columns, injuctions and grouts, and reinforced earth (See Chapter 11 for reinforced earth).

22.13 DYNAMIC COMPACTION AND CONSOLIDATION

1. Vibroflotation. The vibroflot is a cylindrical steel tube (probe) (35 to 45 cm in diameter and about 150 cm long and weight about 20 kN) containing water jets at top and bottom and equipped with a rotating eccentric weight in the lower part, which develops a horizontal vibratory motion. The vibroflot (Brown 1977) is sunk into the soil by a combination of vibration and jetting high pressure water through the lower jets. On reaching the desired depth, the water flow is reduced and diverted to the jets at top of the probe. The probe is raised to the surface in successive small increments (typically in 30 cm lifts) compacting the surrounding material by the vibration process. The upward flow of water maintains a channel around the probe which is continuously backfilled with suitable granular material as the probe is lifted.

The method is applicable to compaction of loose sands with less than 20% fines for depths upto 30 m and is particularly effective for submerged deposits. The effectiveness of compaction diminishes with increasing amount of silt, gravel or cemented sands. Compaction holes should be on a grid pattern spaced at about 2 m. The spacing may be somewhat larger in coarse sand.

2. Terra-probe. The terra-probe consists of a heavy open ended pipe pile (called terra-probe, about 14 m long and upto 80 cm in diameter) and a vibratory pile driver. The vibrator and probe unit are suspended above the ground surface and then vibrated to penetrate the sand to the desired depth (upto 20 m) and then extracted. Compaction is effected by the continuous vibration during penetration and extraction. The process is repeated over a grid pattern.

Somewhat similar to vibroflation, the method is applicable to compaction of loose deposits of sand. Saturated soil conditions are best for maximum effect, although dry sand can also be compacted. Compared to vibroflot, a terra-probe compacts much more rapidly but the extent of compaction achieved per probe is significantly less. A terra-probe is ineffective above 5 m depth (Mitchell 1977).

3. Compaction piles. A compaction pile is a displacement pile driven with the object of compacting a loose sand. Compaction sand piles are formed by driving a hollow casing with a detachable base plate. A sand charge is introduced into the casing which is progressively withdrawn compacting the backfill at the same time. Both the process of driving the casing and the lateral displacement by the compacted sand pile densify the surrounding sandy material. Spacings of compaction piles are generally 1.2 to 1.5 m (Lee et al 1983). Compared to vibroflot, compaction piles can compact sand with a relatively higher percentage of fines. The method is most effective in saturated, loose sandy soils. The maximum effective depth of treatment has been 20 m (Mitchell 1977). A method of driving the casing by a vibrator and forming the compaction pile is known as 'vibro-compozer method' (Dastidar 1985).

4. **Blasting.** Deep compaction of cohesionless soils can be effected by buried explosive charges. Shock waves produced by the explosion cause the loose structure to liquefy and result into a more compact structure.

5. **Heavy tamping.** Heavy tamping consists in dropping a heavy mass (2 to 40 t) from a height of 4 to 35 m on the surface (Tsytovich et al 1974, Dunn et al 1980). Enough energy gets generated to compact the soil to great depths (15 to 20 m). The method was originally used to compact granular soils or fill materials. Now heavy tamping is used for compressing fine grained soils also and the technique is known as 'dynamic consolidation' (Menard and Broise 1975, Dastidar 1985). Compression of cohesive soils is thought to occur due to liquefaction and rapid dissipation of increased pore pressure through radial fissures produced by impact.

22.14 PRECOMPRESSION

The technique of improving soil properties by precompression consists in preloading the soil prior to the construction of the intended structure. The preload is generally in the form of an earthfill exerting pressure equal to or greater than the value which will be applied by the structure. The preload is left in place long enough to induce settlement and then it is removed. This procedure over-consolidates the soil. If preloading produces a pressure in excess of the subsequent foundation pressure, the additional load of the fill is called a 'surcharge' and the technique is known as overloading' or 'surcharging'. Overloading is used to decrease the time necessary to achieve the required settlement.

The precompression technique may also take the form of 'staged construction', which is adopted mainly as a means of gradually increasing the sheer strength of a soft clay which would otherwise be inadequate to support the proposed embankment without failure (Pilot 1981).

Vertical drains. On thick deposits of soft clay or when the permeability is exceptionally low, preloading alone may take long periods of time to bring about significant compression. In such circumstances, vertical drains are adopted to accelerate consolidation. Vertical drains alone or used in conjunction with preloading offer an efficient technique of ground improvement (See Chapter 14).

Preloading and surcharging, (with or without sand drains) can be applied to accelerate consolidation of normally consolidated soft clays, silts, organic deposits and also to sanitary landfills and dredged material.

22.15 ELECTRO-OSMOSIS

When a direct current (d.c.) potential is applied to a moist fine grained soil, the pore water is induced to flow from the positive electrode (anode) to the negative electrode (cathode) and at the same time, electrolysis occurs with the anions and cations of the dissolved salts moving towards the anode and cathode respectively. This process is known as 'electro-osmosis'. The part of the electro-osmosis process by which pore water moves under the influence of an electrical potential is called 'electro-drainage', and the part of the process by which the ions migrate under the electrical potential is called 'electro-injection or electro-chemical treatment.'

1. **Electro-drainage.** The object of electro-drainage is to diminish the moisture content of soft, fine grained soils (silts, clayey silts, and fine clayey silty sands) which cannot be successfully drained by other methods. The reduction in moisture content results in the improvement of undrained shear strength and stiffness of the soil deposit. Fine soil particles are surrounded by an electrical double layer, the outer layer having a concentration of positive ions (See Chapter 1). Upon application of a d.c. voltage between the two electrodes, the positive ions are attracted to the cathode and repelled by the anode and thus are set in motion towards the cathode. The free water in the soil

prores within the outer layer of the particle double layer is also carried along to the cathode by viscous flow. By making the cathode a well, the water can be removed by pumping. The flow of water from anode to cathode sets up negative pore pressures in the soil, with a corresponding increase in effective pressure, since the total pressure is constant. This will lead to consolidation of soil. The coefficient of electro-osmotic permeability, or the rate of flow for electro-drainage, does not vary greatly with soils and may be assumed approximately as 5×10^{-5} cm per sec for a gradient of 1 volt per cm (Casagrnde Leo, 1949, 1952).

2. Electro-injection. During electrolysis the anode either dissolves or is fed with appropriate ion solution which brings about a chemical change in the soil and increase in strength without changing the soil volume or structure (Pilot 1981). This electro-chemical treatment can be applied for insitu strengthening of soil under existing structures without any disturbance.

Electro-osmosis is commonly applied to improve the stability of cuts and trenches. It has also been used to increase the bearing capacity of friction piles (used as anode in the process) (Casagrande Leo 1953).

22.16 LIME PILES AND COLUMNS

Fine silts and silty clays with high moisture contents ($> 50\%$) can be rapidly stabilized by forming quicklime piles and columns.

1. Quicklime piles. Lime piles are formed by compacting quicklime (CaO) in boreholes (Chao and Chin 1963). The piles are about 30 cm in diameter and 10 m long spaced at about 1.0 m. Soil improvement takes place in two stages: on coming in contact with soil quicklime absorbs water equal to about 32% of its weight, and consequently the volumetric expansion of the hydrated lime exerts a lateral pressure of the order of about 1250 kPa causing radial consolidation of soil (Lee et al 1983). There is also a soil-lime interaction by slow diffusion of silica into the lime. The initial cost of the installation equipment is high and special care is needed to handle quicklime.

2. Quicklime columns. As alternative to quicklime piles in the use of 'lime columns' which are formed by insitu mixing of quicklime in proportions of about 10   percent of the dry weight of soil (Broms and Boman 1978, 1979).

To form the lime column, a 0.5 m wide auger fitted with a mixing tool shaped as a giant 'dough mixer or egg beater', is screwed down into the ground to the required depth. The direction of rotation is then reversed and the auger is slowly pulled up. At the same time quicklime is forced out into the soil with compressed air through holes at the level of the mixing tool which mixes lime and soil. Since the blades of the mixing tool are slightly inclined, the stabilized soil also gets compacted. The columns are 0.5 m in diameter and upto 10 m deep. They are normally installed in a triangular pattern with spacing varying from about 1 to 2 metres (Linden-Alimak 1981).

The slaking of lime results in an instant shear strength increase of clay due to reduction in moisture content. The strength of the column increases due to long term chemical reactions. Having higher strength, higher permeability and lower compressibility than the soft, surrounding soil, a lime column acts both as a reinforcing element in the ground and as a vertical drain. Lime columns can be used in soft clays for various purposes, e.g. (i) increasing bearing capacity of poor soils, (ii) reducing settlements, (iii) accelerating the rate of settlement. and (iv) improving the stability of slopes, trenches and deep excavations.

22.17 STONE COLUMNS

An improvement in the bearing capacity and reduction in settle-ment of soft clay deposits can be effected by the installation of stone columns. The method consists in forming vertical holes in the ground which are filled with compacted crushed stone, gravel and sand, or a mixture of these granular materials to form columns or piles, termed as 'stone columns'. Soil improvement occurs in two ways: (i) the stone columns provide strength reinforcement to the soil, and due to their relatively high modulus a large portion of the load applied to the ground surface is transferred to them, (ii) the columns act as vertical drains and accelerate consolidation. Stone columns have been used upto about 20 m depths, and upto about 1.0 m diameter. There are several methods of constructing stone columns (Dastidar 1985, Datye 1982), two of them are described below.

1. Vibro-replacement method. A borehole is drilled to the required depth with a vibroflot. Using water jets when the vibroflot is gradully lifted, crushed stone 20 to 75 mm size or even upto 100 mm size, (Lee et al 1982) is poured down the hole and is vibrated and compacted. Compaction is continued until the lateral resistance to displacement of the soil by the stones is fully developed. The resulting column is fairly irregular: The method is not recommended for fine grained soil with undrained strength less than 20 kPa (Thornburn 1975).

2. Boring and ramming method. It is a non-displacement method in which a cased or uncased borehole is first drilled. The borehole can be made by 'bailer and casing method' or by 'rotary boring'. If bentonite mud is used to stabilize the borehole, it should be pumped out and replaced with water. Stone chips upto 75 mm size or gravel and sand are poured into the hole to fill it upto 0.75 to 1.0 m and rammed with a special hammer (2 t for 75 cm diameter column). The casing is gradually withdrawn, further charge of stones is poured into the hole and compacted. The operations are repeated to complete the column.

22.18 GROUTING

A 'grout' is a stabilizing material of fluid consistency which is pumped into soil pores (and fissures) with the purposes of (i) reducing permeability, (ii) increasing shear strength and (iii) preventing excessive settlement. The process of forcing a grout into the soil, under pressure, is known as 'grouting' or 'injection'. Grouting is also adopted for rock fissures and defects to reduce seepage of water.

1. Types of grouts. Grouts can be divided into two classes, depending on their rheological behaviour:

    (1) Suspension or particulate grouts

    (2) Solution or non-particulate grouts

The commonly used materials for suspension grouts are: (i) cement, (ii) clay, (iii) bitumen, and (iv) combinations of cement-sand-clay, fly ash and other chemicals. Solution grouts are the solutions of chemicals (two-shot solutions or one-shot solutions) which react with one another in soil pores and stabilize the soil.

2. Groutability. The suitability of a grout to penetrate soil or rock is commonly expressed as the 'groutability ratio'. For successful grouting of soils.

$$\frac{(D_{15}) \text{ soil}}{(D_{85}) \text{ grout}} > 25 \qquad (22.1)$$

and for rock fissures,

$$\frac{(D)_{fissure}}{(D_{max}) \text{ grout}} > 5 \qquad (22.2)$$

Generally grouting can be used for soils with permeability greater than 10^{-3} cm/s. Suspension grouts are suitable for soils of permeability greater than 10^{-2} cm/s and the solution grouts for soils of permeability upto 10^{-3} cm/s. Grouting of soils with permeability less than 10^{-3} cm/s is rather difficult, expensive and time consuming.

3. Cement grouting. Portland cement is commonly used for grouting fissured rocks, gravels and coarse sands. The water cement ratio is usually 2:1, somtimes 10:1, by volume. Where considerable quantities of grout are accepted, inert materials such as sand, rockflour, or clay are also added. A typical mixture of cement, sand or clay and water is 1:5:4 by volume. The setting time of a cement grout can be accelerated by adding calcium chloride or retarded by gypsum. Addition of bentonite clay to cement grouts upto about 6% of dry weight of cement increases plasticity of the grout, minimizes 'bleed' and prevents segregation of coarser inert fillers. Cement and fly ash grout is also used. Fly ash is a cheap material. It increases the ease of pumping by its spherical particles and reacts pozzolonically with lime liberated by cement. Fly ash also improves the resistance to attack by sulphates in fills containing a high proportion of cinders (Scott 1974).

Cement grouts gradually solidify forming an impervious barrier, e.g. beneath dams, and also forming a strong, resistant and unerodable foundation support.

4. Clay grouting. Clay bentonite mixtures, with a flocculating agent such as aluminium sulphate to cause coagulation after injection, are mainly used to reduce permeability. Clay grout does not contribute much to the strength properties of the soil. Clay grout can penetrate finer sand compared to cement grout, but it may be removed by vigorous groundwater flow.

5. Solution grouts. Solution grouts are normally used for injecting finer-grained granular materials (sand or silt sized) which will not accept the generally cheaper suspension grouts. A commonly used solution grout consists of solutions of silicic acid (sodium silicate, water glass, $Na_2. nSiO_2 + H_2O$) and calcium chloride. It is injected by the 'two shot' method, known as the 'Joosten method.' A solution of sodium silicate (alone or with some other chemical) is first injected as the injection pipe is advanced by short stages. The pipe is then gradually withdrawn and a second reagent, e.g. a strong solution of calcium chloride, is injected. When calcium chloride (an electrolyte, a salt) is added to silicic acid, it coagulates depositing a stable gel of silica in the pores of the soil. Because of the immediate and quick reaction of these two chemicals forming a precipitation of calcium silicate gel within the soil pores and thus binding the soil particles, the method is applicable also in the presence of groundwater in the soil where k is high (between 1.5×10^{-2} cm/sec to 1×10^{-1} cm/sec). Using a concentrated solution of sodium silicate (e.g. 7 parts of silicate and 3 parts of water, Caron et al 1975) a hard silica gel is obtained which not only seals the soil and checks seepage but also strengthens the soil. The gel is considered stable. Owing to the high viscosity of the concentrated silicate solution, injection pipes are usually spaced at 60 cm centre to centure (Jumikes 1971)

When only permeability reduction is the objective, a very diluted silicate solution is used, e,g. one part of silicate for 9 parts of water (Caron et al 1975). In dilute solutions, the setting time of silica gel is relatively long and the materials may be mixed at the surface and injected in a single shot. Dilute solutions result in a soft gel. There are many other formulations, mostly patented, which can be injected into the ground as one fluid (single shot). These single shot solution grouts consist of two solutions, each of one or more chemicals, which when mixed together have a controlled, delayed reaction ranging from a few minutes to several hours. The process of impermeabilization and strengthening of soil and other geological formations by using solution grouts is also called 'chemical solidification'.

Laboratory studies on dune sand treated with a solution of silicate, formamide and water indicate that the silicate formamide forms a strong bond which is stable in water (Ohri et al 1983). The significant factors influencing the grouted dune sand are grout-mix composition, loading rate and curing. A solution of silicate and formamide in 30:9 mix ratio is more suitable for moist conditions and of 50:9 mix for submerged conditions of sand. The strength of sillicate-formamide stabilized sands decreases as the loading rate decreases, A strength reduction of nearly 50% is observed as the strain rate drops from about 0.7% per sec to about 0.03% per sec.

6. Compaction grouting. Compaction grouting, also known as 'displacement grouting' consists of using a thick grout (e.g. groutlike cement and sand), which, when forced into the soil, displaces, compacts and consolidates the surrounding material about a central core of grout. These grouts are too thick to permeate the soil. Compaction grouts have been used beneath structures to prevent further settlement, and if possible, to raise the building or slab to its former level. The technique is primarily suited for soft, highly compressible silts and sandy silts. (Graf 1969, Brown and Warner 1973).

23

Site exploration and sampling

23.1 OBJECTIVES AND METHODS

Site exploration is required mainly:

(1)   To determine the properties of soil and ground which affect the design and safety of structure, e.g. strength, compressibility and hydraulic characteristics.

(2)   To determine the extent and properties of the material to be used for construction.

(3)   To determine the groundwater conditions.

The investigations for the above objectives are usually carried out in three stages: (i) collection of available information, (ii) preliminary reconnaissance and (iii) detailed exploration. The principal objects of the 'detailed exploration' are as follows:

(1)   To determine the sequence, thicknesses and lateral extent of the soil strata, and where necessary, the level of bedrock.

(2)   To obtain disturbed and undisturbed samples for identification and laboratory testing.

(3)   To identify the groundwater conditions.

(4)   To carry out field tests to know the insitu soil properties.

The direct methods adopted for detailed exploration include excavation, boring and collection of samples and field testing. Field tests depend on the soil properties to be obtained. These include insitu shear tests, penetration tests, pressuremeter test, permeability tests and bearing tests. Indirect 'geophysical methods' are also used to supplement the information obtained from the above mentioned direct methods of exploration. Excavation, boring and sampling, and geopysical methods are described in this chapter.

23.2 EXCAVATION AND BORING METHODS

Trial pits and boreholes are the means of exposing and reaching the soil strata to be examined or tested. Trial pits are used for shallow depths. Borings are generally used for greater depths (greater than 3 m) or when difficult groundwater conditions are met. The borehole diameter generally varies from 50 mm to 250 mm depending on the type of investigation, size of required samples and the type of available equipment. A borehole remains open in firm soils by arching, but in soft clays and in sands below water table, it is kept open by inserting a steel tube, called 'casing' or by filling the hole with a drilling mud, a viscous suspension of bentonite in water.

1. Trial pits. Excavation of trial pits (and trenches) is a simple and reliable method of exploration but it is limited to shallow depths, say about 3 m. Pits and trenches have a great advantage over borings, in that strata can be visually examined, large samples can be easily collected, or tests may be made insitu. Excavation is usually carried out manually. Where there is slightest risk of collapse, suitable methods of timbering and side support should be adopted.

2. Hand and power augers. Hand augers can be used to excavate boreholes to depths of around 6 m in favourable types of soils which have sufficient cohesion to stand unsupported in an unlined borehole and are free from coarse gravel and other obstructions. Two common types of hand augers are the post-hole auger (Fig. 23.1) with diameters upto 200 mm, and the small helical or spiral auger of about 50 mm diameter. The auger is rotated and pressed down into the soil by means of a T-handle on the upper rod. When the blades are loaded with all the soil that can be held, the auger is withdrawn and the soil is removed. As the hole progresses downwards, extension rods are added to the auger.

For deeper explorations, power operated (mechanical) augers may be used. They are generally of the short flight or continuous flight screw augers. Small portable power augers which can be transported and operated by two persons, are suitable for making 10—15 m deep boreholes ranging from 75 mm to 300 mm in diameter. A continous flight screw auger consists of rods with

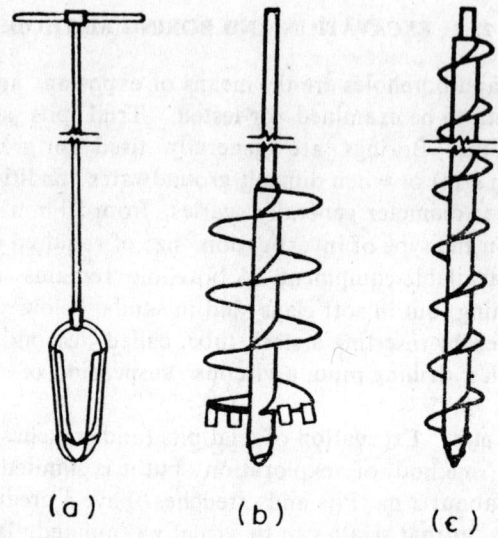

Fig. 23.1 (a) Post-hole auger, (b) short flight screw auger. (c) continous flight screw auger.

a helix covering the entire length. The cuttings of the soil rise to surface along the helix, obviating the necessity of withdrawal. As the auger advances, additional auger sections are added. Continuous flight augers can be used to depths of 30 m or more depending on the equipment available and conditions encountered.

Obviously the soil samples obtained by augers are badly mixed but are sufficient for identification and classification. If carefully done, augering gives the least disturbance of any boring method and undisturbed samples can be obtained after withdrawing the auger from the borehole. Continuous flight augers with hollow stems (75 to 150 mm diameter) are also available. During boring, the hollow steam is closed at the bottom by a plug fitted to a rod running inside the stem. At any depth the rod and the plug may be removed from the hollow stem to allow undisturbed samples to be taken.

3. Bailer and auger boring. Bailer and auger boring, also called 'percussion boring', can be used in widely differing soil types, above and below the water table. Boreholes are usually 150-200 mm diameter and upto about 30 m deep but may extend to 50 m in good ground conditions. The principal tools are as follows (Fig. 23.2).

(a) *Bailer or shell*. This is a heavy steel tube fitted with a cutting edge and a non-return flap valve at the bottom end. It serves as a sand auger.

(b) *Clay cutter*. This is similar to the bailer, but without the flap valve.

(c) *Chisel and drill bits*. A variety of patterns are available to break up hard materials.

(d) *Clay auger*. This is of tubular shape with side and bottom slots.

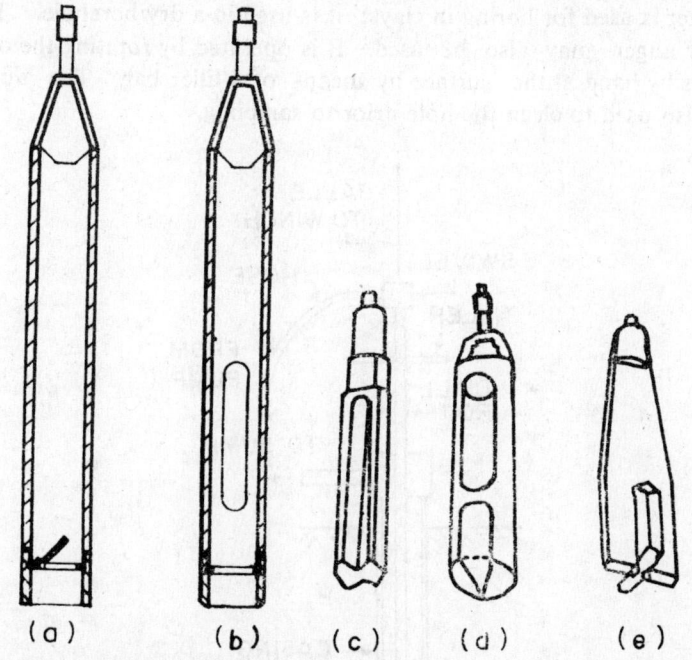

Fig. 23.2   Tools for percussion boring (a) bailer, (b) clay cutter, (c) drill bit,
(d) clay auger, (e) chisel

The 'boring rig' may be a 3-legged tripod or a 4-legged derrick carrying a light steel cable which passes over a pulley block on top of the rig. The cable is worked usually by a motor operated winch, (manual operation may also be used for shallow . boring). An appropriate tool is attached to the bottom of solid drill rods. The tool and the rods are carried by the steel cable and are alternately raised and dropped to bore the hole. A casing (lengths of steel pipe screwed together) is generally required to support sides of the bore-hole, which is driven into the hole.

For boring through sandy strata, the bailer attached to the string of drill rods is repeatedly dropped. The loosened soil forms a slurry with water which is introduced into the borehole. Below the water table, slurry is formed with groundwater. The slurry collects in the bailer which is raised to the surface, when full, and emptied. In hard or dense soil, a chisel is used to break up the material which is brought to the surface in the form of slurry. The clay cutter is used for boring in clays; it is used in a dry borehole. The clay auger may also be used. It is operated by rotating the drill rods by hand at the surface by means of a tiller bar. The auger is also used to clean the hole prior to sampling.

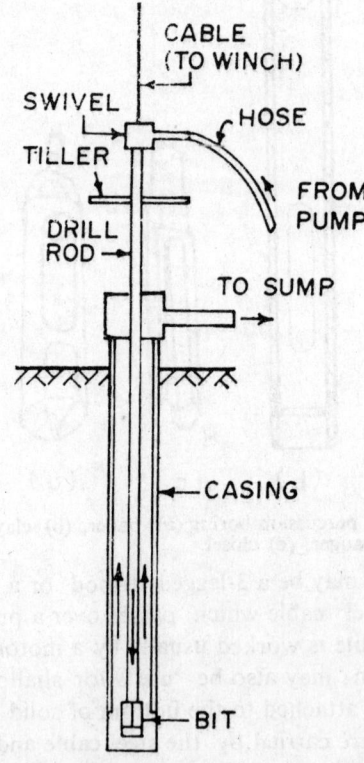

Fig. 23.3 Wash boring

4. Wash boring. In wash boring (Fig. 23.3), water is pumped through a string of hollow drill rods called the 'wash pipe'. Water issues under pressure through narrow holes in a chisel attached to the lower end of the wash pipe. The wash pipe is worked up and down or rotated by hand. The soil is loosened and broken up by the water jets and the up-and-down movement and rotation of the chisel. The water carries the soil up the annular space between the wash pipe and the side of the borehole. The washings are allowed to settle in a pond or tank at the ground level and the fluid is recirculated or disharged to waste as desired. The borehole is generally cased. Drilling mud (bentonite suspension) may be used in place of water, eliminating the need for casing.

Wash boring is a cheap and rapid method of boring in suitable soils (specially uniform sands or clays) and the disturbance of the soil below the bottom of the borehole is generally not much. The washings, however, are so disturbed and mixed up as to be valueless, even for identification.

5. Rotary drilling. Although primarily intended for rocks, rotary drilling is also used in soils. A drilling bit, attached to the bottom of a string of hollow drill rods, is rotated at a high speed with a downward thrust by a power operated rig. The bit may be either a 'cutting bit' or a 'coring bit'. If the object is only to drill a hole, the cutting bit is used, and the method is known as 'open hole drilling'. If a core or sample is also to be recovered, the coring bit is used; the coring bit is fixed to the bottom of a core barrel which in turn is carried by the drill rods. On rotation with downward thrust, the coring bit cuts an annular space in the material and an intact core enters the barrel, to be removed as a sample. This method is known as 'core drilling' and is used in rocks and hard clays. Water or drilling mud is continuously pumped down the drill rods, which emerges under pressure through holes in the bit or barrel. The drilling fluid cools and lubricates the bits and carries up the cuttings to the surface. Drilling mud also supports the sides of the hole where casing is not used.

Rotary drilling is a fast method of boring and the disturbance of the soil below the borehole is slight. Depending on ground conditions and the equipment, 35-60 mm diameter (or larger) holes can be drilled to great depths. The method is not suitable in soil containing high percentage of gravel or larger particles as they tend to rotate under the bit and are not broken up (Craig 1983).

23.3 LAYOUT AND DEPTH OF BOREHOLES

The location, number and depth of borings or pits depend very much both on the nature and variability of soil strata, and on the form and extent of the works. Initially borings should be widely spaced and then intermediate borings made as required, so that sections may be drawn with reasonable accuracy. In uniform soils, borings may be 30 m to 100 m apart or more but spacings of 10 m or less may bε required in very erratic conditions.

For individual, small buildings (plan area less than 300 m²) on uniform soils, at least three borings (not in one line) should be made and for larger buildings at least five borings (one at each corner and one at the middle, and not more than 30 m apart) should be made. Spacing should be decreased for irregular conditions. Whenever possible the boreholes should be sunk close to the proposed foundations but outside their outlines. For small foundations, trial pits should not be located on or close to the intended foundation position because large and deep trial excavation would weaken the ground.

The depth of borings should include all strata liable to be significantly affected by the structure and its construction. Exploration should extend below all strata which would contribute to significant settlement or which might have inadequate shear strength for the support of foundations. In case of foundations, the general rule is to explore upto the 'significant depth' (See Chapter 13) provided there is no weak stratum below this depth which would cause unacceptable settlement. The significant depth is taken to be the level at which either (i) the additional pressure induced by the foundation does not exceed 20% of original, effective overburden pressure, or (ii) the additional pressure is about 20% of the value at the foundation level. The additional pressure reduces to 20% of the value at the foundation level at a depth of about 1.5 B for square or circular foundation and at a depth of about 3 B for strip foundation. Thus the depth of exploration below the foundation level should be at least 1.5 times the width B of the isolated footings (clear spacing ⩾ 4 B) and of rafts. Where the footings are closely spaced so that there is overlapping of pressure zones (clear spacing less than 2 B) the minimum depth of exploration should be 1.5 times the length of footings in case of rectangular footings. For closely spaced rows of footing, exploration should extend to depth equal to 4 times the footing width.   A large piled foundation may be treated as an equivalent raft at a depth of two-thirds of the length of the piles and exploration is done upto 1.5 times the width of the equivalent raft below the raft level.

Where foundations are taken down to solid rocks, drilling should penetrate at least 3 m in rock to confirm that it is bedrock and not a large boulder.

On completion of investigation all boreholes and trial pits should be backfilled with compacted medium cohesive soil. If boreholes penetrate water bearing ground which will be subsequently excavated, it is important to backfill the boreholes with concrete or well rammed puddled clay.

23 4 GROUNDWATER OBSERVATIONS

Determination of groundwater levels and their fluctuations is an important part of any site exploration, and in particular where deep excavations are to be carried out. Elevations at which drilling water is lost, or at which water under excess pressure is met, should be recored. Water levels before and after insertion of casing, where used, should also be noted. When water is met during boring, the borehole should be deepened by 0.5 m and the water level should also be lowered by about 0.5 m (by removing it with a bailer). The borehole should be left for sufficient time for the water to rise to its final value. This time interval depends on the type of water bearing strata. While in sands and gravels 30 to 45 minutes may suffice for the water level to stabilize, at least a 24 hour period should be allowed in silts. Water level cannot stabilize in such short intervals in clays unless pervious seems are present. Although not reliable, the 24-hour water level should be recorded in clays also.

An accurate method of obtaining the groundwater level and observing its fluctuations is to install one or more stand-pipes in boreholes. A simple standpipe consists of a PVC (or steel) tube, 25 mm or 50 mm diameter, with perforations at its lower end. The tube is inserted into the boreholes and packed around with gravel along the perforated portion over which a puddle clay seal is provided. The borehole is than backfilled to prevent access to rainwater. The tube is also fitted with a cap. Groundwater levels should be monitored over as long a period as possible. In irregular groundwater conditions (confined water bearing strata, artesian water, etc.), piezometers installed in each permeable layer may be required.

Water level in boreholes filled with drilling mud is recorded after inserting a casing pipe perforated at its lower end and bailing out the drilling mud.

When groundwater samples are to be collected for chemical analysis, it is important to ensure that they are not diluted or contaminated. They should be taken immediately water is struck in a pit or borehole. If samples are required during progress of boring, the borehole should be pumped or bailed dry and samples taken from water which collects by seepage.

23.5 SOIL SAMPLING AND DISTURBANCE

1. **Effect of disturbance.** Disturbance may be defined as a change in fabric and moisture content of a soil which alters its physical properties. The physical properties of cohesive soils, which can be affected by disturbance are (i) compressive strength (drained and undrained), (ii) modulus 'E' from compression tests (tangent or secant), (iii) shear modulus (iv) strain at peak strength, and (v) C_c, c_v, σ_p', c' and ϕ'. Disturbance does not affect all of these properties to the same degree and in the same manner and that the degree of disturbance varies in various parts of the sample. (Osterberg and Murphy 1979). In case of silty sands. disturbance affects the dynamic shear strength. But the drained shear strengh of sands does not appear to be sensitive against disturbance of soil fabric if density is unchanged (Mori and Koreeda 1979).

2. **Types of samples.** There are two main types of samples which can be taken from trial pits or boreholes.

(a) *Disturbed (representative) sample.* It is a sample which preserves the particle size distribution of the insitu soil but in which the soil structure is significantly or completely disturbed and the moisture content may also differ from the insitu value. Disturbed samples are required mainly for identification and classification tests or for determining the properties of remoulded soil.

(b) *Undisturbed sample.* It is a sample which represents as closely as practicable the true insitu structure and moisture content of the soil. Undisturbed samples are required mainly for shear strength, consolidation and permeability tests.

3. Recovery of disturbed samples. Disturbed samples can be excavated from trial pits or obtained from boring tools, e.g. from augers and the clay cutter. A bailer sample will be deficient in fines and is unsuitable for use as a disturbed sample. Thick wall, split tube samplers, e.g. the sampler used for the standard penetration test (Section 15.8), can also be used to obtain disturbed samples from boreholes. Small samples (about 500 g) are usually put in glass or plastic jars, tins or small polythene bags. Larger samples (5—50 kg) are put in large polythene bags or tins. Samples for natural moisture content determination should be sealed in air tight containers; all containers to be completely filled so that there is negligible air space above the sample.

4. Causes and avoidance of disturbance. It is not possible to obtain a completely undisturbed sample since the very act of sampling must disturb the soil to some extent. The process of boring, driving and withdrawal of the sampler and the stress relief in soil are the main causes of sample disturbance.

Percussion boring (bailer boring) causes considerable disturbance of soil immediately below the bottom of the borehole. This soil should either be removed with an auger before sampling or the upper part of the sample should be rejected.

Driving disturbance in sampling depends on the design features of a sampler and the manner it is driven. The design features relating to sample disturbance are usually expressed in terms of "area ratio A_r", "inside clearance C_i" and "outside clearance C_o" defined as follows (See Fig. 23.4):

$$A_r = \frac{D_w^2 - D_c^2}{D_c^2} \times 100 \ (\%) \tag{23.1}$$

$$C_i = \frac{D_s - D_c}{D_c} \times 100 \ (\%) \tag{23.2}$$

$$C_o = \frac{D_w - D_t}{D_t} \times 100 \ (\%) \tag{23.3}$$

where D_w = outside diameter of the cutting edge which enters the soil during sampling

D_o = inside diameter of the cutting edge of sampler

D_s = inside diameter of the sample tube

D_t = outside diameter of the sample tube

The area ratio is approximately equal to the ratio of the volume of soil displaced by the sampler to the volume of the sample. Other factors being equal, the lower the area ratio, the lower is the sample disturbance. For undisturbed sampling, area ratios of 13 percent or less are acceptable, but the values of 10 percent or less are preferred (U.S. WES practice, Marcuson and Franklin 1979). Samplers with small area ratios (less than about 13%) are called *thin wall samplers*, and those with higher area ratios *thick wall samplers*. For larger diameter samplers (> 100 mm), area ratio upto about 25 percent is in use. The inside clearance reduces friction between the sample and the inside wall of the tube and also allows elastic expansion of the sample on its entry into the sampler. A larger inside clearance will, however, lead to structural disturbance and density reduction in dense soils. A small inside clearance, 0.5 to 1.0%, is normally used (maximum permissible value 3.0%). Smooth finish of the tube and use of oil also reduce wall friction. Low outside wall friction is achieved by provision of the outside clearance, usually between zero to 2 percent. Another important design feature is the provision of a non-return valve in the drive head of a sampler; it allows free exit of water and air during driving and helps sample retention during withdrawal by creating a suction. Fitting of extension pieces at each end of a sampling tube is another way of avoiding disturbane of the top and bottom of the sample. Soil contained in these extension pieces is carefully removed by hand and only the central portion of the sample within the main sample tube is retained for testing.

The *sample recovery ratio*, defined as the ratio of the length of sample retained in the sampler to the depth of penetration, is an important measure of disturbance, specially for clay soils. For an ideal, undisturbed sample, the recovery ratio should be equal to or slightly less than 1.0.

Method of forcing the sampler into the ground has a considerable influence on sample disturbance. A fast, uniform penetration by hydraulic or mechanical jacking (e.g. a system of ropes and pulleys) produces little disturbance, whereas driving dynamically by means of a drop weight induces considerable disturbance. It is imortant not to overdrive the sampler since this compresses the contents. The base of the sample is detached from the ground by rotating the sampler. This causes some disturbance at the lower part of the sample, which should be rejected.

Maintenance of a water balance in the borehole is equally important for undisturbed sampling. During sampling, the water head in the borehole should correspond to the piezometric pressure of porewater in the soil at the level of sampling, This may require extending the casing above ground level or using bentonite slurry instead of water to balance high piezometric pressures.

5. Recovery of undisturbed samples. Undisturbed samples are usually obtained by penetrating thin wall samplers into boreholes, although sample recovery is the best from open excavations, if available.

(a) *Sampling in open excavation.* In an open trial pit or trench, a hand cut, block sample can be easily obtained. A block sample is trimmed in advance and a wooden box or other suitable container, with lid and bottom removed, is kept around the protruding sample. The space between the sample and the box is packed with suitable packing material, e.g. moist sawdust. The top of sample is covered with alternating layers of molten wax and paper and saw dust and the lid is fitted. The sample is cut at the bottom and turned upside down with the container. The bottom side is also sealed and covered with a lid.

Thin wall samplers, or short length cutters, are also easy to be forced into open excavations for obtaining undisturbed samples. When full, the sampler is removed by excavating soil all around and further handled as for sampling in boreholes.

(b) *Sampling in boreholes.* A variety of thin wall sampling devices are available for obtainig undisturbed samples from boreholes. The two types of samplers in common use are the 'open drive samplers' and the 'piston samplers'. (See next section). In general, piston samplers are better and can be used in almost all soils. The following sampling procedures are adopted.

(c) *Sampling with open drive samplers.* Before sampling, clean out the hole by removing all loose and disturbed material. Where casing is used, clean the hole to just below the casing so that sample may be taken *below* the bottom of the casing. Maintain the water level in the hole at or above groundwater level. With the sampler resting on the bottom of the hole, push the sampler into the soil by a continuous and rapid motion, without impact or twisting. Record the depth of penetration of the sampler. Do not push further than the length provided for the soil sample, allowing about 75 mm in the tube for cuttings and sludge. If instead of pushing, the sampler has to be driven by a hammer, record the weight, height and number of blows of the hammer. It is preferable to use a 'down-the-hole hammer' or a jarring link, i.e. the rods delivering the impact at the adaptor head of the sampler (See Fig. 23.4). Wait for about 5 minutes, if necessary, to allow the development of sufficient adhesion inside the tube. Turn the tube at least two revolutions to shear off the sample at the bottom. Pull the sampler out of the hole.

(d) *Sampling with piston sampler.* Seal the lower end of the sampler by keeping the piston at its lowest position and lower the sampler in a cleaned hole to rest at the bottom of the hole. Release the sampling tube from the piston and clamp the piston rod so that the piston remains stationary at the bottom level of the borehole. Push the sampling tube past the piston by a continuous rapid motion until the sampler head meets the top of the piston, or to atleast 90 percent of the effective sampling length. Lock the piston and the sampling tube together, the piston now being at the top end of the tube. Shear the sample at its bottom by giving rotations. Withdraw the tube and the piston together.

(e) *Frequency of sampling.* Samples are normally taken at each change of strata or at intervals not more than 1.5 m, whichever is less (IS: 10108).

(f) *Sampling of sands.* Fine sands and silty sands below the water table can be sampled with piston samplers, if these soils have some cohesion and a core catcher made of spring leaves is fitted just above the cutting edge. Cohesionless sands can be sampled by a piston sampler used with drilling mud (IS : 8763). The coating of the drilling mud at the cutting edge keeps the sample intact during withdrawal. Another is the compressed air technique (IS: 8763) in which compressed air is used to keep the groundwater separated from the sample in order to avoid dispersion of sampled sand. After sampling, the sampler tube is withdrawn into a bell where groundwater has been displaced by compressed air through a continuous pumping process.

(g) *Handling and transport of samples.* On withdrawal of the sampler from the borehole, it is detached from the drill rods and the adaptor head is removed. The length of the sample obtained in the sampling tube is measured and from the known depth of penetration the sample recovery ratio is calculated. For a sample acceptable as undisturbed, the recovery ratio should not be less than 95 percent (IS: 2132, 10108). The cutting shoe, if used, is also removed. Soil is reamed at both ends of the sampling tube and the ends are coated with layers of molten wax. Thin metal discs are usually inserted at the ends before sealing with wax. The seal thickness should not be less than about 25 mm. The ends are suitably capped and taped to prevent breakage of seals. The sample tubes are labelled with the following information: name of project, number of boring and that of sample, depth of sampling, top and/or bottom end of the sample, and date of sampling. The sample tubes should be protected from direct sunshine, shock, etc. while at the site. The samples should be taken to the laboratory with minimum of delay. During transport, they should be protected with suitable resilient packing material to reduce shock, vibration and disturbance.

If testing is to be delayed, the sampling tubes are stored in a humid room. When required, the end seals are removed and the sample is suitably extended with minimum disturbance. It is important that during extraction the movement of the sample relative to the tube is in the same direction as during sampling.

23.6 TYPICAL SAMPLERS

The principal types of samplers can be grouped under three categories: (i) open drive samplers, (ii) piston samplers, and (iii) rotary samplers. The open drive samplers can be 'thick wall' or 'thin wall' samplers depending on the degree of disturbance caused by them during sampling. The piston samplers and the rotary samplers are the thin wall samplers. Samplers used in India are described by Bhandari and Datye (1979).

1. Open drive samplers. These samplers are driven or pushed into soil with their lower ends open and hence the name 'open drive.' They are also known as 'open tube' samplers. They may be in the form of a 'solid tube', 'split tube' or 'split tube with liner'.

(a) *Thick wall solid tube samplers.* The sampler (Fig. 23.4) consists of a steel tube with a screw thread at each end. A cutting shoe is fitted to one end of the tube and a sampler head (adaptor) to the other end. The sampler is connected to a string of drill rods through its adaptor head. The sampler head also incorporates a non-return valve (ball check valve) which allows water and air to escape as the sample fills the tube and helps to retain the sample as the sampler is withdrawn. The sampler may be driven dynamically

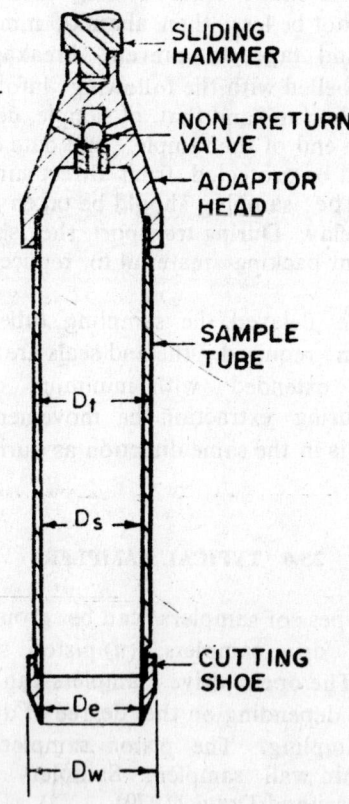

Fig. 23.4 Open drive 100 mm diameter sampler

by means of a drop weight (hammer) or pushed statically by hydraulic
or mechanical jacking. Driving by drop weight is more common.
A 'drive head' or 'anvil' with guide or 'jar length' is attached to the
top of the drill rods, which extends above the borehole. The drop
weight falls through a fixed height on the anvil. An alternative
better arrangement is to use a sliding hammer giving blows directly
to the adaptor head of the sampler.

The internal diameter may range from 35 mm to 100 mm and the
length is 450 mm or 600 mm; the more popular length being 450
mm. The widely used sample tube has an internal diameter of 100
mm (nominal) and a length of 450 mm and the area ratio is
approximately 30 percent. Although the area ratio is quite high,
the sampler is used for undisturbed sampling in cohesive soils; three
38 mm diameter undisturbed specimens can be prepared from a
section of the 100 mm diameter sample for triaxial or unconfined
compression testing. For sampling sands, a spring core catcher is
fitted between the tube and the cutting shoe. The 35 mm diameter
sampler is mainly used in sandy soils along with the standard
penetration test.

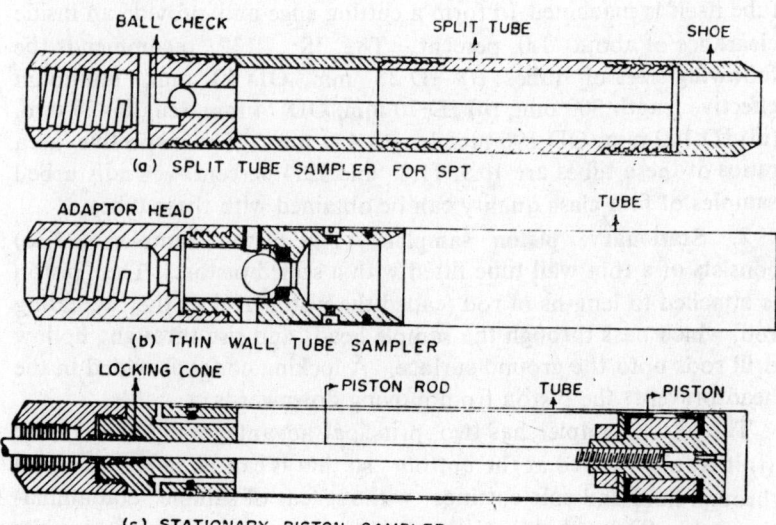

Fig. 23.5 Split tube and thin wall samplers

(b) *Thick wall split tube sampler.* The sampler (Fig.23.5a) consists of a tube which is split longitudinally into two halves. The two halves of the tube can be separated when the cutting shoe and the adaptor head are detached and the sample can be easily removed. The popular dimensions are 35 mm internal diameter, 50 mm external diameter and 600 mm length. This sampler is used for the standard penetration test, (Section 15.8). The sampler causes considerable disturbance to the soil since the area ratio is approximately 100 percent.

A modified version has an internal liner — a thin wall brass or steel tube. After sampling, the liner with sample inside is taken out by opening the sampler. Its ends are sealed in a manner similar to a thin wall sampler and then it is transported to the laboratory. With the development of thin wall samplers, the split tube with liner has declined in popularity.

(c) *Thin wall tube sampler.* The sampler (Fig. 23.5b) consists of a thin wall tube (seemless steel or brass tube, also known as Shelby tube) secured to an adaptor head containing a ball check valve. The head is threaded to receive drill rods. The sampler does not employ a separate cutting shoe. The lower end of the tube itself is machined to form a cutting edge and provide an inside clearance of about 1.0 percent. The IS: 2132 recommends the following sizes of tubes: (i) ID 38 mm, OD 40 mm, minimum effective length 300 mm, (ii) ID 70 mm, OD 74 mm, length 450 mm, (iii) ID 100 mm, OD 106 mm length 450 mm. The respective area ratios of these tubes are 10.9, 11.8 and 12.4 percent. Undisturbed samples of first class quality can be obtained with these tubes.

2. **Stationary piston sampler.** The piston sampler (23.5c) consists of a thin wall tube fitted with a sealed piston. The piston is attached to lengths of rod (called the piston rod or the actuating rod) which pass through the sample head and rise through hollow drill rods upto the ground surface. A locking cone provided in the head prevents the piston from moving downwards.

The piston sampler has two principal advantages (Acker 1974): (i) it is fully sealed at the bottom so that it can be safely lowered through fluid and soft cuttings without fear of sample contamination, and (ii) by holding the piston stationary and pushing the sampler downwards, the top of the sample is completely prevented from any distortion pressure at the top. Thus a much more effective vacuum seal is maintained than with the ball check valve, which helps in better retention of the soil in the tube. The sampler is generally used for soft clays and may also be used for saturated silts and silty sands with some cohesion.

3. Rotary sampler. The rotary sampler, commonly known as a core barrel, is used in connection with rotary drilling. There are two types of core barrels, double tube barrel and triple tube barrel. The double tube barrel is generally used for drilling through boulders or sound rock. The triple tube barrel is used for sampling softer materials, such as hard clay. Such a barrel consists of a liner (the innermost tube), an inner or holding tube and an outer tube. The outer tube rotates while the inner tube and the liner remain stationary. The liner which contains the core (sample) is detachable.

Sampling with core barrels is expensive but a very practical method of taking continuous core in stiff to hard clays and cemented silty sands and sands.

23.7 GEOPHYSICAL METHODS

Geophysical method may be used to explore large areas much more rapidly and economically than is possible by borings. Under certain conditions they are useful to determine stratification of soils and rocks, define the limits of granular borrow areas and large organic deposits and yield general information on subsurface conditions including the depth to groundwater. Because of the numerous limitations to the information obtained by these methods, it is always necessary to spot check the results against data obtained by other direct methods of exploration. The two geophysical methods which are most suitable for Civil Engineering explorations are the 'seismic refraction method' and the 'electrical resistivity method.'

1. Seismic refraction method. The method is based on the fact that shock waves have different velocities in different types of materials. The velocity is faster in denser and more consolidated materials. When artificial impulses are generated either by detonation of explosives or a mechanical blow (usually with a heavy hammer) at ground surface or in a hole at shallow depth, three kinds of waves are produced: longitudinal (compression) waves (called primary or P waves), shear waves (called secondary or S waves) and surface waves (called Rayleigh or R waves). In general, only longitudinal waves are observed. They are classified as 'direct', 'reflected' or 'refracted' waves. Waves travelling in approximately straight lines from the source of impulse to the surface are the direct waves, those which are turned back when they meet a boundary separating media of different seismic velocities are called reflected waves, and those which undergo a change or a bending in their direction on meeting such a boundary are called the refracted waves. The technique based on seismic reflection is suitable for locating deep formations (deeper than approximately 300 m) and is more appropriate to mineral prospecting. In Civil Engineering, the 'seismic refraction' method has been used for subsurface investigations to depths of approximately 300 metres (Lowe and Zaccheo 1979).

The equipment consists of a source of generating shock waves (a detonator and explosive charge or a heavy hammer and metal plate), sensitive vibration transducers (called geophones) and a time measuring device called a seismograph. The seismograph is connected electrically both to the detonator or hammer and to the geophones. At the instant of wave generation the seismograph starts the timing mechanism which stops when the first wave reaches the geophone with the time interval being recorded in milliseconds.

There are two possible arrangements of the geophones. In one set-up, waves are generated on a particular spot and a number of geophones are laid along a linear distance (course) from that spot. In another set-up, a single geophone is fixed in position and a series of detonators or impacts are produced along a line at increasing distances from the geophone.

Consider a two-layer deposit (Fig. 23.6) where the lower stratum is denser and thus has higher seismic velocity. When shock waves are generated the direct waves travel through the surface material in the direction of the geophone. Other waves travel in a downward direction and are refracted on striking the interface with the material of higher seismic velocity. At a certain critical angle, one particular wave will travel along the top of the lower stratum, parallel to the interface, as shown in Fig. 23.6. This critical refracted wave produces continually a weak head wave towards the surface, which is picked up by the geophone. Only the first wave reaching the geophone is recorded. Close to the source, the first wave to arrive will be the direct wave because the refracted wave must travel further. When the distance between the source and the geophone exceeds a certain value (depending on the thickness of the upper stratum), the first arrival will be the refracted wave. This is because the refracted wave, although it travels longer, partly passes through a stratum of higher seismic velocity.

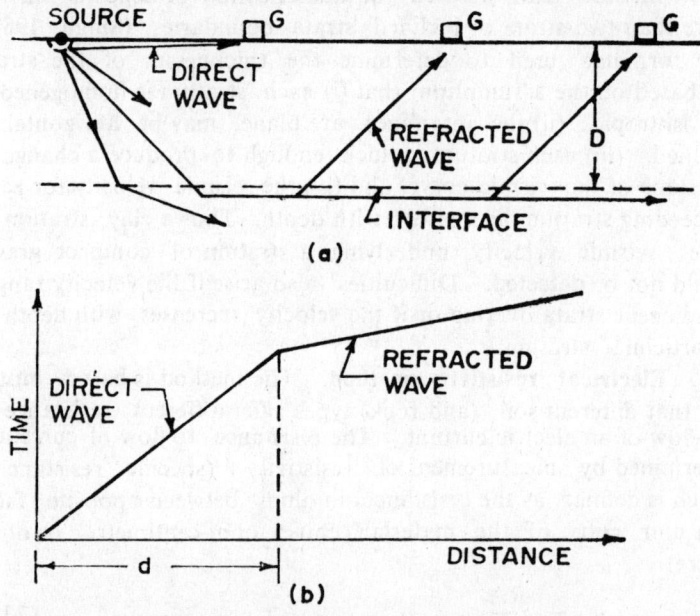

Fig. 23.6   Seismic refraction method

The time of 'first arrivals' is plotted against the distance between source and geophone to give a seismic graph, as shown in Fig. 23.6. The graph is interpreted to know the velocities of wave propogation in the layers and the depth to a layer. For a two-layer deposit, the graph changes its slope at a certain distance 'd'. If the source-geophone spacing is less than d, the distance of the point of break in the curve (called the *critical distance*), the direct wave reaches the geophone in advance of the refracted wave; and at source-geophone spacings greater than d, the first arrival is the refracted wave. The slopes of the two portions of the curves represent the seismic velocities (V_1 and V_2) of the upper and lower strata respectively. The general types of soils or rock can be determined from a knowledge of these velocities. The depth D of the interface between the two strata (provided the thickness of the upper stratum is constant) can be estimated by the relation:

$$D = \frac{d}{2}\left(\frac{V_2-V_1}{V_2+V_1}\right)^{1/2}$$
(23.4)

The method can be used for identification of deposits having more than two strata or inclined strata boundaries (Singh 1981). The formulae used to determine the thicknesses of the strata are based on the assumption that (i) each stratum is homogeneous and isotropic, (ii) the interfaces are plane, may be horizontal or inclined. (iii) each stratum is thick enough to produce a change in the slope of the seismic curve, and (iv) the seismic velocity for each succeeding stratum increases with depth. Thus a clay stratum of lower seismic velocity underlying a stratum of compact gravel would not be detected. Difficulties also arise if the velocity ranges of adjacent strata overlap or if the velocity increases with depth in a particlular stratum.

2. Electrical resistivity method. The method is based on the fact that different soil (and rock) types offer different resistance to the flow of an electric current. The resistance to flow of current is determined by measurement of 'resistivity ρ (specific resistance)', which is defined as the resistance in ohms between opposite faces of a unit cube of the material (units: ohm-centimetre or ohm-metre).

$$\rho = \frac{RA}{L}$$
(23.5)

where R is the resistance in ohms (ratio of electrical potential E to current I) of a block of conducting material of length L and cross sectional area A. The resistivity of an isotropic, homogeneous medium is constant. When the resistivity varies through a medium, the measured value is termed the 'apparent or mean resistivity'.

The resistivity of soil particles and of pure groundwater is high. The flow of current through a soil is mainly through electrolytic action which depends on the concentration of dissolved salts in the pore water. The resistivity of a soil, therefore, depends primarily on moisture content and the concentration of dissolved salts. The resistivity will decrease as both the moisture content and salt content increase.

One of the usual methods consists in driving four equally spaced electrodes into the ground in a straight line (Fig 23.7). A d.c. or a very low frequency a.c. current I of known magnitude is passed between the two outer (current) electrodes A and D, producing an electrical field within the soil. The potential drop is then measured between the two inner (potential) electrodes B and C. The apparent resistivity is given by:

$$\rho = 2\pi a \frac{E}{I} = 2\pi a R \qquad (23.6)$$

where   a = electrode spacing

The apparent resistivity is a weighted average of the true resistivities of the materials to a depth equal to the electrodes spacing 'a', the material close to the surface having a greater weightage.

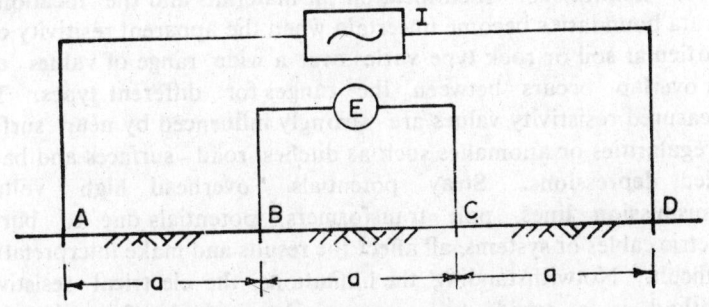

Fig. 23.7 Electrical resistivity method

Two procedures are adopted for exploration, namely 'resistivity profiling' and 'resistivity sounding'.

(a) *Resistivity profiling*. This procedure is used for investigation of lateral variation of soil types. The four electrodes with a constant spacing are moved as a unit either along a single line or along several parallel lines and a series of readings is taken. Apparent resistivity is plotted against the centre position of the four electrodes, to natural scales. Contours of equal resistivity can be plotted for the area traversed. Resistivity profiling is used particularly for delineating boundaries of pervious deposits (sand and gravel deposits) and for locating faults and steeply dipping contacts between different materials.

(b) *Resistivity sounding.* This procedure is used to get information on the variation of subsurface materials with depth. For this purpose, the centre of the electrodes is maintained at a fixed point and the (equal) spacing of the electodes is increased for each successive reading. The spread is thus symmetrical about the central position. With increase of spacing the depth of material affecting the apparent resistivity increases and the changes in material are reflected in the measured values of resistivity. Resistivity sounding can be used to know the sequence of high and low resistivity zones, to estimate the depth to sand and gravel, bedrock or water bearing strata, and to estimate the thickness of strata.

The sounding data are plotted preferably on a log-log graph which can be interpreted by comparing with a set of 'standard curves' or by other available methods (Singh 1981).

(c) *Limitations.* Identification of materials and the location of strata boundaries become uncertain when the apparent resitivity of a particular soil or rock type varies over a wide range of values and an overlap occurs between the ranges for different types. The measured resistivity values are strongly influenced by near surface irregularities or anomalies such as ditches, road surfaces and back-filled depressions. Stray potentials, overhead high voltage transmission lines, pole transformers, potentials due to buried electric cables or systems, all affect the results and make interpretation difficult. Notwithstanding the limitations, the electrical resistivity method is a rapid and economical method of exploration. The method is not considered to be as reliable as the seismic method. For reliability, the method must always be used in conjunction with borings.

23.8 EXPLORATION LOGS

The information obtained from any subsurface exploration should be documented in the form of exploration logs, called the borehole logs. A borehole log (table) contains, for example, the following columns of information: (i) depth below ground surface, (ii) thickness of layers, (iii) water level, (iv) graphical symbol of the soil type, (v) description of soil, (vi) position, type and number of sample, and (vii) results of insitu tests, if performed, etc. This log also includes title and location of the project/job, boring number, method of boring and sampling, surface elevation of the boring and date of work. The borehole logs are ultimately incorporated in a comprehensive report on the investigations.

———————

24

Machine foundations

A body possessing mass and elasticity is capable of vibration. The vibratory motion may be periodic or aperiodic. When the motion repeats itself in equal intervals of time, it is called periodic, and if it does not repeat in equal intervals of time, it is called *aperiodic* or random. The full sequence of a periodic quantity occurring during a period is called *cycle*. Cycle is thus the basic or unit motion which is repetitive. The *period* T is the time required for one cycle of motion or one complete vibration. The number of cycles of motion per unit time is called the *frequency* f. Frequency is thus the reciprocal of the period T:

$$f = \frac{1}{T} \text{ cycles/unit time} \tag{24.1}$$

Frequency in terms of cycles per second cps is also called 'Hertz', abbreviated as Hz.

1. Free vibration Consider an elementary body of mass 'm' affixed to a rigid support by a spring (Fig. 24.1) which is pulled downwards a distance 'z_0' and released. Due to restoring force in the spring, the body will start vibrating in the vertical direction within the range $\pm z_0$ on either side of the equilibrium position. The motion may continue indefinitely if there is no internal friction or external force acting to slow it down. The vibration of a system which is displaced from its equilibrium position by a single application of an external force and left free to vibrate is called "free vibration".

The maximum up or down displacement $\pm z_0$ of the body from the equilibrium position is called the *amplitude* (symbol A).

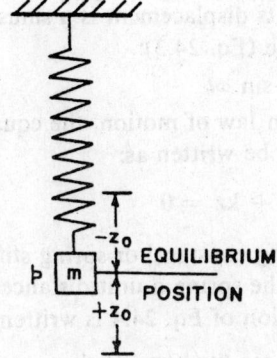

Fig. 24.1 Free vibrating spring mass system

The actual vertical motion of the body may be graphically represented as the projection on the vertical line of a point 'a' rotating along the circumference of a circle of diameter $2z_0$ (Fig. 24.2).

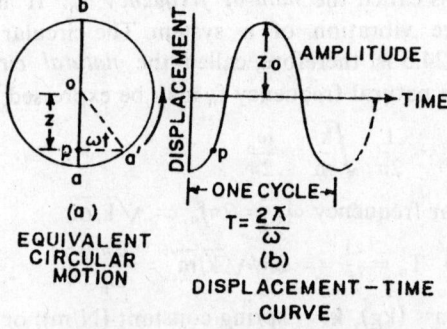

Fig. 24.2 Simple harmonic motion

As the motion is periodic, the point 'a' rotates at a uniform 'angular velocity' ω along the circle. The angular velocity expressed in radian per second, is called *circular or angular frequency*. The period T for one complete cycle is thus $2\pi/\omega$ and

$$f = \frac{1}{T} = \frac{\omega}{2\pi} \text{ (cycles per second)} \tag{24.2}$$

The displacement z can be plotted as a sinusoidal function of time (Fig. 24.2b). A rectilinear vibrating motion is called *simple harmonic motion* if its displacement is a sinusoidal (or cosinusoidal) function of time (Eq. 24.3):

$$z = z_0 \sin \omega t \tag{24.3}$$

Using the Newton law of motion, the equation of motion for the above case may be written as:

$$m \frac{d^2z}{dt^2} + kz = 0 \tag{24.4}$$

where k is the 'spring constant' or spring stiffness, i.e. the force required to deflect the spring a unit distance.

The general solution of Eq. 24.4 is written as:

$$z = c_1 \sin (\sqrt{k/m} \cdot t) + c_2 \cos (\sqrt{k/m} \cdot t) \tag{24.5}$$

where c_1 and c_2 are arbitrary constants determined by initial conditions. Regardless of the values of c_1 and c_2, z is a harmonic function with a circular frequency $\sqrt{k/m}$ ($= \omega$) (cf Eq. (24.3).

The frequency with which a body vibrates naturally (without outside interference) when subjected to the application of an external force is called the *natural frequency* f_n. It is the frequency of free vibration of a system. The circular frequency $\sqrt{k/m}$ in Eq. 24.5 is, therefore, called the *natural circular frequency* ω_n. The natural frequency f_n may be expressed as:

$$f_n = \frac{1}{2\pi} \sqrt{\frac{k}{m}} = \frac{\omega_n}{2\pi} \tag{24.6}$$

Natural circular frequency $\omega_n = 2\pi f_n = \sqrt{k/m}$ (24.7)

Natural period $T_n = \frac{1}{f_n} = 2\pi/\sqrt{k/m} = \frac{2\pi}{\omega_n}$ (24.8)

Units: m = mass (kg), k = spring constant (N/m); or mass (t) and k (kN/m)

The natural frequency may also be expressed as:

$$f_n = \frac{1}{2\pi} \sqrt{\frac{k}{m} \frac{g}{g}} = \frac{1}{2\pi} \sqrt{\frac{k \cdot g}{W}} = \frac{1}{2\pi} \sqrt{\frac{g}{\delta}} \tag{24.9}$$

where δ = static (elastic) deflection = $\frac{W}{k}$

Expressing δ in cm, g = 980 cm/s^2, W in newton N and k in N/m,

$$f_n \simeq \frac{5}{\sqrt{\delta}} \text{ cps} \tag{24.10}$$

2. Damped free vibration In practical problems there is nearly always dissipation of energy or *damping* which causes the amplitude of free vibration to decrease with each successive cycle. The equation of motion of damped free vibration is written as:

$$m \frac{d^2z}{dt^2} + c \frac{dz}{dt} + kz = 0 \qquad (24.11)$$

The second term in Eq. 24.11 represents the damping force. The quantity 'c' is the 'damping coefficient' for viscous damping, which is the ratio of the damping force to the velocity, (i.e. damping force = c × velocity). The minimum viscous damping which allows a displaced system to return to its original position without further oscillation is called 'critical damping'. The motion becomes aperiodic. The value of c for this condition is called *critical damping coefficient* c_c. The ratio of the damping coefficient c to critical damping coefficient c_c is called *damping ratio* or *damping factor* D:

$$D = c/c_c \qquad (24.12)$$

3. Forced vibration Forced vibration occurs when a system is continually excited by some periodic force F(t). The external time dependent force F(t) can be of two types:

(a) *Constant force amplitude excitation* represented as;

$$F(t) = F_0 \sin \omega_1 t \qquad (24.13)$$

where F_0 = amplitude (magnitude) of force − a constant quantity with an exciting circular frequency ω_1

Punch presses or impact machines (hammers) generate constant force (but time dependent) excitation.

(b) *Variable force amplitude excitation*, or rotating mass excitation represented as:

$$F(t) = (m_1 e \omega_1^2) \sin \omega_1 t \qquad (24.14)$$

where m_1 = reciprocating or unbalanced rotating mass

   e = displacement (in case of reciprocating type) or eccentricity (in case of rotating type)

   ω_1 = frequency of motion (operating frequency)

Here the force amplitude (magnitude) is $m_1 e \omega_1^2$ which varies with ω_1. Reciprocating machines or those with unbalanced rotating mass generate frequency dependent forced vibration.

4. Resonance The condition of maximum increase in amplitude of a vibrating system is called 'resonance'. The frequency corresponding to the maximum amplitude of a system subjected to forced vibration is called the *resonant frequency* f_r.

If the force F_0 is applied statically, the static displacement (amplitude) is given by:

$$z_s = F_0/k \tag{24.15}$$

Because of the frequency of the exciting force, the actual amplitude of displacement z_0 is given by:

$$z_0 = Nz_s \tag{24.16}$$

where N is the dynamic magnification factor given by:

$$N = \frac{1}{\sqrt{(1 - r^2)^2 + (2Dr)^2}} \tag{24.17}$$

where r is the *frequency ratio*, ratio of operating (exciting) frequency ω_1 (or f_1) to natural frequency ω_n (or f_n).

When the operating frequency is equal to natural frequency, i.e. $\omega_1/\omega_n = r = 1$ for the undamped system (D = 0), z_0 increases indefinitely (N → ∞). Resonance is said to occur at $r = 1$.

For a damped system (D > 0), the frequency ratio at which the magnification factor is maximum (i.e, z_0 is maximum) is obtained by differentiating the N-function with respect to r and setting the first derivative equal to zero. Thus in case of constant force, the frequency ratio (r_m) for maximum amplitude is given by:

$$r_m = \left(\frac{\omega_1}{\omega_n}\right)_m = \sqrt{1 - 2D^2} \tag{24.18}$$

that is, for damped system, resonance occurs at r < 1. The maximum value of magnification factor (N_r) at resonance is given by:

$$N_r = \frac{1}{2D\sqrt{1 - D^2}} \text{ (at resonant frequency)} \tag{24.19}$$

24.2 TYPES OF MACHINES AND FOUNDATIONS

Based on their operating frequency, the machines may be classified as follows:

(a) Low speed: frequency, less than 300 rpm.

(b) Medium speed: frequency, 300-1000 rpm.

(c) High speed: frequency, greater than 1000 rpm.

Considering their structural form, the machine foundations are, in general, of the following types.

(a) Block foundation consisting of a pedestal of concrete on which the machine rests.

(b) Box or caisson foundation consisting of a hallow concrete block supporting the machine.

(c) Wall foundation consisting of a pair of walls supporting the machine on their top.

(d) Framed foundation consisting of vertical columns and a top horizontal frame work which forms the seat of essential machinery.

Low speed machines producing impact and periodic forces (e.g. forge hammers, presses, low speed reciprocating engines and compressors) are generally supported on block foundations having a large contact area with soil. Medium speed machines (e.g. reciprocating diesel and gas engines) also have, in general, block foundations resting on springs or suitable elastic pads. High speed and rotating type of machines (e.g. internal combustion engines, electric motors and turbogenerator machines) are generally mounted on framed foundations. Other high speed machines are placed on block foundations.

24.3 DESIGN REQUIREMENTS

For satisfactory performance, machine foundations should satisfy the following static load and dynamic load criteria.

(a) *Static load criteria* There should be neither shear failure nor excessive settlement. The stresses on soil under the combined effect of static and dynamic loads should be within permissible limits. For preliminary design, the allowable pressure on soil due to static loads alone may be taken as 40% of the corresponding safe bearing capacity[1]. The IS code[2] specifies that the soil stress should not exceed 80% of the safe stress under static loading.

(b) *Dynamic load criteria* (1) There should be no resonance. To avoid resonance, the natural frequency (or the resonant fre-

---

1. Srinivasulu, P. and Vaidyanathan, C. 1976. Handbook of Machine Foundations. Tata McGraw Hill, New Delhi.

2. IS: 2974-I. Code of Practice for Design and Construction of Machine Foundations, Part 1, Foundations for Reciprocating Type Machines, BIS, New Delhi.

quency) of the machine-foundation-soil system must be either very large or very small compared to the operating speed of the machine. In case of low speed machines, the natural frequency should be high, at least twice the operating frequency, i.e. the frequency ratio $r (= f_1/f_n)$ should be less than 0.5. Natural frequency can be increased (i) by increasing base area or reducing total static weight of the foundation, (ii) by increasing modulus of shear rigidity of the soil by compaction, grouting or injection, and (iii) by using piles to provide the required foundation stiffness. For high speed machines, the natural frequency is kept low. not higher than one-half of the operating value, i.e. $r \geqslant 2$. This can be achieved by increasing weight of the foundation. (2) The amplitudes at both the resonant and operating frequencies should be within permissible limits. (3) There should be proper vibration and shock isolation.

24.4 DESIGN METHODS

The methods used for designing machine foundations can be grouped as follows: (i) empirical and semi-empirical methods, (ii) methods considering soil as a spring, and (iii) method considering soil as a semi-infinite elastic mass (elastic half-space approach) and its equivalent lumped parameter method. Considering the lumped parameter method as the preferred one, this method is described in the next section.

24.5 LUMPED PARAMETER METHOD

The soil is assumed to be homogeneous, isotropic, and semi-infinite elastic material with the foundation resting on the surface of soil. The soil parameters needed for design are the bulk density or unit weight γ, Poisson's ratio μ and the dynamic shear modulus G. The various vibration parameters of mass, damping and flexibility (spring effect) are assumed to be lumped together in each category as *single* 'equivalent mass' m 'equivalent damping constant' c and 'equivalent spring constant' k. Because of this lumping together of the variables in each category, the method is called the 'lumped parameter method'. Using the equivalent lumped values of m, c and k, the equation of motion for vertical vibration may written as follows:

$$m \frac{d^2z}{dt^2} + c \frac{dz}{dt} + kz = F(t) \qquad (24.20)$$

The computations for finding the basic design data of resonant frequency and amplitude of vibration for vertical vibrations are made as follows:

(1) Combined mass of machine and foundation.

$$m = W/g \qquad (24.21)$$

where $g =$ acceleration due to gravity

(2) Equivalent radius r_0 of foundation base for shapes other than circular.

$$r_0 = \sqrt{A_b/\pi} \qquad (24.22)$$

where $A_b =$ area of base

(3) Spring constant k.

$$k = \frac{4Gr_0}{1 - \mu} \qquad (24.23)$$

where $G =$ shear modulus

$\mu =$ Poisson's ratio

(4) Undamped natural frequency f_n.

$$f_n = \frac{1}{2\pi} \sqrt{k/m} \qquad (24.24)$$

(5) Dimensionless mass ratio B_z.

$$B_z = \frac{1 - \mu}{4} \cdot \frac{m}{\rho r_0^3} = \frac{1 - \mu}{4} \cdot \frac{W}{\gamma r_0^3} \qquad (24.25)$$

(6) Damping ratio D.

$$D = 0.425/\sqrt{B_z} \qquad (24.26)$$

(7) Resonant frequency f_r.

$$f_r = f_n \sqrt{1 - 2D^2} \text{ (constant force excitation)} \qquad (24.27)$$

$$f_r = \frac{f_n}{\sqrt{1 - 2D^2}}$$

(variable force rotating mass excitation) \qquad (24.28)

(8) Amplitude (magnitude) of unbalanced force at resonance:

$$F_0 \quad \text{or} \quad m_1 e \omega_r^2$$

(9) Static foundation displacement z_s.

$$z_s = F_0/k \quad \text{(const. force)} \tag{24.29}$$

$$z_s = \frac{m_1 e}{m} \quad \text{(variable force)} \tag{24.30}$$

(10) Magnification factor at resonace N_r.

$$N_r = \frac{1}{2D \sqrt{1 - D^2}} \tag{24.31}$$

(11) Resonant (maximum) amplitude z_r.

$$z_r = z_s N_r \tag{24.32}$$

Effect of embedment Embedment increases resonant frequency and reduces amplitude of vibration, because the static stiffness of soil spring increases with depth. The spring constant k (Eq. 24.23) is then multiplied by the embedment factor n_z, i.e.

$$k = \frac{4Gr_0}{1 - \mu} n_z \tag{24.33}$$

where $\quad n_z = 1 + 0.6 (1 - \mu) (h/r_0)$ $\qquad$ (24.34)

$\quad$ h = depth of embedment

Examples 24.1—24.5

24.1 Find the natural frequency of a foundation block if its elastic deflection is estimated to be 8 mm.

Eq. 24.10: $\quad f_n = \dfrac{5}{\sqrt{\delta}} \text{ cps} = \dfrac{5}{\sqrt{0.8}} = 5.59 \text{ cps}$

24.2 Find the magnification of the amplitude of vibration of a machine at its operating frequency of 5 Hz, if the damping factor is 0.5. The natural frequency of the machine is 2.5 Hz.

$\qquad$ Frequency ratio r $\; = f_1/f_n = 2$

$\qquad$ Damping factor D $= 0.5$

$\qquad$ Magnification factor $N = \dfrac{1}{\sqrt{(1 - r^2)^2 + (2Dr)^2}}$

$\qquad\qquad = 0.277$

24.3 A foundation block of weight 25 kN rests on a soil for which the stiffness may be assured as 200 kN/m. The damping factor is 0.5. It is vibrated vertically by an exciting force of 2.5 sin 30 t (kN) (F_0 sin ωt). Find the natural frequency, natural period and natural circular frequency. Find also the amplitude of vertical displacement.

$$\delta = W/k = \frac{25}{200} = 0.125 \text{ cm}$$

$$f_n = \frac{5}{\sqrt{0.125}} = 14.1 \text{ cps}$$

$$T_n = 1/f_n = 0.071 \text{ s}$$

$$\omega_n = 2\pi f_n = 88.6 \text{ rad/s}$$

Alternately $\omega_n = \sqrt{\dfrac{k \text{ (N/metre)}}{m \text{ (kg)}}}$

$$= \sqrt{\frac{200 \times 10^3 \times 10^2}{(25 \times 10^3)/9.8}}$$

$$= 88.5 \text{ rad/s}$$

Frequency ratio $r = \dfrac{\omega}{\omega_n} = \dfrac{30}{88.5} = 0.34$

$$D = 0.5$$

$$N = \frac{1}{\sqrt{(1 - r^2)^2 + (2Dr)^2}} = 1.055$$

Static deflection $z_s = \dfrac{F_0}{k} = \dfrac{2.5}{200} = 0.0125 \text{ cm} = 0.125 \text{ mm}$

Amplitude $z_0 = Nz_s = 1.055 \times 0.125 = 0.13 \text{ mm}$

24.4 A vertical acting hammer generates an unbalanced force F_0 of amplitude 9 kN. The total mass of the hammer and its block foundation is 72t. The base of the foundation is 4.8 m × 2.5 m in plan. The soil has $\rho = 1.92$ t/m³, shear modulus $G = 19.6 \times 10^3$ kN/m² and $\mu = 0.33$. Find the resonant frequency and the amplitude of the forced vibration if the operating frequency is 200 rpm.

Area of base $A_b = 4.8 \times 2.5 = 12 \text{ m}^2$

$$r_0 = \sqrt{A_b/\pi} = 1.954 \text{ m}$$

$$k = \frac{4Gr_0}{1 - \mu} = \frac{4 \times 19.6 \times 10^3 \times 1.954}{1 - 0.33}$$

$$= 228.647 \times 10^3 \text{ kN/m}$$

$$f_n = \frac{1}{2\pi} \sqrt{\frac{k}{m}} = \frac{1}{2\pi} \sqrt{\frac{228.647 \times 10^3}{72 \text{ (t)}}}$$

$$= 8.9688 \text{ cps} = 538 \text{ rpm}$$

$$B_z = \frac{1-\mu}{4} \cdot \frac{m}{\rho r_0^3} = \frac{0.67}{4} \cdot \frac{72}{1.92 \ (1.954)^3}$$

$$= 0.842$$

$$D = 0.425/\sqrt{B_z} = 0.463$$

$$f_r = f_n \sqrt{1 - 2D^2} \qquad \text{(const. force excitation)}$$

$$= 538 \sqrt{1 - 2 \ (0.463)^2} = 406.6 \simeq 407 \text{ rpm}$$

$$F_0 = 9 \text{ kN}$$

$$z_s = F_0/k = 9/228.647 \times 10^3 = 3.94 \times 10^{-5} \text{ m}$$

$$N_r = \frac{1}{2D \sqrt{1 - D^2}} = 1.218$$

$$z_r \text{ (at resonance)} = z_s \ N_r = 4.799 \times 10^{-5} \text{ m}$$

$$= 0.048 \text{ mm}$$

24.5 A single cylinder engine of weight 12.5 kN rests on a 6 m × 1.5 m block of concrete of weight 680 kN. The engine operates at 1000 rpm and generates a frequency dependent unbalanced force. From the manufacturer data, the quantity $m_1 e = 0.65$ N-s^2 (newton-second^2). The soil data are: $\mu = 0.3$, $G = 17.5 \times 10^3$ kPa and $\gamma = 19$ kN/m^3.

Find the resonant frequency and amplitude of vibration at resonance. Also find amplitude at operating speed.

$$\text{Total mass m} = \frac{12.5 + 680}{9.8} = 70.66 \text{ t}$$

$$r_0 = \sqrt{\frac{6 \times 1.5}{\pi}} = 1.69 \text{ m}$$

$$k = \frac{4Gr_0}{1-\mu} = 169 \times 10^3 \text{ kN/m}$$

$$f_n = \frac{1}{2\pi} \sqrt{k/m} = \frac{1}{2\pi} \sqrt{\frac{169 \times 10^3}{70.66}} = 7.78 \text{ cps}$$

$$= 466.8 \text{ rpm}$$

$$B_z = \frac{1-\mu}{4} \cdot \frac{W}{\gamma r_0^3} = \frac{(1-0.3) \ 692.5}{4 \times 19 \times (1.69)^3}$$

$$= 1.32$$

$$D = 0.425/\sqrt{B_z} = 0.37$$

$$f_r = \frac{f_n}{\sqrt{1 - 2D^2}} \text{ (variable force excitation)}$$

$$= 9.129 \text{ cps} = 547.74 \text{ rpm}$$

$$\omega_r = 2\pi f_r = 2\pi \times 9.129 = 57.36 \text{ rad/s}$$

$$z_s = \frac{m_1 e}{m} = \frac{0.65 \times 10^{-3}}{70.66} = 9.199 \times 10^{-6} \text{ m}$$

$$= 9.2 \times 10^{-6} \text{ m}$$

$$N_r = \frac{1}{2D\sqrt{1 - 2D^2}} = \frac{1}{2 \times 0.37\sqrt{1 - (0.37)^2}}$$

$$= 1.455$$

$$z_r = z_s N_r = 13.386 \times 10^{-6} \text{ m}$$

At operating speed $r = \dfrac{f_1}{f_n} = \dfrac{1000}{466.8} = 2.14$

$$N' = \frac{r^2}{\sqrt{(1 - r^2)^2 + (2Dr)^2}}$$

(variable force excitation)

$$= 1.17$$

z_0 at operative speed $= z_s \times N'$

$$= 9.2 \times 10^{-6} \times 1.17$$

$$= 10.76 \times 10^{-6} \text{ m } (< z_r)$$

25

Design of wells and rings

25.1 I.I.T. BOMBAY (M.O.T.) METHOD OF DESIGNING WELL FOUNDATIONS

Under a research scheme sponsored by the Ministry of Surface Transport (Road Wing) (M.O.T.), Govt. of India, extensive tests on large scale model wells, 1.2 m dia × 5.25 m high, surrounded by clay and resting on clay, sand, soft rock and hard rock were carried out at the Geotechnical Engineering Section of the I.I.T. Bombay. The results have been reported in Final reports of 4 parts (1983-1988), suggesting guide lines for the design of well foundations surrounded by clay. These guide lines are empirically based on model test results. These guide lines are illustrated by typical design examples is Sections 25.2 to 25.5 taken from the I.I.T. reports with minor modifications for conversion into SI units by the author, and with due acknowledgement to I.I.T. Bombay. Only typical parameters of well design have been worked out rather than the complete design, so that the principles of design may be understood.

25.2 WELL SURROUNDED BY CLAY AND RESTING ON HARD ROCK

Example 25.1

The dimensions of the well and the set of forces given in the I.I.T. report (Part -4) are for the well under pier No. 2 of the second Hooghly bridge across the Hooghly river in Calcutta.

Diameter of well (B) = 20.6 m, depth of embedment (D) = 17.53 m, steining thickness = 1.5 m, resultant applied vertical load (V) = 2.486 × 10⁵ kN, resultant applied horizontal load

(H) $= 1.394 \times 10^4$ kN, applied moment at base $(M_a) = 5.813 \times 10^5$ kN-m.

Undrained cohesion of clay $(c_T) = 120$ kN/m^2, $\phi_T = 0$, dry density $(\gamma_d) = 16$ kN/m^3, moisture content (m) = 33%. Compressive strength of rock $= 25 \times 10^3$ kN/m^2 (assumed ultimate bearing capacity q_f).

(1) *Embedment ratio and H/V ratio*
$$D/B = 17.53/20.6 = 0.85$$
$$H/V = (1.394 \times 10^4)/(2.486 \times 10^5) = 5.61\%$$

(2) *Vertical load shared by base and sides of well* From Fig. 25.1, the ratio of load shared by base V_b to applied vertical load V for D/B = 0.85 is given by extrapolating the curves upwards as:
$$V_b/V = 0.905$$

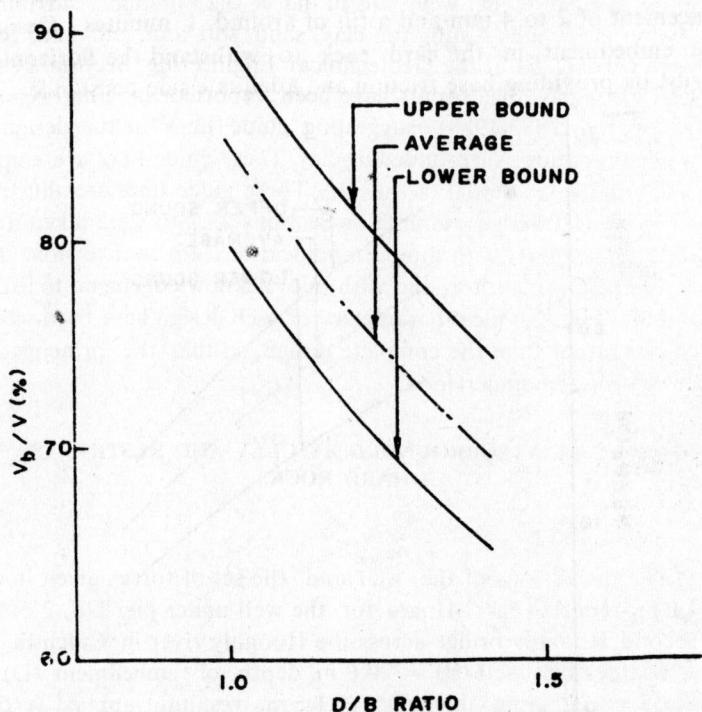

Fig. 25.1 Vertical load shared by base versus D/B ratio (hard rock)

or $\qquad V_b = 0.905 \times 2.486 \times 10^5 = 2.25 \times 10^5$ kN

Load shared by sides $= V_s = V - V_b$

$\qquad\qquad\qquad = 0.236 \times 10^5$ kN

(3) *Moment shared by base and sides of well* From Fig. 25.2, the ratio of moment shared by base M_b to applied moment M_a for $D/B = 0.85$ is given as:

$$M_b/M_a = 0.24$$

or $\qquad M_b = 0.24 \times 5.813 \times 10^5$

$\qquad\qquad = 1.395 \times 10^5$ kN-m

Moment shared by sides M_s

$\qquad = M_a - M_b = 4.418 \times 10^5$ kN-m

Note: The major part of the horizontal thrust and moment was observed to be taken by the sides under a horizontal displacement of 2 to 4 mm and a tilt of around 1 minutes. Design the embedment in the hard rock to withstand the horizontal thrust by providing base friction and adequate side resistance.

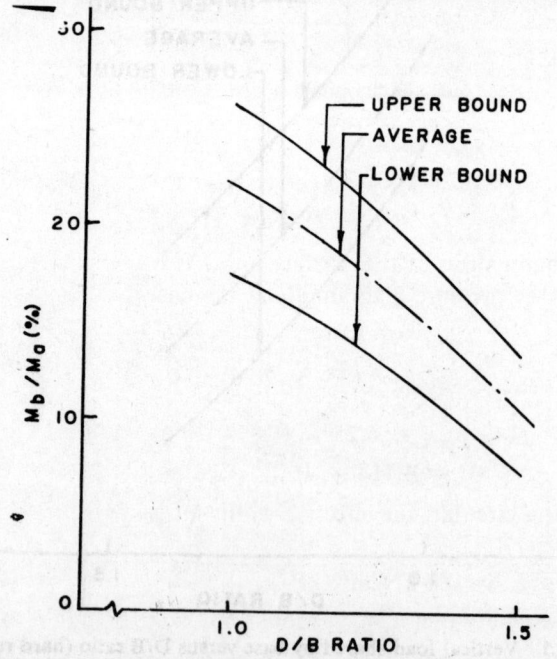

Fig. 25.2 Moment shared by base versus D/B ratio (hard rock)

(4) *Position of centre of rotation* The centre of rotation is at the base for all D/B ratios.

(5) *Pressure distribution on sides of well due to side moment and maximum stress at surface* On the rear side of the well (opposite to the horizontal loading side), the contact pressure distribution may be taken as rectangular upto a depth $2c_T/\gamma$ and then decreasing linearly upto the centre of rotation (which is at the base). (Fig. 25.3).

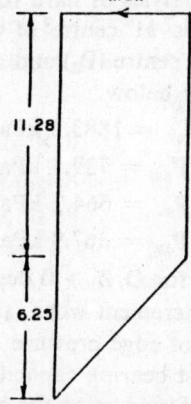

Fig. 25.3 Rear side contact pressure distribution (hard rock)

Bulk density $\gamma = \gamma_d (1 + m) = 16 \times 1.33$
$$= 21.28 \text{ kN/m}^2$$
$$2c_T/\gamma = (2 \times 120)/21.28 = 11.28 \text{ m}$$

The maximum stress at the surface p_{max} is found by taking moment of the pressure diagram about the base and equating it to M_s.

$$p_{max} \times 11.28 \times (0.9 \times 20.6) \times \left(6.25 \times \frac{11.28}{2}\right)$$
$$+ \tfrac{1}{2} p_{max} \times 6.25 \times (0.9 \times 20.6) \times (\tfrac{2}{3} \times 6.25)$$
$$= M_s = 4.418 \times 10^5$$

Note: Being circular, the effective width of well is taken as (0.9 × diameter)

$$p_{max} = 162 \text{ kN/m}^2 \text{ (kPa)}$$

To have the safety factor, p_{max} should not exceed $3c_T$, i.e. $3 \times 120 = 360 \text{ kN/m}^2$.

(6) *Contact pressure at base and safety against ultimate bearing capacity*

Average pressure

$$P_{av} = \frac{V_b}{A} = \frac{2.25 \times 10^5}{\pi (20.6)^2/4}$$

$$= \frac{2.25 \times 10^5}{333.29} = 675 \text{ kPa}$$

It is found from the analysis of hard rock that for $D/B = 0$ and $H = 0$ the pressures at centre of well (P_0), at $3B/16$ from centre (P_1), at $3B/8$ from centre (P_2) and at the edge of well (P_3) are related to P_{av} as given below.

$$P_0 = 2.79 \, P_{av} = 1883.2 \text{ kPa}$$

$$P_1 = 1.095 \, P_{av} = 739.1 \text{ kPa}$$

$$P_2 = 0.984 \, P_{av} = 664.2 \text{ kPa}$$

$$P_3 = 0.545 \, P_{av} = 367.9 \text{ kPa}$$

The pressure increases for $D/B > 0$ depending on H/V ratio.

If ΔP is the pressure increment when there is an overturning moment, the total value of edge pressure is $P'_3 = P_3 + \Delta P$ and the safety factor against bearing capacity will be $q_f/(P_3 + \Delta P)$. The safety factor is also found using the average value P_{av}.

$$F = q_f/P_{av} = (25 \times 10^3)/615 = 37$$

(7) *Tilt of well* The tilt (displacement) of well at the surface (δ) is given by:

$$\delta = Q_{max}/K_h.$$

where Q_{max} = maximum horizontal force per unit width at the surface

K_h = horizontal subgrade reaction

The angle of tilt θ is then given by:

$$\theta = \delta/y$$

where y = distance from surface to the point of rotation

$$Q_{max} = \frac{1.394 \times 10^4}{(0.9 \times 20.6) \times 17.53} = 42.89 \text{ kPa}$$

Assuming $K_h = 4000 \text{ kN/m}^3$

$$\delta = 42.89/4000 = 0.0107 \text{ m}$$

$$y = 17.53 \text{ m (base of well)}$$

$$\theta = 0.0107/17.53 = 1 \text{ in } 1638$$

25.3 WELL SURROUNDED BY CLAY AND RESTING ON SOFT ROCK

Example 25.2

The dimensions of the well, set of forces and the properties of surrounding clay are the same as in Ex. 25.1. The soft rock has $\gamma_d = 19.1 \text{ kn/m}^3$ and $q_f = 2500 \text{ kPa (kN/m}^2)$

(1) $D/B = 0.85$ and $H/V = 5.61\%$

(2) From Fig. 25.4, $V_b/V = 0.9$

$$V_b = 0.9 \times 2.486 \times 10^5 = 2.237 \times 10^5 \text{ kN}$$
$$V_s = V - V_b = 0.249 \times 10^5 \text{ kN}$$

(3) From Fig. 25.5, $M_s/M_a = 0.9$

$$M_s = 0.9 \times 5.813 \times 10^5 = 5.232 \times 10^5 \text{ kN-m}$$
$$M_b = M_a - M_s = 0.58 \times 10^5 \text{ kN-m}$$

(4) The centre of rotation is at the base for all D/B ratios.

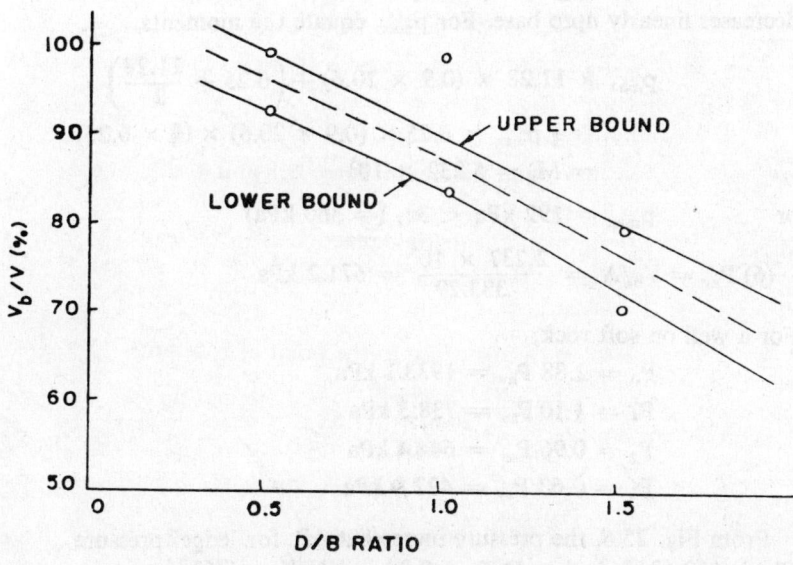

Fig. 25.4 Vertical load shared by base versus D/B ratio (soft rock)

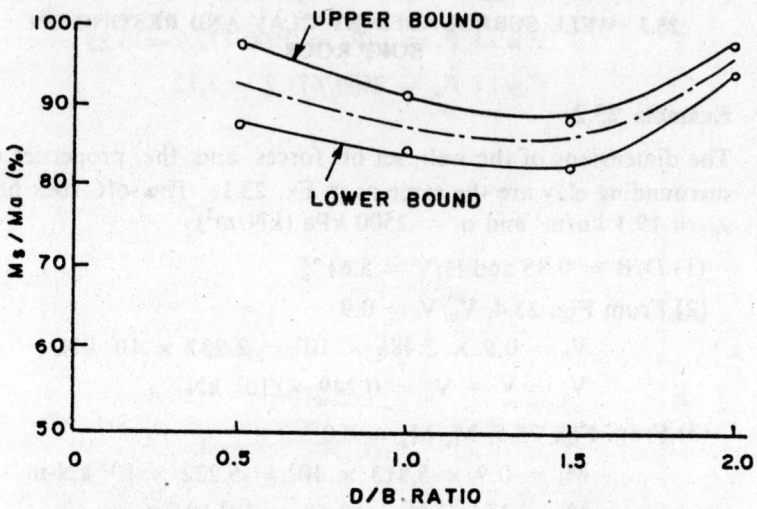

Fig. 25.5 Moment shared by sides versus D/B ratio (soft rock)

(5) Pressure distribution on sides and maximum stress at surface
Similar to well on hard rock (Ex. 25.1), the rear side pressure
distribution is rectangular upto depth $2c_T/\gamma = 11.28$ m and then
decreases linearly upto base. For p_{max}, equate the moments,

$$p_{max} \times 11.28 \times (0.9 \times 20.6) + \left(6.25 + \frac{11.28}{2}\right)$$
$$+ \tfrac{1}{2} p_{max} \times 6.25 \times (0.9 \times 20.6) \times (\tfrac{2}{3} \times 6.25)$$
$$= M_s = 5.232 \times 10^5$$

or $\qquad p_{max} = 192$ kPa $< 3c_T (= 360$ kPa$)$

(6) $P_{av} = V_b/A = \dfrac{2.237 \times 10^5}{333.29} = 671.2$ kPa

For a well on soft rock,

$$P_0 = 2.88\ P_{av} = 1933.1 \text{ kPa}$$
$$P_1 = 1.10\ P_{av} = 738.3 \text{ kPa}$$
$$P_2 = 0.96\ P_{av} = 644.4 \text{ kPa}$$
$$P_3 = 0.63\ P_{av} = 422.9 \text{ kPa}$$

From Fig. 25.6, the pressure increment ΔP for edge pressure
(P_3) is 350 kN/m^2 when $D/B = 0.85$ and $H/V = 5.61\%$

Then the value of P_3' when there is an overturning moment is $P_3 + \Delta P = 422.9 + 350 = 772.9$ kPa

$$F \text{ w.r.t. } P_3 = q_f/P_3 = 2500/772.9 = 3.23$$

$$F \text{ w.r.t. } P_{av} = 2500/671.2 = 3.72$$

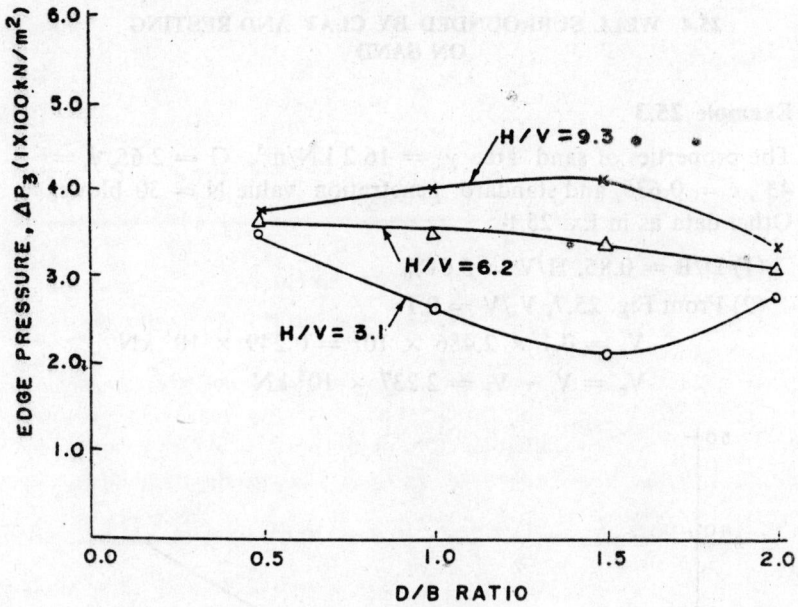

Fig. 25.6 Pressure increment versus D/B ratio (soft rock)

(7) $Q_{max} = 42.89$ kPa (same as in Ex. 25.1)

Assume $K_h = 1000$ kN/m³

$$\delta = 42.89/1000 = 0.043 \text{ m}$$

$$\theta = \delta/y = 0.043/17.53 = 1 \text{ in } 408$$

(8) *Depth of maximum moment* The moment is maximum at the level of zero shear (y_0 from top). The side pressure distribution is rectangular upto 11.28 m (Fig. 25.3).

$$p_{max} \times y_0 \times (0.9 \times B) = H$$
$$192 \times y_0 \times (0.9 \times 20.6) = 1.394 \times 10^4$$
$$y_0 = 3.92 \text{ m}$$

25.4 WELL SURROUNDED BY CLAY AND RESTING ON SAND

Example 25.3

The properties of sand are: $\gamma_d = 16.2 \text{ kN/m}^3$, $G = 2.65$, $\phi = 43°$, $e = 0.635$, and standard penetration value $N = 30$ blows. Other data as in Ex. 25.1.

(1) $D/B = 0.85$, $H/V = 5.61\%$

(2) From Fig. 25.7, $V_s/V = 0.1$

$$V_s = 0.1 \times 2.486 \times 10^5 = 0.249 \times 10^5 \text{ kN}$$
$$V_b = V - V_s = 2.237 \times 10^5 \text{ kN}$$

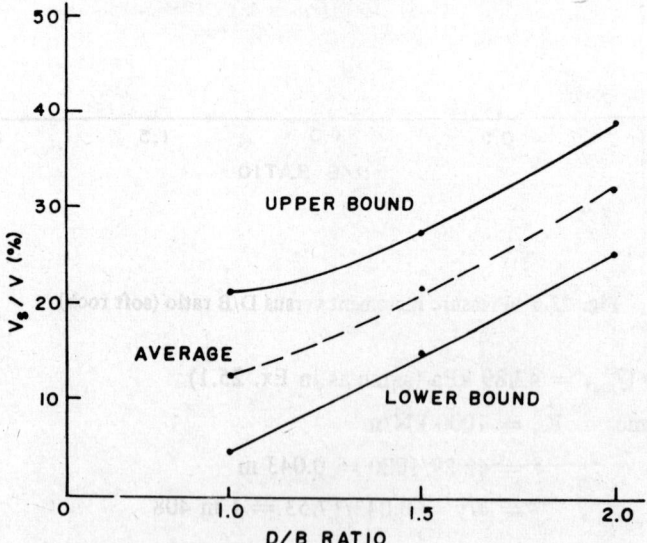

Fig. 25.7 Vertical load shared by sides versus D/B ratio (sand)

(3) From Fig. 25.8, $M_b/M_a = 0.37$

$$M_b = 0.37 \times 5.813 \times 10^5 = 2.151 \times 10^5 \text{ kN-m}$$
$$M_s = M_a - M_b = 3.662 \times 10^5 \text{ kN-m}$$

(4) The centre of rotation depends on D/B and H/V. For $D/B = 0.85$ and $H/V = 5.61\%$,

$$h_0/D = 0$$

where h_0 = height of centre of rotation above base. Centre of rotation is at the base.

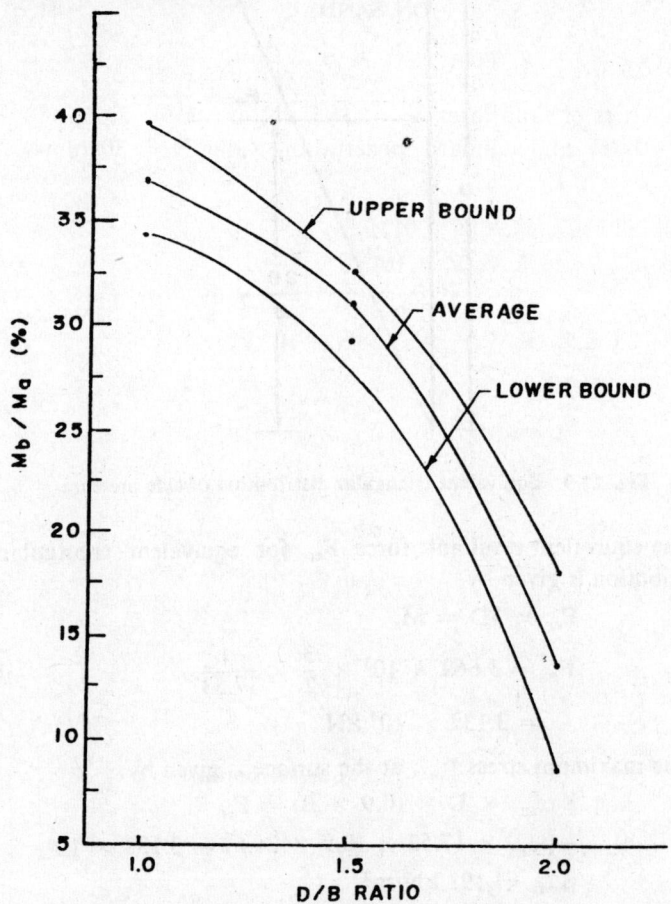

Fig. 25.8 Moment shared by base versus D/B ratio (sand)

(5) Find the ratio of moment above centre of rotation (M_{ab}) on the rear side to moment taken by sides M_s. This ratio depends on D/B ratio. But in this example, $h_0 = 0$, i.e. the centre of rotation is at the base and therefore $M_{ab} = M_s$.

The resultant pressure distribution (of soil reaction) is assumed triangular as shown in Fig. 25.9.

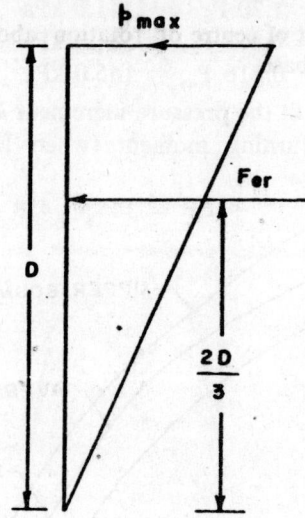

Fig. 25.9 Equivalent triangular distribution of side pressure

The equivalent resultant force F_{er} for equivalent triangular distribution is given by

$$F_{er} \times \tfrac{2}{3}D = M_s$$

or

$$F_{er} = 3.662 \times 10^5 \times \frac{3}{2} \times \frac{1}{17.53}$$

$$= 3.133 \times 10^4 \text{ kN}$$

The maximum stress P_{max} at the surface is given by:

$$\tfrac{1}{2} p_{max} \times D \times (0.9 \times B) = F_{er}$$

$$\tfrac{1}{2} p_{max} \times 17.53 \times (0.9 \times 20.6) = 3.133 \times 10^4$$

or

$$p_{max} = 193 \text{ kN/m}^2$$

$$> 3c_T (= 360 \text{ kN/m}^2), \text{ O.K.}$$

(6) $P_{av} = V_b/A = \dfrac{2.237 \times 10^5}{333.29} = 671.2 \text{ kPa}$

For a well resting on sand,

$$P_0 = 1.90\ P_{av} = 1275.3 \text{ kPa}$$
$$P_1 = 1.70\ P_{av} = 1141.0 \text{ kPa}$$
$$P_2 = 0.867\ P_{av} = 581.9 \text{ kPa}$$
$$P_3 = 0.216\ P_{av} = 145.0 \text{ kPa}$$

From Fig. 25.10, the pressure increment ΔP for edge pressure P_3 due to overturning moment (when D/B = 0.85, H/V = 5.61%) is 450 kPa

$$P_3' = P_3 + \Delta P = 145 + 450 = 595 \text{ kPa}$$

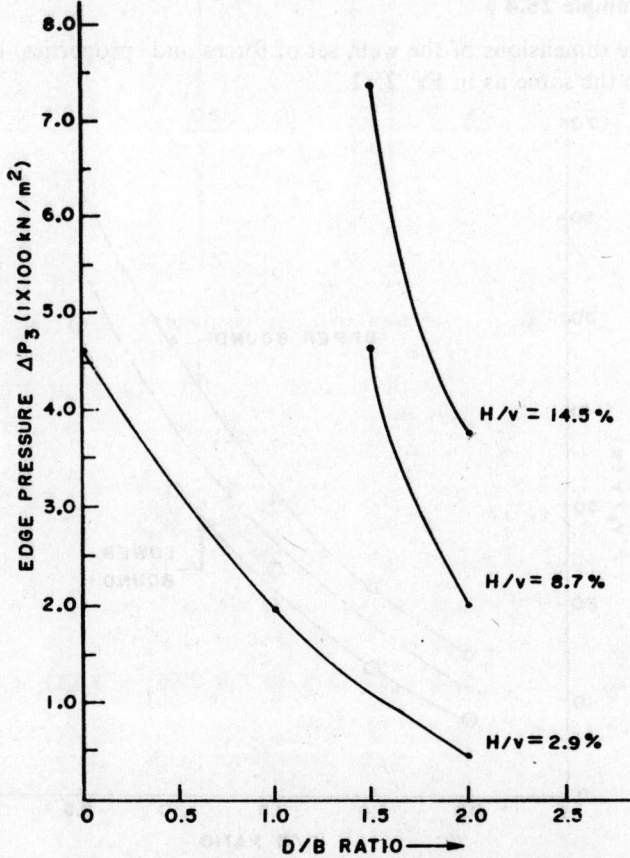

Fig. 25.10 Pressure increment versus D/B ratio (sand)

q_f of sand $= 5.4\, N^2B + 16\,(100 + N^2)\, D\ \text{kg/cm}^2$
$\qquad\qquad$ (IS: 3955 − 1967)
$\qquad\quad = (5.4 \times 30^2 \times 20.6) + 16\,(100 + 30^2) \times 17.53$
$\qquad\quad = 380596\ \text{kg/cm}^2 = 380.6\ \text{t/m}^2$
$\qquad\quad \simeq 3806\ \text{kPa}$

$$\text{F w.r.t. } P_3' = \frac{3806}{595} = 6.40 > 2.0 \qquad \text{(O.K.)}$$

$$\text{F w.r.t. } P_{av} = \frac{3806}{671.2} = 5.67$$

25.5 WELL SURROUNDED BY CLAY AND RESTING ON CLAY

Example 25.4

The dimensions of the well, set of forces and properties of clay are the same as in Ex. 25.1.

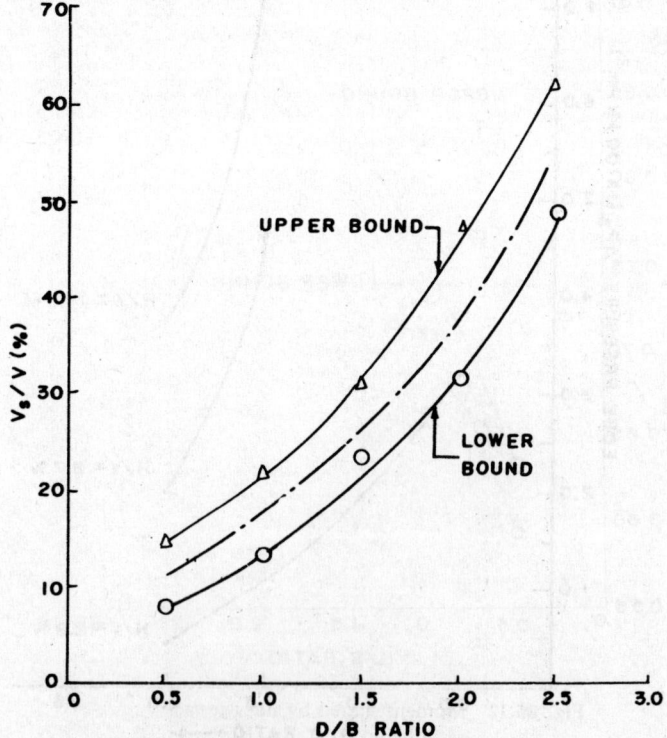

Fig. 25.11 Vertical load shared by sides versus D/B ratio (clay)

(1) $D/B = 0.85$, $H/V = 5.61\%$

(2) From Fig. 25.11, $V_s/V = 0.16$

$$V_s' = 0.16 \times 2.486 \times 10^5 = 0.398 \times 10^5 \text{ kN}$$
$$V_b = V - V_s = 2.088 \times 10^5 \text{ kN}$$

(3) From Fig. 25.12, $M_s/M_a = 0.68$

$$M_s = 0.68 \times 5.813 \times 10^5 = 3.953 \times 10^5 \text{ kN-m}$$
$$M_b = M_a - M_s = 1.86 \times 10^5 \text{ kN-m}$$

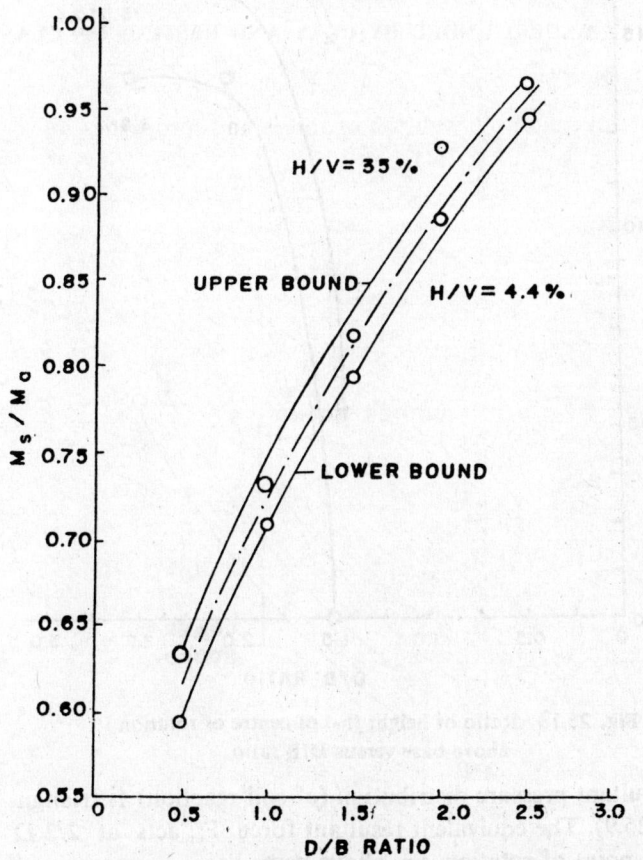

Fig. 25.12 Moment shared by sides versus D/B
ratio (clay)

(4) From Fig. 25.13, $h_0/D = 0$ for $D/B = 0.85$ and $H/V = 5.61\%$, i.e. the centre of rotation is at the base and $M_{ab} = M_s$.

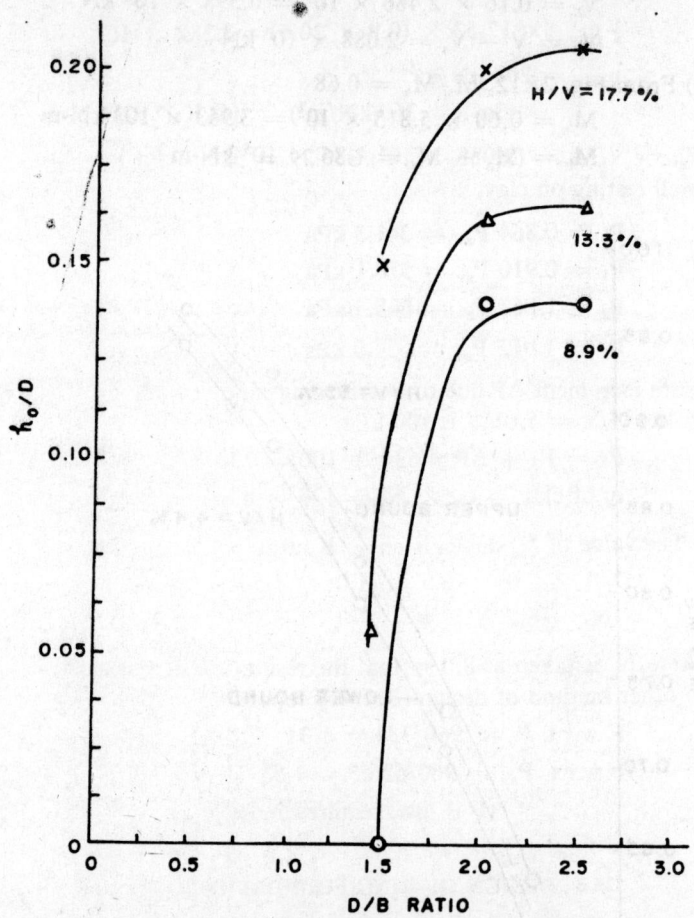

Fig. 25.13 Ratio of height (h_0) of centre of rotation above base versus D/B ratio

The resultant pressure distribution (of soil reaction) is triangular (Fig. 25.9). The equivalent resultant force F_{er} acts at $2/3$ D above the point of rotation, i.e. above base.

$$F_{er} = M_s \bigg/ \frac{2D}{3} = (3.953 \times 10^5 \times 3)/(2 \times 17.53)$$
$$= 3.382 \times 10^4 \text{ kN}$$

The maximum stress p_{max} at surface is given by:

$$\tfrac{1}{2} p_{max} \times D \times (0.9 \times B) = F_{er}$$

$$\tfrac{1}{2} p_{max} \times 17.53 \times (0.9 \times 20.6) = 3.382 \times 10^4$$

or

$$p_{max} = 208 \text{ kN/m}^2$$

$$< 3c_T \ (= 360 \text{ kN/m}^2) \quad (\text{O.K.})$$

(6) $P_{av} = V_b/A = (2.088 \times 10^5)/333.29 = 626.5$ kPa

For a well resting on clay,

$$P_0 = 0.864 \ P_{av} = 541.3 \text{ kPa}$$

$$P_1 = 0.910 \ P_{av} = 570.1 \text{ kPa}$$

$$P_2 = 1.067 \ P_{av} = 668.5 \text{ kPa}$$

$$P_3 = 1.012 \ P_{av} = 634.0 \text{ kPa}$$

Pressure increment ΔP due to overturning moment when $D/B = 0.85$ and $H/V = 5.61\%$ is 100 kPa.

$$P_3 = P_3 + \Delta P = 634 + 100 = 734 \text{ kPa}$$

$$q_f \text{ of clay} = c_T \times N_c = 120 \times 8 = 960 \text{ kPa}$$

Note. The value of N_c depends on D/B ratio.

$D/B =$	0	0.5	1.0	1.5	2.0	2.5
$N_c =$	6.2	7.1	7.7	8.1	8.4	8.6

But here N_c is taken as 8.0 so that the results could be compared with other method of design (Ref. IIT reports).

$$\text{F w.r.t. } P_3 = 960/734 = 1.31$$

$$\text{F w.r.t. } P_{av} = 960/626.5 = 1.53$$

(F is low, requires redesign)

25.6   DESIGN OF RING FOUNDATION

Ring foundations are often used for tower type structures. Their settlement and inclination can be calculated by the method suggested by Egorov et al[*]. The method is applicable to net foundation pressures in the range of 200 to 400 kPa.

---

[*]Egorov, K.E., Konovalov, P.A. Kitaykina, O.V., Salnikov, L.F., and Zinovyev, A.V. 1977, Soil deformation under circular footing. Proc. 9th ICSMFE, Tokyo, Vol. 1, pp. 489-492.

If r_1 is the inner radius of the ring foundation and r_2 is the outer radius and q is the *net* foundation pressure, the significant depth Z of compressible layer below the foundation is first calculated as:

$$Z = r_2 \text{ (for sands)} \tag{25.1}$$

$$Z = 2r_2/3 \text{ (for clays)} \tag{25.2}$$

For mixed soils, Z is suitably interpolated.

The compressible layer Z is suitably divided into a number of small, homogeneous sub-layers (say 3 layers) and the average settlement S of the foundation is calculated from the following relation:

$$S = \frac{2r_2 qM}{m_e}\left[\frac{K_1}{E_1} + \frac{K_2 - K_1}{E_2} + \frac{K_3 - K_2}{E_3} + \ldots\right] \tag{25.3}$$

where M is a coefficient accounting for concentration of stress in the sub-layer (Table 25.1), m_e is a coefficient depending on foun-

Table 25.1

VALUES OF COEFFICIENT M

$2Z/r_2$	M
$0 - \leqslant 0.5$	1.5
$0.5 - \leqslant 1$	1.4
$1 - \leqslant 2$	1.3
$2 - \leqslant 3$	1.2
$3 - \leqslant 5$	1.1
Greater than 5	1.0

dation width $(r_2 - r_1)$ (Table 25.2) and K is dimensionless coefficient for the sub-layer (Fig. 25.14). E is the (total) deformation modulus of the sublayer.

Table 25.2

VALUES OF COEFFICIENT m_e

$(r_2 - r_1)$ (m)	m_e
$5 - \leqslant 10$	1.2
$10 - \leqslant 15$	1.35
Greater than 15	1.5

The coefficient K is obtained from Fig. 25.14 in which 'n' is the radii ratio (r_1/r_2) and the factor 'm' is obtained for the sub-layer as:

$$m = h_1/r_2, \ h_2/r_2, \ etc. \tag{25.4}$$

where h_1, h_2 = depths of bases of the respective sub-layers below the foundation

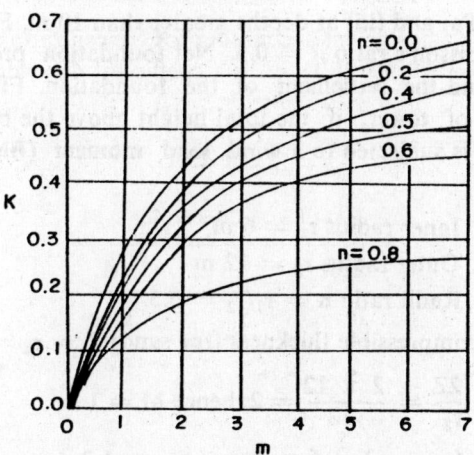

Fig. 25.14 Value of coefficient K (Egorov)

To find the inclination (tilt) of the foundation if it is subjected to a moment (BM), e.g. due to wind pressure on the tower, the weighed average of the deformation modulus E_a is first calculated. The inclination θ is then given by:

$$\theta = \frac{1 - \mu^2}{m_e \, E_a} \, K_m \, \frac{BM}{r_2^3} \tag{25.5}$$

where μ = Poisson's ratio

        K_m = dimensionless coefficient (Table 25.3)

Table 25.3

VALUES OF COEFFICIENT K_m (H = HEIGHT OF TOWER)

H/r_2	0.25	0.5	1.0	2.0	> 2.0
K_m	0.26	0.43	0.63	0.74	0.75

Example 25.5

The ring foundation of a water tower is 12 m in internal diameter and 24 m in external diameter. The base of the ring is at a depth of 2 m below the ground surface in a sandy soil. The (total) deformation modulus of sand is estimated as follows:

(i) 0 to 6 m depth, E_1 = 24,000 kPa (ii) 6 m to 10 m depth, E_2 = 30,000 kPa, and (iii) at depths greater than 10 m, E_3 = 32,000 kPa. The Poisson's ratio μ = 0.3. Net foundation pressure q = 200 kPa. Find the settlement of the foundation. Find also the inclination of tower, if the total height above the base level is 18 m and it is subjected to a wind load moment (BM) of 200 kN-m.

Inner radius r_1 = 6 m

Outer radius r_2 = 12 m

Radii ratio n = r_1/r_2 = 0.5

Significant compressible thickness (for sand) Z = r_2 = 12 m

$$\frac{2Z}{r_2} = \frac{2 \times 12}{12} = 2, \text{ hence } M = 1.3$$

$$(r_2 - r_1) = 6 \text{ m, hence } m_e = 1.2$$

The compressible layer 12 m thick, is divided into three sublayers, each 4 m thick, as shown in Fig. 25.15.

The 'm' value for each layer is calculated and the value of coefficient K is read from Fig. 25.14 for n = 0.5.

First layer $m_1 = h_1/r_2$ = 4/12 = 0.33, $K_1 \approx 0.075$

Second layer $m_2 = h_2/r_2$ = 8/12 = 0.66, $K_2 \approx 0.15$

Third layer $m_3 = h_3/r_2$ = 12/12 = 1.0, $K_3 \approx 0.18$

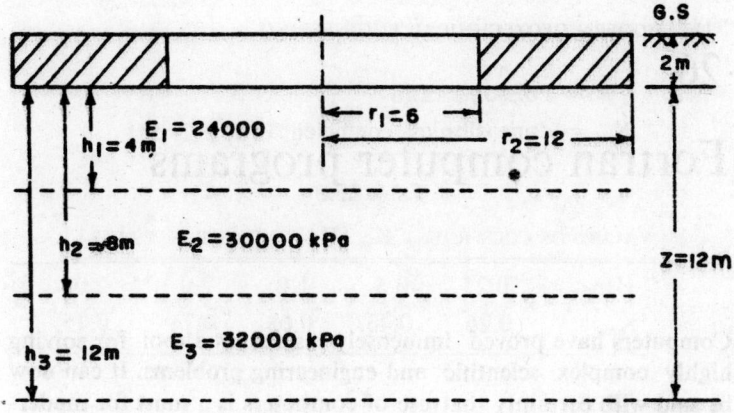

Fig. 25.15 Ring foundation and the division of compressible thickness

The average settlement S is given by:

$$S = \frac{2r_2qM}{m_e}\left[\frac{K_1}{E_1} + \frac{K_2 - K_1}{E_2} + \frac{K_3 - K_2}{E_3}\right]$$

$$= \frac{2 \times 12 \times 200 \times 1.3}{1.2}$$

$$\times \left[\frac{0.075}{24.000} + \frac{0.075}{30,000} + \frac{0.03}{32,000}\right]$$

$$= 0.034 \text{ m} = 3.4 \text{ cm}$$

$$H/r_2 = 18/12 = 1.5$$

$$K_m \text{ (Table 25.3)} = \frac{0.63 + 0.74}{2} = 0.685$$

Weighted average value of E,

$$E_a = \frac{(4 \times 24,000) + (4 \times 30,000) + (4 \times 32,000)}{12}$$

$$= 28,667 \text{ kPa}$$

Wind load moment (BM) = 2000 kN-m

$$\theta = \frac{1 - \mu^2}{m_e E_a} K_m \frac{BM}{(r_2)^3}$$

$$= \frac{1 - (0.3)^2}{1.2 \times 28667} \times 0.685 \times \frac{2000}{(12)^3}$$

$$= 2.09 \times 10^5 \text{ radians}$$

26

Fortran computer programs

Computers have proved immensely useful as a tool for solving highly complex scientific and engineering problems. It can now be said with certainty that use of computers is a must for modernising scientific study program. The list of various fields where computers are used now-a-days is so vast that it is rather impossible to mention all of them. In scientific and engineering fields, many designs which required guessed solutions till now due to very time-consuming repetitive complex computations are now being solved through computer without any delay and with more accuracy. The computer has entered in almost every field of modern society.

This chapter is meant for illustrating the use of computers for solution of geotechnical engineering problems. It may be noted, however, that a computer cannot provide the solution to any problem until required algorithm has been fed to it through a program prepared in a language acceptable to it. Consequently, the technique of converting a few geotechnical problems into the computer language has been illustrated and its application shown through worked examples.

Example 26.1 (Basic properties)

26.1 The mass M of an undisturbed sample is 1250 g and its volume V is 630 cm^3. On oven drying the sample weighs 1102 g (M_s). The average relative density (specific gravity) G of soil is 2.68. Find the bulk density ρ, natural moisture content w, dry density ρ_d, void ratio e, degree of saturation S_r and the air content n_a. What will be the density and unit weight on full saturation? (Ref. Ex. 2.3).

Example 26.2 (Permeability)

26.2 A sand deposit is made up of three horizontal layers, each 3 m thick. The permeability of the top and bottom layers is 2×10^{-4} cm/s and that of the middle layer is 3.2×10^{-2} cm/s. Find the equivalent permeabilities in the horizontal and vertical directions and their ratio (Ref. Ex. 6.22)

Example 26.3 (Consolidation)

26.3 A specimen of normally consolidated undisturbed clay, 75 mm in diameter and 24 mm high, is tested in a consolidometer under pressures as given in Table 14.2. The equilibrium height at each stage of loading and void ratio are also recorded in the table. Calculate the values of coefficient of volume change m_v. Relative density of the soil $G = 2.68$ and dry mass of specimen at end of test is $M_s = 120.89$ g. (Ref. Ex. 14.2)

Example 26.4 (Bearing capacity)

26.4 A 2.5 m square footing is located in dense sand, at a depth of 1.5 m, the shear strength parameters being $c' = 0$, $\phi' = 38°$. Determine the ultimate bearing capacity for the following water table positions: (a) above ground surface ($- 0.5$ m), (b) at ground surface (0.0 m), (c) at 1 m below ground surface, (d) at footing level, (e) at 0.5 m below the footing, (f) at a depth equal to B below the footing, and (g) at a depth greater than B below the footing. The moist unit weight of sand above the water table is 18 kN/m^3 and the saturated unit weight is 20 kN/m.^{3} For $\phi = 38°$, $N_c = 61.4$, $N_q = 48.9$ and $N_y = 58.9$ (Ref. Ex. 15.4)

Example 26.1

```
C          BASIC PROPERTIES OF SOIL.
           IMPLICIT REAL (A-Z)
           OPEN(UNIT=66,FILE='SOIL.DAT',
          1STATUS='OLD')
           OPEN(UNIT=63,FILE='ROIL.DAT',
          1STATUS='NEW')
           READ(66,*)M,V,MS,G,RHOW
           WRITE(63,10)M,V,MS,G,RHOW
10         FORMAT(///1X,'Given data -'////1X,
          1'Bulk mass',15X,'=',F7.6,
          2' tonnes'/1X,'Bulk volume'
          3,13X,'=',F7.6,'  cum.'/1X,
          4'Dry weight',14X,'=',F7.6,
          5' tonnes'/1X,'Sp.Gr.',15X,'=',
          6F7.2/1X,'Density of water',3X,
          7'=',F7.2,' t/cum.'////)
           RHO=M/V
           w=(M-MS)/MS
           w100=w*100.
           RHOD=RHO/(1+w)
           E=(G*RHOW/RHOD)-1
           SR=w*G/E
           NA=(1-SR)*E/(1+E)
           RHOST=(G+E)/(1+E)*RHOW
           GMST=RHOST*9.8
           WRITE(63,20)RHO,w100,RHOD,E,SR,
          1NA,RHOST,GMST
20         FORMAT(1X,'Results-'////1X,'Bulk',
          1' Density',12X,'=',F7.2,2X,
          2't/cum.'/1X,'Moisture content'
          3,3X,'=',F7.2,'  %'/1X,'Dry',
          4' density',13X,'=',F7.2,2X,
          5't/cum.'/1X,'Void ratio',14X,
          6'=',F7.2/1X,'Degree of satura',
          7'tion',4X,'=',F7.2/1X,'Air co',
          8'ntent',13X,'=',F7.2/1X,'Satu',
          9'rated density',7X,'=',F7.2,
          1'  t/cum.'/1X,'Saturated unit',
          2' weight',3X,'=',F7.2,
```

```
    3'  kN/cum.')
    STOP
    END
```

Given data -

Bulk mass	=.001250	tonnes
Bulk volume	=.000630	cum.
Dry weight	=.001102	tonnes
Sp.Gr.	= 2.68	
Density of water	= 1.00	t/cum.

Results-

Bulk Density	= 1.98	t/cum.
Moisture content	= 13.43	%
Dry density	= 1.75	t/cum.
Void ratio	= 0.53	
Degree of saturation	= 0.68	
Air content	= 0.11	
Saturated density	= 2.10	t/cum.
Saturated unit weight	= 20.55	kN/cum.

Example 26.2

```
C         PERMEABILITY OF LAYERED SOIL IN
C         HORIZONTAL & VERTICAL DIRECTIONS.
          DIMENSION AK(10),H(10)
          OPEN(UNIT=7,FILE='PER',
         1STATUS='OLD')
          OPEN(UNIT=8,FILE='REP',
         1STATUS='NEW')
          READ(7,*) N
          READ(7,*)(AK(I),H(I),I=1,N)
          WRITE(8,10)(I,H(I),AK(I),I=1,N)
10        FORMAT(1X,'Given data -'///1X,
         1'Layer   Thickness   Permeability'
         2/1X,' No.   of  layer',4X,
         3'(m/sec.)'/8X,'(m.)'///3(2X,
         4I2,6X,F5.2,7X,F6.4/))
          TH=0.0
          DO 20 I=1,N
20        TH=TH+H(I)
          AKX=0.0
          HBK=0.0
          DO 30 I=1,N
          AKX=AKX+AK(I)*H(I)/TH
30        HBK=HBK+H(I)/AK(I)
          AKZ=TH/HBK
          AKXBKZ=AKX/AKZ
          WRITE(8,40)AKX,AKZ,AKXBKZ
40        FORMAT(10X,'Results -'///10X,
         1'kx   =',F10.4,'cm/sec'/10X,
         2'kz   =',F10.4,'cm/sec'/10X,
         3'kx/kz=',F10.4)
          CLOSE(UNIT=7,STATUS='SAVE')
          CLOSE(UNIT=8,STATUS='SAVE')
          STOP
          END
```

Given data -

Layer No.	Thickness of layer (m.)	Permeability (m/sec.)
1	3.00	0.0002
2	3.00	0.0020
3	3.00	0.0002

Results -

k_x = 0.0108 cm/sec

k_z = 0.0003 cm/sec

k_x/k_z = 36.1125

Example 26.3

```
C         COEFFICIENT OF VOLUME CHANGE
C         DETERMINATION.
          IMPLICIT REAL (M)
          DIMENSION SIG(10),H(10),E(10),
         1DSIG(10),DH(10),DE(10),MVH(10),
         2MVE(10),LINE(40)
          OPEN(UNIT=33,FILE='CMV.DAT',
         1STATUS='OLD')
          OPEN(UNIT=44,FILE='RMV.DAT',
         1STATUS='NEW')
          DATA LINE/40*'_'/
5         FORMAT(1X,40A1)
          READ(33,*)N
          READ(33,*)(SIG(I),H(I),E(I),
         1I=1,N)
          DO 10 I=1,N-1
          DH(I)=H(I)-H(I+1)
          DE(I)=E(I)-E(I+1)
          DSIG(I)=SIG(I+1)-SIG(I)
          MVH(I)=DH(I)/(H(I)*DSIG(I))
10        MVE(I)=DE(I)/((1.+E(I))*DSIG(I))
          WRITE(44,15)LINE
15        FORMAT(1X,'Results -'//1X,40A1
         1/1X,'SIGMA',7X,'H',9X,'DH'
         2,9X,'MV'/1X,'(kPa)',5X'(mm.)'
         3,5X,'(mm.)',5X,'(sqm/kN)')
          WRITE(44,5)LINE
          DO 20 I=1,N-1
          WRITE(44,25)SIG(I),H(I)
20        WRITE(44,30)DH(I),MVH(I)
          WRITE(44,25)SIG(N),H(N)
          WRITE(44,5)LINE
25        FORMAT(1X,F5.1,5X,F5.2)
30        FORMAT(21X,F5.2,5X,F9.7)
          WRITE(44,35)LINE
35        FORMAT(////1X,40A1/1X,'SIGMA',7X,
         1'E',5X,'DE',9X,'MV'/1X,'(kPa)'
         2,25X,'(sqm/kN)')
          WRITE(44,5)LINE
```

```
          DO 40 I=1,N-1
          WRITE(44,45)SIG(I),E(I)
40        WRITE(44,50)DE(I),MVE(I)
          WRITE(44,45)SIG(N),E(N)
          WRITE(44,5)LINE
45        FORMAT(1X,F5.1,5X,F5.3)
50        FORMAT(21X,F5.3,5X,F9.7)
          CLOSE(UNIT=33,STATUS='SAVE')
          CLOSE(UNIT=44,STATUS='SAVE')
          STOP
          END
```

Results -

SIGMA (kPa)	H (mm.)	DH (mm.)	MV (scm/kN)
100.0	19.55		
		0.87	0.0004450
200.0	18.68		
		1.99	0.0005327
400.0	16.69		
		1.89	0.0002831
800.0	14.80		

SIGMA (kPa)	E	DE	MV (scm/kN)
100.0	0.915		
		0.085	0.0004439
200.0	0.830		
		0.195	0.0005328
400.0	0.635		
		0.186	0.0002844
800.0	0.449		

Example 26.4

```
C          BEARING CAPACITY OF SOIL WITH
C          WATER TABLE AT DIFFERENT DEPTHS
C          BY TERZAGHI'S FORMULAE.
           DIMENSION DW(10)
           OPEN(UNIT=11,FILE='TRZ.DAT',
          1STATUS='OLD')
           OPEN(UNIT=12,FILE='RTZ.DAT',
          1STATUS='NEW')
           READ(11,*)N,B,D,COH,PHI,ANC,ANQ,
          1ANGM,GMBD,GMST,GMW
           WRITE(12,25)
20         FORMAT(23X,' (water table'/25X,
          1'above ground'/25X,'surface)'/)
25         FORMAT(///1X,'Results -'////1X,
          1'Case   Depth of water',' Ult',
          2'imate bearing'/1X,'no.',3X,
          3'table below',5X,'capacity of',
          4' soil'/7X,'ground surface',7X,
          5'(kPa)'/12X,'(m.)')
           DO 70 I=1,N
           READ (11,*)DW(I)
           DWA=DW(I)
           IF(DW(I).LT.0.0) DWA=0.0
           IF (DWA.GE.(D+B)) GO TO 55
           GMSUB=GMST-GMW
           IF(DWA-D) 30,45,50
30         SGSUB=(DWA*GMBD+(D-DWA)*
          1(GMST-GMW))/D
           IF(DWA.GT.0.0)GO TO 40
35         QF=COH*ANC+SGSUB*D*(ANQ-1.)+0.4*
          1GMSUB*B*ANGM+GMST*D
           GO TO 60
40         QF=COH*ANC+SGSUB*D*(ANQ-1.)+0.4*
          1GMSUB*B*ANGM+DWA*GMBD+(D-DWA)*
          2GMST
           GO TO 60
45         QF=COH*ANC+GMBD*D*(ANQ-1.)+0.4*
          1GMSUB*B*ANGM+GMBD*D
           GO TO 60
```

```
50        GMAV=((DWA-D)*GMBD+(B+D-DWA)*
        1GMSLB)/B
          GF=CCH*ANC+GMBD*D*ANG+0.4*GMAV*.
        1B*ANGM
          GO TO 60
55        GF=CCH*ANC+GMBD*D*ANG+0.4*B
        1*GMBD*ANGM
60        WRITE(12,65)I,DW(I),GF
65        FORMAT(2X,I2,6X,F7.2,5X,F10.3/)
          IF(DW(I).LT.0.0) WRITE(12,20)
70        CONTINUE
          CLOSE(UNIT=11,STATUS='SAVE')
          CLOSE(UNIT=12,STATUS='SAVE')
          STOP
          END
```

Results -

Case no.	Depth of water table below ground surface (m.)	Ultimate bearing capacity of soil (kPa)
1	-0.50	1363.650
		(water table above ground surface)
2	0.00	1363.650
3	1.00	1735.270
4	1.50	1921.090
5	2.00	2012.964
6	4.00	2390.500
7	4.50	2390.500

27
Soil engineering experiments

EXPERIMENT 1

MOISTURE CONTENT DETERMINATION
1.A OVEN-DRYING METHOD

Object

To determine the moisture content of a soil sample by the oven-drying method. (IS: 2720-2)

Apparatus

Drying oven, controlled at 105 to 110 °C. Sample containers, for fine-grained soils: heat resistant glass weighing bottles, 3.5-cm in diameter, 7.5-cm high with ground glass stoppers, or non-corrosive metal containers 5-cm in diameter and 2-2.5-cm high with tight fitting lids; for coarse-grained soils: cylindrical metal tins of about 0.5-kg capacity with air tight lids. Balance, accurate to 0.01 g for fine-grained soils, and 0.1 g for coarse-grained soils. Desiccator with fresh calcium chloride or, preferably, anhydrous silica gel.

Procedure

1. Place a sample of soil (about 30-50 g of fine-grained soils, 250-300 g of coarse-grained soils) in an appropriate container and weigh along with its lid/stopper (M_1).

2. Place the container, with the lid (or stopper) removed, in the oven and dry for 16 to 24 hours (until constant weight) at a controlled temperature of 105 to 110 °C.

3. After drying, remove the container from the oven, replace the lid and cool the container and contents in a desiccator and weigh (M_2).

If a desiccatar is not available, the dry sample should be cooled in its container with the lid in place and weighed as soon as it is cool enough to handle.

4. Clean and dry the container and weigh (M_3), if its weight is not known.

Calculation

The moisture content w is given by:

$$w = \frac{M_1 - M_2}{M_2 - M_3} \times 100 \text{ (per cent)}$$

Precaution

Use a well designed oven as the temperature variations may lead to errors due to over heating. Higher drying temperatures may burn the organic material and change the composition of mineral grains. Soils containing gypsum should be dried at not more than $80\,^{\circ}C$ and possibly for a longer time. (Gypsum loses its water of crystallisation at high temperatures, which will affect the water content determination.)

Observation Sheet 1.A

MOISTURE CONTENT BY OVEN-DRYING

Test No.			1	2
1. Container No.			25	
2. Mass cont. + wet soil	(M_1)	(g)	75.88	
3. Mass cont. + dry soil	(M_2)	(g)	66.26	
4. Mass water present	(M_w)	(g)	9.62	
5. Mass container empty	(M_3)	(g)	21.60	
6. Mass dry soil	(M_s)	(g)	44.66	
7. Moisture content	(w)	(%)	21.5	

I.B PYCNOMETER METHOD

Object

To determine the moisture content of a soil sample using a pycnometer.

Apparatus

Pycnometer. Balance accurate to 1 g. Glass rod about 30-cm long.

Procedure

1. Weigh a clean and dry pycnometer with its cap (M_1).

2. Place a sample of wet 'soil [(200 to 500 g) into the pycnometer weigh (M_2).

3. Remove the screw-top and add water to the soil (pycnometer about half full). Mix thoroughly with glass rod to remove entrapped air. Continue stirring and add more water. Finally replace the screw-top and fill the pycnometer flush with the hole in the conical top. Remove any remaining air by shaking or rolling the pycnometer on the work table holding one finger over the hole at the top. If forth collects under the top, remove it carefully and fill with water flush to the top. Dry the outside and weigh (M_3).

4. Empty the pycnometer, clean it and refill with clean water flush with the top hole. Wipe dry the outside and weigh (M_4).

Calculation

The moisture content w of the soil sample is given by:

$$w = \left[\frac{M_2 - M_1}{M_3 - M_4} \times \frac{(G - 1)}{G} - 1 \right] \times 100 \text{ (per cent)}$$

Precaution

Care should be taken in removing entrapped air, as the accuracy of the results depend upon the thoroughness with which air is removed. If a vacuum pump or a water-jet pump is available, it should be used for removing air bubbles. The screw-top and the bottle should each have an engraved mark so that the top may be screwed on to the same extent every time keeping the capacity constant.

Observation Sheet 1B

MOISTURE CONTENT BY PYCNOMETER

Specific gravity of soil = 2.67

Test No.		1	2
1. Mass pyc. empty (M_1)	(g)	548	
2. Mass pyc. + wet soil (M_2)	(g)	758	
3. Mass pyc. + soil + water (M_3)	(g)	1560	
4. Mass pyc. + water (M_4)	(g)	1437	
5. Moisture content (w)	(%)	6.8	

EXPERIMENT 2

SPECIFIC GRAVITY DETERMINATION

2.A DENSITY BOTTLE METHOD

Object

To determine the specific gravity of soil particles using a density bottle. (I S: 2720-3 Sec. 1)

Apparatus

Density bottle of 50 or 100-ml capacity with perforated stopper. Balance accurate to 0.001 g. Drying oven. Vacuum desiccator. Thermometer.

Procedure

1. Weigh the clean, dry and cool density bottle with its stopper (M_1).

2. Place about 10 g (in 50-ml bottle), 20 g (in 100-ml bottle), of ovendried and pulverised sample of soil and weigh (M_2). If air-dried soil is used, dry the contents of the bottle at the end of the test to get weight of dried soil.

3. Cover the soil with distilled water and leave for a suitable soaking period, (which should be for at least 12 hours for clays).

4. Add more distilled water until the bottle is about half full. Remove the entrapped air by subjecting the contents to a partial vacuum (not exceeding 10-cm of mercury). This may be done by the use of a vacuum desiccator or a bell jar or by direct connection to an aspirator or vacuum pump. Apply the vacuum for a period of 15 to 30 minutes.

5. Fill the bottle completely with distilled water, wipe dry and weigh (M_3). Measure the temperature (T_t) of the contents.

6. Empty the bottle, rinse it thoroughly, refill with distilled water at the same test-temperature (T_t), wipe dry on the outside and weigh (M_4).

7. Repeat the test twice more.

Calculation

The specific gravity G of soil based on water at test temperature T_t is given by:

$$G \text{ (at } T_t) = \frac{M_2 - M_1}{(M_2 - M_1) - (M_3 - M_4)}$$

The reported result is based on water at 27 °C which in given by:

$$G \text{ (at 27 °C)} = G \text{ (at } T_t) \times \frac{\text{Sp. gr. of water at } T_t, \text{°C}}{\text{Sp. gr. of water at 27 °C}}$$

If the reported specific gravity is to be based on water at 4 °C, it is given by:

$$G \text{ (at 4 °C)} = G \text{ (at } T_t) \times \text{Sp. gr. of water at } T_t \text{ °C}$$

Precaution

Be careful in removing all the entrapped air, otherwise the results will be low. Partial vacuum should be applied at a slow rate (Note). Be careful in weighings and use the same balance throughout the test.

Note— The author has witnessed an accident (1957) where a vacuum desiccator with a completely full 50-ml density bottle inside burst into pieces soon after applying a heavy vacuum at the full speed of a powerfull vacuum pump.

OBSERVATION SHEET 2.A

SPECIFIC GRAVITY BY DENSITY BOTTLE

(i) Test temperature T_t = 28 °C

(ii) Sp. gr. of water at T_t = 0.99626

(iii) Sp. gr. of water at 27 °C = 0.99654

(vi) Correction factor K $= \dfrac{0.99626}{0.99654} = 0.9997$

Test No.	1	2	3
1. Density bottle No.	5		
2. Mass density bottle (M_1) (g)	16.705		
3. Mass bottle + dry soil (M_2) (g)	25.573		
4. Mass bottle+soil+water (M_3) (g)	74.215		
5. Mass bottle + water (M_4) (g)	68.605		
6. Sp. gr. of soil at T_t °C	2.722		
7. Sp. gr. of soil at 27 °C	2.72		

2.B FLASK METHOD

This method is similar to the density bottle method. A measuring flask of 500 or 250-ml capacity is used for the test. The flask should have one graduation mark at the 500/250-ml level and an adaptor for fitting to the vacuum line (e.g , a water jet pump). About 80—100 g of soil is taken for the test, which may be oven-dried or at natural water content. If soil at natural water content is used, the dry mass of soil solids ($M_s = M_2 - M_1$) is obtained by drying and weighing the contents of the flask after taking the weight observation of flask + soil + water (M_3, Ref. Expt. 2.A). It is preferable with clay soils that they are first mechanically dispersed (15 minutes of stirring with a high speed stirrer) and then transferred to the flask for further observations. Weighings are accurate to 0.01 g in case of 250-ml flask and to 0.05 g in case of 500-ml flask.

OBSERVATION SHEET 2.B

SPECIFIC GRAVITY BY MEASURING FLASK

Capacity of flask = 250 ml (approx).

Test No.		1	2
1. Flask No.		3	
2. Mass flask + dry soil (M_2)	(g)	137.97	
3. Mass flask empty (M_1)	(g)	67.97	
4. Mass dry soil ($M_2 - M_1$)	(g)	70.00	
5. Mass flask + soil + water (M_3)	(g)	360.54	
6. Mass flask + water (M_4)	(g)	316.74	
7. ($M_3 - M_4$)		43.82	
8. ($M_2 - M_1$) − ($M_3 - M_4$)		26.18	
9. Specific gravity of soil $= \dfrac{(4)}{(8)}$		2.68	

2.C   PYCNOMETER METHOD

Object

To determine the specific gravity of soil particles using a pycnometer.

Apparatus

Pycnometer fitted with a conical brass screw-top with a sharp edged hole, 6-mm in diameter, at its apex. Balance, accurate to 1 g. Drying oven. Glass rod.

Procedure

Proceed as in Expt. 1. B taking about 300 g of oven-dried sample of soil in place of wet soil. Repeat the test twice more and obtain the average of the results.

Calculation

The specific gravity is given by:

$$G = \frac{M_2 - M_1}{(M_2 - M_1) - (M_3 - M_4)}$$

OBSERVATION SHEET 2.C

SPECIFIC GRAVITY BY PYCNOMETER

Test No.		1	2
1. Pycnometer No.		1	
2. Mass pyc. + dry soil (M_2)	(g)	848	
3. Mass pyc. empty (M_1)	(g)	548	
4. Mass dry soil ($M_2 - M_1$)	(g)	300	
5. Mass pyc. + soil + water (M_3)	(g)	1624	
6. Mass pyc. + water (M_4)	(g)	1436	
7. ($M_3 - M_4$)		188	
8. ($M_2 - M_1$) − ($M_3 - M_4$)		112	
9. Specific gravity of soil $= \dfrac{(4)}{(8)}$		2.68	

EXPERIMENT 3

PARTICLE SIZE DETERMINATION

3.A COMPLETE MECHANICAL ANALYSIS

Object

To determine the particle-size distribution of a soil containing particles ranging from gravel to clay by mechanical analysis.

Apparatus

(a) Set of IS sieves: 37.5, 19, 9.5, 4.75, 2.0 and 1.0 mm sieves, and 600, 425, 300, 212, 150 and 75 (or 63) microu sieves with lids and receivers. Balances accurate to 1 g, 0.1 g and 0.01 g respectively. Brush for cleaning sieves. Basins and trays. Drying oven. Mortar with rubber pestle or wooden mallet. Polyethylene wash bottle, 500-ml capacity. Stop watch. Distilled or demineralised water.

(b) Density hydrometer, calibrated at 27 °C to read in g/ml, range 1.030 to 0.995 g/ml. Two 1000-ml measuring cylinders without lip (also called the sedimentation cylinders), about 7-cm in diameter and 33-cm high marked at 1000-ml volume. Dispersing agent solution (distilled water containing 38 g of sodium hexametaphosphate 'calgon' plus 12 g of sodium carbonate per litre of solution). Thermometer. High speed mechanical stirrer (5000 RPM).

(c) If pipette method is used: 10-ml sampling pipette fitted on a control carriage. Glass weighing bottles. Balance, accurate to 0.001 g. One 1000-ml measuring cylinder without lip (the author prefers it over a 500-ml boiling tube). Constant temperature (27 °C) bath. Rest as above in (a) and (b).

Procedure

(a) Separation of Gravel Fraction

1. Oven-dry a soil sample. Break up the soil particle aggregations using the mortar and rubber pestle. Reduce in quantity by quartering or pouring through a riffle box.

2. Weigh a representative sample accurate to 1 g or 0.1 g depending on the quantity (Note). Keep it in a basin, cover with water and allow it to soak for some time. Puddle the sample. Transfer the soil slurry to the 2-mm sieve and wash with a jet of water so that the

fraction retained on the sieve (gravel and coarse sand fraction) is clean and, all fine material, such as silt or clay, is removed with wash water. When the sample is larger than that can be handled at one time on the sieve, wash the sample in portions.

3. Transfer the washed material retained on the sieve to a container and dry in the oven. Dry also the wash water containing the fraction passing the 2-mm sieve. The fraction retained on the sieve is termed *gravel and coarse sand fraction* and is used for coarse analysis. The fraction passing is used for fine analysis and sedimentation test.

Note. The quantity of sample to be taken for the test depends upon the maximum size of particle present. The quantity should be such that the size of the portion retained on the 2-mm sieve is according to the following schedule (Table 1):

Table 1

Nom. dia. of largest particle sieve size (mm)	Approx. min. mass (g)
9.5	500
19	1000
37.5	3000

Also the size of the dried portion passing the 2-mm sieve should be at least about 125 g for sandy soils and 75 g for silt and clay soils.

Upto 1-kg size of sample, weighing should be accurate to 0.1 g, and beyond it, an accuracy of 1 g will do.

(b) Coarse Sieve Analysis

1. Weigh the dried material retained on the 2-mm sieve and divide it into different fractions by resieving through the set of sieves, 37.5/19/ 9.5/4.75 and 2-mm. The portion retained on each sieve is weighed.

Hand sieving is quite convenient with these sieves. First take the sieve with the largest opening (which should be somewhat larger than the biggest gravel particle), place the dried material over the sieve with a receiver underneath. Shake until no more material passes. Transfer the material obtained in the receiver to the next smaller sieve, which is shaken in turn, and thus the operation of sieving through all the sieves is completed.

(c) Preparation of Sample for Fine Sieve Analysis and Sedimentation Analysis

1. Weigh about 50 to 150 g (accurate to 0.01 g) from the oven-dried

and broken sample passing the 2-mm sieve (Ref. Step a-3). (The quantity of sample to be taken depends upon the soil type. It should be sufficient to yield about 35 g of fraction passing the 75-micron sieve for use in sedimentation analysis).

2. Place the sample in a 250-ml beaker and cover with 100-ml of dispersing agent solution. Add a little distilled water if necessary. Stir until the soil is thoroughly wetted. Allow to soak for at least 18 hours.

3. At the end of the soaking period, transfer the soil-water slurry from the beaker to the 75-micron sieve. Wash with distilled water. Oven-dry the material retained on the 75-micron sieve and make a fine sieve analysis. The soil-water suspension passing the 75-micron sieve is used for sedimentation analysis.

(d) Fine Sieve Analysis

1. Stack a set of fine sieves in descending order of aperture size, with the largest at top, followed by the receiver at the bottom. The set may consist of 1.0-mm, 600, 425, 300, 212, 150 and 75-micron sieves.

2. Place the oven-dried material retained on the 75-micron sieve (Step c-3) in the topmost sieve. Cover with lid and vibrate the assembly in a sieve shaker for 5 minutes. If mechanical sieve shaker is not available, do hand sieving with individual sieves as indicated in Step b-1 starting with the sieve having the largest opening.

3. Weigh (accurate to 0.01 g) the portion retained on each sieve.

(e) Sedimentation Analysis (Hydrometer Method)

1. Before starting the test, calibrate the hydrometer with the measuring cylinder for determining effective depth corresponding to hydrometer reading (Expt. 3.D).

2. *Determination of Composite Correction (C) and Meniscus Correction (C_m)*. Fill a measuring cylinder upto 1000-ml mark with distilled water containing dispersing agent in the same proportion as is used in the sedimentation test. Insert the hydrometer and after a short interval read the hydrometer at the top of the meniscus formed on the stem. Also read at the bottom of meniscus. As the hydrometer is calibrated to read the specific gravity of water as 1.000 for a temperature of 27 °C, the difference between the top meniscus reading and 1.000 is known as the *composite correction* C due to dispersing agent, meniscus and temperature. This may be positive or negative. With the concentration of dispersing agent recommended, it is generally negative. i.e., it is to be subtracted. The difference between the top and bottom meniscus readings is called the *meniscus correction* C_m and it is positive. The contents of the measuring cylinder should be at the same temperature which will prevail for the soil suspension at the beginning of the test.

3. Transfer the soil-water suspension passing the 75-micron sieve (Step c-3) to the cup of a mechanical stirrer. Stir for a period of 1 minute. Immediately after dispersion, transfer the suspension to the 1000-ml measuring cylinder. Use a fine jet of distilled water for all washings. Add enough distilled water to bring the level to the 1000-ml mark.

4. Using the palm of the hand on the open end of the cylinder, turn the cylinder upside down and back for a number of times.

5. After shaking, allow the cylinder to stand and start the stop watch simultaneously. Carefully insert the hydrometer and read the top of the meniscus at 0.5, 1, and 2 minutes (total elapsed time since the cylinder was set down). Remove the hydrometer, clean on the outside and float it in the second cylinder containing distilled water and dispersing agent in the same concentration. Record the temperature of the suspension.

6. Take further readings of the suspension at 5, 10, 15 and 30 minutes, and 1, 2, 4, 8 and 24 hours, after the start of the test. (These timings are only suggestions. Record the exact time of observation each time).

7. Insertion and withdrawal of the hydrometer should be done carefully, taking about 10 seconds over each operation, so as to avoid unnecessary disturbance. After each removal, the hydrometer should be cleaned with a dry cloth and placed in the second cylinder. Alternatively, it may be floated and thus washed in a third cylinder containing clean distilled water, before placing in the second cylinder for recording composite correction. The hydrometer should be inserted about 20-25 seconds before the reading is due to approximately the depth it will have when the reading is taken.

8. Check and record the temperature of the suspension and the composite correction at the time of taking each reading.

9. **Precaution.** Keep both the cylinders at the same temperature. Minimise temperature variations. To avoid unsymmetrical heating and consequent convection currents within the liquid, keep the cylinders out of direct sunlight and away from any local source of heat. The average temperature prevailing during a period should be used for calculation purposes for the reading taken at the end of that period. Preferably use a constant temperature water bath.

(f) Sedimentation Analysis (Pipette Method)

1. Before starting the test, determine the exact capacity of the pipette (V_p) including the hole in the tap by weighing the amount of water to fill it.

2. Same as Step e-3.

3. Keep the measuring cylinder in the water bath at the controlled temperature of 27 °C and leave for some time until the suspension attains this temperature.

4. When the test is to be commenced, take the cylinder out of the bath, closing the open end of the cylinder with the palm of the hand, shake the cylinder thoroughly by turning the cylinder upside down and back.

5. Allow the cylinder to stand in the bath and start the stop watch simultaneously. Take the first sample from a depth of 10 cm below the surface at 0.5 minute after the start of the test. Further samples are taken at 1.5, 3, 5, 10, 15 and 30 minutes, then at 1, 2, 4, 8 and 24 hours (total elapsed time after the start of sedimentation). Transfer each sample to a numbered weighing bottle and dry to constant mass at 105-110 °C.

6. For taking the sample, lower the pipette to touch the surface of the suspension. Note the scale reading. Lower further to a depth of 10 cm, Draw the liquid up into the pipette and a small excess into the safety bulb by applying a small vacuum by sucking on the side spout above the three-way tap. Close the tap. Lift the pipette above the cylinder. By turning the three-way tap the other way, drain off the surplus sample of the safety bulb. Wash with distilled water from the top funnel. Transfer the soil suspension to a numbered weighing bottle. Allow some distilled water to wash further the inside of the pipette into the bottle. Lower the pipette into the suspension about 20 seconds before the sample is due to be taken. Approximately 10 seconds are allowed for lowering the pipette and another 10 seconds for taking the sample.

Calculation

1. **Sieve Analysis.** During the coarse sieve analysis, the weight retained on respective sieves divided by the original dry weight taken for the test and multiplied by 100 gives the percentage retained on that sieve. From the percentages retained, the cumulative per cent retained and cumulative per cent finer (passing any sieve) are calculated. Cumulative per cent retained or finer is the per cent retained or passing any sieve if the whole of the material is sieved through that sieve eliminating all the coarser sieves above it. During the fine sieve analysis and sedimentation analysis, only a portion (M_s) of the dried material passing the 2-mm sieve is used. Therefore, during the fine analysis, the per cent finer (cumulative) is first calculated with respect to M_s then it is reduced on the basis on the entire sample. If M' is the total dry mass originally taken at the start of the test, M is the mass (cumulative) passing the 2-mm sieve, the per cent finer N with respect

to entire sample is given by: $N = N' \dfrac{M'}{M}$ where: N' is the per cent finer calculated on the basis of M_s which was taken from the fraction passing the 2-mm sieve.

2. Hydrometer Test. In recording all hydrometer readings, one is subtracted and the remaining digits are multiplied by 1000, thus 1.025 is read as 25. If R_h' is the hydrometer reading at the top of meniscus at elapsed time t minutes, and C_m the meniscus correction, the effective depth H_e corresponding to $R_h = R_h' + C_m$ is read from the calibration chart. The particle size (D in mm) is given by:

$$D = K \sqrt{\dfrac{H_e}{t}}$$

where $K = \sqrt{\dfrac{0.3\,\eta}{g\,(G-1)}}$, η being the viscosity of water at test temperature in poise units and G the specific gravity of soil particles. g is the gravitational acceleration (9.8 m/s^2). The value of K depends upon the temperature and specific gravity. Its value can be taken from Table 2.

The percentage of soil particles finer than D still remaining in the suspension after elapsed time t is given by: $N' = \dfrac{100\,G}{M_s\,(G-1)}\,R$ where: $R =$ corrected hydrometer reading

$$= R_h + C$$

M_s = mass of dry soil taken for sedimentation analysis from the portion passing 2·mm sieve

The per cent finer N on the basis of the entire soil sample is given by:

$$N = N' \times \dfrac{M'}{M}$$

where: M' is the cumulative mass passing 2-mm sieve and M is the mass of total sample.

3. Pipette Test: The particle size (D in mm) is given by:

$$D = K \sqrt{\dfrac{H_e}{t}},$$

where: $H_e = 10$ cm

and the per cent finer N' is given by the formula:

$$N' = \dfrac{M_D - \dfrac{m}{V}}{M_s/V} \times 100$$

where: M_D = mass per ml of soil solids obtained by oven-drying the samples taken at various elapsed times

m = mass of dispersing agent added to the total suspension (5g, if 100 ml of dispersing agent solution is added, Step c-2).

V = volume of suspension (1000 ml)

N' is then reduced to N as explained above.

4. The results of mechanical analysis are plotted in the form of a particle size distribution curve, with per cent finer as ordinate on an arithmetic scale and particle size as abscissa on logarithmic scale.

Note. In the above procedure, the sedimentation test is recommended on soil fraction passing a 75-micron sieve. Therefore the specific gravity G for use in the calculations should be determined by taking an independent soil sample which is finer than a 75-micron sieve.

Table 2

VALUES OF FACTOR K

$$(1 \times 10^{-4})$$

Temp (°C)	Specific Gravity						
	2.55	2.60	2.65	2.70	2.75	2.8	2.85
16	148	146	144	141	139	137	136
17	146	144	142	140	138	136	134
18	144	142	140	138	136	134	132
19	143	140	138	136	134	132	131
20	141	139	137	134	133	131	129
21	139	137	135	133	131	129	127
22	137	135	133	131	129	128	126
23	136	134	132	130	128	126	124
24	134	132	130	128	126	125	123
25	133	131	129	127	125	123	122
26	131	129	127	125	124	122	120

27	130	128	126	124	122	120	119
28	128	126	124	123	121	119	117
29	127	125	123	121	120	118	116
30	126	124	122	120	118	117	115
31	124	122	121	119	117	115	
32	123	121	120	118	116	114	
33	122	120	118	116	115	113	
34	121	119	117	115	113	112	
35	119	117	116	114	112	111	

OBSERVATION SHEET 3.A (b)

COARSE SIEVE ANALYSIS

(i) Total mass of dry sample (M) = 1000 g

(ii) Mass retained on 2-mm sieve = 110 g

(iii) Mass passing 2-mm sieve (M') = 890 g

S. No.	IS Sieve (mm)	Mass retained (g)	% retained	Cumulative % retained	Cumulative % passing (N)
1	37.5	0	0	0	100
2	19	0	0	0	100
3	9.5	22	2.2	2.2	97.8
4	4.75	38	3.8	6.0	94.0
5	2	50	5.0	11.0	89.0

Gravel fraction (cumulative per cent retained on 4.75-mm sieve) = 6 per cent

3.B   SIEVE ANALYSIS ALONE

Object

To determine the particle-size distribution of a cohesionless soil containing no clay fraction by sieve analysis alone.

Procedure

Take an oven-dried and broken sample. Separate it into two fractions, plus 2-mm and minus 2-mm sieve. Run a coarse sieve analysis on the plus 2-mm fraction and fine sieve analysis on whole of the minus 2-mm fraction. The mass for fine sieve analysis should be about 200 g. Weigh the material retained on each sieve and calculate the per cent finer and plot the gradation curve.

Note 1. If the soil contains clay fraction and sieve analysis alone is required, wash the soil on the 2-mm sieve kept over the 75-micron sieve. Oven-dry the material retained on each sieve and run the coarse and fine sieve analysis on the respective samples.

OBSERVATION SHEET 3.A (d)

FINE SIEVE ANALYSIS

(Passing 2-mm sieve)

(i) Mass original dry sample (M) = 1000 g

(ii) Mass fraction passing 2-mm sieve (M') = 890 g

(iii) Mass sample (M_s) taken from fraction minus 2-mm for fine analysis = 60 g

S. No.	IS Sieve	Mass retained (g)	% retained	Cumulative % retained	Cumulative % passing (N')	Cumulative % passing (whole) $N=100$ M'/M
1	2-mm	Nil	Nil	Nil	100	89.0
2	1-mm	3.78	6.3	6.3	93.7	83.5
3	600-micron	10.80	18.0	24.3	75.7	67.5
4	425-micron	7.89	13.1	37.4	62.6	55.7
5	212-micron	2.59	4.3	41.7	58.3	51.8
6	150-micron	1.34	2.2	43.9	56.1	50.0
7	75-micron minus	1.70	3.0	46.9	53.1	47.3
8	75-micron	31.9				

1. Gravel fraction [from Sheet 3.A (b)]   = 6%
2. Coarse sand (4.75 − 2 mm) = 94.0 − 89.0   = 5%
3. Medium sand (2 mm − 425 micron) = 89.0 − 55.7 = 33.3 %
4. Fine sand (425 − 75 micron) = 55.7 − 47.3   = 8.4%
5. Silt and clay (minus 75 micron)   = 47.3%

OBSERVATION SHEET 3.A(e)

SEDIMENTATION: HYDROMETER ANALYSIS

(i) Mass of dry sample taken from fraction minus 2-mm $(M_s) = 60$ g

(ii) Specific gravity of soil particles minus 75 micron (G) $= 2.70$

(iii) Meniscus-correction $(C_m) = +0.5$

Date	Time	Elapsed time (t) (min)	Hyd. reading (Rh')	Temp.	Composite correction (C)	Rh = Rh' + Cm	Eff. depth (He)	Factor (K) (1 × 10⁻⁴)	Particle size (D) (mm)	R = Rh' + C	% finer (N') based on Ms	% finer (N) based on whole
5.4.65		0.5	21.0	30 °C	−2.50	21.50	11.9	120	0.0584	18.5	49.0	43.6
		1	20.5			21.0	12.1	120	0.0417	18.0	47.7	42.5
		2	18.5			19.0	12.9	120	0.0305	16.0	42.4	37.8
		5	12.5			13.0	15.1	120	0.0209	10.0	26.5	23.6
		10	12.0			12.5	15.4	120	0.0149	9.5	25.2	22.4
		15	11.0	30 °C		11.5	15.7	120	0.0123	8.5	22.5	21.8
	9.45	30	10.5	30 °C	−2.50	11.0	15.9	120	0.00874	8.0	21.2	20.6
	10.15	60	9.5	30 °C	−2.50	10.0	16.3	120	0.00624	7.0	18.5	16.5
	11.15	120	8.5	30 °C	−2.50	9.0	16.7	120	0.00448	6.0	15.9	14.1
	14.30	315	6.25	31 °C	−2.25	6.75	17.5	119	0.00280	4.0	10.6	9.3
	16.30	435	5.75	32 °C	−2.00	6.25	17.7	118	0.00240	3.75	9.9	8.8
6.4.65	8.30	1395	5.00	30 °C	−2.50	5.50	18.0	120	0.00135	2.5	6.6	5.9

OBSERVATION SHEET 3.A(f)
SEDIMENTATION: PIPETTE ANALYSIS

(i) Mass of dry sample taken for test $(M_s) =$

(ii) Specific gravity of soil particles minus 75 micron $(G) =$

(vi) Volume of pipette $(V_p) =$

(iv) Amount of dispersing agent in the suspension $(m) =$

(v) Effective depth $(H_e) = 10$ cm

Date	Time	Elapsed time (t) (min)	Temp.	Factor (K)	Particle size (D) (mm)	Bottle No.	Bottle + dry mass	Mass of bottle	Mass of solids per ml (M_D)
							% finer (N') based on M_s		% finer (N) based on whole

Note 2. If little or no gravel is present, the preliminary division of the soil by the 2-mm sieve is omitted. Sieving is started with the sieve having the opening slightly larger than the biggest soil particle.

3.C SEDIMENTATION ANALYSIS ALONE

Object

To determine the particle-size distribution of a cohesive soil containing little or no sand fraction by sedimentation analysis alone.

Procedure

Take about 35-40 g of oven-dried sample. Suitably disperse (by adding dispersing agent and stirring), make a suspension and fill to the top mark in the measuring cylinder. Perform the sedimentation analysis using either hydrometer or pipette.

After the sedimentation test, wash the suspension through a 75-micron sieve to separate sand fraction, if any. Dry the retained material and weigh. Calculate the percentage finer.

Note. An air-dried sample may be used for the test. A representative sample is kept for determining hygroscopic moisture. If M is the air-dried mass taken for sedimentation test and w the water content (ratio). the dried mass M_s to be used for calculating per cent finer is given by:

$$M_s = \frac{M}{1 + w}$$

3.D CALIBRATION OF HYDROMETER

Object

To plot a calibration curve between hydrometer reading and effective depth for a given hydrometer and sedimentation cylinder.

Apparatus

Hydrometer. 1000-ml sedimentation cylinder with which the hydrometer will be used. Balance accurate to 0.01 g. Vernier calipers. calibrated in metric units, or millimetre scale.

Procedure

1. Measure the volume of the hydrometer (V_h), either by partly filling the 1000-ml cylinder and observing the increase in volume of the water when the hydrometer is immersed in it, or by weighing the hydrometer in grams which approximately gives the volume in cubic centimeters.

2. Find the sectional area A of the 1000-ml sedimentation cylinder by measuring the distance (cm) between two graduations (say 100 and 1000) and dividing the volume included between the graduations by the distance between the graduations.

3. Measure the height of the bulb h, i.e., the distance from the neck to the bottom of the bulb.

4. Measure the distances H from the neck of the bulb to the various major graduations R_h on the stem.

Calculation

The effective depth H_e corresponding to any hydrometer reading R_h is given by:

$$H_e = H + \tfrac{1}{2}\left(h - \frac{V_h}{A}\right)$$

P lot a graph between R_h and H_e.

OBSERVATION SHEET 3.D

CALIBRATION OF HYDROMETER

(i) Hydrometer No. 6919

(ii) Sedimentation cylinder No. 15

(iii) Volume of hydrometer (V_h) = 75 cm^3

(iv) Sectional area of cylinder (A) $= \dfrac{800}{27.1} = 29.5$ cm^2

(v) Height of bulb (h) = 16.5 cm

(iv) Constant $\frac{1}{2}\left(h - \dfrac{V_h}{A}\right) = 6.98$ cm

Hydrometer graduation (R_h)	Distance from neck of bulb to graduation (H − cm)	Effective depth (H_e − cm)
30	1.9	8.88
25	3.7	10.68
20	5.5	12.48
15	7.4	14.38
10	9.3	16.28
5	11.2	18.18
0	13.1	20.08
−5	15.0	21.98

CONSISTENCY LIMITS DETERMINATION

4.A LIQUID LIMIT TEST

Object

To determine the liquid limit of a soil using the Casagrande Liquid Limit Apparatus.

Apparatus

Liquid limit apparatus consisting of a brass cup and a carriage mounted on a hard rubber or micarta block. Grooving tools with 1-cm gauge handle: ASTM and Casagrande (BS) types. Glass plate about 40 cm square or a porcelain evaporating dish about 12-cm in daimeter. 425-micron sieve. Spatula. Balance accurate to 0.1 g. Containers for moisture content samples. Oven, 105 to 110 °C. Stop watch.

Procedure

1. By means of the grooving tool gauge and the adjustment plate, adjust the cup of the liquid limit apparatus to give a drop of exactly 1 cm on the point of contact on the base.

2. Take about 150 g of an air-dried soil sample passing 425 micron sieve and mix thoroughly with distilled water to give a stiff and uniform paste. Leave the soil for a suitable maturing time which may extend upto 24 hours for heavy clays.

3. Place a portion of the paste in the cup, level off with a spatula the top surface symmetrically to give a maximum depth of 1 cm. Cut a uniform, straight groove by drawing firmly a grooving tool through the soil paste along the diameter through the centre of the hinge.

(Note. The groove is a V-shaped gap, 2-mm wide at bottom, 13.6-mm at top and 10-mm deep with the ASTM tool, and 2-mm wide at bottom, 11-mm at top and 8-mm deep with the Casagrande tool.

4. Turn the handle at a rate of 2 revolutions per second and count the number of blows until the two parts of the soil come in contact at the bottom of the groove along a distance of about 13-mm (0.5 in). The groove should be closed by a flow of the soil and not by slippage between the soil and the cup. If no flow but slippage occurs, the test is rejected.

5. Record the number of blows at which the groove closes. Remove about 15 g of the soil forming the edges of the groove that flowed together and determine the water content by oven-drying.

6. Transfer the remaining soil in the cup to the main soil sample on the glass plate (or the evaporating dish) and mix thoroughly after adding a small amount of water. Clean the cup and the grooving tool. Repeat steps 3, 4 and 5.

7. The test should always proceed from the drier to the wetter condition of the soil. In the first test, the moisture content is so adjusted that the groove closes in 35 to 40 blows. The test is repeated at least three times more adding increments of distilled water at the beginning of each stage, thus giving at least four sets of readings in the range of 15 to 40 blows. Each additional increment of water added to the soil is mixed in for at least 5 minutes.

Calculation

A 'flow curve' is plotted on a semi-logarithmic graph representing the number of blows as abscissa on the logarithmic scale and the corresponding moisture content (per cent) as ordinate on the arithmetic scale. The flow curve is drawn as a straight line as nearly as possible through a maximum number of plotted points. The water content corresponding to 25 blows is read off as the liquid limit w_L of the soil.

The flow index I_f or the slope of the curve can be determined from the relation:

$$I_f = \frac{w_1 - w_2}{\log_{10} \dfrac{n_2}{n_1}}$$

where w_1 = moisture content corresponding to n_1 blows

w_2 = moisture content corresponding to n_2 blows

If the flow curve is extended at either end so as to intersect the ordinates corresponding to 10 and 100 blows, the numerical difference in moisture contents at 10 and 100 blows gives directly the flow index.

OBSERVATION SHEET 4.A

LIQUID LIMIT TEST

Test No.	1	2	3	4
1. Container No.	60	61	62	63
2. No. of blows	41	26	19	14
3. Mass cont. + wet soil (g)	34.71	35.62	38.41	38.14

4. Mass cont. + dry soil	(g)	32.52	33.33	35.20	34.80
5. Mass moisture present	(g)	2.19	2.29	3.21	3.34
6. Mass container empty	(g)	23.10	24.03	22.60	22.30
7. Mass dry soil	(g)	9.42	9.30	12.60	12.50
8. Moisture content	(%)	23.3	24.7	25.5	26.7
9. Liquid limit (from graph)	(%)		24.8		

4.B PLASTIC LIMIT TEST

Object

To determine the plastic limit of a soil and also to calculate the plasticity index.

Apparatus

Porcelain evaporating dish, about 12-cm in diameter or a glass plate for mixing soil and water. Spatula, about 2-cm wide. Ground-glass plate, about 20 cm by 15 cm, for rolling of threads. Balance accurate to 0.01 g. Containers for moisture content samples. Oven. Rod, 3-mm in diameter. 425-micron IS sieve.

Procedure

1. Mix thoroughly about 30 g of soil passing a 425-micron sieve with distilled water in the evaporating dish or on the glass plate until it is plastic enough to be shaped into a small ball.

Note. It is often convenient to allow the soil used in the liquid limit test to dry in air to reach this consistency. It may be necessary to suitably mature the soil.

2. Take about 10 g of the plastic soil mass. Form a ball of it and then roll into a thread with the fingers on the ground-glass plate. (A sheet of paper may also be used as a surface for rolling). When a diameter of 3-mm is reached, remould the soil again into a ball.

3. Repeat this rolling and remoulding process until the thread starts just crumbling at a diameter of 3-mm. Keep the crumbled threads for moisture content determination.

4. Repeat the test twice more with fresh samples and calculate the plastic limit w_P as the average of the three moisture contents.

Plasticity Index

After determining the liquid limit w_L and the plastic limit w_P, the plasticity index I_P is calculated from the equation:

$$I_P = w_L - w_P$$

When either the plastic limit or the liquid limit cannot be determined, the plasticity index is reported as 'Non-plastic' (NP). (Note. In case of sandy soils, the plastic limit test should be performed first to know whether the soil is non-plastic.) When the plastic limit is equal to or greater than the liquid limit, the plasticity index is reported as 'zero'.

The 'toughness index', I_T, if also required, can be calculated from the equation:

$$I_T = \frac{I_P}{I_f}$$

where I_f = flow index as determined in Expt. 4.A

The other indices, consistency index I_C and liquidity index I_L, can also be calculated for a soil whose natural moisture content w is known:

$$I_C = (w_L - w)/I_P$$
$$I_L = (w - w_P)/I_P$$

Precaution

Rolling should be gentle and straight. Undue pressure, or oblique rolling may result in mechanical destruction of soil threads. Soil must be well mixed and kneaded and the threads must not be-fissured before the test begins. If soil has not crumbled due to decrease in moisture content only (but due to fissures or negligent rolling), the 3-mm crumbled thread can further be rolled.

OBSERVATION SHEET 4.B

PLASTIC LIMIT TEST

Test No.		1	2	3
1. Container No.		31	32	33
2. Mass cont. + wet soil	(g)	25.94	25.10	26.02
3. Mass cont. + dry soil	(g)	25.48	24.70	25.49
4. Mass moisture present	(g)	0.46	0.40	0.53
5. Mass container empty	(g)	23.10	22.56	22.74
6. Mass dry soil	(g)	2.38	2.14	2.75
7. Moisture content	(%)	19.3	18.7	19.3
8. Average plastic limit	(%)		19.1	

4.C SHRINKAGE LIMIT TEST

Object

To determine the shrinkage limit of a soil.

Apparatus

Circular shrinkage dish of porcelain or stainless steel with flat bottom, about 3.5-cm in diameter and 1.5-cm high. Large porcelain evaporating dish flat bottom (2 Nos.), about 12-cm in diameter or flat-bottom stainless steel dish with lip about 14-cm in diameter and 2-cm high. Small porcelain evaporating dish, about 7.5-cm in diameter, to be used for mixing soil and also to be used as container for weighing mercury. Glass plate with prongs. Plain glass plate, about 7.5 cm by 7.5-cm. Glass cup, about 5.5-cm in diameter and 3-cm high with level and smooth-ground top rim or a stainless steel cup of same dimensions. Balance accurate to 0.01 g. Spatula. Straight edge. Mercury. Oven. 425-micron IS sieve. Desiccator.

Procedure

1. Mix about 50 g of soil passing a 425-micron sieve with distilled water in an amount sufficient to fill the voids completely and to make the soil pasty enough to be readily worked into the shrinkage dish without the inclusion of air bubbles. (In friable soils, this amount of water is equal to or slightly greater than the liquid limit; for plastic soils, it may be about 10 per cent more than the liquid limit.)

2. Take an empty, clean shrinkage dish and coat the inside with a thin layer of vaseline or other grease, place in the centre of the dish an amount of wet soil equal to about one-third the capacity of the dish and tap the dish on a firm surface so that the soil flows to the edges of the dish. (The firm surface should be properly cushioned by several layers of blotting paper or similar material.) Repeat this procedure until the dish is completely full and excess soil stands above its edge. Strike off the top surface with a straight edge and properly clean the outside of the dish.

3. Weigh the shrinkage dish immediately full of wet soil. Allow it to dry in air until the colour of the soil pat turns from dark to light. Then dry in an oven at 105-110 °C. Cool the dish with dry soil in a desiccator.

4. Weigh the shrinkage dish with the dry soil pat.

5. Clean and dry the shrinkage dish and determine its mass when empty.

6. Also weigh an empty porcelain dish (small size) which will be used for weighing mercury. This dish will be known as mercury weighing dish.

7. Keep the shrinkage dish in a large porcelain dish (or stainless steel dish), fill it to overflowing with mercury and remove the excess by pressing the plain glass firmly over the top of the dish. Transfer

the contents of the shinkage dish to the 'mercury weighing dish' and weigh.

8. Place the glass cup in a large dish, fill it to overflowing with mercury, remove the excess by pressing the glass plate with three prongs firmly over the top of the cup.

9. Wipe the outside of the glass cup to remove any adhering mercury, then place it in another larger dish which is clean and empty. Place the dry soil pat on the surface of the mercury and submerge it under the mercury by pressing with the glass plate with prongs.

10. Transfer the mercury displaced by the dry soil pat to the 'mercury weighing dish' and weigh.

Calculation

The shrinkage limit w_s is calculated from the equation:

$$w_s = \left[w - \frac{V - V_d}{M_S} \right] \times 100 \text{ per cent}$$

where: w = moisture content (ratio) of wet soil pat

M_s = mass of dry soil pat (g)

V = volume of wet soil pat (cm^3)

V_d = volume of dry soil pat (cm^3)

The volumes V and V_d are obtained by dividing the two masses mercury by its density (13.6 g/cm^3) (Steps 7 and 10)

Precaution

Be careful that no air is trapped when the shrinkage dish is being filled with wet soil or when the dry soil pat is being forced under mercury.

Alternative Procedure

When the specific gravity of soil is known, the following procedure may be adopted:

1. Make a smooth, round-edged pat of wet soil. The pat can be made either in a small shrinkage dish or in any other small mould, or it may be a specimen cut from an undisturbed wet sample. Dry it in an oven and then cool it in a desiccator.

2. Weigh the dry soil pat after cooling it in the desiccator.

3. Immerse the dry soil pat in mercury and weigh the displaced mercury (Steps 8, 9 and 10 the previous procedure).

Calculation

The shrinkage limit w_s is given by:

$$w_s = \left(\frac{V_d}{M_s} - \frac{1}{G} \right) \times 100 \text{ per cent}$$

where V_d = volume of dry soil pat (cm³)

M_s = mass of dry soil pat (g)

G = specific gravity of soil

Note 1. In the above formulae, the unit weight of water γ_w has been taken as unity.

2. The shrinkage ratio and the volumetric shrinkage, if required, can be calculated. (See Chapter 2).

OBSERVATION SHEET 4.C
SHRINKAGE LIMIT TEST

Test No.		1	2
1. Shrinkage dish No.		A	
2. Mass dish + wet soil pat	(g)	58.72	
3. Mass dish + dry soil pat	(g)	50.58	
4. Mass water present (2) − (3)	(g)	8.14	
5. Mass shrinkage dish empty	(g)	19.98	
6. Mass dry soil pat (M_s)	(g)	30.60	
7. Moisture content (w)	(ratio)	26.6	
8. Mass weighing dish + mercury filling shrinkage dish	(g)	322.18	
9. Mass weighing dish empty	(g)	56.82	
10. Mass mercury	(g)	265.36	
11. Vol. wet soil pat $(V) = \frac{(10)}{13.6}$	(cm³)	19.50	
12. Mass weighing dish + displaced mercury	(g)	273.54	

13. Mass weighing dish empty	(g)	56.82	
14. Mass mercury displaced	(g)	216.72	
15. Vol. dry soil pat $(V_d) = \dfrac{(14)}{13.6}$ (cm³)		15.95	
16. Shrinkage limit $(w_s) = \left[w - \dfrac{V - V_d}{M_s} \right] \times 100$		15.0	

Alternative Procedure

Test No.		1	2
1. Sp. gr. of soil (G)		2.70	
2. Mass dry soil pat (M_s)	(g)	28.43	
3. Mass weighing dish + mercury	(g)	260.82	
4. Mass weighing dish empty	(g)	56.82	
5. Mass mercury displaced	(g)	204.00	
6. Vol. dry soil pat $(V_d) = \dfrac{(5)}{13.6}$ (cm³)		15.00	
7. Shrinkage limit $(w_s) = \left[\dfrac{V_d}{W_d} - \dfrac{1}{G} \right] \times 100$		15.8	

PERMEABILITY DETERMINATION

Object

To determine the permeability of a soil using the Jodhpur Permeameter (designed by the author).

Apparatus

The Jodhpur Permeameter, comprising the following: 0.3-litre permeameter mould, 50 cm^2 in internal cross-sectional area (7.98-cm in dia.) and 6-cm high, fitted with two side-studs. The mould has a recess on the top end for seating the top perforated plate. Top cap fitted with water inlet nozzle and air release valve and having an inside seating for housing a sealing gasket of rubber or other suitable material. Perforated base plate. Top perforated plate. Dynamic compaction base plate. Static compaction flanged end-plugs, 2 Nos., 2.5-cm and 3-cm high Compaction collars, 2 Nos., 2.5-cm and 3-cm high, to fit the top and bottom ends of the 0.3-litre mould respectively. Split collar, 2-cm high. 2.5-kg Dynamic Ramming Tool (2.5-kg DRT) (See Jodhpur Mini-Compactor Test, Expt. 7.D). 0.3-litre core cutter and dolly, the cutter is 6-cm high and 50-cm^2 in cross-sectional area. Centering ring for cutter. Rod tamper. Assembly bolts with nuts, 4 Nos. Bottom tank with side outlet. Constant head tank (a glass tube, 5-cm in diameter and 100-cm long) complete with stopper and air intake tube. Falling head stand pipes (2 Nos.); the constant head tank and stand pipes can be all mounted on a common stand. (Note). A constant head tank can be built on the air-intake-tube principle also with a metallic container or an aspirator bottle. Alternative arrangements for the supply of water to the permeameter at constant head can be used). Sealing washer for top cap, copper wool pad, fine-mesh gauzes (7.8-cm in dia.), filter papers (7.8-cm in dia.), set of spanners, funnel, pinch cocks, flexible tubing, Plate 5.1 shows some of the parts of the permeameter.

4.75-mm and 2.0-mm IS sieves. Mixing basin. Straight edge or trimming knife. Balance accurate to 1.0 g. Stop watch. Graduated measuring cylinder. Beaker. Apparatus for water content determination.

Plate 5.1

Specimen Preparation

Both remoulded and undisturbed specimens can be tested, the former for studying the seepage of water through remoulded soil masses. e.g., rolled earth dam, and the latter in case of seepage through natural soil formations. Remoulded specimens can also be used for studying the variation of permeability with respect to some other physical property of soil, such as voids ratio. The Jodhpur Permeameter can be considered suitable for testing clayey, silty and sandy soils passing a 2.0-mm IS sieve. Composite soils with gravel fraction finer than 4.75-mm sieve can also be tested, provided the gravel fraction is not more than about one-third of the total soil. Larger permeameters are recommended for still coarser soils.

(a) Remoulded Specimen (Statically Compacted)

1. Take a representative sample of soil and raise its moisture content to the desired value. Keep it stored in an air-tight container for a suitable maturing time.

2. Apply thinly a little grease on the inside of the mould. Attach the 3-cm collar to the bottom end of the mould and the 2.5-cm collar to the top end. Support the assembly over the 2.5-cm end-plug with 2.5-cm collar resting on the split collar kept around the 2.5-cm end-plug.

3. Weigh a sufficient quantity of wet soil (M) at known moisture content (w) to give the desired density (ρ_d) when compressed to the volume (V) of the mould (300-cm³):

$$M = V\rho_d (1 + w)$$

Pour the weighed quantity of soil into the mould, tamping the soil by hand during the process of pouring. Insert the top end-plug (3-cm) into the 3-cm collar. Keep the entire assembly in a press (compression machine). Remove the split collar. Compact the specimen by pressing both the plugs until their flanges touch the collars. Release the pressure after a rest period of about ½ minute. Plate 5.2 shows the mould assembly for static compaction.

4. Remove the 3-cm plug and the collar and fix the perforated base plate to the mould after placing a fine mesh gauze or filter paper on the surface of specimen. Turn the mould upside down. Remove the 2.5-cm plug and collar. Place the top perforated plate and fix up the top cap after inserting the sealing gasket. Insert the additional two bolts through the base plate and the top cap and further tighten. The assembly is shown in Plate 5.3.

(b) Remoulded Specimen (Dynamically Compacted)

1. Take about 800 g of pulverised soil and increase its moisture con-

tent to the desired value, say, to the optimum moisture content as determined by the Jodhpur Mini-Compactor Test (Expt. 7.D). Store the soil in an air-tight container for a suitable maturing time.

2. Lightly grease the 0.3-litre mould on the inside, fix it upside down on the dynamic compaction base plate and weigh. Attach the 3-cm collar to the other end.

3. Compact the wet soil in the mould in two equal layers, each layer being given 15 uniformly distributed blows of the 2.5-kg DRT. The surface of the first layer should be scarified to a depth of about 3 mm before placing the second layer. The second layer on compaction should project not more than about 5 mm into the collar. (See Expt. 7.D for compaction procedure). Remove the collar, trim off excess soil and weigh to get the weight of compacted specimen.

4. Place a fine mesh gauze or filter paper on the surface of specimen and then fix the perforated base plate to the mould. Turn the assembly upside down and detach the compaction base plate. Place another fine mesh gauze on top of specimen and over that the top perforated plate. Fix the top cap with sealing gasket inside. Further tighten the top cap and base plate by the additional two bolts.

Note 1. The internal dimensions of the permeameter mould are the same as those of the compaction mould of the Jodhpur Mini-Compactor. The compaction specified above duplicates the Standard Proctor compaction very closely. As a preliminary test, the optimum water content of soil can be determined by compaction in the permeameter mould itself.

Note 2. The number of blows of the 2.5-kg DRT and the moulding moisture content can be varied to simulate particular field conditions.

Specimens of increasing densities can be prepared by increasing the blows per layer and thus the variation of permeability with change in voids ratio can be studied.

Note 3. Compared to static compaction by which specimens of predetermined density can be prepared, dynamic compaction gives specimens requiring density calculations after compaction. If specimens of predetermined density are required by dynamic compaction as well, a weighed quantity of wet soil to give the desired density is compacted to a 300-cm³ volume in the mould fixed on the dynamic compaction base plate. The number of blows (per layer) of the 2.5-kg DRT are to be decided by trial.

Note 4. The rod tamper is used for light compaction and for surface dressing. The depth of the head of the tamper is the same as the depth of collar, i.e., 3 cm.

Note 5. Specimens of cohesionless soils are prepared by vibration in the mould fixed upside down on the compaction base plate.

Note 6. The 0.3-litre core cutter can be used for obtaining remoulded specimens of soil compacted in the field or in larger moulds like the Proctor or the CBR mould. It is assembled between the perforated base plate and top cap as described below under the preparation of undisturbed specimen.

(c) Undisturbed Specimen

1. Weigh the 0.3-litre cutter empty. Obtain a specimen by pushing it gently with dolly attached on top into an undisturbed block of soil. Cut the soil flush with the cutting edge. Remove the dolly and trim off excess soil. Weigh the cutter full of specimen. Determine the water content of the cuttings.

2. Place the centering ring for the cutter on the perforated base plate with 4 holes of the ring coinciding with those of the base plate. Insert the 4 hexagonal bolts through the 4 holes of the base plate and the ring, the base plate resting on the heads of the bolts. Cover the perforations of the base plate with a wire gauze or filter paper. Keep centrally the cutter containing the specimen on the base plate with cutting edge downwards. Place a wire gauze or filter on top of specimen. Insert the copper wool pad in the top cap and the sealing gasket on its seating. Fix the top cap over the cutter by tightening the nuts of the 4 bolts passing through the cap. Plate 5.4 shows the assembly of the 0.3-litre core cutter.

Plate 5.4

3. An alternative method is to keep an undisturbed specimen in the permeameter mould instead of taking it in the 0.3-litre cutter. The specimen should be obtained independently by suitable sampling tubes or cutters not larger than about 6.5-cm in diameter. The specimen is cut exactly to a height of 6 cm and placed centrally in the mould fixed

to the perforated base plate, the perforations of which are covered with a wire gauze or filter paper. A mixture of 10 per cent dry, powdered bentonite and 90 per cent fine sand by weight is tamped in the annular space around the specimen. The top perforated plate and cap are then fixed as usual.

Test Procedure

(a) Saturation

Saturate the specimen by one of the following methods:

1. Place the permeameter mould (or core cutter) assembly with the bottom tank in a vacuum desiccator. Open the air release valve. Fill the desiccator partly with water until it stands well above the top cap. (The water inlet nozzle should remain submerged.) Apply a small vacuum (say 5 to 8-cm of mercury) and maintain it for such a time until air bubbles are seen rising from the specimen. Increase the vacuum *in steps* with suitable rest periods to a maximum of at least 70-cm of mercury. The process of evacuation should be carried out slowly and carefully so that the air trapped in the soil does not bubble out violently. After full saturation, take the permeameter out of the desiccator with the bottom tank full of water. Close the air release valve.

2. Place the permeameter mould assembly in the bottom tank. Fill the tank slowly with water and at the same time connect the water inlet nozzle of the top cap to a vacuum line. Keep the air valve closed. Allow the specimen to saturate by upward flow of water under a small vacuum of 5 to 8-cm of mercury. A small glass tube inserted in the vacuum line may serve as an indicator to watch the air bubbles coming out of the specimen. Carry on the whole operation slowly. High vacuums may cause piping or disturb the specimen.

Note. Vacuum can be created by an aspirator bottle, a filter pump (water-jet suction pump) or a vacuum pump. The constant water-head tank also can be used for this purpose. The tank is kept on the ground and the permeameter with bottom tank is placed on an adjacent table. The constant head tank is filled with water. A small air intake tube (instead of the usual long air intake tube), is inserted through the top stopper of the constant head tank. The air intake tube is connected to the water inlet nozzle of the top cap. Water is allowed to slowly flow out of constant head tank through the bottom outlet. This operation will draw water from the bottom tank through the specimen under a suction head equal to the difference in elevations of the bottom of the air intake tube and the bottom outlet of the constant head tank. The suction head can be varied by lowering or raising the air intake tube within the constant head tank.

3. Allow a specimen of clean sands to saturate without the application of any vacuum but by the upward flow of water through the specimen with water standing in the bottom tank above the top surface of the specimen. The air valve is kept open during saturation and is closed afterwards. (Note. The bottom tank is deep enough to submerge the top cap. Water should be filled slowly in the tank.)

(b) Constant Head Test

1. Remove the stopper with air intake tube from the constant head tank and fill the tank *completely* by allowing the water to enter upwards through the flexible tubing attached to the bottom outlet of the tank. Lower the air intake tube inside the tank and fix the stopper.

2. Connect the outlet tubing of the constant head tank to the water inlet nozzle of the permeameter kept in the bottom tank, the bottom tank being full with water upto its outlet. Adjust the hydraulic head to the required value by raising or lowering the air intake tube within the head tank, or by adjusting the relative elevations of the permeameter and the head tank.

Note. The head is the difference between the elevations of the bottom end of the air intake tube and the free water level in the bottom tank. It remains constant until the water level in the constant head tank does not drop below the bottom end of the air intake tube.

3. Collect from the bottom tank outlet and weigh in grams (or measure) the quantity of water flowing through the specimen for a convenient time interval. Repeat the test twice more under the same head for the same time intervals and calcuate the average rate of flow. Record the temperature of flowing water.

Note 1. If a study state of flow does not exist, erratic results are obtained. Flow observations should be recorded only after establishing a steady state by a preliminary continued flow of water through the specimen.

Note 2. See Precaution-1 for hydraulic gradient.

(c) Falling Head Test

1. Determine the inside cross-sectional areas of the stand pipes (if not known) by weighing (or measuring) the quantity of water contained in known lengths of the stand pipes. Mark the positions of the desired initial and final heads.

2. Connect the permeameter kept in the bottom tank to a suitable stand pipe filled with water. Record the time interval required by the water level to fall in the stand pipe from the initial head to the final head.

3. Refill and repeat the test for at least two more times and record the time intervals for the drop in head from the same initial to final values. Calculate the average time interval. Also calculate the average test temperature by recording the temperature of water in the stand pipe and the bottom tank.

Note. The head is measured from the water level in the bottom tank to the water level in the stand pipe.

Calculation

The coefficint of permeability k (cm/sec) is calculated from equations:

(i) Constant head test; $k = \dfrac{QL}{Ath}$

(ii) Falling head test: $k = 2.3 \dfrac{aL}{At} \log_{10} \dfrac{h_1}{h_2}$

where: Q = total quantity of flow in time t (ml)

t = time interval (sec)

h = constant head (cm)

h_1 = initial head (cm)

h_2 = final head (cm)

L = length of specimen (6 cm)

A = area of specimen (50 cm^2)

a = area of stand pipe (cm^2)

The test temperature, specimen voids ratio and the hydraulic gradient for the constant head test should be reported. The permeability may be converted to the standard temperature of 27 °C.

Precaution

1. Coarse-grained soils should be tested under a low hydraulic gradient to ensure laminar flow conditions for the validity of the Darcy law. For such soils, the gradient may be from 0.2 to 0.3 for a loose state of compactness and from 0.3 to 0.5 for a dense state. The coarser the soil, the lower should be the gradient. If a test is conducted conforming to specific problems involving turbulent flow, the hydraulic gradient should be the same as in the actual problem.

2. The sealing gasket of the top cap, water inlet nozzle and air valve should be leak-proof. If some soil or the wire gauze rides on the recessed seating of the top perforated plate provided in the 0.3-litre mould, the top perforated plate may protrude beyond the top edge of

the mould and the gasket will not remain leak-proof.

3. Clean top water (or preferable distilled water) is used for the test. It is necessary to reduce the air content of water, otherwise, there is danger of air getting released within the specimen, which gets deposited in the pores and reduces the permeability.

Air free water can be obtained by boiling or by spraying it into a partial vacuum. The vacuum tank should be a *stout* cylindrical tank capable of resisting external pressures approaching atmospheric pressure. In constant head test where larger quantities of water are required, an alternative procedure may be to use water at a slightly higher temperature (about 5°) above that of the specimen. When water flows through the specimen, it cools slightly and its capacity for dissolved air is increased. Another method of removing air is to run the water through a porous fine-grained material finer than the material used for testing in the permeameter, before it enters the permeameter. The head tank with air intake tube should be replaced with some alternative arrangement for a constant-head supply of water, when the possibility of air-saturation of water due to air intake tube is to be completely avoided.

4. Perfect sealing of the sides of top rubber stopper and the air intake tube is essential for a successful operation of the constant head tank. If there is leakage, the air intake tube will not remain empty and water will rise through it.

5. No air should get entrapped in the supply line to the permeameter, when flow is started for the test.

OBSERVATION SHEET 5
PERMEABILITY TEST

(a-1) Statically Compacted Specimen

1. Specimen size:	(i) Length	(cm)	6
	(ii) Area	(cm^2)	50
	(iii) Volume	(cm^3)	300
2. Required dry density		(g/cm^3)	1.52
3. Moulding moisture content		(%)	4
4. Mass of soil to be compacted		(g)	475
5. Specific gravity			2.68
6. Voids ratio			0.76

(a-2) Dynamically Compacted/Vibrated Specimen

1. Specimen size:	(i) Length	(cm)	6
	(ii) Area	(cm²)	50
	(iii) Volume	(cm³)	300
2. Mass mould + base plate + specimen		(g)	3,080
3. Mass mould + base plate		(g)	2,573
4. Mass specimen		(g)	507
5. Bulk density		(g/cm³)	1.69
6. Moulding moisture content		(%)	air dry
7. Dry density		(g/cm³)	1.67 (air dry)
8. Specific gravity			2.67
9. Voids ratio			0.58

(a-3) Core-Cutter Specimen

1. Specimen size:	(i) Length	(cm)	
	(ii) Area	(cm²)	
	(iii) Volume	(cm³)	
2. Mass cutter + specimen		(g)	
3. Mass core-cutter		(g)	
4. Mass specimen		(g)	
5. Bulk density		(g/cm³)	
6. Moisture content		(%)	
7. Dry density		(g/cm³)	
8. Specific gravity			
9. Voids ratio			

(b-1) Constant Head Test

Type of specimen: Vibrated coarse sand

1. Hydraulic head	(cm)	4.8
2. Hydraulic gradient		0.8
3. Time interval	(sec)	300
4. Quantity of flow: (i) I test	(ml)	456
(ii) II test	(ml)	455
(iii) III test	(ml	457
5. Average quantity of flow	(ml)	456
6. Coeff. of permeability	(cm/sec)	3.8×10^{-2}
7. Test temperature	(°C)	32

(b-2) Falling Head Test

Type of specimen: Statically compacted silty-sand

1. Area of stand pipe	(cm^2)	0.785
2. Initial head	(cm)	36
3. Final head	(cm)	16
4. Time interval: (i) I test	(sec)	88
(ii) II test	(sec)	86
(iii) III test	(sec)	87
5. Average time interval	(sec)	87
6. Coeff. of permeability	(cm/sec)	8.75×10^{-4}
7. Test temperature	(°C)	32

CONSOLIDATION TEST

Object

To determine the relationship between verticle pressure and voids ratio of a soil and also the coefficient of consolidation using the Jodhpur Consolidometer.

Apparatus

The Jodhpur Consolidometer comprising a *consolidation cell* and a *loading machine*. The cell may be either a floating ring or a fixed ring type.

    (a) The Jodhpur floating ring cell (used in this experiment) comprising the following: Specimen ring, 50-cm^2 in internal cross-sectional area and 24-mm high. Guide rings (2 Nos.). Porous stones (2 Nos.) 79-mm in diameter and 12-mm thick. Pressure pad. Steel ball. Water trough. Specimen cutter. 3-cm extension collar. 3-cm flanged plug. Base plate. Glass plates (2 Nos). Thin rubber sleeve, about-10 cm in diameter and 10-cm long.

    (b) Loading machine, capable of loading the specimen upto a pressure of 10 kg/cm^2, comprising a counter-balanced-loading beam (lever ratio 1:10) with a weight hanger. Set of slotted weights: 0.5 kg (4 Nos.) 1, 2 and 5 kg (1 No., each), 10 kg (4 Nos.). Dial gauge, accurate to 0.01 mm, with graduations preferably increasing in the anticlockwise direction.

The Jodhpur Consolidation cells can be used with other types of loading machines also.

    (c) Balance, accurate to 0.01 g. Stop watch. Mixing basin. Trimming knife, wire saw and spatula. Apparatus for water content and specific gravity determinations.

Specimen Preparation

Both undisturbed and remoulded specimens, fully or partly saturated, can be tested.

(a) Undisturbed Specimen

Weigh the specimen ring and the two glass plates. Attach the cutter to one end of the specimen ring and the extension collar to the other, and place vertically the assembly on the undisturbed soil sample from which the test specimen is to be cut. (Note. The undisturbed sample or block should be at least 10-cm in diameter or a 10-cm cube). Trim excess material with a knife close to the cutting edge of the cutter. Press the ring assembly gently downwards with minimum of disturbance until the soil protrudes into the extension collar above the specimen ring by about 5 mm. Remove the extension collar and trim off excess soil flush with the top end. Cut soil at the level of the cutting edge of the cutter. Place the specimen ring with its top end on a glass plate. Remove the cutter and trim off the excess soil. Cover this face of the specimen also with the second glass plate. Weigh the specimen ring with the specimen inside and covered with glass plates. Keep a sample out of the cuttings for initial water content determination.

(b) Remoulded Specimen

1. **Saturated specimen.** Weigh the specimen ring along with the two glass plates. Place the specimen ring on one of the glass plates. Press remoulded saturated soil into the specimen ring wih a spatula, care being taken to avoid trapping air during the process. Finish flush with the top end of the ring and cover with the second glass plate. Weigh the ring with specimen and glass plates.

2. **Statically compacted specimen.** Mix an appropriate quantity of water to about 350 g of pulverised soil to bring its water content to the desired value. Mature the soil for a suitable time. Keep a sample for moisture content determination.

Keep the specimen ring on the base plate. Attach the extension collar. Weigh a sufficient quantity of wet soil at known moisture content to give the desired density when compressed to a volume of 120 cm³. Pour the soil into the specimen ring. Insert the flanged plug in the extension collar. Compress the assembly in a suitable press or compression machine. Remove the plug and collar and cover the specimen with glass plates.

3. **Dynamically compacted specimen.** Weigh the specimen ring and the two glass plates. Place the ring on the base plate. Attach the extension collar.

Weigh a sufficient quantity of wet soil at known moisture content to give the desired density when compacted to a thickness 3 cm in the 50-cm² ring (i.e., to a volume of 150 cm²). Place the soil in the ring and compact by the 2-5-kg DRT (See Expt. 7D) or by any other suitable tool to a thickness of 3 cm. Remove the collar and trim off 6 mm of excess soil. Cover the specimen with glass plates and weigh.

Note. The Jodhpur Mini-Compactor (Expt. 7.D) can also be used for preparing dynamically compacted specimen. After compaction, the specimen is pushed in the specimen ring and trimmed flush with the ends. Similarly, statically compacted specimens can also be prepared in the 'swelling pressure mould' as described in Expt. 12. If the specimen ring slips down from the specimen when supported on the bottom porous stone during the test, the fixed ring cell should be used.

Test Procedure

1. Soak the porous stones in water and then wipe away any excess water. Place one stone in the central seating of the water trough. Attach guide rings to both ends of the specimen ring and place it gently on the porous stone inside the trough. Place the other porous stone on top of the specimen and then over it place the pressure pad and the steel ball.

2. If a test is to be conducted at the initial water content of a partially saturated specimen, use a thin rubber sleeve about 10-cm in diameter and 10-cm in length, to cover the specimen ring, porous stones and the pressure pad. Tie the top end of the sleeve to the pressure pad by putting a rubber band around the sleeve at the groove provided in the pressure pad. If water is to be filled in the trough at the beginning of the test, do not cover the specimen with a rubber sleeve.

3. Mount the consolidation cell with specimen in the loading machine. Slowely bring the loading beam to bear lightly on the steel ball placed over the pressure pad. Adjust the beam to a level position indicated by the spirit bubble-tube fixed to the beam. (Note. The bubble should, in fact, be slightly off-centre before the application of any load increment so that it will be level when loaded.) Adjust and read the dial gauge.

4. Apply the first load to exert a pressure intensity of 0.1 kg/cm^2 on the specimen and record dial readings at elapsed times indicated in Step 5. If a saturated specimen is under test or if a specimen is to be saturated at the beginning of the test, fill water in the water trough immediately after the application of the load. (Note. If on adding the pressure gauge indicates specimen swelling, immediatly increase water, the dial intensity to $0.2 - 0.5$ kg/cm^2 or, if necessary, to a still higher value. counteract swelling.)

If a rubber sleeve covered specimen is under test; place a saturated cloth in the trough surrounding the rubber sleeve and keep it wet until the specimen is saturated by flooding with water near the end of the test (Step 8.)

5. After the application of first pressure increment (Stop 4), record the dial readings according to either of the following schedules of total elapsed times: (i) 0.25, 1.00, 2.25, 4.00, 6.25, 9.00, 12.25, 16.00, 20.25, 25. 36 and 49 minutes, and 1, 2, 4, 8, 10 and 24 hours; (ii) 0.25,

0.5, 1, 2, 4, 8, 15, and 30 minutes, and 1, 2, 4, 8, 10 and 24 hours. The primary consolidation is generally complete within 24 hours, but if necessary, extend the readings until the completion of consolidation. Before recording the final dial reading, level the beam by bringing the spirit bubble to the centre. (Note. In the auto-level model, the machine is fitted with an electrically operated, automatic levelling device.)

6. After the final dial reading under the first applied pressure of 0.1 kg/cm^2, increase the pressure to 0.2 kg/cm^2 and repeat Step 5. Consolidate the specimen further under applied pressures of 0.5, 1, 2, 4 and 8 kg/cm^2 (or under 10 kg/cm^2 as well, if desired) and take time-dial readings as in Step 5.

7. When the objective is to eastablish only the pressure-voids ratio relationship, the time-dial readings may be omitted. Apply the pressure increments at 24-hour (or longer) intervals and record only the final dial readings for each pressure increment. If the rate of consolidation is to be determined under a particular pressure increment, take time-dial readings for that increment of pressure only.

8. If water was not filled in the trough at the beginning of the test, fill it after the final consolidation has been reached for maximum applied pressure. Before filling water, remove the wet cloth from the trough and the rubber band from the pressure pad. Pull the loose rubber sleeve upwards, catch it together in a clip on one side and leave it in-place. (Water may be added at other pressures when the effect of saturation is to be observed.) Record the dial reading after 24 hours.

9. Having completed the consolidation of specimen under the maximum pressure and when water in the trough fully surrounds the specimen, reduce the applied pressure to zero, and allow the specimen to take up water and swell. After 24 hours, record the final dial reading corresponding to the end of the test. (If desired, load may be decreased in instalments and time-swelling readings also taken.)

10. Remove the consolidation cell from the loading machine. Take out the specimen ring with the specimen inside. Keep the ring with the specimen in a container and dry in an oven for determining final water content and weight of soil solids.

Calculation

The final voids ratio corresponding to each pressure increment is calculated. A graph is plotted between pressure as abscissa on a logarithmic scale and voids ratio as ordinate on an arithmetic scale. The coefficient of consolidation is calculated from time-consolidation data using either the square root of time fitting method or the logarithm of time fitting method.

The dry density of the specimen at any stage of the consolidation test is calculated from the following relation:

$$\rho_d = \frac{M_s}{AH}$$

where M_s = mass of the dry specimen obtained at the end of the test

   A = cross-sectional area

   H = height of specimen

The initial moisture content w_1 at the beginning of the test is given by:

$$w_1 = \frac{M_1 - M_s}{M_s}$$

where M_1 = mass of wet specimen at the beginning of the test

As a check on the value of w_1 so calculated, initial water content is also determined by drying a sample of cuttings obtained during specimen preparation.

The degrees of saturation at the beginning and at the end of test are given by:

$$S_r = \frac{w_1 \, G}{e_1}$$

$$S_r = \frac{w_{11} \, G}{e_{11}}$$

where: w_1 and e_1 = moisture content and voids ratio at the beginning of test

   w_{11} and e_{11} = moisture content and voids ratio at the end of test

The moisture content, at equilibrium under the maximum applied pressure, of the specimen submerged in water is given by:

$$w = \frac{H_w \, A}{M_s}$$

where H_w = height of water in the specimen under applied pressure
   $= H - H_s - H_a$

   H_a = height of air in the specimen, assumed the same for maximum load condition and the expanded condition under zero load at the end of the test

   H_s = height of solids

Precaution

Specimen preparation should receive special care. Bad preparation. leads to inconsistent results. Any disturbance to an undisturbed specimen greatly affects the consolidation results. Porous stones get clogged with use. Their surfaces should be kept clean by washing with acid and hot water. Sometimes, it may be necessary to slightly grind the surfaces. It is better to use filter papers over the stone surface to prevent entry of soil particles into them, specially if stone has a coarse-grained surface. The loading lever should be brought to a level position before recording the final dial reading each time. For accurate determination of final moisture content at the end of the test, the specimen should be removed from the consolidometer only when all swelling is complete. Accurate determination of the final mass of the specimen is very important. All calculations depend upon the value of the final mass and should there be an error in final weighing, the whole work done for the test would be wasted.

OBSERVATION SHEET 6

ONE-DIMENSIONAL CONSOLIDATION OF SOIL

(a) Beginning of Test

1. Specimen ring:	(i) Diameter	(cm)	7.98
	(ii) Height (H_1)	(cm)	2.4
	(iii) Area (A)	(cm^2)	50
	(iv) Volume (V)	(cm^3)	120
2. Mass ring + glass plates + specimen		(g)	496.81
3. Mass ring + glass plates		(g)	287.40
4. Mass wet specimen (M_1)		(g)	209.41
5. Bulk density $= \dfrac{M_1}{V}$		(g/cm^3)	1.74
6. Moisture content (of cuttings)		(%)	10.8
7. Specific gravity (G)			2.7

(b) End of Test

1. Mass container + ring + wet specimen	(g)	656.85
2. Mass container + ring + dry specimen	(g)	627.84

3. Mass water	(g)	29.01
4. Mass container + ring	(g)	439.18
5. Mass dry specimen (M_s)	(g)	188.66
6. Moisture content (w_{11})	(%)	15.4
7. Height of solids (H_s) $= \dfrac{M_s}{GA}$	(cm)	1.395

(c) Pressure and Voids Ratio

Applid pressure σ' (kg/cm²)	Final dial reading $(10^{-2}$ mm)	Dial change ΔH (mm)	Specimen height $H = H_1 \pm \Delta H$ (mm)	Height of voids $H - H_s$ (mm)	Voids ratio $e = \dfrac{H - H_s}{H_s}$	Remarks
0	100		24.00	10.05	0.720	
		−0.05				
0.1	105		23.95	10.00	0.716	
		−0.05				
0.2	110		23.90	9.95	0.713	
		−0.73				
0.5	183		23.17	9.22	0.661	
		−1.27				
1.0	310		21.90	7.95	0.570	
		−1.19				
2.0	429		20.71	6.76	0.485	
		−1.04				
4.0	533		19.67	5.72	0.410	
		−0.93				
8.0	626		18.74	4.79	0.343	
		−1.01				
8.0	627		18.73	4.78	0.342	Water added at
		−0.19				

10.0	646	+1.31	18.54	4.59	0.329	end of period
0	515		19.85	5.90	0.423	

(d) Density and Degree of Saturation

Particular (1)	Beginning of Test (2)	End of Test (submerged specimen) (3)	Under maximum pressure (submerged specimen) (4)
1. Mass dry specimen (M_s) (g)	188.66	188.66	188.66
2. Height of specimen (H) (cm)	2.400	1.985	1.854
3. Dry density $= \dfrac{M_s}{AH}$ (g/cm^3)	1.57	1.90	2.03
4. Height of solids (H_s) (cm)	1.395	1.395	1.395
5. Moisture content (w) (%)	11.0$_a$	15.4$_b$	11.9$_g$
6. Height of water (H_w) (cm)	0.415$_c$	0.580$_c$	0.449$_f$
7. Height of air (H_a) (cm)	0.590$_c$	0.010$_d$	0.010$_e$
8. Degree of saturation (%) $= \dfrac{wG}{e}$ or $\dfrac{H_w}{H - H_s}$	41.3	98.4	97.7

Note (a) $\omega = \dfrac{M_1 - M_s}{M_s}$;

(b) Obtained from Table-b;

(c) $H_w = \dfrac{M_s \cdot w}{A}$;

(d) $H_a = H - H_s - H_w$

(e) Same as in column 3;

(f) $H_w = H - H_s - H_a$

(g) $w = \dfrac{AH_w}{M_s}$

(e) Time-Consolidation Data

The Table is of the following form:

Applied pressure kg/cm^2	Date	Clock time	Elapsed time	Dial reading (10^{-2} mm) Remarks

EXPERIMENT 7

COMPACTION TESTS

7.A STANDARD PROCTOR COMPACTION TEST

Object

To determine the relationship between moisture content and density of a soil and to determine the optimum moisture content to give maximum dry density by the Standard Proctor Test. (IS Light Compaction Test)

Apparatus

(a) Cylindrical mould 10-cm in diameter, 12.73-cm in height and 1000-ml in capacity, fitted with detachable base plate and collar. Rammer, 2.6-kg in mass and 31-cm in drop.

(b) Balance accurate to 1 g. Palette knife. Straight edge. Mixing basin. 19-mm and 4.75-mm IS sieves. Apparatus for moisture content determination. Sample extractor, if available.

Procedure

1. Weigh the empty mould attached to the base plate but without collar.

2. Take about 3 kg of air-dried and pulverised soil passing a 19-mm sieve. Mix it thoroughly with a small quantity of water. Cover it and leave for a maturing time of at least 5 minutes to permit proper absorption of water.

Note. The test was originally restricted to material passing a 4.75-mm sieve but it can be extended to material passing a 19-mm sieve.

The quantity of water to be added for the first test depends upon the probable optimum moisture content for the soil. For coarse-grained soils, a start can be made with an initial 4 per cent moisture content and for fine-grained soils with 10 per cent; these value are to be taken as rough indications only. As a guide, the plastic limit of a soil may be cosidered to approximately represent the optimum moisture content.

3. Attach the collar to the mould fixed on the base plate. Thoroughly remix the matured soil and compact it into the mould in three equal layers by giving to each layer 25 uniformly distributed blows

of the rammer falling from the full drop over the soil. The last compacted layer should project not more than about 6-mm into the collar. During compaction, rest the mould on a uniform rigid foundation, e.g., a concrete floor or a cube of concrete weighing not less than 100 kg.

Note. The compactive energy for the Light Compaction test with 2.6-kg rammer in the 1000-ml mould is 6,045-kg-cm per 1000-cm^3 of soil.

4. Remove the collar and trim off the soil level with the top of the mould. Weigh the mould full of compacted soil with base plate attached.

5. Remove the soil from the mould and keep a representative sample from the centre of the compacted specimen for moisture content determination.

6. Break up with hand the soil removed from the mould, remix with the remainder of the original sample. Raise its moisture content by 1 to 2 per cent, mix thoroughly, and after allowing some maturing time repeat Steps 3-5. Continue this series of tests until there is either a decrease or no change in the wet compacted weight of the soil in the mould.

Note. A separate and new sample should be used for each compaction test if the soil material is fragile enough to be reduced in particle size due to repeated compaction and also in case of clays where it is difficult to incorporate water. Clay samples at different increasing moisture contents should be matured in covered containers for at least 12 hours before making the compaction tests.

Calculation

The bulk density ρ and the corresponding dry density ρ_d for the compacted soil during any compaction test are respectively given by:

$$\rho = \frac{M}{V} \text{ (g/cm}^3) \text{ and } \rho_d = \frac{\rho}{1 + w} \text{ (g/cm}^3)$$

where   M = Mass of wet compacted specimen (g)
        w = moisture content (ratio)
        V = volume of mould, 1000 ml

A compaction curve is plotted between moisture contents as abscissae and the corresponding dry densities as ordinates. The values of the maximum dry density and the corresponding optimum moisture content w_0 are read from the peak of the curve.

Plotting of Zero Air Voids Line

The theoretical relation between the moisture content and the dry density corresponding to zero air voids condition is plotted from the equation:

$$\rho_d = \frac{G\rho_w}{1 + wG}$$

Precaution

Proper breaking and thorough mixing of sample with water before compaction is very important. If water is inadequately mixed with soil and if a suitable maturing period is not allowed, especially for more plastic soils, small nodules of a hard fine-grained soil tend to act like pieces of aggregate and thus alter the effective grading of the soil which affects the test results. Lower densities and high optimum moisture contents are observed to be obtained, if the mould is not placed on a solid foundation but is placed on a wooden table during compaction. Test results also tend to be inaccurate, if the total compacted depth of the specimen is not kept nearly the constant throughout the various tests.

7.B MODIFIED PROCTOR COMPACTION TEST
(IS Heavy Compaction Test)

The apparatus required, in general, is the same as used for the Standdard Proctor test (Expt. 7.A), except that a bigger rammer, 4.9 kg in mass and 45-cm drop is used.

The soil is compacted into the mould in 5 equal layers (instead of 3) by giving 25 full blows of the rammer to each layer. The rest of the procedure is the same as in Expt. 7.A. The optimum moisture content and maximum dry density are obtained by plotting a compaction curve between moisture content and dry density.

Note 1. The compactive energy for the Heavy Compaction test with 4.9-kg rammer in the 1000-ml mould is 27,500 kg-cm per 1000 cm³ of soil.

7.C COMPACTION IN LARGE-SIZE MOULD

For compacting soil containing coarse fraction upto 37.5 mm size, a mould bigger than the Standard Proctor mould (Expt. 7.A) is used. The rammers are the same as used in Expts. 7.A and 7.B. Both the Standard and Modified Proctor tests can be performed, for which the details are as follows. A fresh sample should be taken for each test with increasing water content.

1. **Standard Proctor (or Light Compaction) test :** Mould, 15 cm in

diameter, 12.73 cm in height and 2250-ml in capacity. Rammer, 2.6-kg in weight and 31-cm in drop. Number of layers = 3. Blows per layer = 55.

2. **Modified Proctor (or Heavy Compaction) test** : Mould, 15 cm in diameter and 12.73 cm in height. Rammer 4.9 kg in weight and 45 cm in drop. Number of layers = 5. Blows per layer= 55.

OBSERVATION SHEET 7.A/7.B/7.C

COMPACTION TEST (STANDARD PROCTOR)

(i) Size of mould   = 10 cm dia, × 12.73 cm height

(ii) Capacity of mould = 1000 ml

(iii) Rammer: 2.6 kg × 31 cm

(iv) No. of layers   = 3

(v) Blows per layer   = 25

(a) Density Determination

Test No.	1	2	3	4	5
1. Mass mould + soil (g)					
2. mass empty mould (g)					
3. Mass compacted soil (w) (g)					
4. Bulk density (ρ) (g/cm³)					
5. Dry density (ρ_d) (g/cm³)					

(b) Moisture Content Determination

1. Container No.	5	6	7	8	9
2. Mass cont. (g) + wet soil	271.33	267.02	272.66	287.38	331.03
2. Mass cont. (g) + dry soil	256.82	249.13	251.83	262.57	279.41
4. Mass water (g) present	14.51	17.89	20.83	24.81	31.62
5. Mass container (g) empty	96.91	94.41	96.33	97.37	96.69

6. Mass (g) dry soil	159.91	154.72	155.50	165.20	182.72
7. Moisture (%) content (w)	9.1	11.6	13.4	15.0	17.3
Proctor Needle penetration resistance (if determined)					

(c) Result (from graph): Maximum dry density =

Optimum moisture content =

7.D JODHPUR MINI-COMPACTOR TEST

Object

To determine the relationship between moisture content and density of soil and to determine the optimum moisture content to give maximum dry density using the Jodhpur Mini-Compactor.

Apparatus

The Jodhpur Mini-compactor comprising the following: Compaction mould, 6-cm high, 50-cm^2 in internal cross-sectional area (7.98 cm in dia.) and 300-ml in capacity. Collar, 3-cm high. Base plate. 2.5-kg. Dynamic Ramming Tool (2.5 kg DRT). The 2.5 kg DRT consists of a 2.5 kg drop weight which falls freely through a height of 25 cm on a foot, 4 cm diameter and 7.5 cm high. The foot has two circular grooves around it at distances of 1.5-cm and 4.5 cm from the top. These grooves help in judging the thickness of the compacted layers during the test.

4.75 mm IS sieve. Mixing basin. Straight edge. Cutting knife. Balance accurate to 1 g. Apparatus for moisture content determination.

Procedure

1. Weigh the mould with the base plate but without collar.
2. Take six (or eight) 800-g samples of air-dried and pulverised soil passing a 4.75 mm sieve. Add varying amounts of water to the samples to give moisture contents ranged on either side of the expected optimum value. Keep the samples covered for a suitable maturing period.
3. Attach the collar to the 0.3-litre mould fixed to the base plate. Keep the assembly on a solid base, e.g., a concrete floor. Thoroughly

remix and compact each sample into the mould, in two equal layers each layer being given 15 blows from the 2.5-kg DRT. (Note. If a compacted specimen is to be used for other tests, such as for permeability, consolidation or shear test, the top of the first layer should be scarified to a depth of about 3 mm before placing the second layer.)

The blows should be uniformly distributed over the surface of each layer. The first blow should be given in the centre of the mould. The grooves on the foot of the DRT will indicate whether the soil is sufficient for the layer. If necessary, add a little soil. While giving the remaining 14 blows, the foot should always touch the side of the collar. The top of the second compacted layer should project not more than about 5-mm into the collar.

4. Remove the collar. Cut off the excess soil and level with the top of the mould. Weigh the mould full of compacted specimen with base plate attached.

5. Remove the soil from the mould and keep a representative sample (at least 40-50 g) from the centre of the compacted specimen for moisture content determination.

Calculation

The bulk density ρ and the corresponding dry density ρ_d of the compacted specimen are respectively given by: $\rho = \dfrac{M}{300}$ and

$$\rho_d = \frac{\rho}{1 + w} \ (g/cm^3)$$

where M = mass of wet compacted specimen (g)

  w = moisture content (ratio)

The values of the maximum dry density and the optimum moisture content are read from the peak of the compaction curve plotted between moisture contents as abscissae and the corresponding dry densities as ordinates.

Note. The values of maximum dry density and optimum moisture content obtained by the Jodhpur Mini-Compactor test are very nearly the same as those obtained by the Standard Proctor Compaction test.

OBSERVATION SHEET 7.D

JODHPUR MINI-COMPACTOR TEST

  (i) Capacity of mould = 300 ml

  (ii) No. of layers = 2

(iii) **Rammer** = 2.5 kg DRT

(iv) **Blows per layer** = 15

(a) Density Determination

Test No.		1	2	3	4	5
1. Mass mould + soil	(g)	3047	3083	3101	3106	3090
2. Mass mould empty	(g)	2459	2459	2459	2459	2459
3. Mass compacted soil	(g)	588	624	642	647	631
4. Bulk density (ρ)	(g/cm³)	1.96	2.08	2.14	2.16	2.10
5. Dry densty (ρ_d)	(g/cm³)	1.80	1.87	1.89	1. 88	1.80

(b) Moisture Content Determination

	5	6	7	8	9
1. Container No.	5	6	7	8	9
2. Mass cont + wet soil (g)	271.33	267.02	272.66	287.28	311.03
3. Mass cont. + dry soil (g)	256.82	249.13	251.83	262.57	279.41
4. Mass water present (g)	14.51	17.89	20.83	24.81	31.62
5. Mass container empty (g)	96.91	94.41	96.33	97.37	96.69
6. Mass dry soil (g)	159.91	154.72	155.50	165.20	182.72
7. Moisture content (w) (%)	9.1	11.6	13.4	15.0	17.3

(c) Result (from graph)   (i) Maximum dry density = 1.89 g/cm³

(ii) Optimum moisture
content = 13.4 %

7.E PROCTOR NEEDLE TEST

Object

To establish the moisture content—penetration resistance relationship of a fine-grained soil as determined by the Proctor needle apparatus.

Apparatus

Proctor needle apparatus (also known as the Proctor soil penetrometer) with a set of standard needle points having end sectional areas of 6, 5, 3, 2, 1.5, 1.0, 0.5 and 0.25 cm². Rest as for the Standard Proctor Test, Expt. 7.A.

Procedure

1. Compact the soil in the mould as described in the Proctor test, (Expt. 7.A). Remove the collar, strike off excess soil and weigh. After taking the weight of the compacted specimen, measure the penetration resistance, before removing the specimen from the mould.

2. Attach a *suitable* needle point to the penetrometer. Keep the mould full of campacted soil between the feet and hold the penetrometer over the specimen in a vertical position. Drive the needle point near the centre into the specimen gradually and firmly at a rate of 12.5-mm per sec to a depth of not less than 75-mm. Record the penetration resistance. (Before pushing in, the sliding ring in the calibrated stem type apparatus and the dial indicator in the hydraulic type, should read zero.) Repeat twice more and take the average of the readings. Penetrations should be near the centre and should not interfere with one another.

3. Repeat the test on each soil specimen compacted with increasing water content in the compaction test.

Calculation

The penetration resistance is calculated by multiplying the average penetrometer reading by the reciprocal of the end area of the needle point.

A curve is plotted between moisture contents as abscissae and the corresponding penetration resistances as ordinates. This curve usually has the x-axis (moisture contents) common with the compaction curve.

EXPERIMENT 8

FIELD DENSITY AND VOIDS RATIO DETERMINATION

8.A SAND REPLACEMENT METHOD

Object

To determine the dry density of natural or compacted soil *in situ* by sand replacement method.

Apparatus

Sand pouring cylinder mounted above a pouring cone and separated by a valve or shutter. Calibrating container. Tray with central circular hole. Chisel and scoop for excavating hole. Balance accurate to 1 g. Apparatus for moisture content determination. Uniformly graded, dry, clean sand, preferably passing a 600-micron sieve and retained on the 300-micron IS sieve.

Calibration of Apparatus

(a) Determination of the moisture of sand required to fill the pouring cone of the cylinder Fill the pouring cylinder with sand upto a height about 2 cm below the top and weigh to get the total initial mass (M_1). (Note. The initial mass is to be maintained constant throughout the tests for which the calibration is used.) Run sand out of the cylinder equal in volume to that of the calibrating container. Place the cylinder over a plain surface, open the shutter and allow the sand to run out and fill the cone. Close the shutter when no further movement of sand takes place in the cylinder. Collect the sand filling the cone and weigh (M_2). Take average of three readings.

(b) Determination of the bulk density of sand ρ_{sand}. Determine the capacity of the calibrating container, if not known, either by measuring the inside dimensions of the container or by weighing the quantity of water required to fill the container. Fill the pouring cylinder with sand to the constant mass M_1 and place it concentrically on the top of the calibrating container. Open the shutter and allow the sand to run out. When no further movement of sand takes place in the cylinder, close the shutter and weigh the cylinder (M_3). Take average of three readings.

Measurement of Soil Density

1. Fill the cylinder to the initial constant mass M_1. Weigh the empty tray.

2. Clean and level off about 50-cm square of the site where density determination is to be made. Place the tray on the prepared surface. Excavate the test hole to the desired depth within the circular hole of the tray and carefully collect all excavated material into the tray.

3. Weigh the tray containing the excavated soil. Keep a representative sample (at least 100 g) for moisture content determination.

4. Carefully place the sand pouring cylinder directly over the excavated hole to cover it concentrically. (The tray is removed before placing the cylinder.) Open the shutter. When sand stops running out, close the shutter, remove the cylinder and weigh.

Calculation

The mass of sand required to fill the calibrating container (M') is given by:

$$M' = M_1 - M_3 - M_2$$

where M_1 = mass of cylinder + sand before pouring into calibrating container (g)

M_3 = mass of cylinder + sand after pouring into calibrating container (g)

M_2 = mass of sand in cone

The bulk density of sand ρ_{sand} in g/cm^3 is given by:

$$\rho_{sand} = \frac{M'}{\text{Vol. of container in ml}}$$

The mass of sand required to fill the excavated hole is given by:

$$M'' = M_1 - M_4 - M_2$$

where M_1 = mass of cylinder before pouring sand into hole (g)

M_4 = mass of cylinder after pouring sand into hole (g)

M_3 = mass of sand in cone (g)

$$\text{In } \textit{situ} \text{ volume of excavated soil} = \frac{M'}{\rho_{sand}} \text{ (cm}^3\text{)}$$

$$\text{In } \textit{situ} \text{ bulk density of soil} \quad \rho = \frac{M}{M'/\rho_{sand}} \text{ (g/cm}^3\text{)}$$

where M = mass of excavated soil (g)

$$\text{In } \textit{situ} \text{ dry density of soil } \rho_d = \frac{\rho}{1+w} \text{ (g/cm}^3\text{)}$$

Precaution

The sand used should be dry and clean, as otherwise, its bulk density considerably varies. Closely graded sand gives better results. The bulk density of sand is found to decrease with decrease in the depth of the calibrating container. The bulk density of sand also decreases with the decrease in the level of sand in the pouring cylinder. Therefore, the depth of the calibrating container used should be approximately equal to the depth of the hole to be excavated. Also, the mass of the cylinder filled with sand should be kept constant at the beginning of each test, including the calibration test. Since the dry density of soil varies appreciably from point to point, a number of determinations should be made to get the average.

OBSERVATION SHEET 8.A

DRY DENSITY BY SAND REPLACEMENT METHOD

(a) Calibration of Apparatus

1. Initial mass of cylinder + sand (M₁)	(g)		10.543
2. Mass sand in pouring cone (M₂)	(g)		435
3. Capacity of calibrating container	(ml)		1,178
4. Mass cylinder + sand after pouring in the calibrating container (M₃)	(g)		8,388
5. Mass sand to fill container (M') = M₁ − M₃ − M₂	(g)		1,720
6. Bulk density of sand (ρ_{sand}) = $\frac{M'}{(3)}$	(g/cm³)		1.46

(b) Density of Soil in Place

1. Mass tray + excavated soil	(g)		4,067
2. Mass tray empty	(g)		1,741
3. Mass excavated soil (M) = (1) − (2)	(g)		2,326
4. Mass cylinder + sand after pouring sand in the hole (M_4)	(g)		8,395
5. Mass sand in hole + cone = $M_1 − M_4$	(g)		2,148
6. Mass sand in hole $M'' = M_1 − M_4 − M_2$	(g)		1,713
7. Volume of hole $(V) = \dfrac{M''}{\rho_{sand}}$	(cm³)		1,172
8. Bulk density of soil $(\rho) = \dfrac{M}{V}$	(g/cm³)		1.98
9. Moisture content (w)	(ratio)		0.08
10. Dry density of soil $(\rho_d) = \dfrac{\rho}{1 + w}$	(g/cm³)		1.83

(c) Moisture Content Determination
Use the Table of sheet 1.A

8.B CORE CUTTER METHOD

Object

To determine the dry density of natural or compacted soil *in situ* by core cutter method.

Apparatus

Core cutter apparatus consisting of a steel cutter, 10 cm in diameter and 13-cm high, a 2.5-cm high steel dolly and a rammer. Balance accurate to 1 g. Palette knife. Straight edge. Spade. Pickaxe, Trowel. Apparatus for water content determination. (Note. As an alternative, the 0.3-litre cutter of the Jodhpur Permeameter can also be used for density determination.)

Procedure

1. Measure the inside dimensions of the cutter and calculate its volume, if not already known.
2. Weigh the cutter without dolly.
3. Clean the top soil on the site. Place the dolly over the cutter

and ram it gently into the soil until about 1 cm of the dolly protrudes above the surface.

4. Dig out the cutter containing the soil from the ground. Remove the dolly and trim off any soil extruding from the ends.

5. Weigh the cutter full of soil and keep a representative sample for water content determination.

6. Calculate the dry density of the soil sample by knowing its mass, volume and moisture content.

Precaution

The cutting edge of the cutter should be kept sharp. The cutter cannot be used on stony or cohesionless soils. Care should be taken not to rock the cutter during the process of ramming.

OBSERVATION SHEET 8.B

DRY DENSITY BY CORE CUTTER METHOD

(a) Density Determination

1. Mass cutter + soil	(g)	2,994
2. Mass cutter empty	(g)	1,002
3. Mass wet soil (M) = (1) − (2)	(g)	1,992
4. Volume of cutter (V)	(cm³)	1,021
5. Bulk density $(\rho) = \dfrac{M}{V}$	(g/cm³)	1.95
6. Moisture content (w)	(ratio)	0.10
7. Dry density $(\rho_d) = \dfrac{\rho}{1 + w}$	(g/cm³)	1.77

(b) Moisture Content Determination
Use the Table of Sheet 1.A.

8.C   WATER DISPLACEMENT METHOD

Object

To determine the dry density of a sample of natural or compacted soil by water displacement method.

Apparatus

A cylindrical metal container fitted with an overflow outlet tube in the upper half; a rubber tube with clip is attached to the outlet tube. Balance accurate to 1 g. Paraffin wax of known specific gravity. Equipment for melting wax. Cutting knife or wire saw for trimming specimen. Graduated measuring cylinder of 1000-ml capacity. Apparatus for moisture content determination.

Procedure

1. Trim to more or less regular shape a small specimen from a larger sample of soil and weigh. The size of the specimen to be prepared depends upon the size of the sample of soil available and the size of the metal container.

2. Cover the specimen with a thin layer of paraffin wax either by directly dipping the specimen in molten wax or by applying wax with a brush. Give two coats of wax; the second coat being applied after the first has hardened. Weigh the waxed specimen.

3. Fill the metal container to above the overflow level. Allow the excess water to run off through the overflow outlet by releasing the clip on the rubber tube fitted to the outlet. Immerse the waxed soil specimen in water. Release the clip and measure the volume of displaced water by running it into a measuring cylinder or by weighing it in grams. After the overflow of displaced water, check that the specimen is fully immersed, otherwise repeat with a smaller specimen.

4. Remove the specimen from water, wipe dry on the outside and divide it down the centre. Take a representative sample for moisture content determination.

Calculation

The volume of the specimen is calculated from the equation:

$$V = V_w - \frac{M_2 - M_1}{\rho_P}$$

where V = volume of specimen (ml)

V_w = volume of water displaced (ml)

M_1 = mass of specimen (g)

M_2 = mass of waxed specimen (g)

ρ_P = density of paraffin wax (g/ml),

(ρ_P is about 0.908 g/ml, it may be determined by any of the usual methods)

The bulk density and the dry density are respectively given by:

$$\rho = \frac{M_1}{V} \text{ and } \rho_d = \frac{\rho}{1 + w}$$

Note. The volume of a waxed specimen can be measured by displacement of water in a partly filled measuring cylinder itsel', if the size of the specimen is small enough to be put in the cylinder.

Alternatively, a small cutter of known volume, such as the 0. -litre cutter of the Jodhpur Permeameter, can be used to cut a specimen from a larger sample of soil. By knowing the weight, volume and moisture content of the cut specimen, the dry density is calculated.

Precaution

Wax should only be just molten. Care should be taken to see that any cavities on the surface of the specimen do not form air bubbles during waxing or immersion of the specimen.

OBSERVATION SHEET 8.C

DRY DENSITY BY WATER DISPLACEMENT METHOD

(a) Density Determination

1. Mass waxed specimen (M_2)	(g)		2,444
2. Mass soil specimen before applying wax (M_1)	(g)		2,385
3. Mass wax used $= M_2 - M_1$	(g)		59
4. Density of wax	(g/ml)		0,098
5. Volume of wax used	(ml)		65
6. Volume of water displaced (V_w)	(ml)		1,219
7. Volume of specimen (V) = (6) − (5)	(cm³)		1,154
8. Bulk density (ρ) $= \dfrac{(2)}{(7)}$	(g/cm³)		2.07
9. Moisture content (w)	(ratio)		0.15
10. Dry densitry (ρ_d)	(g/cm³)		1.8

(b) Moisture Content Determination
Use the Table of Sheet 1.A

8.D VOIDS RATIO DETERMINATION

The voids ratio of natural or compacted soil is calculated from the known values of the dry density and the specific gravity of soil. The dry density ρ_d may be determined by any one of the methods described in Expts 8.A, 8.B or 8.C. The specific gravity G of soil particles is determined as given in Expt. 2. Sometimes an assumed value of G is taken. The voids ratio e can then be calculated from:

$$e = \frac{G\rho_w}{\rho_d} - 1$$

where: ρ_w = density of water which may be taken as g/ml

EXPERIMEMT 9

SHEAR STRENGTH DETERMINATION

9.A SHEAR BOX TEST

Object

To determine the shear strength parameters (c_u, ϕ_u) (c_T, ϕ_T) of a soil by undrained, direct shear test.

Apparatus

Square shear box for testing 6-cm square and 2-cm thick specimens complete with porous stones (2 Nos.), plain grids (2 Nos.), and pressure pad. Alternatively, author's circular shear box for testing 2.4 cm thick specimens of 50-cm² cross-sectional area, complete with stones, grids and pressure pad. (Plate 9.1 — Circular shear box)

Shearing machine fitted with a geared jack to give constant rate of strain and a proving ring complete with dial gauge. Loading yoke. Set of weights to give loading intensity upto 3 kg/cm² (300 kP$_a$) on the specimen. Dial gauge of 0.01 mm accuracy for measuring vertical displacement of specimen.

Balance accurate to 0.1 g. Stop watch. Apparatus for specimen preparation and moisture content determination.

Specimen Preparation

Both undisturbed and remoulded specimens can be tested. For the 6-cm square box, the undisturbed specimen is first obtained by a cutting ring, 2-cm high and 10-cm in diameter. It is pushed out of the ring with a wooden dolly and then cut in the form of a 6-cm square piece. Remoulded specimens can be prepared by first compacting the soil in suitable large-size moulds and then cutting the specimens, 6-cm square and 2-cm thick, from the compacted soil. Alternatively, the soil can be compacted in the shear box itself.

For the 50-cm² circular shear box, undisturbed specimens are obtained by the specimen ring, 50-cm² in area and 2.4-cm high (See parts of author's consolidometer. Expt. 6). Dynamically compacted specimens are prepared by the Jodhpur Mini-Compactor (Expt. 7.D). Statically compacted specimens are prepared in the 50-cm² mould as described in Expt. 12.

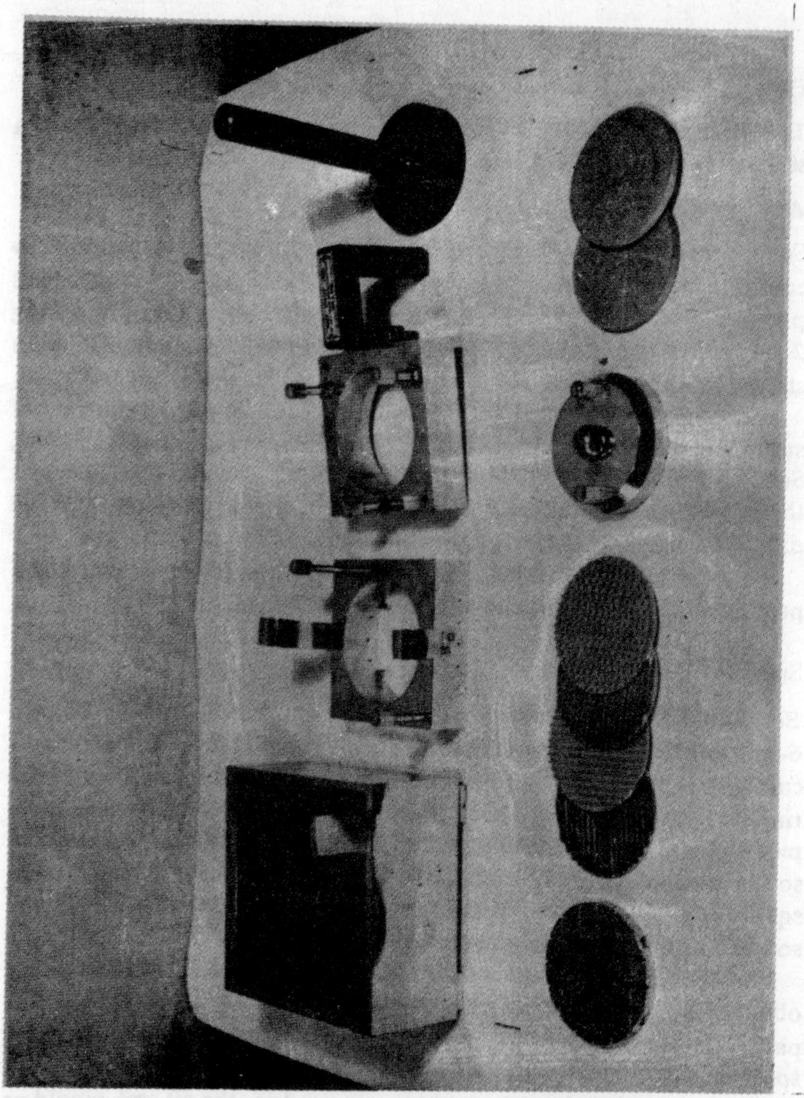

Plate 9.1

Specimens of cohesionless soils should be prepared by filling a constant weight of soil directly in the box.

Test Procedure

1. Fix the upper part of the box to the lower part by tightening the locking screws. Place a porous stone in the box and a plain grid on the stone, keeping the serrations of the grid at right angles to the direction of shear. Place the specimen carefully in the box. Place the upper grid on top of the specimen with serrations at right angles to the direction of shear. Place the upper porous stone on the grid and the pressure pad on the stone. (Note. It may be necessary in many cases to submerge the specimen with water for preventing moisture-loss from the specimen during the test).

2. Mount the shear box with specimen inside on the shearing machine and adjust so that the upper part touches the proving ring. Bring forward the jack to bear up against the box container, which will be indicated by a slight movement of the proving ring dial gauge. Set the dial gauge to read zero.

3. Set the loading yoke on top of the pressure pad. Fix the vertical displacement dial gauge to read zero in contact with the top of yoke. Apply the desired normal load (Note 1.) and record any vertical displacement in the specimen.

4. Remove the locking screws. Raise the upper part of the box slightly relative to the lower part by turning the spacing screws. Afterwards, slacken off the spacing screws.

5. Shear the specimen at a suitable constant rate of strain (Note 2.) Record readings of the proving ring dial gauge and the vertical dial gauge at every half minute. Continue the test until the specimen fails. The specimen is assumed to fail when the proving ring dial gauge begins to recede after recording the maximum, or at a shear displacement of approximately 15 per cent of the length (diameter) of the specimen, if no definite maximum is obtained and the same readings continue for increasing values of strain.

6. Repeat the test on identical specimens under increasing normal loads. At least three tests should be carried out.

Note 1. The intensity of normal loads can be in the order of $\frac{1}{4}$, $\frac{1}{2}$, 1, 2, and 3 (or 4) kg/cm^2. (25, 50, 100, 200 and 300 kP$_a$).

Note 2. For undrained test, the rate of shear displacement may vary from 1.0 to 2.5 mm per minute. The normal rate may be suggested as 1.25 mm per minute.

Note 3. For drained tests, perforated grids are used in place of plain grids. The rate of shear displacement may vary from 0.005 to

0.02 mm per minute for clays and from 0.2 to 1.0 mm per minute for sands. The lower the permeability, the slower is the rate of shear displacement.

Calculation

A graph is plotted between shear displacement (or strain) as abscissa and shear stress (or shear force) as ordinate for each test. The shear stress at failure is called the shear strength under a particular normal load. Another graph is plotted between the normal stress (or normal load) as abscissa and the shear strength (or shear force at failure) as ordinate, from which the values of c_u and ϕ_u are determined. Sometimes the graph between shear strain and stress may be omitted, and the shear strength can be directly known from the proving ring dial gauge readings.

OBSERVATION SHEET 9.A

SHEAR BOX TEST

(Type of test: Undrained)

(a) Specimen Data

(i) Length (or dia.) = 6 cm   (iii) Thickness = 2 cm
(ii) Area = 36 cm²   (iv) Volume = 72 cm³

Test No	1	2	3	4
1. Initial mass (g)	158.4	158.4	158.4	
2. Initial bulk density (g/cm³)	2.2	2.2	2.2	
3. Initial moisture content (%)	16	16	16	
4. Final moisture content (%)	—	15.8	15.5	

(b) Shear Force and Deformation

(i) Proving ring constant: 1 div = 0.368 kg
(ii) Mass of loading yoke = 3.5 kg
(iii) Rate of shear = 1.25 mm/min
(iv) Test No. = 2
(v) Normal load (inclusive of yoke) = 18 kg

Elapsed time (min)	Shear displacement (mm)	Proving ring dial (div)	Shear force (kg)	Vertical dial (div)	Vertical dial change (10^{-2} mm)
0	0	0	0	550	Before load
0	0	0	0	530	After load
$\frac{1}{2}$	—	15	4.62	526	—4
1	1.25	21	6.47	526	—4
$1\frac{1}{2}$	—	26	8.00	525	—5
2	2.50	29	8.93	523	—7
$2\frac{1}{2}$		31	9.55	522	—8
3	3.75	33	10.16	521	—9
$3\frac{1}{2}$		34	10.47	520	—10
4	5.00	35	10.78	519	—11
$4\frac{1}{2}$		35	10.78	519	—11
5	6.25	35	10.78	519	—11

Note. Use as many tables as the number of tests.

(c) Normal Stress and Shear Strength

Test No.	Normal load (kg)	Shear force at failure (kg)	Normal stress (kg/cm^2)	Shear strength (kg/cm^2)	Shear parameters from graph
1	9	8.00	0.25	0.222	$c_u = 0.167$ kg/cm^2
2	18	10.78	0.50	0.299	$\phi_u = 15°$
3	36	15.40	1.0	0.428	

9.B TRIAXIAL COMPRESSION TEST

Object

To determine the shear strength parameters (c_u, ϕ_u) (c_T, ϕ_T) of a soil by undrained, triaxial compression test.

Apparatus

Triaxial cell. Solid end-caps and pressure cap for specimen. Apparatus for preparation of specimen, such as, sampling tube, sample extractor, split mould, sand former, membrane suction tube. Rubber membrane and sealing rings.

Constant-rate-of-strain compression machine fitted with a proving ring complete with dial gauge. Dial gauge accurate to 0.01 mm for measuring axial deformation (strain). Apparatus for building cell pressure consisting of a water reservoir, compressor or pump, pressure gauge and connecting tube for the cell. Stop watch. Balance accurate to 0.1 g. Apparatus for moisture content determination.

Procedure

1. Obtain an undisturbed or remoulded specimen. Determine its bulk density and also its initial moisture content from cuttings. (Note. The specimen may be of any one of the following sizes: 3.8-cm in diameter by 7.6-cm high; 4-cm in diameter by 8-cm high.

2. Enclose the specimen with the rubber membrane. Mount it on the base of the triaxial cell with solid end caps placed on either end of the specimen. Place also the pressure cap on top. Seal the membrane on to the caps with rubber rings. Carefully assemble the cell with piston of the top cap raised in the upward position.

3. Place the triaxial cell in the compression machine. Admit water in the cell with air release valve open until water escapes from the valve, which is then closed. Raise the hydrostatic pressure to the required amount and keep it constant till the end of test.

4. By hand operation of the compression machine, lower the proving ring to rest on the cell piston and to move it slightly downwards. Set the proving ring dial gauge to read zero, after subtracting the reading as was necessary to move the piston against the friction force and the cell pressure.

5. Move the piston down by hand to rest on the pressure cap and bring the proving ring again in contact with the piston. Mount and adjust the strain dial gauge to read zero.

6. Start the axial loading at a constant rate of strain of approximately 2 per cent per minute (or 1.25 mm per minute). Take readings of the stress and strain dial gauges for every 0.5-mm deformation upto 7.5-mm and, thereafter, for every 1.0-mm deformation. Continue the test until a recession of axial load is observed or three consecutive equal load values are obtained. Sketch the failed specimen. Weigh and determine the moisture content of the whole specimen.

7. Repeat the test at least on two more identical specimens with increasing cell pressure.

Calculation

Assuming the volume of specimen to remain constant and the area of specimen to increase uniformly with decrease in length, the deviator strees is calculated on the basis of changed area at various strains for each test. (See calculations of Expt. 9.C). A graph is plotted between strain as abscissa and deviator strees as ordinate, from which the maximum value of deviator stress is read. A Mohr circle is drawn for each test at failure. A common tangent to the Mohr circles gives the failure envelope from which the parameters c_u and ϕ_u can be determined.

Note. For a detailed treatment of triaxial testing, See author's book Soil Engng, in Theory and Practice Vol. 2. Geotechnical Testing and Instrumentation.

OBSERVATION SHEET 9.B
TRIAXIAL COMPRESSION TEST

(Type of test: Undrained)

(a) Specimen Data

   (i) Diameter = 3.75 cm

   (ii) Length   = 7.5 cm

   (iii) Area   = 11.04 cm^2

   (iv) Volume   = 82.8 cm^3

Test No.	1	2	3	4
1. Initial mass (g)	167.3	167.3	167.3	
2. Initial bulk density (g/cm^3)	2.02	2.02	2.02	
3. Initial moisture content (%)	22.5	22.5	22.5	
4. Final moisture content (%)	22.2	22.2	22 0	

(b) Compression Test

   (i) Proving ring constant: 1 div = 0.310 kg

   (ii) Rate of strain = 1.25 mm/min

   (iii) Test No. = 2

   (iv) Cell pressure = 2.1 kg/cm^2

Axial deformation (mm)	Proving dial (div)	Additional axial load (kg)	Specimen area (cm^2)	Deviator stress (kg/cm^2)	Axial strain (%)
0	0	0	11.04	0	0
0.5	13	4.03	11.10	0.363	0.67
1.0	22	6.82	11.20	0.608	1.33

1.5	31	9.61	11.25	0.855	2.00
2.0	37	11 47	11.35	1.010	2.66
2.5	42	13.02	11.40	1.144	3.33
3.0	48	14.88	11.50	1.293	4.00
3.5	52	16.12	11.55	1.396	4.66
4.0	57	17.67	11.65	1.518	5.33
4.5	61	18.91	11.75	1.610	6.00
5.0	64	19.64	11.83	1.660	6.66
5.5	68	21.08	11.90	1.775	7.33
6.0	70	21.70	12.00	1.810	8.00
6.5	73	22.63	12.10	1.875	8.66
7.0	75	23.25	12.18	1.910	9.33
7.5	78	24.18	12.26	1.970	10.00
8	80	24.80	12.36	2.005	10.66
9	83	25.75	12.55	2.050	12.00
10	86	26.66	12.74	2.090	13.33
11	90	27.90	12.95	2.155	14.70
12	92	28.52	13.15	2.165	16.10
13	95	29.45	13.35	2.206	17.33
14	95	29.45	13.57	2.178	18.67
15	95	29.45	13.80	2.136	20.00
16	96	29.76	14.05	2.120	21.33
17	97	30.07	14.30	2.100	22.67
18	97	30.07	14.53	2.065	24.00
19	99	30.69	14.80	2.075	25.33
20	101	31.31	15.05	2.080	26.67

Note. Use as many tables as the number of tests.

(c) Principal Stresses and Shear Parameters

Test No.	Cell Presure (kg/cm²)	Deviator stress at failure (kg/cm²)	Major stress (kg/cm²)	Shear parameters (from graph)
1	1.4	2.10	3.50	$c_u = 1.07$ kg/cm²
2	2.1	2.20	4.30	$\phi_u = 0$ degree
3	2.8	2.14	4.94	

9.C UNCONFINED COMPRESSION TEST

Object

To determine the unconfined compressive strength and the undrained shear strength of a clayey soil with $\phi_u \cong$ zero.

Apparatus

Unconfined compression tester fitted with flat-face seatings for specimen, a 100-kg proving ring complete with dial gauge and a deformation dial gauge of 25-mm travel and 0.01-mm accuracy. Sampling tube, 3.8-cm in internal diameter, 4-cm in outside diameter and 15 to 20-cm in length. Split mould, 3.8-cm in diameter and 7.6 cm long. sample extractor screw type or simply a piston (dolly) with flat end. Balance accurate to 0.1 g. Trimming knife. Apparatus for moisture content determination. (Alternatively, the sampling tube is 4-cm in internal diameter and the split mould is 4-cm in diameter and 8-cm long.)

Procedure

1. Obtain a sample by the sampling tube. (Note. Samples can be undisturbed or remoulded. Undisturbed samples can be obtained directly from the ground or can be cut from larger blocks or samples obtained by other suitable tools. Remoulded samples can be first prepared in a larger mould and then cut by the sampling tube or can be directly prepared in the split mould.)

2. Lightly oil the inside of the split mould and weigh. Push the sample out of tube into the split mould. Cut off flush with the ends. and weigh the mould with specimen. Keep the cuttings for moisture content determination. After slightly pushing the specimen in the split mould, open the mould and take out specimen. (Note. The sample should always be pushed in the sampling tube or the mould along the same direction in which it enters the tube or the mould.)

3. Place the specimen centrally on the lower seating and move down the upper seating to touch the top of specimen. Set the proving ring and deformation dial gauges to read zero. Load the specimen by turning the handle at a rate of 4 revolutions per minute (at a rate of strain of approximately 5 per cent per minute). Record the proving ring dial gauge readings corresponding to every millimeter deformation of the specimen. Continue the test until the load dial gauge recedes after recording a definite maximum or upto a vertical deformation of 2 cm, whichever occurs earlier. Sketch the failure pattern of the specimen.

Calculation

The axial strain ϵ, changed cross-sectional area A at any strain and the corresponding axial stress σ are calculated from the equations:

$$\epsilon = \frac{100 \, \Delta L}{L_1}$$

$$A = \frac{100 \, A_1}{100 - \epsilon} = \frac{V_1}{L_1 - \Delta L}$$

$$\sigma = \frac{Q}{A}$$

where: ΔL = change in length of specimen

L_1 = initial length of specimen

A_1 = initial area of specimen

V_1 = initial volume of specimen

Q = compressive load

A graph is plotted between axial strain (or deformation) as abscissa and axial stress as ordinate. The maximum stress is termed the unconfined compressive strength q_u. If no maximum occurs, the stress at 20 per cent strain is termed q_u. The undrained shear strength c_u may be taken as $\frac{1}{2}q_u$ in case of saturated intact clays.

OBSERVATION SHEET 9.C

UNCONFINED COMPRESSION TEST

(a) Specimen Data

1. Specimen:	(i) Diameter	(cm)	3.75
	(ii) Area	(cm²)	11.04
	(iii) Length	(cm)	7.5
	(iv) Volume	(cm³)	82.8
2. Mass split mould + specimen		(g)	484.2
3. Mass split mould		(g)	318.4
4. Mass specimen		(g)	165.8
5. Bulk density		(g/cm³)	2.0
6. Moisture content		(%)	25

(b) Compression Test

(i) Proving ring constant: 1 div = 0.1432 kg

Axial deform-ation (mm)	Proving ring dial (div)	Compres-sive load (kg)	Specimen length (mm)	Specimen area (cm²)	Compres-sive stress (kg/cm²)	Axial strain (%)
0	0	0	75	11.04	0	0
1	30	4.3	74	11.20	0.394	1.33
2	55	7.9	73	11.35	0.695	2.66
3	70	10.0	72	11.50	0.870	4.00
4	90	12.9	71	11.65	1.108	5.33
5	110	15.8	70	11.83	1.335	6.66
6	130	18.6	69	12.00	1.550	8.00
7	150	21.5	68	12.18	1.770	9.33
8	160	22.9	67	12.36	1.850	10.66
9	175	25.0	66	12.55	1.990	12.00
10	184	26.4	65	12.74	2.070	13.33
11	198	28.4	64	12.95	2.190	14.70
12	206	29.5	63	13.15	2.250	16.00
13	216	31.0	62	13.35	2.320	17.33
14	222	31.8	61	13.57	2.345	18.67
15	230	33.0	60	13.80	2.390	20.00
16	230	33.0	59	14.05	2.350	21.33
17	232	33.2	58	14.30	2.320	22.67
18	240	34.4	57	14.53	2.360	24.00
19	250	35.8	56	14.80	2.420	25.33
20	265	38.0	55	15.05	2.520	26.67

Result: (i) Failure strain = 20 per cent

(ii) Unconfined compressive strength = 2.39 kg/cm²

EXPERIMENT 10

CALIFORNIA BEARING RATIO DETERMINATION

10.A TEST ON LABORATORY COMPACTED SPECIMEN

Object

To determine the California Bearing Ratio (CBR) of a soil by testing laboratory compacted specimen.

Apparatus

(a) Alternatively. CBR mould, 15-cm in diameter and 17.5-cm high, with detachable perforated base plate. Collar, 5-cm high. Displacer disc, 14.8-cm in diameter and 4.77-cm thick, with a removable handle (Note. With this mould and disc, the test specimen is 15-cm in diameter, 12.73-cm in height and 2,250 cm³ in volume.) Standard plunger, 5-cm in diameter (19.635-cm² cross-sectional area). One annular and several slotted surcharge weights, 2.5-kg each. Compaction rammers, 2.6-kg weight with 31-cm drop, 4.9-kg weight with 45-cm drop.

(b) Loading machine with a capacity of at least 5 tonnes, fitted with a calibrated proving ring to which the standard plunger shall be attached. Micrometer dial gauge, 25-mm travel accurate to 0.01 mm for measuring penetration. Soaking tank. Swelling gauge consisting of a perforated plate with an adjustable extension stem, bridge for holding dial gauge, and dial gauge accurate to 0.01 mm. IS sieves, 4.75 mm and 19 mm. Straight edge. Filter paper, 15-cm in diameter. Mixing basin. Balance accurate to 1 g. Apparatus for water content determination.

Procedure

Note. Test specimens are prepared in the laboratory by either dynamic or static compaction. The dry density for remoulding should be equal to either the field density, or if the subgrade is to be compacted, to a value at which the subgrade will be compacted, i.e, equal to either Standard Proctor (light) compaction or Modified Proctor (heavy) co-

compaction. The moulding water content should be that which, it is estimated. the soil will have during construction. If the penetration test is to be performed on unsoaked specimen, themoulding water content should be the same as the equilibrium water content which the soil is likely to attain subsequent to the construction of the pavement.

(a) Preliminary Work

1. Before preparing a specimen for the CBR test, perform a separate, preliminary compaction test in the CBR mould (Expt. 7.C) using the Standard of Modified Proctor compaction, as the case may be, and determine the optimum moisture content maximum dry density can be fixed on the basis of Expts 7.A, 7.B, or 7.D.

2. Take about 5 to 6 kg of soil (greater quantity required in case of granular material) passing a 19-mm sieve. Discard the coarse aggregate retained on the 19-mm sieve and replace it by an equal portion of the original material passing the 19-mm sieve but retained on the 4.75-mm sieve. Alternatively, soil material upto (40 mm) size can be used directly for the test. Add enough water to raise the water content to the desired value say optimum moisture content). It is necessary to know the hygroscopic water present in the air-dried sample before adding additional water. Mix thoroughly and keep a representative sample for moisture content determination. Compact by either dynamic or static compaction as follows.

(b) Dynamic Compaction

1. Weigh the empty mould without base plate and collar. Keep the displacer disc on the base plate and a filter paper over the disc. Fix the mould to the base plate, with displacer disc inside the mould, and attach the top collar.

2. Compact the wet soil in the mould using either light or heavy compaction. For light compaction, compact the soil in 3 equal layers, each layer being given 55 uniformly distributed blows of the 2.6-kg rammer. For heavy compaction, compact the soil in 5 equal layers, each layer being given 55 blows of the 4.9-kg rammer. The top layer should project not more than about 10-mm in the collar.

3. Remove the collar and trim off excess soil in level with the top of the mould. Detach the base plate and remove the displacer disc also. Weigh the mould full of compacted specimen to get to the wet mass of the specimen. By knowing the moulding moisture content, the dry density of the compacted specimen can be calculated.

4. Place a filter paper on the perforated base plate. Fix the mould upside down to the base plate, so that the surface of the specimen

which was down wards in contact with the displacer disc during compaction is now turned upwards on which the penetration test is performed. The specimen is then ready for penetration test (if unsoaked test is to be performed) or for soaking.

(c) Static Compaction

1. Take about 5 to 6 kg of soil and increase its moisture content to the desired value w (Step a-2).

2. Weigh in grams the required amount of wet soil (W), having a moisture content w, to give the desired dry density ρ_d (g/cm³), as calculated by the equation:

$$W = \rho_d (1 + w) V$$

where: V = volume of compacted specimen
         = 2,250 cm³, for 15-cm dia. mould

3. Fix the mould on the base plate. Place a filter paper inside. Pour the weighed amount of wet soil into the mould. Tamp the soil by hand during the process of pouring and make the top surface roughly level. Put a filter paper on top and insert the displacer disc. Keep the mould assembly in a compression machine (hydraulic press) and press the disc down in level with the top of the mould. Release the pressure, after allowing a little rest period. Remove the disc. The specimen is then ready for soaking or for penetration test, if soaking is not desired.

(c) Soaking Specimen

Weigh the mould with base plate and specimen. After keeping a filter paper on top of the specimen, place the perforated top plate with adjustable stem over the specimen. Apply a surcharge equivalent to the expected pavement. (Note. Each 2.5-kg weight is approximately equivalent to 7 cm of construction. Surcharge is applied in multiples of full weight of 2.5-kg Minimum surcharge is 5 kg.) Immerse the whole assembly in water, so as to allow free access of water to both the top and bottom of specimen. Mount the dial gauge bridge on the mould and set the dial gauge to read zero in contact with the stem of the perforated plate. Allow the specimen to soak under constant water level for 4 days or until the rate of swelling has reduced to almost zero. (A short immersion period may be permissible for soils taking up water readily.) Record the swelling every 24 hours. At the end of soaking, take out the mould and allow it to drain downwards for 15 minutes it may be necessary to tilt the specimen for removing free water. Remove the surcharge weights, the perforated top plate and the filter paper. Weigh the specimen to know the weight of water absorbed.

(e) Penetration Test

1. Place the mould in the testing machine. Apply surcharge over the specimen equivalent to the weight of pavement (See note of Step d). If the specimen has been soaked previously, the surcharge should be the same as used during soaking test. (Note. In case annular surcharge weights are available, all the weights should be placed before seating the plunger, otherwise, the first annular weight should be placed before seating the plunger and the remaining weights (slotted) can be placed after seating the plunger).

2. Seat the plunger on the surface of specimen under a load not exceeding 4 kg. Set the load and penetration measuring dial gauges to read zero.

3. Apply load to the plunger so that the rate of penetration remains constant at 1.25 mm per minute. Record the load readings at the following penetrations: 0.5, 1.0, 1.5, 2.0, 2.5, 3.0, 4.0, 5.0, 7.5, 10 and 12.5 mm.

4. When the penetration test is complete, take a sample for moisture content determination from the top 3-cm layer of the specimen at the seat of the plunger.

Calculation

A smooth curve is drawn between the penetration of plunger, plotted as abscissa, and the load on plunger, plotted as ordinate. If the initial portion of the curve is concave upwards, the corrected origin point is shifted to the point of intersection of a tangent drawn to the curve at the point of greatest slope with the penetration axis. Corrected test loads corresponding to 2.5-mm and 5-mm penetrations are read from the curve. The test load divided by the corresponding standard loads and expressed as a percentage give the CBR at 2.5-mm and 5-mm penetrations. The greater of the values is taken for design purposes. Usually, the CBR value at 2.5-mm penetration is greater. If it is not so, the test is repeated and the 5-mm value adopted, only when identical results follow.

The standard loads for the plungers are given in Table 3.

The exansion ratio during the soaking test is determined from the expression:

$$\text{Expansion ratio} = \frac{R_f - R_i}{H_i} \times 100$$

Table 3
STANDARD LOADS

Plunger 5 cm in diameter

Penetration (mm)	Load (kg)
2.5	1370
5.0	2055
7.5	2630
10.0	3180
12.5	3600

where: R_f = final dial gauge reading (mm)

R_i = initial dial gauge reading (mm)

H_i = initial height of specimen (mm)

OBSERVATION SHEET 10A

CALIFORNIA BEARING RATIO TEST

(a-1) *Dynamic Compaction*

(i) Rammer: 4.9 kg. × 45 cm

(ii) No. of layers = 5

(iii) Blows per layer = 55

(iv) Moulding moisture content = 10%

1. Mass mould + compacted specimen	(g)		13,010
2. Mass mould empty	(g)		7,855
3. Mass compacted specimen	(g)		5,155
4. Vol. of specimen	(cm³)		2,250
5. Bulk density	(g/cm³)		2.29
6. Dry density	(g/cm³)		2.08

(a-2) *Static Compaction*

1. Dry density required	(g/cm³)	
2. Moulding moisture content	(%)	
3. Wt. wet soil compacted	(g)	

(b) *Soaking Test*

(i) Mass mould + base plate + specimen before soaking

= 15,720 g

(ii) Mass mould + base plate + specimen after soaking
$$= 15,845 \text{ g}$$

(iii) Mass water absorbed = 125 g

(iv) Surcharge weight = 5 kg.

Date/Time	4.8.91 8.00	5.8.91 8.00	6.8.91 8.00	7.8.91 8.00	8.8.91 8.00	9.8.91 8.00
Dial gauge reading (mm)	0.00	1.05	1.46	1.71	1.88	1.89
Total expansion (mm)		1.05	1.46	1.71	1.88	1.89
Final expansion ratio	$\frac{1.89}{127} \times 100 = 1.5$ per cent					

(c) *Penetration Test* (Unsoaked/soaked)

(i) Proving ring constant: 1 div = 4.32 kg

(ii) Plunger: 5 cm in dia

(iii) Surcharge weight = 5 kg

(iv) Test-end of specimen: displacer disc end/flush end

Penetration (mm)	Proving ring dial (div)	Load on plunger (kg)	Corrected load (kg)	Standard load (kg)	CBR (%)
0.0	0				
0.5	31	134			
1.0	69	298			
1.5	88	381			
2.0	98	424			
2.5	106	458	480	1370	35
3.0	113	489			
4.0	123	531			
5.0	130	561	570	2055	27.8
7.5	142	614			
10.0	150	648			

(d) *Moisture Content after Penetration Test*

$$w_{11} = 12.5 \text{ per cent}$$

(use the Table of Sheet 1.A)

10.B TEST ON FIELD SPECIMEN

The CBR test can be run on undisturbed field specimens obtained from an existing subgrade soil in its natural condition or from field compacted subgrades. A CBR mould 15-cm in diameter and only 12.73-cm high, is used for taking the specimen. A steel cutting shoe is fitted to the mould which is lightly oiled inside and out and is gently pushed into the ground. The ground should be trimmed ahead away from around the cutting edge to help the mould just slide over the sample so formed. Light tappings may be applied on the mould for driving. The mould full of soil is dug out, ends trimmed flush and covered with wax coating and then transported to the laboratory.

In the laboratory, wax coating is removed and the specimen is weighed to get its bulk density. Base plate is fixed on one end and a collar on the other end. After keeping the necessary surcharge weights, penetration test is performed. A representative sample from the tested end is then taken for moisture content determination. If soaking and subsequent testing is desired, the specimen is reweighed, soaked and then tested on the other end which was not used for unsoaked penetration test.

EXPERIMENT 11

NORTH DAKOTA CONE TEST

Object

To determine in situ the bearing value of a subgrade by the North Dakota Cone apparatus.

Apparatus

The North Dakota Cone apparatus complete with weights so that the cone may be loaded to 4.5, 9, 18 and 36 kg. (*Note*. The weights are inclusive of the weight of the shaft and the cone.)

Procedure

1. Scrap level the test site. Set the apparatus in place and lower the cone to just touch the surface of the ground. Lock the shaft and take the initial reading on the scale.

2. Load the cone to 4.5 kg. Unlock the shaft slowly to prevent impact and allow the cone to penetrate for 1 minute. Lock the shaft and note the penetration reading.

3. Repeat Step 2 with cone loaded to 9, 18 and 36 kg.

Calculation

The bearing value is given by:

$$q_c = \frac{Q}{\pi(\rho \tan 7°45')^2} = \frac{Q}{0.058 \, \rho^2}$$

where: q_c = bearing value (kg/cm^2)

Q = load on cone (kg)

ρ = (corrected) penetration of cone (cm)

Note. Theoretically, neglecting friction, the penetration under 4.5 kg is half of the penetration under 18 kg; and the penetration under 9 kg is half of the penetration under 36 kg. Actually it is never so, because of the rounding of the tip of the cone and also because the cone may have slightly penetrated or may be slightly above the surface at the start of the test. A correction is, therefore, added to or subtracted from all the readings so that the penetration ρ_9 under 9 kg becomes half of the penetration ρ_{36} under 36 kg. The correction is thus given by:

$$C = \rho_{36} - 2\rho_9$$

While determining the mean bearing value, it is usual to omit the reading corresponding to the 4.5-kg load.

OBSERVATION SHEET 11

NORTH DAKOTA CONE TEST

Load (kg)	Scale reading (cm)	Penetration ρ' (cm)	Corrected penetration ρ (cm)	Bearing value (kg/cm²)
0	8.65			
4.5	7.05	1.60	1.95	
9	6.45	2.20	2.55	23.9
18	5.35	3.30	3.65	23.3
36	3.90	4.75	5.10	23.9

Correction = 4.75 − 2 × 2.20 = + 0.35 cm	Mean bearing value = 23.7 kg/cm²

EXPERIMENT 12

SWELLING PRESSURE DETERMINATION

Object

To determine the swelling pressure of an expansive soil.

Apparatus

(a) Swelling pressure apparatus comprising the following: swelling pressure cylindrical mould. 50 cm² in internal cross-sectional area (7.98 cm in dia.) and 6-cm high.* Perforated base plate. Porous stones (2 Nos), 79-mm in diameter and 12-mm thick. Perforated pressure pad. Compaction end-plugs, 2 Nos, 12-mm and 24-mm high. Water trough. Load frame fifted with a 1000-kg proving ring complete with dial gauge, a load transfer piston and a strain measuring dial gauge of 0·01-mm accuracy. (All submersible parts of the apparatus are of non-corrodible material) Figure 7 shows the author's apparatus.

(b) Balance accurate to 0.1 g. Filter paper, 79 mm in diameter. Mixing basin. Apparatus for moisture content determination.

Procedure

1. Mix an appropriate quantity of water to about 350 g pulverised soil to bring its moisture content to the desired value and keep it in air-tight container for a maturing period of at least 18 hours. Afterwards, keep a sample for moisture content determination.

2. Weigh a sufficient quantity of wet soil at known moisture content to give the desired density when compressed to a test specimen, 2.4-cm in thickness, 50-cm² in area and 120-cm³ in volume.

3. Keep the swelling pressure mould on the 12-mm end-plug with bottom end of the mould resting on the flange of the plug. Pour the weighed quantity of wet soil into the mould after placing a filter paper on the end-plug. Press and roughly level the soil with hand, place a filter paper on the soil and insert the 24-mm end-plug. Form the specimen by pressing in the top plug by means of a compression machine or other suitable press. Keep the plugs pressed for a while before releasing the pressure. Remove the plugs. Fix the mould to the perforated base

*Same internal dimensions as those of the mould of Jodhpur Mini-Compactor (Expt. 7.D).

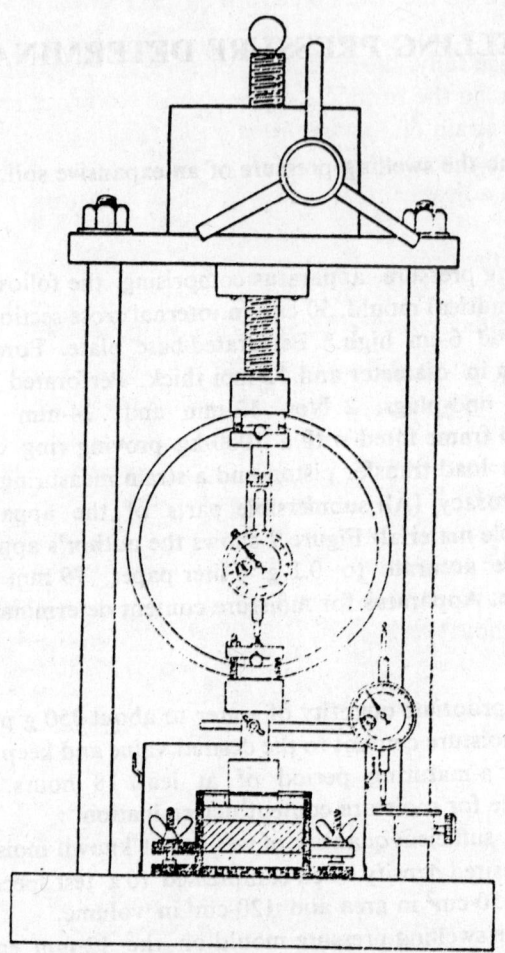

Author's Swelling Pressure Apparatus

plate with a porous stone inserted in the bottom end of the mould. Place the other porous stone on top of the specimen inside the mould and over it the perforated pressure pad.

Note. Remoulded specimens can alternativley be prepared by the Jodhpur Mini-Compactor. The compacted sample is ejected out of the mould, from which a 2.4-cm thick specimen can be cut. Specimens can also be prepared from undisturbed or field compacted blocks of soil, for which the author's specimen rings, 2.4-cm high and 50-cm² in area (See Expt. 6), can be used. The end-plugs can also be pressed in under dynamic blows of the 2.5-kg DRT.

4. Keep the specimen mould inside the water trough and place the assembly centrally on the base of the load frame. Apply a seating load

of 3.5 kg and set the load and strain measuring dial gauges to read zero.

5. Submerge fully. the specimen, top stone and pressure pad by filling water in the trough, and keep them so throughout the test. As soon as the strain dial gauge measures a swelling by 1 to 2 divisions, bring it to read zero by moving the handle on top of the load frame. (Note. Such adjustment is necessitated often during the first day of of the test.)

6. Record the proving ring dial gauge readings at total elapsed times of 0.25, 0.5, 1, 2, 4, 8, and 24 hours, and thereafter, every 24 hours until the swelling pressure becomes essentially constant.

(Note The test should be continued for at least 4 days. Increase the pressure on the specimen, if the strain dial gauge indicates swelling. After the first one-hour reading, this adjustment should be done about 20 minutes before the pressure-reading time and also in between, whenever necessary. It may be necessary to leave the strain dial gauge to read short of zero as the increase in pressure may cause consolidation.)

7. After the completion of pressure observations, keep a sample of soil for moisture content determination. (*Note.* It may be preferable to dry the whole specimen.)

Calculation

The load (kg) indicated by the proving ring divided by 50 to get the swelling pressure in kg/cm² for any observation time. A graph is then plotted between observation time as abscissa and pressure as ordinate. The maximum, constant pressure exerted is termed the swelling pressure. The degree of saturation on completion of test is also determined.

OBSERVATION SHEET   12

SWELLING PRESSURE TEST

(*a*) *Specimen Data*

1. Hleight of specimen	(cm)	2.4
2. Area of specimen	(cm²)	50
3. Volume of specimen	(cm³)	120
4. Initial moisture content	(%)	12
5, Dry density	(g/cm³)	1.75
6. Mass wet soil compacted	(g)	235

(*b*) *Swelling Pressure*

(i) Proving ring constant: 1 div = 0.447 kg

Date	Clock time	Elapsed time (hr)	Proving ring dial (div)	Load (kg)	Pressure (kg/cm²)
21.8.65	8.00	0	0	—	—
	8.15	0.25	64	28.6	0.57
	8.30	0.5	96	42.9	0.86
	9.00	1.0	131	58.6	1.17
	10.00	2	177	79.0	1.58
	12.00	4	206	92.0	1.84
	16.00	8	240	107.1	2.14
22.8.65	8.00	24	243	108.6	2.17
23.8.65	8.00	48	245	109.5	2.19
24.8.65	8.00	72	246	110.0	2.20
25.8.65	8.00	96	246	110.0	2.20

(c) **Final Moisture Content and Degree of Saturation**

1. Final moisture content	(%)	20.8
2. Dry density	(g/cm³)	1.75
3. Specific gravity		2.78
4. Voids ratio		0.59
5. Degree of saturation	(%)	98

REFERENCES

Acker III, W.L. 1974. Basic Procedures for Soil Sampling and Core Drilling. Acker Drill Co Inc, Scranton,, Pa, USA.

Aitchison, G.D., Russam, K. and Richards, B.G. 1965. Engineering concepts of moisture equilibrium and moisture changes in soils. Moisture Equilibria and Moisture Changes in Soils beneath Covered Areas (Symposium). Butterworths, Sydney, Australia, 7-21.

Altmeyer, W.T. 1955. Discussion of 'Engineering properties of expansive soils'. Proc ASCE, Vol 81, Separate No 658, March.

Arnold, M. 1980. Prediction of footing settlements on sand. Ground Engng, UK, March, 40-49.

ASCE. 1941. Pile driving formulas. Proc ASCE, May, 67 (5): 853-866.

Audric, T. and Bouquier, L. 1976. Collapsing behaviour of some loess soils from Normandy. Quaterly J Engng Geol, 9:265-278. (See also Bell 1981).

Baguelin, F., Jezequel, J.F. and Shields, D.H. 1978. The Pressure-meter and Foundation Engineering. Trans Tech Publications, Germany.

Barron, R.A. 1947. Consolidation of fine grained soils by drain wells. JSMF Div, ASCE, Vol 113, 1948.

Barton, N. 1978. The shear strength of rock and rock joints. Publ 119, NGI, Oslo.

Barton, N. and Choubey, V. 1978. The shear strength of rock joints in theory and practice. Publ 119, NGI, Oslo. Also in Rock Mechanics, 10:1-54, Springer-Verlag, 1977.

Barton, N., Lien, R. and Lunde, J. 1975. Engineering classification of rock masses for the design of tunnel support. NGI Publ 196. Also in Rock Mechanics, Vienna, Austria, 6(4): 189-236, Dec 1974.

Bedse, V.M., Dater, P.C., Kalantri, B.N. and Ambadkar, R.M. Bridge foundations on weathered rock. Indian Highways, IRC, New Delhi, Oct, 12(10): 22-30.

Bell, A.L. 1915. The lateral pressure and resistance of clay and the supporting power of clay foundations. Proc Instn Civil Engrs,

199: 233-272.

Bell, F.G. 1978. Introduction. Chapter 1 in Foundation Engineering in Difficult Ground, (ed) F.G. Bell. Newness-Butterworths, London.

Bell, F.G. 1981. Engineering Properties of Soils and Rocks. Butterworths, London.

Berezantzev, V.G., Khristoforov, V.S. and Golubkov, V.N. 1961. Load bearing capacity and deformation of piled foundations. Proc 5th Intern Conf SMFE, Paris, 2, 11-15.

Bhandari, R.K. and Datye, K.R. 1979. Soil sampling practice in India. State of the Art on Current Practice of Soil sampling. Proc Intern Symp Soil Sampling, Singapore, July, pp 157-170.

Bhati, R.S. 1982. Diaphragm walls for bridges — pre-trench method using bentonite. IRC J, New Delhi, 43-2: 217-252.

Bieniawski, Z.T. 1976. Rock mass classification in rock engineering. Proc Symp Exploration for Rock Engng. (ed) Z.T. Bieniawski, Vol 1, A.A. Balkima, Rotterdam, Holland, pp 97-106.

Bieniawski, Z.T. 1978. Determining rock mass deformability, experience from case histories. Intern J Rock Mech and Mining Sc, Oct, 15(5): 237-241.

Bishop, A.W. 1955. The use of the slip circle in the stability analysis of slopes. Geotechnique, London, March, 5(1): 83-117.

Bishop, A.W. 1958. Test requirements for measuring the coefficient of earth pressure at rest. Brussels Earth Conf.

Bishop, A.W. 1960. The measurement of pore pressure in the triaxial test. Proc Conf Pore Pressure and Suction in Soils, London, March, pp 38-46. Butterworths, London, 1961.

Bishop, A.W. 1966. The strength of soils as engineering materials. Geotechnique, Vol 16, London.

Bishop, A.W. and Morgenstern, N.R. 1960. Stability coefficients for earth slopes. Geotechnique, London, 10(4): 129-150.

Bjerrum, L. 1954. Geotechnical Properties of Norwegian Marine clays. Geotechnique, London, 4(2): 49.

Bjerrum, L. 1963. Discussion on compressibility of soils. Proc European Conf SMFE, Wiesbaden, Vol 2:135-137. (Also NGI Publication No 98).

Bjerrum, L. and Eggestad, A. 1963. Interpretation of loading tests on sand. Proc European Conf SMFE, Wiesbaden, 1: 192-203.

Bond, D. 1961. Influence of foundation size on settlement. Geotechnique, London, June, 11(2): 121-143.

Boussinesq,˙ J.V. 1885. Application des Potentiels a L'etude de L'équilibre et du Mouvement des Solides Elastiques. Gauthier-Villars, Paris.

Bowles, J.E. 1979. Physical and Geotechnical Properties of Soils. McGraw Hill Book Co., New York.

Bowles, J.E. 1983. Foundation Analysis and Design. McGraw Hill.

BRE. 1977. Building Construction. BRE Digests, 2/e, Building Res Est, UK.

Brewer, R. 1964. Fabric and Mineral Analysis of Soils. John Wiley, New York.

Briaud, J.L. and Meyer, B. 1983. Insitu tests and their application in offshore design. Proc Conf Geotech Practice in Offshore Engng, Univ of Texas at Austin, April, (ed) S.G. Wright, ASTM, 1983.

Broms, B.B. 1975. Landslides. In Foundation Engineering Handbook, (eds) H.F. Winterkorn and H.Y. Fang, Van Nostrand Reinhold Co, New York.

Broms, B. and Boman, P. 1978. Stabilization of Soil with Lime Columns. Design Handbook, Dept of Soil and Rock Mechanics, Royal Inst of Technology, Stockholm.

Broms, B. and Boman, P. 1979. Lime columns — a new foundation method. JGE Div, ASCE, 105 (GT4): 539-556.

Brown, D.R. and Warner, J. 1973. Compaction grouting. JSMF Div, ASCE, 99 (SM8): 589.

Brown, E. 1977. Vibroflotation compaction of cohesionless soils. JGE Div, ASCE, 103 (GT12): 1437-1451.

Brown, E.T. 1981. Rock characterization, testing, and monitoring, Inter Soc Roc Mech, Suggested Methods, Pergamon Press, 211 p.

BS: 1377. 1975. Methods of Test for Soils for Civil Engineering Purposes. BSI, London.

Cannon, F.W. 1976. Fabrics in Civil Engineering. Civil Engng, London, March, pp 39-42.

Caquot, A. 1934. Equilibre des Massifs a Frottement Interne. Gauthier-Villars, Paris.

Carman, P.C. 1956. Flow of Gases through Porous Media. Academic Press, New York.

Caron, C., Herbst, T.F. and Cattin, P. 1975. Injections. Chapter in Foundation Engineering Handbook, (eds), H.F. Winterkorn and H.Y. Fang, Van Nostrand, Reinhold Co, New York.

Carter, M. 1983. Geotechnical Engineering Handbook. Pentech Press, London.

Casagrande, A. 1932. Research on the Atterberg limits of soils. Public Roads, Oct, USA.

Casagrande, A. 1936. The determination of the preconsolidation load and its practical significance. Proc 1st Intern Conf SMFE, Cambridge, Mass, 3: 60-64.

Casagrande, A. 1937. Seepage through dams. J New England Water Works Assoc, June, pp 131 170. Also Pub 209, Graduate School of Engng, Harvard Univ. Also in Contributions to Soil Mechanics, 1925-40, Bonston Soc Civil Engrs, 1940.

Casagrande, Leo. 1949. Electro-osmosis in soils. Geotechnique, London, 1 (3): 159-177.

Casangrade, Leo. 1952. Electro-osmotic stabilization of soils. J Boston Soc Civil Engrs, Jan, 39:51-83.

Casagrande, Leo. 1953. Review of the past and current work on electro-osmotic stabilization of soils. Harvard Soil Mech Ser No 45, Dec.

CBRI. 1978. Handbook on Under-reamed and Bored Compaction Pile Foundation. Central Bldg Res Inst, Roorkee.

Cedergren, H.R. 1977. Seepage, Drainage and Flow Nets. John Wiley and Sons, New York.

CGS. 1985. 1978. Canadian Foundation Engineering Manual, 2/e (1985). Canad Geotech Soc, Montreal. Bi-Tech Publishers Ltd, Vancouver, BC.

Chan, H.R. and Kenney, T.C. 1973. Laboratory investigation of permeability ratio of New Liskeard varyed soil. Canad Geotech J, 10: 453—472.

Chao, K.H. and Chin, K.Y. 1963. The study of improving bearing capacity of Tapei silt by using quickline pile. Proc 2nd Asian Conf SMFE, Tokyo, 1:387-389.

Chen, F.H. 1965. The use of piers to prevent uplifting of lightly loaded structures founded on expansive soils. Proc 1st Intern

Res and Engng Conf on Expansive Clay Soils. Texas A and
    M Press, College Station, Tex, pp 152-171.

Chen, F.H. 1975. Foundation on Expansive Soils. Elsevier
    Scientific Pub Co., Amsterdam.

Christian, J.T. and Carrier, W.D. III, 1978. Janbu, Bjerrum and
    Kjaernsli's chart reinterpretted. Canad Geotech J, 15(1):
    124—128.

Clemence, S.P. 1985. Collapsible soils: indentification, treatment
    ment and design considerations. Current Practices in
    Geotechnical Engineering, (eds) A. Singh and R.C. Joshi, Geo-
    Environ Academia, Jodhpur, 1:17-33.

Clemence, S.P, Finbarr, A.O., 1981. Design considerations for
    collapsible soils. JGE Div, ASCE, March, 107 (GT3): 306-317.

Collins, K. and McGown, A. 1974. The form and functions of
    microfabric features in variety of natural soils. Geotechnique,
    London, 24(2).

Cornfield, G.M. 1975. Sheet pile structures. Chapter in Foundation
    Engineering Handbook, (eds) H.F. Winterkorn and H.Y. Fang,
    Van Nostrand Reinhold Co., New York.

Coulomb. C.A. 1976. Essai sur une application des r'egles des
    maximis et minimis a quelques problemes de statique relatifs a
    l'architecture. Mem Acad Roy Pres Div Savants, Paris, 7:38.

CP:2002.1981. Code of Practice for Site Investigations, BS 5930
    BSI, London.

Craig, R.F. 1983. Soil Mechanics. Van Nostrand Reinhold (UK)
    Co Ltd, England.

Culmann, K. 1866. Die Graphische Statik. Zurich.

Dalmatav, B.I., 1962. Major Principles Governing Design and
    Construction of Foundations according to the Practice in the
    Soviet Union. (ed) J. Narain. Univ of Roorkee. Refer to
    Foundation Beds for Buildings and Structures, Building Code
    Regulations, Part II, Section B, Chapter 1, Code of Design
    Practice (BC and R II-B. 1-62), Moscow, 1966.

Darcy, H. 1856. Les Fontaines Publiques de la Ville de Dijon,
    Dalmont, Paris.

Das, B.M. 1983. Advanced Soil Mechanics. Hemisphere Publishing
    Corporation, New York; McGraw-Hill Book Co., New York.

Dastane, N.G. 1972. A Practical Manual for Water Use Research
    in Agriculture. Navabharat Prakashan, Poona.

Dastidar A.G. 1985. Treatment of weak soils — an Indian perspective. Current Practices in Geotech, Engng, Vol 1:179-232. (ed) A. Singh. Geo-Environ Academia, Jodhpur.

Datya, K.R. 1982. Simpler techniques for ground improvement. 4th IGS annual lecture. IGJ, New Delhi, Jan, 12(1): 1-82.

De Beer, E. and Martens, A. 1957. Method for computation of an upper limit for the influence of the heterogeneity of sand layers in the settlement of bridges. Proc, 4th Intern Conf SMFE, London, 1: 275-282.

De Beer, E. and Vesic, A. 1958. Etude experimentale de la capacite portante du sable sons des fondations directes etablies en surface. Annales des Travaux Publics de Belgigue, 59 (3): 5-58.

De, P.L. 1978. Foundations in Poor Soils including Expansive Clays. Overseas bldg notes, BRS, Garston, England, No 179, April.

De Ruiter, J. and Beringun, F.L. 1979. Pile foundations for large North Sea structures. Marine Geotechnology, 3(3): 267-314.

Deere D.U. 1968. Geological considerations. Chapter in Rock Mechanics in Engineering Practice, (eds) Stagg M.G. and Zienkiewiez, O.C. Wiley, pp 1-19.

Deere, D.U., Hendron, A.J. Patton, F.D., Cording, E.J. 1967. Design of surface and near-surface construction in rock. Proc 8th Symp Rock Mech, AIME, Minnosota. pp 237-302.

Desai, M.D. 1972. Applicability of Dynamic Cone Test to the Problems of Civil Engineering. Ph.D. Thesis, South Gujarat Univ, Surat.

DGI. 1985. Code of Practice for Foundation Engineering. Bull No 36, Danish Geotechn. Inst, Copenhagen. (1st edition 1978, Bull No. 32)

Dudley, J.H. 1978. Review of collapsing soils. JSMF Div, ASCE, 96 (SM 3): 925-947.

Dupuit, J. 1963. Etudes Theoriques et Pratiques sur le Mouvement des Eaux a travers les Terrains Permeables. Corilian — Goeury, Paris, 2/e.

Fellenius, W. 1936. Calculation of the stability of earth dams. Trans 2nd Congr Large Dams, Washington, D.C. 4: 445.

Flaate, K. 1964. An investigation of the validity of three pile driving formulae in cohesionless materials. NGI Pub No 56, 11-12.

Fookes, P.G. and Horswill, P. 1970. Discussion on 'The load deformation behaviour of the Middle Chalk at Mundford, Norfolk' — article in 'Insitu Investigations in Soils and Rocks.' BGS, London, 53-57.

Fox, E.N. 1948. The mean elastic settlement of a uniformly loaded area a depth below the ground surface. Proc 2nd Intern Conf SMFE, Rotterdam, 1: 129-132.

Garlanger, J.E. 1973. Prediction of downdrag load at the Cutler Circle bridge. Symp on Downdrag of Piles, Massachusetts Institute of Tech, USA.

Gibbs, H.J. and Holtz, W.G. 1957. Research on determining the density of sands by spoon penetration testing. Proc 4th Intern Conf SMFE, London, 1: 35-39.

Gibbs, H.J. and Bara, J.P. 1967. Stability problems of collapsing soils. JSMF Div, ASCE, 93 (SM 4): 577-994.

Gibson, R.E. and Morgenstern, N.R. 1962. A note on the stability of cuttings in normally consolidated clays. Geotechnique, London, 12(3): 212-216.

Gilboy, G. 1933. Hydraulic fill dams. Proc 1st Intern Congr Large Dams, Stockholm.

Golubkov, V.M. 1950. Niesushchaja sposobnost svajnykh osnovanji, GISL, Moscow.

Graf, E.D. 1969. Compaction grouting techniques and observations. JSMF Div, ASCE, 95 (SM3): 1151.

Grant, R., Christian, J.T. and Vanmarcke, E.W. 1974. Differential settlement of buildings. JGE Div, ASCE, Sept, 100(GT9): 973-991.

Greenwood, J.R. 1983. A simple approach to slope stability, Ground Engng, Foundation Publications Ltd., Brentwood, Essex, England, May, 16(4): 45-48.

Grim, R.E. 1959. Physico-chemical properties of soils: clay minerals. JSMF Div, ASCE, 85 (SM2).

Grim, R.E. 1962. Applied Clay Mineralogy McGraw Hill, New York.

Grim, R.E. 1968. Clay Mineralogy. McGraw Hill, New York 2/e.

Gromko, G.J. 1974. Review of expansive soils. JGE Div, ASCE, 100(GT6): 667-687.

Hansen, J.B. 1970. A revised and extended formula for bearing capacity. Bull No 28 Danish Geotech Inst, Copenhegen.

Harr, M.E. 1962. Groundwater and Seepage. McGraw Hill, New York.

Hazen, A. 1911. Discussion on 'Dams on sand foundations by A.C. Koenig'. Trans, ASCE, 73:199.

Head, K.H. 1982 Manual of Soil Laboratory Testing, Vol. 2: Permeability, Shear Strength and Compressibility Tests. Pentech Press, London.

Herrero, O.R. 1980. Universal compression index equation. JGE Div, ASCE, 106 (GT 11): 1179-1200, Nov. Discussions: Vol. 109, Oct 1983 and Vol 109, May 1983.

Herrin, M. and Mitchell, H. 1961. Lime soil mixtures. Bull 304, Natl Acad of Sciences, Natl Res Counc, HRB, Washington, DC.

Hiley, A. 1925, A rational pile driving formula and its application in practice explained. Engineering, Vol 129:657.

Hiley, A. 1930. Pile driving calculations with notes on driving forces and ground resistance. The Structural Engr, London, July-Aug, Vol 8.

Hillel, D. 1971. Soil and Water: Physical Principles and Processes Academic Press, New York.

Hirschfeld, R.C. 1963. Stress deformation and strength characteristics of soils. Harvard Univ (Unpublished). 87 pp.

HMSO 1952. Soil Mechanics for Road Engineers. Road Res Lab, UK.

Hoek, E and Bray, J.W. 1974, 1977. Rock Slope Engineering. Instn Mining & Metallurgy, London, 1/e(1974), 2/e(1977).

Hoek, and Brown. E.T. 1980. Empirical strength criterion for rock masses. JGE Div, ASCE, Sept, 106 (GT9): 1013-1035.

Holden, J.C, 1974. Penetration testing in Australia. Proc European Symp Penetration Testing, Stockholm, 1: 155-162.

Holtz, W.G. 1959. Expansive clays—properties and problems. Colorado School of Mines Quarterly, Oct, 54(4): 89-125.

Holtz, R.D. and Kovacs, W.D. 1981. An Introduction to Geotechnical Engineering. Prentice Hall. New Jersey.

Hunt, C.B. 1972. Geology of Soils, their Evaluation, Classification and Uses. W.H. Freeman and Co, San Francisco.

Hunter, J,H. and Schuster, R,L. 1968, Stability of simple cuttings in normally consolidated clays. Geotechnique, 13: 372-378.

Hutchinson, M.T., Daw, G.P., Shotton P.G. and James, A.H. The properties of bentonite slurries used in diaphragm walling and their control. Proc Conf on Diaphragm walls and Anchoroges. Instn Civ Engrs, pp. 33-40.

Hvorslev, M.J. 1951, Time lag and Soil Permeability in Ground Water Observations. Bul No 36, USWES, Vicksburg, Miss.

IRC: 37-1984. Guidelines for the Design of Flexible Pavements. IRC, New Delhi.

IRC: 45-1972, Recommendations for Estimating the Resistance of Soil below the Maximum Scour Level in the Design of Well Foundations of Bridges. IRC, New Delhi.

IRC: 78.1983. Standard Specifications and Code of Practice for Road Bridges, Section VII — Foundations and Substructures. IRC, New Delhi.

IRC: 1978a. State of the Art: Compaction of Earthwork and Subgrades. Spl Rep IRC HRB, New Delhi.

IRC: 1978b. Specification for Road and Bridge Works. IRC, New Delhi.

IRC: 1981. Recommended Practice for Lime Fly Ash Stabilized Soil Sub-bases in Pavement Constructions (Draft — under preparation). Soil Engng. Committee of IRC, New Delhi.

IRC. 1983. Recommended Design Criteria for the use of Soil-lime Mixes in Road Construction. Under preparation. See also IRC: 51-1973 (The earlier recommendations)

IRC: 1985. Pocket book for High way Engineers. IRC, New Delhi.

IS: 1904.1966. Code of Practice for Structural Safety of Buildings: Foundations. ISI, New Delhi.

IS: 1948.1970. Classification and Identification of Soils for General Engineering Purposes. ISI, New Delhi.

IS: 2132.1972. Code of Practice for Thin Walled Tube Sampling of Soils. ISI, New Delhi.

IS: 2720-3.1980. Determination of Specific gravity. Section 1: Fine grained soils. Section 2: Medium and Coarse grained Soils. ISI, New Delhi.

IS:2720-5. 1985. Determination of Liquid and Plastic Limits. ISI, New Delhi.

IS: 2720-6-1978. Determination of Shrinkage Factors. ISI, New Delhi.

IS: 2720-7.1980. Determination of Water Content — Dry Density Relation using Light Compaction. ISI, New Delhi.

ISI: 2720-8.1983. Determination of Water Content — Dry Density Relation using Heavy Compaction. ISI, New Delhi.

IS: 2720—14. 1983. Determination of Density Index (Relative Density) of Cohesionless Soils. ISI. New Delhi.

IS:2720-16.1979. Laboratory Determination of CBR. ISI, New Delhi.

IS: 2720-28. 1974. Determination of Dry Density of Soils in-place by the Sand Replacement Method. ISI, New Delhi.

IS: 2720-29.1975. Determination of Dry Density of Soils in-place by the Core Cutter Method. ISI, New Delhi.

IS: 2720-30.1980. Laboratory Vane Shear Test. ISI, New Delhi.

IS: 2720-38.1976. Compaction Control Test (Hill Method). ISI, New Delhi.

IS: 2720-40.1977. Determination of Free Swell Index of Soils. ISI, New Delhi.

IS: 2911-1.1964. Code of Practice for Design and Construction of Pile Foundations, Part 1 Load Bearing Concrete Piles. ISI, New Delhi.

IS: 2911-3.1973. Code of Practice for Design and Construction of Pile Foundations, Part 3, Under-Reamed Pile Foundation, ISI, New Delhi.

IS: 2911-8. Code of Practice for Design and Construction of Pile Foundations, Part 8, Load Tests on Piles. ISI, New Delhi.

IS: 3955-1967. Code of Practice for Design and Construction of Well Foundations. ISI, New Delhi.

IS: 4434-1978. Code of Practice for Insitu Vane Shear Test for Soils. ISI, New Delhi.

IS: 4968-1.1976 Method for Subsurface Sounding for Soils, Part 1, Dynamic Method using 50 mm Cone without Bentonite Slurry. ISI, New Delhi.

IS: 4968-2.1976, Method for Subsurface Sounding for Soils, Part 2, Dynamic Method using Cone and Bentonite Slurry, ISI, New Delhi.

IS: 8009-1.1976. Code of Practice for Calculation of Settlements of Foundations. Part 1, Shallow Foundations subjected to Symmetrical Static Vertical Loads. ISI, New Delhi.

IS: 8783.1978. Guide for Undisturbed Sampling of Sands. ISI, New Delhi.

IS: 9198.1979. Specifications for Compaction Rammer for Soil Testing. ISI, New Delhi.

IS: 10074.1982. Specifications for Compaction Mould Assembly for Light and Heavy Compaction Tests for Soils. ISI, New Delhi.

IS: 10108.1982. Code of Practice for Sampling of Soils by Thin Wall Sampler with Stationary Piston. ISI, New Delhi.

Jain, G.S., Gupta, S.P. and Prakash, C. 1969. Construction of under-reamed piles in ground with high water table. Civil Engng Construction and Public Works J, 11(2):23-27.

Jaky, J. 1948. Pressure in soils. 2nd Intern Conf SMFE, Rottardam.

Janbu, N. 1953. An energy analysis of pile driving using non-dimensional parameters. Annales Institut Technique du Batiment et des Travoux Publics, No. 63-64, 352-360 (also NGI Publ No 3).

Janbu, N. 1954. Stability analysis of slopes with dimensionless parameters. Harvard Soil Mech Series, No. 46, 81 p.

Janbu, N. 1954b. Application of composite slip surfaces for stability analysis. Proc. European Conf on Stability of Earth slopes, Stockholm, Vol 3: 43-49.

Janbu, N. 1968. Slope stability computations. SMFE, Rep No 23.8, 66 p. Tech Univ, Trondheim.

Janbu, N. 1969. An advanced method of slope stability analysis. Lecture manuscript, Univ of California.

Janbu, N. 1980. Critical evaluation of the approaches to stability analysis of landslides and other mass movements. Proc Intern Symp Landslides, New Delhi, April, 2: 109-128.

Janbu, N., Bjerrum, L. and Kjaernsli, B. 1956. Norwegian Geotech Inst (NGI), Oslo, Publ 16.

Jennings, J.E. and Knight, K. 1957. The additional settlement of foundations due to collapse of sandy soils on wetting. Proc 4th Intern Conf SMFE, Aug, 1: 316-319.

Jennings, J.E. and Kerrich, J.E. 1962. The heaving of buildings and the associated economic consequences, with particular reference to the Orange Free State Goldfields. Civil Engr in South Africa, 4(11).

Johnson, A.i. 1962. Methods of measuring soil moisture in the field. US Dept of Interior, Geological Survey Water Supply

Paper 1619-U. (As quoted by Hillel 1971).

Johnson, L.D. Overview for design of foundations on expansive soils. Miscellaneous paper GL-79-21, USAEWES, Vicksburg, Miss.

Johnson, L.D. and Snethen, D.R. 1979. Prediction of potential heave of swelling soil. Geotech Testing J, ASTM, Sept, 1(3): 117-124.

Jumikis, A.R. 1962,1965. Soil Mechanics. Van Norstrand, New York (1962). Affiliated East-West Press, New Delhi (1965).

Jumikis, A.R. 1971. Foundation Engineering. Intext Educational Publishers, Scranton.

Kapre, B.S. and Kulkarni, R.P. 1972. One point method of liquid limit test. Bul ISI, New Delhi, 24(9): 395-398.

Katti, R.K. 1979. Search for solutions to problems in black cotton soils. First IGS annual lecture. IGJ, New Delhi, 9(1): 1-82.

Kenney, T.C. 1959. 'Discussion': Geotechnical properties of glacial lake clays. JSMF Div, ASCE, 85 (SM3): 67-79.

Kerisel, J. 1965. Vertical and horizontal bearing capacity of deep foundations in clay. Proc Symp on Bearing Capacity and Settlement of Foundations, Duke Univ, Durham, NC. April.

Kezdi, A. 1974. Handbook of Soil Mechanics, Vol 1, Soil Physics Akademiai Kiado, Budapest.

Knight, K. 1963. The origin and occurrence of collapsing soils. Proc 3rd Reg Conf SMFE, Aprica, 1: 127-130.

Koerner, R.M. 1984. Construction and Geotechnical Methods in Foundation Engineering. McGraw Hill Book Co, New York.

Komornik, A. and David, D. 1969. Prediction of swelling pressure of clays. JSMF Div, ASCE, Jan, 95 (SM1): 209-225.

Kondner, R.L. 1963. Hyperbolic stress-strain response: cohesive soils. JSMF Div, ASCE, Feb, 89 (SM1): 115-143.

Kozeny, J. 1927. Ueber Kapillare Leitung des Wassers im Boden. Wien, Akad Wiss, 136 (Pt 2a): 271.

Ladd, C.C. 1971. Strength parameters and stress strain behaviour of saturated clays. Res Rep R 71-23, Soils publications 278. Dept Civil Engng, MIT. 280 pp.

Ladd, C.C., Foote, R. Ishhara, K. Schlosser, F. and Poulos, H.G. 1971. Stress deformation and strength characteristics. State-of-the-art report. Proc 9th Intern Conf SMFE, Tokyo. 2: 421-494.

Lade, P.V., Lee, K.L. 1976. Engineering Properties ot Soils. Report, UCLA-ENG-7652, 145 pp. (Ref. Holtz and Kovacs 1981).

Lambe, T.W. 1953. The structure of inorganic soils. Proc ASCE, 79, Separate 315, Oct.

Lambe, T.W. 1958a. The structure of compacted clays. JSMF Div, ASCE, 84 (SM2): 1654-1 to 1654-34.

Lambe, T.W. 1958b. The engineering behaviour of compacted clays. JSMF Div, ASCE, 84(SM2): 1655-1 to 1655-35.

Lamb, T.W. 1960. Compacted clay: structure. Trans ASCE, 125: 682-717.

Lambe, T.W. 1967. stress path method. JSMF Div, ASCE, 93 (SM 6): 309-311.

Lambe, T.W. and Whitman, R.V. 1973, 1979. Soil Mechanics. Wiley Eastern Pvt Ltd, New Delhi. SI version (1979), Wiley, New York.

Lambe, T.W., Marr, W.A. 1979. Stress path method: 2/e. JGE Div, ASCE, 105 (GT6): 727-738.

LCPC. 1971. Essai Pressiometerique Normal. Modes operative due Laboratoire Central des Ponts et Chaussees, Dunod, Paris.

LCPC. 1976. Reinforced Earth. Laboratoire Central des Pont et Chaussees, Paris.

Lee, K.L., 1970. Comparison of plane strain and triaxial tests on sand. JSMFE Div, ASCE, 96 (SM3): 901-923.

Lee, I.K., White, W. and Ingles, O.G. 1983. Geotechnical Engineering. Pitman Books Ltd, London.

Leflaive, E. 1985. Geotextiles: their rationale and future. Geotextiles and Geomembranes, 2(1): 23-30. Elsevier Applied Sc Publishers, England.

Leliavsky, S. 1955. Irrigation and Hydraulic Design, Vol 1, Chapman and Hall, London.

Leonards, G.A. 1976. Estimating consolidation settlements of shallow foundations on overconsolidated clays. Sp Rep 163, Transportation Res Bd, USA, pp 13-16.

Lio, S.S.C. and Whitman, R.V. 1986, Overburden correction factors for SPT in sand. JGE, ASCE, March, 112(3): 373-377.

Linden-Alimak. 1981. Alimak Lime Column Method. Brochure, Linden-Alimak, AB, Skelleftea, Sweden.

Little, A.L. 1969. The engineering classification of residual tropical soils. Proc Speciality Session-Engng Properties of Lateritic Soils, 7th Intern Conf SMFE, Mexico, Bangkok 1: 1-10.

Londe, P. 1972. The Mechanics of Rock Slopes and Foundations. Rock Mechanics Res Report 17, Imperial College of Science and Technology, April, London.

Louis, C. 1974. Rock hydraulics. Chapter in Rock Mechanics, (ed) L. Muller, Springer Verlag, Wien, New York.

Lowe III, J. and Zaccheo, P.F. 1979. Subsurface exploration and sampling. Chapter in Foundation Engineering Handbook, (eds) H.F. Winterkorn and H.Y. Fang, Van Nostrand Reinhold Co, New York.

Marcovici, M. and Ciuraru, Y. 1963. Rapid field moisture control using the Proctor needle. Proc 2nd Asian Regional Conf SMFE, Japan, 1: 157-160.

Marcuson, W.F. III and Bieganousky, W.A. 1976. Laboratory Standard penetration tests on fine sands. JGE Div, ASCE, June, 103 (GTG): 565-568.

Marcuson, W.F. and Franklin, A.G. 1979. State of the art of undisturbed sampling of cohesionless soils. State of the Art on Current Practice of Soil Sampling. Proc Intern Symp on Soil Sampling, Singapore, July, pp 57-71.

Mathur, B.L. 1982. Effect of Water Table on Bearing Capacity of Model Footings on Dune Sand. M.E. Thesis, Univ Jodhpur.

Mc Carthy, D.F. 1977. Essentials of Soil Mechanics and Foundations. Reston Pub Co Inc, Virginia

Means, R.E. and Parcher, J.V. 1963. Physical Properties of Soils. Charles E. Merril Books Inc, Columbus, Ohio.

Menard, L. 1956. An Apparatus for Measuring the Strength of Soils in Place. M.Sc. Thesis, Univ of Illionois, Urbana.

Menard, L. 1975. The interpretation of pressuremeter test results. Sols-soils, Paris, 7(26). Also printed as brochure, D. 60. AN, on Interpretation and Application of Pressuremeter Test Results. Techniques Louis Menard, Longjumeau, France.

Menard, L. and Broise, Y. 1975. Theoretical and practical aspects of dynamic consolidation. Geotechnique, London, 25(1):3-18.

Merrit, F.S. and Gardner, W.S. 1983. Geotechnical Engineering. Section 7 in Standard Handbook for Civil Engineers, (ed) F.S. Merritt, McGraw Hill Book Co, 3/e.

Mesri, G. Godlewski, P.M. 1977. Time and stress-compressibility interrelationship. JGE Div, ASCE, 193(GT5): 417-430.

Meyerhof, G.G. 1955. Influence of roughness of base and ground water conditions on the ultimate bearing capacity of foundations. Geotechniqne, London, Sept, 5(3): 227-242.

Meyerhof, G G. 1956. Penetration tests and bearing capacity of cohesionless soils. JSMF Div, ASCE, Jan 82 (SM1): 19 p.

Meyerhof, G.G. 1963. Some recent research on bearing capacity of foundations. Canad Geotech J, Sept, 1(1): 16-26.

Meyerhof, G.G. 1965. Shallow foundations. Proc ASCE, 82 (SM-2): 21-31.

Meyerhof, G.G. 1976. Bearing capacity and settlement of pile foundation, JGE Div, ASCE, March, 102 (GT 3): 137-227. Closure by G.G. Meyerhof, JGE Div, Vol 103, Sept, 1979.

Meyerhof, G.G. 1983. Scale effects of ultimate pile capacity. JGE Div, ASCE, June, 109(6):797-806.

Miller. R.H. and Low, P.F. 1963. Threshold gradient for water flow in clay systems. Proc Soil Science Soc Amer, 27(6): 605-609.

Michigan SHC. 1985. A Performance Investigation of Pile Driving Hammers and Piles. Final report, Michigan State Highway Commission, Lansing, Michigan, 338 pp.

Mitchell, J.K. 1976a. Fundamentals of Soil Behaviour. John Wiley and Sons, New York.

Mitchell, J.K. 1976b. Introduction: Survey of methods of soil improvement. Continuing Education in Engng. Univ of California, Berkeley. (See Lee et al 1983).

Mitchell, J.K. 1977. Soil improvement methods and applications. Prepared for Utah State Univ Spl Course for U.S. Soil Conservation Engrs, Logan, Utah.

Mitchell, J.K. and Younger, J.S. 1967. Abnormalities in Hydraulic Flow through Fine Grained Soils. Special Tech Pub 417, ASTM, pp 106-141,

Mitchell, J.K. and Katti, R.K. 1981. Soil improvement State-of-art report (preliminary), 10th Intern Conf SMFE, Stockholm, General reports, p 264.

Mitchell, R.J. 1983. Earth Structures Engineering. Allen and Unwin, Inc, Boston.

Mogi, K. 1966. Pressure dependence of rock strength and transition from brittle fracture to ductile flow. Bull Earth Res Inst, Tokyo, 44: 215-232.

Moh, Zo-Chief. 1969. Engineering properties of lateritic soils. Proc 7th Intern Conf SMFE, Mexico, Speciality session, Vol 3, pp 453-454.

Mohan, D., Jain, G.R.S. and Sharma, D. 1966. Bearing capacity of multi-under reamed bore piles. Proc 3rd Reg Conf SMFE, Haifa, Israeil.

Mori, H. and Koreeda, K. 1979. State of the art report on the current practice of sand sampling. State of the Art on Current Practice of Soil Sampling. Proc Intern Symp Soil Sampling, Singapore, July, pp 73-93.

NASRA. 1972. Prediction of Soil Moisture Conditions for Pavement Design. National Assoc of Australian State Road Authorities Paper R 27, Proc 7th Conf Aust Road Res Bd, 7(Part 8): 172-202, 1974.

Nagaraj, T.S. and Jayadeva, M.S. 1981. Re-examination of one-point methods of liquid limit determination. Geotechnique. London, 31(3): 413-425.

Nagaraj, T.S. and Jayadeva, M.S. 1983. Critical reappraisal of plasticity index of soil. JGE, ASCE, July, 109(7): 994-1000.

Nagaraj, T.S. and Murthy, B. R S. 1986a. A critical reapppraisal of compression index equations. Geotechnique, London, 36)(1).

Nagaraj, T.S. and Murthy, B.R.S. 1986b. Prediction of compressibility of overconsolidated uncemented soils. JGE, ASCE, April, 112(4): 484-488.

Nakase, A. 1966. Contribution to the $\phi_u = 0$ analysis of stability. Port and Harbour Res Inst Nagase, Yokosuka, Japan, Soils Div, Rep 1, pp 1-64.

Narahari, D.R. and Rao, B.G. 1979. Skirted-soil-plug foundation. 6th Asian Regional Conf on Soils, Singapore, IV/13:319-322.

NAVFAC. 1982. (a) Soil Mechanics, NAVFAC DM-7.1. (b) Foundations and Earth Structures, NAVFAC DM-7.2. Dept of Navy, Naval Facilities Engng Command, Alexandria, VA, USA. Reprinted by Scientific Publishers, Jodhpur, 1985.

Newmark, N.M. 1942. Influence charts for computation of stresses in elastic foundations. Bull Univ Illinois Engng Expt Stn, Urbana, Illinois, No 338, p 28.

NGI. 1972. Bearing Capacity of Shallow Foundation on Cohesionless Soils. Interim report, Sept, Norwegian Geotech Inst (NGI), Oslo.

Nishida, Y. 1956. A brief note on compression index of soils. J SMF Div, ASCE, July, 82 (SM 3), Proc paper 1.27.

Norman, L.E.J. 1958. A comparison of values of liquid limit determined with apparatus with bases of different hardnesses. Geotechnique, London, 8(1): 70-91.

Northey, R.D. 1969. Engineering properties of loess and other collapsible soils. Speciality session rep, Proc 7th Intern Conf SMFE, Mexico, 3: 445-452.

O'Connor, M.J. and Mitchell, R.J. 1977. An extension of the Bishop and Morgenstern slope stability charts. Canad Geotech J, 14(1): 144-155.

Ohri, M.L., Singh, A. and Mathur, R.C. 1981. Interference studies of footings on sand. Proc Symp on Engng Behaviour of Coarse Grained Soils, Boulders and Rocks, IGS, Hyderabad, Dec, 1:369-373.

Ohri, M.L., Singh, A. and Malhotra, R.N. 1983. Behaviour of silicate stabilized dune sand. Proc Indian Geotech Conf-83, IIT Madras, Dec. 21-24, pp V-7 to V-12,

Ohri, M.L., Singh, A. and Goel, J.K. 1985. Dune sand stabilization with U.F. resin sodium silicate-lime—fly ash. Proc. Indian Geotech Conf (IGC-85), Rorkee, Dec, 1: 215-219.

Olsen, H.W. 1965. Deviations from Darcy's law in saturated clays. Proc Soil Sci Soc Am, pp 135-140.

Olsen, H.W. Simultaneous fluxes of liquid and charge in saturated kaolinite. Proc, Soil Sci Soc, Am, 33(3): 338-344.

Olson, R.C. and Flaate, K.S. 1967. Pile driving formulas for friction piles in sand. JSMF Div, ASCE, 93 (SM6): 279-296.

O'Neill, M.W. and Ghazzaly, O.I. 1977. Swell potential related to building performance. JGE, ASCE, Dec, 103 (GT 12): 1363-1379.

O'Neill, M.W, and Poormoayed, N. 1980. Methodology for foundations on expansive clays. JGE, ASCE, Dec, 106 (GT 12). 1345-1367.

Osterberg, J.O. and Murphy, W.P. 1979. State of the art of undisturbed sampling of cohesive soils. State of the Art on

Current Practice of Soil Sampling. Proc Intern Symp of Soil Sampling, Singapore, July, pp 43-50.

Parry, R.H.G. 1977. A direct method of estimating settlements in sands from SPT values. Proc Conf Midlands SMFE Soc, Birmingham, England, pp 29-37.

Parry, R.H.G. 1978. Estimating foundation settlements in sand from plate bearing tests. Geotechnique, London, March, 8(1) 107-118.

Patton, F.D. 1966a. Multiple Modes of Shear Failure in Rock and Related Materials, Thesis, Univ of Illinois.

Patton, FD 1966b. Multiple modes of shear failure in rock. Proc 1st Congr Intern Soc Rock Mech, Lisbon, 1: 509-513.

Peck, R.B. 1969. Deep excavations and tunnelling in soft ground. Proc 7th Intern Conf SMFE, Mexico (State of the art volume).

Peck, R.B., Hanson, W.E. and Thornburn, T.H. 1974. Foundation Engineering, 2/e, John Wiley and Sons, Inc.

Perloff, W.H. and Baron, W. 1976. Soil Mechanics, Principles and Applications. Ronald Press Co., New York.

Pilot, G. 1981. Methods of improving the engineering properties of soft clay. Chapter in Soft Clay Engineering, (eds), E.D. Brand and R.P. Brenner, Elsevier Scientific Publishing Co, Amsterdam.

Poncelet, J.V. 1840. Memoire sur la stabilite des revetments et de leurs fondation. Memorial de l'officier du genie, Vol 13, Paris.

Poulos, H.G. 1981. Pile foundations subjected to vertical loading. Symp Geotech Aspects of Coastal and Offshore Structure, Bangkok, Dec 14-18. Also, in Geotechnical Aspects of Coastal and Offshore Structures, (eds) Yudhbir and A.S. Balasubramanniam, A.A. Balkema, Rotterdam, 1983.

Poulos, H.G. and Davis, E.H. 1974. Elastic Solutions for Soil and Rock Mechanics. John Wiley and Sons, New York.

Pariklonskii, V.A. 1949. Gruntovedenye, G.J.G.L. Moscow. Cited by Wilun and Starzewski, 1975.

Proctor, R.R. 1933. Fundamental principles of soil compaction. Engng News Record, Vol 3, Nos 9,10,12 and 13.

Rangnathan, B.V. and Satyanarayana, B. 1965. A rational method of predicting swelling potential for compacted expansive soils. Proc 6th Intern Conf SMFE, Montreal, 1: 92-96.

Rankine, W.J.M. 1857.   On the stability of loose earth.   Phil Trans Royal Soc, Vol 147.

Rao, B.G. and   Bhandari,   R.K. 1977.   Use of skirted footing in expansive clay.   Proc 1st National Symp Expansive Soils, HBTI, Kanpur, Dec, 19-21.

Rawlins, S.L. and Dalton, F.M. 1967.   Pyschrometric measurement of soil water potential without precise temperature control. Proc Soil Sci Soc, Am 31:297-301.

Raymonds,   G.P.   and   Wahls,   H.F.   1976.   Estimating one-dimensional consolidation including secondary compression of clay loaded from overconsolidated to normally consolidated state.   Sp Rep 163, Transportation Res Bd, Washington, pp 17-23.

Rebhann, G. 1871.   Theorie des Erddruckes under Futtermauern. Vienna.

Resal, J. 1910.   La Poussee des Terres.   Paris.

Richards, B.G.   1968.   Review of measurement of soil water variables and flow parameters.   Proc 4th Conf Australian Road Res Board, 4(Part 2): 1843-1861.

Richards, B.G. 1971.   Psychrometric techniques for field measurement of negative pore pressure in soils.   Proc Australia New Zealand Conf Geomechanics, Melbourne, 1: 387-394.

Richards, B.G. 1974.   Behaviour of unsaturated soils.   Chapter 4 in Soil Mechanics—New Horizons, (ed) I.K. Lee, Newnes-Butterworths, London.

Robertson, P.K. Campanella, R.G. and Wightman. 1983. SPT-CPT correlations. JGE, ASCE, Nov, 109(11): 1449-1459.

Robertson,   P.K., and   Campanella,   R.G. 1985.   Liquefaction potential of sands using the CPT.   JGE, ASCE, Mar, 111(3): 384-403.

Rosenak, S. 1966.   Soil Mechanics.   B.T.   Batsford Ltd, London.

SAA. 1978.   Piling Code, AS 2159.   Standard Association for Australia, Sydney.

Sanglerat, G. 1972.   The Penetrometer and   Soil Exploration. Elsevier Pub Co, Amsterdam.

Sastry, M.V. 1981.   Bridge foundations, bearing capacity, criteria for fixing depth of foundations   and different types of foundations.   Indian Highways, IRC, Dec, 9(12): 68-78.

Sawaguchi, S. and Takahashi, K. 1976. Influence charts for circular arc method. Rep Port and Harbour Res Inst, Nagase, Yokosuka, Japan, March, 15(1): 3-18.

Saxena, R.K. 1971. Well foundations for road bridges. JIRC, New Delhi, Nov, 34(2):391-435.

Schmertmann, J.H. 1953. Estimating the true consolidation behaviour of clay from laboratory test results. Proceedings, ASCE, Vol 79, Separate 311

Schmertmann, J.H. 1955. The undisturbed consolidation behaviour of clay. Trans ASCE, No 120, Paper 2775, pp 1201-1233.

Schmertmann, J.H. 1970. Static cone to compute static settlement over sand. JSMF Div ASCE, 96 (SM-3): 1011-1043.

Schmertmann, J.H., Hartman, J.P. and Brown, P.R. 1978. Improved strain influence factor diagrams. JGE Div, ASCE, 104 (GT 8); 1131-1135.

Schofield, R.K. 1935. The pF of the water in soil. Trans 3rd Intern Cong Soil Sci, Vol 2, pp 37-48.

Schofield, A. and Wroth, P. 1968. Critical State Soil Mechanics. McGraw Hill.

Scott, C.R. 1974. An Introduction to Soil Mechanics and Foundations. Applied Sci Pub Ltd, London.

Seed, H.B., Woodward, R.J. and Lundgren, R. 1962. Prediction of Swelling Potential for Compacted Clays. JSMF Div, ASCE, June, 88 (SM 3): 53-87.

Seed, H.B., Idriss, I.M. and Arango, I. 1983. Evaluation of liquefaction potential using field performance data. JGE, ASCE, March, 109(3): 458-482.

Sherard, J.L., Decker, R.S. and Ryker, N.L. 1972. Piping in earth dams of dispersive clays. Proc Speciality Conf of Performance of Earth and Earth-Supported Structures, ASCE, June, Vol 1, Part I, pp 589-626.

Sherard, J.L., Dunnigan, L.P. and Decker, R.S. and Steele, E.F. 1976a. Pinhole test for identifying dispersive soils. JGE Div, ASCE, 102 (GT 1): 69-85.

Sherard, J.L., Dunnigan, L.P. and Decker, R.S. 1976b. Identification and nature of dispersive soils. JGE Div, ASCE, 102 (GT4) 287-301.

Sherwood, P.T. and Ryley, M.D. 1970. An Investigation of a cone penetrometer method for the determination of the liquid limit.

Geotechnique, London, 20(2): 203-208.

Simons, N.E. and Menzies, B.K. 1977. A Short Course in Foundation Engineering Newnes-Butterworths, London.

Singh, A. 1964. Development of some new apparatus and procedures for testing soils. 4th Annual Number, Instn Engrs (India) Rajasthan Centre, Jaipur, Feb, pp 10-22.

Singh, A. 1965. Instrumentation for going metric in soil engineering testing. J Ind Soc SMFE, New Delhi, April, 4(2): 207-226.

Singh, A. 1966. The determination of swelling pressure of expansive clays. Invention Intelligence. Invention Promotion Board, New Delhi, Feb, 1(7): 5-8.

Singh, A. 1972. A practical method of estimating settlement of foundations on cohesionless soils. Symp Modern Trends in Civil Engng, Roorkee, Nov, 1: 43-47. Discussion, 2: 77-80.

Singh, A. 1981. Soil Engineering in Theory and Practice, Vol. 2, Geotechnical Testing and Instrumentation. Asia Pub House, Bombay, and APT Book Inc, New York.

Singh, A. 1987. Soil Engineering in Theory and Practice, Vol. 1, Fundamentals and General Principles, Jaisingh & Mehta Publishers, Pvt, Ltd, Bombay. 3/e. (in press).

Singh, A. and Punmia, B.C. 1965. A new laboratory compaction device and its comparison with the Proctor Test. Highway Res News, HRB, Washington, Feb, No 17: 37-42.

Singh, A., Punmia, B.C. and Phatak, M.B. 1972. Study of uplift pressures below apron with downstream cut-off founded on two layer media. IGJ, IGS, New Delhi, 2(4): 267-290.

Singh, A, Punmia, B.C, and Ohri, M.L. 1973. Interference between adjacent square footings on cohesionless soil. IGJ, IGS, New Delhi, 3(4): 275-284.

Singh, A, Ohri, M.L. and Grover, G.S. 1976. Interference between circular footings on sand. Proc 1st Ind Conf Desert Technology, Jodhpur, Nov. Trans Ind Soc Desert Technology and Univ Centre Desert Studies, 1(2): 159-162.

Singh, A. and Maheshwari, K.C. 1977. A study of friction angle and critical voids ratio of dune sand. Proc 1st Indian Conf on Desert Tech, Jodhpur, Nov 1976. Trans Ind Soc Des Tech and Univ Centre Des Studies, Jodhpur, July, 2(1): 17-22.

Singh, A., Ohri, M.L. and Moorthy, G.G.K. 1. ?? Use of shirted foundations on dune sand. Proc Conf Construction Practices and Instrumentation in Geotech Engng, Dec 1:275-280.

Singh, A., Punmia, B.C. and Ohri, M.L. 1985. Desert Soils. In Commemorative Volume on Indian Contribution to Geotechnical Engng. IGS, New Delhi.

Singh, A., Punmia, B.C. and Ohri, M.L. 1986. Geotechnical behaviour of dune sands. Current Practices in Geotech Engng, monograph on Sc and Engng of Desert Soils. (eds) A. Singh and M.L. Ohri, Geo-Environ Academia, Jodhpur, Vol 3: 165-212.

Sivaram, B. and Swamee, R.K. 1977. A computational method for consolidation coefficient. Soils and Foundations, Japanese Soc SMFE, Tokyo, May 17(2): 48-52.

Skempton, A.W. 1951, The bearing capacity of clays. Proc British Bldg Res Congress, 1: 180-189.

Skempton, A.W. 1953a. The colloidal activity of clays. Proc 3rd Intern Conf SMFE, Zurich, Vol 1: 57-61.

Skempton, A.W. 1953b. Soil mechanics in relation to geology. Proc Yorkshire Geol Soc, 29, Part 1, No 3, pp 33-62.

Skempton, A.W. 1954. The pore pressure coefficients A and B. Geotechnique, London, Vol 4: 143-147.

Skempton, A.W. 1959. Cast-in-situ bored piles in London clay. Geotechnique, Vol 9: 153-173.

Skempton, A.W. and Northey, R.D. 1952. The sensitivity of clays. Geotechnique, London, 3(1): 30-53.

Skempton, A.W., Yassin, A.A. and Gibson, R.E. 1953. Theorie de la force portante des pieux dans le sable. (Theory of the bearing capacity of piles in sand). Ann Inst Tech Bati Travaux Pubs, 63-64, 285-290. See also Skempton (1953): Discussion, 3rd Intern Conf SMFE, Zurich, Vol 3, p 172.

Skempton, A.W. and Bjerrum, L. 1957. A contribution to the settlement analysis of foundations on clay. Geotechnique, 7(4): 168-178.

Skempton, A.W. and Hutchinson, J. 1969. Stability of natural slopes and embankment foundations. Proc 7th Intern Conf SMFE, State of the art volume, Mexico, pp 291-340.

Smart, P. 1969. Soil structure in the electron microscope. Proc Intern Conf Struct Solid Mech Engg Div I, Wiley Interscience, New York, pp 249-255.

Smith, G.N. 1978. Elements of Soil Mechanics for Civil and Mining Engineers. Granade Publishing Ltd, UK, 4/e.

Snethen, D.R. 1979. Technical Guidelines for Expansive Soils in Highway Subgrades. US Army Engineer Waterways Experiment Station, Vicksburg, Miss, June, Rep No FHWA-RD-79-51.

Soderblom, R. 1969. Salt in Swedish Clays and its importance for quick clay formation. Swedish Geotech Proceedings, No 22, p 63.

Sorensen, T. and Hansen, B. 1957. Pile driving formulae, an investigation based on dimensional considerations and a statistical analysis. Proc 4th Intern Conf SMFE, London, Vol 2: 61-65.

Southwell, R.V. 1946. Relaxation Methods in Theoretical Physics. Oxford Univ Press.

Sowers, G.F. 1962. Shallow foundations. Chapter in Foundation Engineering, (ed) G.A. Leonards, pp 525-632. McGraw Hill Book Co Inc.

Sowers, G.B. and Sowers, G.F. 1970. Introductory Soil Mechanics and Foundations. Collier Macmillan International, Inc, New York. 3/e.

Talwar, D.V., Singh, A. and Saran, S. 1987. Reinforced earth, design principles and applications. In Current Practices in Geotech Engng, (eds) A. Singh and M.L. Ohri, Vol 4, Geo-Environ Academia, Jodhpur.

Taylor, D.W. 1937. Stability of earth slopes. J Boston Soc Civil Engrs, 24(3).

Taylor, D.W. 1942. Research on consolidation of clays. MIT Dept of Civ and Sanit Engng, No 82.

Taylor, D.W. 1948. Fundamentals of Soil Mechanics. John Wiley and Sons, New York.

Terzaghi, K. 1925. Erdbaumechanik. F. Deuticke, Vienna.

Terzaghi, K. 1936. The shearing resistance of saturated soils. Proc 1st Intern Conf, SMFE, 1: 54-56.

Terzaghi, K. 1943. Theoretical Soil Mechanic. John Wiley & Sons, New York.

Terzaghi, K. 1950. Mechanism of landslides. In Application of Geology to Engineering Practice, Berkey Volume, Geol Soc of America, pp 83-123.

Terzaghi, K. and Peck, R.B. 1967.   Soil Mechanics in Engineering Practice,   John Wiley and Sons, New York, 2/e.

Thompson, M.R. 1964. Lime reactivity of Illinois soils as it relates to compressive strength.   Ph. D. Dissertation Univ of Illinois, Urbana.

Thornburn, S. 1963.   Tentative correction chart for the standard penetration test in non-cohesive soils.   Civil Engng Public Works Rev, 58: 752-753.

Thompson, M.R. and Robnett, Q.L. 1976.   Pressure injected lime for swelling soils.   Transportation Res Rec 568, TRB, Washington, DC, pp 24-34.

Thornburn, S. 1975.   Building structures supported by stabilized ground.   Geotechnique, London, 25(1): 83-94.

Thornburn, S. 1978.   Discussion on 'Estimating foundation settlements in sand from plate bearing tests, by R.H.G. Parry'. Geotechnique, London, Dec, 28(4): 481.

Tomlinson, M.J. 1977.   Pile Design and Construction Practice. Viewpoint Publications, Cement and Concrete Association, London.

Tomlinson M.J. 1980. Foundation Design and Construction. Pitman Books Ltd., London, 4/e.

Trofimenkov, J.B. 1974.   Penetration testing in USSR.   State-of-the art rep, Proc European Symp on Penetration Testing, Stockholm, 1: 147-154.

Tsytovich, N. 1976. Soil Mechanics (Concise course), Mir Publishers, Moscow.

Tuma, J.J. and Hady, M.A. 1973.   Engineering Soil Mechanics Prentice Hall, Inc, New Jersey.

USBR. 1973. Design of Small Dams. USBR, Washington, D.C. 2/e.

USBR. 1974. Earth Manual. USBR, Washington, D.C. 2/e.

USSR Code. 1962.   Foundation Beds for Buildings and Structures. Building Code and Regulations, Part II, Section B, Chapter 1, Code of Design Practice (BC and R, II-B, 1-62), State Committee on Construction of the USSR, May, 1962, Moscow, 1966.

Van der Merwe, D.H. 1975. Contribution to speciality session on 'Current theory and practice for building on expansive clays'. Proc 6th Regional Conf SMFE, Africa. Durban, Vol 2.

Van Olphen, H. 1963.   An Introduction to Clay Colloid Chemistry. Wiley Interscience, New York.

Vesic, A.S. 1967.   A Study of Bearing Capacity of Deep Foundations.   Final rep, Project B-119, Georgia Inst Tech, Atlanta, Georgia.

Vesic, A.S. 1970.   Tests on instrumented piles, Ogeechee River Site. J SMF Div, ASCE, 96 (SM 2): 561-584.

Vesic, A.S. 1973. Analysis of ultimate loads of shallow foundations. J SMF Div, ASCE, Jan, 99 (SM1): 45-78.

Vesic, A.S. 1977.   Design of Pile Foundations.   National Cooperative Highway Res Program Synthesis 42, TRB.   (See also NAVFAC DM-7.2).

Von M Harmse, H.J.   1975.   Discussion on paper — The occurrence of dispersive soil piping in Central South Africa by G.W. Domldson.   Proc 6th Reg Conf SMFE, Africa, Durban, Vol 2.

Wagner, A.A. 1957.   The use of the unifield soil classiffcation system by the Bureau of Reclamation.   Proc 4th Intern Conf SMFE, London, Butterworths, 1: 125-134.

Wawryk, S. 1972.   Deep Foundations. In Practical Aspects of Soil Mechanics, (ed) M. Arnold, Univ of Adelaide, South Australia.

Westergaard, H.M. 1938.   A problem of elasticity suggested by a problem in soil mechanics: soft material reinforced by numerous strong horizontal sheets.   Contributions to the Mechanics of Solids, S. Timoshenko 60th Anniversary Volume, Macmillan Pub Co, New York.

Whitaker, T. and Cooke, R.W. 1961. A new approch to pile testing. Proc 5th Intern Conf SMFE, Paris, 2: 171-176.

Whitaker, T. 1963.   The constant rate of penetration test for the determination of the ultimate bearing capacity of pile. Proc Instn Civil Engrs, London, 26: 119-123.

Whitaker, T. 1976.   The Design of Pilcd Foundations.   Pergamon Press, Onford, 2/e.

Williams, A.A.B. 1958.   Discussion on Jenning and Knight paper 'The prediction of total heave from the double oedometer test'. Trans S Afr Instn Civ Engrs, 8 (6): 123-124.

Wilun, Z. and Starzewski, K. 1975.   Soil Mechanics in Foundation Engineering. Vol 1: Properties of Soils and Site Investigations.

Vol 2: Theory and Practice. 2/e. Surrey Univ Press in association with Intern Textbook Co, Ltd, London.

Winterkorn, H.F. 1975. Soil stabilization. Chapter in Foundation Engineering Handbook, (eds) H.F. Winterkorn and H.Y. Fang, Van Nostrand Reinhold Co, New York.

Winterkorn, H.F. and Fang, H.Y. 1975. Soil technology and engineering properties of soils. Chapter 2 in Foundation Engineering Handbook, Van Nostrand Reinhold Co., New York.

Wright, S.G. 1969. A Study of the Slope Stability and the Undrained Shear Strength of Clay Shales. Ph. D. Thesis, Univ California, Berkeley, 348 pp.

Wright, S.G., Kulhawy, F.H., and Duncan, J.M. 1973. Accuracy of equilibrium slope stability analysis. JSMF Div, ASCE, Oct, 99 (SM10): 783-792.

Wroth, C.P. and Wood, D.M. 1978. The correlation of index properties with some basic engineering properties of soils. Canad Geotech J, Ottawa, May 15(2): 137-145.

Yong, R.N. and Sheetan, D.E. 1973. Fabric unit interation and soil behaviour. Proc Intern Symp Soil Structure, Guthenburg, Sweden, pp 176-183.

Yong, R.N. and Warkentin, B.P. 1975. Developments in Geotechnical Engineering, 5 — Soil Properties and Behaviour. Elsevier Scientific Pub Co, Amsterdam.

Youssef, M.S. el Ramli, A.H. and el Demery, M. 1965. Relationship between shear strength, consolidation, liquid limit, and plastic limit for remoulded clays. Proc 6th Intern Conf SMFE, Montreal, 1: 126-129.

Zeevaert, L. 1972, 1983. Foundation Engineering for Difficult Subsoil Conditions. Van Nostrand Reinhold Co, New York. 1/e 1972, 2/e 1983.

Zienkiewicz, O.C. 1971. The finite element method in engineering science. McGraw Hill, London.

INDEX